Extrusion Processing Technology

Extrusion Processing Technology

Food and Non-Food Biomaterials

Jean-Marie Bouvier
Clextral, Firminy, France

Osvaldo H. Campanella
Purdue University, Indiana, USA

WILEY Blackwell

Registered Office
John Wiley & Sons, Ltd, The Atrium, Southern Gate, Chichester, West Sussex, PO19 8SQ, UK

Editorial Offices
9600 Garsington Road, Oxford, OX4 2DQ, UK
The Atrium, Southern Gate, Chichester, West Sussex, PO19 8SQ, UK
111 River Street, Hoboken, NJ 07030-5774, USA

For details of our global editorial offices, for customer services and for information about how to apply for permission to reuse the copyright material in this book please see our website at www.wiley.com/wiley-blackwell.

Library of Congress Cataloging-in-Publication Data

Bouvier, Jean-Marie.
Extrusion processing technology : food and non-food biomaterials / Jean-Marie Bouvier and Osvaldo Campanella.
pages cm
Includes index.
ISBN 978-1-4443-3811-9 (cloth)
1. Polymers–Extrusion. 2. Food–Extrusion. 3. Biochemical engineering. I. Campanella, Osvaldo. II. Title.
TP156.E98B68 2014
660.6′3–dc23

2013046765

A catalogue record for this book is available from the British Library.

Wiley also publishes its books in a variety of electronic formats. Some content that appears in print may not be available in electronic books.

Cover image: Extrusion machine ©Clextral
Cover design by Meaden Creative

Set in 9.25/11.5pt Minion by SPi Publisher Services, Pondicherry, India

1 2014

Contents

Foreword

Extrusion as a means of forming products has been a mainstay in the food and non-food industries for nearly 100 years and has been used to produce a wide variety of products with a wide range of compositions, densities, structures, and textures. Much work and many publications have, over the years, been devoted to the extrusion of food and non-food biomaterials but few works have successfully brought together the engineering aspects of the extrusion process with the physical and chemical changes made to the product at both the fundamental and practical levels.

The authors of this book, who are truly experts and leading scientists in extrusion technology, have produced a very impressive text that surely will be the standard for many years to come. This work will no doubt be extremely useful not only for academicians but also for practicing engineers. To facilitate a thorough understanding of the process, the authors introduce the concept of the Generic Extrusion Process (GEP) to present in a very comprehensive way the fundamentals of extrusion derived from experience in other industries. The basic functions include melt formation, micromixing and reaction, cooking and texturization, and solid-liquid separation. This concept allows for an understanding of the complexity of the extrusion process, linking science with long-term practical experience.

The book is divided into three parts.

- Part 1: Extrusion equipment design and engineering analysis with emphasis on fundamental engineering principles.
- Part 2: Application of extrusion to food and non-food biomaterials that is demonstrated with numerous case studies.
- Part 3: Emerging extrusion technologies with emphasis on developing sustainable extrusion processes.

Each part is well written and very comprehensive.

It is my belief that this work will stand the test of time and be one of best books on extrusion technology.

Martin R. Okos
Professor, Agricultural and Biological Engineering Department
Purdue University, West Lafayette, Indiana, USA

Acknowledgements

While this book grows out of several years of experience in extrusion processing technology both in academia and in industry, its final format and content reflect the valuable input of several people. My greatest debt is to my friend Professor Maurice Gélus, now retired, who encouraged me to develop research studies in extrusion processing at the Compiègne University of Technology, France. I would like to thank the professors and the graduate students who worked with me and inspired me in my research work in extrusion at the Compiègne University of Technology. They have contributed importantly to the enrichment of the content of this book. I would also like to extend my gratitude to the management of Clextral Company, France, who gave me the opportunity to use my scientific background to the benefit of product and process innovation and development in extrusion, and be confronted with industrial reality. Undoubtedly, my responsibility in Clextral has significantly enriched my experience in the area and has inspired the format and content of this book. My gratitude goes to my colleagues of Clextral who have contributed their valuable experiences in many aspects related to this book. Many thanks also to my friend Professor Osvaldo Campanella, co-author of this book, for the numerous resourcing exchanges we had before and during the completion of this book.

My final debt is to my close family, my wife Colette, and my children, Jérôme, Bérengère and Guillaume, whose love sustained me confidently through the achievement of this long-term project.

Jean-Marie Bouvier

Several teachers have been influential in stimulating my interest in research and encouraging me to continue graduate studies. I would like to thank Professor Ramon Cerro, now at the University of Alabama, who was my first advisor and mentor in my early research career at the University of Litoral, Santa Fe, Argentina. Professor Micha Peleg, my advisor at the University of Massachusetts, who has not only been a scientific mentor, but also a great friend and colleague all these years. In addition to these mentors and colleagues, I would also like to extend my gratitude to students who helped me to advance in the extrusion area. My thanks also go to colleagues who have shared with me knowledge and experience in extrusion and related areas, and Guibing Chen, Zhengjun Xue, Stanley Harlow, and Brenda Bowen who have contributed to create some figures in the book. Professor Jean-Marie Bouvier, co-author of this book, Clextral and his personnel gave me the opportunity to spend a sabbatical leave with them, initiate the book project and gain an enriching professional and personal experience. My gratitude also goes to Purdue University for their support over all these years and especially during the time I was working on this book.

Last but not least, I would like to thank my wife Estela, daughters Carolina and Cecilia and mother Leda. Without their never-ending support and encouragement this book would not have been completed. I would like to dedicate this book to them for always bringing joy to my life, and also to the memory of my father Osvaldo and sister Marta who have brought joy to my life during the times I shared with them.

Osvaldo H. Campanella

1 Generic Extrusion Processes

Extrusion, one of the most important innovations of the 20th century, is often presented as a model of scientific and technology transfer between different processing industries, such as the polymer and plastics, food and feed, and paper-milling industries in particular. Although the first technical designs of screw extruders were introduced in the latter years of the 19th century, extrusion processing really established itself approximately 60 years later, with the development in the plastics industry of polymer-based materials. It was later successfully exploited by the industries that processed plant biopolymers and has developed into a widespread extrusion processing culture over the past 80 years.

The purpose of this introductory chapter is to give a brief historical overview of the emergence of screw extruders and of the extrusion processing culture that owes its existence to the remarkable transfers of technology from polymer processing to food and feed processing and to paper milling.

1.1 A history of extrusion processing technology

Extrusion is commonly defined as the operation of forming and shaping a molten or dough-like material by forcing it through a restriction, or die. This operation is extensively applied in many processes as a batch or continuous operation. While adhering strictly to this definition, the understanding and analysis of extrusion are quite simple and straightforward.

Extrusion processing technology relies on a continuous process operation which uses extruders to handle process functions such as the transport and compression of particulate components, melting of polymers, mixing of viscous media, cooking of polymeric or biopolymeric materials, product texturization and shaping, defibering and chemical impregnation of fibrous materials, reactive extrusion, fractionation of solid–liquid media, etc. Extrusion processing technology is highly complex and in-depth descriptions and discussions are required in order to provide complete understanding and analysis of this subject.

Extrusion processing technology uses two different equipment designs: the single screw extruder and the twin screw extruder. Within each design, there are various engineering options, which depend upon the equipment manufacturers and/or the processing requirements. Readers who are interested in a complete historical review of the development of extruders should refer to the excellent and well-documented book by White (1990). This introductory chapter focuses mainly on single screw extruders and intermeshing co-rotating twin screw extruders.

1.1.1 The introduction of screw extruders

Single screw and intermeshing co-rotating twin screw extruders appeared over the last 30 years of the 19th century. An 1871 US patent to Sturges (1871) presents a single screw machine for pumping and forming soap. Gray (1879) developed a single screw extruder for processing and extruding gutta percha with specific application to wire coating. In fact, the rubber industry was an early user of screw extrusion machinery for the continuous compounding of rubber. Throughout the period from 1880 to 1930, though no notable industrial application emerged, there was significant mechanical engineering activity that brought remarkable improvements to the designs of the single screw extruder (segmented screws, threaded and grooved liners, pin barrel designs, steam-heated barrels, etc.).

In 1869, the first patent for a fully intermeshing co-rotating twin screw extruder was granted to Coignet (1869). This patent described a machine called a malaxator, which pumped and processed artificial stone paste. There was real development of the intermeshing co-rotating

Extrusion Processing Technology: Food and Non-Food Biomaterials, First Edition. Jean-Marie Bouvier and Osvaldo H. Campanella.

twin screw extruder design in the late 1930s, when Colombo (1939) proposed an advanced design which was manufactured by the Italian company Lavorazione Materie Plastiche (LMP). LMP machines were deployed extensively to the polymer industry over the period between 1940 and 1956, through manufacturing licenses (to Clextral, formerly CAFL and then Creusot-Loire in France, to R.H. Windsor in England and to Ikegai Iron Works in Japan), and through exportation sales, to IG Farbenindustrie in Germany, for example.

The introduction of thermoplastics in the 1930s gave a boost to the development of extrusion processing technology. In Germany in 1939, Paul Troester Maschinenfabrick (PTM) built an electrically heated, air-cooled single screw extruder, with automatic temperature control and variable screw speed. This was the precursor to modern automated extrusion technology for the emerging plastics industry. In following years, various ancillary equipments were introduced to meet the process requirements of the polymer-processing industry (venting, breaker plates, screen packs, co-extrusion dies, film blowing, etc.).

As the polymer industry burgeoned, along with the complexity of polymer formulations, single screw extruders showed real processing limitations in their ability to efficiently mix, compound and pump polymer melts of specific characteristics. Then, following World War II and at the request of plastics manufacturers, twin screw extrusion technology was introduced, adding significant value to the process. For example, in July 1956 CAFL delivered its first intermeshing co-rotating twin screw extruder (72.5 mm interaxis) to the French Pechiney company (Pennaroya factory of Noyelles-Godault in France), for manufacturing plastic pipes. At the same time, at the request of Bayer, the German company Werner and Pfleiderer made the first prototype of the ZSK 83 intermeshing co-rotating twin screw extruder. Figure 1.1 shows a vertical intermeshing co-rotating twin screw machine supplied by CAFL in 1963, for nylon and tergal spinning.

It must be noted that screw extruders played a determinant role in the spectacular growth of the polymer-processing industry (plastics and rubber) between 1940 and 1960, when world production of plastics exploded from 300,000 to 12 million metric tonnes per year (Utracki, 1995). And extrusion technology really established its processing potential with the development of the polymer-processing industry, where, together with the availability of reliable machinery, flexible and productive processing could be carried out at a competitive cost-to-performance ratio.

Figure 1.1 Intermeshing co-rotating twin screw extruder designed in 1963 for nylon and tergal spinning.
Source: Reproduced with permission of Clextral, France.

1.1.2 The generic extrusion process concept

The success of extrusion processing technology in the polymer industry resulted in a close relationship between extrusion equipment designers and plastics manufacturers. Both industries pragmatically combined their efforts and skills to make the use of screw extruders technically and economically viable at an industrial level, in order to produce marketable plastic materials.

The term "use of screw extruders" means extrusion processing. This term combines all the relevant technical expertise, know-how and knowledge that allow the process developers to properly operate screw extruders in order to produce the targeted materials with the expected functional and end-use properties. In other words, extrusion processing completes the synthesis, revealing the technical harmony between equipment engineering and product design. This technical harmony, which has revealed itself in the numerous extrusion applications, has led to a few generic processes whose characteristics are worth presenting and analyzing. At this point, the authors wish to introduce the generic extrusion process (GEP) concept as one way of manifesting the synthesis and revealing the fundamentals of extrusion processing technology.

Currently, the GEP concept is fundamental to the presentation and analysis of the extrusion processing culture in a comprehensive way, and is important in achieving the following aims:

- understanding the complexity of extrusion processing technology
- allowing extrusion practitioners to be less dependent on long-term experience when carrying out experimental developments
- linking extrusion science and the long-term experience, the lack of which often dramatically places a limit on the optimal use of screw extruders
- creating the potential for generating new ideas for extrusion applications.

The GEP includes the basic process functions which govern the performances of screw extruders and resulting products in certain groups of extrusion applications. They are presented in the following sections and take into account the industrial application of extrusion processing technology in different fields.

1.1.3 Extrusion technology in the polymer-processing industry

The technical synergy which occurred between extrusion equipment manufacturers and the polymer-processing industry led to the first main generic extrusion process, GEP I, "Thermomechanical Plasticating of Polymers and Polymer Melt Forming." Its process functions are applied in a vast number of industrial applications, to process polymeric materials into marketable products. In the extrusion world, this initial generic extrusion process can essentially be considered as the "big bang" of extrusion processing technology, since it has inspired numerous scientists, process engineers, and product manufacturers, and has led to outstanding process and product innovations.

GEP I involves continuous melting, mixing, and forming of polymeric mixtures (polymer resins with various additives) in screw extruders, in order to adjust the processability as well as the physical and chemical characteristics of the resulting compounds, or blends, for future uses. The energy required for this process is provided by heat which can be either transferred by conduction and/or convection between the barrel wall and the material, and/or created by viscous dissipation of mechanical energy into thermal energy. Hence the notion of thermomechanical plasticization of polymers, which has proved to be very efficient, in terms of processing time-temperature history, compared to all other types of alternative processes. The operation can be performed within a short residence time, making it possible to minimize damage to polymers and to preserve their physical and functional properties.

GEP I has many different applications, from the manufacture of simple compounds (polymers compatible with plasticizers and neutral fillers), to speciality compounds and alloys (polymer compounds with active fillers and functional additives). Single screw extrusion is the ideal solution for the production of basic and simple compounds while intermeshing co-rotating twin screw extrusion is better suited to speciality compounds which require a relatively high level of mixing in the extruder and flexibility of the process.

The polymer-processing industry also contributed to the emergence of a second generic extrusion process, GEP II, called "Thermomechanical Micromixing and Reactive Extrusion." This came about in the early 1940s when added-value polymer blends appeared on the market to satisfy specific properties with a broader range of characteristics, depending on the end-uses of the resulting materials. Polymer blend technology aims to combine several components (polymers, highly reactive compatibilizers and impact modifiers; occasionally fillers, fibers, etc.), the end-result being materials characterized by a controlled chemical constitution and morphology. Intermeshing co-rotating twin screw extruders are particularly well suited to continuous reactive processing of polymer blends, as they allow extensive variation in thermal and mechanical energy inputs, control of the residence time and the application of efficient micromixing (intense mixing at the molecular level, due to the intermeshing zone of the screw profile). Thus, the time-temperature-shear history applied in this extruder design shows a satisfactory ability to handle chemical reactions in viscous media. It favorably combines chemical kinetics with the flow and thermal properties of the reactive components and products, in order to control the strong interrelation between processing, material morphology, and product performance.

GEP II is well known and recognized in the polymer-processing industry. It made a significant contribution to the development of the polymer blend industry from the 1960s (polymerization, polymer grafting, cross-linking, etc. applied almost exclusively to synthetic polymers), where extrusion technology offered high added-value processing. As an example, Figure 1.2 shows a two-stage

Figure 1.2 Two-stage reactive extrusion equipment designed in 1966 for polymerization. Source: Reproduced with permission of Clextral, France. For color detail, please see color plate section.

reactive extrusion process applied to continuous rubber polymerization, as well as to nylon and polyester polymerization. It combines two intermeshing co-rotating twin screw extruders in series (one vertical and one horizontal machine). This assembly was designed in 1966, and several units were supplied to the polymer-processing industry.

Today, reactive extrusion is highly active in the technology of blending synthetic polymers, for processing high-performance blends in particular. Besides polymer blending, there are currently very few industrial applications. Among these are the chemical modification of biopolymers such as milk casein converted into caseinates (neutralization of the carboxylic acid functions of casein by use of alkalis) and chemically modified starch for food and non-food uses (hydrolysis, etherification, and esterification). Also, it has been successfully applied to convert diphasic solid–liquid media in the organic chemical industry (e.g. saponification), as well as in the paper-milling industry (e.g. alkaline hydrolysis of lignocellulosics, peroxidic bleaching of cellulose pulps).

Undoubtedly, the role of reactive extrusion can only increase over the coming years with the emergence of biomaterials whose morphology and characteristics can be more precisely tailored to end-use requirements. It can also be used to convert biomass into intermediate components (green chemistry), or to substitute existing batch processes with a significant improvement of processing sustainability.

1.1.4 Extrusion technology in the food- and feed-processing industry

The food and feed industry uses screw extrusion technology extensively (both single screw and intermeshing co-rotating twin screw extruders), in particular in two different fields: the cereal-processing industry and the oilseed-processing industry. To date, there has been no real technological synergy between these two industries from the equipment and process engineering standpoints. The cereal-processing industry uses extrusion technology in the second transformation of cereals, which consists of converting cereal flours by kneading, cooking, forming, and texturizing functions to produce both ready-to-eat (RTE) food products and functional ingredients. The cereal-processing industry, due to efficient technology transfer, has benefited considerably from the

(a)

(b)

Corn curl snacks
(made on single screw extruder)

Crispy flat bread and co-filled flat bread
(made on intermeshing co-rotating twin screw extruder)

Figure 1.3 The former extrusion-cooked food products. Source: Reproduced with permission of Clextral, France. For color detail, please see color plate section.

polymer-processing industry in applying extrusion technology to its numerous processes.

The oilseed-processing industry instead uses extrusion technology in the initial transformation of oilseeds which consists of fractionating its main components: the seed hulls, the vegetable oil, and the proteins. This industry has developed specific designs of single screw extruders for solid–liquid separations.

1.1.4.1 Extrusion technology in the cereal-processing industry

Because of the importance of continuous large-scale processing in the pasta industry, it is worth mentioning the development of single screw extruders for continuous pasta pressing in the mid-1930s. The aim is to mix at low shear the ingredients (mainly semolina and water) and pump the resulting dough into a forming die to produce different types of pasta products – long and short pasta. Continuous pasta processing is referred to as cold extrusion (slightly over the ambient temperature), which is now widely used at the start of continuous pasta production lines. Cold extrusion is not covered in this book, as it is extensively described and discussed in many documents and books dealing with pasta processing.

Intermediate/high-shear, hot-extrusion processing appeared almost simultaneously in the cereal-processing industry, due to the remarkable technology transfer of GEP I applied to native cereal-based biopolymers. Harper (1981) presented a well-documented historical review of the deployment of extrusion technology in the cereal-processing industry. A review of the key steps follows.

General Mills Inc. (USA) was the first to introduce a single screw extruder in the processing line of RTE cereals in the late 1930s. The extruder was only used to shape a precooked cereal dough which was subsequently dried and flaked or puffed. In this case, the process functions of the extruder were quite similar to those of pasta pressing. The first expanded, extrusion-cooked product was made in 1936; it was the former corn-based snack product which was commercially exploited by Adams Corporation (USA) from 1946. Figure 1.3a shows the former extrusion-cooked food product (corn curl snack) made using single screw extrusion technology. In the 1950s came the development of dry, expanded extrusion-cooked pet foods. In the 1960s the development of expanded, extrusion-cooked RTE cereals should be mentioned, as well as the introduction of textured vegetable protein (extrusion-cooked soy-based raw materials). All these applications used single screw extruders to continuously process cereal-based raw materials.

Intermeshing co-rotating twin screw extrusion technology was introduced in the late 1960s, spurred on by Clextral and by the industrial and commercial development of crispy flat bread in Europe, which was the result of a collaboration between the CTUC (food technical centre in France), the BSN Group (now the Danone Group), and Clextral. Figure 1.3b shows the former extrusion-cooked food product (crispy flat bread and co-filled flat bread) made using twin screw extrusion technology. From the early 1970s, secure in the potential of twin screw extrusion technology, Clextral successfully promoted intermeshing co-rotating twin screw extrusion technology worldwide to the major segments of the cereal-processing industry (snacks, RTE cereals, textured vegetable proteins, food ingredients such as precooked flours, encapsulated flavors, pet foods, etc.). Then, from the early 1980s, other manufacturers joined Clextral to supply twin screw extruders to the cereal-processing market. It must be noted that the former extrusion-cooked food products (corn curl snack and crispy flat bread) are still on the market today, and they lead their market segment.

Another user of screw extruders is the fish feed industry. In the 1970s, it adopted extrusion technology (single screw and intermeshing co-rotating twin screw extruders) to mix, cook, and texturize aquafeed pellets according to the various needs of fin and shell fish, making it possible to intensify aquaculture with high productivity, while preserving the health of both animals and environment. In this field, large extrusion units are used for producing the main supply of aquafeed pellets. As an example, Figure 1.4 shows a large intermeshing co-rotating twin screw extruder (200 mm screw diameter), able to process approximately 25 metric tonnes of feed mix per hour, which gives approximately 30 metric tonnes of finished aquafeed pellets per hour (after drying and fat coating, in the case of typical salmon feed, for example).

Nowadays, the cereal-processing industry is the second most important user of extrusion technology, owing to an efficient technology transfer of GEP I. Cereals are composed of natural plant polymers or biopolymers, mainly starch polymers (the fractions of amylose and amylopectin), which are neutral semi-crystalline macromolecules, and polymers combining natural amino acids (proteins), which are amphoteric macromolecules. When applying GEP I to native biopolymers with water as the plasticizer, thermomechanical melting of biopolymers occurs, which causes irreversible physicochemical modifications, making them easily digestible by human and animal digestive systems. This has given rise to a new method of continuous cooking of starchy and proteinaceous foodstuffs called high-temperature, short-time (HTST) extrusion cooking which favors mechanical action for the conversion of starches and proteins and consequently offers a unique alternative to traditional hydrothermal cooking. Extrusion cooking of starchy and proteinaceous materials is intended to give the resulting products a specific texture (on the macromolecular and macroscopic level) for a sensory and functional purpose (Guy, 2001; Mercier et al., 1989).

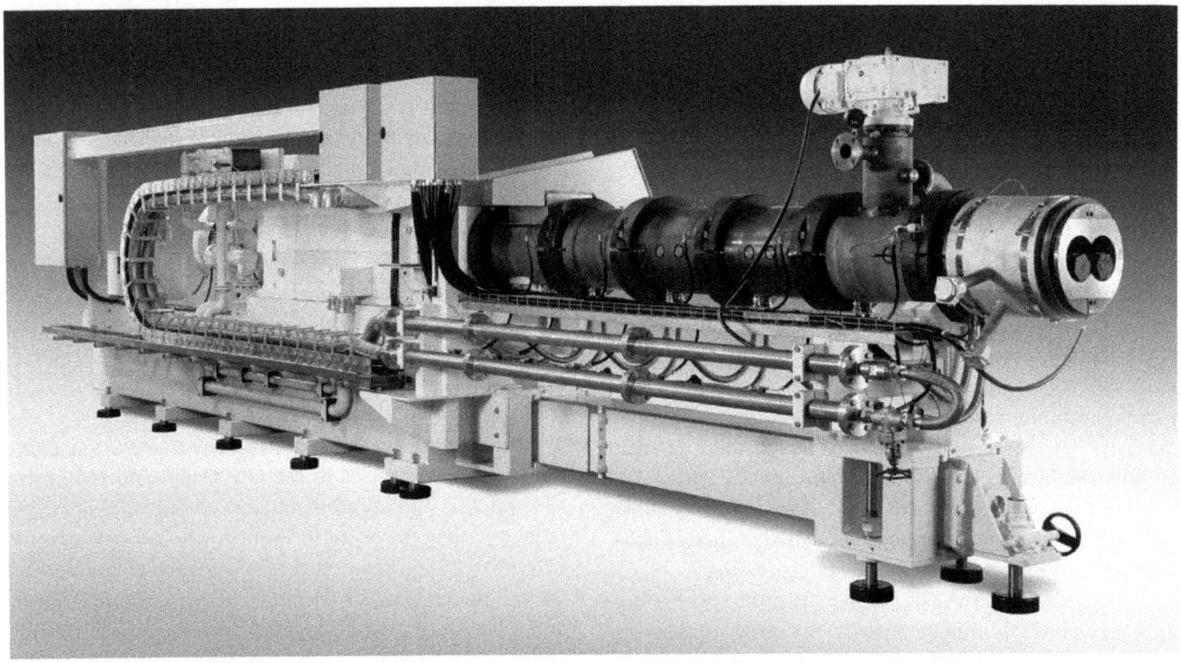

Figure 1.4 Large intermeshing co-rotating twin screw extruder (200 mm screw diameter) used in the aquafeed processing industry. Source: Reproduced with permission of Clextral, France.

The cereal-processing industry subsequently developed an innovative GEP III called "Thermomechanical Cooking and Food Product Texturization." It was immediately adopted by the food and feed industry for processing starch- and protein-based raw materials, resulting in crispy, textured human-food products (snacks, breakfast cereals, textured vegetable proteins, etc.) and textured animal food products (pet foods and aquafeed pellets in particular).

Both single screw and intermeshing co-rotating twin screw extruders are used. Single screw extrusion is generally used for simple formulations and easy-to-control product characteristics, while intermeshing co-rotating twin screw extrusion is better suited to mixtures with several ingredients and products which require greater flexibility and control of quality. Both single screw and twin screw extrusion technologies are now well established in their respective markets within the cereal-processing industry, on the basis of technical and economic criteria.

1.1.4.2 Extrusion technology in the oilseed-processing industry

The first transformation of oilseeds aims to fractionate their main components by use of a continuous process with the following unit operations in series: dehulling, grinding, thermal cooking, mechanical pressing, and solvent extraction (Laisney, 1984). Mechanical pressing is mainly required for high fat content oilseeds such as sunflower seeds, rapeseeds, cotton seeds, etc.

Mechanical pressing consists of separating part of the oil from the seeds. Originally made by using batch hydraulic presses, the move to continuous mechanical pressing was particularly challenging for the oilseed-processing industry. This became possible when Anderson (USA) created the first continuous single screw press in 1904. Thanks to the efforts of Anderson and a number of other suppliers, this became really popular in the oilseed industry following World War II. Continuous twin screw pressing of oilseeds has been investigated fairly recently, in the 1990s (Bouvier & Guyomard, 1996), but has not yet been industrialized. Though it has been applied in a different industrial field (the paper-milling industry), the first intermeshing co-rotating twin screw press was introduced by Clextral in late 1970s (see section 1.1.5).

The design of single screw presses is similar to that of single screw extruders. The main difference is in the barrel of the single screw press which has openings to collect the oil when it is expressed from the seeds through the pressing effect of the rotating screw. The proteinaceous meal is then extruded at the exit of the screw-barrel assembly.

Continuous single screw pressing has been deployed to other agro-food processing industries to eliminate liquid aqueous juices from various agro-materials such as sugar beet pulp, alfalfa, etc., thus increasing the dryness of those materials at high energy efficiency.

The introduction of continuous screw pressing for solid–liquid separation led to GEP IV, called "Thermomechanical Pretreatment and Solid–Liquid Separation." This process makes the most of the mechanical energy input to generate pressure along the screw and compress plugs of material from which the liquid exudes, while the screw design makes it possible to fine-tune both the pressure distribution and the free volume and to convey the plugs forward. As the barrel has filtering sections where pressurization occurs, continuous fractionation of the liquid is obtained.

GEP IV is most important for the processing industries that deal with biomaterials for food and non-food uses. In fact, it offers a real opportunity to potentially save energy in solid–liquid separation processing, by eliminating water in its liquid state and significantly reducing the amount of water when water-aided extraction is required, leading to environmentally friendly processing. Therefore, GEP IV is a worthwhile and relevant subject for presentation and analysis in this book.

1.1.4.3 An emerging, innovative generic extrusion process

Extrusion technology is still spreading through new processing industries. Engineers and researchers continue to come up with new ideas and invent new processes and products in all fields, both to improve the performances of existing processes and to satisfy new needs expressed by society.

In view of this, there is one innovative GEP that needs to be mentioned. It is presently in the development phase and should reach industrial maturity in the coming years. This new process concerns the field of instant porous powders produced from liquid or from liquid–solid concentrates for food and non-food uses. Instant powder production technology must be able to handle the dehydration of the starting material and the texturization of the resulting powder. The most widespread technologies and processes used are spray drying, drum drying, and freeze drying. One characteristic of these technologies and processes is their generally high investment and operating costs (notably energy), sometimes prohibitively so. Because of its capability to deal with viscous media, to texturize and form mono- and multiphase mixtures, intermeshing co-rotating twin screw extrusion technology is a valuable aid in handling instant powder products and "porosifying" (creating porosity), more cost-effectively than traditional processes and technologies.

The advantages and potential of twin screw extrusion in processing and functionalizing porous powders have led Clextral and its partners to take a hands-on interest in the field in recent years (Bouvier et al., 2006). The current R&D investigations have lead to GEP V, called "Thermophysical Micromixing and Material Porosification." It is also called the "Extrusion-Porosification" process, by its inventors.

This process aims to use the mixing efficiency of intermeshing co-rotating twin screw extrusion technology to mix gases into viscous media, in order to produce aerated foams which lead to porous particles when later dried. This process was invented during consideration of the requirements of the dairy industry for improved performances of the spray-drying process (to reduce operating costs and increase process flexibility) in the production of powders from milk and dairy co-products. However, the dairy industry aside, the process is conceptually generic, as it can be applied to various types of powders (food and non-food) which require easy and quick rehydration for use. For this reason, the authors have dedicated a full chapter in this book to this emerging new extrusion process in the field of porous powder texturization.

1.1.5 Extrusion technology in the paper-milling industry

The paper-milling industry is the third most significant user of extrusion technology, owing to two consecutive technology transfers, one from the polymer-processing industry and the other from the food- and feed-processing industry. It adopted extrusion technology quite recently, in the late 1980s, to promote environmentally friendly continuous processing of lignocellulosic materials.

Actually, the adventure began at the end of the 1970s when process engineers and technologists attempted to use intermeshing co-rotating twin screw extrusion

Figure 1.5 Intermeshing co-rotating twin screw extruder (240 mm screw diameter) used in the paper-milling industry. Source: Reproduced with permission of Clextral, France.

for cellulose fractionating, to take advantage of the continuous nature of the extrusion process and minimize the consumption of utilities (water, energy, and chemicals). This represented a real technological breakthrough for the paper-milling industry, and was the result of the collaboration between Clextral and the Centre Technique du Papier (Grenoble, France), which played a decisive role in the success of this invention. At present, Clextral is the only twin screw extrusion equipment manufacturer in the world that is capable of handling this application. Figure 1.5 shows a Clextral intermeshing co-rotating twin screw extruder for the paper-milling industry.

The extrusion fractionating process includes several unit operations in two intermeshing co-rotating twin screw extruders in series, which use two generic extrusion processes presented earlier, GEP II and GEP IV, as follows.

- In the first extruder, mechanical defibering of lignocellulosics and chemical impregnation of the plant substrate are carried out. This unit exploits the principles of reactive extrusion (GEP II) in a solid–liquid medium to digest the lignin component and bleach the cellulose pulp, making it possible to significantly decrease the consumption of chemicals and reduce the total processing time.
- In the second extruder, washing and fiber refining of the cellulose pulp take place through continuous solid–liquid separation. This unit exploits the principles of mechanical expression (GEP IV), through the ability of the intermeshing co-rotating twin screw extruder to generate pressure build-up sections, thus creating a series of dynamic plugs which cause the separation of the liquid (liquor) from the solid fibrous substrate (cellulose). Mechanical expression allows very effective washing to be achieved with small amounts of water, which leads to an important saving of water, compared with traditional cellulose-washing processes.

Because of the technical and economic environment for cellulose fractionation, twin screw extrusion technology is used for speciality pulps from annual plant by-products such as cotton-based pulp for bank notes and security papers, flax- and hemp-based pulp for cigarette papers and filter papers, and cotton-based pulp for printing and writing paper, etc.

1.2 Factors driving the development of extrusion processing technology

Given the major applications that were developed and successfully industrialized in the 20th century, it appears that the adoption of extrusion technology in the processing industries was driven by the following three factors.

- Process productivity (continuous, flexible, versatile processing)
- Product innovation and functionality
- Environmentally friendly processing

1.2.1 Process productivity

Extrusion is a continuous process. This is a determinant characteristic in industrial processing, particularly when dealing with viscous media. Actually, early developments and industrialization of screw extruders were driven by continuous processing, such as pasta screw pressing, oilseed screw pressing, and polymer processing. And continuous extrusion processing has been always valuable when promoting extrusion-processed materials at an industrial level.

Extrusion is a flexible and versatile process, due to the modularity of the screw configuration and the wide variations of independent process variables, which make it possible to make extensive adjustments to the processing conditions (energy inputs, residence time, temperature profile, on-line addition of ingredients, etc.), depending upon process requirements and characteristics of the materials. It should be noted that screw extruders can handle a wide variety of media (various viscosities and rheological behaviors and both single- and multiphase mixtures), process functions (intense mixing, chemical reactions, cooking, physical separation, cooling, degassing, etc.) and product characteristics (forming, shaping, texturizing, etc.). Continuous, flexible, versatile processing is globally linked to process productivity. This has been particularly well investigated and exploited in the following fields: polymer processing, oilseed processing, feed processing, and cost-competitive manufacturing.

1.2.2 Product innovation and functionality

Extrusion technology has brought determinant value to processing industries in several areas by permitting material and product innovation, in particular through the following:

- physical and chemical fine-tuning of functional properties of the end-products, including polymer blending, modified starches, flavor encapsulation, etc.
- physical fine-tuning of end-use properties of the finished products such as structured polymer-based materials (such as multilayer films, reinforced composites, etc.), textured food and feed cereal-based products (crispy flat

bread, multigrain snacks, crispy-crunchy cereals, chewy pet treats, aquafeeds, etc.).

The growth, in both volume and value, of extrusion-processed materials and products (for food and non-food uses) in the business-to-business and the business-to-consumer markets indicates how defining the impulse of extrusion technology was in the dynamism of product innovation in various processing industries.

1.2.3 Environmentally friendly processing

Continuous extrusion is characterized by low processing volumes and relatively low residence times, compared to traditional batch processing. In other words, the extrusion processing volume is restricted in comparison with the processing volume of current agitated vessels. As a matter of fact, the rate of physical and chemical modifications that occur in extrusion processing is usually far higher (heat, mass and momentum transfers, reaction rates, for example) due in particular to significant mechanical energy inputs. This places extrusion processing technology in the process intensification class, leading to economies in energy, raw materials and consumables, a reduced equipment footprint and reduced investment costs compared to non-intensified processing, thus minimizing the impact of the processing industries on the environment.

The positive impact of extrusion processing technology on the environment became particularly noticeable from the 1980s, when promoting this technology in the aquafeed-processing and paper-milling industries. This factor will certainly become increasingly important over the coming years and decades, given the trend driven by sustainable development, which should increasingly encourage process intensification in processing industries.

1.3 The industrial and economic importance of extrusion processing technology

From the early designs of screw extruders in the latter years of the 19th century until now, important industrial activity and business have been generated in two main fields:

- *extrusion equipment engineering*; that is, companies which deal with equipment design and manufacturing
- *extrusion process engineering*; that is, processing industries which deal with product design and manufacturing.

Extrusion-related industries really started to develop significant business in the 1930s, when processing industries had to create the appropriate technologies for processing viscous polymer-based materials. Since then, screw extruders have made important contributions to the growth of those industries, which have continuously benefited from the engineering input of equipment manufacturers.

Nowadays, owing to the successful technology transfers that occurred over the 20th century, three main processing industries use extrusion technology in their production units:

- polymer and plastics industry
- food and feed industry
- paper milling industry.

Though there are no established statistics, the industrial impact of extrusion processing technology can be qualitatively assessed. There are several thousand screw extruders presently in operation, with approximate distribution as follows:

- two-thirds of screw extruders in the polymer-processing industry
- approximately 30% of screw extruders in the food- and feed-processing industry (which includes both the cereal-processing and oilseed-processing industry)
- and the balance (less than 5% of screw extruders) in the paper-milling industry and in miscellaneous processing industries such as the chemical (fine chemicals, minerals) and pharmaceutical industries, etc.

The industrial application of extrusion processing technology has generated no less than 100 applications. Some of the major applications used in processing industries include the following.

1.3.1 In the polymer and plastics industry

Plastic films for bags, buildings; plastic pipes for water, gas, drains; plastic tubing for automobiles, control cable housing, medical use; plastic insulated wire for the home, automobiles, appliances, electric power distribution; plastic-coated paper for film cartons, meat packaging, moisture barriers; plastic sheeting for formed products, lighting; plastic profiles for sliding and storm windows in the home; polymer alloys, masterbatches, fiber-reinforced composites for compression, injection; energy materials, plastics recycling and reprocessing, modified biopolymers (starches, proteins), and much more.

1.3.2 In the food and feed industry

3D snacks, snack pellets, 3D breakfast cereals, pellet-to-flakes cereals, co-filled cereals, crispy flat bread, textured

Table 1.1 Extrusion processing technology: synthesis.

Generic extrusion process	Major process functions	Screw extruders used	Relevant applications concerned
I	Thermomechanical polymer melting Mixing Material forming	SSE and TSE	Polymer compounding Plastics shaping/forming Masterbatches Fiber-reinforced composites
II	Thermomechanical micromixing Chemical reaction Cooling	TSE	Polymer blends Chemically modified biopolymers Hydrolysis of lignocellulosics and cellulose pulp bleaching
III	Thermomechanical plant biopolymer melting Mixing and cooking Product texturization/shaping	SSE and TSE	Snacks and breakfast cereals Crispy flat bread Pet foods and aquafeeds Textured proteins
IV	Thermomechanical pretreatment Solid–liquid separation	SSE and TSE	Oil separation from oilseeds Dewatering of agro-products
	+ Material refining	TSE	Cellulose pulp washing and refining
V	Thermophysical micromixing (gas–liquid) Cooling Product porosification	TSE	Instant porous powders (food and non-food products)

SSE, single screw extruder; TSE, intermeshing co-rotating twin screw extruder.

vegetable proteins, encapsulated flavors, casein-to-caseinates, dry pet foods, pet treats, aquafeeds, oilseed fractionation, sugar beet dewatering, alfalfa dewatering, etc.

1.3.3 In the paper milling industry

Cellulose pulping for speciality papers (bank notes, security papers, cigarette papers, filter papers, etc.), wood pulping for horticultural uses, etc.

Table 1.1 gives a synthesis of extrusion processing technology. For each of the generic extrusion processes I–V, the main process functions applied in the extruders are reviewed, together with the screw extruders used and the resulting major applications.

1.4 Contents and structure of the book

Industrial applications of extrusion technology have led to the establishment of a rich extrusion processing culture which has emerged progressively from continuous interaction between practitioners and scientists. This is particularly true where extrusion processing media are based on synthetic polymers. As this field has been a former user of extrusion technology, it has had the advantage of extensive scientific investigation and modeling, making it possible to predict process performances. Polymer extrusion processing applied to synthetic polymers has led to many relevant scientific and engineering developments, and the resulting information given in numerous books is particularly rich and well documented.

However, biomaterials extrusion processing (for food and non-food uses) is relatively new for most of the applications, and it suffers from a lack of written information. In addition, this field tends to be more segmented and application oriented. In biomaterials extrusion processing, an accumulation of long-term experience generally precedes the scientific developments, since it usually deals with complex raw materials whose characteristics are not sufficiently known. A multidisciplinary approach to extrusion processing must then be applied, for which the complete prediction of the process performances is not yet available.

The purpose of this book is to provide an engineering analysis of extrusion processing technology dedicated to biomaterials, for food and non-food uses. The content has been inspired both by existing applications of extrusion technology and by emerging applications that are mainly the result of sustainable development. The objective of this work is to broach the distance between long-term experience in extrusion and the science of extrusion, in order to create a link between the world of extrusion practitioners and that of extrusion scientists in the field of biomaterials. For this purpose, the generic extrusion process concept (as described in section 1.1.2) has been applied in order to present and discuss the extrusion processing culture and to organize the framework of the book content.

Apart from this introductory chapter, the book is composed of three main parts and nine chapters.

Part I: extrusion equipment and extrusion engineering

Chapter 2 reviews the basics of the design and configuration of extrusion equipment. It is dedicated mainly to single screw extruders and intermeshing co-rotating twin screw extruders, as well as to ancillary equipment which is relevant to extrusion processing.

Chapter 3 gives an engineering analysis of extrusion processing (fluid mechanics and transport phenomena) in close relation to the process functions which are presented and discussed in Part II through the generic extrusion processes. Of great interest to extrusion practitioners is the comparison between single screw and twin screw extrusion, which is included at the end of the chapter.

Part II: generic extrusion processing and industrial applications

Chapter 4 presents and analyzes GEP I, focusing on extrusion processing of bio-based plastics. Resulting applications are included, such as extrusion processing of biodegradable materials and fiber-reinforced composites.

Chapter 5 presents and analyzes GEP II, focusing on reactive extrusion applied to the functional modification of biomaterials for food and non-food uses.

Chapters 6 and 7 are concerned with GEP III, focusing on extrusion cooking of plant biopolymers and methods relating to quality analysis of extrusion-cooked food products. Relevant existing and emerging applications are included.

Chapter 8 deals with the presentation and analysis of GEP IV, focusing on existing applications of screw extrusion pressing. Relevant industrial applications are considered.

Part III: the prospects for extrusion processing technology

Chapter 9 presents and discusses the emerging GEP V, focusing on bio-based instant porous powders. Although this process has not yet been applied at an industrial level, its potential in terms of industrialization is discussed.

Chapter 10 presents the process intensification concept, and discusses the ability of extrusion technology to intensify specific process functions and, in doing this, to meet the requirements of sustainable development in the processing industries.

References

Bouvier JM, Guyomard P (1996) Method and installation for continuous extraction of a liquid contained in a raw material. EP 0914243.

Bouvier JM, Maller G, Scott M, Moffatt D (2006) Twin screw extrusion: texturization of porous powders. Congrès Scientifique International Clextral, 18–19 October, pp. 277–303.

Coignet F (1869) US Patent 93035.

Colombo R (1939) Italian Patent 370578.

Gray M (1879) British Patent 5056.

Guy R (2001) *Extrusion Cooking: technologies and applications*. Cambridge: Woodhead Publishing.

Harper JM (1981) *Extrusion of Foods*, vol.1. Boca Raton: CRC Press.

Laisney J (1984) *L'Huilerie Moderne*. Paris: Compagnie Française pour le Développement des Fibres Textiles.

Mercier C, Linko P, Harper JM (1989) *Extrusion Cooking*. St Paul: American Association of Cereal Chemists.

Sturges JD (1871) US Patent 114063.

Utracki LA (1995) History of commercial polymer alloys and blends (from a perspective of the patent literature). Polymer Engineering and Science 35(1): 2–17.

White JL (1990) *Twin Screw Extrusion: technology and principles*. Munich: Hanser Publishers.

2 Extrusion Equipment

It is worth describing the design of extrusion equipment in this chapter, particularly hardware components, which have a direct impact on extrusion processing performance. Single screw and intermeshing co-rotating twin screw extruders are considered exclusively, as they both support the generic extrusion processes which are presented and analyzed in the following chapters.

Screw extruders are composed of four main parts, which are generally supported by a frame (Figure 2.1).

- The kinematics (motor and gearbox, in particular), which delivers the mechanical power required by the extrusion process.
- The screw-barrel assembly, in which the material is converted.
- The die assembly, through which the converted material is shaped, formed, or textured, depending upon the targeted products.
- The central operating cabinet or "brain" of the equipment, which monitors equipment operation (control of independent variables, automation, equipment security, display, etc.).

Equipment parts are presented in this chapter, with emphasis on the hardware of the screw-barrel assembly, the design of which determines the quality of the product obtained in the extrusion process. Screw extruders are associated with ancillary equipment the role of which is to suitably preprocess raw materials upstream, or postprocess products downstream; sometimes they manage on-line the addition or removal of process-related components. Thus, most important ancillary equipments required for traditional extrusion processing are also presented.

The aim of this chapter is not to provide a complete and exhaustive mechanical engineering background on the hardware of extruders and ancillary equipment, but rather to point out those engineering characteristics which are important for successful extrusion processing. Readers who are interested in a detailed description of the hardware of extruders should refer to the excellent and well-documented books published by Martelli (1983) and Rauwendaal (2001).

2.1 Extruders

Figure 2.1 shows a schematic view of an extruder and its major parts, all supported by a frame. This can be either a fixed frame or a ready-to-open frame which allows quick, easy access to the screw configuration for cleaning, checks, and maintenance, in particular. The frame opening can be handled either manually or with electrical assistance. Ready-to-open frames are offered by a few suppliers of intermeshing co-rotating twin screw extruders, for the food and feed industry, as well as for the paper-milling industry, as this design does offer recognized operating and economical added value through improved process productivity and flexibility.

The characteristics and functions of extruder parts with respect to their major functions in extrusion processing are now described.

2.1.1 The kinematics of extruders

The kinematics of extruders is composed of a motor drive system and a gear reducer, the designs of which determine the torque-speed domain covered by the extrusion equipment. The torque-speed domain is fundamental to defining the extruder's performance in terms of process productivity and product conversion.

2.1.1.1 The gear reducer

The gear reducer transmits the power supplied by the motor to the shank of the screw, the screw shaft for single screw extruders, or the two screw shafts for intermeshing co-rotating twin screw extruders. The gear reducer also

Extrusion Processing Technology: Food and Non-Food Biomaterials, First Edition. Jean-Marie Bouvier and Osvaldo H. Campanella.

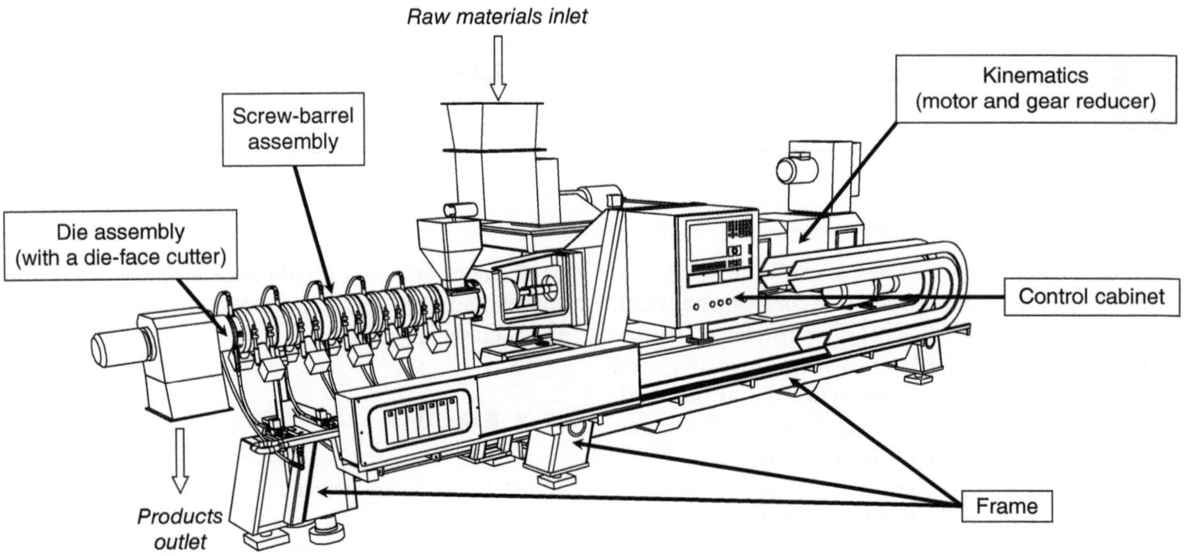

Figure 2.1 Schematic view of an extruder with the major parts. Source: Reproduced with permission of Clextral, France.

has to match the relatively low speed of the screw(s) to the high speed of the motor drive. The mechanical design of the gear reducers defines the torque-speed domain of the extruder and therefore, the maximum torque and screw speed available for extrusion processing. It should be noted that the efficiency of gear reducers is high, generally greater than 95%.

The design of a twin screw extruder gear reducer is more complex, as it must symmetrically divide the torque between each screw. It is composed of two stages: the first stage reduces the speed of the motor drive (1200 to 2000 rpm) to the final speed of the screws (generally 100 to 1200 rpm for modern intermeshing co-rotating twin screw extruders), while the second stage distributes torque between the two screw shafts. The design and manufacturing of high-performance gear reducers require a high level of expertise due to the tightness of the interaxis of the exit shafts.

Manufacturers have tremendously improved the design and performance of gear reducers over the past 30 years. This is clearly illustrated by the development of the standard maximum torque density-speed characteristics of intermeshing co-rotating twin screw extruders, which progressed from about 2.5 N.m/cm^3-250 rpm in the 1970s, to 12 N.m/cm^3-1200 rpm in the 1990s. This allowed end-users to significantly improve the competitiveness of extrusion processes through process productivity and product functionalities. Over the same period of time, gear reducers have achieved a high level of reliability which today ensures long lifetimes, over 50,000 hours for most extrusion applications, provided end-users rigorously adhere to the appropriate maintenance and lubrication procedures.

2.1.1.2 The motor drive

The motor drive system allows the torque and screw speed to be fine-tuned according to process requirements, i.e.:

- it maintains a constant screw speed
- it varies the screw speed over a relatively wide range
- it operates at constant torque over the range of screw speed.

Two main motor drive systems are available: DC and AC. Currently, the DC motor drive system is the one most commonly used for both single screw and intermeshing co-rotating twin screw extruders, as it covers a very wide range of screw speed from zero to the nominal speed, and accurately controls the exit torque which is strictly proportional to the intensity of the current delivered to the motor. This system is simple and cost-effective. The main drawback to using DC motors is maintenance of the brushes and commutator.

The AC motor, together with the frequency drive, is increasingly used, as its performance has significantly improved in recent years. It is reliable and, unlike the DC motor, it requires very low maintenance.

Though it is rarely used on regular extruders, it is worth mentioning the hydraulic drive system. It generally consists of a constant-speed AC motor driving a hydraulic pump which drives a hydraulic motor. This system can be controlled quite well, and has a satisfactory level of efficiency and reliability. It is sometimes used in critical environments (explosive environments, for example), or in situations where a very compact extrusion processing unit is required, as the pump assembly can be physically separated from the extruder.

2.1.1.3 Process operation

In practice, most single screw and intermeshing twin screw extruders operate at constant torque over the range of screw speeds. Thus, torque-speed characteristics can be used to determine power-screw speed characteristics by using the well-known relationship $P = CM_dN$, where P is the power (expressed in W), M_d the screw torque (expressed in N.m), N the screw speed (expressed in revolutions per minute, or rpm), and C a constant ($C = 2\pi/60$).

If the torque is consistent with the screw speed, then the power is directly proportional to the screw speed. Thus, the maximum power of the drive is utilized when the motor is running at full speed. Mechanical power consumption is mainly determined by the screw configuration, process throughput, and rheological characteristics of the material in the screw-barrel assembly. Figure 2.2 demonstrates the evolution of the normalized mechanical power, *PN*, consumed by an intermeshing co-rotating twin screw extruder, as a function of the productivity index, *IP*. *IP* is the ratio of the volumetric throughput to the free volume of the screw-barrel assembly, for three different material viscosities assuming a Newtonian behavior. Note that *PN* and *IP* are independent of the size of the extruder. Figure 2.2 was based on calculated data obtained from LUDOVIC® computer software at constant screw configuration (Della Valle et al., 1993; Sciences Computers Consultants, 2013). As the calculation assumed Newtonian behavior of the material, the figure shows a linear relationship between *PN* and *IP*, the slope of which increases when material viscosity increases. Values of *PN* at $IP = 0$ correspond to the power consumed with zero throughput where the extruder barrel theoretically contains some material in the filled sections; in this case, *PN* at $IP = 0$ is as high as the material viscosity. Obviously, the capacity of the extruder is governed by screw speed with low-viscosity materials (in the case of feed applications, pet food and aquafeed extrusion processing, for example), while it is governed by torque with high-viscosity materials (in the case of thermoplastic polymer extrusion processing, for example).

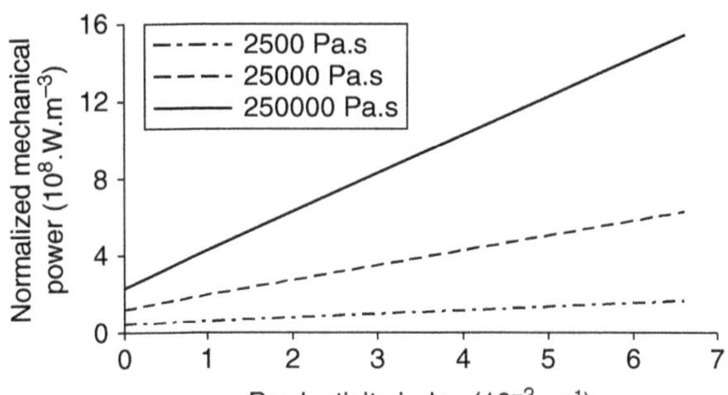

Figure 2.2 Normalized mechanical power as a function of productivity index (calculations made from LUDOVIC® software).

It is worth mentioning that a lot of consideration must be given to the kinematics as it has a significant effect on extrusion processing performance. Firstly, the torque-speed characteristics of the kinematics determine the productivity of the extrusion equipment. Secondly, the ability of the kinematics to consistently manage the screw speed, that is, maintain a constant screw speed in a stationary state and vary the screw speed over a relatively wide range, determines both the flexibility and stability of extrusion processing.

2.1.2 The screw-barrel assembly

Mechanical power is delivered to the screw-barrel assembly via a thrust bearing box which, in the case of single screw extruders, corresponds to the point where the screw shank, or the screw shaft, connects to the output shaft of the gear reducer; in the case of twin screw extruders, this is where the screw shafts connect to the output shafts of the gear reducer.

The screw-barrel assembly is the central part of an extruder. It receives the main stream of raw materials at the feed port and then it physically and chemically converts them using thermal and mechanical energy, according to product requirements. Its design requires special consideration as it contributes to process performance and the quality of targeted products.

There are two main categories of screw-barrel assembly.

- The "monobloc" screw-barrel assembly, which consists of a single-piece, solid screw surrounded by a cylindrical

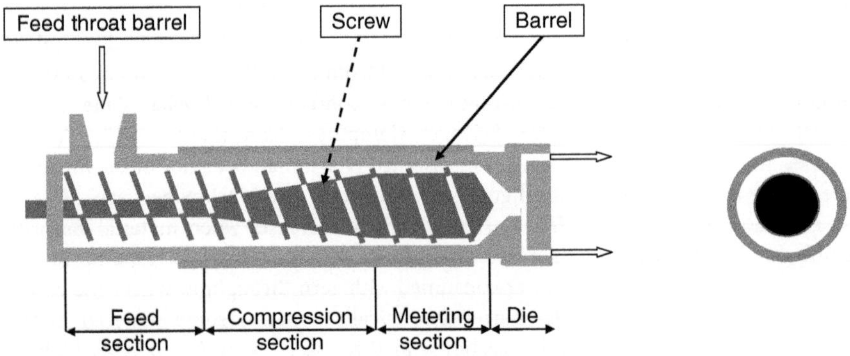

Figure 2.3 Schematic view of a one-stage 'monobloc' screw-barrel assembly with operating sections.

barrel. This is the simplest and most traditional design for the single screw extruder screw-barrel assembly.

- The "modular" screw-barrel assembly, which consists of segmented worm screw elements and a barrel. This design is found on advanced single screw extruders, and is used as standard on intermeshing co-rotating twin screw extruders.

2.1.2.1 The "monobloc" design

Figure 2.3 presents a simple version of a single-piece, single-flighted worm screw of a conventional plasticating single screw extruder. It shows three distinct geometrical sections with a channel having variable depth and constant pitch.

- The first section, called the feed section, has deep flights, in order to offer a high transportation capacity for solid and particulate raw material (solid powders and particulates). Its function is to convey the material down the screw.
- The following section, called the compression section, is where the material is compressed under the compressive effect of the screw (in a channel of decreasing depth). At the same time, the material is heated by interparticular friction and conductive heat transfer until melting occurs. In this section, the material changes from a solid particulate state into a melt state (viscous fluid). This is the longest section, and is essential for correct operation of the extruder, as it must be designed in such a way as to completely melt polymeric materials. The melting mechanism of polymeric materials is analyzed in Chapter 4.
- The third section, called the metering section, is where the material is ideally in a molten state. This is a pumping section in which the pressure needed to convey and feed the molten material through the die opening is built up. But it can also be used to fine-tune conversion of the

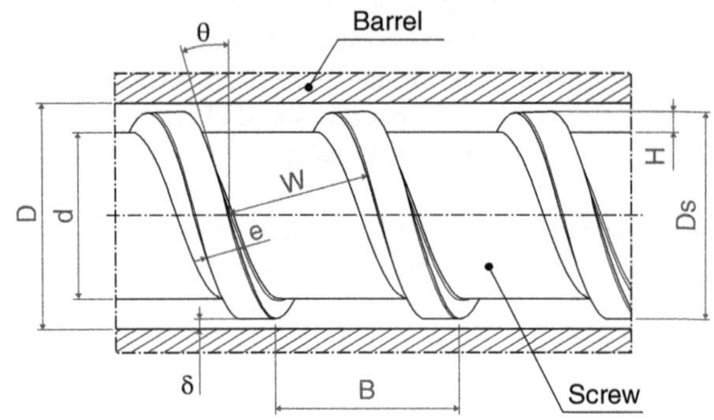

- Internal diameter of the barrel: D
- Screw root diameter: d
- Screw pitch: B
- Flight width: e
- Screw clearance: δ
- Actual screw diameter: $Ds = (D - 2\delta)$
- Actual height of screw flight: $H = (Ds - d)/2$
- Channel width: $W = B.\cos\theta - e$
- Length of channel in one complete turn of the helix: $Z = B/\sin\theta$
- Peripheral speed of the screw: $V = \pi.Ds.N \approx \pi.D.N$

Figure 2.4 Geometry of screw-barrel assembly of single screw extruder (monobloc design).

material. In fact, the metering section is characterized by shallow flights which generate high shearing conditions. The mechanical energy dissipated allows the material to be converted into a rheological state which is compatible with satisfactory processing and forming. The melt flow mechanism of molten polymers is analyzed in Chapter 3.

The monobloc screw-barrel assembly is traditionally defined by geometrical parameters, which are presented in Figure 2.4.

The screw pitch determines the helix angle θ, i.e. the angle which the flight makes with a plane normal to the axis of the screw. In fact, this angle θ varies with the distance to the screw axis but as the channel depth is much lower than the screw diameter, a unique value can be considered, such as: $tan\theta = B/\pi Ds \approx B/\pi D$. The helix angle, θ, is important as it determines the performance of the extruder screw, particularly the output. Several authors have modeled the material flow in the different sections of the screw. Indeed, the models allow the optimum helix angle to be determined, depending upon the type of screw section and the state of the material. In the solids conveying section (the feed section), where the material is in a quasi-solid state (particulates or powders), the solid flow depends on friction forces between the material, the barrel and the screw. It has been demonstrated that the output is proportional to the screw speed, *N*, and it depends on the geometry of the feed section (*D, d, H, e*) and friction coefficients. For a given extruder, assuming that friction coefficients between the material and screw and the material and barrel are equal, optimum output corresponds to a helix angle close to 20° for most polymeric materials (Tadmor & Klein, 1970, pp. 49–66). In the melt conveying section (the metering section), where the material is in a molten state, the melt flow depends on the rheological characteristics of the polymeric melt. For a Newtonian melt under isothermal conditions, it has been demonstrated that maximum output corresponds to a helix angle of 30° (Tadmor & Klein, 1970, pp. 221–223). Standard plasticating extruders are equipped with screws having a helix angle of 17.66°, which corresponds to the square pitch design almost universally adopted by extruder manufacturers.

The barrel is the cylinder that surrounds the screw. It has a feed throat in the feed section, where the material is introduced into the screw channel. The feed throat is generally water cooled, to prevent the temperature of the material from rising too soon, thereby preventing material from sticking to the barrel wall and causing restriction of the flow into the extruder. Thus, an effective cooling capacity and efficient thermal barrier between the feed throat section and the downstream barrel are important requirements for optimizing the output of the screw-barrel assembly in the feed section.

In the feed section, maximum output is determined by the friction of material particulates against the barrel wall. This is promoted by the grooved feed throat barrel. Grooves can be either straight or helical over the feed section. The depth of the grooves varies with the axial distance. Depth is greatest at the inlet and reduces to zero at the end of the feed section. The cylindrical barrel which surrounds the compression and metering sections is designed to ensure temperature regulation.

The standard monobloc screw-barrel assembly of a single screw extruder is designed with the following general characteristics.

- Total length: 15D to 30D ($15 < L/D < 30$; L is the total axial distance of the barrel)
- Length of the feed section: 4D to 8D
- Length of the metering section: 5D to 10D
- Single flighted screw
- Helix angle: 17.66°
- Channel depth ratio (channel depth in the feed section divided by channel depth in the metering section): 2 to 4
- Grooved feed throat barrel

2.1.2.2 The "modular" design

Figure 2.5 illustrates the principles of the modular design of a segmented screw-barrel assembly. It is composed of several screw elements in series, surrounded by several barrel modules in series. The screw elements are assembled on splined shafts (Figure 2.6). In fact, the splined shafts transmit the torque required to turn the screw, and it is important that they have sufficient mechanical strength to withstand the stresses generated by the conveying process in the extruder. Thus, the design of the splined shafts is of great importance, as this greatly determines the capacity and reliability of the screw extruders, particularly in the case of intermeshing co-rotating twin screw extruders.

The number and design characteristics of screw elements and barrel modules depend upon process requirements. For example, Figure 2.5 shows a typical modular configuration of a twin screw extruder used as a cooker extruder in the cereal-processing industry. This configuration has five independent barrel modules, including the feed throat barrel, and several screw elements in series, the design of which determines the process functions of the operation sections.

The screw-barrel assembly modular design exists for both single screw and intermeshing co-rotating twin screw extruders. It should be noted that modular single screw extruders are used mostly in the food and feed industry, whilst modular intermeshing co-rotating twin screw extruders are used in all industrial fields. These machines have removable screw and barrel elements. The screw profile can be modified by changing the sequence of the screw elements along the shaft; similarly, the sequence of barrel elements can be configured

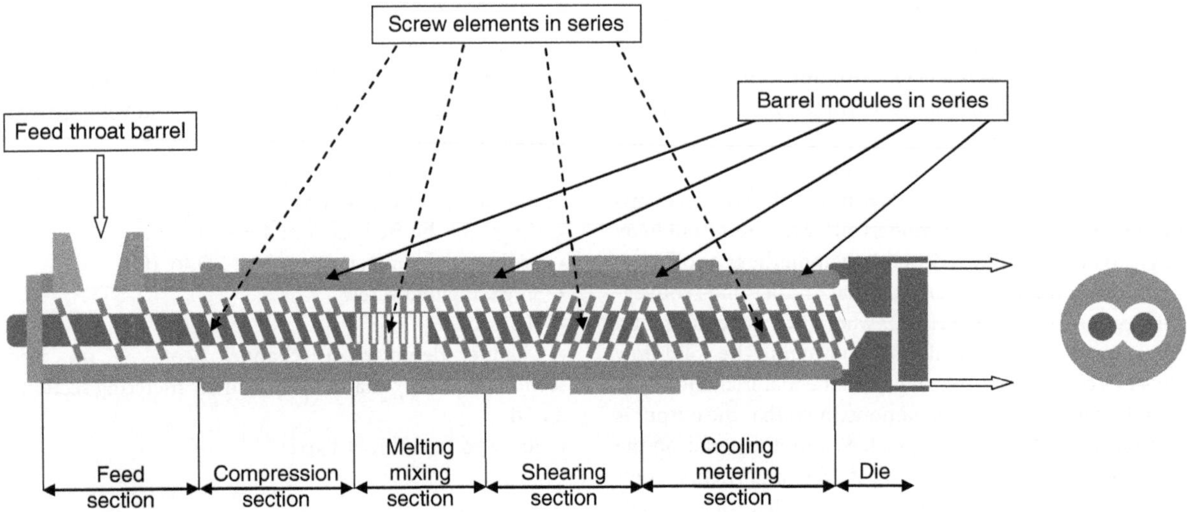

Figure 2.5 Schematic view of a 'modular' screw-barrel assembly with operating sections.

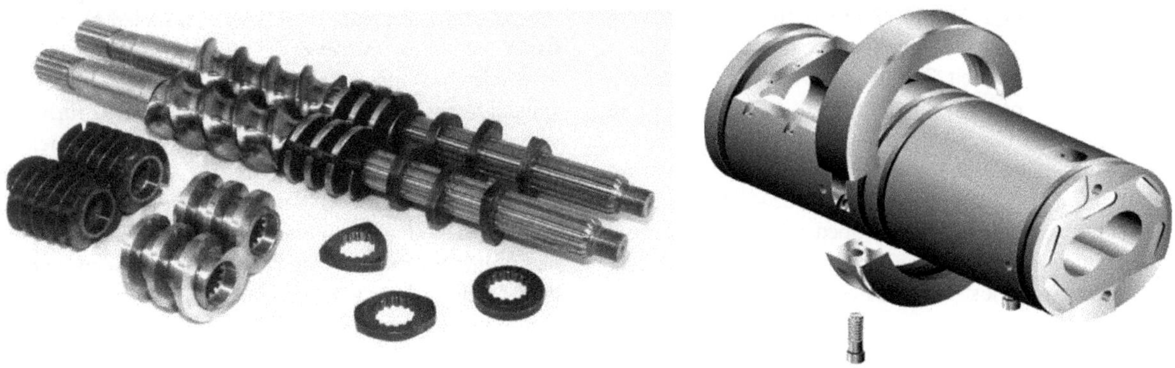

Figure 2.6 Splined shafts, modular screws and barrels of an intermeshing co-rotating twin screw extruder. Source: Reproduced with permission of Clextral, France.

according to the screw profile. Thus, the modular design offers a high level of flexibility and versatility in configuring the screw-barrel profile, particularly for those processes which require more than one process function in the same extruder.

As shown in Figure 2.5, the modular design of screw-barrel assembly has a wider range of operating sections than the monobloc design.

- A feed section with large pitch screw elements.
- A compression section, in which the screw elements are of decreasing pitch. In contrast to the monobloc screw, note that in the modular design, the root of the screw is constant in diameter, hence changing the pitch of worm flights to compress the material in this section.
- At the output of the compression section, the densified material enters a so-called processing section, which has various subsections in series where special dedicated screw elements determine the features according to the required goal: for example, steam locks for shearing in single screw extruders, counterthreads (with reverse pitches) for shearing, or mixing disks for intense mixing, in intermeshing co-rotating twin screw extruders. In general, the first processing subsection is dedicated to polymer melting, which is analyzed in Chapter 4. The following processing subsections deal with molten polymers. The intensity of shearing and mixing is then governed by the geometric characteristics of the screw elements. In the processing section, the barrel modules

are configured according to the screw profile, in order to consistently handle the temperature profile, any possible additive injections, venting, etc. The characteristics of screw elements and barrel modules will be presented and discussed in section 2.2.

- A metering section, with positive conveying screw elements to feed the processed material through the forming die. The melt flow mechanism in this section is analyzed in Chapter 3.

2.1.2.3 Specificities of the intermeshing co-rotating twin screw-barrel assembly

Twin screw extruders are generally classified according to two main criteria (Figure 2.7):

- *the screw rotation direction*: the screws are co-rotating if the two screws rotate in the same direction, whilst the screws are counter-rotating if the screws rotate in the opposite direction
- *the position of the screws in relation to one another*: intermeshing screws in which the flight of one screw penetrates to a greater or lesser degree the channel of the other (the center-to-center distance of the screws, also called interaxis, is significantly lower than the sum of the radii of the two screws); non-intermeshing screws in which the screws are tangent and do not penetrate each other's threads (the screw interaxis is larger than the sum of the radii of the two screws). The conveying process in non-intermeshing twin screw extruders is similar to that of a single screw extruder.

This paragraph is dedicated to the intermeshing co-rotating twin screw extruder. It is worth mentioning conjugated and non-conjugated screws. Conjugated screws have flights with the same shape and dimension of, and slightly fit the channels of the other screw with low clearance (Figure 2.8a). Non-conjugated screws are those in which the flight does not fit perfectly into the channel of the other screw and leaves material flows all around (case of trapezoidal flight; see Figure 2.8b).

The screw-barrel assembly of intermeshing co-rotating twin screw extruders is defined by the geometrical parameters presented in Figure 2.9.

The helix angle which the flight makes with a plane normal to the axis of the screw, called ϕ, is defined as

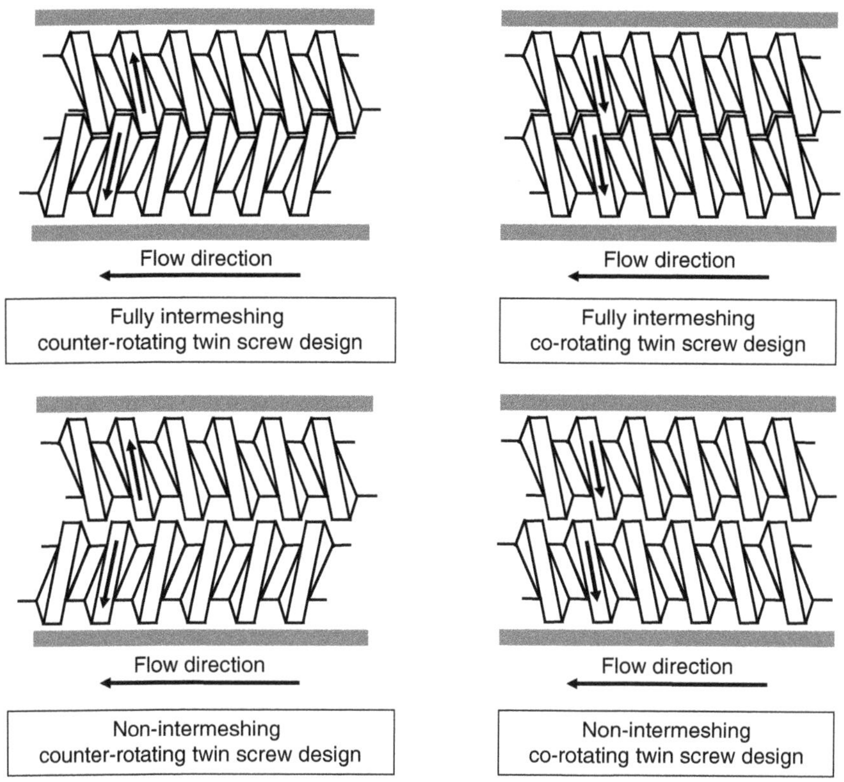

Figure 2.7 Basic designs of twin screw extruders.

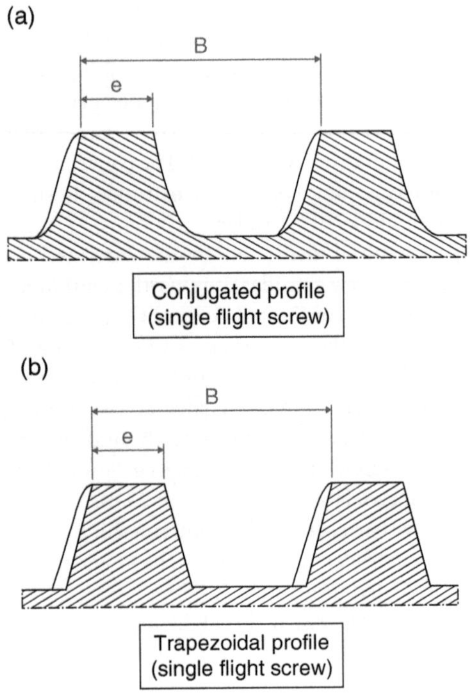

Figure 2.8 Schematic view of conjugated screws (a) and non-conjugated screws (b).

follows: $tan\phi = B/\pi Ds$. The helix angle determines the pitch of the screw. Unlike the single screw extruder, in which the pitch is relatively large and channel depth relatively low, in the twin screw extruder the pitch is much shorter and the channel depth much greater. Hence the ϕ angle is smaller; it varies from 15° in large pitch screws (in the feed section) to 6–8° in low pitch screws (in the pumping section).

Screw elements can be single flight or double flight (rarely more) according to the number of parallel channels along the screw (Figure 2.10). Single-flight screws have the largest channel cross-section and are useful when large conveying capacity is required, such as in the feed section. Double-flight screws have a smaller channel cross-section and are used when rigorous mixing and high output are required.

2.1.3 The die assembly

When examining all relevant industrial applications of extrusion processing technology, the profile of extrusion

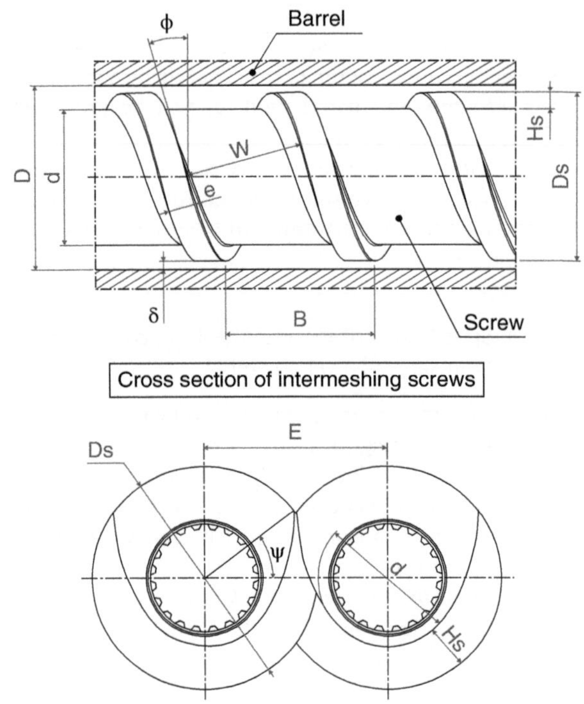

- Internal diameter of the barrel: D
- Screw root diameter: d
- Screw pitch: B
- Flight width: e
- Screw clearance: δ
- Actual screw diameter: Ds = (D − 2δ)
- Actual height of screw flight: H = (Ds − d)/2
- Channel width: W = [(B.cosϕ)/n − e]; where n is the number of flights
- Center-to-center distance of the screws: E = (D–H)

Figure 2.9 Geometry of screw-barrel assembly of intermeshing co-rotating twin screw extruder.

processes varies with respect to the use of dies. In fact, there are applications where no die is required (in the case of cellulose extrusion pulping). But applications where a die system is required are numerous. For most applications, the die system is rather sophisticated and adds significant value to the extrusion process (e.g. in polymer extrusion processing and cereal extrusion processing).

The aim of this section is to present and discuss the classification of die systems in extrusion processing. The main types of die/cutting systems will then be presented, particularly those which are relevant to biomaterials extrusion processing. For detailed information on

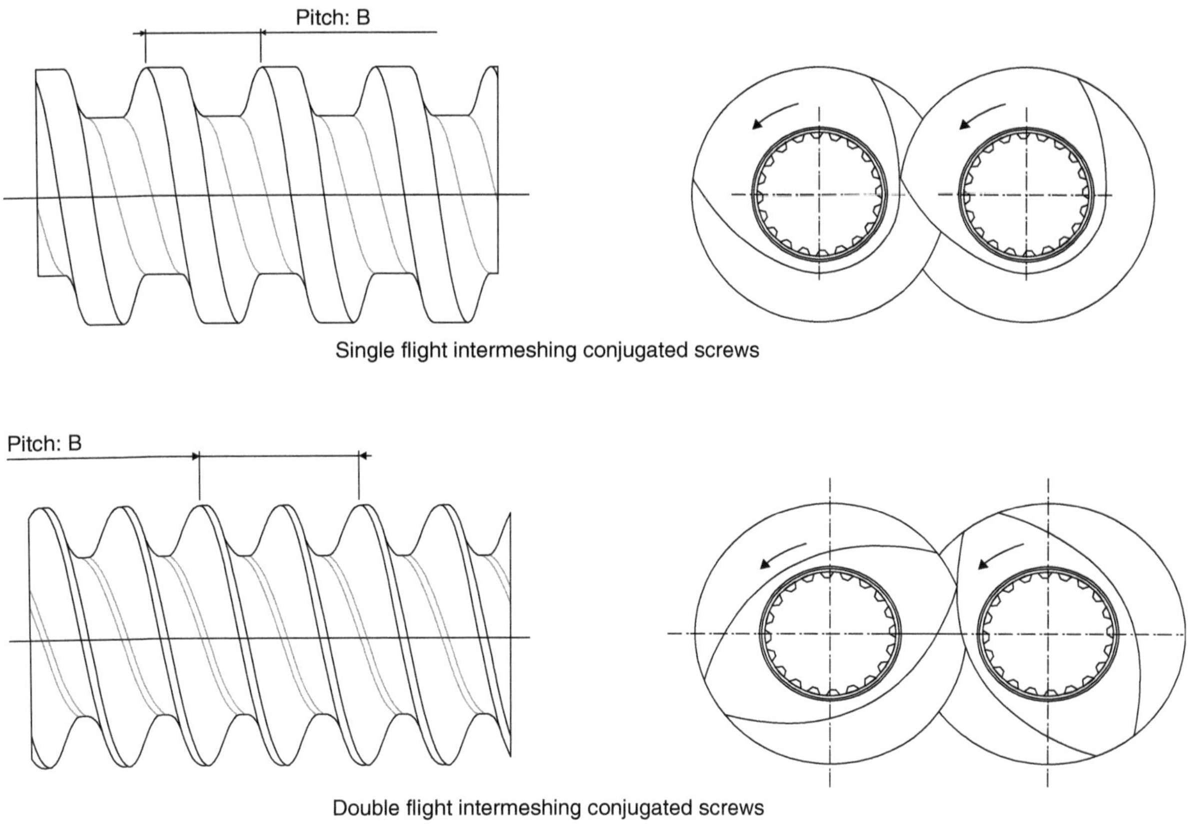

Figure 2.10 Schematic view of single- and double-flight intermeshing co-rotating screws. Source: Reproduced with permission of Clextral, France.

design relating to extrusion dies, readers should refer to Michaeli's book (2003).

2.1.3.1 Die system classification

Die system classification is shown in Table 2.1. This classification is based on the actual configuration of industrial extrusion processes. Four different categories of dies can be distinguished.

- *Category I.* No die is present ahead of the screw-barrel assembly. The cellulose extrusion pulping process is an example of this category, in which the screw-barrel assembly continuously delivers the converted material to the next unit operation, with no use of a die system.
- *Category II.* Flaking die ahead of the screw-barrel assembly. The die collects the converted material and produces flakes out of the extruder. A few reactive extrusion processes belong to this category, such as the casein-to-caseinate extrusion process. This is also the case for most continuous screw pressing processes (oilseed screw pressing, screw pressing for agricultural products such as sugar beet pulp, alfalfa, etc.). In this category, the die has low process and product added value as its role simply consists of flaking the converted material with no significant effect on process performance or physical and functional properties of the products.
- *Category III.* Forming die ahead of the screw-barrel assembly. The die produces shaped products to facilitate further product processing and/or to determine the properties of end-products. As mentioned in Table 2.1, this is the case for most extrusion processes for thermoplastic polymers and a few cereal extrusion processes. In this category, the die has high product added value, as it converts a shapeless polymeric melt into well-shaped plastic products, often with tight dimensional specifications and functional characteristics. The forming die can be combined with die-face cutting or downstream cutting, depending upon the application.

Table 2.1 Die system classification in extrusion processing.

Category of die system	Product/process added value	Type of cutting associated	Relevant applications concerned
Category I: No die	None (open discharge screw channel)	None	Cellulose extrusion pulping
Category II: Flaking die	Low	Die-face cutting	Casein-to-caseinate extrusion processing
		None	Oilseed screw pressing
		None	Screw pressing of agricultural products
Category III: Forming die	High	Die-face cutting	Extrusion compounding (thermoplastic raw materials; pellet-to-flake cereals; 1D snacks pellets; encapsulated flavors)
		Downstream cutting (strand pelletizing)	Extrusion compounding (plastics raw materials; pellet-to-flake cereals)
		Downstream cutting	Extrusion sheeting (plastics processing; crispy flat bread; 2D & 3D snacks pellets)
		Downstream cutting	Blown film extrusion (plastics processing)
		Downstream cutting	Piping/tubing extrusion (plastics processing)
		Downstream cutting	Co-extrusion (plastics processing; co-filled cereals; pet treats)
Category IV: Texturizing die	High	Die-face cutting	Directly expanded food products (in particular, snacks, breakfast cereals; textured vegetable proteins; dry pet foods; aquafeed pellets)
		Downstream cutting	High moisture protein fibration (meat analogs)

- *Category IV.* Texturizing die ahead of the screw-barrel assembly. The purpose of this die is to produce shaped and textured products. As presented in Table 2.1, this is the case for most cereal extrusion processes. In this category, the die has an additional feature in that it converts a shapeless biopolymeric melt into food products with well-defined shapes and textures. As in category III, the texturizing die can be combined with die-face cutting or downstream cutting, depending upon the application.

Categories III and IV deal with multicomponent polymeric melts derived from either synthetic polymers or plant polymers. Polymeric melts can be homogeneous, such as formulated mixes in a molten state, but they may also be multiphase mixes, such as polymeric melts mixed with solid fillers, or solid fibers. Polymeric melts demonstrate enormous complexity by exhibiting non-Newtonian flow behavior, with remarkable viscoelasticity dictated by their micro- and macrostructures. It should be noted that the rheological behavior of polymeric melts is not well described over a wide range of flow conditions, particularly in the case of bio-based polymers.

Nevertheless, the purpose of the die assembly is to distribute polymeric melts with a uniform velocity when they exit the die, in order to obtain appropriate product definition downstream. Product definition mainly results from the flow properties of the melt, the die geometry, the flow rate through the die, and the temperature field in the die. Lack of knowledge regarding the flow properties of polymeric melts, combined with the complexity of die geometry, means that it is often difficult to predict and optimize material flow in the die assembly. This is even more so when dealing with biomaterials where the lack of flow properties is more critical compared with synthetic polymers. Though there are useful prediction methods and software packages, it should be noted that die design is often based on experimental trial-and-error procedures, relying essentially on the designer's experience and the know-how of extrusion practitioners, particularly when processing biomaterials.

From a design standpoint, the die assembly is an extension of the screw-barrel assembly of the extruder, thanks to a suitable die plate which ensures the transition between both parts. From a process standpoint, the screw-barrel assembly and die assembly are fully interdependent. In general, the screw-barrel assembly physically and chemically converts the material into a shapeless melt with suitable rheological properties, and the die assembly forms the melt and processes its micro- and macrostructures according to the expected shape and texture of end-products.

The next two sections focus on die systems relevant to biomaterials extrusion processing.

2.1.3.2 Forming die systems

Two cases must be envisaged depending upon the cutting system associated with the die: either die-face cutting, where the die and cutter are physically connected, or downstream cutting where the die and cutter are separate.

For detailed technical descriptions of forming die systems, readers should refer to the handbooks edited by Hensen (1997) and Giles et al. (2005).

2.1.3.2.1 Forming die combined with die-face cutting

Extrusion compounding is the main example of this die/cutting configuration. Extrusion compounding is one of the oldest processes used in polymer extrusion processing (polymer reinforcing, polymer filling, masterbatches, polymer blends, etc.), as it is extensively used to produce various kinds of intermediate products, from low-grade to high-grade compounds.

Polymer compounding is usually the final step in manufacturing plastic raw materials for the plastics molding industry. A multihole die delivers a polymeric melt and a fast rotating knife, almost touching the die, cuts the melt into particles, uniform in shape and size, which makes them more suitable for further processing. Two main methods are used: water ring pelletizing and die air cooling pelletizing. They are both robust, reliable techniques that can be easily adapted to biomaterials compounding, owing to their specific rheological and physical properties.

The extrusion compounding process has been adapted to food extrusion processing, to produce pellet-to-flake cereals, 1D snack pellets and encapsulated flavors, in particular. For these examples, die air cooling pelletizing is required. Figure 2.11 shows a die-face cutting system which is currently used in 1D snacks pellet extrusion processing.

For detailed information on mixing and compounding polymers, readers should refer to Manas-Zloczower's book (2009).

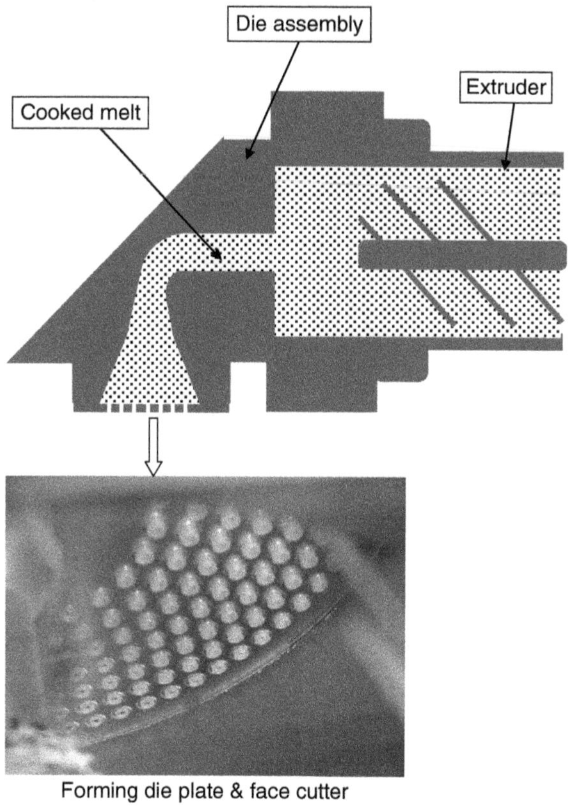

Figure 2.11 Die-face cutting system used in 1D snacks pellet extrusion processing. For color detail, please see color plate section.

2.1.3.2.2 Forming die combined with downstream cutting

There are many forming techniques in the polymer processing industry which use this die/cutting configuration to produce various formed plastic products, such as sheets, films, bags, coated wires, pipes, tubes, etc. The aim of this section is to present some of these techniques.

Technique I: extrusion compounding
It is worth mentioning again extrusion compounding through strand pelletizing where the die extrudes multiple strands which, after cooling in a 3–4 meter long water trough, are cut into cylindrical pellets, uniform in size (approximately 3 mm long and 3 mm in diameter). Strand extruding and pelletizing is a simple and straightforward process. Figure 2.12 shows a sketch of traditional strand pelletizing in polymer compounding (Automatik Plastics Machinery GmbH, 2010). This forming technique is also used quite extensively in some food extrusion

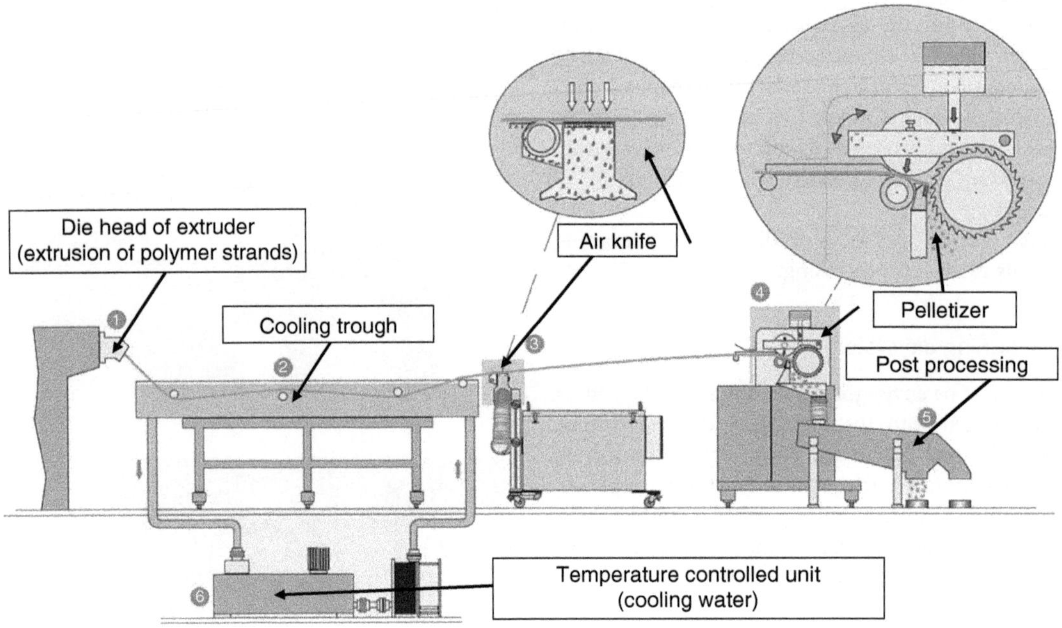

Figure 2.12 Strand pelletizing in polymer compounding. Source: Reproduced with permission of Automatik Plastics Machinery GmbH, Germany.

cooking processes such as pellet-to-flake cereal extrusion processing and pet treats extrusion processing.

Technique II: film and sheet extrusion

Film and sheet extrusion relies on continuous extrusion of a polymeric melt through a horizontal die. The extruded product is called a sheet if it is more than 0.25 mm thick and a film if it is equal to or less than 0.25 mm (ASTM definition). The die design is of critical importance to ensure that a product is manufactured with uniform thickness and exit velocity distributions over the entire width of the die exit.

The most widely used die designs are the T-type die, fish-tail die, and coat-hanger die. The T-type die is simple and economical but the distribution of the polymer melt is not very uniform. It is more suitable for low-viscosity polymers. The coat-hanger die (Figure 2.13) is commonly used in sheet extrusion, as it delivers a uniform flow rate across the die exit. The die cavity typically consists of a large manifold to distribute the melt across the die, and adjustable features such as a choker bar and flexible die lips to accommodate process variations by altering the melt flow within the die cavity, according to possible changes in temperature, material, and/or flow rate. Choker bar and flexible lip adjustments are based on several bolts which allow local deformations of the flow channel along the width of the die. Choker bar adjustment causes a change in the depth of the flow channel, and thus allows adjustment of the flow distribution in the die. Flexible lip adjustment allows local adjustment of the final land gap of the die exit and hence fine-tuning of product thickness.

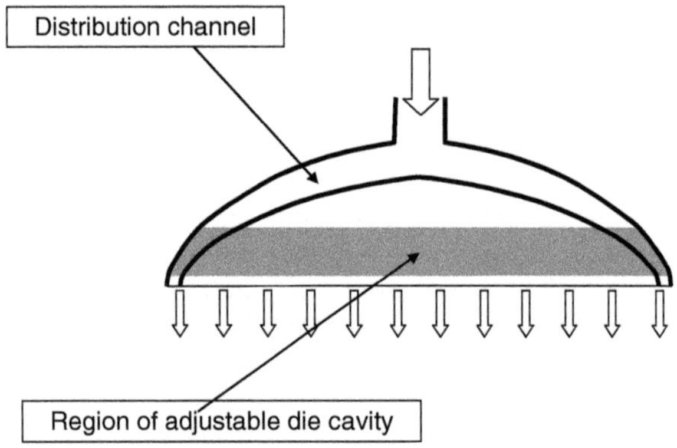

Figure 2.13 A coat-hanger die.

Suitable sheeting die design is of primary importance in order to operate high-capacity processing lines and

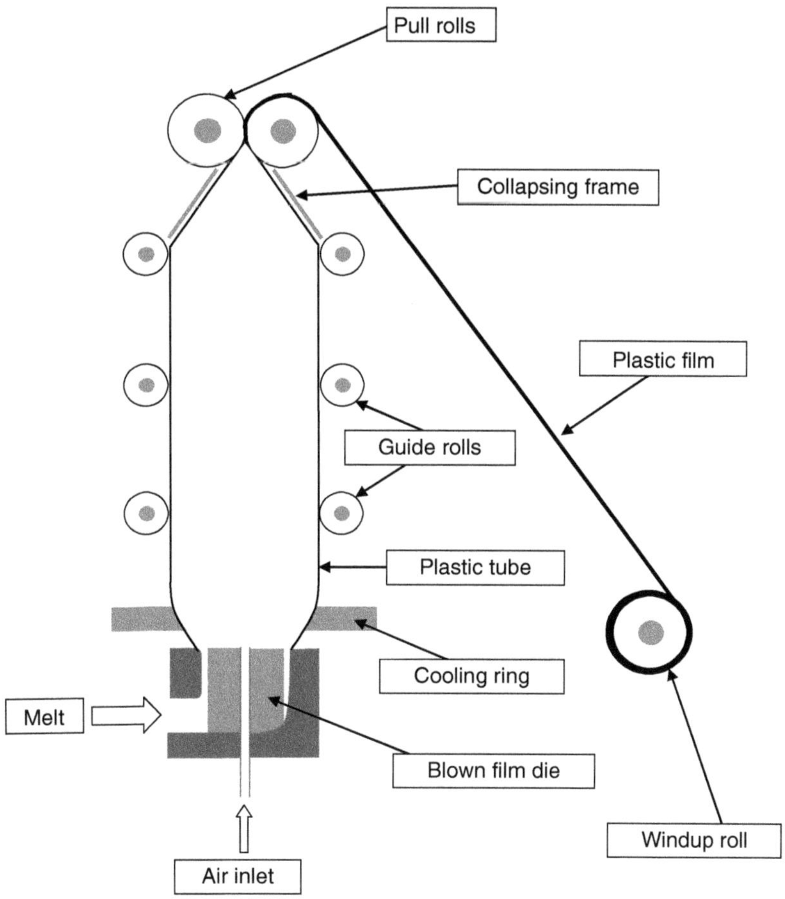

Figure 2.14 Blown film extrusion.

produce high-quality films or sheets. There have been many scientific and technical articles dealing with the design and operation of sheeting dies. Dies with uniform exit velocity distributions are well covered but far less consideration has been given to pressure drop. In addition, the ability of dies to accommodate variations in material properties has not been considered. For a detailed analysis of sheeting dies, readers should refer to Qi Wang's dissertation (2007) and the paper of Reid et al. (2003).

Film and sheet extrusion techniques are extensively used in plastics processing, as well as in two major applications relating to cereal extrusion processing: crispy flat bread and 2D and 3D snack pellets.

Technique III: blown film extrusion

Blown film extrusion is the most common method for making plastic films, particularly for the packaging industry. Figure 2.14 shows a typical process configuration for blown film extrusion. The extruder feeds the polymer melt into the blown film die, which extrudes it through a circular opening as a thin tube. Air is injected through a hole in the centre of the die, and the pressure causes the extruded melt to expand into a bubble; constant pressure is maintained to ensure uniform thickness of the film. The bubble is pulled upwards from the die and a cooling ring blows air onto the film, to reduce the temperature inside the bubble while maintaining its diameter. Once it is sufficiently cooled, the tube is then collapsed through nip rollers and flattened into a double-layer film. Blown film extrusion is highly flexible, as the same die can make films with different widths and sizes without significant trimmings.

The most common die used in blown film extrusion is the spiral mandrel die, in which the polymer melt is divided into a number of spiraling channels with the depth of the channels reducing in the direction of flow. The use of spiraling channels allows good distribution of the polymer melt exiting the die, resulting in films with satisfactory wall thickness uniformity. Thickness uniformity can be improved by making the die rotate.

For more detailed information on blown film extrusion, readers should refer to Cantor's book (2006).

Technique IV: piping and tubing extrusion

As a general rule, small-diameter products of less than about 10 mm are referred to as tubing and large-diameter products are referred to as pipe. Tubing is often extruded on cross-head dies while pipe is often extruded with in-line pipe dies.

In cross-head dies, the direction of the inlet flow is perpendicular to the outlet flow. As the polymer melt flows into the die, it splits over the core tube and recombines below. Thus, there is a weld line where the melts rejoin after flowing around the core tube. The length of the core tube is adjusted to enable the polymer to cure sufficiently after the point of weld line formation. The curing process is affected by time, temperature, and pressure. After the 90° turn in the die, the polymer melt flows through the annular flow channel where it adopts the shape of the final die land. A slight internal air pressure may be applied through the center of the core tube in order to maintain the internal diameter, or to prevent collapse of the tubing.

In in-line dies, the center line of the die is in line with the center line of the extruder. The core tube is supported by a number of spider legs, usually three. Although the spider legs are thin, there is a slight disruption of the velocity profile when polymer melts flow through. After the spider support, the polymer recombines, forming a weld line. Thus, the spider support should be far enough away from the die exit to enable the polymer to heal sufficiently.

Technique V: co-extrusion

Co-extrusion is a generic technical term meaning simultaneous extrusion of at least two streams through a single die. In the die, the streams are joined together so that they form distinct, well-linked parts resulting in a single extruded product. Co-extrusion can be applied in the previously described plastic forming extrusion techniques such as film, sheet, blown film, piping, and tubing extrusion. Each co-extruded stream is a polymer melt which gives the end-product specific properties. For example, multilayer packaging films result from the co-extrusion of two or more polymers in a film die to form a film with well-linked layers, thus offering effective moisture resistance, gas barrier properties, etc. thanks to a suitable combination of different polymers. The design of co-extrusion dies depends upon the differences in flow properties of the various melt streams. When the flow properties are similar, a simple, compact die can be used. When the differences in flow properties are great, multimanifold combining dies with thickness control are used, to minimize interface distortion.

Co-extrusion is also effective for food extrusion processing, to combine different textures and flavors in a single product, such as co-filled extruded foods where a soft flavored paste (chocolate-based, fruit-based) is included in a crispy cereal-based shell; pet treats (bone analogs, for example). Figure 2.15 shows a co-extrusion die assembly used in food extrusion processing.

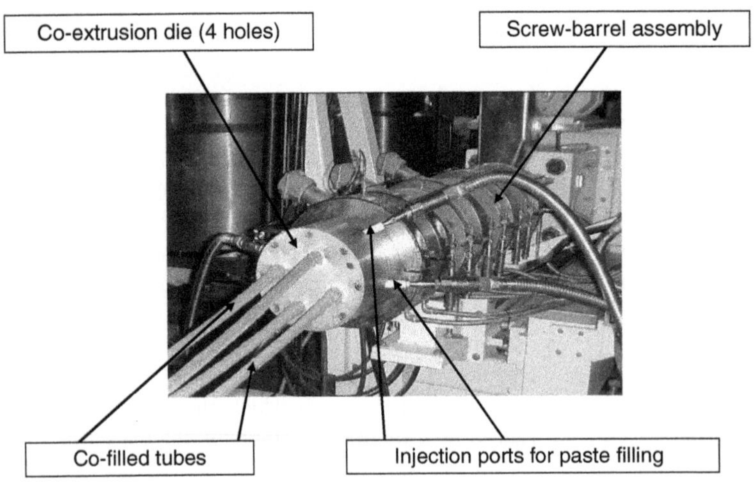

Figure 2.15 Four-exit co-extrusion die assembly for the production of co-filled extruded foods. Source: Reproduced with permission of Clextral, France.

2.1.3.3 Texturizing die systems

2.1.3.3.1 Texturizing die combined with die-face cutting

This configuration refers mainly to the directly expanded extrusion cooking process which is used extensively in cereal extrusion processing. Figure 2.16 shows a traditional die assembly combined with a face cutter, used in cereal extrusion cooking. In this process, the die collects the extrusion-cooked melt in a central feed die plate and feeds it evenly into a distributing die plate which has several holes or inserts. The number of holes or inserts may vary extensively, from less than 50 in the production of directly expanded snacks and cereals to several thousands in the production of directly expanded aquafeed pellets. The holes or inserts have different shapes and dimensions depending upon the specifications of the targeted products. Figure 2.17 shows the functional parts of a die assembly for the directly expanded extrusion cooking process. The central feed die plate and the distributing die plate are designed to ensure an even flow in each hole, and therefore a uniform linear velocity, while the design of the final hole or insert determines the final shape and dimensions of the extruded food product.

In this process, shaping and texturizing phenomena strongly interact, as product shaping is affected by a steam-induced expansion phenomenon which occurs at the die exit. The design of the hole or insert must take into account the extent of the expansion phenomenon, which is tricky when non-circular geometries are targeted (in the case of snacks and breakfast cereals). For example,

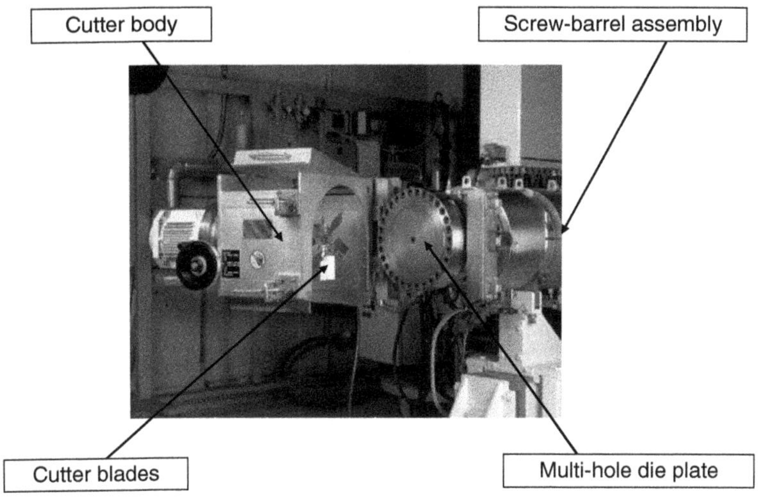

Figure 2.16 Die-face cutter assembly for the production of directly expanded extrusion-cooked foods. Source: Reproduced with permission of Clextral, France.

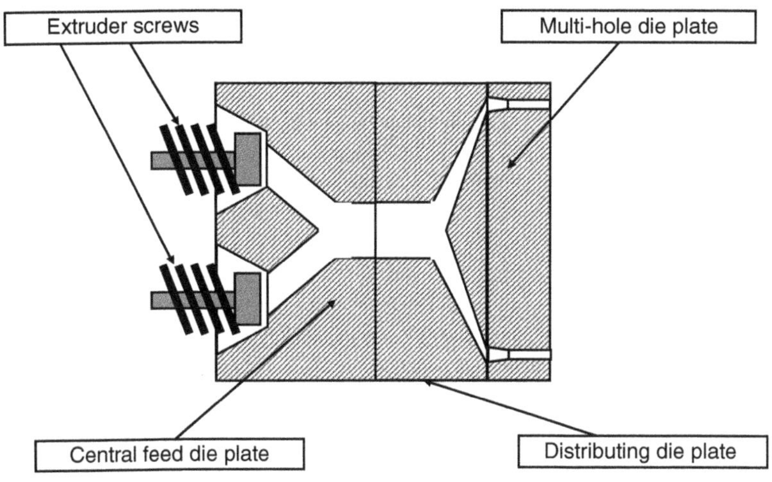

Figure 2.17 Die assembly for directly expanded extrusion-cooking process. Source: Reproduced with permission of Clextral, France.

a cross-shaped insert is used to produce a directly expanded square-shaped product.

Melt expansion at the die exit also generates a cell structure in the product matrix, the organization of which determines specific textural properties. Thus, understanding the expansion mechanism is crucial in the extrusion cooking process, in order to be able to define the right flow conditions in the die and ensure the expected product shaping and texturization. Die expansion mechanism is presented and discussed in Chapter 6.

2.1.3.3.2 Texturizing die combined with downstream cutting

One major example of this configuration is protein extrusion processing, i.e. the high moisture protein fibration process. In this process, the screw-barrel assembly produces a cooked protein-based melt and feeds it into a long, temperature-controlled cooling die, the purpose of which is to texturize the protein melt in order to mimic the texture of animals' muscles. Indeed, when cooling the melt in the die, the streamlines which are developed by the laminar flow tend to generate fibers when the liquid melt is converted into a gel. The high moisture protein fibration process is presented and discussed in Chapter 6.

2.1.4 The central operating cabinet

Though some screw extruders are still operated manually, most extrusion processing units are operated automatically. In general, the central operating cabinet is composed of the following items: the terminal operator, which is the interface between the extrusion processing machinery and the operator, and the programmable logic controller (PLC); the power cabinet, which supplies suitable electrical power to all parts of the extruder and its ancillary equipment, is also included.

The PLC and terminal operator allow:

- the extruder hardware to be controlled (the main drive, safety devices, screw speed, temperature regulation of the screw-barrel assembly, etc.), including ancillary equipment. Nowadays, extruder hardware control is closed-loop, PID-monitored (proportional + integral + derivative). The variable to be controlled is measured and this information is sent to a control unit from which a signal is sent to an actuator that adjusts the process so that the control variable is as close as possible to the setpoint (desired value)
- the development of different response variables to be checked (temperatures, die pressure, etc.)
- the creation and use of procedures for automatic start-up and shutdown of the extruder, to optimize process start-up and shutdown
- operating parameters to be displayed in real time.

The purpose of the PLC and terminal operator is also to control the extrusion process in relation to targeted products. As is the case in many processing systems, extrusion processing is governed by independent process variables (operating variables) such as flow rate, screw speed, barrel temperature, mix composition, etc. and the response of the extrusion processing system is determined by dependent process variables such as product temperature, product pressure at the die, specific mechanical energy, product properties (density, texture), etc. In fact, the operation of extrusion processing systems relies on the relationship between independent and dependent process variables. In practice, the operator focuses on those response variables which give the best representation of the targeted process and product, and on the manipulation of operating variables according to his or her experience. The operating variables can be adjusted without control, the process then operating as an open-loop control system. The problem with open-loop control systems is that they tend to drift with time and be too operator dependent. Thus, closed-loop systems are increasingly used to control extrusion processing, in order to ensure high process performance and product consistency. However, this requires the use of advanced and reliable process/product sensors which allow dependent process variables and product properties to be measured (ideally on-line). Undoubtedly, significant development in this area will occur over the next few years.

2.2 Extruder screw-barrel configurations

Generally, optimal screw-barrel assembly of screw extruders results from a decision procedure which takes into account three important criteria.

- Ability to operate at the highest possible throughputs, with acceptable product quality.
- Ability to combine several process functions in one extruder, to add value to extrusion processing and minimize the time-temperature history applied to polymeric materials.
- Ability to operate with minimum wear rate. In fact, as screw extrusion is used in such diverse areas as plastics

processing, chemistry, food and feed processing and paper milling, in relatively aggressive environments and conditions, the selection of materials for screw and barrel parts, as well as the way the screw-barrel assembly is configured, must be carefully considered.

The aim of the following section is to discuss in more detail different options relating to the aforementioned criteria.

2.2.1 Single screw extruders

2.2.1.1 Monobloc screw-barrel configurations

The simple standard version of the monobloc screw-barrel assembly (as described in section 2.1.2.1) is referred to as a one-stage processing configuration, where the material is melted (in the compression section) and the melt pumped into the die (in the metering section). Thus, all combined, the compression and metering sections constitute one single processing section, which is symbolized by the pressure build-up curve shown in Figure 2.18.

The one-stage processing configuration has a rather limited processing capability. In addition, the melt obtained in a screw-barrel configuration such as this is generally heterogeneous due to non-uniform melting in the compression section. This tends to restrict the use of the monobloc screw-barrel configuration when efficient laminar mixing is required. Thus, significant technical improvements have been made by extruder manufacturers to overcome these process limitations. For example, vented monobloc single screw extruders have been designed to continuously extract low molecular weight volatiles (monomers, oligomers, solvents, moisture, etc.) from the molten polymers. In this case, the barrel of the extruder is equipped with one opening (vent port) through which volatiles are extracted while the screw is designed so that there is no positive pressure in the polymer under the vent port, to prevent the polymer from escaping through the vent port. The resulting screw-barrel assembly leads to a two-stage processing configuration (Figure 2.19) which adds to the flexibility and versatility of monobloc single screw extruders. However, the processing capability of such extruders is still limited to low devolatilization yields (less than 5% volatiles). Extra vent ports can be added to increase the yield but the length of the screw, as well as the length of the L/D ratio of the extruder, can become a problem (difficult screw handling, screw deflection, etc.). When substantial amounts of volatiles need to be removed, the twin screw extruder is more competitive from a process and cost standpoint.

It is worth mentioning technical improvements aimed at generating mixing through the incorporation of

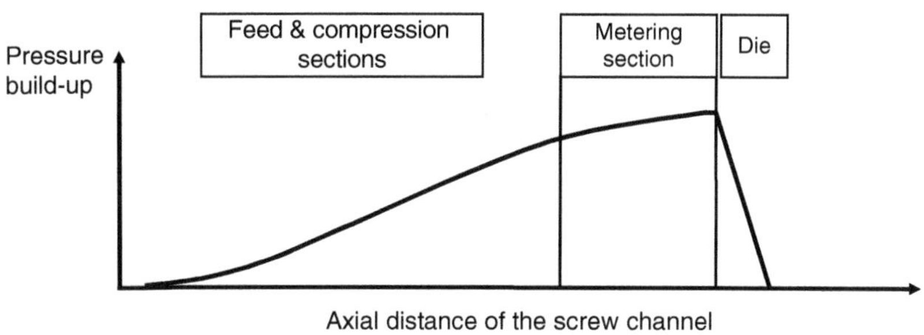

Figure 2.18 Simplified graph for pressure build-up in a single screw extruder.

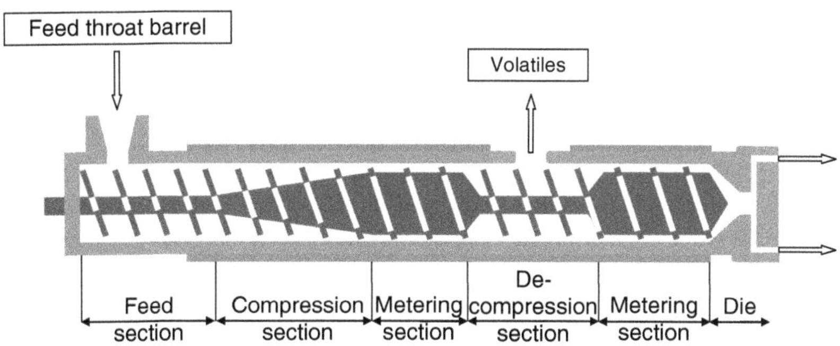

Figure 2.19 Schematic view of a two-stage 'monobloc' single screw-barrel assembly with operating sections.

specialized screw designs in the metering section, or at the front of the screw. Depending on the quality of mixing required, either distributive or dispersive mixing elements are commonly added. Distributive mixing concerns the blending of different polymers and the incorporation of additives such as fillers, dyes, chemicals, etc. in a polymer matrix, while dispersive mixing is required when agglomerates need to be broken down with substantial shear action. The diversity of mixing elements that have been used on extruder screws is huge; readers who are interested in a detailed description of mixing elements which can be used with monobloc single screw-barrel configurations should refer to Rauwendaal's book (2001; see pp. 491–525).

2.2.1.2 Examples of monobloc single screw extruders

There are many examples of single screw extruders equipped with a monobloc screw-barrel assembly. The single-stage processing configuration, described in section 2.1.2.1, is extensively used in the polymer-processing industry for basic extrusion operations such as feeding molding machines. In the previous section, a two-stage processing configuration was presented for polymer devolatilization.

Apart from these two important examples, inspired in particular by the polymer-processing industry, there are a few other monobloc single screw extruders which have been developed by the food processing industry. Below are three relevant examples.

- *Pasta single screw extruders.* Although cold forming extrusion is not included in this book, it is worth mentioning pasta extruders. They are characterized by a deep flight screw and smooth barrel, which make them ideal for processing, at a low shear rate, moistened semolina flour and pressing it through a die with little or no cooking.
- *Collet single screw extruders.* These are characterized by a very short screw-barrel configuration (L/D ≈3), shallow channel depth, multiple flight screw and grooved barrel, which enable them to operate with a very high shear rate (150–200 s^{-1}), high screw speed (300–450 rpm), low residence time (6–8 seconds), and high temperature (160–180°C). These extruders are designed to extrusion-cook low-moisture, cereal-based raw materials (12–15% moisture content); consequently, energy input comes only from the viscous dissipation of the mechanical energy. Collet extruders are mainly used to produce directly expanded snacks.
- *The turbo single screw extruder*, from Schaaf. This is characterized by a short screw-barrel configuration which delivers the cooked material into a turbine-assisted mixing chamber for intense shearing and mixing. The original configuration allows high temperature and short time processing to be applied, and uniform melt to be obtained. However, it must be noted that the efficiency of the extrusion process makes the extruder parts very susceptible to wear. The turbo extruder is mainly used to cook and texturize cereal-based directly expanded products (snacks and breakfast cereals, in particular).

2.2.1.3 Modular screw-barrel configurations

Despite technical improvements, monobloc screw configurations show real limitations: limited process functions, process flexibility and versatility, in particular. As mentioned in section 2.1.2.2, the introduction of modular screw configurations lead to significant technical improvements, which increase the process capability of related single screw extruders. In fact, owing to the possibility of incorporating a variety of screw elements (single- and double-flight screws, mixing lobes, shear locks of various diameters) and barrel modules (grooved and spiral-walled barrels) into the screw configuration, the modular design enables extruder users to locally vary the degree of shear and mixing. The modular design also allows the residence time to be adjusted by adapting accordingly the L/D ratio of the screw-barrel assembly, and sections to be allocated specific process functions (such as venting, mixing, cooking, cooling, etc.). This contributes to significant improvement of process flexibility and the capability of modular single screw extruders compared with those equipped with monobloc screw-barrel configurations.

The best example of the aforementioned modular single screw extruder design is that proposed by Wenger Manufacturing Co. (USA). The Wenger range of single screw extruders is particularly well designed and recognized for feed extrusion processing.

In this category of modular screw-barrel configurations, single screw presses must also be mentioned. In fact, single screw presses are equipped with a splined shaft upon which screw elements of various pitches and channel depths are assembled. The different screw elements are usually separated by straight or conical rings, to periodically squeeze the material. The barrel of such presses is also modular, as it has closed and filtering barrel modules to collect the expressed liquid product from the solid–liquid raw material.

Modular single screw extruders are still subject to process constraints due to the intrinsic design of the single screw concept which leads to poor mixing and heat transfer, non-uniform melting and shearing, and imprecise temperature control, in particular. In addition, the increase in process functions along the screw-barrel assembly tends to significantly decrease extruder capacity, which makes it less cost-effective.

2.2.2 Intermeshing co-rotating twin screw extruders

Twin screw extruders offer the most valuable process performances such as positive displacement type transfer, effective mixing and heat transfer, uniform melting, and effective temperature control, in particular. As described in section 2.1.2.2, the design of these machines is fully modular, as they have removable screw and barrel elements. This ensures a very high level of process flexibility when optimizing the screw-barrel configuration for a particular application.

Table 2.2 shows standard modular screw elements and kneading blocks available for intermeshing co-rotating screw extruders. In fact, twin screw extruder manufacturers can offer different screw element and kneading block designs in order to meet as accurately as possible the required time-shear history in each process section of the screw configuration. The table presents the main characteristics of these screw elements and kneading blocks, as well as their typical process functions in the screw configuration.

Table 2.3 focuses on the mixing effects of kneading blocks. In fact, depending upon the design of mixing disks (monolobe, bilobe, and thickness) and the way they are assembled to make up kneading blocks by adjusting the staggering angle, various degrees of shearing and mixing can be obtained which vary extensively in terms of intensity and efficiency, from distributive to dispersive mixing. For instance, the greater the thickness of the mixing disks making up the kneading blocks, the better the dispersive mixing and the lower the distributive mixing and vice versa. Kneading blocks derived from a neutral 90° staggering angle arrangement favor distributive mixing whilst dispersive mixing is enhanced by using kneading blocks derived from left-handed staggering angle configuration.

Thus, the different ways of selecting the design and length of kneading blocks and screw elements, and of configuring them together enable extrusion practitioners to adapt the screw profile with a large degree of flexibility according to process and product requirements.

In addition to standard screw and kneading/mixing element designs, twin screw manufacturers may also offer very specific designs, resulting from advanced process know-how. This is particularly the case for manufacturers who have a strong background in extrusion process engineering combined with long-term experience in widely applying extrusion processing technology. Innovative screw elements often allow extrusion users to optimize the functional properties of extruded products.

Table 2.4 presents standard, modular barrel elements for an intermeshing co-rotating twin screw extruder. Their process functions are also specified. It should be noted that the variety of barrel designs allows extrusion users to feed various types of materials (solids, liquids, gases) or remove volatiles, at different stages of the screw-barrel configuration. In addition to barrel module standard designs, twin screw manufacturers may offer specific designs for particular extrusion process applications. For example, in cellulose extrusion pulping, Clextral offers filtering barrel modules for solid–liquid separation.

The huge variety of screw and barrel elements allows the process designer to create an almost infinite number of configurations. By way of example, Figure 2.20 shows the typical screw-barrel configuration of an intermeshing co-rotating twin screw extruder for polymer compounding. This configuration has the following characteristics.

- Number of barrel elements: 10
- Number of screw elements: 41
- L/D ratio: 40
- Number of sections in series: seven, where melting and mixing sections are composed of kneading blocks combined with a short reverse screw element; the other sections are all set up with positive conveying screw elements with different pitches
- Number of processing stages: three (three-stage processing configuration)

The pressure profile down the barrel length, as shown in Figure 2.20, clearly indicates that the screw-barrel assembly is partially filled, as the pressure is close to zero over a large length of the extruder. In fact, pressure is developed only in sections equipped with restrictive screw elements (melting section, mixing section) and before the die. In general, the pressure is highest in the melting section, where the melt is colder and more viscous. In transport sections, there is no pressure development, hence low shear extent.

The pressure profile down the barrel length on intermeshing co-rotating twin screw extruders can be varied to allow different process functions (mixing, shearing, venting, partial cooling, on-line feeding of ingredients), which can be designed in series according to process purposes. Such a succession of processing zones in series provides a high level of process flexibility when compared

Table 2.2 Standard modular screw elements and kneading blocks for an intermeshing co-rotating twin screw extruder.

Type of screw elements or kneading blocks	Main characteristics and process functions
Conjugated, double flight, right-handed pitch 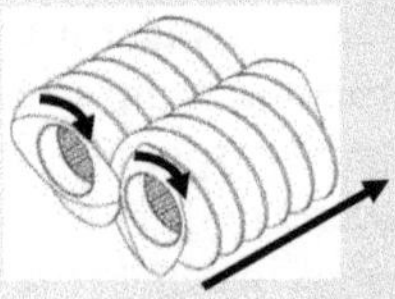	• Various pitches and lengths. Low shearing. High free volume • Used in feed sections (primary and secondary) and transport sections (such as degassing and cooling)
Conjugated, single flight, right-handed pitch  Trapezoidal, single flight, right-handed pitch 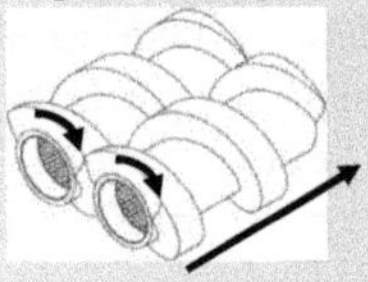	• Various pitches and lengths. Low shearing. High pumping capacity • Used in compression, transport (pressure build-up) and metering section (discharge pressure build-up)
Conjugated, double flight, left-handed pitch 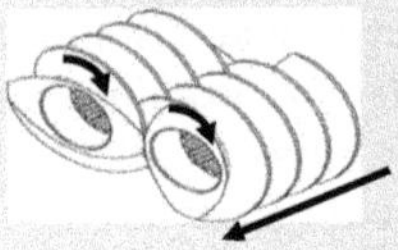	• Various pitches and lengths. Intense shearing. High material retention • Used in melting and shearing sections. Efficient to create a seal before a degassing section
Conjugated, single flight, left-handed pitch 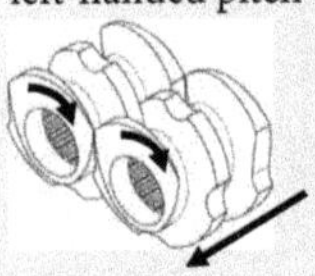 Conjugated, double flight, left-handed pitch 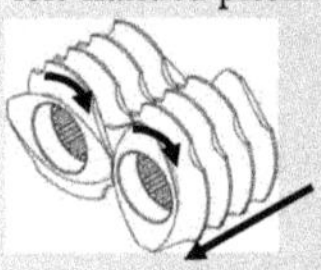	• Various pitches and lengths. Peripheral openings. Intense shearing. High material retention • Used in melting and shearing sections (such as cooking, defibering, pressing, etc.)
Monolobe kneading disks 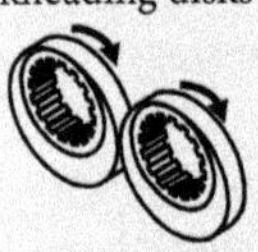 Monolobe kneading blocks, right-handed 	• Intense shearing. High mixing capacity. Intermediate material retention • Used in mixing and shearing sections (particularly with high viscosity materials) • Generate elongational flows to improve dispersive mixing
Bilobe kneading disks 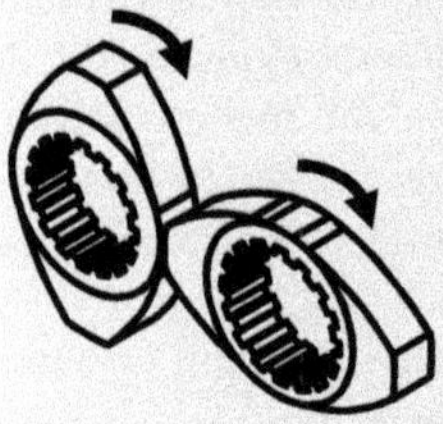 Bilobe kneading blocks, right-handed 45° staggering 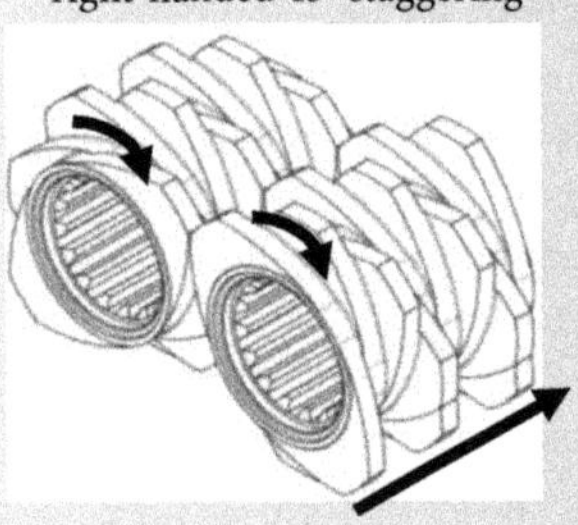	• Various disk thicknesses. Blocks with various lengths and staggering angles. Medium shearing and mixing. Intermediate material retention • Used in melting, mixing and shearing sections

Table 2.2 (continued)

Type of screw elements or kneading blocks	Main characteristics and process functions
Bilobe kneading blocks, neutral 90° staggering 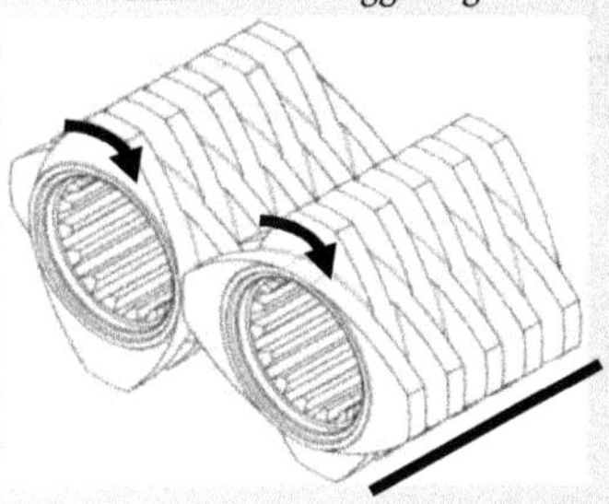	• Various disk thicknesses. Blocks with various lengths. Intense shearing and/or mixing. High material retention • Used in melting, mixing (distributive and dispersive mixing) and shearing sections (cooking, reactive extrusion, etc.)
Bilobe kneading blocks, left handed 45° staggering 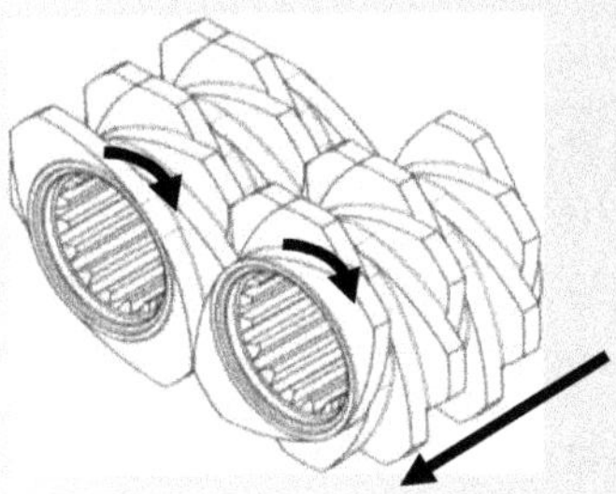	• Various disk thicknesses. Blocks with various lengths and staggering angles. Intense shearing and mixing. High material retention • Used in melting, mixing and shearing sections • Efficient to create a seal before a vacuum section

↘: Direction of screw rotation.
↗↙↗: Flow directions induced by screw elements or kneading blocks (positive, negative or neutral).
Source: Reproduced with permission of Clextral, France.

with single screw extrusion processing. Thus, multistage processing configuration (from three to five) designs on intermeshing co-rotating twin screw extruders used in various industrial applications such as polymer blending, reactive extrusion, food extrusion cooking, and cellulose extrusion pulping are very common.

The best example of the aforementioned design of modular intermeshing co-rotating twin screw extruders is that offered by the Clextral Company (France). Clextral's range of twin screw extruders is particularly suitable for technical and process specifications required for biomaterial extrusion processing, for food and non-food uses. In fact, Clextral extrusion technology benefits from long-term experience in this area, and is particularly well suited to the specific technical and process challenges relating to biomaterials.

2.2.3 Screw-barrel configuration and wear

Single screw and twin screw extrusion processing leads to substantial wear of screw-barrel assemblies under normal processing conditions, due to the fact that these technologies have screws rotating inside a fixed barrel.

In extrusion processing, wear is an important issue as it seriously affects the performance of extruders from an economical and technical standpoints. In fact, the operating cost of extrusion units is as high as the wear rate, due to downtime when replacing worn parts and the cost of replaced parts. Wear also increases the screw clearance, δ, which tends to reduce extruder throughput and modify product characteristics. This is particularly critical in single screw extrusion processing where increasing the screw clearance leads to incomplete melting, temperature non-uniformities,

Table 2.3 Trends of the mixing effect of kneading blocks in intermeshing co-rotating twin screw extruders.

Type of kneading block	Design profile	Distributive mixing	Dispersive mixing
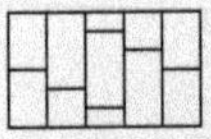	Bilobe kneading blocks, **right-handed staggering angle**	+	+ +
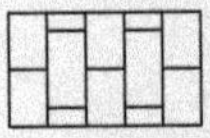	Bilobe kneading blocks, **neutral 90° staggering angle**	+ + + +	+ + +
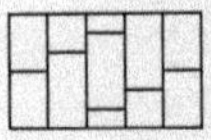	Bilobe kneading blocks, **left-handed staggering angle**	+ +	+ + + +
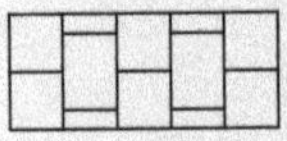	Bilobe kneading blocks, neutral 90° staggering angle, **large kneading disks**		
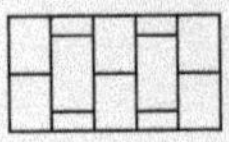	Bilobe kneading blocks, neutral 90° staggering angle, **medium kneading disks**	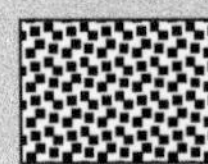	
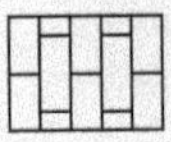	Bilobe kneading blocks, neutral 90° staggering angle, **short kneading disks**		

pressure fluctuations, poorer mixing and heat transfer, and reduced throughput. In twin screw extrusion processing, the effect of wear on extruder performance is less critical, as the consequences of screw clearance increase on processing can be compensated by screw speed adjustments over a large extent of processing conditions, in order to retain the nominal capacity of extruder and product characteristics.

Optimization and consistency of extruder performance are achieved by minimizing the rate of wear in extrusion processing, which requires an understanding of wear mechanisms.

2.2.3.1 Wear mechanisms

Basically, the rate of wear depends on three main factors:
- the composition of the mix to be processed; for example, the presence of minerals or aggressive chemicals in the mix composition tends to increase the rate of wear
- the material properties of the screw and barrel elements; these are defined by the metallurgical composition, and by the physical and mechanical characteristics of the metal alloys
- the screw configuration; for example, the compression ratio of the screw in single screw extrusion and the design and L/D ratio of processing stages in twin screw extrusion may all significantly affect the rate of wear.

To reduce the rate of wear of screw-barrel configurations, it is paramount to understand wear mechanisms which occur in extrusion processing. Five mechanisms of wear can be distinguished.

2.2.3.1.1 Abrasive wear

Abrasive wear occurs through deformation, microcutting and/or lifting material off the surface of a mechanical part through friction under the action of hard particles. This type of wear is revealed by scratches, local polishing and the tearing off of chips, with some modifications to the surface layers of the materials. In the case of three-body abrasion with hard particles such as extrusion processing, one solution is to increase the hardness of the surfaces (of screw and barrel elements) to be protected, the third body being the particles generated by metal wear or already present in the material to be processed (minerals, fillers, for example). Particle size and form, the angle of impact, and displacement speed of the extruded load in relation to the surface are parameters which influence the type of wear. Abrasive wear usually occurs in the feed and compression sections of the screw-barrel assembly where there is a pressure build-up while the polymer material is not completely melted. It should be noted that strain hardening or

Table 2.4 Standard modular barrel elements for an intermeshing co-rotating twin screw extruder.

Type of barrel module	Process function
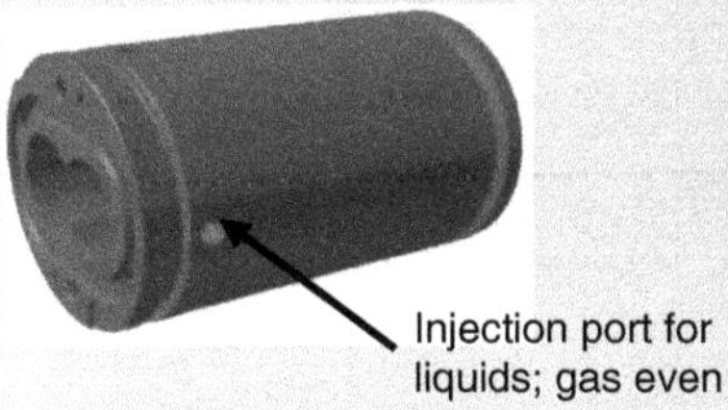	**CLOSED BARREL MODULE** **Used in all sections when no solids feeding, nor products removal, is required. It allows liduids, gas even, to be added on-line.** **It is combined with all types of screw elements**
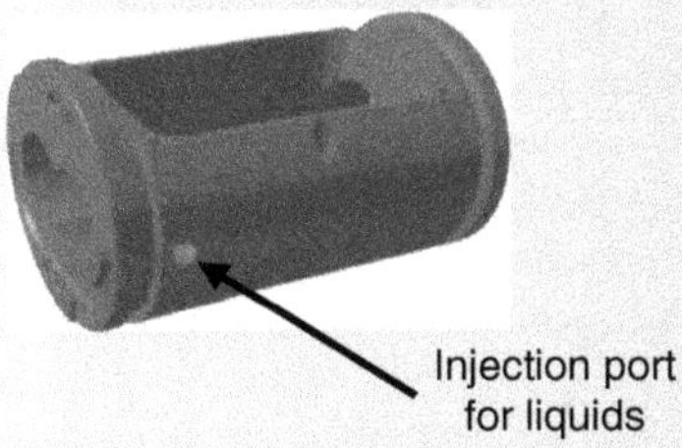	**FEEDING BARREL MODULE** **Used in primary and secondary feeding sections, to feed raw materials (powders, flours, granulates, etc.). It can be used for degassing. It allows liquids to be added on-line. It is usually combined with large forward pitch screw elements**
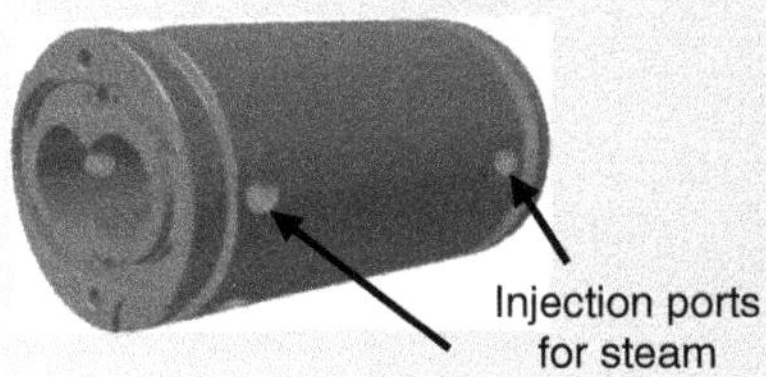	**STEAM INJECTION BARREL MODULE** **Used in sections where direct steam injection is needed (thermal energy input). It is usually combined with forward pitch screw elements**
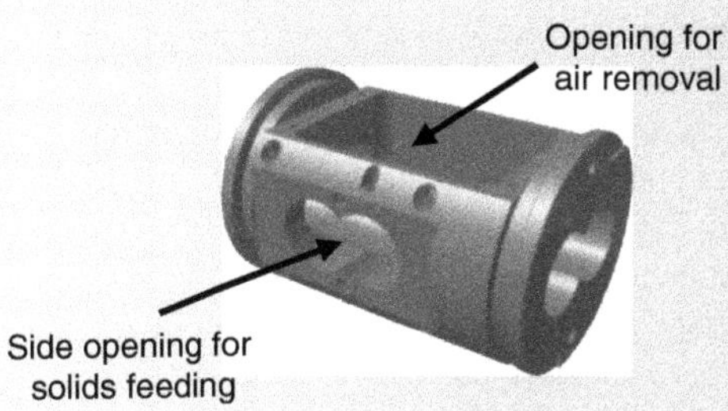	**SIDE FEEDING BARREL MODULE** **Used in side feeding sections to feed solids on-line (powders, flours, fibers, etc.). It is usually combined with relatively large forward pitch screw elements**

Source: Reproduced with permission of Clextral, France.

precipitation hardening does not improve abrasive wear resistance since the microcutting process already yields maximum local strain hardening.

2.2.3.1.2 Adhesive wear

Adhesive wear occurs with metal-to-metal contact under high stresses (e.g. screw tip to barrel, screw flight to screw flight, screw tip of one screw to channel bottom of the other screw). The surfaces of the screws and barrel parts after machining, even finishing, still have some roughness. When parts come into contact and forces are applied to those parts, local adhesion can form at points of contact, causing cold welding. Adhesive wear varies according to the different material combinations in sliding contact. In adhesive wear, degradation takes the form of strong plastic deformations with fissuring and the emission of very hard microscopic fragments which lead to an increase in abrasive wear. The rule is to prevent two similar or identical materials from coming into contact.

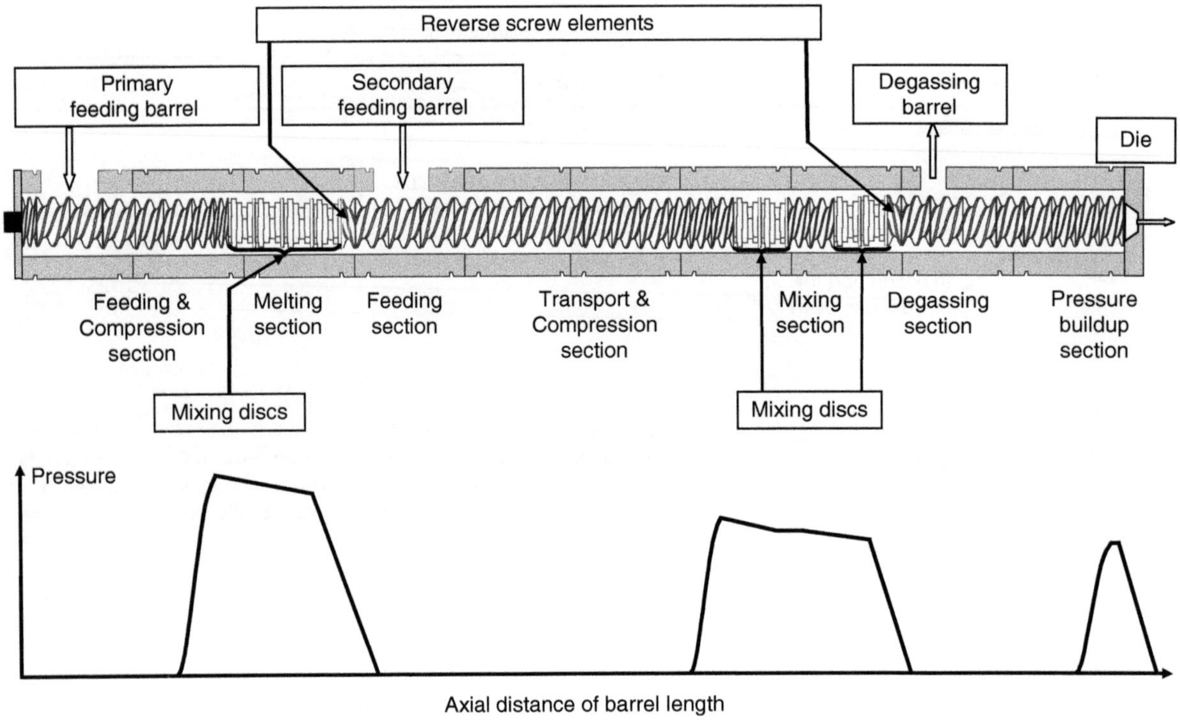

Figure 2.20 Classic screw-barrel configuration of an intermeshing co-rotating twin screw extruder in polymer compounding; hypothetical pressure profile. Source: Reproduced with permission of Clextral, France.

The best means of combating adhesive wear is to use appropriate lubrication, where a fluid or solid film plays the role of a dissipative interface between the first two bodies. Under normal extrusion processing conditions, melted polymeric materials or solid–liquid media play this role which explains why few adhesive wear phenomena are observed, provided that material hydrodynamics in the screw-barrel configuration is well established, which depends on the screw profile. Where the lubricant film is broken, metal particles are generated and a third body may form. This allows metal-to-metal contact to occur; abrasive particles are then introduced into the lubricant which may cause abrasive problems.

2.2.3.1.3 Surface fatigue wear

Surface fatigue wear is caused by repeated normal and tangential loads due to screw rotation, which lead to microcracks and pitting, causing microscopic and macroscopic metal particles to come away from the surface of the screw and/or barrel parts. Prevention methods are: (i) decreasing contact stresses using suitable geometry; (ii) reducing friction and adhesion phenomena; (iii) using materials with as few faults as possible and increasing their strength.

Surface fatigue wear is characterized by long induction times and a relatively large depth of penetration. In extrusion processing, the majority of mechanical parts required by contact actions may contain cracks, which are sometimes present when they come off the production line. However, a crack does not constitute a problem unless it spreads. A mechanism comprising cracked parts may continue to operate without any problems once the cracks stop spreading.

2.2.3.1.4 Corrosive wear

Corrosion is defined as the set of chemical and mechanical interactions contributing to the removal of solid body material in a tribological contact (Kaczorowski, 2002). Corrosive wear occurs when a chemical reaction attacks at least one part of the screw-barrel configuration, converting it into a friable one. In extrusion processing, corrosive wear usually occurs in combination with abrasive wear and the synergy between both mechanisms can cause wear far in excess of what would be expected based on their individual contribution.

2.2.3.1.5 Cavitation-induced wear

Cavitation-induced wear is caused by localized and repetitive overpressure pulses due to the implosion of vapor cavities within a liquid. The phenomenon of cavitation is the successive appearance and disappearance of vapor cavities caused by flow at a given temperature within a liquid. When the localized pressure suddenly drops below the vapor pressure, the liquid undergoes localized tearing, encouraging the formation of vapor cavities. The cavities interact with the flow (enlargement, coalescence) and when they are in a high-pressure area, they implode and generate relatively high overpressure spikes in the form of shock waves. Almost hemispherical holes appear outside the friction area or on its edges and then the whole surface is eroded. Destruction, which is often catastrophic, is very rapid and loud. The best way to prevent wear caused by cavitation is to prevent depressions in the screw-barrel configuration.

In extrusion processing, cavitation-induced wear may occur in modular screw-barrel configurations, when abnormal process conditions are applied such as an inappropriate screw profile combined with direct steam injection or an unsuitable pressure profile in decompression sections of the screw-barrel configuration. Obviously, preventing cavitation-induced wear requires a screw profile which is suitable for the process functions of the extruder.

Most of the time, overall wear is due to several processes or forms of degradation, which occur at the same time and, more rarely, in one well-defined and identifiable process. The effect of these simultaneous actions is often more significant than the sum of the effects of the different processes acting separately. This is sometimes called "overadditivity."

In extrusion processing, abrasive and corrosive wear are the most important wear mechanisms which occur simultaneously under normal processing conditions. Surface fatigue wear may also contribute significantly. Even though each mechanism is well known individually, the wear rate induced by their combined action is often unpredictable. Cavitation-induced wear may also occur when applying inappropriate extrusion processing conditions (inappropriate screw profile and process parameters, in particular). Therefore, precise measurement of wear rates combined with visual observation of worn parts are absolutely vital, to allow technical experts to make a diagnosis and find a suitable solution to wear phenomena.

2.2.3.2 Solutions to wear

In a well-designed extruder, the majority of wear should be restricted to screws, as they can be replaced and rebuilt more easily than the barrel. And it should be pointed out that screw replacement is definitely less expensive than barrel replacement.

Extruder manufacturers offer screw and barrel parts with varying degrees of hardness. For a given metal, barrel parts are made of harder material than that of screw parts, which allows screw wear to develop faster than barrel wear. The adjustment in the difference of hardness between both parts is a result of rigorous engineering analysis and long-term experience.

Rebuilding extruder screws is very common and follows a well-controlled procedure. Screws can be successfully rebuilt 3–5 times, which allows significant savings to be made when replacing worn screw parts; savings generally range from 20% to 40% depending upon the type of extruder and screw-barrel configuration.

If the difference in hardness between screw and barrel parts is suitable, barrel replacement occurs much less frequently than screw replacement. Rebuilding worn barrels is usually much more difficult than rebuilding screws. In most cases, the worn barrel is replaced by new one but as an alternative, extruder manufacturers can allow the worn barrel to be reused by converting it into a liner holder. This is a reliable solution which can be implemented providing that the internal diameter of the worn barrel does not exceed a maximum value according to the technical recommendations of equipment manufacturers. Then, the worn barrel can be honed to a larger diameter, and a new barrel liner can be inserted into the larger diameter barrel, which then becomes a liner holder. This solution, which consists of combining a modified liner holder with a new barrel liner, allows the cost of barrel replacement to be reduced by 30–60%, compared with a completely new barrel. Figure 2.21a shows a schematic view of a barrel liner to be inserted into a liner holder. Barrel liners can be made of various metals.

A large number of materials are available for screw and barrel parts. The simplest standard solution is to use carbon steel material which, after machining, receives a hardening treatment through nitriding. Hardened nitrided surfaces have an effective hardened depth of up to approximately 1 mm but the hardness of the nitrided layer decreases as the depth from the surface into the body of the material increases. This solution can be implemented for screw and barrel parts while, as mentioned above, a difference in hardness between both parts

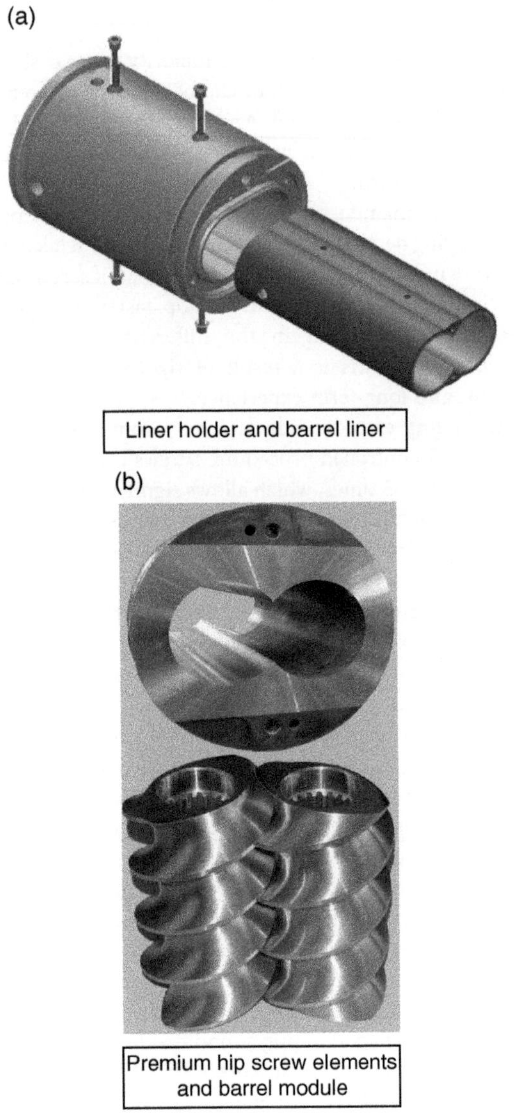

Figure 2.21 (a) Schematic view of combined liner holder and barrel liner. (b) Premium HIP-based screw elements and barrel module. Source: Reproduced with permission of Clextral, France.

is rigorously respected. Today, nitrided screws and barrels for extruders are technically reliable and cost-effective in numerous industrial applications such as polymer extrusion processing and food and feed extrusion processing. It should be noted that this solution provides effective resistance to abrasive wear, but much less resistance to corrosive wear as nitriding treatment significantly reduces the resistance of nitrided materials to corrosion. With regard to fatigue surface wear, nitrided screws and barrels provide effective resistance to wear.

Equipment manufacturers and material suppliers have made real progress in offering advanced, premium solutions to globally reduce the contribution of wear to the operating costs of extrusion processing units. For a given metallurgical formulation, the difficulty in material development for screws and barrels lies in combining a material which is as hard as possible (to reinforce resistance to abrasive wear, in particular) with good material resilience (to reinforce resistance to surface fatigue wear, in particular). One solution consists of combining, in one piece of metal, two different, strongly bonded materials: a resilient material and a hard material. For example, a screw element would be composed of a central core made of resilient material (carbon steel or stainless steel) which would be the screw root of the final screw, and an external shell made of hard material (cobalt-based, nickel-based) in which the screw flight would be machined, both the resilient central core and hard external shell being closely bound. The most reliable manufacturing process for producing such bimetallic materials is the so-called hot isostatic compression process (HIP). The HIP produces materials with uniform hardness. Figure 2.21b presents premium HIP-based screw elements and a barrel module.

Hot isostatic compression process-based materials allow premium screw elements and barrel modules to be produced which can be successfully used in industry from a technical and economic standpoint. As extrusion processes have different degrees of sensitivity to wear mechanisms, depending upon the composition of the raw materials to be processed, the HIP offers real flexibility in terms of adapting the required materials and metallurgical formulations, in order to create an optimum solution and minimize the wear rate as much as possible. Screw and barrel parts made from HIP-based materials are significantly more expensive than nitrided parts. Nevertheless, they are increasingly used in biomaterials extrusion processing (food and feed extrusion processing, cellulose extrusion pulping) as they last much longer than nitrided parts. In fact, owing to the longer lifetimes of HIP-based screws and barrels, combined with reduced downtime due to parts replacement, premium solutions become more cost-effective than standard nitrided solutions. It should be noted that screw-barrel configurations with modular designs do offer the advantage that just worn parts can be replaced, rather than the whole screw or barrel, which leads to considerable reductions in wear-related costs.

Screw-barrel assembly wear is complex as it is a multidimensional phenomenon. As stipulated in section 2.2.3.1, the wear rate of extruder screw elements and barrel modules results from the combination of two or three different wear mechanisms, mainly abrasive wear, corrosive wear, and surface fatigue wear. In the case of mature extrusion processing applications, where long-term experience has been accumulated by equipment manufacturers and extruder users, the synergy of wear mechanisms is well known and metallurgical solutions for screws and barrels are generally successful.

Technical issues may arise when dealing with new extrusion processing applications or when new processing conditions are applied in mature applications. In fact, in extrusion processing, the development of wear solutions depends not only on the right metallurgical solution for screws and barrels, but also on two other important factors, i.e. the composition of raw materials and the screw configuration. For a given raw material, the composition may vary significantly. This is particularly the case with biomaterials for which variations in the elementary composition (lipids, minerals, for example) can significantly affect the extent of wear mechanisms. The second factor, i.e. the screw configuration, is far more difficult to manage in terms of wear issues. It must be borne in mind that the material acts as a lubricating film which prevents metal-to-metal contact (screw-to-barrel contact, or screw-to-screw contact in the case of twin screw extrusion), providing the hydrodynamics in the screw-barrel assembly is well developed. This is particularly sensitive to the screw profile according to the rheological behavior of the extruded material.

As biomaterials will be increasingly important in the world of extrusion of tomorrow, it is worth noting that pertinent metallurgical solutions for minimizing wear in extrusion processing will result from very close dialog between metallurgists and process engineers. Equipment manufacturers who have long-term experience in both fields are more able to successfully handle such situations than pure specialists in metallurgy or material engineering.

2.3 Ancillary equipment

High-performance screw extruders require suitable, reliable ancillary equipment, the purpose of which is to provide significant process assistance to the extruders. Three main categories of ancillary equipment can be distinguished depending upon their position in the extrusion processing unit: upstream, on-line, and downstream ancillary equipment (Figure 2.22). The process function of ancillary equipment is to pretreat raw materials (e.g. preconditioners), feed raw materials and additives (e.g. feeders, pumps, side feeders), supply thermal energy (e.g. steam injection kits), remove volatiles (e.g. degassing kits, vent stuffers), and postprocess extruded products (e.g. cutters, gear pumps, pinching/stamping machines). All the ancillary equipment listed above is presented in this section. Of course, this list is not exhaustive but it includes ancillary equipment which is at least required for biomaterials extrusion processing.

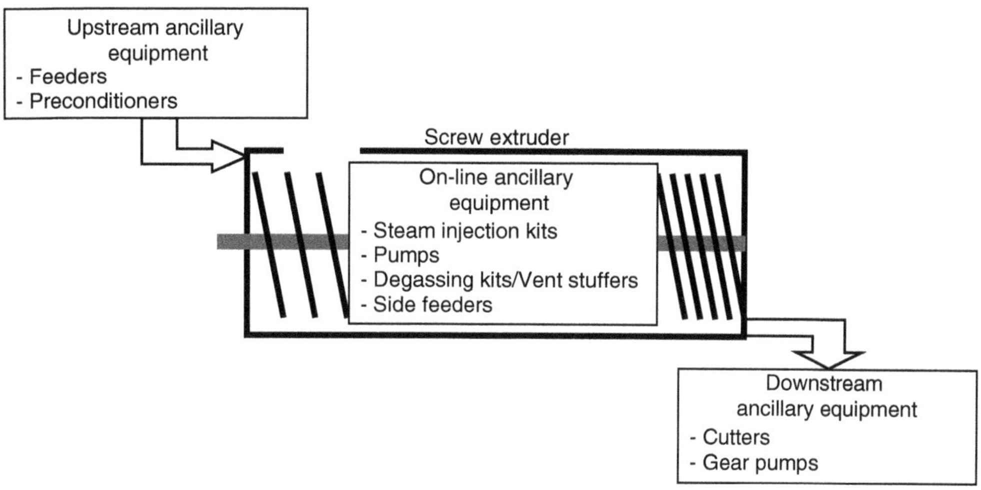

Figure 2.22 Main ancillary equipment used in extrusion processing.

2.3.1 Upstream ancillary equipment

2.3.1.1 Feeders

In the majority of industrial applications relating to polymer extrusion processing as well as food and feed extrusion processing, the raw materials fed into a screw extruder consist of powders, pellets, and granules, in a dry state. It is of paramount importance that the raw materials are fed consistently and at regular intervals, in order to obtain a uniform feed rate and prevent output fluctuations. In food and feed extrusion processing, it is common to start with a premix of various ingredients which may have different particle sizes; of course, this process context requires specially designed hoppers to prevent segregation of ingredients.

Two categories of feeders are used to feed powders, pellets or granules into screw extruders: volumetric and gravimetric.

2.3.1.1.1 Volumetric feeder (Figure 2.23a)

This is designed to provide a constant volumetric flow rate of materials. It is assumed that the density of feed materials does not change over time, so that a constant volume of feed results in a constant mass flow rate. The most common volumetric feeder is the single screw feeder where the volumetric feed rate is proportional to the screw speed. Of course, the operational efficiency of these feeders depends on the ability of the feed materials to flow into and fill the rotating screw. It is suitable for free-flowing materials. With flooding materials (such as soy flour and casein powder, for example) as well as with sticky materials, the use of intermeshing, self-wiping twin screw feeders is recommended since they convey these types of raw materials more consistently than single screw feeders. Volumetric feeders operate through calibration and are open loop in nature. Feeding accuracy depends on the uniformity and bulk density of the raw materials. Feeding accuracy generally ranges from 2% to 8%: the wide range depends upon the extent of changes in product characteristics such as density, temperature, humidity, etc. For materials which have a uniform density and which are free-flowing, volumetric feeders are the simplest and least expensive feeders. We should also mention volumetric belt feeders, for which the feed rate is controlled by the speed of the belt, the material being fed onto the belt by a hopper.

2.3.1.1.2 Gravimetric feeders (Figure 2.23b)

These are designed to deliver a uniform mass flow rate of materials, which is independent of density. The two most

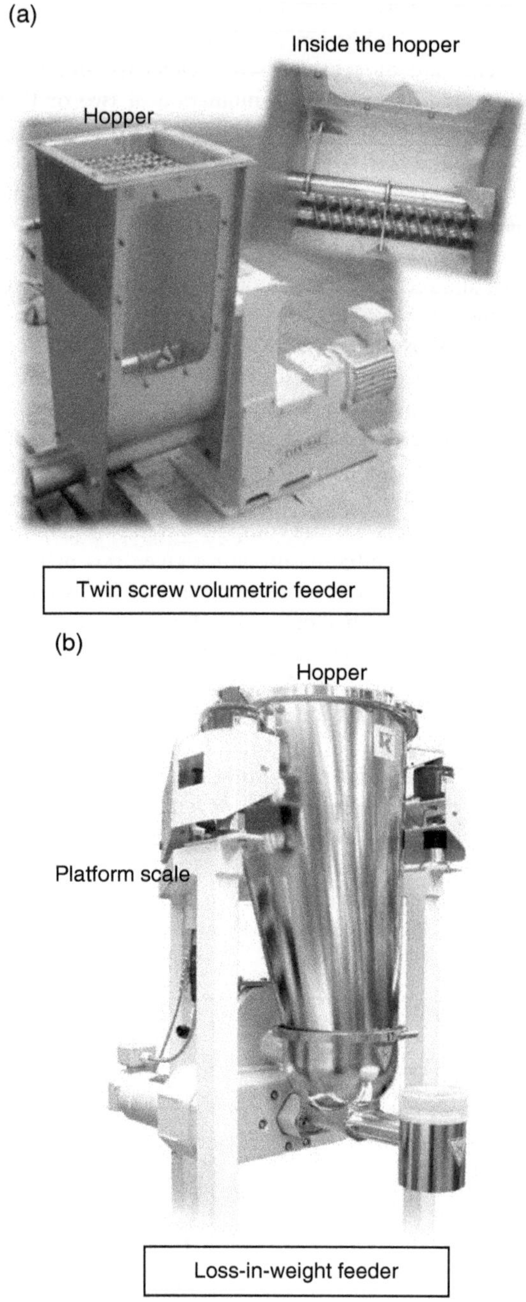

Figure 2.23 (a) Volumetric twin screw feeder. Source: Reproduced with permission of Clextral, France. (b) Loss-in-weight feeder. Source: Reproduced with permission of K-Tron, Switzerland.

common gravimetric feeders are the loss-in-weight and weigh-belt feeders.

Loss-in-weight feeder principle

The feeder consists of a screw-aided discharge unit (single screw, intermeshing self-wiping twin screw) and a hopper containing the material to be fed. It is placed on a platform scale or suspension scale weighing system. The feeder functions as follows.

The weight of the feeding device and hopper is electronically tared. The bulk material is discharged from the hopper via the feeding device and the resultant weight loss per unit of time is determined by the weighing and control system. This actual weight loss per unit of time is compared to a desired value based upon a desired continuous feed rate. Any difference between the actual and desired weight loss per unit of time results in correction of the speed of the feeding device. When the hopper content reaches a predetermined minimum weight level, weight loss control is briefly interrupted and the hopper is refilled. During refill, sensed weight is not useful for controlling the feed rate. Therefore, the simplest solution consists of freezing the screw speed of the feeding device until refill is complete. But an advanced solution is available, i.e. K-Tron (K-Tron, 2010). In fact, K-Tron Smart Refill technology memorizes recent gravimetric performance at various hopper weights and varies screw speed to compensate for changes in material density during refill, maintaining accurate performance and enabling a smoother transition to normal operation. During the refill period, weight is increasing and the controller regulates the speed of the feeding device based on the historical weight and speed information that was determined during the previous weight loss cycle.

Owing to its operating principle, loss-in-weight feeding requires continuous weighing of the feeder, and the feeder cannot be constantly resupplied with material as it must be periodically refilled. Without any basis for weight-based control during this brief but periodic refill phase, maintaining momentary feeder accuracy becomes a concern. A traditional method consisted of maintaining a constant metering speed throughout the refill phase; that is, a speed corresponding to the screw speed associated with gravimetric control just prior to entering the refill phase. If, for example, metering speed averaged 60 rpm just prior to the system sensing the need to refill the supply hopper, screw speed was maintained at 60 rpm for the full duration of the refill operation. After refill was completed, the material had settled, and the feeder sensed an appropriately declining system weight, the feeder then returned to gravimetric operation where metering speed once again became the control parameter.

The problem with the traditional approach is that in most cases bulk density within the metering zone (the area from which process material is drawn) increases as material rises in the hopper. When metering speed is held constant throughout refill, an overfeed condition is then experienced. When gravimetric control is restored after refill, the feeder senses the high-density condition and abruptly reduces screw speed. Laboratory tests and field experience involving many hundreds of types of material have shown that, in practical terms, headload-related loss-in-weight overfeeding may range anywhere between 1% for relatively constant density materials to 10–15% for powders and other materials whose density can vary substantially. The preferred solution to this problem begins with determining the relationship between system weight and material density within the metering zone of the feeder. If this relationship is known, metering speed does not need to be constant throughout refill, but may be smoothly reduced as the hopper is refilled, in order to avoid overfeeding and preserve feeder performance throughout the refill phase. To ensure the highest possible feeder accuracy throughout the refill phase, K-Tron has developed a concept called the refill array to enable metering speed to be gradually adjusted during refill in order to accurately counterbalance the effects of variations in material density occurring within the metering zone as hopper weight increases.

During refill, metering speed is determined by an array of indices called feed factors. These values correspond to material density and its mechanical behavior within the feeder. They are computed and stored during the entire gravimetric feeding phase immediately preceding each refill. On the basis of increasing hopper weight during refill, material density within the metering zone may be inferred, and a metering speed corresponding to its feed factor array value may be used. In this way, gravimetric feeding accuracy during the brief refill period may be maintained and, once refill is complete, smooth resumption of gravimetric operation is ensured.

The loss-in-weight feeder principle is most accurate when using a high-resolution, fast-response, vibration-proof weighing system combined with self-tuning controls. With these advanced technical solutions, K-Tron loss-in-weight feeders generally guarantee +/−1% accuracy. Therefore, loss-in-weight feeders are the most accurate devices for a wide range of feed materials to be fed into screw extruders.

Weigh-belt feeder principle

The feeder operates by continuously weighing a moving bed of material on its short conveyor, and controlling belt speed in order to obtain the desired flow rate at discharge. Taring or zeroing is a major concern since both the belt and material are weighed, and any error in tare produces a repetitive and systematic error in the feed rate. Sources of potential changes in tare include belt wear, impregnation of material into the belt, and adherence of material to the belt. Changes in belt weight due to material build-up are inevitable, and the use of a belt scraper at discharge and elsewhere within the feeder minimizes but, for many materials, cannot eliminate the concern. Thus, periodic taring has historically been required.

Early efforts to address the issue of periodic taring included partial automation of the taring procedure. Upon user demand, the empty belt feeder was cycled through a single belt revolution and a single tare value correction was automatically generated. To account for variations in belt weight along the length of the belt (important in lower belt speed applications), an indexing feature was later added to allow belt weight to be measured and recorded centimeter by centimeter along the length of the belt. During operation, these indexed belt segment tare values were applied in order for the corresponding belt segment to pass over the weighing section.

While these measures simplified the taring process and made it more accurate, taring remained a burdensome and periodic requirement, and tare drift was still a problem. To resolve these difficulties and fully automate the taring process, K-Tron has developed its exclusive Continuous Auto-Tare feature (K-Tron, 2010). When adding a second weight sensor upstream of the material inlet, where no bed of material is present, taring can be accomplished accurately, automatically and continuously online without emptying the feeder. This real-time, fully indexed taring procedure eliminates belt weight variation concerns regardless of their cause, and ensures the highest possible weigh belt feeding accuracy.

As for feeding lignocellulosics, the physical state of raw materials (cotton combers, cotton linters, straws from annual plants, etc.) differs from that of raw materials encountered in polymer extrusion processing and cereal extrusion processing. In fact, these raw materials are fiber based; they are large, highly compressed bales in which long bast-fiber strands are entangled. Feeding such raw materials into an extruder is technically challenging. The bales are opened, bast-fiber strands are partially disentangled and fed into the extruder by use of a specially designed feeder which aims to ensure a constant mass flow rate delivery of raw materials into the feeding section of the extruder. Clextral has developed reliable feeding technology associated with twin screw extrusion processing technology for feeding lignocellulosics raw materials (Clextral, 2013).

2.3.1.2 Preconditioners

Preconditioners were soon introduced by the cereal extrusion processing industry, to preheat and prehumidify dry, cereal-based feed mixes prior to extrusion cooking, by mixing them with steam and water. Although pressurized preconditioners would bring about obvious processing advantages, such as quicker moisture diffusion and higher discharge temperatures, for example, preconditioners generally operate under atmospheric pressure conditions; this offers simpler machinery design and full continuous processing.

Two main types of preconditioners can be distinguished: single shaft in a cylindrical chamber and double shaft in an octagonal chamber (Figure 2.24). Shafts are equipped with adjustable paddles which, when combined with the rotation of the shafts, allow the feed materials to be mixed with steam and water injected into the chamber of the preconditioner. In single-shaft preconditioners,

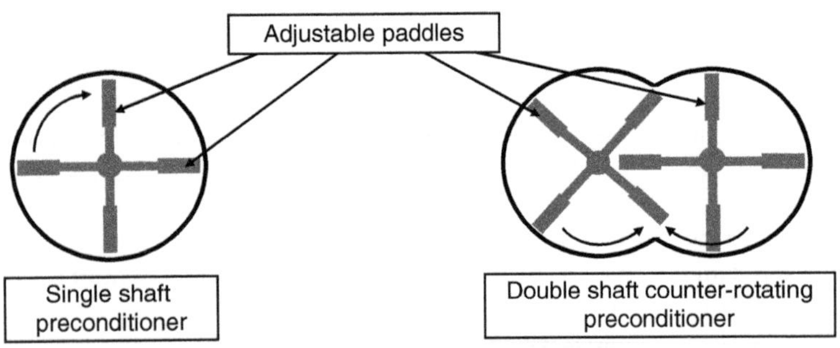

Figure 2.24 Single- and double-shaft paddle preconditioners.

mixing efficiency is significantly lower than that observed in double-shaft preconditioners. In fact, double-shaft preconditioners have an intermeshing counter-rotating upwards design, which tends to 'fluidize' the bed of powdery materials, hence ensuring intense mixing in the preconditioner. Double-shaft preconditioners are much more common in modern cereal extrusion-processing units.

The motion of powder particles is ensured by paddles that are threaded, bolted, or welded onto the shafts. Paddle design varies depending upon the manufacturer: beater-type paddles (flat, rectangular shaped) are commonly used, as they are simple and operate effectively – the simpler, the better. Paddle configuration is important as it has a significant effect on the residence time and filling ratio in the preconditioner. Paddles are fixed at regular intervals along the shafts through a helical design to globally generate a forward flow of the powdery material. Paddle angle results from the orientation of the paddle width with respect to the shaft axis direction. An angle of 0° is obtained when the paddle width is parallel to the shaft axis (Figure 2.25); this position is called a neutral pitch and offers very good radial mixing, but no axial mixing. An angle of +45° would favor forward flow while an angle of −45° would induce reverse flow of the powdery material. Forward and reverse positions, or forward and reverse pitches, would affect both radial and axial mixing in the preconditioner. The effect of the paddle configuration on mixing and flow behavior has been discussed by Levine (1995). Varying the paddle angle in any direction makes it possible to vary the intensity of the axial flow of particles in the preconditioner, and to combine axial and radial mixing components. Paddle configuration results from the arrangement of neutral, forward, and reverse sections to ultimately optimize time history, filling ratio, and mixing efficiency in the preconditioner.

There are various optional designs for double-shaft preconditioners.

- The intermeshing counter-rotating double-shaft preconditioner, the semi-interacting chambers of which have identical diameters (symmetrical chambers or octagonal chamber), as shown in Figure 2.24. This design is commonly used and well recognized in the cereal-processing industry. Figure 2.26 shows such a preconditioner.
- The intermeshing counter-rotating double-shaft preconditioner, the semi-interacting chambers of which have different diameters (non-symmetrical chamber). This design also offers the possibility of applying differential shaft speeds.
- Dual single-shaft preconditioners in series, the chambers of which do not interact. These are composed of two chambers. The first single-shaft chamber operates at high shaft speed to enable the powdery material to be well mixed with water and steam. This first section delivers the mix into a second single-shaft chamber underneath, which operates at low screw speed, allowing moisture to penetrate the particles of powdery material.

As shown in Figure 2.26, double-shaft preconditioners are highly versatile and flexible and offer the possibility of adding liquid water from the top (for material moistening) and injecting steam from the bottom (for material heating); both fluids are injected in the first part of the preconditioner. Downstream, it is possible to add various liquids and slurries, which enter the mix formulation. The preconditioning chamber is usually made of stainless steel. It has door openings for process inspection, cleaning, etc. The gearbox allows the shaft speed to vary according to process specifications.

Preconditioners contribute considerably to the precooking of cereal-based raw materials, in combination with the next extrusion-cooking operation. Consequently,

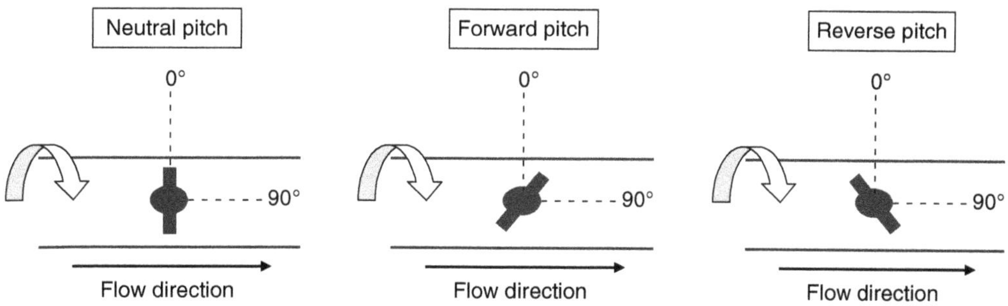

Figure 2.25 Flow direction versus paddle configurations in preconditioners.

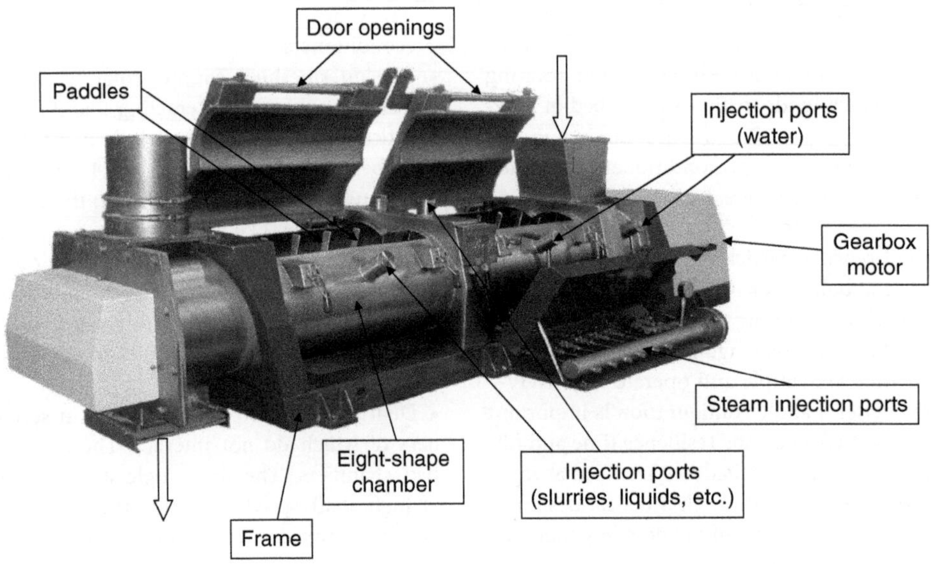

Figure 2.26 An intermeshing counter-rotating double shaft preconditioner. Source: Reproduced with permission of Clextral, France.

due to their importance in the whole extrusion-cooking process, preconditioners are further analyzed and discussed in Chapter 6.

2.3.2 On-line ancillary equipment

2.3.2.1 Steam injection kits

In feed extrusion processing applications, steam must be added to the screw-barrel assembly of the extruder and/or the preconditioner, if there is one, in order to provide the extrusion-cooking process with thermal energy. The steam must be dry and its mass flow rate fully controlled, which is technically challenging. Extruder manufacturers can offer self-contained, reliable and flexible steam injection kits of varying sizes, which allow the factory's steam supply network (6–8 bar) to be converted to meet extrusion process requirements (dry steam, 1.5–2 bars; controlled flow rate).

Steam injection kits are generally characterized by the following components: steam inlet connection, steam trap for condensate recovery, flow meter, control valve with electronic regulation, steam pressure probe, steam temperature probe, and steam outlet connection. They are thermally insulated and all parts are made from stainless steel. Flow rates should range from a few kg per hour for pilot plant units to 2.5–3 tons per hour for large feed extrusion processing units.

2.3.2.2 Pumps

In most extrusion processing applications, it is essential to be able to add fluids (e.g. low-viscosity liquids in particular: water, organic chemicals, acids, alkalis, etc.) to the screw-barrel assembly. As seen in Table 2.4, barrel modules are equipped with suitable injection ports for this purpose. In this case, pumps are key ancillary equipment in the extruder environment. It should be noted that flow rates of liquids added to the extruder are relatively low: from just a few litres per hour to a few hundred litres per hour.

Since the density of liquids is almost constant over the relevant temperature range, volumetric metering devices are primarily considered. A constant volume of liquid is then delivered into the extruder. Among the different categories of pumps available on the market, metering pumps and rotary pumps are generally used in extrusion processing. Piston or membrane metering pumps are perfectly suited to the targeted flow rates. They can operate up to a temperature of approximately 150°C and 300 bars of back pressure. They can be made from various materials such as carbon steel, stainless steel, elastomeric polymers, etc. Different designs of piston and membrane pumps are available, which provide different levels of flow rate accuracy and regularity. Peristaltic pumps are also used, as they are simple and easy to maintain but they deliver pulsated flow rates of liquids.

Lobe pumps and screw pumps may be used for specific applications where intermediate-viscosity liquids or liquids with solid particulates are processed, in food and feed extrusion processing in particular.

2.3.2.3 Degassing kits. Vent stuffers

In extrusion processing, it is often necessary to remove volatiles from the extruder's screw-barrel assembly (for example, polymer devolatilization in polymer extrusion processing; water flashing off for quick melt cooling in food and feed extrusion processing). This is called degassing.

As seen in sections 2.2.1 and 2.2.2, this requires appropriate screw-barrel configurations which need to be connected with dedicated ancillary equipment to handle removed volatiles (organic vapors, steam, in particular). Atmospheric degassing results from liquid flash-off at atmospheric pressure, while vacuum degassing is carried out at a pressure lower than atmospheric pressure, which allows degassing yield to be more effective. In both cases, the degassing barrel is connected to a volatile collecting pipe, in order to facilitate transportation of volatiles to the volatile treatment station of the factory or a water-cooled condenser. In the case of vacuum degassing, a vacuum pump must be supplied, together with a condenser, which protects the pump from product contamination. Thus, these different hardware components allow the degassing kit to be defined according to process requirements.

When degassing from an extruder barrel, the melt which travels along the screw channel may escape from the screw-barrel assembly through the vent port. For a given screw profile, this tends to occur with high throughputs. To prevent the melt from escaping and still maintain high extrusion throughput, it is possible to fit screw-aided equipment on top of the vent port. This screw-aided equipment is commonly called a vent stuffer. Its purpose is to ensure that the melt remains at all times inside the screw-barrel assembly, while allowing the volatiles to flow freely through the screw assembly. Single screw and intermeshing twin screw vent stuffers are offered by equipment manufacturers. Vent stuffers are used in industrial polymer extrusion processing, as well as in feed extrusion processing (pet treats and aquafeed applications, in particular). Figure 2.27 shows the principle of vent stuffing,

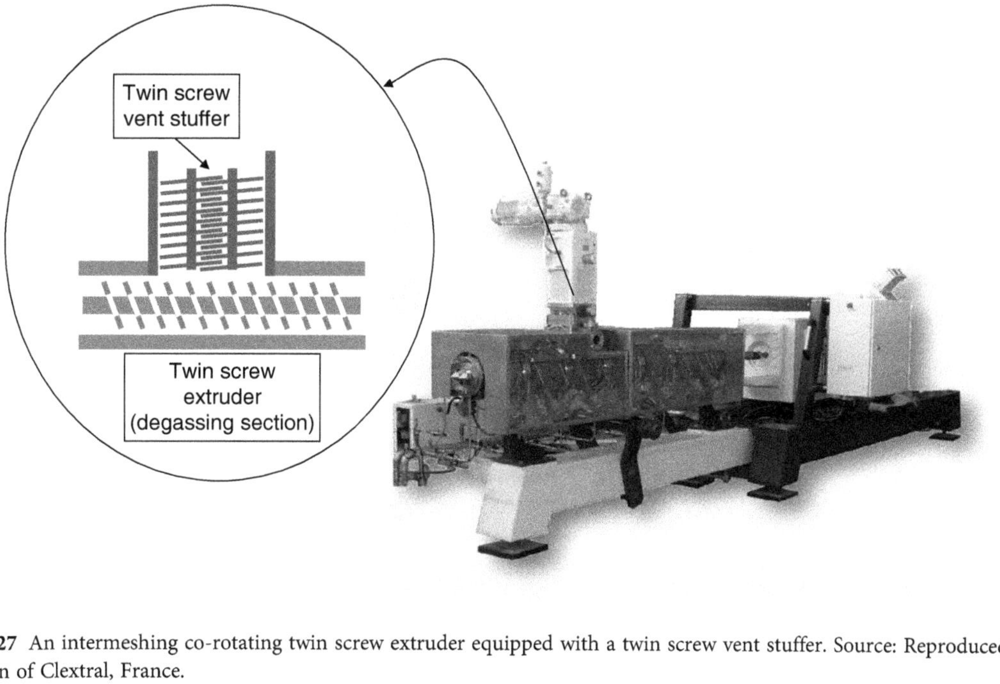

Figure 2.27 An intermeshing co-rotating twin screw extruder equipped with a twin screw vent stuffer. Source: Reproduced with permission of Clextral, France.

and a picture of an intermeshing co-rotating twin screw extruder equipped with an intermeshing twin screw vent stuffer; this assembly is used in pet treats extrusion processing.

2.3.2.4 Side feeders

In polymer extrusion processing (polymer blending in particular), as well as in some food and feed extrusion processing applications, solid materials such as minerals, resins, reinforcements, fibers, precooked cereal flours, etc. must be added to the screw-barrel assembly of the extruder. As a result of these additives being added, two main methods can be envisaged: gravity feeding through a feeding barrel module (secondary feeding) and side feeding. Side feeding provides higher addition rates than gravity feeding.

Side feeders are either single or intermeshing twin screw feeders. Intermeshing twin screw feeders (co-rotating, counter-rotating) are more commonly used. The screw profile is equipped with self-wiping, constant forward pitch screws that positively convey additives into the main stream of the extruder. The barrel is generally composed of a feeding barrel module and an outlet barrel module which are normally not cooled or heated. The screw and barrel are nitrided, but premium wear or corrosion-resistant elements are optionally available. The side feeder normally operates in a starve-fed mode for additive rate control; the advantage of starve feeding is that additives are fed in the proper ratio with regard to the formulation composition. As shown in Figure 2.28, the screw-barrel assembly and kinematics (gearbox and motor) of the side feeder are supported by a fixed frame. The entire piece of equipment can finally be mounted on wheels, which allow the side feeder to be easily transported.

2.3.3 Downstream ancillary equipment

2.3.3.1 Cutters

Two types of cutters can be distinguished, as per section 2.1.3: die-face cutters, which are physically connected to the die assembly of extruders, and downstream cutters which are separate from the die assembly.

2.3.3.1.1 Die-face cutters

This section focuses on die-face cutters combined with texturizing dies in cereal extrusion processing (section 2.1.3.3.1). Die-face cutters are high-speed rotary cutters centrally mounted on the die plate of the extruder die assembly. Die-face cutters are generally composed of:

- an interface plate which is fixed to the die plate of the die assembly
- a cage which holds the extruded products after cutting
- a blade holder which holds various types of blades depending upon the types of products to be die-face cut
- a blade adjustment system for tight, accurate adjustment of the cutting blades and the face of the die plate
- a motor which controls the rotation speed of the blade holder
- safety tools to ensure operator protection
- a frame which supports all parts of the cutter.

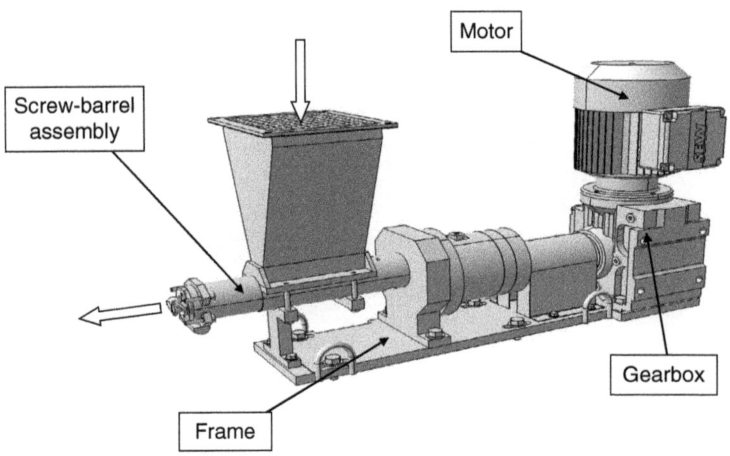

Figure 2.28 An intermeshing twin screw side feeder. Source: Reproduced with permission of Clextral, France.

A lot of consideration must be given to blade design, particularly for food and feed extrusion processing applications when die-face cutting directly expanded products. In fact, the final shape of directly expanded products depends to a large degree on the angle of attack and clearance angle of the blade (Figure 2.29a). If the angle of attack is small, the blade tends to chop the extrudate; when the angle of attack is large, the blade has a sliding motion which enhances clean cutting. The blade width should be small so that the trailing edge does not modify the continuous flow of the extrudate. It should be noted that the design of cutter blades requires long-term experience in extrusion processing in order to find a solution adapted to product characteristics.

Equipment manufacturers are constantly improving the design of die-face cutters, to allow extruder users to reduce downtime (allowing dies to be changed quickly, blades to be checked and replaced without stopping the extruder) and preserve product quality (minimizing water condensation and product damage thanks to special cage design). Figure 2.29b shows a die-face cutter recently developed by Clextral, which satisfies these process and product requirements.

Figure 2.29 (a) A typical blade holder and sketches of blade cutting. (b) A modern die-face cutter. Source: Reproduced with permission of Clextral, France.

2.3.3.1.2 Downstream cutters

Three different cutting systems can be mentioned to illustrate this category.

- The strand pelletizing cutter for polymer compounding. As seen in Figure 2.12 (section 2.1.3.2.2), the strand pelletizer is separated from the die by a cooling trough. The strand pelletizer is a compact piece of equipment. Its active part is the cutting head, equipped with wear-resistant cutting tools made of various materials (e.g. stainless steel, tungsten carbide, ceramic, diamond, etc.), depending upon the composition of the thermoplastics compounds. This equipment is robust and reliable, and can be easily adapted to any biomaterial-based compound.
- "The chipstick" cutter for downstream cutting of extrusion-cooked products. Formerly introduced for cutting directly expanded, extrusion-cooked potato-based strands to produce crispy sticks (so-called "chipsticks"), this simple cutter allows accurate cutting to be achieved through a spiral action with a special hood to reduce multiple cutting of the product. It enables multiple strands to be cut to provide sticks uniform in length. The chipstick cutter is well suited to products which require straight cutting and show low resistance to cutting. It is effective at cutting various types of strands from crispy, directly expanded products (crispy snacks) to soft non-expanded products (snack pellets, pet treats).
- The pinching cutter for downstream pinching of co-filled extruded foods (see Figure 2.15, section 2.1.3.2.2). The equipment is designed to combine two main process functions: (i) flattening the co-filled ropes exiting the die; (ii) forming individual co-filled products from continuous ropes. Thus, the pinching cutter is composed of a belt conveyor which has pressing

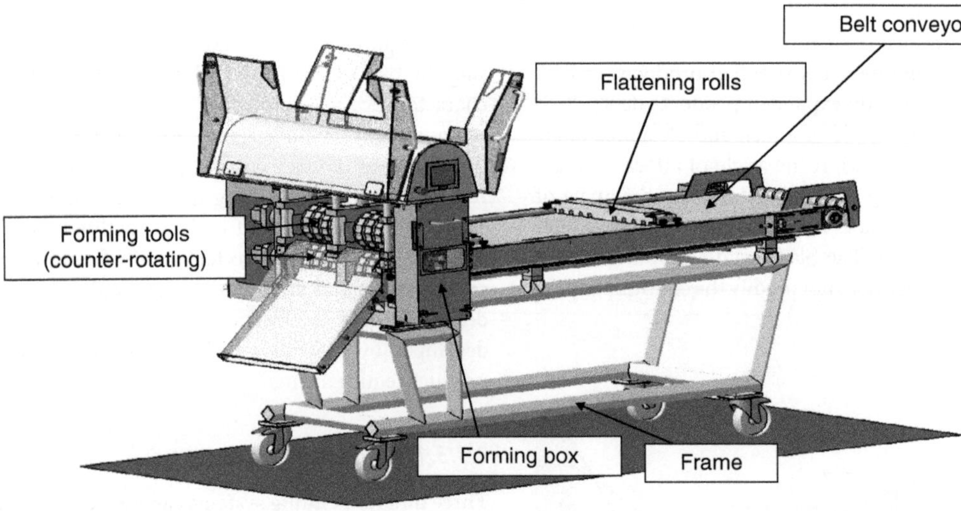

Figure 2.30 Schematic view of a pinching cutter. Source: Reproduced with permission of Clextral, France.

rolls (gentle pressing action). The belt and rolls aim to regulate the linear speed of co-filled ropes (up to about 100 m/minute). The belt conveyor delivers the ropes to a forming box which has two counter-rotating tool holders; tool holders make shapes (pillow shapes, for example) and individual pieces of co-filled products (25 × 25 mm pillows, for example) through a pinching action rather than a pure cutting action. Tool holders can accept various types of tools, which makes the equipment very flexible and versatile. Processing parts are supported by a frame and the whole piece of equipment is mounted on wheels, which allow the pinching cutter to be easily transported. Figure 2.30 shows an example of a flexible, versatile pinching cutter.

2.3.3.2 Gear pumps

Gear pumps are often used in polymer extrusion processing, where they are effective in assisting screw extruders and allow pump-assisted extrusion units to operate consistently with higher capacity and product quality.

Gear pumps are rotary volumetric pumps which are designed to pump highly viscous materials at a relatively high pressure. As shown in Figure 2.31, the active section of such pumps is composed of a drive gear shaft and a driven gear shaft; the gears have a self-cleaning configuration. The rotary action of the gears allows suction, transport, and discharge of polymer melts to be carried out at controlled pressure and constant throughput. The design of the gears is of primary importance as it determines process consistency and efficiency (operating pressure, shear extent, throughput constancy, in particular).

The use of gear pumps in combination with screw extruders adds considerable value in plastics extrusion processing through process optimization (increased throughput, reduced pressure fluctuations, lower time-temperature history, lower energy consumption, reduced extruder screw wear) and product quality improvements (lower thermal degradation of polymer melts, better dimensional uniformity of finished products). In gear pump-assisted extrusion processing, screw extruders focus on polymer feeding, plasticizing and mixing, while gear pumps aim to build pressure and consistently feed die systems, at constant pressure and throughput. As seen in Figure 2.32, gear pumps are then placed between the screw extruder and the die system. This type of arrangement is commonly used in most plastic-forming configurations presented in section 2.1.3.2.2, such as film and sheet extrusion, blown film extrusion, piping and tubing extrusion, co-extrusion, etc. Hence, gear pump-assisted

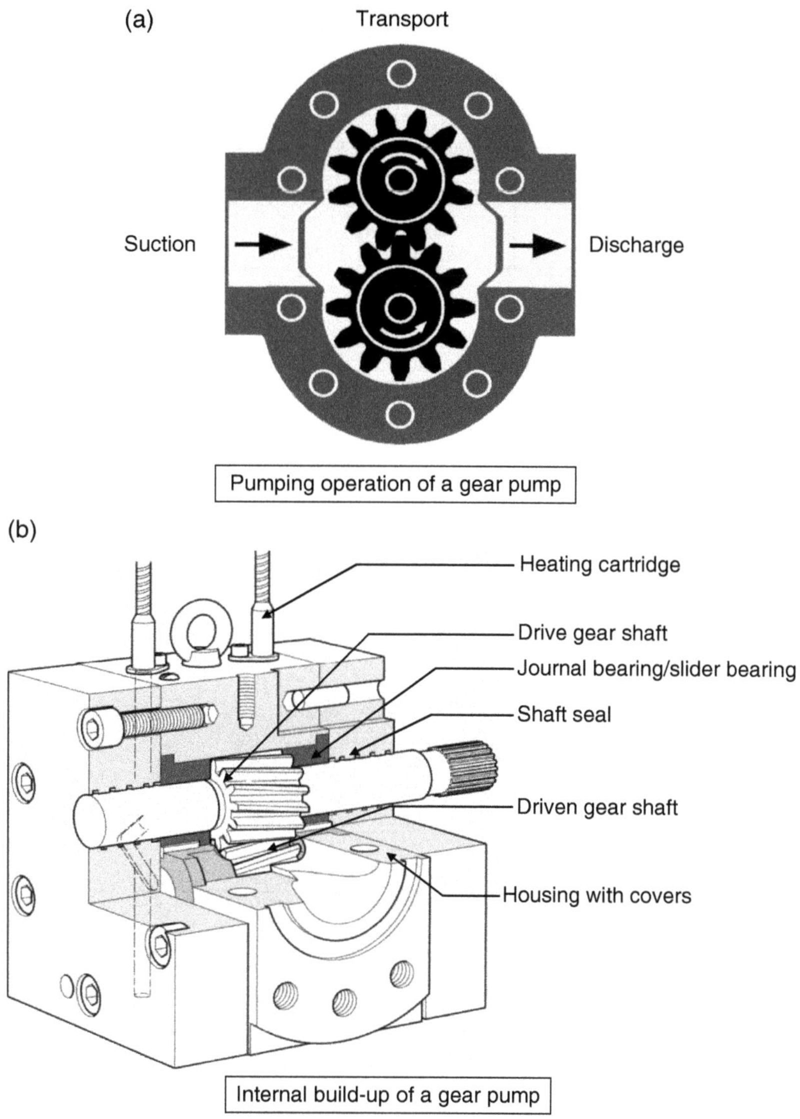

Figure 2.31 (a) Pumping operation of gear pump. (b) Internal build-up of gear pump. Source: Reproduced with permission of Maag Pump Systems AG, Switzerland.

extrusion processing is significantly more competitive than extrusion processing with no gear pump assistance, from a technical and economical standpoint.

Maag Pump Systems AG, a gear pump equipment manufacturer, offers a wide range of gear pumps which can be associated with screw extrusion for thermoplastics extrusion processing (Maag Pump Systems AG, 2010). These gear pumps allow a wide range of operating conditions to be covered: melt viscosity up to 30,000 Pa.s, temperature up to 350°C, suction pressure up to 120 bars, discharge pressure up to 350–500 bars.

As biomaterials are generally thermal-sensitive polymers, the use of gear pumps in biomaterials extrusion processing should allow operation with a lower time-temperature history, with the advantage of preserving the quality of processed products.

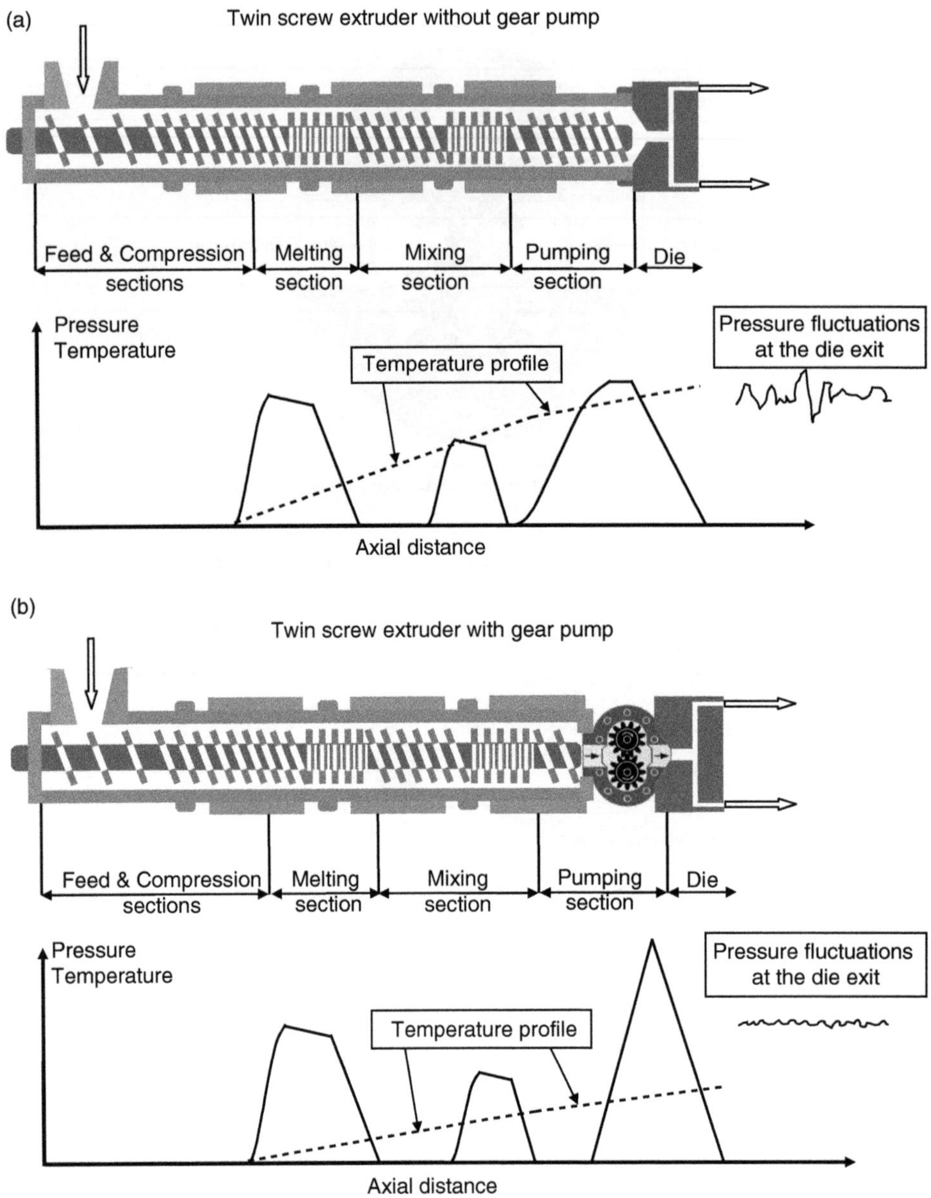

Figure 2.32 Typical configurations of extrusion processing units: (a) without gear pump; (b) with gear pump. Source: Reproduced with permission of Maag Pump Systems AG, Switzerland.

References

Automatik Plastics Machinery GmbH (2010) Private communication: Strand pelletizing in polymer compounding. Information available through: www.automatikgroup.com/.

Cantor K (2006) *Blown Film Extrusion: an introduction*. Munich: Carl Hanser Verlag.

Clextral (2013) Private communication: Feeding technology for long bast fibre materials. Information available through: www.clextral.com/.

Della Valle G, Barres C, Plewa J, Tayeb J, Vergnes B (1993) Computer simulation of starchy products transformation by twin-screw extrusion. Journal of Food Engineering 19: 1–31.

Giles HF, Wagner JR, Mount EM (2005) *Extrusion: the definition processing guide and handbook*. New York: Plastics Design Library, William Andrew Inc.

Hensen F (1997) *Plastics Extrusion Technology*. Munich: Carl Hanser Verlag.

Kaczorowski D (2002) Usures d'un acier inoxydable austénitique dans de l'eau à haute pression et haute température. PhD dissertation, Ecole Centrale de Lyon, France.

K-Tron (2010) Private communication: Loss-in-weight feeding principle and weight-belt feeding principle. Information available through: www.ktron.com/.

Levine L (1995) Of paddle mixers and preconditioners. Cereal Foods World 40(6): 452–453.

Maag Pump Systems AG (2010) Private communication: Basics of gear pumps and basics of compound pelletizing. Information available through: www.maag.com/.

Manas-Zloczower I (2009) *Mixing and Compounding of Polymers: theory and practice*, 2nd edn. Munich: Carl Hanser Verlag.

Martelli FG (1983) *Twin-Screw Extruders: a basic understanding*. New York: Van Nostrand Reinhold.

Michaeli W (2003) *Extrusion Dies for Plastics and Rubber: design and engineering computations*, 3rd edn. Munich: Carl Hanser Verlag.

Qi Wang (2007) Advanced analysis and design of polymer sheet extrusion. PhD dissertation, University of Missouri, Columbia, USA.

Rauwendaal C (2001) *Polymer Extrusion*, 4th edn. Munich: Carl Hanser Verlag.

Reid JD, Campanella OH, Corvalan CM, Okos MR (2003) The influence of power-law rheology on flow distributions in coathanger manifolds. Polymer Engineering and Science 43(3): 693–703.

Sciences Computers Consultants (2013) LUDOVIC: Logiciel d'Utilisation de DOubles VIs Corotatives, version 5.2. Information available through: www.scconsultants.com/.

Tadmor Z, Klein I (1970) *Engineering Principles of Plasticating Extrusion*. Polymer Science and Engineering Series. Florida: Robert E. Krieger.

3 Extrusion Engineering

As discussed in Chapter 1, extrusion processing technology has developed rapidly during the past 70 years with applications being continually expanded to new areas of the polymer processing industry. Through these developments, process engineers are faced with many extrusion-related problems or questions such as:

- setting up screw configuration and extrusion variables according to product objectives
- maximizing extruder productivity
- solving flow instabilities in an extruder
- explaining abnormal wear development in an extruder
- specifying the control parameters for an extruder
- scaling up an extrusion process from laboratory experiments
- designing an optimum extrusion system.

Due to the complexity of both the extrusion process and the properties of the materials being converted as well as associated transitions, answers to many of these questions and concerns have been primarily of an empirical nature rather than responses based on the analysis of the system using theoretical bases. Extrusion processing technology has been developed using trial-and-error approaches involving costly experimental runs, which coupled with a smart sense of observation have led to many successful applications in various industrial sectors. Comprehensive understanding on transport phenomena in screw extruders should overcome this empirical approach and lead to the optimization of existing processes as well as boosting development of new extrusion processes.

This chapter focuses on the engineering analysis of thermomechanical processing in screw extruders, with emphasis on the flow analysis in single screw extruders which has benefited from numerous relevant experimental and theoretical studies in the past 40 years and provides the basics of extrusion engineering.

3.1 Thermomechanical processing in screw extruders

3.1.1 Process configuration of single screw extruders

As explained in Chapter 2 (see section 2.1.2.1), single screw technology typically consists of a one-piece screw with a channel of variable depth and constant pitch, or a splined shaft that accepts modular screw sections with a constant channel depth and variable pitch. In regard to the process functions, the screw-barrel assembly is normally composed of four sections in series (Figure 3.1).

- The feed section, with a high transport capacity (deep channel or large screw pitch) for solid and particulate raw material (solid powders and particulates).
- The compression section, where the material is compressed and densified under the compressive effect of the screw (in a channel of decreasing depth or pitch). During transport in this section, the material is heated by friction and conductive heat transfer through heated barrels until melting occurs. In this section, the material changes from a solid particulate state to a continuum state (viscous fluid). The conversion from a powder to a fluid melt, which is known as plastication, is described in Chapter 4 (see section 4.2).
- The metering section, with screw elements having a small channel depth or pitch. In this section the material is sheared under laminar conditions. The mechanical energy dissipated during shearing is transformed into thermal energy (increasing the temperature of the material) and also used to physically and chemically modify the material, and to bring it to a rheological state compatible with satisfactory structure forming and product shaping. In this section, there is also a pumping action, under which the pressure needed to convey the material

Extrusion Processing Technology: Food and Non-Food Biomaterials, First Edition. Jean-Marie Bouvier and Osvaldo H. Campanella.

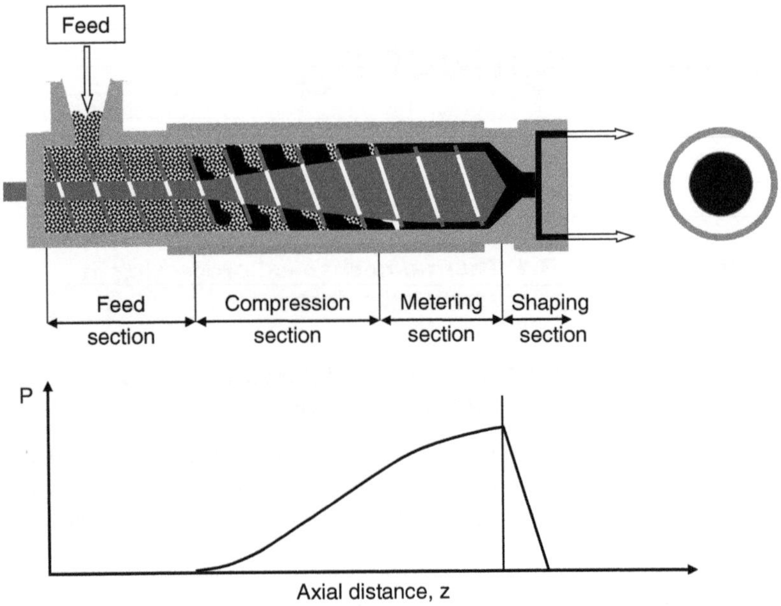

Figure 3.1 Thermomechanical processing in a monobloc single screw extruder.

through the die opening is built up. As illustrated by the pressure profile down the barrel length (see Figure 3.1), this section operates pretty much as the only effective processing section of the single screw extruder.

- The shaping section, where the melt, which has been physically and chemically transformed in previous sections, passes through a die or die head to shape the product. For instance, extrusion compounding processes consist of extruding strands of the material for further pelletization, while in the extrusion-cooking process, the material at the exit of the die is expanded by evaporation of the water present in the product in liquid form due to the high pressure existing inside the extruder. During expansion and water evaporation evaporative cooling promotes hardening of the product because its temperature reaches or is lower than the product glass transition temperature. Aspects of product texturization in extrusion cooking are discussed in Chapter 6, whereas the glass transition phenomenon is discussed in Chapter 4.

In the metering section of single screw extruders, the flights of the rotating screw convey the viscous material down the barrel. The adhesion of the extrudate at the barrel wall prevents its turning with the screw, allowing the positive pitch angle of the flights to push it along the barrel. This is a drag flow, and its velocity is directly proportional to the screw speed. In the single screw extruder, the material is conveyed by friction forces whose efficiency depends mainly on the friction of the material with the barrel walls. If friction does not exist, the material turns with the screw to the detriment of the shearing rate that can reach a value of zero. The differential velocity between the screw and the barrel is essential to promote the drag flow. On the other hand, if the material adheres to the barrel, maximum shearing will occur with the existence of a true drag flow. To smooth the drag flow, it is a common practice to reduce the friction coefficient of the extruded material with the screw (by polishing its surface) and to increase the friction coefficient with the barrel (by including grooves on the barrel walls).

As drag flow moves the material forward in the screw channel of the metering section, a positive gradient pressure is produced which generates a pressure flow opposite to the drag flow. This pressure flow depends on the pressure gradient and thus on the die restriction downstream. Pressure flow causes a reduction in the net flow of the material exiting the extruder. Thus, the net volumetric throughput of a single screw extruder is a combination of drag and pressure flows and the operational throughput of the extruder is dictated by the screw speed, die pressure, and rheological properties of the flowing material. The coupling of these variables limits the operating range and flexibility of the extruder. The operational throughput is a global response of the characteristics within the screw-channel assembly. Flow characteristics in the screw channel of the metering section are analyzed thoroughly in section 3.2 of this chapter, as it provides tools to determine the mixing efficiency, convective heat transfer, residence time distribution, and distribution of shear and extensional strains. Actually, these different variables significantly affect the extent and quality of the material that flows into the screw channel.

3.1.2 Process configuration of intermeshing co-rotating twin screw extruders

As discussed in Chapter 2, modular twin screw technology uses two splined shafts that can hold screw elements with a constant channel depth, and various pitches and designs such as reverse pitches for enhanced shear, mixing disks for intense mixing, large positive pitches for venting, etc. As illustrated in Figure 3.2, the screw-barrel assembly of modular, intermeshing co-rotating twin screw extruders has typically similar functions to those of the monobloc single screw extruder but due to the characteristics of the modular screw design, it enables incorporation of several process functions in series. In regard to the process functions, the screw-barrel assembly is normally composed of the following sections (see Figure 3.2):

- The feeding and conveying section with large pitch screw elements.
- The compression section, where the screw elements have decreasing pitch.
- The metering section, or processing section, which usually comprises several process functions in series. At the output of the compression section, the densified material enters the processing section, where specially dedicated screw elements may perform functions such as melting, intense mixing, shearing, degassing/venting, cooling, pumping, etc. The intensity of shearing and mixing is governed by the geometric characteristics of special screw elements, such as those described in Chapter 2 (see Table 2.2). In addition, screw elements can be single flighted or double flighted (rarely triple flighted), which are counted as the number of parallel channels along the screw. In comparison to double-flight screw elements, for the same channel depth and pitch, single-flight screw elements have less conveying capacity, higher pressure build-up, and a more limited range of variation of the residence time. In addition, double-flight screw elements have the advantage of applying a more uniform shearing rate within the screw channel, which is a positive factor for achieving uniform processing of the material.
- The shaping section, which generally functions in the same manner as for single screw extruders.

As seen in Figure 3.2, the pressure profile down the barrel length on intermeshing co-rotating twin screw extruders can be varied to allow for different process functions, which are designed in series according to the process

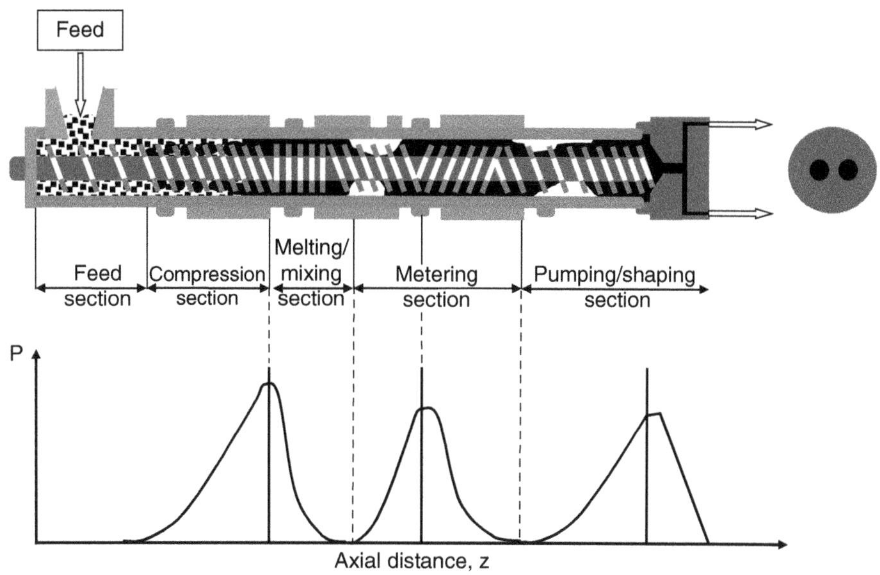

Figure 3.2 Thermomechanical processing in intermeshing, co-rotating twin screw extruders.

objectives. Such a succession of processing zones brings a high level of process flexibility for this type of extruder. It is worth noting that intermeshing co-rotating twin screw extruders are not totally filled. In fact, the screw-barrel assembly shows filled characteristics in pressure build-up zones whereas it is partially filled in conveying zones comprised between processing zones.

3.1.3 Processing specificities

Extrusion can be regarded as a combination of unit operations that simultaneously transports, mixes, heats, and shapes polymeric and non-polymeric materials, often under conditions of extremely high temperature and shear. As such, during the extrusion process the processed material is subjected to thermomechanical treatments that have a profound influence on its molecular structure and properties. Although the implications of molecular transformations on the properties of extruded materials could be of a different nature during the extrusion of plastics and the extrusion of bio-based polymers, in both areas there has been great deal of interest in characterizing the molecular properties of the extruded material. During the extrusion of polymeric plastics, the interest lies in forming a melt that facilitates its shaping but having minimum degradation of the polymeric molecules. Numerous studies can be found in the area of plastics, mainly assessing changes in the molecular weight of the material and the presence of compounds that may result from deleterious reactions, e.g. oxidation reactions, which negatively affect the properties of the final extruded product (Capone et al, 2007).

During the extrusion of bio-based polymers, great consideration has been given to the fate of key macromolecules forming most of these biomaterials, notably polysaccharides and proteins. Non-starch polysaccharides, for example food gums, which are largely used in food and non-food formulations, have received less attention. Although they are not macromolecules, lipids and other small molecules such as oligosaccharides and mono- and disaccharides (e.g. sugars) and minerals (e.g. salt) are also important in the extrusion of bio-based polymers given the physicochemical interactions that these components have with key biomacromolecules which are of relevance to the quality of the final product in terms of nutrition and texture for foods and specific functional properties for non-food materials.

Figure 3.3 illustrates the different approaches followed in the extrusion of plastic polymers and bio-based

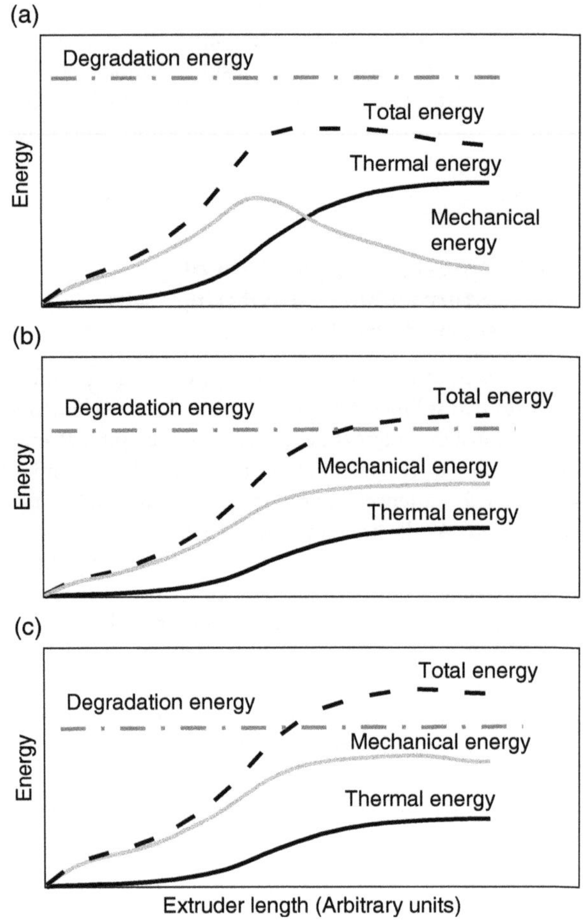

Figure 3.3 Different approaches used for the extrusion of plastic polymers and biomaterials in terms of thermal and mechanical energies profiles in the process. (a) Energy profile used for extruding plastic polymers. (b) Energy profile used for extruding biopolymers with minimum molecular degradation, e.g. extrusion of protein-based formulations. (c) Energy profile for extruding polymers where molecular degradation is sought, e.g. starch-based products with enhanced digestion.

polymers. Essentially, the energy supplied to the material being extruded is composed of a thermal energy which is applied by a variety of heating technologies, most of which are described in Chapter 2. The other type of energy supplied is the mechanical energy that is applied by the rotation of one or two screws inside a barrel while the extruded material is conveyed from the feed to the shaping section. The different screw profiles used in extrusion, described in Chapter 2, are generally designed to provide increased mechanical energy when the product moves through the barrel from the entrance to the exit of the extruder. In plastics extrusion, the increase in mechanical energy is intentionally avoided and instead the screw profile is designed to have a decreasing input of mechanical energy when the material approaches the shaping section. Thus, the total energy supplied to the product is kept below the energy needed to degrade its polymeric structure (Figure 3.3a) and avoid deleterious effects in the quality of the final product such as surface quality, optical transparence, or a decay of the overall mechanical and electrical performances (Brent Strong, 1996; Capone et al, 2007).

The approach used in the extrusion of biomaterials is different because degradation of the bio-based polymers is often necessary and sought to enhance some of its functional properties in the final product, e.g. better digestion of starch, better cellular structure and texture. The structure of these extruded bio-based materials is related to many of their rheological properties, notably viscoelasticity (Della Valle et al, 1998). These intended changes in the structure of the biomaterials make necessary the use of screw profiles that can input a mechanical energy that continuously increases from the feed to the metering section (Figure 3.3b, c). Thus, in addition to the multiplicity of unit operations that extruders provide, they are also considered as reactors where a number of physical and chemical transformations take place; the type of extruder has a large influence on those transformations of key raw materials such as starch and proteins (Kokini, 1993; Vergnes & Berzin, 2010). Extrusion conditions also significantly affect the thermomechanical treatment applied to the raw materials during the process, which in turn translates into the properties of the final product.

In order to systematically study the very complex process of extrusion, the large number of parameters involved in the process can be separated into three different groups:

- process parameters
- system parameters
- product parameters.

Process parameters include variables associated with the extruder equipment which are discussed in Chapter 2 and include screw configuration, barrel temperature, and type of extruder as well as parameters associated with the product formulation, notably moisture content, raw material characteristics such as the type of flour, its particle size distribution, and operational variables such as screw speed, feed rate, and die-head configuration. Process parameters have influence over the system parameters which are mainly associated with the process and include material residence time distribution in the extruder and mechanical and thermal energy input during the process, which can be estimated via one variable such as the barrel temperature or by a combination of variables to yield a parameter such as the specific mechanical energy (SME). The SME concept is defined in Chapter 6 (see section 6.1.1.4).

The system parameters in turn affect the product parameters that are associated with the physicochemical properties of the product. Therefore, product parameters are of importance because they are associated with the product quality. Among them, and related to foods, can be included texture, taste, color, expansion, and solubility. For both plastics and food products, these parameters are mainly associated with the structural changes to which the extruded material is subjected during the extrusion process. The interrelationship among the group of parameters and the material structural changes are seen as happening in an intensive reaction process (later discussed in Chapter 10), which can be described by a system analytical model proposed for the extrusion process (Brümmer et al, 2002; Meuser & Van Lengerich, 1984). Although the system analytical model approach has been used to describe structural changes of starches, its application to other biomacromolecules and plastics macromolecules may follow a similar methodology.

Figure 3.4 schematically shows a list of variables that can affect the extrusion process separated by groups of parameters identified by Meuser and collaborators. The interrelationship between these variables and their effects on the extrusion process are also illustrated in the figure. Variables grouped in the process parameters such as the raw material characteristics and extruder operational conditions can be regarded as the input to the system analytical model that yields variables such as the SME, residence time distribution (RTD), temperature, pressure, and viscosity of the melt, which form the system parameters. As discussed above, the quality of the product is associated with product parameters which among others include product expansion, bulk density, texture, and product morphology. It is also important to note that extrusion can be affected by external disturbances (system disturbances) such as ambient conditions and equipment wear, notably in the screws, aspects that have been discussed in Chapter 2.

Although the system analytical model is of a great conceptual value, the large number of parameters involved in extrusion and the complexity of the interactions occurring during the process make it very difficult to establish relationships that could help to understand the effect of process parameters on product quality. Because of this

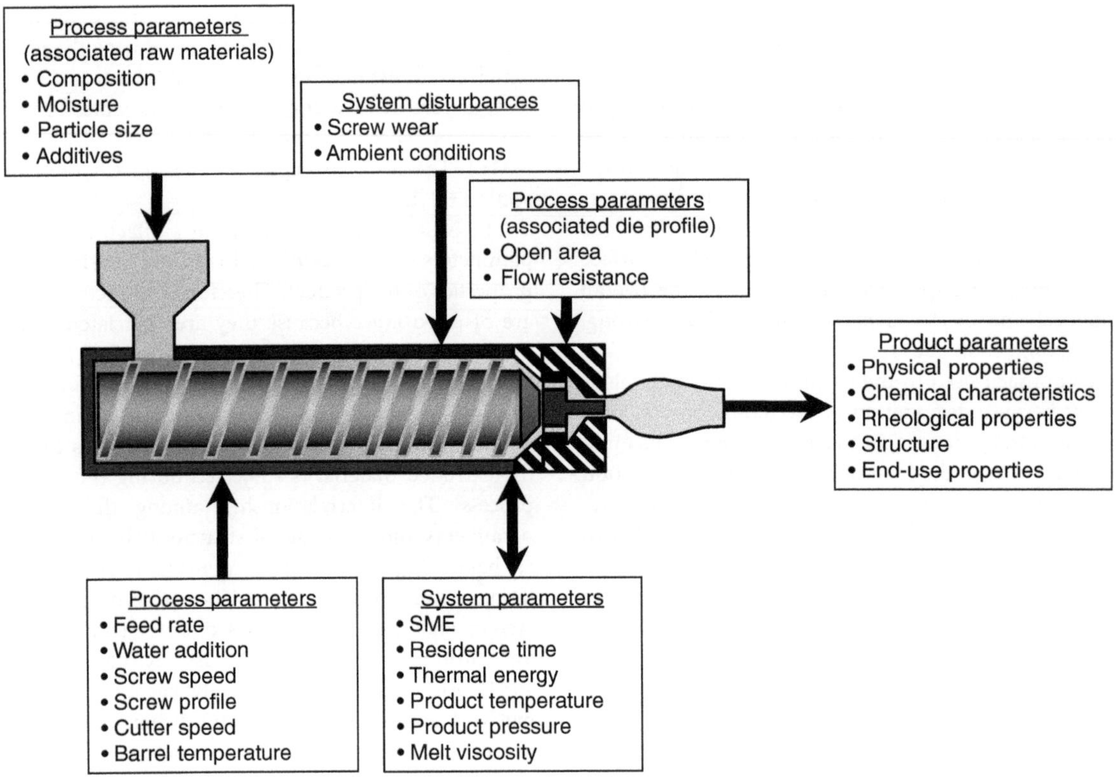

Figure 3.4 Interrelationship between variables involved in the extrusion process. Source: Adapted from Harper 1989 and Bouvier 2008.

difficulty, simplified approaches for research and applied food extrusion have been adopted in the past, in which the effects of a few variables were studied at a time. This approach has diverted the study into different directions, which focus on only a part of the whole extrusion process. By defining key variables and establishing the basis of the interactions between these and product quality, progress towards modeling the extruder performance as well as the development of relationships between input and output variables can be made.

For modeling purposes, in terms of fluid flow and energy transfer in the extrusion process, the rheology/viscosity of the melt within the extruder barrel is of crucial importance. If the melt rheology is known, it helps not only in the understanding of the extrusion process in terms of effects of formulations and extrusion conditions on the process performance, but also in the design of an extrusion control system. The rheology of the melt is the result of many of the transformations undergone by the extruded material, which due to the intensive thermal and mechanical energy input occur in a time of the same order of magnitude as the extruder residence time. Thus, it is important to describe physicochemical transformations that occur in extrusion, with an emphasis on the effects that these changes have on the properties of biomaterials such as viscosity, other thermomechanical properties and the extrusion process itself. More detailed description focusing on the chemical and physical changes of these materials is given in the following chapters.

The purpose of the following section of this chapter is to analyze and discuss thermomechanical processing in screw extruders, in single screw extruders in particular, using the basics of extrusion science and engineering available in the literature.

3.2 Thermomechanical flow in screw extruders

3.2.1 Modeling approaches

The first question that arises regarding modeling the extrusion process is what is the reason for obtaining a descriptive and representative model of the complex extrusion operation? There are many reasons for modeling the extrusion process, which include (Kohlgrüber, 2008):

- increasing process understanding
- simulating and creating qualitative "what if" scenarios, to calculate inaccessible process variables
- simulating system or critical process conditions
- scaling processes up or down
- saving on pilot and industrial costs
- saving training costs
- promoting process monitoring and control.

Figure 3.5 illustrates the interrelationships between extrusion modeling and the different areas involved in an extrusion process. It also clearly shows that extrusion modeling can establish an alternative, probably less expensive, link between a planned process and results of that process including the transformation of raw material into high-quality finished products. Extrusion modeling can also serve as a link between small pilot plant and large production trials. In reference to Figure 3.5, it is noteworthy that the feasibility of a model to describe a process must be assessed by a validation step that can apply to both the small pilot plant and production scales.

The second question arising in regard to the benefits of modeling is related to the feasibility of obtaining a very rigorous model that can describe the flow of a material that is thermomechanically modified in the complex geometry of the extruder. Thus, in addition to including the complexity of the equipment geometry, the model should take into account the changing properties of the material during the process. Property changes of extruded materials are further described and discussed in following chapters.

A related question to the modeling approach is how rigorous the model needs to be. In this respect, the approach to developing a model can vary according to specific needs. It can be a simple model relating dimensionless or dimensioned process parameters. At the other extreme, there is a very general and rigorous model that does not need parameters beyond the properties of the material being extruded as well as the processing conditions and equipment geometry. A different modeling approach combining these two concepts would be that of a semi-empirical model linking experimental results and theory.

Kohlgrüber (2008) classifies the models by dimensions. Models of dimension 0 are simple mass and energy balances that look at what is going into and what is coming out from the extruder in terms of matter and energy. These models are described by simple overall balances of matter and energy without considering smaller details.

3.2.1.1 Dimension 0 models

Let us consider a complex process that involves transformation from a powdered solid to a melt, venting and degassing the material, shaping and expansion (in the case of an extrusion-cooking process, for instance). A macroscopic material and energy balance to the system will give:

$$\dot{P}_{shaft} + \dot{q}_{barrel} = \Delta\dot{U} + Q\Delta p + \dot{q}_{loss} + \Delta\dot{H}_{vap} + \Delta\dot{H}_{melting} \tag{3.1}$$

where $\dot{P}_{shaft}$ is the power in the extruder motor, $\Delta\dot{U}$ the rate of internal energy change in the product, p is the pressure, $\Delta\dot{H}_{vap}$ the rate of enthalpy change involved in the

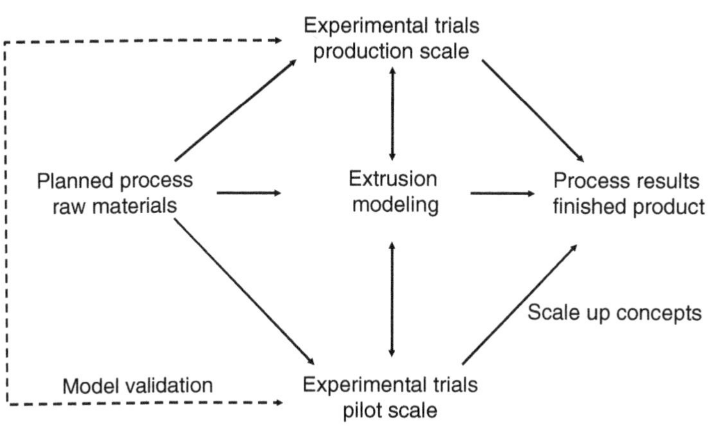

Figure 3.5 The extrusion modeling approach. Interrelationship between activities associated with extrusion modeling.

vaporization of the plasticizer/solvent (mostly water in the extrusion of most biomaterials) in the degassing zone and the evaporation during the expansion of the product, $\Delta\dot{H}_{melting}$ the rate of enthalpy necessary for the transformation of the feed from a powdery solid to a melt, all expressed in kW, and Q is the volumetric flow rate through the extruder (m^3/s). $\dot{q}_{barrel}$ and $\dot{q}_{loss}$ are the rate of heat transferred through the extruder barrels and the heat loss, respectively, both in kW.

Although the model represented by Eq. 3.1 is a simple algebraic equation, it can provide a qualitative description of the process. It clearly demonstrates that all the energy input through the power consumed in the extruder motor and the external heat transferred through the barrel is used to convey and thermomechanically transform the material during the product transit from the feed to the die. Equation 3.1 also provides qualitative information and trends of some of the variables that are relevant to the extruder operation, notably the temperature. A typical and simple calculation derived from this equation is the temperature rise in an autogeneous extruder ($\dot{q}_{barrel} = 0$), which with the assumption of negligible heat losses as well as enthalpies for melting and solvent evaporation in the degassing section simplifies Eq. 3.1 to:

$$\Delta\dot{U} = \dot{P}_{shaft} - Qdp \qquad (3.2)$$

The rate of internal energy change in the product due to the heating from the extruder inlet temperature, T_{in}, to the temperature of the product before leaving the die, T_{out}, can be calculated from the mass flow rate of product in the extruder, $\dot{m}$, an average heat capacity for the product, c, and the temperature difference $T_{out} - T_{in}$ as:

$$\Delta\dot{U} = \dot{m}c(T_{out} - T_{in}) \qquad (3.3)$$

Changes in the properties of the extruded material are discussed in following chapters, where it becomes apparent that the transformations of the product from a powder state in the feed to a melt state at the die are accompanied by changes in the thermomechanical properties such as heat capacity c. Assuming a constant specific heat could result in an oversimplification if a rigorous model describing the extrusion operation more quantitatively is sought. However, results using this approximation provide useful trends that help to understand underlying mechanisms in the process.

By substituting Eq. 3.3 into Eq. 3.2 and with further rearranging, the following equation can be obtained:

$$T_{out} - T_{in} = \frac{e}{c} - \frac{\Delta p}{\rho c} \qquad (3.4)$$

where $e = \frac{\dot{P}}{\dot{m}}$ is a specific energy, in kW.h/kg, defined as the drive power, and $\frac{\dot{m}}{Q} = \rho$ is the product density. Equation 3.4 predicts that the increase of product temperature in the extruder, or in a section of the extruder, is related to the drive power and to the pressure difference between the inlet and outlet locations of the section being considered. It is important to note that for most extrusion processes, an examination of the right side of the equation shows that the first term is considerably larger than the second term (Kohlgrüber, 2008) and for some conditions will be the prevalent term for the calculation of the product temperature increase in a given section of the extruder. If considered, and depending on the type of screw profile used, the second term could be positive or negative. For those cases, changes in product temperatures predicted by Eq. 3.4 are related to the pressure difference. As discussed in Chapter 2, in single screw extruders the pressure increases smoothly from the compression section to the die at different rates that are governed by the screw geometry; whereas at the die there is a sudden decrease of pressure (see Figure 3.1). In consequence, in the compression and metering sections, the pressure difference is positive and the increase of temperature is not as high as that theoretically calculated assuming the sole effect of the driving power. At the die, the pressure difference is negative but there is no input of energy, so Eq. 3.4 predicts a small increase of temperature.

In twin screw extruders the use of specific modular screw profiles creates pressure differences that can be either positive or negative in the different sections of the extruder. As seen in Chapter 2, the use of kneading or mixing disks and the selected staggering angle between them enable the creation of conveying and pressure generation zones, neutral (non-conveying action), or a back feeding action that creates a pressure drop. The conveying action of a set of screws can be defined by the kinematic flow parameter Λ which is defined as (Kohlgrüber, 2008):

$$\Lambda = \frac{Q}{A_1 \cdot N \cdot D^3} \qquad (3.5)$$

A_1 is a dimensionless parameter that provides information on the maximum rate of conveying of a given screw, N is the screw speed, and D the internal diameter of the barrel. The parameter A_1 is defined by the screw geometry and measures the inherent conveying ability of a screw or arrangement of screws (e.g. when mixing paddles are used). It can be experimentally measured or theoretically calculated through a rigorous computer model.

In Eq. 3.5 the relation $\frac{Q}{ND^3}$ is a dimensionless parameter known as the throughput characteristic, which characterizes the ability of the screw or an extruder section to handle a given volumetric flow rate Q. Thus, Eq. 3.5 represents the ratio between the throughput characteristic for a given operating condition and the maximum "inherent" throughput that a screw or an extruder section can handle. That ratio defines the kinematic flow parameter, Λ, which can be larger or smaller than one. Figure 3.6 illustrates the pressure profile that can be generated in different sections of twin screw extruders. Noted in the figure there are positive ($\Delta p > 0$) and negative pressure gradients ($\Delta p < 0$) from which, depending on their magnitude, Eq. 3.4 may predict different changes in temperature. The magnitude of those pressure gradients and thus the temperature changes predicted by Eq. 3.4 can be calculated by the kinematic flow parameter Λ. For filled conveying sections, it is assumed that the inherent maximum throughput, defined by the parameter A_1, has been reached. Thus, if the throughput characteristic in the extruder is greater than the maximum inherent possible flux A_1, the kinematic flow parameter holds the condition $\Lambda > 1$ and a negative pressure gradient $\Delta p < 0$ is established. From Eq. 3.4 that would generate a positive temperature change for the section being considered. Conversely, when the characteristic throughput is lower than the inherent maximum possible, $\Lambda < 1$ and a positive pressure gradient ($\Delta p > 0$) is generated in the section (see Figure 3.6). From Eq. 3.4, the temperature change generated under that condition is lower than that produced by the sole effect of the driving power.

3.2.1.2 1D models

These models provide a view of what is happening in the process along a dimension, which is in the direction of the screw shaft. Thus, they neglect possible variations in any cross-sectional area of the extruder, notably velocity and pressure profiles and temperature gradients. The approach to develope this model is to relate the pressure and drag flow generated in a typical shear test, schematically shown in Figure 3.7. As an illustration, Figure 3.7a shows a typical drag flow created by the relative velocity between a moving and a stationary plate. It should be noted that although the simple shear flow is an oversimplification of the flow generated in the extruder, it provides some qualitative information on the shear stresses and shear rates involved in the process. Figure 3.7b is a schematic of the screw geometry and the generated flow whereas Figure 3.7c describes an approximation of

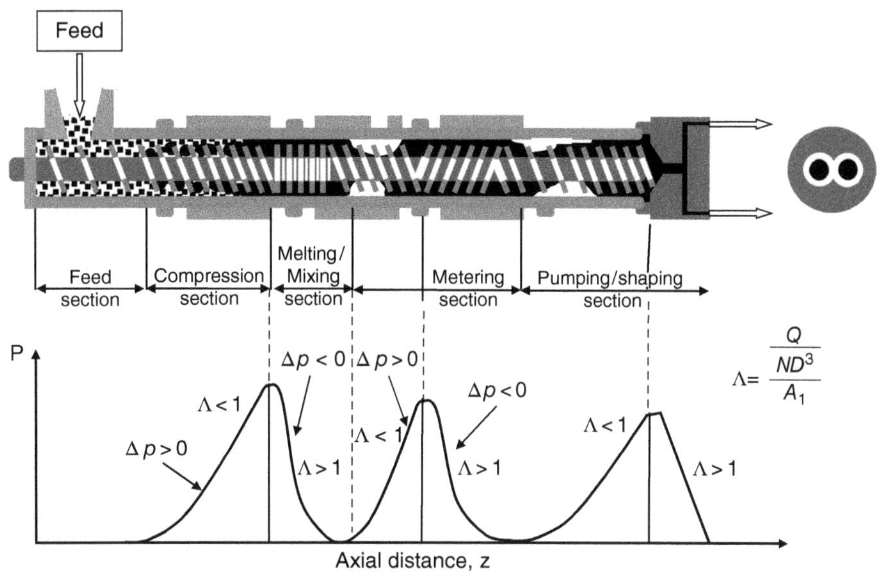

Figure 3.6 Simplified balance of energy based in an order 0 model to qualitatively analyze changes of pressure and temperatures in different sections of a twin screw extruder.

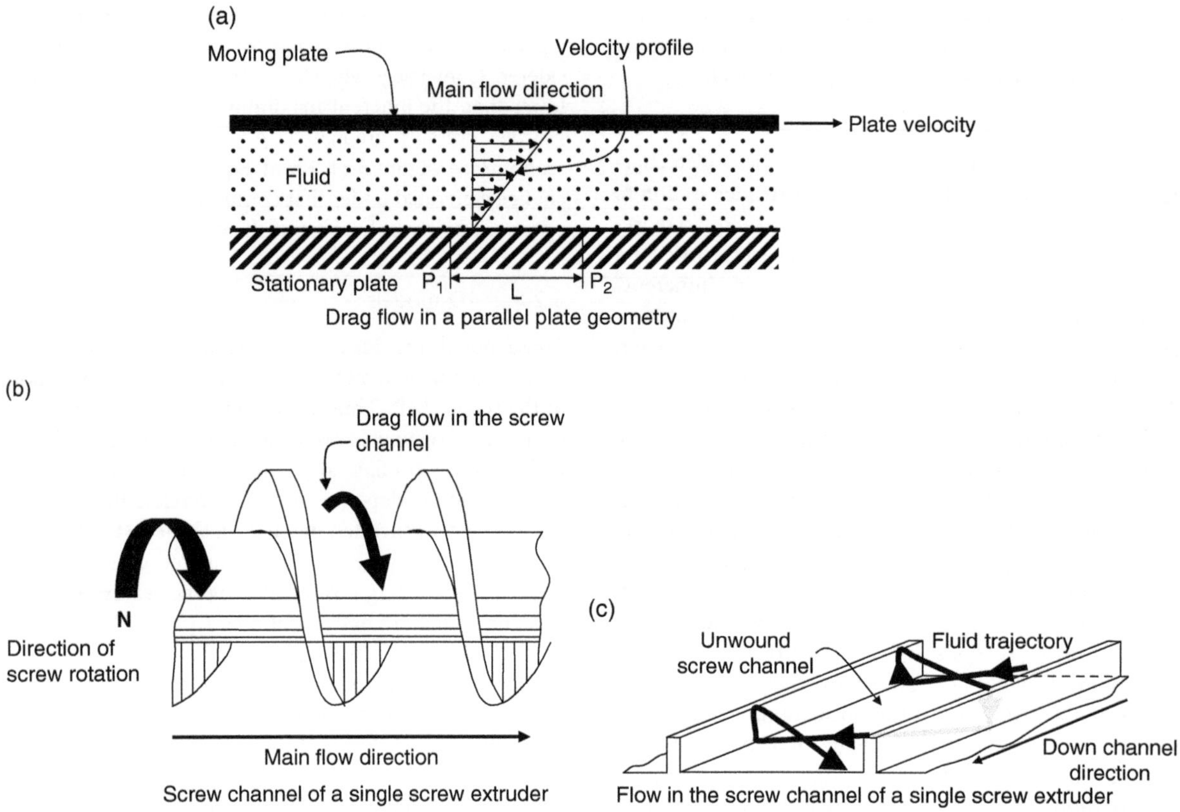

Figure 3.7 Schematic of a shear flow. (a) Shear flow generated by the movement of a plate with a constant velocity to deform a fluid located between this plate and a fixed one. (b) Shear drag flow in the channel of a single screw extruder. (c) Trajectory of fluid particles in the channel of a single screw extruder due to the presence of a shear flow.

the flow of material and the trajectories of the fluid particles in a rectangular co-ordinate system. To be able to approximate the real screw flow to the flow illustrated in Figure 3.7c, seen in rectangular dimensions, the screw channel has to be unwound. The fluid dynamics of the flow in the screw channel has shown that, under laminar flow, the fluid particles have different velocities and do not interact, which leads to a dispersion of residence times and poor mixing. However, these aspects of the flow are ignored in the dimension 1 modeling approach, and only variations in the channel direction are considered. By using fluid mechanics concepts applied to Newtonian flows, it can be demonstrated that for the dimension 1 model there is a linear relationship between the pressure gradient, Δp, generated in a section of the extruder and the volumetric throughput Q (Kohlgrüber, 2008). It can also be demonstrated that for Newtonian materials, the following relationship holds for both single screw and twin screw extruders:

$$\frac{1}{A_1}\frac{Q}{ND^3}+\frac{1}{A_2}\frac{\Delta p \cdot D}{\mu \cdot N \cdot L}=1 \tag{3.6}$$

The parameter A_1 was defined above, whereas A_2 is a new parameter that identifies the hypothetical pressure generation capacity of a screw or screw section when the flow in the extruder is completely restricted (Kohlgrüber, 2008); μ is the Newtonian viscosity of the material and L the length of the screw or the extruder section considered in the analysis. By combining Eqs 3.5 and 3.6, the following equation can be obtained:

$$\Delta p=\mu\frac{NL}{D}\cdot\frac{A_2}{A_1}\left(A_1-\frac{Q}{ND^3}\right)=\mu\frac{NL}{D}\cdot A_2(1-\Lambda) \tag{3.7}$$

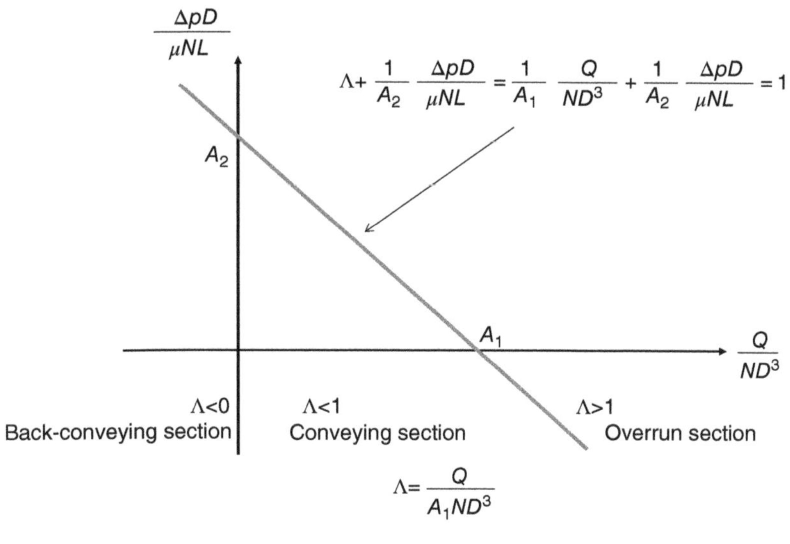

Figure 3.8 Pressure versus volumetric flow characteristics for extruders. Source: Adapted from Kohlgrüber 2008.

A relationship of this type is important because it enables a qualitative estimation of the pressure profile generated in different sections of the extruder. Figure 3.8 illustrates the theoretical relationship between pressure drop and volumetric throughput; it enables the location of regions which provide the sign of the pressure gradient. From either Eq. 3.5 or Eq. 3.7, it can be seen that for points on the theoretical curve with abscissa values such as $\frac{Q}{ND^3} > A_1$, the kinematic flow parameter is $\Lambda > 1$, i.e. there is an overrun situation on the screw or screw section, and Eq. 3.7 shows that the pressure gradient is negative. For points in the theoretical curve with abscissa values such as $0 < \frac{Q}{ND^3} < A_1$, the kinematic flow parameter is $\Lambda < 1$ and Eq. 3.7 predicts that there is a conveying action with a positive pressure gradient in that extruder region. For values of abscissa such as $\frac{Q}{ND^3} < 0$, the kinematic flow parameter is $\Lambda < 0$ and the screw section is back conveying. By estimating the power on the extruder shaft P_s, a linear equation similar to Eq. 3.6 relating the power and the volumetric throughput can be obtained:

$$\frac{1}{B_1}\frac{Q}{ND^3} + \frac{1}{B_2}\frac{P_s}{\mu N^2 D^2 L} = 1 \tag{3.8}$$

and by combining it with Eq. 3.5, the following equation is obtained:

$$P_s = \mu N^2 D^2 L \cdot \frac{B_2}{B_1} \cdot \left(B_1 - \frac{Q}{ND^3} \right) = \mu N^2 D^2 L B_2 \cdot \left(1 - \frac{1}{B_1} \cdot \frac{Q}{ND^3} \right) \tag{3.9}$$

where B_1 and B_2 are characteristic parameters for the screw. Thus, the theoretical energy characteristics of an extruder can be represented as before in a plot of $\frac{P_s}{\mu N^2 D^2 L}$ as a function of $\frac{Q}{ND^3}$. Figure 3.9 indicates that the greatest energy transferred to the extruded material takes place in the back-conveying region and is still positive for values of the throughput characteristic $\frac{Q}{ND^3}$ smaller than the parameter B_1. Over the abscissa it is possible to locate the parameter A_1, which as indicated in Figure 3.9 separates the conveying section where the screw (or the extrusion section being considered) is partially filled from the overrun section in which the screw (or screw section) is completely filled. For the partially filled section, the energy transferred to the product is lower and identified by the parameter B_4 (Figure 3.9).

If the throughput characteristic is larger than the parameter B_1, known as the turbine point, the energy would be negative (Figure 3.9), i.e. it means that it would be transferred from the product to the shaft, a situation that although theoretically possible is practically unrealizable.

3.2.1.3 Rigorous 2D or 3D models

Rigorous 2D or 3D models properly describing the extrusion process with a greater deal of detail can be regarded as tools to see the inside of the extruder or observe the process through a "glass extruder" (Kohlgrüber, 2008). These models rely on a system of partial differential equations

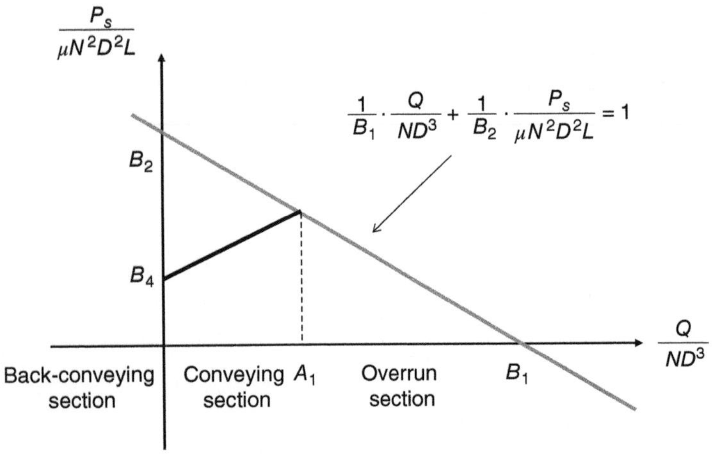

Figure 3.9 Power versus volumetric flow characteristics for extruders. Source: Adapted from Kohlgrüber 2008.

that describe the physics of material flow through a complex geometry, including heat transfer through the barrel and heat generated by friction as well as the chemical and biochemical transformations of the components forming the extruded material during its transit in the extruder channel. Thus, a multiphysics approach describing the fluid mechanics and the energy and mass transfer involved in the extrusion process must be considered.

Each of these models has parameters and properties that are mutually dependent and their solutions require the simultaneous solution of a system of partial differential equations describing a multiphysics model. For example, the viscosity of the product being processed is a material property relevant to the fluid mechanics model, but that strongly depends on the temperature, which is estimated from the model describing the balance of energy. Viscosity is also important because it is involved in the generation of heat by friction. Thus, a rigorous 2D or 3D model must be observed as a multiphysics approach in which all equations useful to study the material flow, energy transfer to the material through the barrel and/or generated by friction, and the chemical and biochemical conversion of the extrudate material components must be solved simultaneously.

In the two preceding chapters, different types of extruders used to process biomaterials and plastics have been discussed. In this section, a mathematical model describing a thermomechanical processing operation in which a material is forced to flow through a shallow channel formed between a rotating screw and a cylindrical barrel, in the case of a single screw extruder, or in the eight-shaped barrel of a twin screw extruder is described. The processed material exits through one or more flow restrictions, the die or die plate, which helps to create a high-pressure environment within the extruder. The high shear environment inside the extruder results in a mixing and shearing action on the material that leads to the generation of heat by friction and a mechanical energy input that transforms the material and changes its physicochemical properties, some of which are described and discussed further in the following chapters. Since the material is transformed by both mechanical and thermal energy, the mathematical model representing this process has to include appropriate equations that describe the flow of material through narrow channels and through the dies as well as equations that describe the energy balance that takes into account the generation of heat by friction, the transfer of that heat to the material and the external heating that can be applied through the extruder barrel. In addition, rheological models relating the stress and the rate of strain for the extruded material should be included in these equations; these rheological models applied to bio-based polymers and plastics are discussed in Chapter 4.

The equations of continuity, momentum, and energy applied to extrusion are essentially general equations describing the laws of conservation of mass, momentum, and energy in flowing systems. These equations are discussed in great detail in specialized transport phenomena books (Bird et al, 2007; Slattery, 1999). They are mathematically expressed as a system of partial differential equations in which the dependent variables are the three components of the fluid velocity vector and the fluid pressure in the momentum

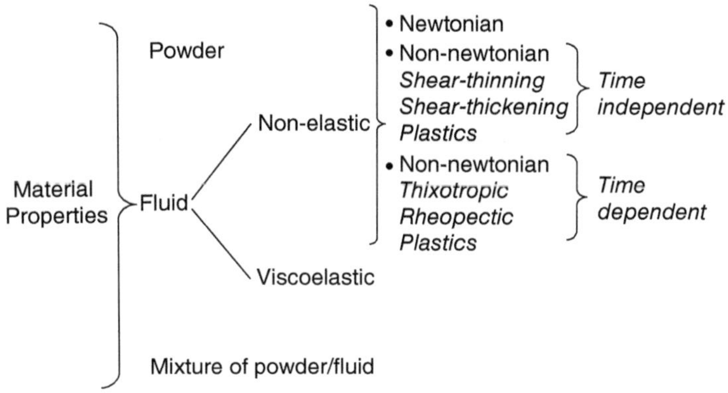

Figure 3.10 Schematic diagram showing the characteristics of materials that are extruded including a classification of their rheological behaviors.

equation and the fluid temperature in the energy equation. For both cases, the dependent variables are functions of position and time. In the event that a chemical reaction is produced inside the extruder when it is used as a chemical reactor, in a process known as reactive extrusion (described in Chapter 5), another partial differential equation (the mass transfer model) is used to estimate the concentration of one or more components as a function of the location and time. Changes in the concentration of some molecular species, e.g. volatiles in the degassing section of a twin screw extruder, can also be estimated by the mass transfer model. The complete simultaneous solution of this system of partial differential equations can be achieved only by using powerful numerical approaches such as the finite difference (FD) or the finite element (FEM) among other methods. Given the complex 3D geometry comprising the extruder mechanical parts, many of them already described in Chapter 2, adaption of these geometries to finite domains is not trivial and requires experience not only in software dedicated to the mechanical design of components (e.g. AutoCad™, ProE™) but also in optimal discretization of the domain representing those complex surfaces. Specialized extrusion software, such as LUDOVIC®, can help to perform modeling work. Examples of results obtained with this particular software are given at the end of this chapter.

It must be noted that a rigorous extrusion model not only depends largely on the thermal and mechanical properties of the materials subjected to the thermomechanical transformations, but also on the external conditions or spatial borders of the model, mathematically known as boundary conditions. In addition, the characteristics and properties of the material being extruded change drastically during the extrusion process and have a large influence on the results obtained from these models. Figure 3.10 illustrates the relevance of material properties to describe the flow and fluid mechanics of the materials in the process and shows how these materials are classified. As described in this chapter (section 3.1), there are different and well-defined sections in an extruder. Close to the feed and before the material is processed in the compression and metering sections, the properties of the material are those of a powdery solid. Flow of powdery materials is difficult to describe because the hypothesis of the continuum, which is essential to represent momentum, energy and mass transport phenomena, can hardly apply to systems which are discrete in nature. Despite that, several approximations have been used and are described in the following section (Hyun et al, 1996).

The flow of non-elastic fluids has been well covered in the extruder modeling literature with very good results. These types of fluids are those existing in the metering section of the extruder and generally denominated melts. Most of the modeling work has been applied to time-independent Newtonian and non-Newtonian fluids, including plastic fluids that depending on the stress applied can exhibit either solid or liquid behavior. Time-dependent fluids are essentially structured fluid materials, whose structure is seriously affected by shear. As shear forces applied during the extrusion process are extremely high, the time dependence of the fluid can be rapidly eliminated in the high shear ambient existing in the metering section. Most of the extruder melts are composed of polymeric molecules that tend to confer elastic properties to the fluid to transform it into a viscoelastic material. Modeling the flow of a viscoelastic material is a complicated task even in simple geometries, so modeling viscoelastic flows in the complicated screw geometry has been done with very limited success, often using unrealistic viscoelastic properties that do not resemble those of true extruded melts. Some of the earlier work in this area can be found in references which have mostly focused on plastic polymers and on the extrusion of dough-like materials (Dhanasekharan & Kokini, 2000; Karapetsas & Tsamopoulos, 2008; Yue & Zhao, 2008).

The other aspect to be considered in suitable models is the characteristic of the extrusion process where the material being extruded passes through the different extruder sections. As discussed in Chapter 2 as well as earlier in this chapter, sections that are partially filled and fully filled (in the case of an intermeshing co-rotating twin screw extruder) can be considered. The existence of a material in the screw having only one phase which can be either a powder or a melt (liquid), or a mixture of a solid and melt phase, or solid/air or melt/air in a partially filled screw can be even considered. Furthermore, the extruder can operate under steady-state or unsteady-state conditions. Most of the simulation work has focused on steady-state conditions.

The solution of a 2D or 3D multiphysics model will yield the main or primary variables (material velocity components, pressure, and temperature) that depend on time and the location on the extruder; thus these variables are considered fields. The primary variables fields can also be used to calculate other secondary variables fields that are relevant to the extrusion process (e.g. shear stress, shear rates fields). Other variables that are calculated as an overall averaged field and are scalar in nature include power, an axial pressure or a defined temperature, e.g. average, maximum or minimum. Lastly, all these calculated variables can be linked with thermal-mechanical degradation of the material, chemical conversion of components, and extruder residence time (Figure 3.11).

Simplified versions of this system of equations that describe the multyphysics model can be obtained using some realistic approximations of the extrusion system. As these approximations change when they are applied to single screw and twin screw extruders, it is convenient to differentiate the models specifically applied to these two systems. First, the general equations will be presented, followed by specific approximations to apply to the two types of extruders known. The other aspect to be considered for the modeling of the extrusion operation is the state of the matter at the different positions of the extruder. That is discussed in Chapter 4 where the transformation of a polymer powder to a melt is estimated in terms of models that predict the conversion of the extruded material and the state of the material at the different locations along the extruder length. Essentially, it is assumed that the state of material in the extruder is solid (powder) or liquid (melt). In the next section, the discussion will focus on the flow of solids in the section of the extruder close to the entrance, i.e. before the metering section.

The general equations are presented only in vector form for the sake of conciseness but components of these equations in rectangular, cylindrical and spherical

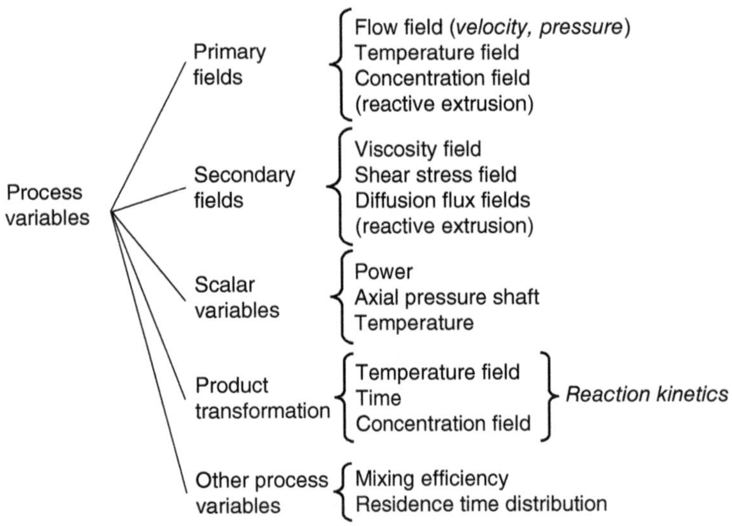

Figure 3.11 List of process variables and parameters of relevance in extrusion. Classification and utilization of those process variables to estimate parameters of relevance to the process.

co-ordinates are given in specialized transport phenomena books (e.g. Bird et al., 2007).

The incompressible and isothermal fluid flow governed by the velocity and pressure of the material within the extruder flow channel can be described by the equations of continuity and fluid motion given below (Tadmor & Klein, 1970):

$$\frac{\partial \rho}{\partial t} + \nabla \cdot \rho \underline{v} = 0 \tag{3.10}$$

$$\frac{\partial (\rho \underline{v})}{\partial t} + \nabla \cdot (\rho \underline{v}) = -\nabla p + \rho \underline{g} + \nabla \cdot \underline{\underline{\tau}} \tag{3.11}$$

where ρ denotes the density (kg/m^3), $\underline{v}$ the velocity vector (m/s), $\underline{\underline{\tau}}$ the stress tensor (Pa), p the pressure (Pa) and $\underline{g}$ a body force like gravity. An extrusion operation can be considered as a steady-state flow of an incompressible and highly viscous fluid. These assumptions allow one to discard all time derivatives and derivatives of the density with position, as well as inertial effects and volume forces such as gravity, which yield the simplified equations:

$$\nabla \cdot \underline{v} = 0 \tag{3.12}$$

$$0 = -\nabla p + \nabla \cdot \underline{\underline{\tau}} \tag{3.13}$$

The above two equations are used to describe the fluid mechanics of a continuous material with liquid-like properties which can be achieved at the end of the melting section as well as in the metering section of the extruder. For numerous applications such as thermomechanical plasticating of polymers or thermomechanical cooking of biopolymers, in the first processing section of the screw extruder the material is flowing in a powder form and in this extruder section the screws are not fully filled so the above equations are not appropriate for describing the flow of that material before the melting section. For the analysis of material flow, it is preferably to separate the analysis into the flow of powders, called solids conveying section, and the flow of melts, called melt conveying section.

3.2.2 Solids conveying section

The flow of particulate solids is important in the feed section close to the extruder inlet where generally a powder is fed by either volumetric or weight loss feeders (discussed in Chapter 2, section 2.3.1). It has been found experimentally that during the conveying of these solid particulates, they consolidate and tend to form a solid plug, which is mainly transported forward by the relative motion of the screw and the barrel surfaces. In practice, the screws are rotating and the barrel does not move. However, it is convenient for the derivation of pertinent equations to assume the opposite. As the analysis is based on relative velocities, results do not change using this assumption. It is also assumed that the velocity of the solid plug is uniform in the screw channel. To induce a solid plug flow, the generation of pressure in the screw channel has to be significant and depends on the degree of fill of the screw. When the transport occurs with no or little powder compaction, i.e. without the presence of a plug flow, the flow is called an Archimedean transport, which generally occurs when the feed flow is smaller than the conveying rate of the compacted solid. For starve feed extruders, an Archimedean transport condition is created intentionally and generally associated with a partially filled screw channel (Rauwendaal, 2001).

Tadmor and Klein (1970) developed equations to model the conveying of solids in the extrusion operations. These equations are based on the interaction of solids surfaces and the resulting friction and do not necessarily apply to the transport of solid powders or pellets, instead they describe the flow of a continuum system such as that found in the metering section of the extruder. However, the equations obtained can provide tools to evaluate trends that could help to understand mechanisms applicable to the flow of solids in the feed section.

Some earlier simplified analyses of the drag flow conveying the solid plug have considered that the drag is caused by the imbalance of forces created by the different friction coefficients between the solid plug and the screw and the barrel (Figure 3.12). This simplified approach is based on experimental observations showing that low friction between the solid plug and the screw and high friction between the solid plug and the barrel favor solids conveying. It has also led extruder manufacturers to develop polished screws of low friction and rugged barrel that promote friction with the conveyed solids. Friction forces can be calculated by multiplying the normal force acting on the surface by the friction factor (Tadmor & Klein, 1970). If a constant pressure p is considered in the solid plug and the contact area between the plug and the barrel and the plug and the screw are designed as A_b and A_s respectively, the forces needed to drag the solid plug would be $F_b = pA_bf_b$ at the barrel and $F_s = pA_sf_s$ at the screw. As shown in Figure 3.12a, the analysis of the

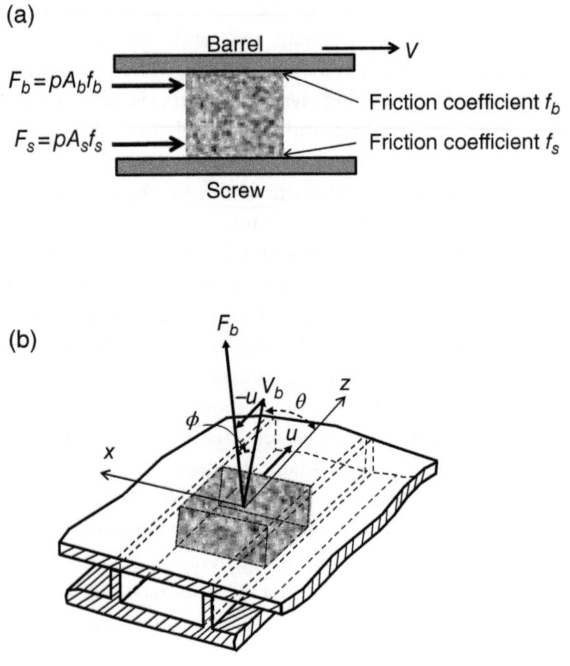

Figure 3.12 Schematic illustrating the drag flow of a solid plug. (a) Drag flow created by a plate moving with a constant velocity *V*. (b) Solid plug flow created in the channel of a single screw extruder.

powder/solid flow in extruders has been intended considering these differences in friction and the resulting forces depicted in the figure, from which several situations can be considered. The optimum condition is when $F_b > F_s$, and the solid plug is accelerated from a starting velocity u to reach the velocity of the plate V, producing the highest conveying action. Conversely, if $F_b < F_s$ the velocity of the solid plug will slow down until flow stops. At the equilibrium condition, the two forces are equal and the solid plug reaches a steady velocity that is lower than the velocity of the top plate (V) with the consequent reduced conveying action.

A more elaborated analysis takes into account the angle of the screw flight. Figure 3.12b illustrates the velocity of the top plate V_b (barrel), which is forming an angle θ with the direction of the channel (z). In addition, the solid plug may have a velocity u in the direction of the channel. The force at the barrel, F_b, which is able to drag the solid plug will have the direction of the velocity that results from the composition of the barrel velocity V_b and the velocity of the solid plug u in the channel direction. The drag force at the barrel, whose magnitude is calculated as $A_b f_b p$ (see

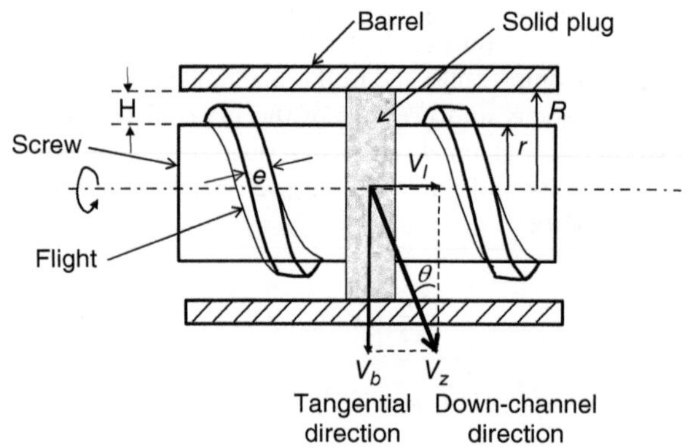

Figure 3.13 Flow of a solid plug in the channel of a single screw extruder.

Figure 3.12a), forms an angle ϕ with the barrel velocity. Thus, the component of the force in the channel direction can be estimated as $A_b f_b p\ cos(\theta + \phi)$. The friction in the screws without considering the flights enables establishment of an equilibrium condition by the following equation:

$$A_b f_b p cos(\theta + \phi) = A_s f_s p \tag{3.14}$$

As illustrated in Figure 3.12b, the angle ϕ is determined by the solid plug velocity u. Being the *cosine* a function that yields values in the range (0,1) for angles in the range $0 < \phi < \pi/2$, Eq. 3.14 provides the condition $A_b f_b > A_s f_s$ to specify the conveying action of the solid plug, which agrees with the experimental observations discussed above.

The real solid flow is, however, more complex than the model derived from the flow depicted in Figure 3.12. To develop a more realistic model, the effects of the flights should be incorporated as well as the channel curvature. This approach is schematically illustrated in Figure 3.13, in which a portion of the solid plug is moving in the curved channel with a velocity that is composed by an axial translation V_l, with a velocity component in the down-channel direction V_Z, which can be calculated by the following trigonometric relationship:

$$V_Z = \frac{V_l}{sin\theta} \tag{3.15}$$

In addition, the velocity of the solid plug at the barrel surface V_b can be obtained from Figure 3.13 as:

$$V_b = \frac{V_l}{tan\theta} \tag{3.16}$$

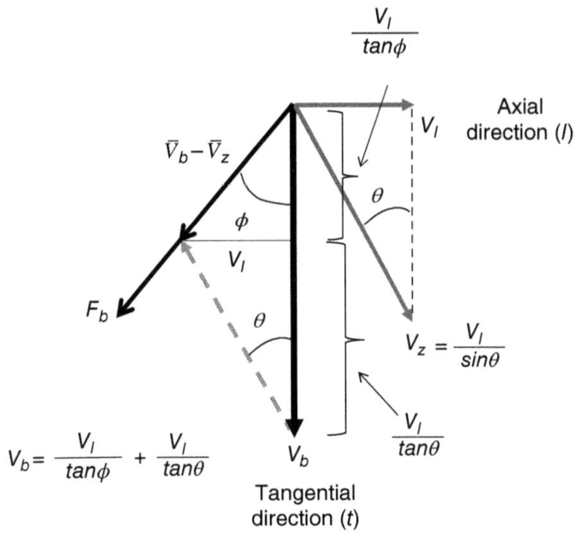

Figure 3.14 Schematic illustrating all velocities involved in the flow of a solid plug in the channel of a single screw extruder.

The friction forces between the barrel and the solid plug depend on the relative velocity between the barrel and the solid plug moving in the down-channel direction, which is graphically determined by the vector difference indicated in the balance of velocities illustrated in Figure 3.14 as $\vec{V}_b - \vec{V}_Z$. The angle that the vector difference forms with the barrel velocity, which is designed as ϕ, has a singular importance because it provides the direction of the friction force acting at the boundary between the barrel and the solid plug. Thus, one way to obtain that angle is to write a balance of forces in the solid. By using geometric relationships schematically illustrated in Figure 3.14, the following equation is obtained:

$$tan\phi = \frac{V_l}{V_b - \frac{V_l}{tan\theta}} \tag{3.17}$$

By rearrangement of the above equation, the velocity of the solid plug in the axial direction can be obtained as:

$$V_l = V_b \frac{tan\phi \cdot tan\theta}{tan\phi + tan\theta} \tag{3.18}$$

With the solid plug velocity component in the axial direction, the flow of the solid plug through the extruder can be estimated. The area to be considered is the cross-sectional area of the solid plug that is calculated from the geometry of the screw channel including the curvature and the width of the flight (see Figure 3.13). Thus the flow of the solid plug is:

$$Q = V_l \int_{d/2}^{D/2} \left(2\pi R - \frac{q_f e}{sin\theta}\right) dR = V_l \left[\frac{\pi}{4}(D^2 - d^2) - \frac{q_f eH}{sin\,\bar{\theta}}\right] \tag{3.19}$$

where q_f is the number of flights in parallel, H, d, D and e are the channel depth, diameters of the screw and the barrel and the thickness of the flight, respectively. Considering that for most of the current extrusion technology $d >> H$, the helix angle and the width of the screw channel are weak functions of the radial distance and can be assumed as average values $\bar{\theta}$ and $\bar{W}$, a simple integration of Eq. 3.19 can be achieved (Rauwendaal, 2001). Average values of the helix angle and the channel width are also assumed in the analyses that involve unrolling the screw channel in a plane channel, which is discussed in the next section. The linear velocity of the barrel can be calculated as $V_b = N\pi D$, where N is the screw speed (in rad/s). Substituting the expression of this velocity component along with Eq. 3.18 into Eq. 3.19 yields:

$$Q = \pi ND \frac{tan\phi \cdot tan\,\bar{\theta}}{tan\phi + tan\,\bar{\theta}} \left[\frac{\pi}{4}(D^2 - d^2) - \frac{q_f eH}{sin\,\bar{\theta}}\right] \tag{3.20}$$

A schematic of the area considered to calculate the flow is illustrated in Figure 3.15 and shows that the term $(D^2 - d^2)$ can be replaced by $4H(D - H)$, which substituted in Eq. 3.20 and after further rearrangement yields:

$$Q = \pi^2 NHD(D-H) \frac{tan\phi \cdot tan\,\bar{\theta}}{tan\phi + tan\,\bar{\theta}} \left[1 - \frac{q_f e}{\pi(D-H) sin\,\bar{\theta}}\right] \tag{3.21}$$

The above equation can be further simplified if written in terms of the average channel width $\bar{W}$ (see Figure 3.15) which can be estimated as $\bar{W} = \frac{\pi(D-H) sin\,\bar{\theta}}{q_f} - e$ to yield the following equation:

$$Q = \pi^2 NHD(D-H) \frac{tan\phi \cdot tan\,\bar{\theta}}{tan\phi + tan\,\bar{\theta}} \left[\frac{\bar{W}}{\bar{W} + e}\right] \tag{3.22}$$

The equation predicts the flow of solids in the extruder as a function of geometric parameters associated with the

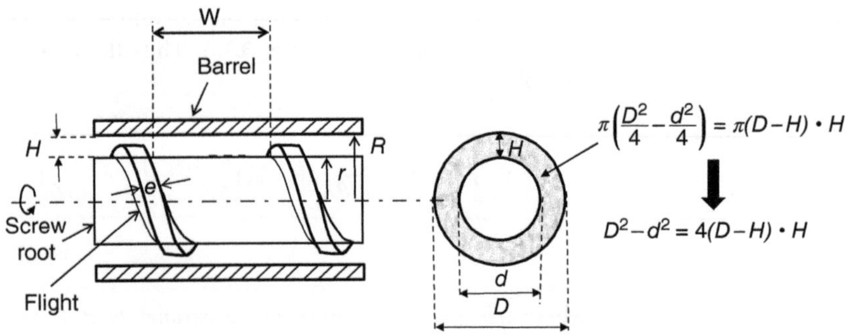

Figure 3.15 Area considered for the flow of the solid plug in the channel of a single screw extruder.

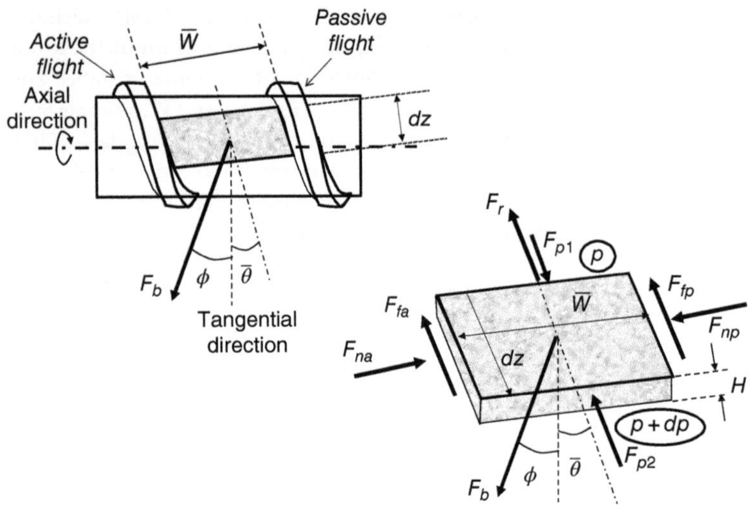

Figure 3.16 Schematics of the balance of forces in a differential element of the solid plug.

extruder equipment, the screw speed, and the angle ϕ. This angle, which is formed by the dragging force acting on the barrel F_b and the barrel tangential velocity V_b, is associated with the characteristics of the barrel and screw surfaces and also the properties of the solid plug. It can be estimated from the corresponding friction coefficients between the solid and the surfaces of the barrel and the screws. It can be also noted that the direction of the dragging force at the barrel surface is similar to that of the vector difference between the barrel tangential velocity V_b, and the down-channel velocity V_Z (see Figure 3.14).

Figure 3.16 schematically illustrates the forces involved in a differential region of the solid plug of thickness dz during its flow through the screw channel. Before establishing forces and momentum balances between these forces to determine unknown variables such as the angle ϕ and an extra normal force acting on the active flight, designed as F^*, it is important to write equations to calculate these forces and express them in two main components: the axial direction and the tangential direction. These equations are given below.

- Dragging force, F_b, acting on the solid plug in a differential area $\bar{W}\cdot dz_b$:

$$F_b = f_b p \bar{W} dz_b \tag{3.23}$$

where p is the pressure in that location and f_b is the friction coefficient between the solid plug and the barrel. The axial direction (indicated by "l" in Figure 3.14) is calculated as:

$$F_{bl} = f_b p \bar{W} dz_b \sin\phi \tag{3.24}$$

whereas in the tangential direction (indicated by "t" in Figure 3.14) is calculated as:

$$F_{bt} = f_b p \bar{W} dz_b \cos\phi \tag{3.25}$$

- Forces on the solid plug in the down-channel direction (direction z): By considering that the solid pressure is uniform in a cross-sectional area but varies in the down-channel direction z, the difference between forces F_{p1} and F_{p2} can be calculated from that pressure gradient as:

$$F_{p2}-F_{p1}=H\bar{W}dp \tag{3.26}$$

which separated in the axial (l) and tangential (t) components yields:

$$\left(F_{p2}-F_{p1}\right)_l=H\bar{W}dp\,sin\,\bar{\theta} \tag{3.27}$$

$$\left(F_{p2}-F_{p1}\right)_t=H\bar{W}dp\,cos\,\bar{\theta} \tag{3.28}$$

- Forces acting on the solid plug on the flight flanks [passive (p) and active (a)]: These forces acting on the passive and active flight flanks can be calculated as:

$$F_{np}=pH\,dz \tag{3.29}$$

$$F_{na}=pHdz+F^* \tag{3.30}$$

The force F^* is an additional unknown force that acts on the active flight flank to push the solid plug forward. It is the result of the drag force acting at the barrel surface in the direction normal to the flight flanks and can be determined by substracting the above two equations:

$$F^*=F_{na}-F_{np} \tag{3.31}$$

and when expressed in their axial (l) and tangential components (t) yields:

$$\left(F_{na}-F_{np}\right)_l=F^*cos\,\bar{\theta} \tag{3.32}$$

$$\left(F_{na}-F_{np}\right)_t=F^*sin\,\bar{\theta} \tag{3.33}$$

- Friction forces acting on the flight flanks (F_{fa} active and F_{fp} passive) and the root of the screw, F_r: These forces expressed in their axial (l) and tangential components (t) yield:

$$F_{fa},l=f_s(pHdz+F^*)\,sin\,\bar{\theta} \tag{3.34}$$

$$F_{fa},t=f_s(pHdz+F^*)\,cos\,\bar{\theta} \tag{3.35}$$

$$F_{fp},l=f_spHdz\,sin\,\bar{\theta} \tag{3.36}$$

$$F_{fp},t=f_spHdz\,cos\,\bar{\theta} \tag{3.37}$$

$$F_{r,l}=f_sp\bar{W}dz\,sin\,\bar{\theta} \tag{3.38}$$

$$F_{r,t}=f_sp\bar{W}dz\,cos\,\bar{\theta} \tag{3.39}$$

The assumption of a steady velocity in the axial direction allows a balance among all forces acting in that direction, written as:

$$F_{b,l}+\left(F_{p2}-F_{p1}\right)_l+\left(F_{na}-F_{np}\right)_l+F_{fa,l}+F_{fp,l}+F_{r,l}=0 \tag{3.40}$$

which after substitutions, rearrangement, and definition of newly associated variables results in an equation that enables the estimation of the unknown force F^*:

$$F^*=\frac{\alpha_1pdz+\alpha_2dp}{cos\,\bar{\theta}-f_ssin\,\bar{\theta}} \tag{3.41}$$

where the associated variables α_1 and α_2 are defined as:

$$\alpha_1=\bar{W}(f_bsin\,\phi+f_ssin\,\bar{\theta})+2Hf_ssin\,\bar{\theta} \tag{3.42}$$

$$\alpha_2=H\bar{W}sin\,\bar{\theta} \tag{3.43}$$

A similar balance supported by an assumed steady state can be established for the momentum of the tangential components of the acting forces with respect to the screw shaft:

$$F_{bt}\cdot\frac{D}{2}-\left(\left(F_{p2}-F_{p1}\right)_t+F_{fa,t}+F_{fp,t}+F_{na,t}+F_{np,t}\right)\times\frac{\bar{D}}{2}-F_{r,t}\cdot\frac{d}{2}=0 \tag{3.44}$$

It can be noted in Figure 3.16 that the tangential and normal forces acting on the flight flanks are distributed on the flank height which is approximately equal to the screw characteristic dimension H. Thus, an average distance from the center plane of the solid plug to the shaft can be defined as $\bar{D}=\dfrac{D+d}{2}$. Equation 3.44 also assumes that the forces acting on the solid plug are constant in the cross-section of the solid plug. By substituting equations for the tangential components of the acting forces and upon rearrangement, the following is obtained:

$$F^*=\frac{\beta_1pdz-\beta_2dp}{\dfrac{\bar{D}}{D}(sin\,\bar{\theta}+f_s\,cos\,\bar{\theta})} \tag{3.45}$$

where the associated geometric constants β_1 and β_2 can be calculated as:

$$\beta_1=\bar{W}\left[f_b\,cos\,\phi-f_s\,cos\,\bar{\theta}\left(\frac{d}{D}+\frac{2H}{\bar{W}}\frac{\bar{D}}{D}\right)\right] \tag{3.46}$$

$$\beta_2 = \bar{W}H(\cos\bar{\theta})\frac{\bar{D}}{D} \quad (3.47)$$

By eliminating F^* from Eqs 3.41 and 3.45, a differential equation is obtained:

$$\frac{dp}{dz} = \frac{\beta_1 - \alpha_1 K}{\alpha_2 K + \beta_2}p \quad (3.48)$$

where K is defined as:

$$K = \frac{\bar{D}}{D}\cdot\frac{\sin\bar{\theta} + f_s\cos\bar{\theta}}{\cos\bar{\theta} - f_s\sin\bar{\theta}} \quad (3.49)$$

Integration of Eq. 3.48 between two different locations where pressures are known yields:

$$\int_{p_1}^{p_2}\frac{dp}{p} = \int_{z_1}^{z_2}\frac{\beta_1 - \alpha_1 K}{\alpha_2 K + \beta_2}dz$$

$$\ln\frac{p_2}{p_1} = \left(\frac{\beta_1 - \alpha_1 K}{\alpha_2 K + \beta_2}\right)(z_2 - z_1) \quad (3.50)$$

Equation 3.50 predicts the pressure increase during the flow of the solid plug provided the geometry of the screw system as well as the screw speed and the angle ϕ are known. Substituting the expressions that define the parameters α_1, α_2, β_1, and β_2 into Eq. 3.50, the following expression for the angle ϕ can be obtained:

$$\cos\phi = K\sin\phi + M \quad (3.51)$$

M can be estimated as:

$$M = \frac{f_s}{f_b}\cdot\left[K\sin\bar{\theta}\left(2\frac{H}{\bar{W}} + 1\right) + \cos\bar{\theta}\left(\frac{2H}{\bar{W}}\frac{\bar{D}}{D} + \frac{d}{D}\right)\right] + \frac{1}{f_b}\frac{H}{z_2 - z_1}\left(K\sin\bar{\theta} + \frac{\bar{D}}{D}\cos\bar{\theta}\right)\ln\frac{p_2}{p_1} \quad (3.52)$$

These equations neglect the effect of the radius of the helix angle and the width of the screw, which is different in the root of the screw (W_s) and in the barrel (W_b). Thus average values are considered for the helix angle and the channel width, which are designed by $\bar{\theta}$ and $\bar{W}$ respectively. Analyses that consider these geometric variables as functions of the radius are given by Darnell and Mol (1956) and Tadmor and Klein (1970). As discussed by Rauwendaal (2001), the uncertainty in the friction coefficients between the solid plug and the screw and the barrel surfaces may overcome the errors introduced in considering average values for these geometric parameters and justify the use of average values.

Equation 3.51 can be solved for sin ϕ to yield:

$$\sin\phi = \frac{\sqrt{K^2 - M^2 + 1} - KM}{K^2 + 1} \quad (3.53)$$

Equation 3.53 enables the calculation of the angle formed by the tangential component of the barrel velocity and the dragging force F_b (see Figure 3.16). Furthermore, the flow Q of the solid plug in the screw channel at the solid powder conveying region can be calculated by Eq. 3.22, provided the pressure drop in the solids conveying region is known. Conversely, if the flow of the solids is known, the pressure drop in the solids conveying section can be calculated by Eq. 3.50.

3.2.3 Melt conveying section

In general, processing technologies aim at converting one component, or a mix of components, into one single product or a mix of products. In that sense, extrusion processing promotes many physical and chemical modifications of these components, setting out what is known as the *processing history* of the product. The final characteristics of the product depend upon that processing history.

In extrusion, the processing history of the product can be quantitatively assessed by the fluid dynamics and the energy transport within the extruder from both experimental and modeling standpoints. It enables the estimation of velocity patterns, distribution of residence times, distribution of the strain applied to the material, temperature, and shear in the extrusion process, which influence the conversion of the components fed into the extruder, and affect the properties of the resulting products. However, only when the hypothesis of the continuum applies to the extruded material the equations of momentum, energy and mass transfers are suitable to describe the non-isothermal flow of the material in the extruder. This corresponds to numerous applications such as those related to thermomechanical plasticating of polymers or thermomechanical cooking of biopolymers, where the extruded material is in a molten state. Chapter 4 (section 4.2) provides a description and analysis of the melting mechanism both in a monobloc single screw extruder and in a modular, intermeshing co-rotating twin screw extruder. The following section provides an engineering analysis to model the momentum and energy transfer in the fully molten state.

3.2.3.1 Single screw extruders

Several assumptions are necessary to develop models describing the fluid mechanics of the flow inside the extruder.

- The process is under steady state, meaning that the flow is independent of time and is fully developed.
- Inertial forces are negligible due to the high viscosity of the melt. Gravity forces are considered negligible.
- The fluid is incompressible.
- There is no slip at the barrel and screw walls.
- The rheological properties of the melt can be described by non-elastic rheological models.

Specifically Newtonian and non-Newtonian power law models are used for the analysis.

As discussed in section 3.2.1.3, the approximated equations resulting from the assumptions of a steady-state process and small Reynolds numbers (i.e. assuming negligible inertia effects) that are used to describe the flow of the melt in a screw-barrel assembly are given by Eqs 3.12 and 3.13. In Eq. 3.13 the stress tensor $\underline{\underline{\tau}}$ incorporates the velocity components of the fluid flow by specific functions that provide the rheology of the material. The simplest case is a fluid exhibiting a Newtonian rheological model, whereas more complex but also more realistic cases include non-Newtonian rheological models which are fully described in Chapter 4. The system described by Eqs 3.12 and 3.13 consists of a system of three partial differential equations for a 2D model and four partial differential equations for a 3D model. The variables to be solved as a function of the position in the extruder are two velocity components and pressure profiles in the 2D model and three velocity components and pressure profiles in the 3D model.

3.2.3.1.1 Newtonian melts, isothermal process

The continuity and motion equations (Eqs 3.12 and 3.13) are applied to solve the flow in the metering zone of the single screw system (Figure 3.1). Extruder diameters in commercial single screw extruders range from 2 cm (laboratory scale) to 75 cm (industrial scale) with axial lengths in terms of length to diameter (L/D) ratios of approximately 24 to 26, including an uncommon design that can go to L/D values as high as 40 and as low as 8

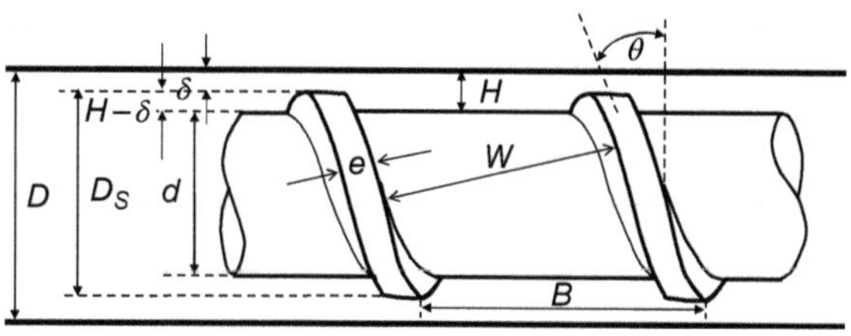

Figure 3.17 Schematic of the screw channel in a single screw extruder, including names of most relevant geometrical variables.

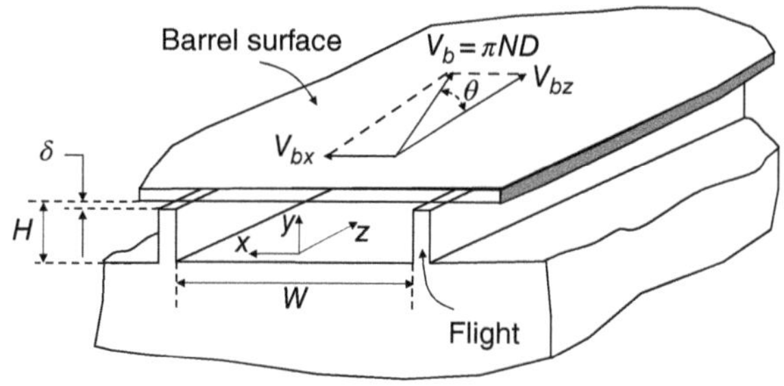

Figure 3.18 Unwound channel of a single screw extruder indicating main geometric characteristics and velocity components.

(Tadmor & Gogos, 2006). A closer look at a region of the screw is illustrated in Figure 3.17. The barrel has a diameter D and the clearance between the tip of the flight and the inner part of the barrel is given by the thickness δ which is of the order of 0.1–0.3% of the barrel diameter; thus the diameter of the screw is $d = D - 2\delta$ and the axial distance of a full turn is B (see Figure 3.17). The thickness of the screw flight is designated by e and the flight forms an angle θ with the vertical line. The design of the screws in most single screw extruders is such that $B = d$ which are known as single-flight screws (Tadmor & Gogos, 2006). The radial distance between the root of the screw and the inner surface of the barrel, designed as H, is known as the channel depth and could be variable or specially designed to increase the shear applied in the metering zone.

In order to solve the fluid mechanics model, the helical channel formed by the screw and the barrel can be unwound to obtain the rectangular prism geometry illustrated in Figure 3.18. Although the actual geometry of the screw is helical and has some curvature, the small thickness δ used in commercial extruders enables to assume a negligible effect of curvature and solve the equation on rectangular co-ordinates. With these assumptions, as illustrated in Figure 3.18, the barrel becomes a flat plate with velocity $V_b = \pi ND$, which can be resolved in a z-axial component $V_{bz} = \pi ND \cos\theta$ and a x-cross-channel component $V_{bx} = \pi ND \sin\theta$. With these assumptions Eq. 3.12 and the x, y and z components of Eq. 3.13 can be written as:

$$\frac{\partial v_x}{\partial x} + \frac{\partial v_y}{\partial y} + \frac{\partial v_z}{\partial z} = 0 \quad (3.54a)$$

$$\frac{\partial p}{\partial x} = \frac{\partial \tau_{xx}}{\partial x} + \frac{\partial \tau_{yx}}{\partial y} + \frac{\partial \tau_{zx}}{\partial z} \quad (3.54b)$$

$$\frac{\partial p}{\partial y} = \frac{\partial \tau_{xy}}{\partial x} + \frac{\partial \tau_{yy}}{\partial y} + \frac{\partial \tau_{zy}}{\partial z} \quad (3.54c)$$

$$\frac{\partial p}{\partial z} = \frac{\partial \tau_{xz}}{\partial x} + \frac{\partial \tau_{yz}}{\partial y} + \frac{\partial \tau_{zz}}{\partial z} \quad (3.54d)$$

For a Newtonian fluid, the above equations are transformed in the following equations (Bird et al, 2007):

$$\frac{\partial p}{\partial x} = \mu\left(\frac{\partial^2 v_x}{\partial x^2} + \frac{\partial^2 v_x}{\partial y^2} + \frac{\partial^2 v_x}{\partial z^2}\right) \quad (3.55a)$$

$$\frac{\partial p}{\partial y} = \mu\left(\frac{\partial^2 v_y}{\partial x^2} + \frac{\partial^2 v_y}{\partial y^2} + \frac{\partial^2 v_y}{\partial z^2}\right) \quad (3.55b)$$

$$\frac{\partial p}{\partial z} = \mu\left(\frac{\partial^2 v_z}{\partial x^2} + \frac{\partial^2 v_z}{\partial y^2} + \frac{\partial^2 v_z}{\partial z^2}\right) \quad (3.55c)$$

For a fully developed 2D flow, the derivatives of the velocity components with respect to the z-co-ordinate can be discarded. The unwound geometry illustrated in Figure 3.18 resembles the geometry of a typical shear flow in which, due to the shallow channel, the fluid pressure does not vary in the direction perpendicular to the channel (direction y) and the velocity in that direction could be discarded. By using those assumptions, the equation of continuity and the equations of motion simplify to:

$$\frac{\partial v_x}{\partial x} = 0 \rightarrow v_x \neq f(x) \quad (3.56)$$

$$\frac{\partial p}{\partial x} = \mu \frac{\partial^2 v_x}{\partial y^2} \quad (3.57)$$

$$\frac{\partial p}{\partial z} = \mu\left(\frac{\partial^2 v_z}{\partial x^2} + \frac{\partial^2 v_z}{\partial y^2}\right) \quad (3.58)$$

The right side of Eq. 3.57 is a function of the co-ordinate y only, whereas its left side indicates that it is only a function of the co-ordinate x. In consequence, for consistency, the left side of Eq. 3.57 must be a constant, so Eq. 3.57 can be readily integrated to give:

$$v_x = \frac{1}{2\mu}\left(\frac{\partial p}{\partial x}\right)y^2 + C_1 y + C_2 \quad (3.59)$$

where C_1 and C_2 are two constants that are determined using the conditions of non-slip at the barrel wall ($y = H$) and the root of the screw ($y = 0$). These boundary conditions are $v_x(y = H) = -V_{bx}$ and $v_x(y = 0) = 0$. Although in single screw extruders the screw is rotating and the barrel is fixed, for convenience the opposite is considered without affecting the final equations because relative velocities are used. Using these boundary conditions, the velocity component of the melt in the direction x results in:

$$u_x(\xi) = \frac{H^2}{2\mu V_{bx}}\left(\frac{\partial p}{\partial x}\right)(\xi^2 - \xi) - \xi \quad (3.60)$$

where $\xi = \frac{y}{H}$ and $u_x = \frac{v_x}{V_{bx}}$. In order to estimate the pressure gradient in the direction x, a condition of no melt flow (no leakage is assumed) in the direction x is applied to give:

$$\int_0^1 u_x d\xi = \int_0^1 \left[\frac{H^2}{2\mu V_{bx}}\left(\frac{\partial p}{\partial x}\right)(\xi^2 - \xi) - \xi\right] d\xi = 0 \quad (3.61)$$

and:

$$\left(\frac{\partial p}{\partial x}\right) = -\frac{6\mu V_{bx}}{H^2} = -\frac{6\mu\pi ND \sin\theta}{H^2} \quad (3.62)$$

By substituting Eq. 3.62 into Eq. 3.60, the following equation is obtained:

$$u_x(\xi) = 2\xi - 3\xi^2 \quad (3.63)$$

Figure 3.19a shows a schematic of the flow in the x-direction of the unwound rectangular channel formed by the root of the screw and the barrel, whereas Figure 3.19b illustrates the melt dimensionless x-component velocity profile described by Eq. 3.63. As noted in these figures, this velocity component promotes a great deal of recirculation and probably mixing in the rectangular channel.

The derivation of the equation to estimate the axial velocity component v_z requires the solution of the partial

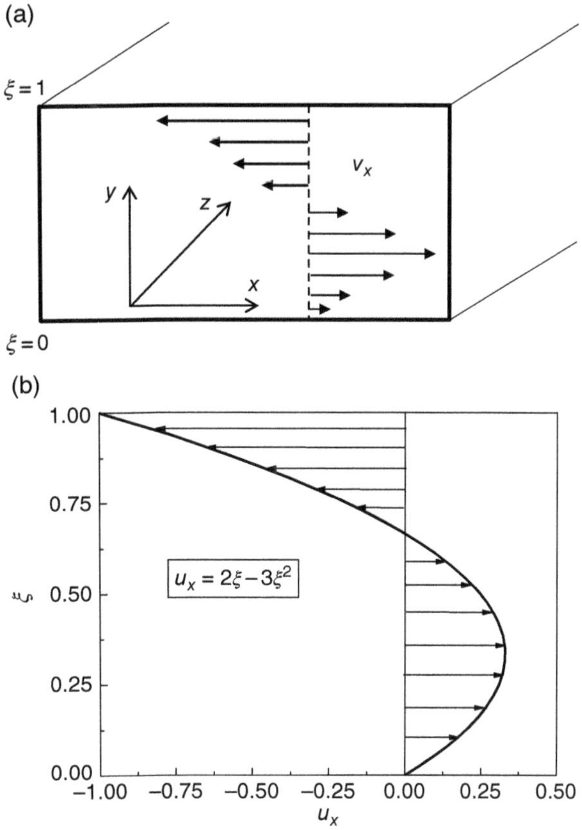

Figure 3.19 (a) Rectangular channel of a single screw extruder showing the profile of the *x*-component of the fluid velocity. (b) Velocity profile of the dimensionless *x*-velocity component as a function of the dimensionless cross-channel direction ξ.

differential Eq. 3.58, which is performed by the classic method of separation of variables and the following boundary conditions:

$$\begin{aligned} v_z &= 0 \quad at \quad \xi = \frac{y}{H} = 0 \\ v_z &= V_{bz} \quad at \quad \xi = \frac{y}{H} = 1 \\ v_z &= 0 \quad at \quad \chi = \frac{x}{W} = 0 \\ v_z &= 0 \quad at \quad \chi = \frac{x}{W} = 1 \end{aligned} \tag{3.64}$$

which yields the following result (Tadmor & Gogos, 2006):

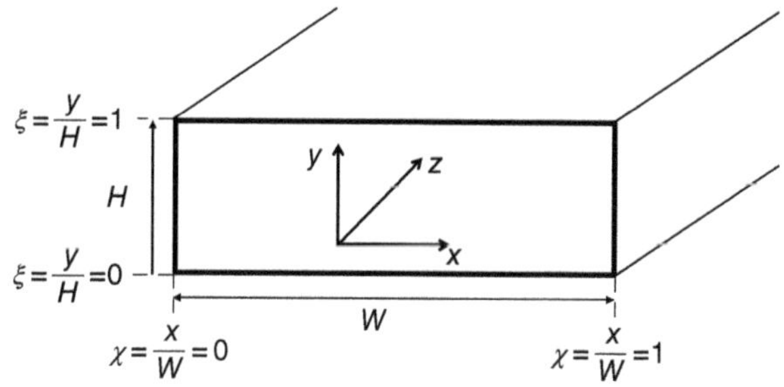

Figure 3.20 Geometry used to solve the model for the flow of melt in the rectangular channel of a single screw extruder.

$$u_z(\chi,\xi) = \frac{4}{\pi}\sum_{i=1,3..}^{\infty}\frac{sinh(i\pi h\xi)}{isinh(i\pi h)}\cdot sin(i\pi\chi) + \left(\frac{H^2}{2\mu V_{bz}}\frac{\partial p}{\partial z}\right)\times \left[\xi^2 - \xi + \frac{8}{\pi^3}\sum_{i=1,3..}^{\infty}\frac{cosh\left(\frac{i\pi(\chi-0.5)}{h}\right)}{i^3 cosh\left(\frac{i\pi}{2h}\right)}\cdot sin(i\pi\xi)\right] \tag{3.65}$$

where $u_z = \frac{v_z}{V_{bz}}$, $\chi = \frac{x}{W}$, and h is a geometric factor defined as $h = H/W$ which for shallow channels is significantly smaller than 1. Equation 3.65 contains terms that contribute to the drag flow and terms that contribute to pressure flow, primarily given by the dimensionless pressure gradient $\left(\frac{H^2}{2\mu V_{bz}}\frac{\partial p}{\partial z}\right)$. The geometry used for the flow model is schematically depicted in Figure 3.20, whereas results of the analytical solution of the model (Eq. 3.65) are illustrated in Figure 3.21 for different screw geometries given by the ratio H/W and different processing conditions given by an imposed constant pressure gradient $\frac{\partial p}{\partial z}$. For the preparation of the figures, a melt viscosity of 10,000 *mPa.s* and a linear velocity V_{bz} of 2 m/s were used. The linear velocity was estimated from realistic processing conditions by assuming a screw speed of 200 rpm and a barrel internal diameter of approximately 10 cm. The figures show contour lines of the axial velocity component in the channel. By comparing Figures 3.21a and 3.21b that were prepared with a nil pressure gradient, i.e. a velocity profile originated solely by the drag flow, it is

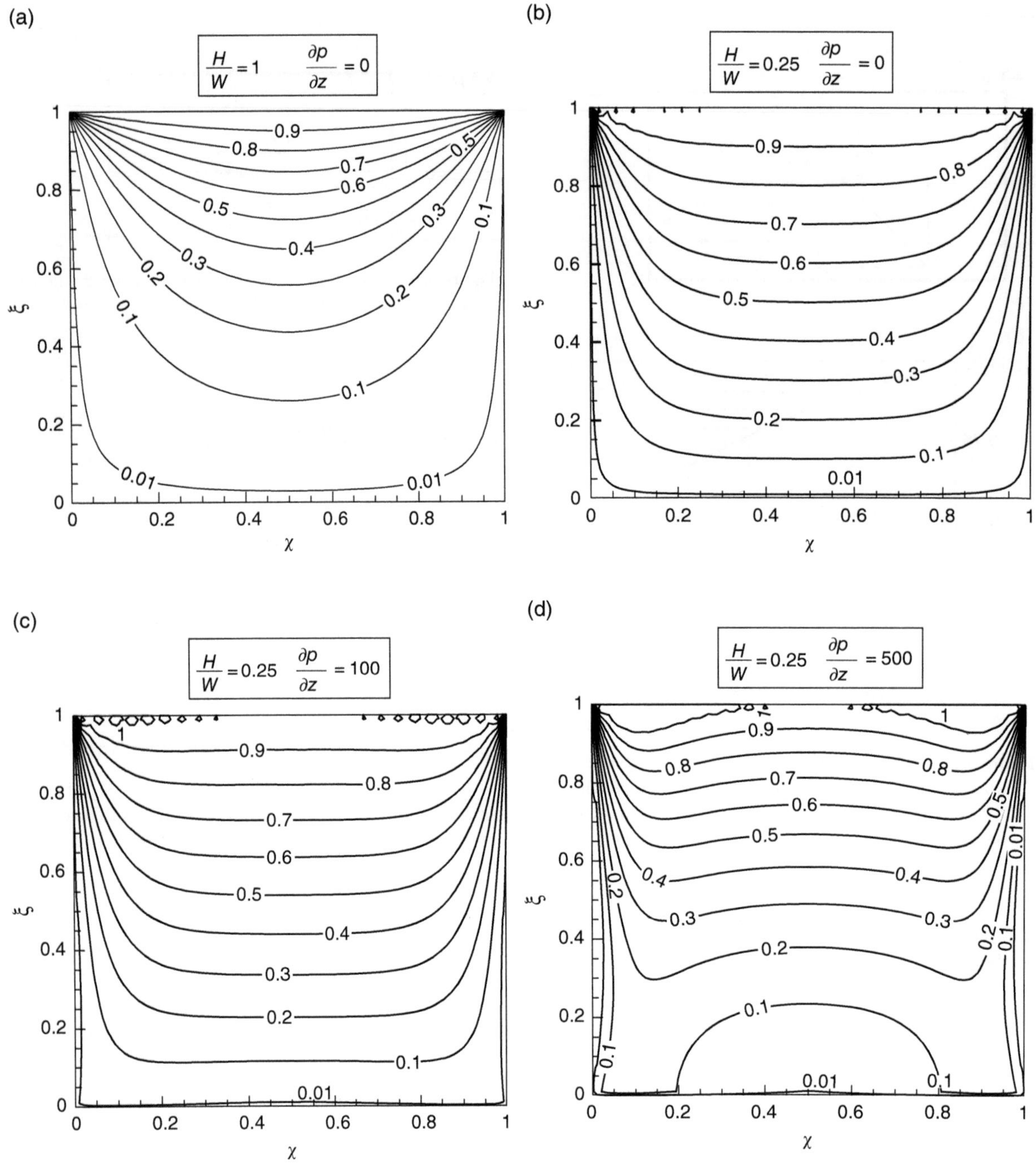

Figure 3.21 Contour plots showing the axial component of the melt velocity as a function of the dimensionless axial and radial co-ordinates for different pressure gradients and ratios H/W. (a) $H/W = 1$, $\partial p/\partial z = 0$. (b) $H/W = 0.25$, $\partial p/\partial z = 0$. (c) $H/W = 0.25$, $\partial p/\partial z = 100$. (d) $H/W = 0.25$, $\partial p/\partial z = 500$.

noted that less deep channels increase the restriction to the flow and less separation in the isolines is observed. Less separation between isolines is associated with larger shear rates, which is evident for the more shallow channels (lower H/W value). The effect of the imposed pressure gradient is illustrated by comparing Figure 3.21b with Figures 3.21c and 3.21d. For small pressure gradients, the contour lines are slightly more flat and a small flow restriction to the flow is observed (Figure 3.21c). For large pressure gradients, the contour lines have a less regular pattern and indicate the possibility of zones with negative axial velocity components (Figure 3.21d).

Although the information depicted in Figure 3.21 provides characteristics of the flow in the channel and areas of potential fluid recirculation that could improve the mixing action in the extruder, its main application is to estimate the flow rate in the extruder as well as pressure profile, power consumption and other main variables of interest for the design of an extrusion process. In particular, the net volumetric flow rate Q could be obtained by integration of Eq. 3.65 as written below (see Figure 3.20):

$$Q = WHV_{bz}\int_0^1\int_0^1 u_z(\chi,\xi)d\xi d\chi \qquad (3.66)$$

Tadmor and Gogos (2006) give the result of the above integration as:

$$Q = \frac{V_{bz}WH}{2}F_d + \frac{WH^3}{12\mu}\left(-\frac{\partial p}{\partial z}\right)F_p \qquad (3.67)$$

where F_d and F_p are two parameters, known as shape factors, associated with the drag and pressure flows respectively and defined by the following equations:

$$F_d = \frac{16W}{\pi^3 H}\sum_{i=1,3,5....}^{\infty}\frac{1}{i^3}tanh\left(\frac{i\pi H}{2W}\right) \qquad (3.68)$$

$$F_p = 1 - \frac{192H}{\pi^5 W}\sum_{i=1,3,5....}^{\infty}\frac{1}{i^5}tanh\left(\frac{i\pi W}{2H}\right) \qquad (3.69)$$

For practical convenience in the use of Eq. 3.67, plots of these parameters as a function of the ratio H/W are illustrated in Figure 3.22. The figure clearly shows that the geometry of the screw has a larger effect on the pressure-driven flow. As observed in the contour velocity profiles given in Figure 3.21, higher channel depth increases the resistance to both drag and pressure-driven flows. A simplified version of Eq. 3.67 frequently useful for analysis of the extrusion process in terms of the total throughput Q is:

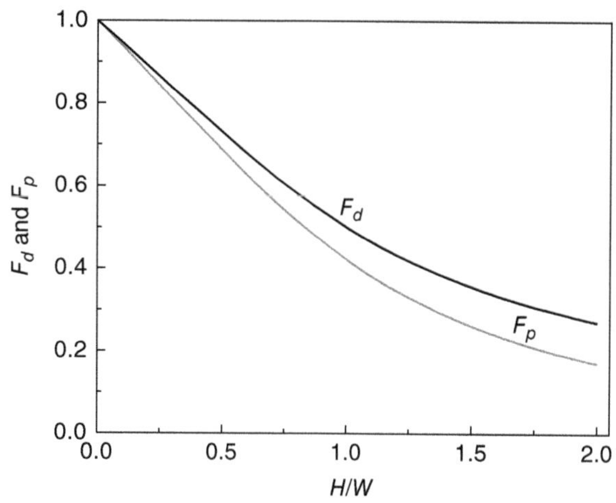

Figure 3.22 Shape factors F_d and F_p defined by Eqs 3.68 and 3.69 as a function of the ratio H/W.

$$Q = Q_d + Q_p \qquad (3.70)$$

On examining Eq. 3.67, it can be seen that depending on the sign and the magnitude of the pressure gradient $\left(-\frac{\partial p}{\partial z}\right)$, the pressure-driven flow can be negative. As discussed in previous chapters and also as schematically shown in Figure 3.1, in single screw extrusion processing the pressure gradient $(\partial p/\partial z)$ is positive and makes the pressure-driven flow negative. Under these conditions Eq. 3.70 shows that $Q < Q_d$. Simulations shown in Figure 3.21 clearly illustrate the effect of the pressure gradient $\frac{\partial p}{\partial z}$ on the profile of the axial component of the melt velocity. Furthermore, for very large pressure gradient, negative axial component velocities can be observed (not shown in Figure 3.21). Tadmor and Gogos (2006) noted that those negative velocity components do not originate a backward flow of the melt through the screws. In fact, the backward flow is along the z-direction but not along the extruder shaft direction. As indicated in Figure 3.23, the melt velocity in the shaft direction can be calculated in terms of the z- and x-directions as:

$$v_l = v_x \cos\theta + v_z \sin\theta \qquad (3.71)$$

Equation 3.70 provides the condition for a zero net flow, i.e. $Q = 0$ or $Q_p/Q_d = -1$. From Eq. 3.67, that ratio can be

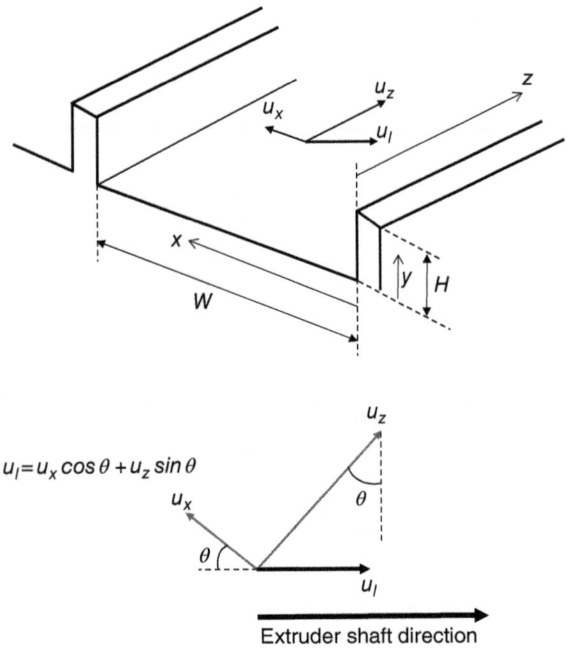

Figure 3.23 Schematics showing the melt velocity components in the cross-channel (x), down-channel (z), and extruder shaft (l) directions.

written in terms of the pressure gradient $\left(\frac{\partial p}{\partial z}\right)$, the geometry of the extruder (W and H, D, and θ), the operating conditions (e.g. the screw speed, N), and the rheology of the melt (μ for a Newtonian fluid) as:

$$\frac{Q_p}{Q_d} = -\frac{H^2}{6\mu V_{bz}}\left(\frac{\partial p}{\partial z}\right)\frac{F_p}{F_d} \qquad (3.72)$$

By substituting Eq. 3.72 into Eq. 3.65, the following equation is obtained:

$$u_z(\chi,\xi) = \frac{4}{\pi}\sum_{i=1,3..}^{\infty}\frac{sinh(i\pi h\xi)}{isinh(i\pi h)}\cdot sin(i\pi\chi) - \left(3\frac{Q_p F_d}{Q_d F_p}\right)\times$$

$$\left[\xi^2 - \xi + \frac{8}{\pi^3}\sum_{i=1,3..}^{\infty}\frac{cosh\left(\frac{i\pi(\chi-0.5)}{h}\right)}{i^3 cosh\left(\frac{i\pi}{2h}\right)}\cdot sin(i\pi\xi)\right] \qquad (3.73)$$

Equations 3.63, 3.73 and 3.71 enable calculation of the velocity component profiles $u_x(\xi)$, $u_z(\chi,\xi)$ and $u_l(\chi,\xi)$, respectively for different processing conditions (mainly specified by the pressure gradient), which are illustrated in Figure 3.24. For the calculations, the values $\theta = \frac{\pi}{4}$ and $\chi = 0.5$ were used. It can be noted that the component x of the melt velocity is only a function of the y (or ξ) direction whereas its z-velocity component is a function of both x and y (χ and ξ); the latter also applies to the velocity component in the direction of the extruder shaft u_l. The top row of Figure 3.24 shows values of the dimensionless velocity components x, z, and l as a function of the dimensionless cross-channel direction ξ for a typical drag flow, i.e. when $Q_p/Q_d = 0$. As observed in the figure, the x-component velocity profile is parabolic whereas the z-component velocity is linear (typical of a pure drag flow). When these melt velocity components are combined as indicated by Eq. 3.71, the velocity component in the direction of the extruder shaft (u_l) results in a parabolic velocity profile. The magnitude of the velocity component $u_l(\chi,\xi)$ is significantly decreased when pressure effects increase (e.g. for $Q_p/Q_d = -0.5$) and eventually reduces to zero when the condition $Q_p/Q_d = -1$ is achieved (third column and third row in Figure 3.24). Equations providing these velocity components profiles can be used to calculate the power shaft required for the extruder operation.

As indicated in Figure 3.25a, the force acting on the barrel surface S can be calculated by the vector product between the unit vector normal to the barrel surface and the stress tensor, which depend on the rheology of the melt by the following equation (Morrison, 2001):

$$\underline{f}(\underline{n}) = \underline{n}\cdot\underline{\underline{\tau}}\, dS \qquad (3.74)$$

and the power applied by the extruder shaft:

$$P_w = \int_S \underline{f}(\underline{n})\cdot\underline{v}\, dS = \int_S \left[\underline{n}\cdot\underline{\underline{\tau}}\right]\cdot\underline{v}\, dS \qquad (3.75)$$

where S is the surface on which the force is acting, in this case the surface of the barrel, and $\underline{n}$ the unit vector normal to the surface. As indicated in Figure 3.25a, the surface of the barrel is WZ, where Z is the length of the unwound channel. By assuming uniform stress components and fluid velocity components along the barrel, Tadmor and Gogos (2006) demonstrated that Eq. 3.75 simplifies to:

$$P_s = \left[\tau_{yz}\big|_{barrel\ surface} V_{bz} + \tau_{yx}\big|_{barrel\ surface} V_{bx}\right] W\cdot Z \qquad (3.76)$$

The components of the shear stress tensor indicated in the above equation can be obtained from the analysis of a

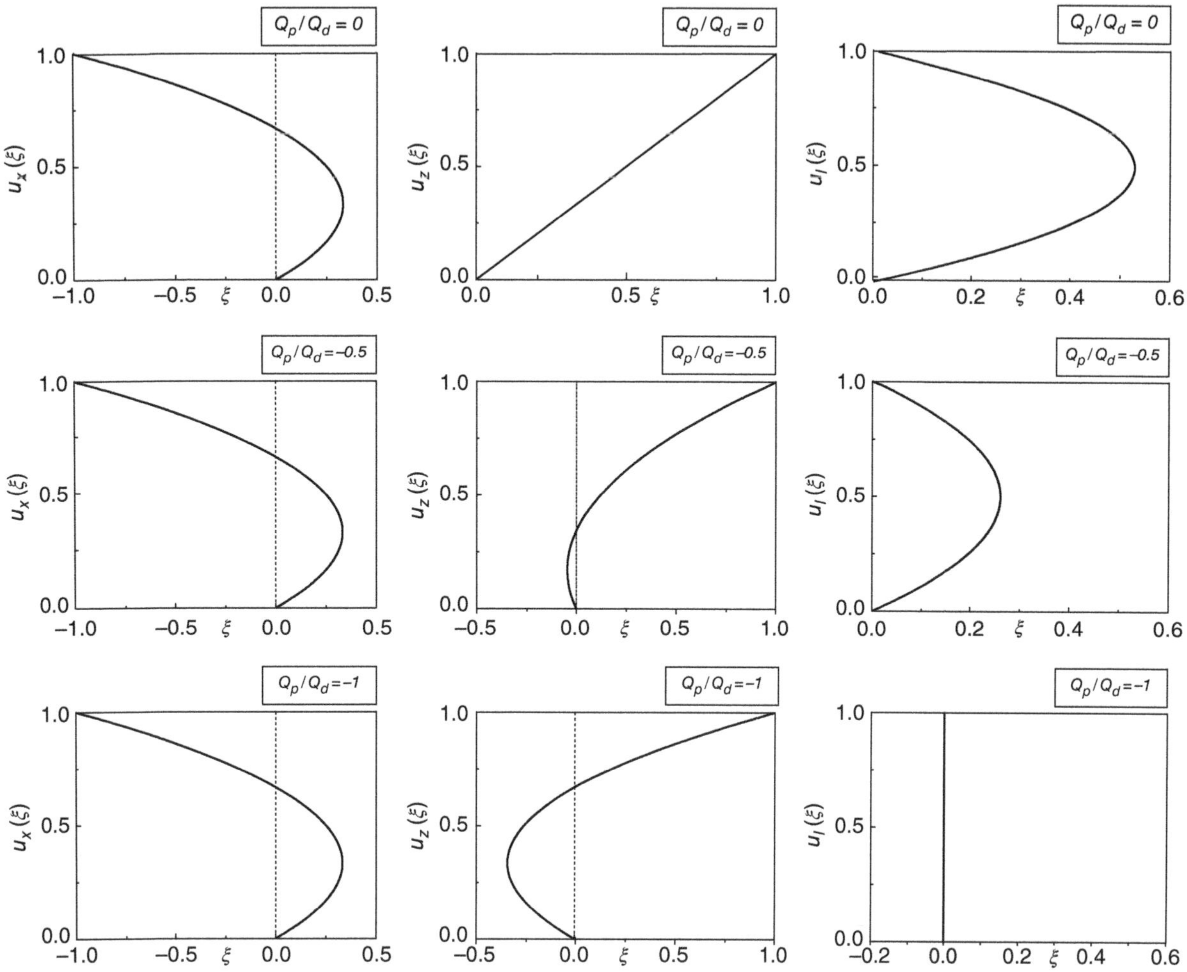

Figure 3.24 Melt velocity component profiles as a function of the dimensionless cross-channel direction at different pressure gradients or Q_p/Q_d (see Eq. 3.72). Velocity components are the cross-channel direction (x), the down-channel direction (z), and the extruder shaft direction (l).

drag flow with an imposed pressure gradient for Newtonian fluids, which is carried out by Tadmor and Gogos (2006) resulting in the following:

$$\tau_{yz}\big|_{barrel\ surface} = \mu\frac{V_{bz}}{H}\left(1-3\frac{Q_p}{Q_d}\right) = \mu\frac{\pi ND\cos\theta}{H}\left(1-3\frac{Q_p}{Q_d}\right) \tag{3.77}$$

$$\tau_{yx}\big|_{barrel\ surface} = \mu\frac{V_{bx}}{H} = \mu\frac{\pi ND\sin\theta}{H} \tag{3.78}$$

Expressing the V_{bz} and V_{bx} components in terms of the screw speed N and the flight angle θ and by substituting Eqs 3.77 and 3.78 into Eq. 3.76, the following is obtained:

$$P_s = \frac{\mu\pi^2N^2D^2}{H}\cdot W\cdot Z\cdot\left[1-3\frac{Q_p}{Q_d}\cos^2\theta\right] \tag{3.79}$$

The above equation expressed in terms of the total flow $\frac{Q_p}{Q_d} = \frac{Q}{Q_d} - 1$ and the length of the screw $L = Z\cdot\sin\bar{\theta}$ is transformed into:

$$P_s = \frac{\mu\pi^2N^2D^2}{H\cdot\sin\bar{\theta}}\cdot W\cdot L\cdot\left[1-3\left(\frac{Q}{Q_d}-1\right)\cos^2\theta\right] \tag{3.80}$$

where $\bar{\theta}$ is the average screw helix angle.

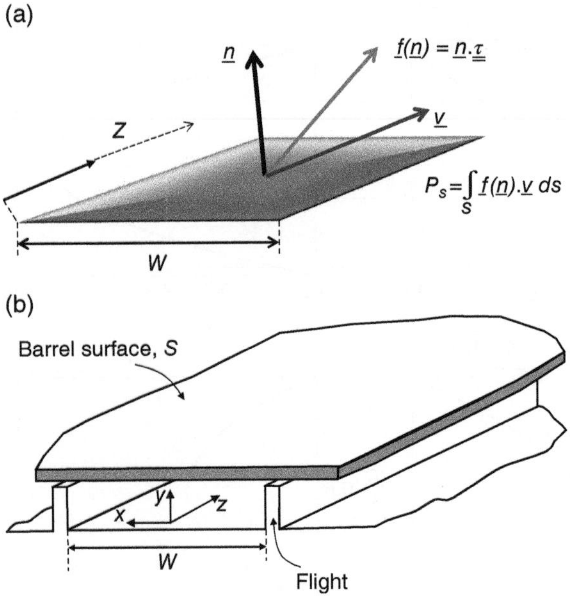

Figure 3.25 Schematics of the geometry involved in calculation of the power on the extruder shaft as a function of the unit vector normal to the barrel surface and the stress tensor acting on that surface.

Although equations predicting velocity profiles in the screw channel have large applicability for calculation of the power required for the process and also to estimate the fluid circulation in the screw channel, which have implications in mixing, the net flow in the extruder under different processing conditions is of key importance for the process and merits an analysis based on the flow equation; for the analysis Eq. 3.67 could be rewritten as:

$$Q = \alpha N - \frac{\beta \Delta p}{\mu} \tag{3.81}$$

where:

$$\alpha = \frac{\pi}{2} D(\cos\theta)\, WHF_d \tag{3.82}$$

$$\beta = \frac{WH^3}{12L} F_p \tag{3.83}$$

$$\left(-\frac{\partial p}{\partial z}\right) \approx \frac{p_o - p_L}{L} = -\frac{\Delta p}{L} \tag{3.84}$$

It is worth analyzing the net flow Q as a function of key processing variables and Figure 3.26 shows, based on Eq. 3.81, a schematic of the net throughput as a function of the screw speed under increasing pressure differences and temperatures. The plot clearly demonstrates that

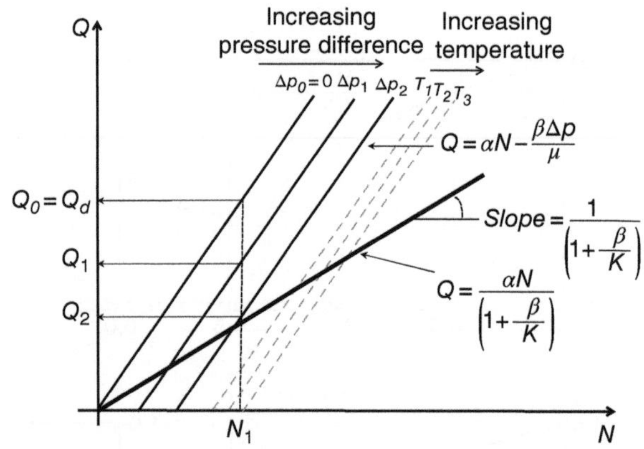

Figure 3.26 Operating characteristics of a single screw extruder showing the net flow as a function of the screw speed considered for different back pressures and temperatures.

for a given screw speed (e.g. N_1 in Figure 3.26), increases in the pressure difference (back pressure) generated in the extruder result in decreases of the net throughput. For example, a decrease from Q_0, for $\Delta p = 0$, to Q_1 is observed when back pressure increases from $\Delta p = 0$ to Δp_1, and so on (for example to yield the flow Q_2 upon a pressure gradien Δp_2). A similar effect can be seen with increases in temperature from T_1 to T_3. This behavior can be explained by the decreasing effect that temperature has on melt viscosity, which increases the pressure flow, resulting in a decrease of the net flow. This analysis considers the applicability of the flow equation until the melt reaches the entrance of the die. Until that point, the flow is characterized by a continuous increase of the pressure along the extruder length (see Figures 2.18 and 3.1). However, at the die the pressure starts to drop. This behavior could be incorporated if the pressure drop in the die can be estimated. For the simple case of a Newtonian melt flowing in a cylindrical die, the Hagen–Poiseuille equation, which is given below (Bird et al, 2007), can be used:

$$Q_{die} = \frac{\pi \Delta p_{die} R_{die}^4}{8 \mu L_{die}} = \frac{K \Delta p_{die}}{\mu} \tag{3.85}$$

where the constant K is a function of the die geometry, which for Newtonian fluids is given by the following equation:

$$K = \frac{\pi R_{die}^4}{8 L_{die}} \tag{3.86}$$

For other types of die geometries such as those schematically illustrated in Figure 3.27 and melt rheologies

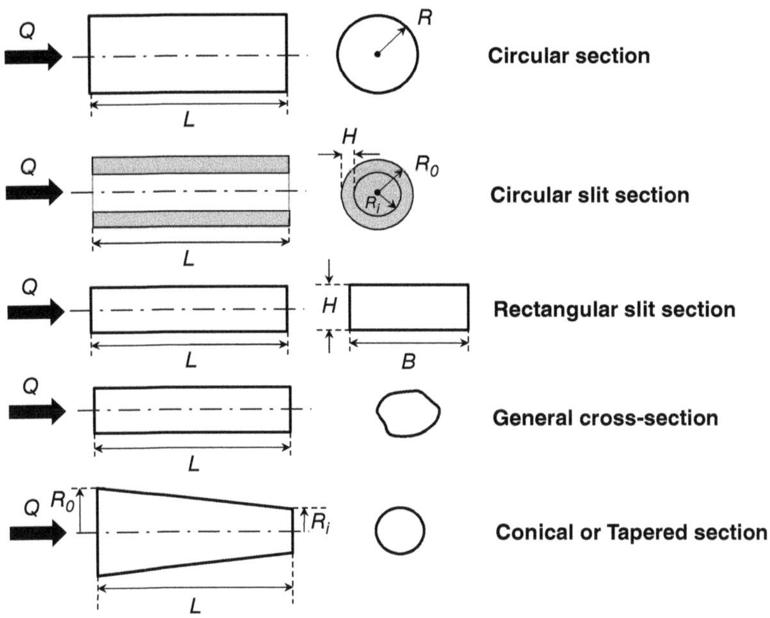

Figure 3.27 Different die arrangements used in extrusion processes along with their relevant geometrical dimensions.

described by the Newtonian and non-Newtonian power law models similar equations have been obtained.

For the power law rheological model, a relationship between the net flow and the pressure drop was established (Michaeli, 2003):

$$Q_{die} = K' \left(\frac{\Delta p_{die}}{m} \right)^{1/n} \tag{3.87}$$

Equations to determine the values of the K and K' constants, known as die conductance, are given in Table 3.1, where m and n are the consistency and the flow indexes corresponding to the power law model. The equations included in Table 3.1 were derived from and apply to certain conditions, some similar to those outlined above, and enumerated below (Michaeli, 2003).

- Isothermal flow.
- Steady-state flow.
- Incompressible and laminar flow; given the high viscosity of melt, that condition is likely to be prevalent in the extrusion operation. In addition, with pressures normally used in extrusion operations, melts can be considered incompressible fluids.
- Inlet and outlet effects are not considered, nor are viscoelastic effects. The incorporation of viscoelastic effects results in models that require complex numerical approaches because analytical solutions are not accessible.
- Non-slippage at the barrel and the screw is assumed.

For geometric conditions in which an analytical equation cannot be obtained, a correction factor, f_p, as a function of the geometric ratio B/H of the die (illustrated in Figure 3.28) is used to estimate the die conductance for Newtonian fluids.

In order to incorporate the pressure drop in the die to the net flow calculations, an integrated form of the continuity equation, which implies that under steady-state conditions the flows in the extruder and the die are the same, can be used:

$$Q = Q_{die} \tag{3.88}$$

Rearranging Eq. 3.85 and substitution in Eq. 3.81 yields the following equation for Newtonian fluids:

$$Q = \frac{\alpha N}{\left(1 + \frac{\beta}{K} \right)} \tag{3.89}$$

Equation 3.89 shows that the net flow varies linearly with screw speed, which is represented in Figure 3.26 by a straight line with slope $\frac{1}{\left(1 + \frac{\beta}{K} \right)}$.

Table 3.1 Equations of to calculate the constant K for different types of die geometries and for Newtonian and non-Newtonian power law fluids.

Die geometry	Newtonian fluid ($\tau = \mu\dot{\gamma}$)	Power law fluid ($\tau = m\dot{\gamma}^n$)
Circular section	$K = \frac{\pi R^4}{8L} \quad \frac{L}{R} \gg 1$	$K' = \frac{\pi n R^{1/n+3}}{2^{1/n} L^{1/n}(1+3n)}$
Circular (annular) section	$K = \frac{2\pi R + H}{12L} H^3 \quad \frac{H}{R} > 0.1$ $K = \frac{\pi R H^3}{6L} \quad \frac{H}{R} \ll 0.1$ (considered as slit) $K = \frac{\pi R_i^4}{8L}\left[(1-\kappa^4) - \frac{(1-\kappa^2)^2}{\ln\frac{1}{k}}\right]$ $\kappa = \frac{R_i}{R_o}$	$K' = \frac{\pi n (R_i + R_o)\cdot(R_o - R_i)^{\frac{1}{n}+2}}{2^{\frac{1}{n}+2}(2n+1)L^{\frac{1}{n}}}$
Rectangular slit section	$K = \frac{B\cdot H^3}{12L} \quad (B \gg H)$ Introduction of correction factor f_p for $B/H \leq 20$ (see Figure 3.28)	$K' = \frac{B\cdot n\cdot H^{\frac{1}{n}+2}}{2^{\frac{1}{n}+2}(2n+1)L^{\frac{1}{n}}} \; for \; \frac{B}{H} > 20$
General cross-section	$K = \frac{B\cdot H^3}{12L}\cdot f_p$ For values of f_p, see Figure 3.28	$K' = \frac{2n\cdot A^{\frac{1}{n}+2}}{(3n+1)\cdot L\cdot P^{\frac{1}{n}+1}}$ A and P are the area and the perimeter of the section respectively
Conical (tapered) section	$K = \frac{\pi R_i^4}{8L}\cdot\frac{3\left(\frac{R_o}{R_i}-1\right)}{1-\left(\frac{R_i}{R_o}\right)^3}$	$K' = \frac{\pi n}{2^{\frac{1}{n}}(3n+1)}\left(\frac{3n\left(\frac{R_o}{R_i}-1\right)}{L\left(1-\left(\frac{R_i}{R_o}\right)^{\frac{3}{n}}\right)}\right)^{1/n}\cdot R_i^{\frac{1}{n}+3}$

Source: Adapted from Michaeli 2003.

It is also important to analyze the behavior of the net flow and how it is affected by pressure and the geometry of the screw elements. Figure 3.29 illustrates schematically the predicted drag flow as a function of the pressure change (die pressure or back pressure) in the extrusion operation at different screw speeds (Figure 3.29a) and different screw channel depth, i.e. H values (Figure 3.29b). The horizontal lines indicate the independence of the drag flow with pressure differences. However, if the pressure flow, which as indicated in Eq. 3.67 is independent of the screw speed and a linear function of the pressure difference, is superimposed on the drag flow, a constant decrease of the net flow at each pressure difference (back pressure) is observed as schematically shown in Figure 3.30. The same behavior results if the pressure flow is superimposed on the drag flow for different channel heights, which is schematically represented in Figure 3.29b.

The effect of the die characteristics based on Eq. 3.85 derived for a Newtonian material is schematically illustrated in Figure 3.31. As noted, it is represented by a straight line of slope $\frac{K}{\mu}$. Either Eq. 3.86 or the relationships

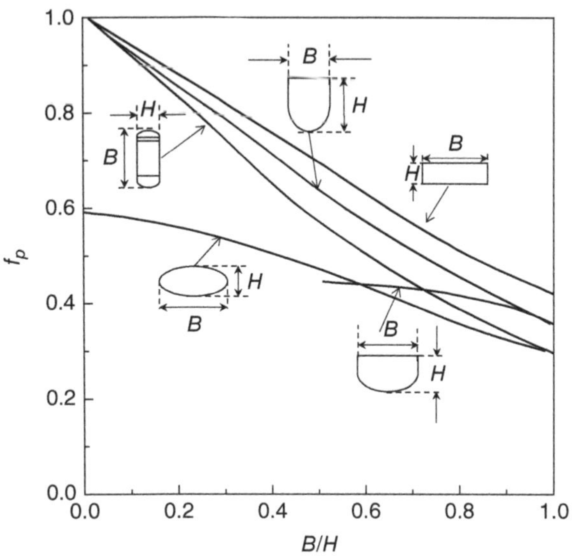

Figure 3.28 Correction factor f_p used to estimate the die conductance of non-regular geometries and Newtonian melts as a function of the die geometrical characteristics.

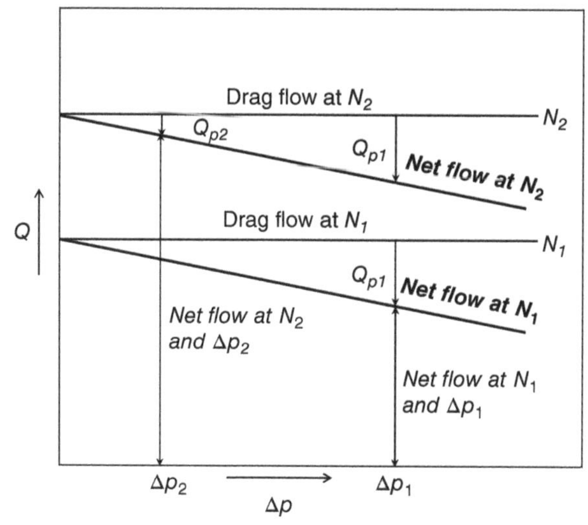

Figure 3.30 Characteristic curves for single screw extruders in terms of net flow as a function of the pressure difference (back pressure).

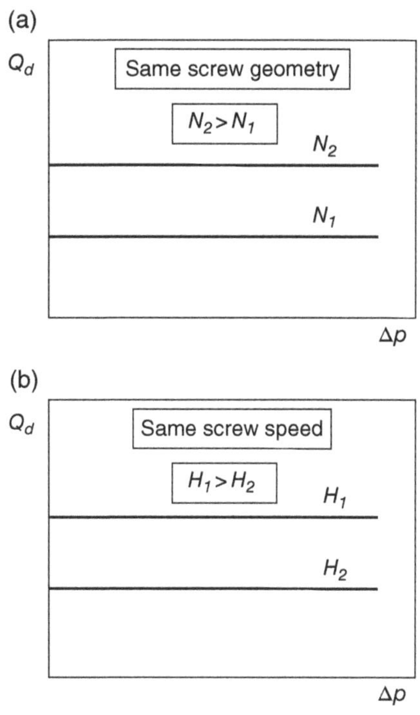

Figure 3.29 Characteristic curves for single screw extruders showing the drag flow as a function of the pressure difference (back pressure). (a) Influence of the screw speed. (b) Influence of the screw geometry (channel depth H).

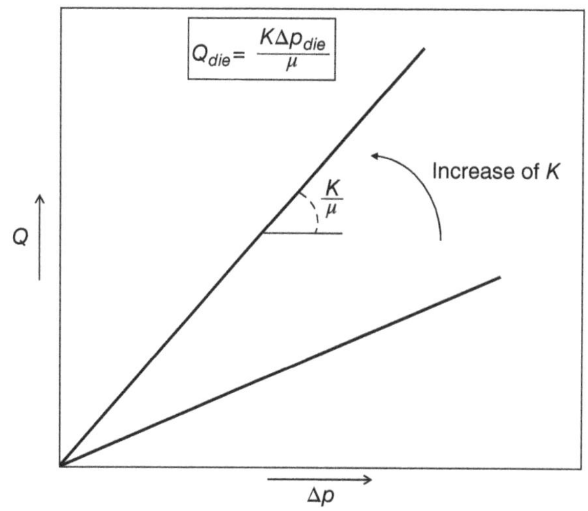

Figure 3.31 Characteristic curves for the flow of a Newtonian melt through a die represented as the net flow as a function of the pressure drop at the die for different dies defined by a conductance K.

for other geometries given in Table 3.1 show that the die conductance K increases with increases of the die radius (flow area) and decreases with the length of the die. In order to locate the steady processing operation in the extruder, it is necessary to take into account the net flow

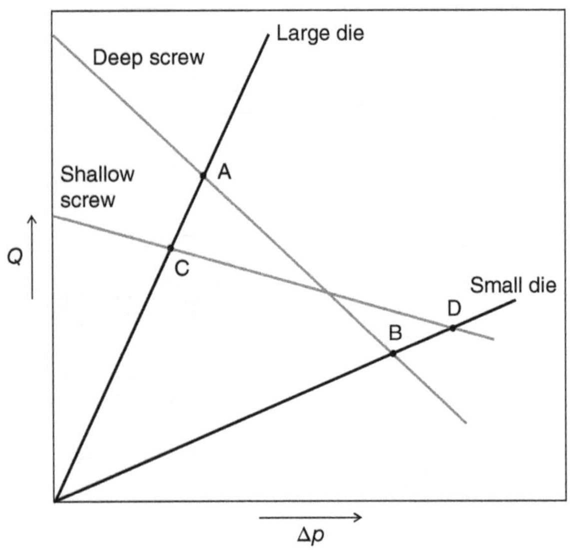

Figure 3.32 Characteristic curves for a single screw extruder operation considering the flow in the screw channel and the flow through the die.

in the screw and at the die, mathematically described by Eqs 3.81 and 3.85 and schematically illustrated in Figures 3.30 and 3.31. The result of combining these two effects can be seen in Figure 3.32 where several operation points such as A, B, C, and D can be identified and are dependent on the processing conditions. The figure clearly shows the increase in pressure and the reduction of the operating net flow with the reduction of the size of the die while the screw profile is maintained; the operating flow and pressure moves from A to B. If the operating points A and C, which result from the same die size but geometrically different screws, are compared it can be readily observed that the use of a deep screw geometry will provide higher net flow and pressure (point A) than the use of a screw with a shallow channel height (point C). Conversely, if the operating points B and D, which result from using two different types of screw and a small size die, are compared it can be observed that the use of shallow screws provides higher net flow and pressure than those obtained using deep screws.

Some of the assumptions used in this theoretical model have been revisited and adjusted according to experimental observations. Li and Hsieh (1996) argued that there are differences in considering the hypothetical situation where the barrel is rotating instead of the more realistic situation assumption where the screw is rotating. The rotation of the flights may influence the melt flow. They solved the model represented by Eq. 3.56 to Eq. 3.58 for Newtonian fluids and isothermal conditions but with the drag flow associated with the rotation of the screw rather than the barrel. Solutions provide a similar flow pattern in terms of a total net flow calculated as the sum of the drag flow and the pressure flow as indicated in Eq. 3.70 but the shape factors for both the drag and pressure flows were different. It is claimed that with the revisited boundary conditions, the model can provide more accurate predictions for geometries in which the ratio H/D is high. Results of the theoretical model were compared with existing experimental data on extrusion of Newtonian materials (Choo et al, 1980; Griffith, 1962).

3.2.3.1.2 Non-Newtonian melts (power law fluids), isothermal process

The assumptions used in this case are similar to those utilized to develop the flow model for Newtonian materials. However, the main difference lies in the fact that the shear stress tensor components are described by a different expression, which for this case is the power law model. The equations of motion in terms of the stress relevant this flow model are still valid and represented by Eqs 3.54b and 3.54d, which for clarity are repeated below including only terms that are relevant to the flow:

$$\frac{\partial p}{\partial x} = \frac{\partial \tau_{yx}}{\partial y} \tag{3.90}$$

$$\frac{\partial p}{\partial z} = \frac{\partial \tau_{xz}}{\partial x} + \frac{\partial \tau_{yz}}{\partial y} \tag{3.91}$$

In order to express the components of the stress tensor in terms of the velocity gradients for the power law model, for convenience, it is preferable to use the Generalized Newtonian Model concept that expresses the stress tensor as (Morrison, 2001):

$$\underline{\underline{\tau}} = \eta(\dot{\gamma})\underline{\underline{\dot{\gamma}}} \tag{3.92}$$

$\eta(\dot{\gamma})$ is a functional expression that can take several forms, among them the power law model described above and in more a greater detail in Chapter 4, which is given by the following equation (Morrison, 2001):

$$\mu_{app}(\dot{\gamma}) = m\dot{\gamma}^{n-1} \tag{3.93}$$

where the functional expression is assumed to be the apparent viscosity of the material μ_{app}, and $\dot{\gamma}$ is the second invariant of the strain rate tensor $\underline{\underline{\dot{\gamma}}}$, a scalar defined as:

$$\dot{\gamma} = \sqrt{\frac{1}{2}\underline{\underline{\dot{\gamma}}} : \underline{\underline{\dot{\gamma}}}} \tag{3.94}$$

The colon : in the above equation indicates a tensor operation known as a tensor scalar product (Morrison, 2001), the strain rate tensor $\underline{\underline{\dot{\gamma}}}$ is defined in terms of the velocity gradient tensor $\nabla \underline{v}$ and its transpose $(\nabla \underline{v})^T$ as:

$$\underline{\underline{\dot{\gamma}}} = \nabla \underline{v} + (\nabla \underline{v})^T \tag{3.95}$$

Forms of the strain rate and the velocity gradient tensors are tabulated for different flows, including drag and pressure-driven flows such as those encountered in extrusion processing (Bird et al, 1987; Morrison, 2001). From the kinematics of the flow assumed in the analysis of a Newtonian fluid in a single screw extruder, the only non-zero velocity gradient components and the resulting stress components are:

$$\tau_{yx} = m\left(\left(\frac{\partial v_x}{\partial y}\right)^2 + \left(\frac{\partial v_z}{\partial y}\right)^2 + \left(\frac{\partial v_z}{\partial x}\right)^2\right)^{\frac{n-1}{2}} \frac{\partial v_x}{\partial y} \tag{3.96a}$$

$$\tau_{yz} = m\left(\left(\frac{\partial v_x}{\partial y}\right)^2 + \left(\frac{\partial v_z}{\partial y}\right)^2 + \left(\frac{\partial v_z}{\partial x}\right)^2\right)^{\frac{n-1}{2}} \frac{\partial v_z}{\partial y} \tag{3.96b}$$

$$\tau_{xz} = m\left(\left(\frac{\partial v_x}{\partial y}\right)^2 + \left(\frac{\partial v_z}{\partial y}\right)^2 + \left(\frac{\partial v_z}{\partial x}\right)^2\right)^{\frac{n-1}{2}} \frac{\partial v_z}{\partial x} \tag{3.96c}$$

The 2D model described by Eqs 3.90, 3.91, 3.96a, 3.96b and 3.96c involves non-linear partial differential equations with no straightforward analytical solution for the velocity profiles and the pressure gradient in the direction x (Rauwendaal, 2001). Some additional assumptions about the flow can simplify this set of equations and make them more amenable to either an analytical solution of a 1D model (Tadmor & Gogos, 2006; Tadmor & Klein, 1970) or the numerical solution of a 2D simplified flow (Steller, 1990).

1D analytical model

Given the practical importance of the flow of shear-thinning/power law melts in a single screw extruder, an analytical solution is obtained for a simplified 1D flow in which only the velocity component in the direction z is considered. For that simplified case, the relevant equation is the following:

$$\frac{\partial p}{\partial z} = \frac{\partial \tau_{yz}}{\partial y} = \frac{d}{dy}\left[m\left(\frac{dv_z}{dy}\right)^{n-1}\frac{dv_z}{dy}\right] \tag{3.97}$$

Using the same dimensionless variables defined above as $\xi = \frac{y}{H}$ and $u_z = \frac{v_z}{V_{bz}} = \frac{v_z}{\pi N D \cos\theta}$ in addition to the two new dimensionless variables suggested by Tadmor and Klein (1970):

$$\Gamma_R = \frac{H}{V_{bz}} \cdot \left(\frac{H}{m} \cdot \left(\frac{\partial p}{\partial z}\right)\right)^s = \frac{H}{V_{bz}}\left(\frac{H}{m} \cdot \left(\frac{\partial p}{\partial z}\right)\right)^{1/n} \quad \textit{and}$$

$$\Gamma_R^n = \frac{H^{n+1}}{m V_{bz}^n} \cdot \left(\frac{\partial p}{\partial z}\right) \tag{3.98}$$

$$\phi = \frac{u_z}{\Gamma_R} \tag{3.99}$$

where $s = 1/n$ and both Γ_R and Γ_R^n are considered dimensionless pressure gradients. Substituting these defined dimensionless variables into Eq. 3.97, the following ordinary differential equation is obtained:

$$\frac{d}{d\xi}\left[\left(\frac{d\phi}{d\xi}\right)^{n-1}\frac{d\phi}{d\xi}\right] = \frac{d}{d\xi}\left[\left(\frac{d\phi}{d\xi}\right)^{n}\right] = 1 \tag{3.100}$$

By integrating the above equation twice, the following is obtained:

$$\phi(\xi) = \frac{1}{s+1}(\xi + C_1)^{s+1} + C_2 \tag{3.101}$$

C_1 and C_2 are two integration constants whose values can be determined using the boundary conditions given by Eq. 3.64. With the boundary condition $\phi = 0$ at $\xi = 0$, an expression for C_2 in terms of C_1 can be found:

$$C_2 = -\frac{1}{s+1}C_1^{s+1} \tag{3.102}$$

and by using the boundary condition $\phi = 1/\Gamma_R$ at $\xi = 1$, an expression that requires a numerical approach to find the constant C_1 is obtained:

$$(1 + C_1)^{s+1} - C_1^{s+1} = \frac{s+1}{\Gamma_R} \tag{3.103}$$

The velocity component in the z-direction can be then estimated from the following equation:

$$u_z(\xi) = \Gamma_R\left[\frac{(\xi + C_1)^{s+1} - C_1^{s+1}}{s+1}\right] \tag{3.104}$$

The constant C_I can be numerically estimated from Eq. 3.103 if the rheological parameters of the melt such as m and n, the processing conditions and the screw characteristics such as N, $\partial p/\partial z$, D, θ, and H are known. As for Newtonian melts, it is also important to estimate the net flow in the extruder which can be calculated by integration of the velocity profile over the thickness of the screw channel. It is worth noting that for this case the z-component of the melt velocity is not a function of the x-direction, hence the flow in the rectangular channel is calculated as (for the name of variables refer to Figure 3.20):

$$Q = W\int_0^H v_z dy = WV_{bz}H\Gamma_R\int_0^1 \frac{(\xi + C_1)^{s+1} - C_1^{s+1}}{s+1} d\xi \tag{3.105}$$

which by substitution yields:

$$Q = \frac{WV_{bz}H\Gamma_R}{s+1}\left[\frac{(1 + C_1)^{s+2} - C_1^{s+2}}{s+2} - C_1^{s+1}\right] \tag{3.106}$$

The numerical solution of Eq. 3.103 to get the constant C_I requires positive values of the constant except for cases in which the exponent $s + 1$ is an integer. Hence, there are conditions, mainly related to the rheological parameter n, under which the solution does not converge and the net flow calculated from Eq. 3.106 cannot be estimated. In addition, an examination of Eq. 3.106 would appear to indicate that when $\Gamma_R \to 0$ the net flow Q is zero. However, for those conditions the function $\frac{(1 + C_1)^{s+2} - C_1^{2+s}}{s+2} - C_1^{s+1}$ yields values so that:

$$lim_{\Gamma_R \to 0}\frac{Q}{WV_{bz}H} = lim_{\Gamma_R \to 0}\Gamma_R\left[\frac{(1 + C_1)^{s+2} - C_1^{s+2}}{s+2} - C_1^{s+1}\right] \approx 0.5 \tag{3.107}$$

The other condition which yields positive real solutions from Eq. 3.103, and thus warranty convergence of the velocity component predicted by Eq. 3.104 for any value of the flow index n, is related to the ratio $\frac{s+1}{\Gamma_p}$. Values of

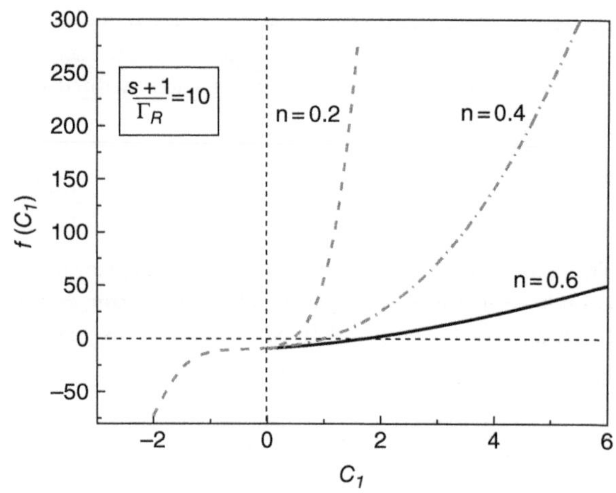

Figure 3.33 Plot of the function $f(C_I)$, given by Eq. 3.108, as a function of the parameter C_I for $(s + 1)/\Gamma_p = 10$ and different values of the flow index n.

the equation $f(C_I)$, written below for clarity, for different values of n and $\frac{s+1}{\Gamma_p} = 10$ are illustrated in Figure 3.33:

$$f(C_1) = (1 + C_1)^{s+1} - C_1^{s+1} - \frac{s+1}{\Gamma_R} \tag{3.108}$$

The figure clearly shows that for $n = 0.2$ and the value $s + 1 = 6$ being an integer, the function $f(C_I)$ yields real values for both positive and negative values of C_I, whereas it does not yield real values for negative values of C_I for $n = 0.4$ and $n = 0.6$, i.e. when the corresponding $s + 1$ values are not integers. It can also be demonstrated that for n values for which values of $s + 1$ are not integers, values of the function $f(C_I)$ will be real numbers only if the following relationship applies:

$$\frac{s+1}{\Gamma_R} \geq 1 \quad or \quad s + 1 \geq \Gamma_R \tag{3.109}$$

The above condition imposes limitations to the range of the dimensionless parameter Γ_R that could be used to predict the net flow by Eq. 3.106. This no-convergence of the solution for certain conditions can be explained by examining Figure 3.34. In the case of a small pressure gradient condition (small values of Γ_R), the velocity gradient illustrated in Figure 3.34a is always positive whereas for large pressure gradient conditions (large values of Γ_R), the velocity gradient can be either positive or negative (Figure 3.34b).

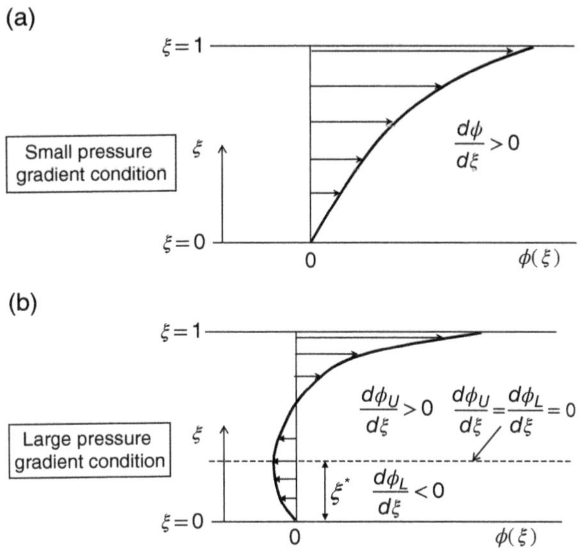

Figure 3.34 Dimensionless down-channel velocity component profiles for (a) small pressure gradients and (b) large pressure gradients.

For cases in which the velocity gradient is negative, i.e. ($d\phi/d\xi < 0$), the solution of Eq. 3.100 may or may not converge to real solutions, depending on the value of n. This convergence problem was overcome by solving the differential equation (Eq. 3.100) in two different regions (Tadmor & Klein, 1970). For the region where the velocity gradient is positive (upper region), the solution, a upper velocity component (ϕ_U), is obtained as described above with a similar boundary condition at $\xi = 1$ but the other boundary condition is at $\xi = \xi^*$ where the upper velocity component reaches a minimum, i.e. when $\dfrac{d\phi_U}{d\xi} = 0$ (see Figure 3.34b). Solution of Eq. 3.100 with the above boundary conditions yields (Tadmor & Klein, 1970):

$$\phi_U(\xi) = \frac{1}{s+1}\cdot\left[(\xi-\xi^*)^{s+1} - (1-\xi^*)^{s+1}\right] + \frac{1}{\Gamma_R} \quad for \quad \xi^* \le \xi \le 1 \tag{3.110}$$

For the region where the velocity gradient is negative (lower region), the following differential equation can be used to get the lower solution ϕ_L:

$$-\frac{d}{d\xi}\left[\left(\frac{-d\phi_L}{d\xi}\right)^n\right] = 1 \tag{3.111}$$

In this case the boundary condition $\phi_L = 0$ at $\xi = 0$ is used in addition to the condition for the minimum observed in the lower velocity component ϕ_L at ξ^* (see Figure 3.34b). The solution to the differential equation (Eq. 3.111) is:

$$\phi_L(\xi) = \frac{1}{s+1}\cdot\left[(\xi^*-\xi)^{s+1} - (\xi^*)^{s+1}\right] \quad for \quad 0 \le \xi \le \xi^* \tag{3.112}$$

In order to use these solutions in the two regions, the value of ξ^* must be known. That value is obtained by considering the continuity of the velocity profile at $\xi = \xi^*$, that is:

$$\phi_U(\xi) = \phi_L(\xi) \quad at \quad \xi = \xi^* \tag{3.113}$$

Thus, substituting Eqs 3.110 and 3.112 in the above equation, an equation to estimate ξ^*, is obtained:

$$(\xi^*)^{s+2} - (1-\xi^*)^{s+1} + \frac{s+1}{\Gamma_R} = 0 \tag{3.114}$$

It must be noted that values of ξ^* have to be obtained using a numerical approach. As before, these velocity profiles can be used to estimate the net flow in the extruder as:

$$Q = W\int_0^H v_z dy = WV_{bz}H\Gamma_R\left[\int_0^{\xi^*}\phi_L(\xi)d\xi + \int_{\xi^*}^{1}\phi_U(\xi)d\xi\right] \tag{3.115}$$

which upon integration gives:

$$\frac{Q}{WV_{bz}H} = \frac{\Gamma_R}{s+1}\left[\frac{(\xi^*)^{s+2} + (1-\xi^*)^{s+2}}{s+2} - (\xi^*)^{s+1}\right] \tag{3.116}$$

Normalized net flows as a function of the dimensionless pressure gradient Γ_R^n for different shear-thinning materials (defined by the flow index n) are illustrated in Figure 3.35. The effect of the material rheology on the net flow is remarkable. However, it should be noted that the dimensionless pressure gradient Γ_R also depends on the flow index n (see Eq. 3.98). But when the normalized net flow is plotted as a function of the dimensionless pressure gradient without rising to the power of n (i.e. Γ_R), a similar behavior is observed except when the dimensionless gradient is around or larger than 10 (Figure 3.36).

Thus, a question arises about the convenience of using either Γ_R or Γ_R^n to describe the relationships between processing parameters and the rheology of the melt. An examination of Eq. 3.87 indicates that a non-linear relationship is established between the net flow and the

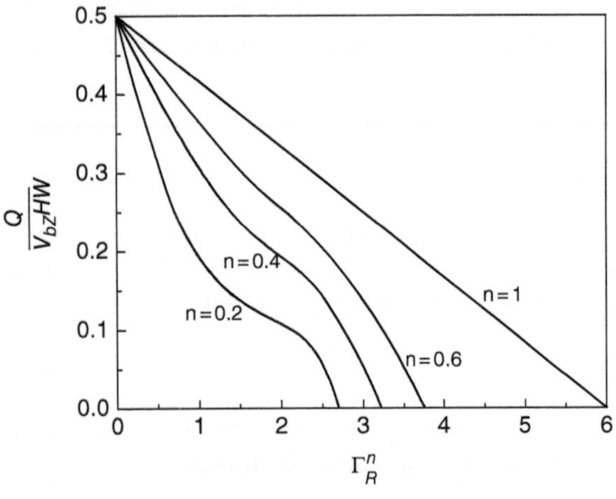

Figure 3.35 Dimensionless flow as a function of the dimensionless pressure gradient Γ_R^n for power law fluids and different flow indexes.

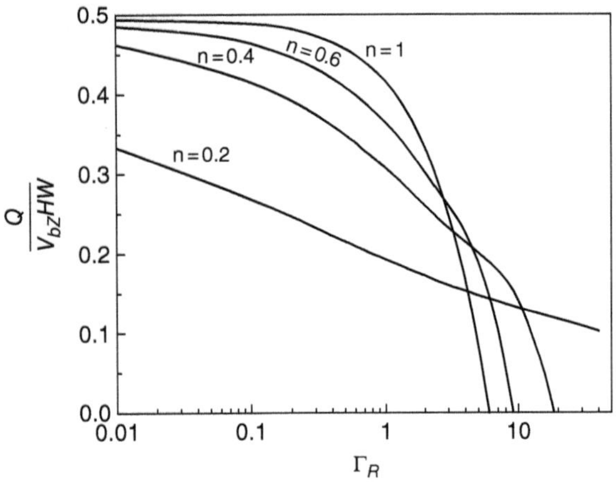

Figure 3.36 Dimensionless flow as a function of the dimensionless pressure gradient Γ_R for power law fluids and different flow indexes.

pressure gradient but it becomes linear if the pressure gradient is raised to the power of flow index n. Thus, the dimensionless pressure gradient Γ_R^n is preferable for further analysis. Figure 3.35 shows that for the same value of the dimensionless pressure gradient, the normalized net flow decreases with decreasing values of the flow index (i.e. when the material is more shear-thinning). This behavior can be explained by a quick observation of Eq. 3.87 which can be applied to the screw channel instead of the die. As in the screw channel the pressure gradient is

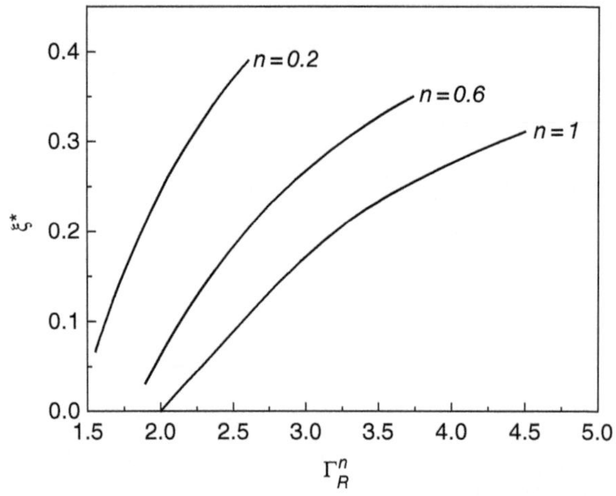

Figure 3.37 Transition cross-channel location (ξ^*) where the dimensionless gradient of the down-channel velocity components in the upper $\phi_U(\xi)$ and lower $\phi_L(\xi)$ regions of the screw channel are equal to zero as a function of the dimensionless pressure gradient and for different flow indexes.

positive, increases of the pressure flow generated by the pressure gradient will oppose the drag flow created by the rotation of the screw and reduce the net flow (see Eq. 3.70). For same values of pressure gradients, lower values of n will produce larger pressure flows, which reduce the net flow. It is also useful to examine values of ξ^* as a function of the dimensionless pressure gradient and the flow index, which is illustrated in Figure 3.37. When the dimensionless pressure gradient Γ_R^n increases, the region in which the z-component velocity is negative expands for all materials, regardless their rheology. The effect of the flow index on the value of ξ^* is related to results shown in Figure 3.35, i.e. lower values of the flow index n decrease the net flow at the same dimensionless pressure gradients in correspondence with an increase of the region where the velocity z-component is negative.

Simplified 2D numerical model

Steller (1990) simplified Eqs 3.90 and 3.91, assuming that for a channel with constant height and width, the non-zero velocity components v_x and v_z only depend on the direction y. These assumptions also simplify the stress tensor components given by Eqs 3.96a, 3.96b and 3.96c, to give the following set of equations:

$$\frac{\partial p}{\partial x} = \frac{\partial \tau_{yx}}{\partial y} \tag{3.117}$$

$$\frac{\partial p}{\partial z} = \frac{\partial \tau_{yz}}{\partial y} \tag{3.118}$$

$$\tau_{yx} = m\left(\left(\frac{\partial v_x}{\partial y}\right)^2 + \left(\frac{\partial v_z}{\partial y}\right)^2\right)^{\frac{n-1}{2}} \frac{\partial v_x}{\partial y} \tag{3.119}$$

$$\tau_{yz} = m\left(\left(\frac{\partial v_x}{\partial y}\right)^2 + \left(\frac{\partial v_z}{\partial y}\right)^2\right)^{\frac{n-1}{2}} \frac{\partial v_z}{\partial y} \tag{3.120}$$

By assuming constant pressure gradients in directions x and z, integration of Eqs 3.117 and 3.118 gives:

$$\tau_{yx} = \left(\frac{\partial p}{\partial x}\right)(y - C_1) \tag{3.121}$$

$$\tau_{yz} = \left(\frac{\partial p}{\partial z}\right)(y - C_2) \tag{3.122}$$

where C_1 and C_2 are integration constants. Combining Eqs 3.119 and 3.120 with Eqs 3.121 and 3.122, the following equations are obtained:

$$\frac{\partial v_x}{\partial y} = \frac{\left(\frac{\partial p}{\partial x}\right)\cdot(y - C_1)}{m^{\frac{1}{n}}}\left[\left(\left(\frac{\partial p}{\partial x}\right)\cdot(y - C_1)\right)^2 + \left(\left(\frac{\partial p}{\partial z}\right)\cdot(y - C_2)\right)^2\right]^{\frac{1-n}{2n}} \tag{3.123}$$

$$\frac{\partial v_z}{\partial y} = \frac{\left(\frac{\partial p}{\partial z}\right)\cdot(y - C_2)}{m^{\frac{1}{n}}}\left[\left(\left(\frac{\partial p}{\partial x}\right)\cdot(y - C_1)\right)^2 + \left(\left(\frac{\partial p}{\partial z}\right)\cdot(y - C_2)\right)^2\right]^{\frac{1-n}{2n}} \tag{3.124}$$

By using similar dimensionless variables like the ones defined before, in addition to newly defined variables given below, Eqs 3.123 and 3.124 can be written in dimensionless form as:

$$C_1^* = \frac{C_1}{H} \quad C_2^* = \frac{C_2}{H} \quad \upsilon = \frac{\partial p}{\partial x}\Big/\frac{\partial p}{\partial z} \quad U_o = \left(\left(\frac{\partial p}{\partial z}\right)\cdot\frac{H}{m}\right)^{1/n}\cdot H \tag{3.125}$$

It is helpful to relate the parameter U_o, which has dimensions of velocity, with the dimensionless pressure gradient Γ_R defined above for the 1D analysis. Since the 2D analysis involves two velocity components, the approach of using the parameter U_o appears more convenient and dimensionless velocity components are defined as:

$$u_x = \frac{v_x}{U_o} \quad \textit{and} \quad u_z = \frac{v_z}{U_o} \tag{3.126}$$

Equations 3.123 and 3.124 in dimensionless form can be written as:

$$\frac{du_x}{d\xi} = \upsilon(\xi - C_1^*)\left[\upsilon^2(\xi - C_1^*)^2 + (\xi - C_2^*)^2\right]^{\frac{1-n}{2n}} \tag{3.127}$$

$$\frac{du_z}{d\xi} = (\xi - C_2^*)\left[\upsilon^2(\xi - C_1^*)^2 + (\xi - C_2^*)^2\right]^{\frac{1-n}{2n}} \tag{3.128}$$

Integration of the above two equations gives analytical solutions in terms of the hypergeometric function known as the Appell function (Exton, 1978). However, it is mathematically cumbersome to use the Appell functions to estimate the parameters C_1^* and C_2^*, so Steller (1990) applied a numerical approach to estimate the net flow. If the function $\Phi(\xi) = \upsilon^2(\xi - C_1^*)^2 + (\xi - C_2^*)^2$ is defined, the following equation is obtained:

$$\int_0^1 \Phi^{\frac{1-n}{2n}}(\xi)\, d\xi = \frac{I_x - \upsilon \cdot I_z}{\upsilon\cdot(C_2^* - C_1^*)} \tag{3.129}$$

where I_x and I_z are two integrals defined as:

$$I_x = \int_0^1 \upsilon\cdot(\xi - C_1^*)\,\Phi^{\frac{1-n}{2n}}(\xi)\, d\xi \tag{3.130}$$

$$I_z = \int_0^1 (\xi - C_2^*)\,\Phi^{\frac{1-n}{2n}}(\xi)\, d\xi \tag{3.131}$$

Integration of Eq. 3.129 plus a number of transformations yields the equations:

$$\frac{n}{n+1}\cdot\left[(\Phi(\xi=1))^{\frac{1+n}{2n}} - (\Phi(\xi=0))^{\frac{1+n}{2n}}\right] = \upsilon\cdot I_x + I_z \tag{3.132}$$

$$n\cdot\left\{\left[(\Phi(\xi=0))^{\frac{1+n}{2n}}(n\upsilon^2(1 - C_1^*) + (2n+1)(1 - C_2^*)\right]\frac{I_x}{\upsilon} - \left[n(1 - C_2^*) - (2n+1)(1 - C_1^*)\right]I_z\right\} = 0 \tag{3.133}$$

The expressions above involve the boundary conditions at $\xi = 0$ and $\xi = 1$ that have been transformed by the function Φ (ξ). The transformation avoids the use of the hypergeometric functions to obtain the integration constants C_1^* and C_2^* which are necessary to obtain the velocity components in the cross-channel (x) and down-channel directions (z) by respective integration of

Eqs 3.127 and 3.128. To complete the mathematical model, it is necessary to calculate the two flow rates associated with the velocity components u_x and u_z in the extruder channel. Flow rate determination has already been discussed for Newtonian melts (see Eqs 3.61 and 3.66). By using similar equations as well as integration by parts along the transformed boundary conditions, the following relationships are obtained:

$$\int_0^H v_x dy = 0 \;\Rightarrow\; I_x - \int_0^1 \xi\left(\frac{du_x}{d\xi}\right)d\xi = 0 \tag{3.134}$$

$$W\int_0^H v_z(y)\,dy = Q \;\Rightarrow\; I_z - \int_0^1 \xi\left(\frac{du_z}{d\xi}\right)d\xi = \frac{Q}{HWU_o} = \bar{Q} \tag{3.135}$$

where $\bar{Q}$ is a dimensionless net flow defined as $\bar{Q} = \frac{Q}{HWU_o}$.

By using the transformed boundary conditions defined above, an equation to estimate the dimensionless net flow in the extruder $\bar{Q}$ can be obtained (Steller, 1990):

$$\bar{Q} = (1 - C_1^*)I_z - (1 - C_2^*)\frac{I_x}{v} \tag{3.136}$$

For the isothermal flow of non-Newtonian power law melts, the set of expressions from Eq. 3.129 to Eq. 3.136 allows the calculation of the net flow $\bar{Q}$ at different extrusion processing conditions and for different melts. This approach clearly shows that more complicated systems including non-isothermal conditions, more complex rheological behavior and geometries, such as that of twin screw extruders, require a numerical approach based on the multiphysics concept used in many computational packages (Ansys™, Comsol™, LUDOVIC®) which are discussed below. Results obtained by solving the above set of equations are represented in terms of the net flow in the extruder as a function of pressure gradient expressed in its dimensionless form Γ_R^n, which is related to the parameter U_o (defined by Eq. 3.125) as:

$$\Gamma_R^n = \left(\frac{U_0}{V_b \cos\theta}\right)^n \tag{3.137}$$

Results obtained from this set of equations are illustrated in Figure 3.38 in terms of the ratio between the net Q and the pure drag flow Q_d as a function of the dimensionless pressure gradient which is calculated by the following equation:

$$\frac{Q}{2Q_d} = (1 - C_1^*)\cos\theta - (1 - C_2^*)\frac{\sin\theta}{v} \tag{3.138}$$

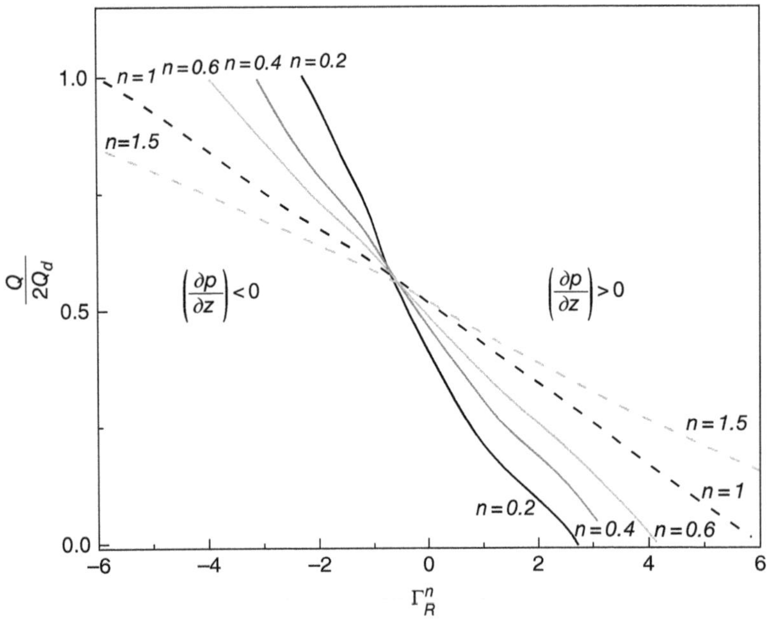

Figure 3.38 Schematic plot showing the ratio between the net and drag flows as a function of the dimensionless pressure gradient for different power law fluids including Newtonian ($n = 1$), shear-thinning ($n < 1$) and shear-thickening ($n > 1$) melts.

Results shown were calculated for a square pitch screw with an angle of $\theta = 17.6°$. They resemble the trend presented in Figure 3.35 for the 1D assumed flow in which for positive pressure gradients, decreases of the flow index produce a decrease of the net flow due to the increase of pressure flow. The behavior is reversed when the pressure gradient is negative. Results are also shown for melts that are dilatant or shear-thickening ($n = 1.5$). For this case when the pressure gradient is negative, a pressure flow with the same direction as the drag flow is generated but the higher viscosities encountered at high shear rates produce a low-pressure flow that does not contribute significantly to the drag flow and net flows are lower than net flows obtained with shear-thinning fluids. Conversely, for positive pressure gradients, the high viscosity of dilatant melts at high shear rates produces low-pressure flow that opposes the drag flow and thus a higher net flow is obtained compared with that of shear-thinning fluids for positive pressure gradients. Other studies using this model include the effect of helix angle and the material rheology on the net flow (Steller, 1990).

A summary of some of the approaches utilized to describe the isothermal flow of non-Newtonian power law fluids for a specific case of a flow index $n = 0.5$ in comparison with results obtained for the Newtonian model is illustrated in Figure 3.39. The largest deviation is observed when an equation similar to Eq. 3.70 is used to predict the net flow. The drag flow is independent of the melt rheology so its calculation does not present problems. However, the pressure flow for a power law fluid is strongly dependent on the melt rheology and when this is added to the drag flow to estimate the net flow, as indicated by Eq. 3.70, significant differences between results of this model and the Newtonian melts are observed (see Figure 3.39). The figure also shows that results obtained by solving the fluid mechanics equation for a power law model do not differ significantly whether the 1D or 2D approaches are used. Simulations represented in Figure 3.39 were performed using a flow index $n = 0.5$ but similar results can be observed if other flow indexes are used.

Other numerical approaches have been used to describe the isothermal flow of power law fluids with some degree of variation in the methodology, but among them, seminal works are those of Griffith (1962) and Kroesser and Middleman (1965). One of the main conclusions arising from those works when comparing the flow of Newtonian melts with, more realistic, flows involving non-Newtonian melts can be seen in Figure 3.40, which is a modification of the schematic figure illustrated in Figure 3.30 for Newtonian flows.

3.2.3.1.3 *Use of simplified models to study the performance of single screw extruders*

It is worth using the engineering analysis on the fluid dynamics of the flow of melts in channels of single screw

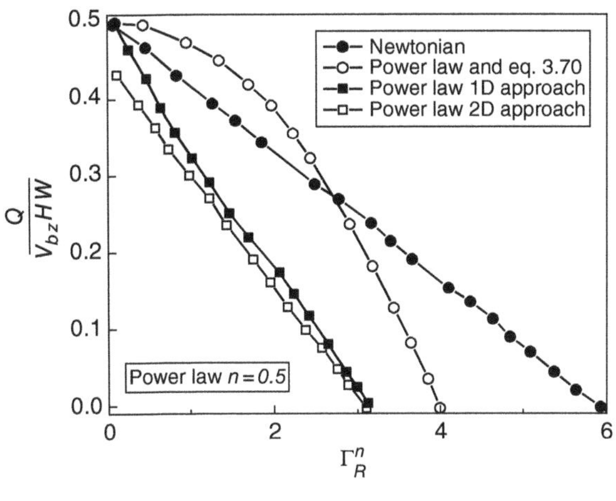

Figure 3.39 Comparison of model results shown as dimensionless net flow versus dimensionless pressure gradient. Behaviors for a Newtonian fluid and power law fluids using 1D, 2D and the approximated approach described above. The flow index used for the simulations was $n = 0.5$.

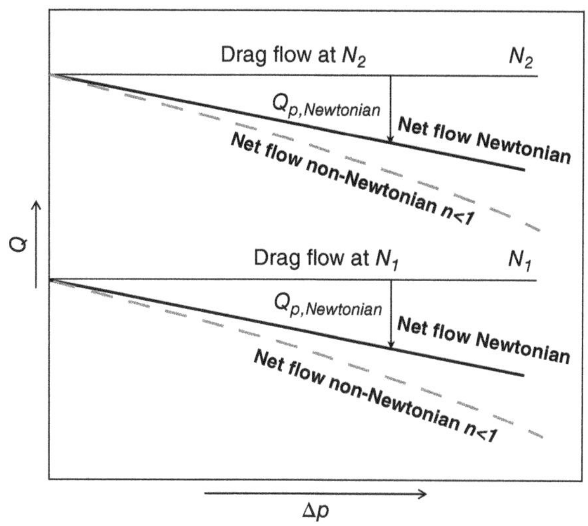

Figure 3.40 Schematic characteristic curves of net flow versus pressure differences (back pressure) for Newtonian and non-Newtonian melts.

extruders under isothermal conditions to describe and understand the performance of these extruders. An aspect of interest in the analysis of extrusion operations using the aforementioned simplified models is the possibility of performing "what if" qualitative analysis of the extrusion process. Some of the analyses that can be incorporated into the isothermal flow of Newtonian and non-Newtonian melts include special screw design or screw profile associated with the metering section, such as: (a) leakage flow through the flight clearance; (b) stepped screw; (c) tapered screw; and (d) venting.

Leakage flow

With new extruders, the flight clearance, which is the clearance between the top of the flight and the barrel wall (indicated by δ in Figure 3.17) is usually small and the extruded material acts like a lubricant that prevents metal-to-metal friction. Stevens and Covas (1995) stated that the leakage flow is produced in the flight clearance as a consequence of the relative motion of the screw flight and the barrel and the pressure gradient existing in the extruder. The first is a form of drag flow created by two parallel faces, which can be calculated by the following equation (Stevens & Covas, 1995):

$$Q_{dL} = \frac{\pi^2 N D^2 \delta \tan\theta}{2} \tag{3.139}$$

which, after several geometric considerations, can be estimated as a percentage of the total drag flow by the following equation:

$$\frac{Q_{dL}}{Q_d} = \frac{\delta}{H\left(1-\frac{e}{B}\right)\cos^2\theta} \tag{3.140}$$

where B is the screw pitch and e is the screw flight width.

The leakage flow due to pressure flow can be estimated by an equation that is derived by assuming a flow originated by a pressure difference in the two extremes of the screw flight. It is important to note that since pressure gradients in the longitudinal and transverse directions exist, their corresponding leakages are created and given by the following expressions: for the longitudinal leakage flow:

$$Q_{pL} = \frac{\pi^2 D^2 \delta^3}{12\mu e \cos\theta} \cdot \frac{\partial p}{\partial z} \tag{3.141a}$$

and for the cross-leakage flow:

$$Q'_{pL} = \frac{\pi^2 D^2 \delta^3 N(B-e)\tan\theta}{2eH^2} \tag{3.141b}$$

The ratio of the leakage pressure flow with respect to the pressure flow of a Newtonian melt is estimated as:

$$\frac{Q_{pL}+Q'_{pL}}{Q_p} = \frac{\pi^2 D^2}{(B-e)e}\left(\frac{\delta}{H}\right)^3 \cdot \frac{1}{\cos^2\theta} + \frac{6\mu\pi^2 D^2 \delta^3 N}{eH^5} \cdot \frac{\sin\theta}{\cos^2\theta \cdot \left(\frac{\partial p}{\partial z}\right)} \tag{3.142}$$

It is important to recognize that leakage flows are calculated per turn and the leakage of successive turns is mostly in series, thus the net flow reduction due to leakage flow should not be estimated by multiplying the flow per turn by the number of turns (Stevens & Covas, 1995). Calculations of these flows using typical single screw extruder configurations and high-viscosity Newtonian materials have shown that leakage flows are relatively low and can be assumed negligible as long as the flight clearance is low. But leakage flow can be significantly high when the flight clearance becomes large, as is the case for worn screws, for instance. The impact of screw wear on the performance of single screw extruders is further discussed in section 3.2.4.

Stepped screw

A typical stepped screw is schematically illustrated in Figure 3.41 that clearly shows a step change in the root diameter of the screw suddenly changing the height of the channel from a large value H_1 to a small value H_2. Hence the values of the flow calculated by Eq. 3.67, assuming F_d and F_p to be approximately equal to 1, will vary accordingly (see Figure 3.41). However, under steady-state conditions, the continuity of the flow $Q_1 = Q_2$ in the two zones is required. Stevens and Covas (1995) reported that with those assumptions, the pressure p along the extruder shaft varies as schematically illustrated in Figure 3.41, and the pressure p is a function of the net flow, which for Newtonian melts can be calculated as:

$$p = K_1\left(0.5 - \frac{Q}{WV_{bz}H_2}\right) \tag{3.143}$$

where

$$K_1 = \frac{12\mu W L_2}{H_2^2}\left(1+\frac{L_1}{L_2}\cdot\left(\frac{H_2}{H_1}\right)^3\right) \tag{3.144}$$

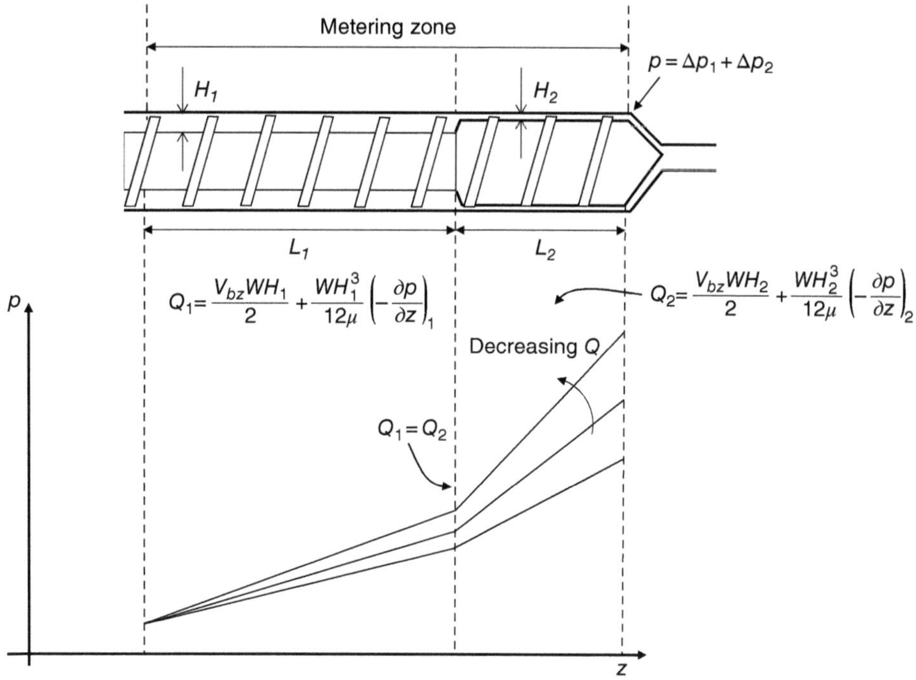

Figure 3.41 Pressure profiles in single screw extruders with stepped screws. Effects of the net flow rate in the pressure profile.

The above equations indicate that die pressure increases with decreases of the net flow Q (see Figure 3.41).

Tapered screw

The diameter of the screw root increases linearly which produces a linear decrease in the channel height as illustrated in Figure 3.42. If as indicated in the figure at $z=0$, $H(z)=H_1$, and at $z=Z_1$, $H(z)=H_2$, the change of channel height with the co-ordinate z can be written as:

$$H(z)=H_1-\frac{H_1-H_2}{Z_1}z \tag{3.145}$$

By applying the derivative with respect to the variable z to the above equation, the independent variable z can be changed to the variable H as:

$$dz=-\frac{Z_1}{H_1-H_2}dH \tag{3.146}$$

By assuming that the coefficients F_d and F_p are approximately equal to 1, Eq. 3.67 can be rearranged to give:

$$\frac{\partial p}{\partial z}=\frac{12\mu V_{bz}}{H^2}\left(\frac{1}{2}-\frac{Q}{WHV_{bz}}\right) \tag{3.147}$$

and by integration from $z=0$, $H=H_1$ where the pressure is $p=0$ to $z=Z_1$, $H=H_2$ where the pressure is p_1, the following equation is obtained:

$$p_1=\frac{6\mu V_{bz}Z_1}{H_1H_2}\left(1-\frac{Q}{WH_2V_{bz}}\cdot\frac{H_1+H_2}{H_1}\right) \tag{3.148}$$

For $z>Z_1$ the height of the channel is constant, and integration of Eq. 3.147 until $z=Z_2$ where pressure is p_2 yields:

$$p_2=\frac{6\mu V_{bz}Z_2}{H_2^2}\left(1-\frac{2Q}{WHV_{bz}}\right) \tag{3.149}$$

Thus, the die pressure can be calculated as the sum of these two pressures to give:

$$p=\frac{6\mu V_{bz}Z_1}{H_1H_2}\left(1-\frac{Q}{WH_2V_{bz}}\cdot\frac{H_1+H_2}{H_1}\right)+\frac{6\mu V_{bz}Z_2}{H_2^2}\left(1-\frac{2Q}{WHV_{bz}}\right) \tag{3.150}$$

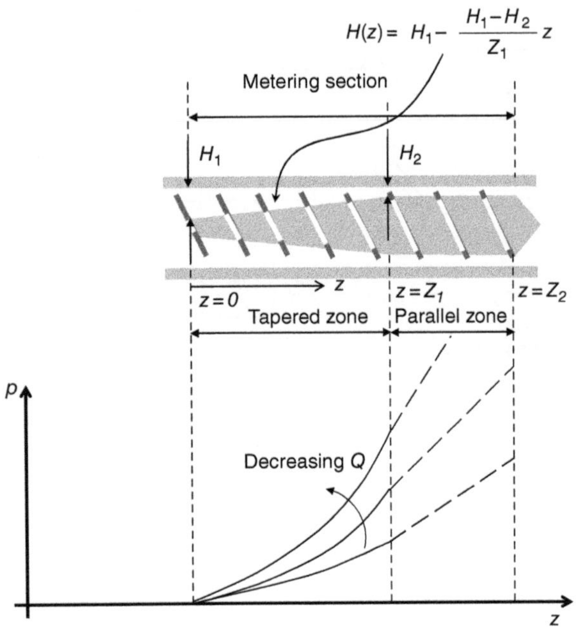

Figure 3.42 Pressure profiles in single screw extruders with tapered channels. Effect of the net flow rate in the pressure profile.

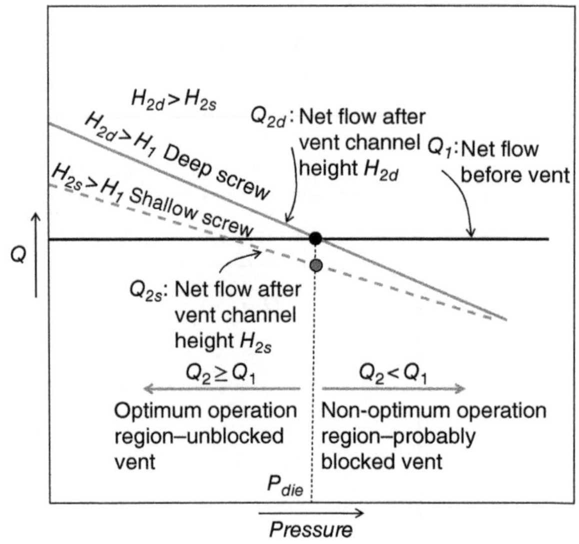

Figure 3.43 Characteristic net flow rate versus pressure difference curves for single screw extruders in the presence of vents. Analysis of the extruder performance considering large (deep) and small height (shallow) screw elements.

The pressure profile of this tapered channel geometry is schematically illustrated in Figure 3.42. The slope of the pressure profile estimated by Eq. 3.147 is not constant but increases with decreases of H following the relationship $1/H^2$ from $p = 0$ until the die pressure calculated by Eq. 3.150. The equation also shows that pressure increases when the net flow decreases.

Venting

The benefits of using venting on extrusion processing can be implemented with more practical advantages in twin screw extrusion for functions such as recovery or elimination of volatiles and water vapor. Use of venting in single screw extruders is also possible but more caution should be used in the design of the screws. With the help of the analytical solutions for Newtonian melts, it is possible to conceptually see some of the constraints imposed on the screw design to minimize blockage in vents. It is important to note that when a venting port is inserted, the gauge pressure in an unblocked vent is reduced to zero. So after the vent, a suitable screw geometry is necessary to raise the pressure in the melt to the required die pressure.

The design of the screw profile required to achieve this pressure increase is crucial in the operation of single screw extruders because if the die pressure is too high, it can cause blockage of the vent. While the venting is unblocked, the zone before the vent port is operating with a zero back pressure, so the net flow in the extruder is provided by the drag flow and only depends on the screw speed and the screw geometry as indicated by Eq. 3.67. That flow is illustrated in Figure 3.43 and labeled Q_1. Conversely, the flow after the vent is acting against the die pressure. As described before (see Figure 3.32), that flow will be of the type shown in Figure 3.43 as Q_2. To keep the vent unblocked, the extruder must operate with a restriction such as $Q_2 \geq Q_1$, so for a given die pressure P_{die} two operation zones can be distinguished which are indicated in Figure 3.43. It can be demonstrated that for increases of pressure in the extruder zone after the vent, the channel height of the screw after the vent, H_2, has to be larger than the channel height before the vent, H_1. However, as noted in the figure, the height of the channel cannot be selected arbitrarily. If for that die pressure a shallow channel height (indicated by H_{2s} in the figure) is selected, the vent would be blocked because the net flow Q_{2s} becomes smaller than the drag flow Q_1.

As demonstrated in this section, although models describing isothermal extrusion of Newtonian and non-Newtonian melts provide analytical relationships useful for analysis like the ones done above, practical extrusion operations are non-isothermal and models representing more realistic processes require numerical solutions. A review of the approaches used to describe these processes is given in the following section.

3.2.3.1.4 Newtonian and non-Newtonian models, non-isothermal process

The energy balance provides the temperature field in the extruded material and incorporates the generation of heat by viscous dissipation in the extrusion process. During the process, heat is transported in and out of the extruded material by convection and conduction whereas the heat generated by viscous dissipation can be introduced as a heat source in the energy balance estimated by the term $\underline{\underline{\tau}}:\nabla\underline{v}$, where : indicates the double tensor product between the stress tensor $\underline{\underline{\tau}}$ and the velocity gradient tensor $\nabla\underline{v}$. This term is a scalar and establishes a link between the fluid mechanics model described above and the rheology of the material, which is covered in more detail in Chapter 4. A balance of energy suitable for the extrusion process can be written as:

$$\rho C\left(\frac{\partial T}{\partial t}+\underline{v}\cdot\nabla T\right)=-\nabla.q+\left(\underline{\underline{\tau}}:\nabla\underline{v}\right) \tag{3.151}$$

where the heat gain or loss per conduction through the barrel wall can be expressed as:

$$\underline{q}=-k\nabla T \tag{3.152}$$

In Eq. 3.151, the first term on the left side of the energy equation represents the accumulation of thermal energy in the product whereas the second term, also in the left side, is the energy transferred by convection due to the flow of the melt. By substituting Eq. 3.152 into Eq. 3.151, the following equation is obtained:

$$\rho C\left(\frac{\partial T}{\partial t}+\underline{v}\cdot\nabla T\right)=\nabla.k\nabla T+\left(\underline{\underline{\tau}}:\nabla\underline{v}\right) \tag{3.153}$$

C and k are the heat capacity and the thermal conductivity of the melt, respectively, and ρ its density. The conversion of mechanical energy into heat in the extruder volume can be estimated as:

$$E_V=\int_V\left(\underline{\underline{\tau}}:\nabla\underline{v}\right)dV \tag{3.154}$$

Since a rigorous derivation including non-isothermal conditions even for Newtonian fluids usually ends in models that require numerical solutions, this section will provide a simple analysis for the simple non-isothermal Newtonian case.

Newtonian melts

For the Newtonian case, analytical solutions were obtained using a viscosity that is dependent on the temperature, which is also a function of the location in the extruder. The problem was solved by assuming only the down-channel direction z component of the melt velocity, which is dependent on the direction y. The non-isothermal condition was expressed as the temperature in the screw channel being dependent on the direction y. Under those assumptions, Eq. 3.54d simplifies to the following:

$$\frac{\partial p}{\partial z}=\frac{\partial}{\partial y}\left[\mu\frac{\partial v_z}{\partial y}\right] \tag{3.155}$$

whereas the temperature dependence of the viscosity is given by the following equation:

$$\mu=\mu_o e^{-A(T-T_s)} \tag{3.156}$$

where μ_o and A are two empirical parameters. Although the relationship used to describe the effect of temperature on the viscosity seems to differ from the expression used in Eq. 4.18 in Chapter 4, both give a similar trend which is an exponential or Arrhenius-like equation that predicts a decrease of viscosity with temperature. For convenience, in the derivation of an analytical solution for the non-isothermal model that relationship is expressed by Eq. 3.156. Substituting Eq. 3.156 into Eq. 3.155 and using the dimensionless variables for the z-velocity component of the flow and the direction y defined above, the following equation is obtained:

$$\frac{d}{d\xi}\left[e^{b\Theta}\frac{du_z}{d\xi}\right]=\frac{H^2}{V_{bz}\mu_o}\left(\frac{\partial p}{\partial z}\right) \tag{3.157}$$

where the parameter Θ is a dimensionless temperature defined as:

$$\Theta=\frac{T-T_s}{T_b-T_s} \tag{3.158}$$

and in terms of the above defined dimensionless temperature the change of the melt viscosity with temperature is expressed as:

$$\mu = \mu_o e^{b\Theta} \tag{3.159}$$

where $b = -A(T_b - T_s)$.

Rather than solving Eq. 3.157 along the energy equation to get the temperature profile (multiphysics model), which would require a numerical approach, a linear temperature profile is assumed with the aim of getting a close solution of this simple non-isothermal problem. Thus, the use of a linear temperature profile yields:

$$\frac{d}{d\xi}\left[e^{b\xi}\frac{du_z}{d\xi}\right] = \frac{H^2}{V_{bz}\mu_o}\left(\frac{\partial p}{\partial z}\right) = \Gamma_1 \tag{3.160}$$

with the following boundary conditions:

$$\begin{aligned} u_z &= 0 \quad at \quad \xi = 0 \\ u_z &= 1 \quad at \quad \xi = 1 \end{aligned} \tag{3.161}$$

where the parameter $\Gamma_1 = \frac{H^2}{V_{bz}\mu_o}\cdot\left(\frac{\partial p}{\partial z}\right)$ is a dimensionless pressure gradient. The solution of Eq. 3.160 is:

$$u_z(\xi) = \Gamma_1 e^{-b\xi}\left(-\frac{\xi}{b} - \frac{1}{b^2} + \frac{C_1}{b}\right) + C_2 \tag{3.162}$$

The integration constants C_1 and C_2 are determined from the boundary conditions as:

$$C_1 = \frac{-b}{\Gamma_1(1-e^{-b})} + \frac{1-e^{-b}(b+1)}{b(1-e^{-b})} \tag{3.163}$$

$$C_2 = \Gamma_1\left[\frac{1}{b^2} - \frac{C_1}{b}\right] \tag{3.164}$$

The set of expressions given by Eqs 3.162 to 3.164 can be used to estimate the z-component velocity profile in the extruder channel. Curves illustrated in Figures 3.44 and 3.45 show this velocity component profile in the channel for a number of different processing conditions. In Figure 3.44 velocity profiles as a function of pressure gradients are shown for a melt in which its viscosity decreases with temperature (negative values for the parameter b). It clearly shows velocity profiles following the same trend as those observed for isothermal processing (refer to Figures 3.24 and 3.34 for Newtonian and non-Newtonian melts, respectively). When the pressure gradient is positive, a pressure flow opposed to the drag flow exists, which, depending on the magnitude of the pressure gradient, can promote negative velocity z-components. Conversely when the pressure gradient is negative, there is a contribution of the pressure flow to the drag flow, hence increasing the melt velocity. Figure 3.44 clearly illustrates that the increase in the magnitude of the pressure gradient

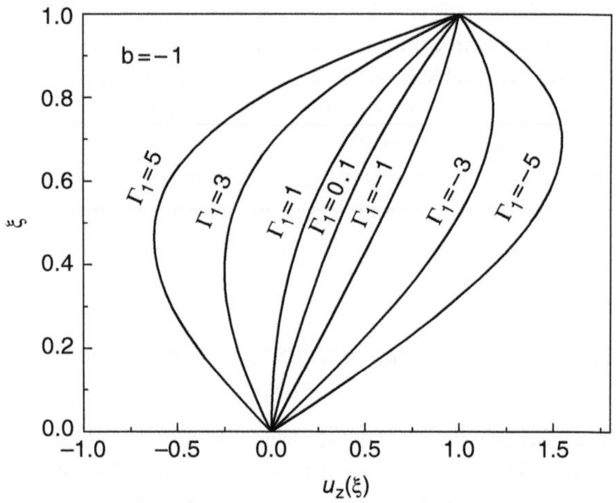

Figure 3.44 Dimensionless z-component of the velocity profile in the screw channel for different dimensionless pressure gradients. Newtonian melt under non-isothermal conditions.

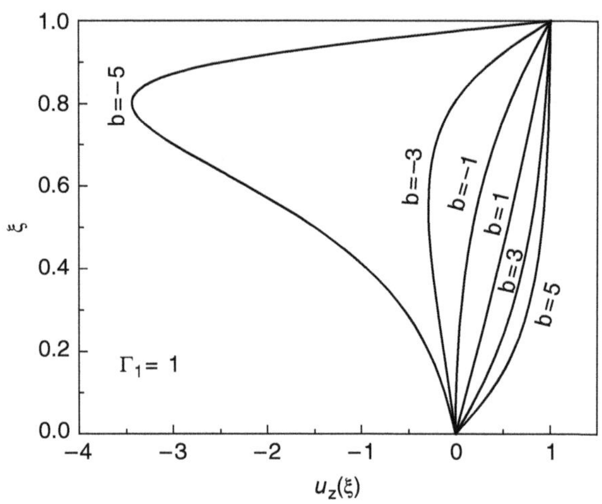

Figure 3.45 Dimensionless z-component of the velocity profile for melts whose viscosities have different dependence with temperature (different values for the parameter b).

promotes both larger negative and larger positive velocities for positive and negative gradients, respectively. Figure 3.45 illustrates the effect of the melt viscosity temperature dependence on the z-component of the velocity profile. Melts whose viscosities are highly dependent and decrease with temperature (large negative values of b) show significant negative velocity components due to the increase of the pressure flow resulting from the large decrease in viscosity. Conversely, melts whose viscosities increase with temperature show positive velocity profiles. Melts that exhibit increases of viscosity with increases of temperature may be difficult to find among usual materials. However, in the metering section of the extruder, materials undergo physicochemical transformations that may tend to increase the viscosity of the melts concurrently with the increase in temperature in the melt.

Once the z-velocity component is known, the volumetric flow rate in the z-direction can also be determined as indicated below (Rauwendaal, 2001):

$$Q = \int_{\xi=0}^{\xi=1} u_z(\xi)\, d\xi \tag{3.165}$$

which yields the following expression:

$$Q = \Gamma_1 \left[\left(e^{-b} - 1\right)\left(\frac{1}{b^2} + \frac{2}{b^3} - \frac{C_1}{b^2}\right) + \frac{2}{b^2} - \frac{C_1}{b} \right] \tag{3.166}$$

Figure 3.46 shows dimensionless flow in the channel as a function of the parameter b for different dimensionless pressure gradient values. To avoid the division by zero, the dimensionless flow for $b = 0$ is obtained from the Newtonian and isothermal approach by integration of Eq. 3.55c which results in the expression $Q = \frac{1}{2}\left(1 - \frac{\Gamma_1}{6}\right)$. As clearly observed in the figure, melts whose viscosities decrease strongly with temperature (large and negative values of b) exhibit lower flow rates, at a given positive pressure gradient, in comparison with the isothermal case (line going through $b = 0$). When the pressure gradient is negative, the opposite is observed (see Figure 3.46). As already explained, the lower viscosities observed at high temperatures which result in larger pressure flow are the reason for the observed behavior for positive and negative pressure gradients. For melts in which viscosity increases with temperature, due for example to physicochemical transformations of the melt, the opposite behavior is observed. If the ratio

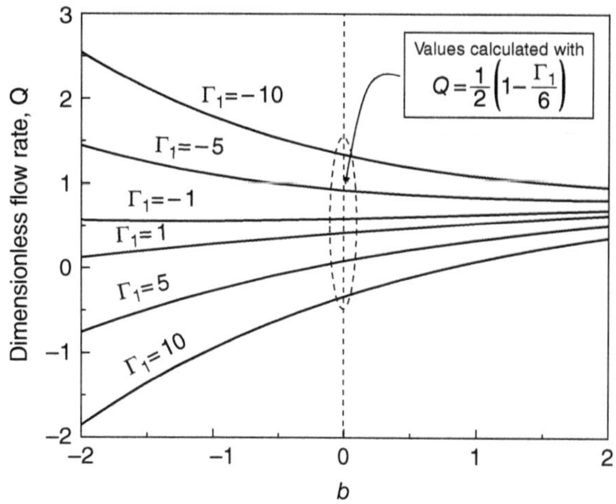

Figure 3.46 Dimensionless flow rate in the channel as a function of the parameter b for different dimensionless pressure gradient values. The melt is assumed to be Newtonian.

between the dimensionless flow obtained under non-isothermal and isothermal conditions is calculated, Figure 3.47 results. The curves intersect at $b = 0$ where the ratio is equal to 1. It must be noted that for negative and low dimensionless pressure gradients, the respective curves are relatively flat whereas large changes in flow rates are observed when the dimensionless pressure gradients are large and positive. As a result, temperature-related extrusion instabilities are more likely to occur at large positive pressure gradients (Rauwendaal, 2001).

The solution presented above for Newtonian melts preassumes a linear temperature profile between the barrel and screw temperatures in the screw channel. However, the temperature profile in the screw channel is not linear and can be determined from the balance of energy (Eq. 3.153) under some specific assumptions. For instance, by considering steady-state, negligible convection and a flow velocity given by the z-velocity component, the balance of energy equation is reduced to:

$$k\frac{d^2T}{dy^2} = \underline{\underline{\tau}} : \nabla \underline{v} \tag{3.167}$$

which through integration of Eq. 3.160 and by using Eqs 3.155 and 3.156 results in:

$$k\frac{d^2T}{dy^2} = \frac{\mu_o V_{bz}^2}{H^2} e^{-b\xi} \Gamma_1^2 (\xi - C_1)^2 \tag{3.168}$$

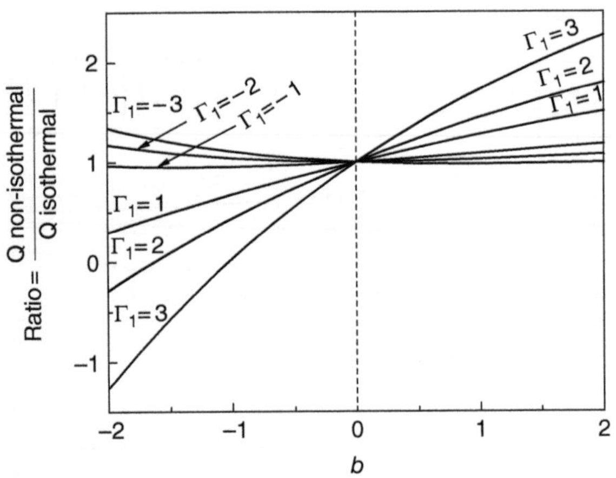

Figure 3.47 Ratio between the dimensionless net flow obtained under non-isothermal and isothermal conditions as a function of the parameter b and at different dimensionless pressure gradients.

The dimensionless Brinkman number, N_{Br}, representing the ratio of the heat generated by friction to the heat conducted by the temperature difference between the barrel and the screw, is defined as:

$$N_{Br} = \frac{\mu_o V_{bz}^2}{k(T_b - T_s)} \tag{3.169}$$

By using dimensionless variables for the direction y, the temperature and the z-velocity component defined above, Eq. 3.168 becomes:

$$\frac{d^2\Theta}{d\xi^2} = -\Gamma_1^2 N_{Br} e^{-b\xi}(\xi - C_1)^2 \tag{3.170}$$

Rauwendaal (2001) solved the above equation using general thermal boundary conditions that prescribe heat flux in the screw and barrel surfaces. These boundary conditions are expressed as:

$$\text{At the screw surface:} \left.\frac{d\Theta}{d\xi}\right|_{\xi=0} = -B_{i,s}\Theta(\xi=0) \tag{3.171a}$$

$$\text{At the barrel surface:} \left.\frac{d\Theta}{d\xi}\right|_{\xi=1} = B_{i,b}[\Theta(\xi=1)-1] \tag{3.171b}$$

where $B_{i,s}$ and $B_{i,b}$ are the Biot numbers for the screw and the barrel, respectively, calculated as:

$$B_{i,s} = \frac{h_s H}{k_s} \quad \text{and} \quad B_{i,b} = \frac{h_b H}{k_b} \tag{3.172}$$

h_s and h_b are the convection heat transfer coefficients at the interface between the screw and the melt and between the barrel and the melt, whereas k_s and k_b are the thermal conductivities of the screw and the barrel, respectively.

Solution of Eq. 3.170 with these boundary conditions is given below by Eq. 3.173, along with the relevant integration constants used (Eqs 3.174a–f):

$$\Theta(\xi) = -\Gamma_1^2 N_{Br} e^{-b\xi}\left[\frac{\xi^2}{b^2} + \left(\frac{4}{b^3} - \frac{2C_1}{b^2}\right)\xi + \frac{C_1^2}{b^2} + \frac{6}{b^4} - \frac{4C_1}{b^3}\right] + A_1\xi + A_2 \tag{3.173}$$

$$A_1 = -B_1 - B_{i,s}B_2 - B_{i,s}A_2 \tag{3.174a}$$

$$A_2 = \frac{-B_3 + B_1 + B_{i,s}B_2 + B_{i,b}B_4 - B_{i,s}B_{i,b}B_2 - B_{i,b}B_1 - B_{i,b}}{-B_{i,b} + B_{i,s}B_{i,b} - B_{i,s}} \tag{3.174b}$$

$$B_1 = \frac{-\Gamma_1^2 N_{Br}(2bC_1 - 2 - b^2C_1^2)}{b^3} \tag{3.174c}$$

$$B_2 = \frac{-\Gamma_1^2 N_{Br}(-4bC_1 + 6 + b^2C_1^2)}{b^4} \tag{3.174d}$$

$$B_3 = \frac{-\Gamma_1^2 N_{Br} e^{-b}(-b^2 + 2b^2C_1 - 2b + 2bC_1 - 2 - b^2C_1^2)}{b^3} \tag{3.174e}$$

$$B_4 = \frac{-\Gamma_1^2 N_{Br} e^{-b}(b^2 + 4b - 2b^2C_1 - 4bC_1 + 6 + b^2C_1^2)}{b^4} \tag{3.174f}$$

In the above equations C_I is evaluated with Eq. 3.163. Figure 3.48 illustrates dimensionless temperature $\Theta(\xi)$ versus the dimensionless co-ordinate ξ for a number of parameters that represent typical extrusion operations. It is important to note that all plots in Figure 3.48 are calculated for melts whose viscosities decrease with temperature ($b = -1$) and for adiabatic conditions at the screw (zero heat flux at the screw, i.e. $B_{i,s} = 0$). The effect of the viscous heating, represented by the Brinkman number, can be observed by comparing Figure 3.48a and Figure 3.48b in which significantly higher temperatures are observed for conditions with large Brinkman numbers (high viscous dissipation). The effect of the dimensionless pressure gradient is also clearly illustrated in all figures, where it is shown that larger

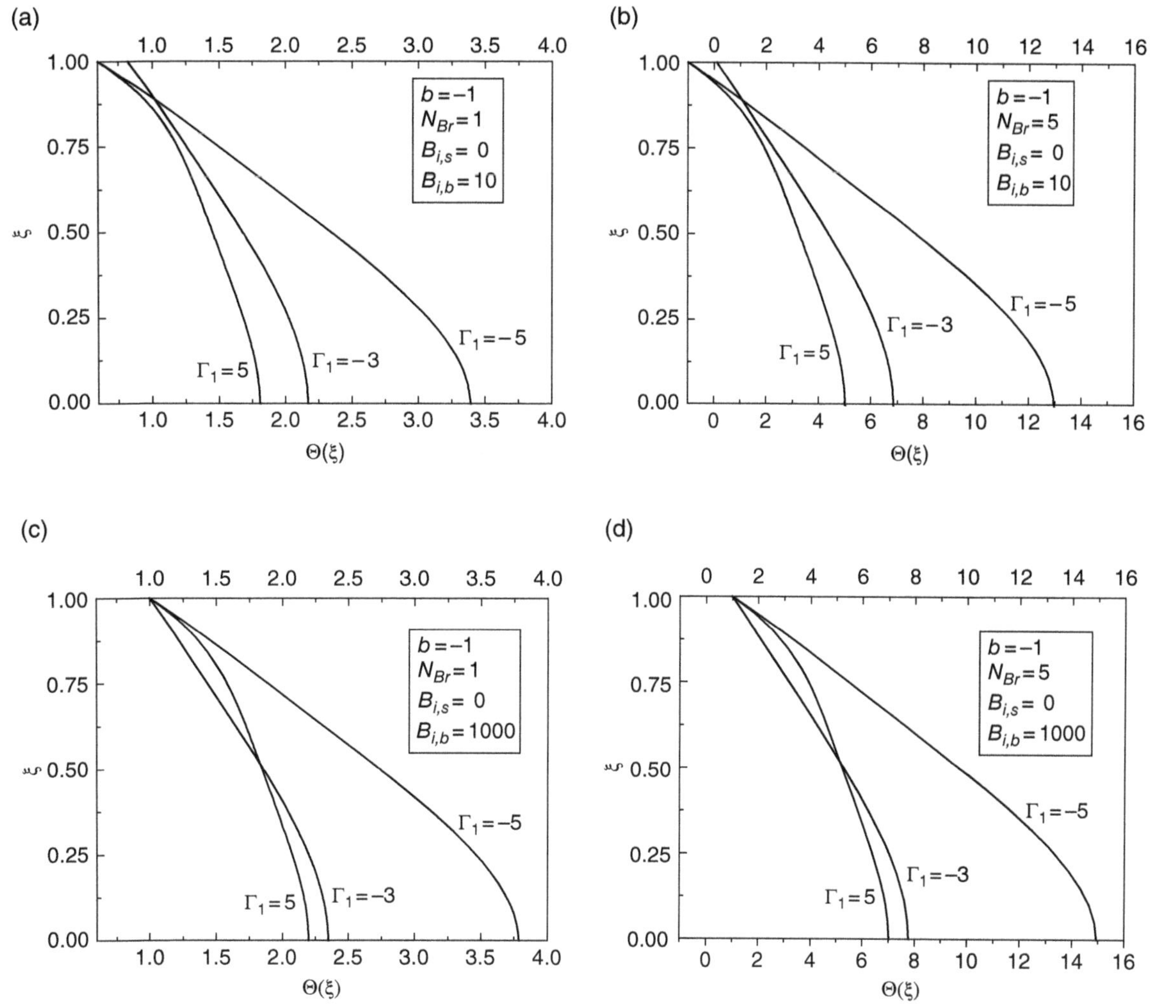

Figure 3.48 Dimensionless temperature versus dimensionless cross-channel direction for a number of parameters representing typical extrusion operations.

negative pressure gradients cause higher temperatures in the melt. The effect of the heating/cooling through the barrel can be observed by comparing Figures 3.48a and 3.48c and also Figures 3.48b and 3.48d where the Biot number at the barrel is changed from 10 to 1000. For larger external heat transfer (large Biot numbers), the melt temperature increases. It is noteworthy to observe in Figure 3.48b that for large viscous heating and low external heating conditions, the barrel temperature can reach or even be smaller than the screw temperature [negative $\Theta(\xi = 1)$]. For large Biot numbers the temperature of the melt approaches the barrel temperature for all conditions [$\Theta(\xi = 1) = 1$].

Non-Newtonian melts (power law fluid)

Development of 3D models for the non-isothermal flow of power law melts in single screw extruders requiring numerical solutions is presented in the literature (e.g. Syrjälä, 1999).

Solutions to models describing non-isothermal extrusion of non-Newtonian melts, specifically the power law model, have been obtained only for simplified cases, which include no heat dissipation, pure drag flow, and considering only the z-velocity component. The effect of temperature on the consistency index m was assumed to follow a similar relationship to that used for the Newtonian viscosity expressed by Eq. 3.159, whereas the flow index n was considered to be independent of temperature.

With these assumptions, the dimensionless forms of the equations of motion (Eq. 3.97) and energy (Eq. 3.167), the latter with the additional assumption of steady state and no convection, reduce to:

$$\frac{d}{d\xi}\left[e^{b\Theta}\left(\frac{du_z}{d\xi}\right)^n\right]=0 \qquad (3.175)$$

$$\frac{d^2\Theta}{d\xi^2}=0 \qquad (3.176)$$

and with the following boundary conditions:

$$\left.\begin{array}{lll} u_z=0 & at & \xi=0 \\ \Theta=0 & at & \xi=0 \end{array}\right\}(\text{screw}) \qquad (3.177a)$$

$$\left.\begin{array}{lll} u_z=1 & at & \xi=1 \\ \Theta=1 & at & \xi=1 \end{array}\right\}(\text{barrel}) \qquad (3.177b)$$

the following solutions for the dimensionless temperature, the z-velocity component and the flow (obtained from Eq. 3.165) are obtained:

$$\Theta(\xi)=\xi \qquad (3.178)$$

$$u_z(\xi)=\frac{1-e^{-\frac{b\xi}{n}}}{1-e^{-\frac{b}{n}}} \qquad (3.179)$$

$$Q=\frac{2n-2b-2ne^{-\frac{b}{n}}}{1-e^{-\frac{b}{n}}} \qquad (3.180)$$

A model incorporating viscous dissipation but still considering only drag flow was presented by Rauwendaal (2001). The model that combines Eqs 3.97 and 3.167, expressed in dimensionless variables, is:

$$\frac{d^2\Theta_1}{d\xi^2}+D_1e^{\Theta_1}=0 \qquad (3.181)$$

The effect of temperature in the power law model is defined by the following equation:

$$m=m_oe^{A(T_o-T)} \qquad (3.182)$$

m_o is the value of the consistency index at the reference temperature T_o. A new dimensionless variable $\Theta_1=\frac{A}{n}(T-T_o)$ is introduced, whereas the constant D_1 is given by:

$$D_1=-\frac{AH^2\tau_o^{\frac{n+1}{n}}}{nk_{melt}m_o^{1/n}} \qquad (3.183)$$

k_{melt} is the thermal conductivity of the melt and τ_o a parameter obtained from the solution of the motion equation (Eq. 3.97), and defined as:

$$\tau_o=v_z(y)\left(\int_0^H m^{-\frac{1}{n}}dy\right)^{-n} \qquad (3.184)$$

The unit of the parameter τ_o is given in pascals if all variables involved in the equation are expressed in the metric system. Solutions of Eqs 3.181 and 3.175 as well as the flow rate were obtained by Rauwendaal (2001) and are given below:

$$\Theta_1(\xi)=ln\left[A_c sech^2(y_1\xi+y_1)\right] \qquad (3.185)$$

$$u_z(\xi)=\frac{(1+e^{-2y_2})(e^{-2y_2}+e^{2y_1})}{(1-e^{2y_1})(e^{-2y_2}+e^{2y_1\xi})}+\frac{e^{-2y_2}+e^{2y_1}}{e^{2y_1}-1} \qquad (3.186)$$

$$\frac{Q}{V_{bz}H}=\frac{e^{-2y_2}+e^{2y_1}}{e^{2y_1}-1}+\frac{(1+e^{-2y_2})(e^{-2y_2}+e^{2y_1})e^{2y_2}}{2y_1(1-e^{2y_1})}\,ln\left[\frac{e^{2y_1}(1+e^{-2y_2})}{e^{2y_1}+e^{-2y_2}}\right] \qquad (3.187)$$

where:

$$y_1=0.5\sqrt{2D_1} \qquad (3.188)$$

and:

$$y_2=0.5\Theta_{1,s}-ln\left(\sqrt{A_c}+\sqrt{A_c-e^{\Theta_{1,s}}}\right) \qquad (3.189)$$

$\Theta_{1,s}=\Theta_1(\xi=0)=\frac{A}{n}(T_o-T_s)$ and $\Theta_1(\xi=0)=\theta_{1,b}=\frac{A}{n}(T_b-T_o)$ whereas the constant A_c can be determined from the following equation:

$$\left(\frac{N_{Na}}{4n}\right)^{\frac{1}{n}}=2A_c(-y_2-y_3)^{\frac{1-n}{n}}\left[\frac{1}{1+e^{2y_2}}-\frac{1}{1+e^{-2y_3}}\right]^{\frac{1+n}{n}} \qquad (3.190)$$

where N_{Na} is the Nahme number and y_3 is a parameter, both defined by the equations given below:

$$N_{Na}=\frac{Am_oV_{bz}^{n+1}}{k_{melt}H^{n-1}} \qquad (3.191)$$

$$y_3 = 0.5\Theta_{1,b} - ln\left(\sqrt{A_c} + \sqrt{A_c - e^{\Theta_{1,b}}}\right) \qquad (3.192)$$

It is clear from the above equations that there is no explicit solution for the constant A_c for given processing and melt conditions such as melt rheology, barrel and screw temperatures, screw speed, and screw geometry. Thus, a numerical solution is required except for simple cases such as $\Theta_{1,s} = \Theta_{1,b} = 0$. Use of other boundary conditions requires the numerical solution of Eq. 3.187 to estimate the constant A_c.

The description and solution of these relatively simple models clearly show that even for very simplified situations, a numerical solution is required. Although the incorporation of temperature effects and the introduction of non-linear rheology were used to approximate the models to more realistic conditions that resemble those existing during the extruder operation, the fact that the physicochemical characteristics of the melt might change during the process was ignored. As mentioned above, the flowing material is not always inert and many physical and chemical reactions involving a modification of the melt may take place, which requires the use of computational multiphysics models which are described in section 3.3.

3.2.3.1.5 Residence time distribution

Using concepts already defined in the area of reactors where it is important to know the residence time distribution of the different components of the reaction in the reactor, the "age" of a particle of the fluid is defined as the time elapsed since that particle enters the reactor (Froment et al, 2011). Thus, a residence time distribution may be visualized as particles of different age in the reactor or a processing vessel like an extruder. The concept of a particle or fluid element was introduced in a series of papers by Danckwerts (Danckwerts, 1957, 1958a, 1958b) who defined it as a small volume with respect to the size of the reactor but still large enough to contain enough molecules to assume that in the volume, the properties of the system are continuous functions. This is also known as the hypothesis of the continuum. Experimental determination of residence time distributions using tracers allows for a visualization of the concept of residence time distribution. If in an instant of time, all fluid elements or particles entering the reactor or the processing vessel are marked and if the exit of those tagged particles is monitored, the residence time distribution of the particles in the processing vessel can be estimated.

Two ideal systems have been assumed in the reactor area: the continuous stirred tank reactor (CSTR), which assumes a perfect mixing behavior and a high dispersion of residence times of the fluid elements, and the continuous plug flow reactor (CPFR) with an assumed perfect radial mixing but no mixing in the axial direction and only a residence time for all the fluid elements. For real systems and imperfect mixing, the tracer concentration-time pattern is very different. At the beginning of the process, a volume of fluid with a lower concentration of tracer is exiting the processing vessel. With time, the concentration of tracer increases until a maximum concentration is reached at the mean residence time in the vessel. After that time, the concentration of the tracer decreases eventually to zero when all the tagged fluid has been washed off by the incoming feed of non-tagged fluid. The concepts of residence time distribution (RTD) and ideal flow reactors are also used and discussed further in this book, in Chapters 5, 6, and 10.

Measuring RTD in single screw extruders has singular importance because it is related to many parameters that provide an assessment of the extrusion performance, such as the degree of mixing and the degree of uniformity in the thermomechanical treatment of the melt. Bruin et al. (1978) determined the trajectory of fluid elements of an isothermally extruded Newtonian fluid and predicted results similar to those qualitatively illustrated in Figure 3.7c. The analysis is based on the cross- and down-channel velocities of the melt. The equation for the cross-channel melt velocity was given by Eq. 3.63 as:

$$u_x(\xi) = 2\xi - 3\xi^2 \qquad (3.63)$$

whereas the down-channel melt velocity can be calculated by Eq. 3.73, which was derived considering it is a function of the co-ordinates x and y (see Eq. 3.58). For the sake of simplicity, it can be assumed that the down-channel velocity is only a function of the co-ordinate y. Hence, the solution of the simplified Eq. 3.58 with the corresponding boundary condition gives a dimensionless down-channel velocity component expressed as (Rauwendaal, 2001; Tadmor & Gogos, 2006):

$$u_z(\xi) = \xi + 3\frac{Q_p}{Q_d}\xi(1-\xi) \qquad (3.193)$$

As this velocity component $u_z(\xi)$ is affected by the operating conditions, given by the ratio Q_p/Q_d, the fluid particle trajectories should be analyzed for different ratios. A particular one is when $Q_p/Q_d = -1$ because for that value both the cross- and down-channel velocities vanish

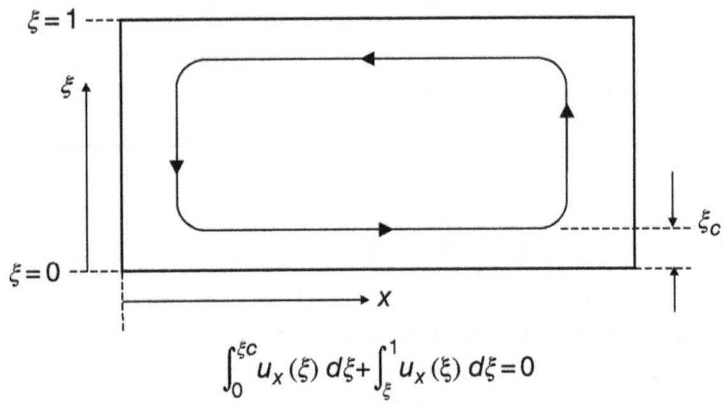

Figure 3.49 Schematic of the path of a fluid particle in the rectangular screw channel of a single screw extruder.

at $\xi = 2/3$. A fluid element located at $\xi > \frac{2}{3}$ will have a negative cross-channel velocity component and a positive down-channel velocity component, whereas the opposite is true for $\xi < \frac{2}{3}$. The path of a fluid particle is schematically illustrated in Figure 3.49 where the negative cross-channel velocity component is moving to the left and down the channel until it comes close to the leading edge of the advancing flight where the fluid element is turning over and starts moving to the right or positive x direction (Tadmor & Klein, 1970). Since the net flow on the cross-channel direction is zero, a relationship can be established to determine the location in the channel where the cross-channel velocity component is positive as:

$$\int_0^{\xi_c} u_x(\xi)\, d\xi = -\int_{\xi_c}^{1} u_x(\xi)\, d\xi \qquad (3.194)$$

which after substitution of Eq. 3.63 into Eq. 3.194 yields:

$$\xi_c^2 - \xi_c^3 = \xi^2 - \xi^3 \qquad (3.195)$$

Tadmor and Gogos (2006) have given solutions of Eq. 3.195 in terms of ξ or ξ_c:

$$\text{For } 0 \le \xi_c \le \frac{2}{3} \quad \xi = \frac{1}{2}\left(1 - \xi_c + \sqrt{1 + 2\xi_c - 3\xi_c^2}\right) \qquad (3.196)$$

$$\text{For } \frac{2}{3} \le \xi \le 1 \quad \xi_c = \frac{1}{2}\left(1 - \xi + \sqrt{1 + 2\xi - 3\xi^2}\right) \qquad (3.197)$$

If trajectories of the fluid elements are determined for the different vertical positions, the residence time of a particle in a cycle is obtained by dividing the channel width by the local cross-channel velocity calculated at that vertical position. The fraction of time that a fluid element spends on the upper part of the channel can be calculated as:

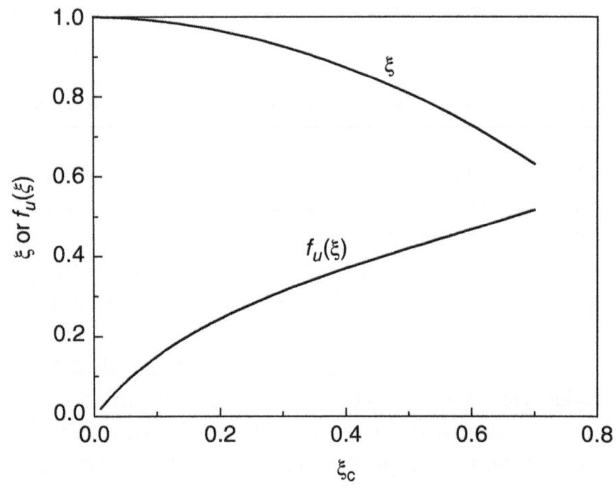

Figure 3.50 Relationship between the upper (ξ) and lower (ξ_c) positions of a fluid particle flowing in a cycle in the cross-channel section of a single screw extruder, and fraction that the particle spends on the upper part of the channel $f_u(\xi)$.

$$f_u(\xi) = \frac{1}{1 - \dfrac{\xi(2-3\xi)}{\xi_c(2-3\xi_c)}} \qquad (3.198)$$

Figure 3.50 illustrates the position of the fluid element in the upper part of the channel, ξ, as a function of its position in the lower part of the channel, ξ_c. Values of the fraction of time that fluid elements spend in the upper channel, $f_u(\xi)$, are also plotted as a function of ξ_c. The figure clearly shows that when the fluid is close to the center of the channel and when $\xi_c \approx \frac{2}{3}$, the fraction of time the fluid element spends in the upper part of the channel increases. To evaluate thermomechanical treatments, the residence time of the fluid elements should be known. It can be evaluated by dividing the length of the extruder barrel by the component of the melt velocity in the direction of the extruder shaft which can be calculated by Eq. 3.71. Assuming a down-channel velocity component solely dependent on the y direction, Tadmor and Gogos (2006) obtained an equation for the residence time of Newtonian fluids:

$$t(\xi) = \left(\frac{1}{3V_b\left(1 + \dfrac{Q_p}{Q_d}\right)\sin\theta\cos\theta}\right) \cdot \frac{3\xi - 1 + \sqrt{1 + 2\xi - 3\xi^2}}{\xi\left(1 - \xi + \sqrt{1 + 2\xi - 3\xi^2}\right)} \qquad (3.199)$$

The minimum residence time is for fluid elements located at $\xi = 2/3$, i.e. at the location where the cross-channel velocity is zero and the fluid element recirculation vanishes; the minimum residence time is calculated as:

$$t_{min} = \frac{3z}{2V_{bz}\left(1 + \frac{Q_p}{Q_d}\right)} \quad (3.200)$$

where $z = l/\sin\theta$ and V_{bz} is the barrel velocity in the down-channel direction ($V_{bz} = V_b \cos\theta$). The mean residence for these conditions can be calculated as:

$$\bar{t} = \int_{t_{min}}^{\infty} tf(t)dt = \frac{4}{3}t_{min} = \frac{2z}{V_{bz}\left(1 + \frac{Q_p}{Q_d}\right)} \quad (3.201)$$

The cumulative RTD function $F(t)$ can be obtained by integration of Eq. 3.198 to yield:

$$F(\xi) = \frac{1}{2}\left(3\xi^2 - 1 + (\xi - 1)\sqrt{1 + 2\xi - 3\xi^2}\right) \quad (3.202)$$

The above equation is used in conjunction with Eq. 3.199 to estimate the value of the classic cumulative RTD function, $F(t)$, which is currently plotted as a function of the dimensionless time, $t/\bar{t}$, where $\bar{t}$ is the average residence time of the fluid in the extruder. For detailed information on RTD functions, readers should refer to Levenspiel (1984).

Figure 3.51 shows experimental results of the RTD $F(t)$ function measured in a single screw extruder using a Newtonian liquid resin as the melt and a radioactive tracer, compared with results obtained from the theoretical model developed for Newtonian fluids using a combination of ideal plug flow and perfect mixing models (Wolf & White, 1976). For the sake of comparison, the figure also illustrates RTD functions for ideal systems such as the CSTR model, the CPFR model, and the RTD function calculated for the flow of a Newtonian fluid in a pipe (Poiseuille flow).

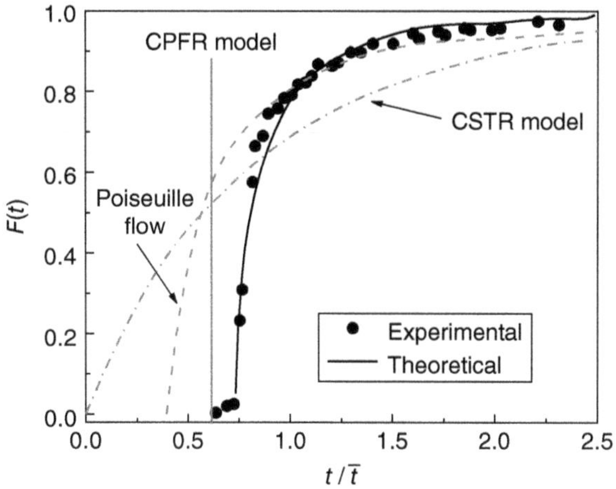

Figure 3.51 Experimental results of the RTD $F(t)$ function measured in a single screw extruder, which are compared with results obtained from a theoretical model for Newtonian fluids that uses a combination of ideal plug flow and perfect mixing model developed by Wolf and White (1976). For comparison, theoretically developed RTD $F(t)$ functions for ideal plug flow and continuous stirred tank reactor models as well as for a Poiseuille flow are also included.

3.2.3.1.6 Strain distribution

The extruded melt is very sensitive to the thermomechanical treatment, which has been described in preceding sections. Thus, it is important to have detailed knowledge not only of the RTD of the melt during the process but also the strain to which the melt is subjected during flow in the channel. As demonstrated, the flow velocity of the melt varies in the extruder channel for both Newtonian and non-Newtonian fluids and also for processes under isothermal and non-isothermal conditions. These different flow velocities cause the fluid particles to follow different trajectories (Figure 3.7c), which also give rise to different shear rates.

Definitions

The shearing model is illustrated in Figure 3.52, in which effects of the shearing action induced by the shear model on the fluid are schematically shown. At time $t = 0$, the fluid is contained between two parallel plates separated by a distance H. The top plate is moved with a constant velocity V whereas the bottom plate is maintained stationary, initiating the shear deformation which is illustrated in the figure for a time $t > 0$. The different fluid elements indicated by black squares are deformed due to the applied shear rate $\dot{\gamma} = \dot{\gamma}_o = \frac{V}{H}$ which in this case is constant (Figure 3.52a). However, the shear rate in the extruder channel is not constant but varies with the position of the fluid element in the channel. Shear and extensional deformations are pertinent to the extrusion process but given the importance of drag flow, typical shear flow and shear strains are further discussed. Figure 3.52b illustrates the deformation of the fluid element created by the simple shear flow. Two fluid elements or particles

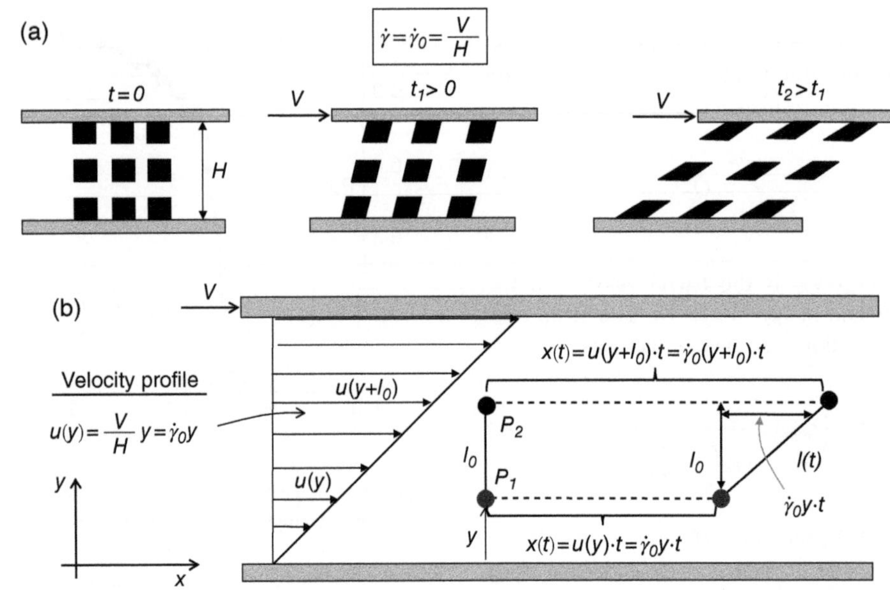

Figure 3.52 Shearing model. (a) Schematic description of the test and fluid particles deformation under this flow. (b) Schematic of the deformation of a fluid element created by the simple shear flow and description of the approach to estimate the strain applied to the fluid.

initially separated by a distance l_o experiencing a shear flow for a given time will be further separated due to the different velocities at the locations y and $y+l_o$. The velocity profile for this simple shear flow is linear and the velocity of the fluid particle located at a distance y from the bottom fixed plate is:

$$u(y)=\dot{\gamma}_o y \tag{3.203}$$

After an elapsed time t, the distances x traveled by the particles P_1 and P_2 can be calculated from the shear flow kinematics and given by the following equations (Figure 3.52b):

$$\text{For particle } P_1\text{: } x(t)=u(y)\cdot t=\dot{\gamma}_o y\cdot t \tag{3.204}$$

$$\text{For particle } P_2\text{: } x(t)=u(y+l_o)\cdot t=\dot{\gamma}_o(y+l_o)\cdot t \tag{3.205}$$

As indicated in Figure 3.52b, the difference between the distances traveled by the two particles can be calculated as $\dot{\gamma}_o y\cdot t$. Thus, after a time t has elapsed, the distance between the two fluid particles changes from l_o to $l(t)$. For long times the separation between the two particles $l(t)$ can be calculated as (Morrison, 2001):

$$l(t)\approx l_o\dot{\gamma}_o t \tag{3.206}$$

and the strain, defined as relative deformation, is calculated as:

$$\gamma(t)=\frac{l(t)}{l_o}=\dot{\gamma}_o t \tag{3.207}$$

Although the concept of shear strain, defined as a relative deformation of a body and applied to simple flows (e.g. shear flow), appears to be straightforward, its measurement and calculation are more involved in complex flows like the ones existing in an extruder. For flows with non-constant shear rates, the shear strain can be calculated as:

$$\gamma(t)=\int_0^t \dot{\gamma}(t')dt' \tag{3.208}$$

For calculation of the shear strain, it is important to use a deformation measure that is applicable to general flows (Morrison, 2001). The method of strain calculation in a 3D flow is schematically illustrated in Figure 3.53 where the trajectory of an element of fluid or fluid particle,

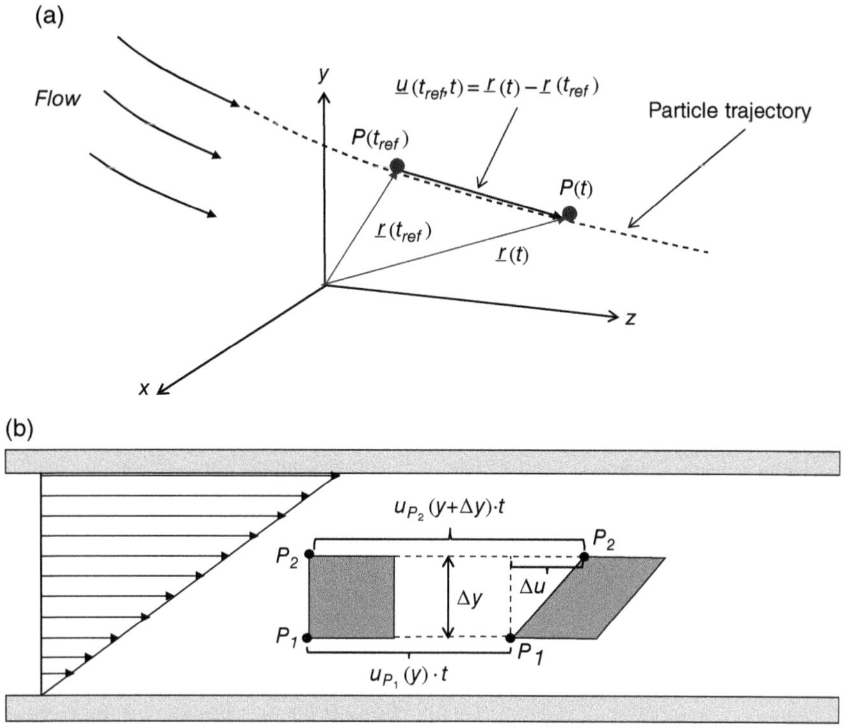

Figure 3.53 (a) Description of the method for strain calculation in a general 3D flow. (b) Application of the approach to calculate strain in a simple pure shear flow.

identified by P, can be followed by the vector position $\underline{r}(t_{ref})$ that locates the particle at a reference time t_{ref} (the reference time for convenience is usually considered as 0). Due to the flow and after a time t has elapsed, the particle moves to a new position that is defined by a new position vector $\underline{r}(t)$. The displacement of the particle P due to the flow is determined by the displacement function $\underline{u}$ defined as:

$$\underline{u} = \underline{r}(t) - \underline{r}(t_{ref}) \qquad (3.209)$$

The ratio of that displacement function, a vector with three non-zero components in a typical 3D flow, to one reference distance or distance vector provides an estimation of the strain applied to that particular fluid particle at a given time and for a particular location on the flow domain. As before, a physical understanding of the shear strain calculation can be obtained if the method is applied to the simple shear flow illustrated in Figure 3.53b. Since the flow only has a velocity component in the x direction, the displacement function u becomes a scalar and it is estimated from the x position of the particles P_1 and P_2 as:

$$\Delta u = u_{P_2}(y+\Delta y)\cdot t - u_{P1}(y)\cdot t = \dot{\gamma}_o(y + \Delta y - y)\cdot t = \dot{\gamma}_o \Delta y \cdot t \qquad (3.210)$$

If Δy is considered as the reference distance, i.e. the initial separation between the particles before the flow has started, the shear strain calculation is given by the following equation:

$$\gamma = \frac{\partial u}{\partial y} = \dot{\gamma}_o t \qquad (3.211)$$

which agrees with the estimation of the shear strain given by Eq. 3.207. Strain estimations for other types of flow such as extensional flow or mixed flows are given by Morrison (2001).

Since the shear rate in the extruder channel is not constant, different fluid particles experience different strains. As indicated by Eq. 3.208, strain is a function of time under which the shear rate is applied, thus the RTD of the different fluid particles plays an important role in the distribution of strains to which the fluid melt is subjected. Residence time distribution and strain distribution histories are related and both should be evaluated for proper estimation of the thermomechanical treatment experienced by the melt in a given extrusion set-up and specific processing conditions.

Strain distribution functions in single screw extruders
As demonstrated for shear flows and indicated by Eqs 3.207 and 3.208, calculation of the strain imposed on the fluid (in either shear or other flows such as the extensional flow) is estimated by the product of the strain rate and the time that the strain rate is applied. For a Newtonian fluid, the corresponding shear rates can be calculated from Eq. 3.63 and the simplified solution used to estimate the RTD function in the previous section (Eq. 3.193):

$$\dot{\gamma}_{yx} = \frac{V_{bx}}{H}\frac{du_x(\xi)}{d\xi} = \frac{2V_b sin\theta}{H}(1-3\xi) \tag{3.212}$$

$$\dot{\gamma}_{yz}(\xi) = \frac{V_{bz}}{H}\frac{du_z(\xi)}{d\xi} = \frac{2V_b cos\theta}{H}\left[1+3(1-2\xi)\frac{Q_p}{Q_d}\right] \tag{3.213}$$

The magnitude of the strain rate tensor calculated by Eq. 3.95 results in:

$$\dot{\gamma}(\xi) = \left(\dot{\gamma}_{yx}^2+\dot{\gamma}_{yz}^2\right)^{1/2} = \frac{V_b}{H}\left\{(4-3\xi)^2 sin^2\theta + \left[1+3(1-2\xi)\frac{Q_p}{Q_d}\right]cos^2\theta\right\}^{1/2} \tag{3.214}$$

Due to the recirculation flow, schematically shown in Figure 3.49, fluid particles are subjected to different shear rates in the upper and lower regions of the channel. By assuming that the shear strains imposed on a fluid element during its transit through the upper and lower regions of the channel are additives, the total strain applied to that fluid particle can be calculated as:

$$\gamma(\xi) = \dot{\gamma}(\xi)f_u(\xi)\cdot t(\xi) + \dot{\gamma}(\xi_c)[1-f_u(\xi)]\cdot t(\xi) \tag{3.215}$$

The first term on the right side of the above equation represents the strain applied to a fluid particle in the upper region of the channel, i.e. for $\xi > \xi_c$ whereas the second term provides the strain applied to the same fluid particle during its transit in the lower region of the channel, i.e. for $\xi \le \xi_c$. After substituting Eqs 3.197, 3.198, 3.199, and 3.214 into Eq. 3.215, an expression for the shear strain distribution in the extruder was obtained (Tadmor & Gogos, 2006):

$$\gamma(\xi) = \frac{1}{3H}\left(\frac{1}{1+Q_p/Q_d}\right)\cdot\left[E_1(\xi,\theta,Q_p/Q_d) + E_2(\xi,\theta,Q_p/Q_d)\right] \tag{3.216}$$

where the functions E_1 and E_2 are given in the following two equations:

$$E_1(\xi,\theta,Q_p/Q_d) = \frac{\frac{2f_u(\xi)}{cos\,\theta}\left[(1-3\xi)^2 + \frac{cotg^2\theta}{4}\left(1+3\frac{Q_p}{Q_d}-6\xi\frac{Q_p}{Q_d}\right)^2\right]^{1/2}}{\xi_c(1-\xi_c)+t_f(\xi)\cdot(\xi-\xi_c)\cdot(1-\xi-\xi_c)} \tag{3.217}$$

$$E_2(\xi,\theta,Q_p/Q_d) = \frac{\frac{(1-f_u(\xi))}{sin\,\theta}\left[4(1-3\xi_c)^2 tan^2\theta + \left(1+3\frac{Q_p}{Q_d}-6\xi_c\frac{Q_p}{Q_d}\right)^2\right]^{1/2}}{\xi_c(1-\xi_c)+t_f(\xi)\cdot(\xi-\xi_c)\cdot(1-\xi-\xi_c)} \tag{3.218}$$

Shear strain as a function of the vertical position ξ in the screw channel calculated from Eq. 3.216 is illustrated in Figure 3.54 for three different running conditions. Calculations are based on a large single screw extruder 2.54 m long, a channel depth of 5 mm, and a 20° helix angle. Running conditions include: Run 1 with $Q_p/Q_d = 0$, Run 2 with $Q_p/Q_d = -0.05$, and Run 3 with $Q_p/Q_d = -0.075$ (Tadmor & Gogos, 2006). For the calculation, the two functions E_1 and E_2 must be evaluated, which require the following parameters:

- $f_u(\xi)$ which is a function of the parameters ξ and ξ_c; it is estimated by (Eq. 3.198)
- the relationship between ξ and ξ_c which is given by Eqs 3.196 and 3.197.

Results for Run 1, which is the pure drag flow ($Q_p/Q_d = 0$), show that the strain is fairly uniform in the channel and the minimum strain coincides with the position $\xi = 2/3$ which, as was shown above, corresponds to the case for which the cross-channel and down-channel velocities are vanishing. For large pressure flows (Run 3), the shear strain is not uniform and dispersive mixing is predicted. When the back pressure increases (and consequently the ratio Q_p/Q_d), the minimum strain changes to regions closer to the barrel surface. The strain distribution function is calculated from Eq. 3.202. For the case of pure drag flow, $\xi = 2/3$ and ξ is related to γ by Eq. 3.215. The increase of the strain in the material with increases in back

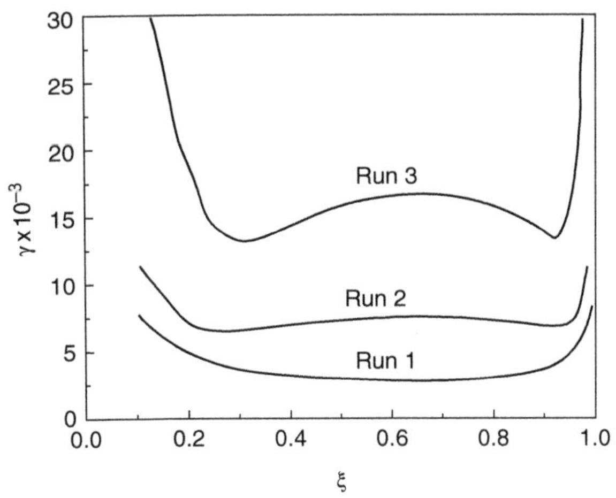

Figure 3.54 Shear strains applied to fluid elements as a function of their dimensionless vertical position in the upper region of a single screw extruder for three different processing conditions: Run 1 is a pure drag flow, Run 2 for $Q_p/Q_d = -0.05$ and Run 3 for $Q_p/Q_d = -0.075$.

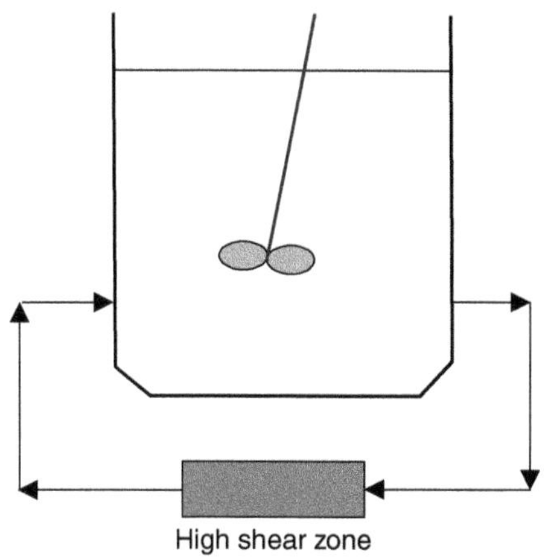

Figure 3.55 Schematics of a perfectly stirred tank with a bypass through a high shear zone to study empirically the residence time distribution of fluid elements in a single screw extruder.

pressure, supported by experimental observations, is also shown in Figure 3.54. For the general case in which pressure flow is present, it is necessary first to find the location with the minimum strain. Once this is done, several methods can be used to get the strain distribution function. Details of those approaches are given by Tadmor and Gogos (2006).

Bigg and Middleman (1974) developed a mathematical model describing the 2D flow of a power law fluid in a single screw extruder. The model was based on a simplified set of momentum equations similar to those presented above (Eqs 3.90 and 3.91) but with the additional assumption of neglecting changes of the stress components in the x direction. A numerical approach was used to determine the velocity components in the cross- and down-channel directions which enabled description of the circulation pattern and trajectories of fluid elements in the extruder channel. The trajectories obtained were similar to those schematically illustrated in Figure 3.7 and depended on the power law parameters, notably the flow index. The velocity profiles in the extruder channel were also used to determine the dimensionless net flow as a function of dimensionless pressure in the extruder that yielded results that are in good agreement with those shown in Figure 3.35. Following a similar approach to the one presented above for Newtonian fluids, with the obtained velocity profile, Bigg and Middleman (1974) were able to determine the cumulative RTD function $F(t)$ and the strain distribution in the extruder as a function of the rheological parameters of the power law model. Theoretical results showed good agreement with experimental results obtained using a pilot single screw extruder.

Although the previous analysis and the existence of some recirculation in the screw channel would be indicating some mixing in this type of extruders, it has been recognized that the single screw extruder does not produce an efficient dispersive mixing, which requires repeated passages (at least 20–30) of fluid elements through high shear regions. The analysis of RTD and strain distribution for isothermal extrusion of Newtonian fluids does not allow for the calculation of parameters that could assess more accurately the mixing degree in the extruder, including both cross-channel and dispersive mixing. Those more in-depth analyses require more sophisticated mathematical models that can be solved solely using numerical methods, some of which are discussed in section 3.3.

It is worth mentioning a simple analysis performed by Tadmor and Gogos (2006) to assess dispersive mixing in a single screw extruder. This is based on the analysis of a well-stirred tank in which a recycling stream passes through a high shear zone (Figure 3.55). The concept is based on the assumption that part of the fluid which passes through the high shear zone contributes to mixing. In single screw extruders, the high shear zone can be the zone between the tip of the screw flight and the barrel wall. The passage distribution function, g_k, which is defined as the fraction of the volume in a circulating batch system that has k passages in the zone of interest, can characterize the performance of the mixing. The passage distribution function can be calculated as:

$$g_k = \frac{\lambda^k}{k!} e^{-\lambda} \quad (3.219)$$

where $\lambda = \frac{t}{\bar{t}}$ and $\bar{t} = V/Q$.

By assuming pure drag flow, the volume and the flow passing over the high shear flight zone located between the tip of screw flight and the barrel wall (for geometric considerations see Figure 3.18) can be determined and the mean residence time $\bar{t}$ for the passage of an element of fluid through that zone calculated as:

$$\bar{t} = \frac{V}{Q} = \frac{\pi \bar{D} H \Delta z}{\pi N D \Delta z \frac{\delta}{2}} = \frac{2H\bar{D}}{ND\delta} \quad (3.220)$$

The residence time or elapsed time of the fluid element in the extruder, using the approximation of plug flow, can be calculated as:

$$t = \frac{\pi L H \bar{D}}{Q} \quad (3.221)$$

Thus the parameter λ is calculated as:

$$\lambda = \frac{t}{\bar{t}} = \frac{\frac{\pi L H \bar{D}}{Q}}{\frac{2H\bar{D}}{ND\delta}} = \frac{\pi N L D \delta}{2Q} = \frac{Q_f}{Q} \quad (3.222)$$

where Q_f is the flow rate passing over the flight (high shear zone) along the extruder.

For example, the passage distribution function can be calculated for a 60 mm screw diameter extruder (D), with a L/D equal to 10/1 ($L = 600$ mm), and a flight clearance (δ) of 0.125 mm running at 100 rpm with a flow of 100 L/h. The parameter λ calculated from Eq. 3.222 gives:

$$\lambda = \frac{\pi \cdot (100/60) \cdot (600x10^{-3})(60x10^{-3})(0.025x10^{-3})}{2 \cdot 100/3600 \cdot 10^{-3}} = 0.424$$

The elements of the passage distribution function are calculated as follows.

- For $k = 0$, $g_0 = e^{-0.424} = 0.654$, that is 65.4% of chances that a fluid element passes 0 time through the flight high shear zone.
- For $k = 1$, $g_1 = \frac{0.424}{1} e^{-0.424} = 0.277$, that is 27.7% of chances that a fluid element passes 1 time through the flight high shear zone.
- For $k = 2$, $g_2 = \frac{0.424^2}{2} e^{-0.424} = 0.059$, that is 5.9% of chances that a fluid element passes 2 times through the flight high shear zone.
- For $k = 3$, $g_3 = \frac{0.424^3}{6} e^{-0.424} = 0.0083$, that is 0.83% of chances that a fluid element passes 3 times through the flight high shear zone.

This simple analysis clearly shows some drawbacks in the mixing degree in single screw extruders.

3.2.3.2 Intermeshing co-rotating twin screw extruders

Intermeshing co-rotating twin screw extruders work in a different way than the single screw extruder. The interpenetration of the screws creates a positive movement of the material although the machine is not filled. The intermeshing co-rotating twin screw extruder is a positive displacement pump which allows a large range of materials behavior to be handled (viscous, oily, sticky, very wet ...) with the same level of pumping efficiency. Therefore, the throughput and screw speed are not linked for a wide range of restrictive conditions, which enables multiple operating points for a given throughput and a large variation of raw materials and formulations. It must be noted that the positive displacement pumping effect of the screws leads to a narrow RTD, and that throughput is independent of die pressure (Harper, 1989).

Unlike the screw channel in the single screw extruder, much of the channel of twin screw extruder is not completely filled. When the flow is restricted by the screw configuration (counter-thread, mixing elements, die), the material accumulates upstream, creating a fully filled working section or processing section (an active area where the material is processed by shearing and mixing), with significant heating produced by viscous dissipation which contributes to the physical and chemical transformation of the material (Janssen, 1989).

The screw configuration may hold several processing sections in series and any flow restrictions in the screw profile or at discharge create a filled section. The succession of processing sections down the barrel length of twin screw extruders permits dedication of process-oriented functions according to the whole process requirements. For example, a melting section, a mixing section, a cooking section, a reaction section, a venting/cooling section, etc. can be easily configured. This ability to design process functions in series by suitable selection of the screw configuration is a key process advantage in extrusion processing.

Figure 3.56 qualitatively illustrates the effects of the main operating variables and the screw configuration on the length of the fully filled process sections of co-rotating twin screw extruders. For a given screw configuration, the length of fully filled sections depends mainly on throughput and screw speed: it would decrease with decreasing throughput (Figure 3.56a) and increasing screw speed (Figure 3.56b) For a given throughput and screw speed, the length of fully filled sections depends on the design and length of the screw elements that generate the flow restriction: it would increase when decreasing the opening area of the restrictive screw element (e.g. the area offered to the fluid flow), or when increasing the length of the restrictive screw element (Figure 3.56c). The discussion on flow characteristics shows how extensive is the flexibility of twin screw extrusion technology which among other factors includes multiple processing sections in series, independence of throughput and screw speed, and multiple operating points. It is worth noting that the extensive capability to organize the screw configuration due to the great diversity of screw designs enables operators to vary and control local shear-time-temperature histories in co-rotating twin screw extruders. This brings tremendous processing advantages (high process productivity and flexibility) when compared to single screw extrusion technology.

Flow analysis in twin screw extruders is much more complicated than that of single screw extruders, and complete modeling of the flow is not available, even if reasonable assumptions are made. A schematic of the flow pattern existing in intermeshing, co-rotating twin screw extruders is qualitatively illustrated in Figure 3.57. Nevertheless, the flow in the C-shaped chambers of the screw-barrel assembly can be described as long as the end effects near the intermeshing zone are neglected, thus, partial modeling of the flow together with experimental observations allow the process performance of twin screw extruders to be discussed. Modeling approaches to analyze engineering aspects of twin screw extrusion technology are described in section 3.3. Simple analyses of the flow of co-rotating twin screw extruders for Newtonian fluids and isothermal conditions have been carried out and similar equations to those used in single screw extruders have been obtained, such as Eq. 3.81, taking the following form:

$$Q = \alpha_t N - \frac{\beta_t \Delta p}{\mu L} \tag{3.223}$$

Calculated drag and pressure flow coefficients, α_t and β_t, are presented in Tables 3.2 and 3.3 for bilobe kneading

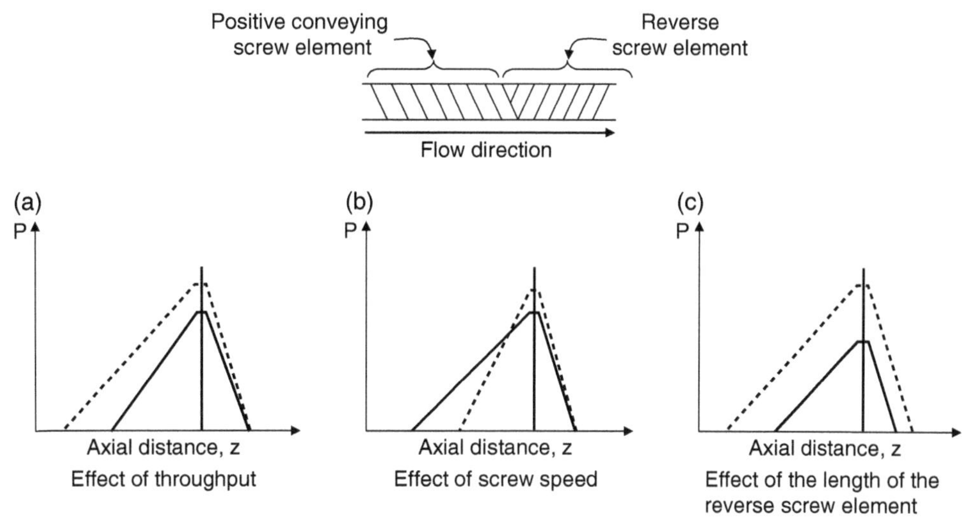

Figure 3.56 Effect of processing parameters on pressure build-up and length of fully filled zones of intermeshing co-rotating twin screw extruders (- - -: high value of the parameter; —: low value of the parameter).

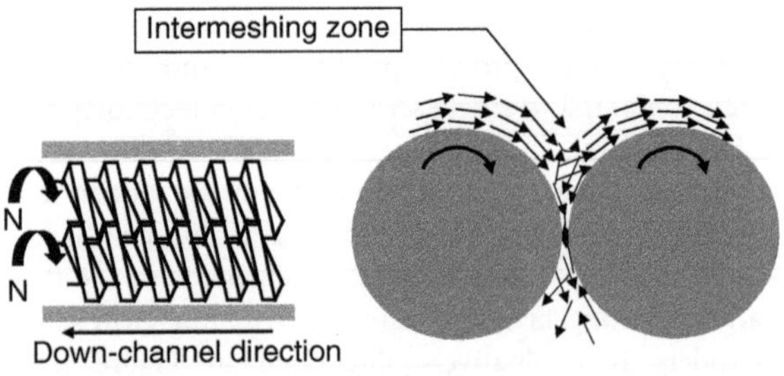

Figure 3.57 Schematics of the flow pattern in the intermeshing zone of co-rotating twin screw extruder.

Table 3.2 Drag flow and pressure flow coefficients for bilobe kneading disks of a 50.8 mm twin screw extruder.

Staggering angle	Disk thickness, mm	α_t, cm^3	β_t, cm^3
30°	12.7	51.5	0.53
45°	6.3	18.7	0.19
45°	12.7	31.1	0.34
45°	25.4	36.4	0.60
60°	6.3	5.7	0.23
60°	12.7	17.9	0.36
60°	25.4	22.9	0.49
90°	12.7	0.0	0.43

Source: Kohlgrüber 2008. Reproduced with permission of Carl Hanser Verlag Publisher, Munich.

Table 3.3 Drag flow and pressure flow coefficients for positive conveying screw elements of a 50.8 mm twin screw extruder.

Helix angle	α_t, cm^3	β_t, cm^3
6.1°	51.5	0.53
18°	18.7	0.19

Source: Kohlgrüber 2008. Reproduced with permission of Carl Hanser Verlag Publisher, Munich.

disks and positive conveying screw elements of a 50.8 mm twin screw extruder.

Considerable work has been devoted to the 3D simulation of the flow of melts in positive conveying screw elements and kneading disks of twin screw extruders (e.g. Ishikawa et al, 2000). Although models and approaches used in these simulations as well as some relevant results are discussed in section 3.3, a summary of the advantageous aspects of twin screw extrusion technology performance is given in this section.

The mixing function is performed effectively in intermeshing co-rotating twin screw extruders, partly because of the interpenetration of the screws (see Figure 3.57) and partly because of the efficiency of mixing elements that are generally full. As a result, the twin screw extruder provides a high level of micromixing (mixing of ingredients at the molecular and meso-scale levels), making it perfectly appropriate as an effective physicochemical reactor. In addition, the increase in the degree of mixing is very favorable for enhanced heat transfer during processing of very viscous materials. Lastly, residence times are significantly less dispersed than in the single screw extruder; in particular, there are no stagnant zones that show the presence of the characteristic "tails" often found in the single screw extruder.

When compared with single screw extrusion, co-rotating twin screw extrusion provides more efficient basic unit operations with improved performance such as mixing degree, heat transfer, RTD, and consequently a more controlled shear-time-temperature history imposed on the product in the screw-barrel assembly. Thus, more homogeneous processed melts can be obtained with very good molecular bindings. This brings valuable advantages in many extrusion applications which involve the processing of biomaterials in both food and non-food fields. Altogether, this enables efficient handling of various raw materials, and development of more consistent material conversion and product quality.

Processing advantages of co-rotating twin screw extrusion technology are further discussed in the following chapters based on practical observations made from relevant generic applications such as reactive extrusion (Chapter 5), thermomechanical cooking of food mixes (Chapter 6), and thermophysical micromixing of gas–liquid mixes (Chapter 9). It is worth mentioning that the unique processing characteristics of twin screw extrusion technology contribute to classifying extrusion processing technology in the class of process-intensifying methods, as discussed in Chapter 10.

3.2.4 Single screw extrusion versus twin screw extrusion

Single screw and twin screw extrusion processing technologies are now well recognized in processing industries as they allow various materials to be handled (from a few compatible ingredients in single screw extrusion to numerous non-compatible ingredients in twin screw extrusion, i.e. going from simple to complex formulated mixes). They allow materials with different physical and chemical transformations to be processed (from simple conversions such as polymer melting to chemical reactions in reactive extrusion), and extensive ranges of end-products to be created (from simple shapes to sophisticated products; from low to high added-value products). Both technologies are easy to operate and maintain, due to the fact that equipment manufacturers have designed efficient training programs to allow newcomers and seasoned operators to properly exploit extrusion equipment and processes. They also have smart upgrading offers which permit existing equipment to be improved significantly.

As seen in the previous sections, both technologies show numerous technical differences, and there are important distinctive characteristics which affect the performance of extrusion processes and the properties of resulting products. When investigating an investment project, end-users need to understand those differences from technical and process standpoints, in order to invest in the right extrusion equipment and to optimize its use and profitability.

Comparison of single screw (SSE) and twin screw (TSE) extrusion technologies is a permanent debate in the extrusion community, particularly in industries discussing processing capability of extrusion equipment or deciding on an investment. When end-users decide to invest in an extrusion processing unit, they have many key questions to ask and the answers serve to explain the advantages and disadvantages of both technologies. Although not an exhaustive list, some relevant questions include the following.

- What are the differences between SSE and TSE, in terms of processing capability and end-products made?
- Can you describe the process advantages of TSE versus SSE?
- For a given process, is TSE technology more productive than SSE?
- What products can TSE produce that SSE cannot?
- What are the operational costs (wear, energy, etc.) of TSE versus SSE?
- Can you justify the capital cost difference between TSE and SSE?
- Why should I buy a TSE rather than a SSE?
- Is my operational staff skilled enough to handle TSE versus SSE?

Promoters of SSE technology may suggest the following benefits.

- Low capital cost.
- Low wear cost.
- Low mechanical energy input, which may allow SSE promoters to claim low energy costs.
- Simple machinery with easy operating and maintenance.

Of course, these arguments are important and lead to valuable advantages but TSE promoters can produce another list outlining the potential advantages of TSE technology over SSE technology.

- High processing flexibility.
- High processing consistency and productivity.
- High product quality.
- Easy adaptation to changes in raw materials, formulations, product specifications.
- In addition, there are product conversions made on TSE that cannot be made on SSE.

Comparing single screw extrusion and twin screw extrusion from an operational standpoint is not an easy task. Although there are very few studies on this subject, it is worth mentioning the studies of Harper (1989) and Bouvier (2008). Extrusion technology is a long-term investment which should allow investors to adapt their production units according to market requirements over 15 years and more. Potential investors should not only account for what can be produced now but also what could be produced several years after the initial investment.

The purpose of this chapter was to review and discuss the process characteristics of both single screw extrusion and twin screw extrusion by use of the basics of extrusion science and engineering which have been developed in the previous sections, and then to propose an objective comparison.

In the world of extrusion processing technology, SSE is the simplest and cheapest technology on the market. For a given processing unit which includes the process capacity, the raw materials and the product to be produced, TSE may cost up to twice as much as SSE. This is significant in the cost structure of final products.

As discussed in section 3.2.3.2, TSE enables the handling of multiple processing sections (such as melting, intense mixing, cooking, venting, cooling, etc.) in one extruder unit which, combined with throughput and screw speed independence and the positive pumping action of the screws, offers substantial operational advantages to product manufacturers in terms of process flexibility and unit productivity. Additionally, TSE has the capability to reach and control the local shear-time-temperature histories required in the different processing sections of the screw profile. This leads to consistent material conversion and product quality.

Process flexibility and unit productivity both have a positive impact on the cost structure of final products.

Of great importance for the comparison of TSE and SSE technologies is the discussion of the mixing capability of the respective equipment. In the metering section of SSE, based on analysis of the thermomechanical flow in the screw channel (see section 3.2.3.1), the flow patterns can be qualitatively illustrated as in Figure 3.58. A helical flow is obtained in the down-channel direction (Figure 3.58a), which demonstrates that each fluid particle has different trajectories and velocities depending upon its position in relation to the center of the channel: the closer to the center, the faster the velocity – which leads to a relatively large dispersion of residence time of the fluid particles in the screw channel. In the cross-section of the channel (Figure 3.58b), the flow pattern indicates that the streamlines of the material show low interaction with each other, which means that the fluid particles do not mix efficiently and so the mixing in the screw channel is rather poor.

Mixing in SSE can be upgraded to some extent by use of cut flight screw elements and increased back pressure but this invariably leads to a significant decrease in extruder capacity. Flow analysis in the SSE channel clearly shows that mixing is rather limited by the laminar flow conditions, and residence times are extensively dispersed. Any flow restriction to enhance mixing, through special screw designs for example, would significantly depress extruder throughput. Low mixing in the screw channel means consequently low convective heat transfer and low mechanical energy input. The diversity of flow velocities means not only dispersion of residence times but also dispersion of shear rates and strain in the screw channel (Harper, 1989); this leads to dispersion of product conversion and temperatures in the material at the end of the metering section, which gives some heterogeneities of the melt properties when exiting the die for product shaping or texturization. These local heterogeneities tend to negatively affect the final quality of products, their functional characteristics and end-use properties.

Unlike the metering section of SSE, as seen in section 3.2.3.2 the mixing function is performed effectively in metering sections of intermeshing, co-rotating TSEs (see Figure 3.57), due to the interpenetration of the screws and the extensive diversity of mixing/shearing elements in the screw profile. As a result, the intermeshing co-rotating TSE provides a high mixing degree (dispersive and distributive mixing), making it more efficient in handling the basic processing conditions such as better convective heat transfer coefficient in viscous media, narrow RTD, and controlled local shear-time-temperature histories. Consistent material conversions as well as homogeneous melts can then be obtained.

Undoubtedly, the mixing capability of intermeshing, co-rotating TSE gives effective advantages over SSE from process and product standpoints. It is considered as the key valuable parameter of TSE technology, which impacts very positively on all dependent variables of extrusion processing (or system parameters as defined in Figure 3.4), the physical and chemical characteristics of the converted material, and ultimately the functional and end-use characteristics of extruded products (or product parameters as defined in Figure 3.4). The high mixing capability of TSE technology and subsequent

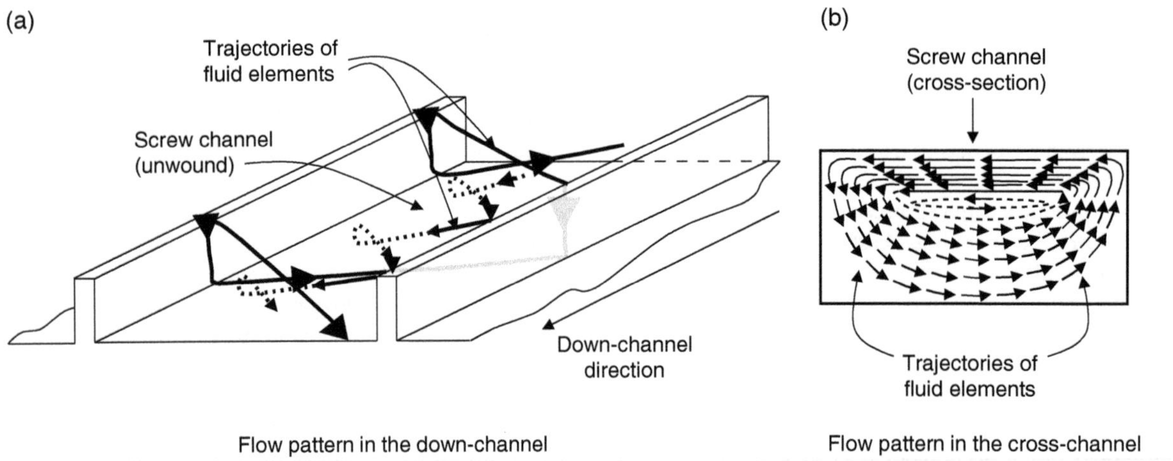

Figure 3.58 Schematics of the flow pattern in the screw channel of the metering section of a single screw extruder.

improvements in processing conditions bring significant benefits from an economical standpoint.

In addition, as described in Chapter 4 (section 4.2.4), it has been observed that polymer melting occurs over shorter lengths of barrel in TSE compared with SSE, leading to better consistency of the extrusion processing of molten materials.

Another aspect of interest is the energy cost of extrusion technologies. For a given extrusion process, it is fair to say that the specific processing energy, which is the energy consumed per kg of finished product, should be similar whichever type of extruder is used. But it must be noted that TSE technology operates over a much larger range of mechanical energy input than SSE technology. Thus, SSE technology usually requires more thermal energy input to compensate for limitations of mechanical energy input. Thus, claims that TSE technology requires more energy for a given extrusion process are inappropriate.

Any comparison between SSE and TSE technologies must include screw wear. As seen in section 3.2.3.1.3, leakage flow in SSEs remains low with new extruders but becomes significant when the flight clearance increases. Although the converting material in the screw channel acts as a lubricant in this clearance to prevent metal-to-metal friction, as the extruder operates the flight clearance increases due to screw wear caused by material abrasion and corrosion. Screw wearing in SSE significantly affects extruder performance, in particular the extruder throughput as defined by Eq. 3.81 taking the following form (Tadmor & Klein, 1970):

$$Q = \alpha . N\left(1 - \frac{\delta}{H}\right) - \frac{\beta . \Delta p}{\mu}(1 + f_w) \qquad (3.224)$$

where $^{\delta}/_{H} \gg f_w$ at small flight clearance; f_w is a dimensionless parameter which accounts for the effect of flight clearance on the pressure flow. Eq. 3.224 shows clearly that as screw wear increases, leakage flow is generated which invariably decreases the extruder throughput until it reaches beyond a tolerable level which depends on process and product requirements. Figure 3.59 illustrates qualitatively the impact of screw wearing on the throughput of SSE and TSE. In addition, the magnitude of screw wear depresses the mixing degree, the convective heat transfer coefficient, and the mechanical energy input which means that as screw wear progresses, the quality of material conversion and the final product decreases. Thus, screw wear causes some inconsistency from process and product standpoints.

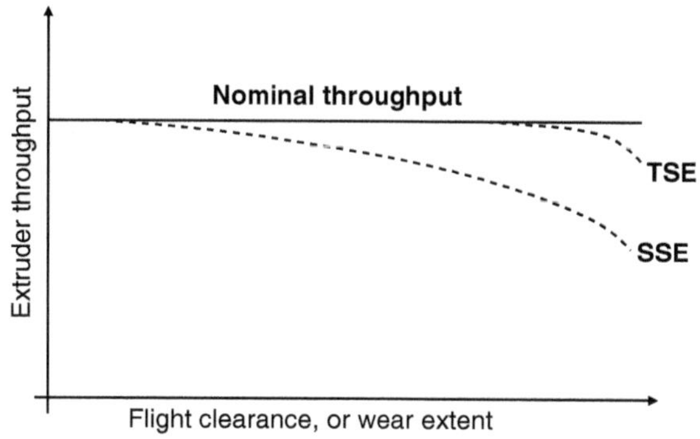

Figure 3.59 Schematics showing the effects of screw wear in single screw extruders (SSE) and twin screw extruders (TSE).

Screw wear also occurs in TSE technology, particularly at the screw restrictions and preceding screw elements. But screw wear does not affect the process performances of TSE in a large range of operations. In fact, as shown in Figure 3.56b, as screw wear increases both throughput and mechanical energy input still remain at the required levels by slightly increasing the screw speed which is throughput independent, in order to maintain the process throughput at the nominal level as well as processing performances for most of the lifetime of screw elements. Figure 3.59 illustrates qualitatively the impact of screw wearing on the throughput of TSE. For a given extrusion process and extruder capacity, though the cost of screw replacement may be slightly higher for TSE compared to SSE, wear cost of SSE is globally higher than that of TSE due to the decrease in extruder capacity and product consistency as screw wear increases over the lifetime of the screw.

The last point which needs to be considered when comparing SSE and TSE technologies is the effect of melt slipping on extruder performance. In SSE, optimum performance of the extruder is obtained when the material adheres to the barrel wall, a condition in which throughput and shearing are maximum. If material adherence to the barrel does not occur, extruder performance will drop dramatically; this is what happens when the material slips at the barrel wall. Slip phenomenon depends on the rheological characteristics of the melt in the metering section of the extruder, and so it depends closely on melt composition. Basically, the melt is usually composed of polymers and various low molecular weight components. Slip phenomenon in the screw channel starts when internal friction between macromolecular segments of polymers decreases. This is due to components which interact with macromolecular segments and interface those segments; in a way they act as "slip inducers." For instance, in feed extrusion processing (pet food as well as aquafeed mixes), slip inducers include moisture, meat slurries, protein hydrolyzates, fat, etc. That is why SSE performance starts to decrease when fat content and meat slurry in formulations are over 8% and 15% respectively, and as the content of those ingredients increases, SSE performance decreases, leading to decrease of throughput, mixing degree, heat transfer coefficient, mechanical energy input, and quality of end-products, and increase of melt heterogeneities. Conversely, with co-rotating TSE, melt slipping does not affect throughput and process parameters considerably, mainly due to the positive pumping of the screws. Thus, the capability of TSE technology to handle consistently slipping effects in extrusion processing enables product manufacturers to maximize the productivity of their extruder when processing slip-sensitive mixes and still perform intense mixing and consistent processing.

Table 3.4 gives a comparative synthesis of SSE and TSE technologies, based on the aforementioned analysis and discussion and on the previous studies published by Harper (1989) and Bouvier (2008). Basically, SSE is cheaper as the initial investment is significantly lower than that of TSE. That is why SSE is classically used for simple extrusion processes and to produce basic products, from low to intermediate added-value products. However, SSEs are rather limited from process and products standpoints. Conversely, TSE technology offers performance consistency as well as process flexibility, mainly due to possibilities to set screw configuration and operating parameters, which enable end-users to minimize operating costs. That is why TSEs are typically used to operate more complex extrusion processes requiring multiple processing sections in one single extruder, to process multicomponent formulations, and to produce an extensive

Table 3.4 Technical and economical comparison between single screw and twin screw extrusion.

Characteristics	Twin screw extrusion	Single screw extrusion
Initial investment cost	Higher investment cost (TSE/SSE = 1.4 to 2)	Smaller investment cost
Process operation		
• Replacement of worn screw parts	Slightly higher cost (TSE/SSE = 1.1 to 1.5)	Slightly lower cost
• Global wear cost	Slightly lower cost	Slightly higher cost (SSE/TSE = 1.1 to 1.3)
• Energy	Large range of mechanical energy input	Low range of mechanical energy input
Process flexibility		
• Processing sections	Multiple	Single
• Operating points	Multiple	Single
• Dependency of process variables	Independence of throughput and screw speed	Dependency of throughput and screw speed
• Recipe flexibility	Large range of raw materials and recipes	Tight specifications of raw materials and recipes
Process performances		
• Throughput consistency	Low sensitivity to recipe composition and wear	Sensitive to recipe composition and wear
• Mixing degree	Good	Poor
• Mechanical energy input	From low to high	Low
• Convective heat transfer	Good	Poor
• Residence time distribution	Relatively low dispersion	Relatively high dispersion
• Shear rate and strain	Uniform	Non-uniform
Melt characteristics		
• Material conversion	Uniform and complete	Non-uniform, may be incomplete
• Melt temperature	Uniform	Non-uniform
• Melt composition	Homogeneous	Heterogeneous
• Physicochemical binding	Good	Incomplete

Source: Adapted from Harper 1989 and Bouvier 2008.

range of products. As screw extruders are long-lasting, when deciding to invest in either a single screw extruder or a twin screw extruder, future investors are encouraged to consider not only short-term objectives but also long-term needs unless the end-user is targeting simple extrusion processing with standard raw materials and simple products both in the short and long term.

It is worth bearing in mind that the extra investment cost of a TSE over an SSE is largely compensated in the mid-term by lower operating costs associated with the ability to produce various upgraded products. Additionally, TSE enables operators to adapt the configuration of extrusion production units very easily, according to the continuous evolution of the extrusion business due to raw materials, equipment upgrading, product quality, product design, etc.

3.3 Thermomechanical extrusion processing: numerical methods

3.3.1 Single screw extrusion

Fenner (1977) reviewed experimental and theoretical work related to the single screw extrusion of polymers. A key aspect in that review indicated the preponderance of theoretical contributions compared with experimental work. It is logically argued that it is not a big challenge to measure overall performance parameters such as output rate, power consumption, pressure and temperature profiles, and specific mechanical energy (SME). However, measuring other parameters such as local temperatures and pressures inside the extruder presents technological and expensive challenges. From that standpoint, mathematical models that truly represent the extruder operation provide tools for comprehensive analysis of the extrusion operation. Models for extrusion processes are mostly based on the principles of continuum mechanics which do not consider interactions at the molecular level. The isothermal case involves solution of the system of partial differential equations represented by Eqs 3.54a, 3.54b, 3.54c, and 3.54d from which some analytical solutions were obtained for some simplified cases including Newtonian and non-Newtonian fluids.

Non-isothermal models involve the inclusion of a partial differential equation that describes the balance of energy during the extrusion processes (Eq. 3.151). Solutions for very simplified conditions were given in previous sections. Fenner (1977) has included mostly literature concerning models describing the extrusion of plastic polymeric materials. Most of the analyses were based on the screw geometry illustrated in Figures 3.17 and 3.18. Although from a rheology standpoint plastic polymers are non-Newtonian fluids exhibiting shear-thinning and viscoelastic effects, most of the models considered in the literature have involved non-elastic non-Newtonian fluids. The power law rheological model was most often used but other models such as the Herschel–Bulkley, Bird–Carreau, Carreau–Yasuda, and Cross models have been incorporated into specialized computer packages that are used to solve extrusion models, e.g. Polyflow (Yamsaengsung & Noomuang, 2010).

For plastic polymeric materials, the effect of temperature on the melt viscosity was characterized by an Arrhenius-type equation (Fenner, 1977). As discussed in reference to Figures 3.3a and 3.3b, in many applications for plastic polymers the extrusion process is designed to produce minimum degradation of the polymeric nature of the extruded material. Thus, the assumption of a melt viscosity that depends on shear rate and temperature is adequated to incorporate the viscous heating phenomenon in these models.

Conversely, biomaterials may be significantly degraded by effects of temperature and shear (see Figure 3.3c), so the assumption of a melt viscosity solely dependent on shear rate, temperature, and to a lesser extent the plasticizer type (water for most bio-based polymers) does not provide a realistic condition for the extrusion of these biomaterials. Hence, the effect of the processing history in terms of shear and thermal energy during the residence of the melt in the extruder must be considered in the rheological model.

Dhanasekharan and Kokini (2003) developed a model to describe the 3D non-isothermal flow of materials whose rheology is influenced by shear rate, temperature, and moisture content as well as the temperature-time and strain history. In the model, a rheological constitutive equation is used, which in addition to shear rate and temperature effects also includes moisture content and a number of empirical parameters related to the physicochemical transformation of the material due to heat (ψ) and shear/strain (ϕ) effects applied in the extrusion process:

$$\eta(\dot{\gamma},T,MC,\psi,\phi) = \left[m_o\dot{\gamma}^{n-1} + \frac{\tau_o}{\dot{\gamma}}\right]\cdot e^{A\left(\frac{1}{T}-\frac{1}{T_o}\right)}\cdot e^{\alpha(MC-MC_o)} \times \left[\left(1+B\left(1-e^{-k_a\psi}\right)\right)^a\right]\cdot\left[1-\beta\left(1-e^{-d\phi}\right)\right] \tag{3.225}$$

The above equation is a modification of the rheological models proposed by Morgan et al. (1989) and Mackey and Ofoli (1990). A similar equation in which the temperature-time and strain history effects are replaced by the SME parameter is discussed in Chapter 4 (section 4.4). The use of SME has some benefits because it represents the amount of mechanical energy applied to the product during the process and its experimental determination is direct.

Similar rheological models have been used by other authors to solve extrusion models with application in the food area (e.g. Chiruvella et al, 1996a; Gopalakrishna et al, 1992).

As discussed above, a single screw extruder can be considered as a processing arrangement with typical operations that simultaneously pump, heat or cool, chemically and physically transform and ultimately form a very viscous material (Chiruvella et al, 1996a). Thus, integrated models that take into account all these phenomena should include the relevant equations that describe each phenomenon. The multiphysics approach to define a model that includes the non-isothermal flow of a material that is physically and chemically modified by the process involves the use of the equation of motion with negligible inertia (Eq. 3.13), which after the approximation of negligible gravity effects results in:

$$\nabla p = \nabla \cdot \underline{\underline{\tau}} \tag{3.226}$$

The above equation is used along with the continuity equation (Eq. 3.54) that represents the microscopic mass balance in a given location of the extruder, and the stress tensor that is calculated by Eq. 3.92, whereas the strain rate tensor and its invariants are calculated by Eqs 3.94, and 3.93.

The balance of energy used by Dhanasekharan and Kokini (2003), which provides the temperature of the melt in the channel, is given by Eq. 3.153, which for a steady-state process results in the following:

$$\rho C\left(\underline{v} \cdot \nabla T\right) = \nabla . k \nabla T + \left(\underline{\underline{\tau}} : \nabla \underline{v}\right) \tag{3.227}$$

Simulations were performed by the commercial software Polyflow (Fluent Inc, New Hampshire, USA) to obtain numerical solutions of the partial differential equation system representing the model employing the finite element method (Reddy & Gartling, 2010). Results obtained were for conditions used in pasta extrusion which are very moderate compared to other extrusion processes aimed at the thermomechanical conversion of biomaterials. However, the objective of this model application was to establish fundamental rules and conditions that can be used to scale up extrusion processes without having to resort to more empirical approaches based on residence time and SME.

Gopalakrishna et al. (1992) used the finite difference method described by Morton and Mayers (2005) to solve a model similar to that described by Dhanasekharan and Kokini (2003). A power law rheological model with parameters changing with temperature and moisture content was considered. Moisture content changes were associated with the gelatinization of starch, and water was assumed to bind to the starch granules to produce starch gelatinization in a reaction considered to follow first-order kinetics. Since the viscosity of the melt was assumed to be affected by the moisture content, a partial differential equation providing the changes of moisture content in the extruder was included in the model. The main governing parameters in the non-isothermal flow were the dimensionless volumetric flow and the dimensionless Peclet number, defined as $\frac{V_{bz}H}{\alpha}$, which provides the importance of the convection in the screw channel over the heat conduction through the melt represented by the value of its thermal diffusivity $\alpha = \frac{k}{\rho, C}$. Other input parameters were related to the rheology of the melt which was dependent on shear rate, temperature, and moisture content. Heat dissipation included in the energy equation (second term on the right side of Eq. 3.205) and reaction rate included in the mass balance equation were represented by two dimensionless numbers, the Griffith number and the reaction number, respectively. Typical non-slippage boundary conditions for the 2D flow were used along a constant temperature in the barrel surface and adiabatic condition on the screw surface. Results were presented in terms of velocity, temperature, and moisture distributions as well as the pressure distribution along the channel. These primary variables were used to calculate other parameters of practical importance in the extrusion operation such as bulk temperature and residence time (Gopalakrishna et al, 1992). Results of the model were also used to evaluate uniformity of temperature in the extruder channel.

Chiruvella et al. (1996a) extended the previous model and included chemical changes in the starch such as starch gelatinization and dextrinization, which are discussed in Chapter 6 (section 6.2.2). The finite difference method was also used for the numerical solution of these mathematical models. Comparison of the effects of physicochemical modification on the starch fraction under different heat transfer conditions was possible using these models. One heat transfer condition included external heat transferred through the barrel whereas the other neglected external heating, and only the heat dissipated by friction was used to transform the starch in the product. Simulations in which the heat used to transform the material is generated by friction while the temperature of the barrel is maintained constant (at 90°C) are illustrated in Figure 3.60. Figure 3.60a shows isotherms in a plane section parallel to the extruder shaft when the barrel and the inlet temperatures are equal (90°C), whereas Figure 3.60b illustrates the corresponding degree of conversion of the starch. A larger degree of conversion is observed closer to the exit of the extruder, i.e. when the distance z increases (Figure 3.60b). As noted in the temperature isotherms and the starch degree of conversion for this isothermal case, the heat dissipated by friction is enough to increase the temperature of the melt temperature for more than 10°C and convert the starch component. Increasing the barrel temperature to 115°C further increases the temperature of the product (Figure 3.60c), favoring more conversion of the starch component (Figure 3.60d).

Processing parameters such as die bulk temperature, die pressure, and starch conversion as a function of the feed rate for different screw speeds are illustrated in Figure 3.61. Conditions were set up for an experimental single screw extruder at relatively high moisture content (30%). Results show a similar trend to that observed by using analytical solutions from more simple models which were discussed in previous

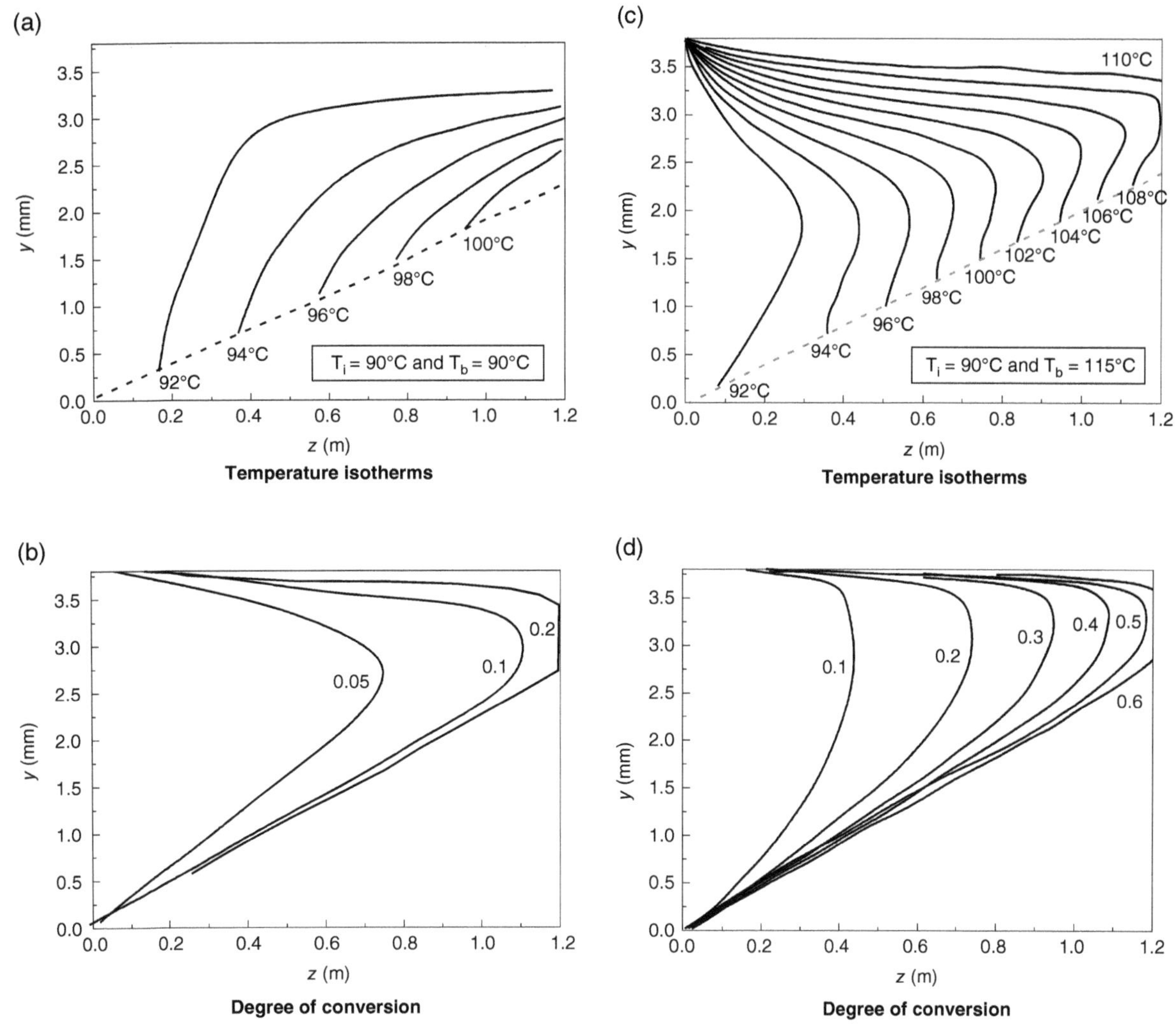

Figure 3.60 Simulations showing isotherm lines in the channel of single screw extruders for different barrel (T_b) and initial (T_i) temperature conditions (*top row figures*). Effects of those conditions on the degree of starch conversion (*bottom row figures*).

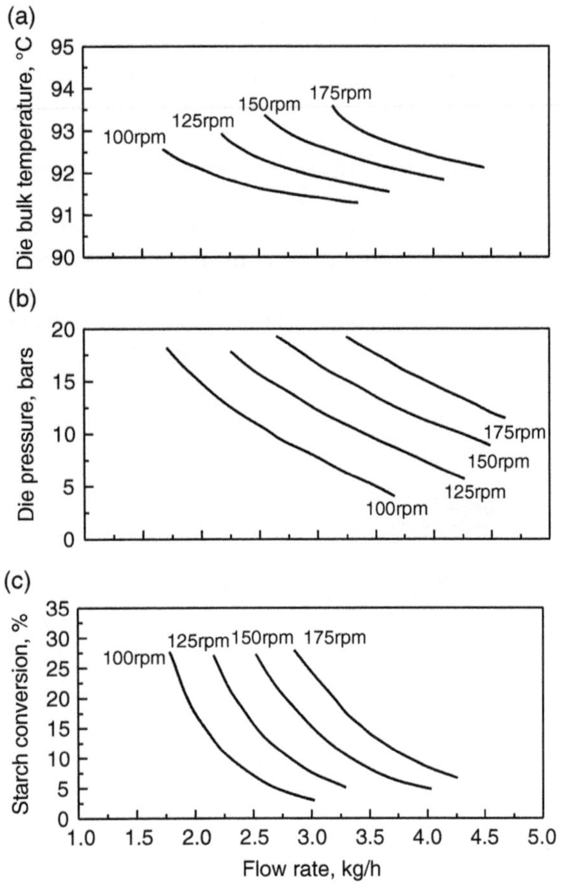

Figure 3.61 Simulations illustrating the effect of processing parameters such as feed rate and screw speed on die bulk temperature, die pressure, and starch conversion.

sections. For a given screw speed, increase of the flow rate reduces the residence time of the material in the extruder and consequently the thermal energy transferred through the barrel, which results in a decrease of the product temperature at the die. At a constant flow rate, an increase in the screw speed significantly increases the shear rate which, as indicated by Eqs 3.153 and 3.154, increases viscous dissipation and consequently the bulk temperature at the die (Figure 3.61a). The decrease of die pressure with increases in feed rate at a constant screw speed, as illustrated in Figure 3.61b for single screw extruders. It is consistent with the necessary increase of the die opening able to handle the increased feed rate. A similar trend is observed from the analytical solutions obtained for Newtonian melts (e.g. Eqs 3.143 and 3.144 for stepped screws) which clearly demonstrates that die pressure decreases with increases of the feed rate or volumetric flow rate. At constant feed rates, increase in screw speed increases the drag flow and the product pressure at the die; that behavior has been observed in Newtonian fluids and for more complicated geometries, for instance for tapered screws (see Eq. 3.150). Of great importance for product quality are the results illustrated in Figure 3.61c where simulations enable comparison of the degree of conversion of starch at different extrusion conditions.

3.3.2 Twin screw extrusion

The development of mathematical models that describe twin screw extrusion processes uses the same basic equations of flow motion and thermal transport given in the previous section (Eqs 3.226 and 3.227), which generate a multiphysics model. The model describing the rheology of the melt in terms of shear rate, temperature, and moisture content as well as changes induced by the thermomechanical treatment is also similar. Two aspects are unique in TSE mathematical modeling: one is related to the more complex screw system formed by the co- or counter-rotating technologies as well as the intermeshing and non-intermeshing features of twin screw systems. The other aspect is related to the multiplicity of screw elements having modular and interchangeable characteristics providing a large variety of functions and options, which complicate the development and optimization of new extrusion projects based solely on experimental approaches.

Mathematical modeling of these processes offers a less expensive alternative to explore various preliminary approaches that can help in process development. Although there are abundant reports describing arrangements used in TSE technology, the intermeshing co-rotating twin screw system is more commonly used for processing biomaterials, so major attention will be given to this system here. Mathematical models include 2D and 3D approaches. In spite of major difficulty, 3D analyses could provide a more precise evaluation of local properties such as temperatures, melt velocities, shear rates, and extension rates as well as their corresponding strains, which as described in previous sections constitute essential tools to quantitatively assess the thermomechanical treatment of the extruded material. Successful extrusion of bio-based polymers relies on a good understanding and control of the thermomechanical treatment.

White et al. (1987) reviewed the different technologies used in TSE systems, including mechanical aspects and modeling. Ishikawa et al. (2000) developed a 3D mathematical model to describe the flow characteristics of a non-Newtonian flow in kneading disks with a staggering angle of 90°, a configuration that enhances mixing. The rheology used was the Carreau model, and the model parameters were assumed to be depending on temperature according to the Arrhenius relationship which is further discussed in Chapter 4. Similar work was carried by Bravo et al. (2004) who developed a 3D mathematical model to describe the flow of a melt in the kneading disks of intermeshing co-rotating twin screw extruders. The melt was modeled with the Carreau rheological model. Determination of the velocity profile allowed the estimation of theoretical residence time distributions of particle tracers computationally introduced in the entrance of the kneading disks. Specifically, results showed how backflow is generated in a kneading disk section as well as the configuration that generates larger backflow to promote mixing. Further details on the mixing characteristics of twin screw extruders are covered in Chapter 5.

Since the application was oriented to the extrusion of plastic polymers, no effects of thermomechanical treatments on the material rheology were considered. The model was validated with pressure and temperature measured under the same configuration and the same operational conditions as the numerical simulations. The experimental set-up used for measuring the pressure and temperature is illustrated in Figure 3.62. Figure 3.62a shows the location of the temperature and pressure probes whereas Figure 3.62b indicates the different sections considered for the analysis. As shown in the figure, three positions, an upper stream section (section AA'), a central section (section BB') and a downstream section (section CC'), were considered. The numerical solution of the mathematical model was done using the finite element method (FEM). Simulations at different screw speeds were considered. A distinctive aspect noted in the simulations was the presence of forward and backward flows due to changes in the pressure gradients. The simulated results were experimentally validated by pressure measurements at different angular positions (Figure 3.63). As illustrated in the figure, there are changes of pressure gradients that may promote better mixing of the material. Temperatures changes with the axial distance at different screw speeds are shown in Figure 3.64. Simulations showed that temperature increased with the screw speed, likely to be due to the higher viscosity dissipation. Discrepancies between results of the model and the experiments at the higher screw speed were attributed to some heat loss by convection and potential overestimation of the viscous dissipation. The use of an accurate rheological model may alleviate those deviations.

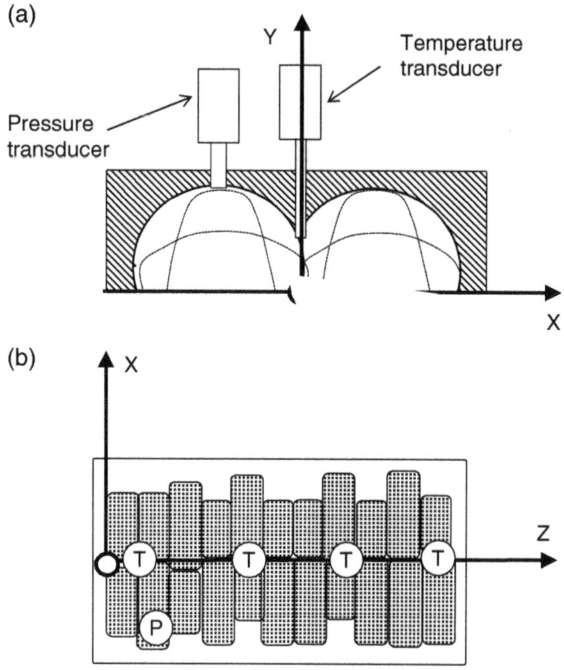

Figure 3.62 Schematic illustrating the location of pressure and temperature sensors to study temperature and pressure fields in a twin screw channel to validate the 3D numerical model developed by Ishikawa et al. (2000). Source: Ishikawa, Kihara & Funatsu 2000. Reproduced with permission of John Wiley & Sons.

Models that describe twin screw extrusion of bio-based polymers were developed by other researchers (Berzin et al, 2010; Chiruvella et al, 1996b; Ficarella et al, 2006; Sastrohartono et al, 1995) using the finite differences method, finite volumes method or commercial packages, e.g. Polyflow, LUDOVIC®, the latter will be briefly discussed in the next section, along the general use of FEM as the numerical tool for solving these mathematical models. Although thermomechanical treatment is important when processing biomaterials, only the effects of temperature and moisture content were considered to affect the assumed non-Newtonian viscosity of the extruded materials. Elastic effects are rarely included in the rheology of extruded materials, and in many cases the viscoelastic models used are not a good representation

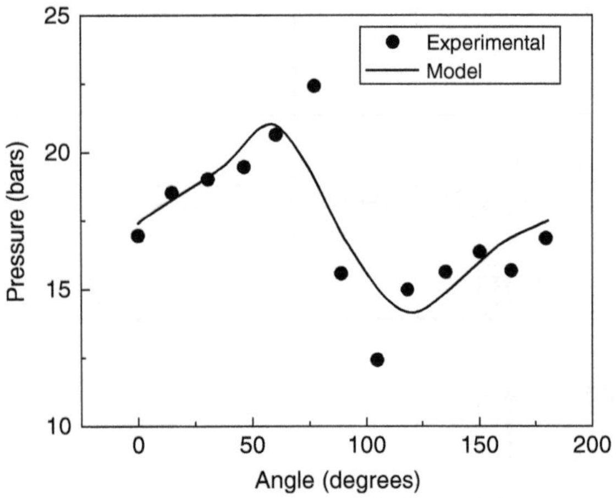

Figure 3.63 Experimentally and simulated results illustrating pressure values during half a turn of the screws. Source: Ishikawa, Kihara & Funatsu 2000. Reproduced with permission of John Wiley & Sons.

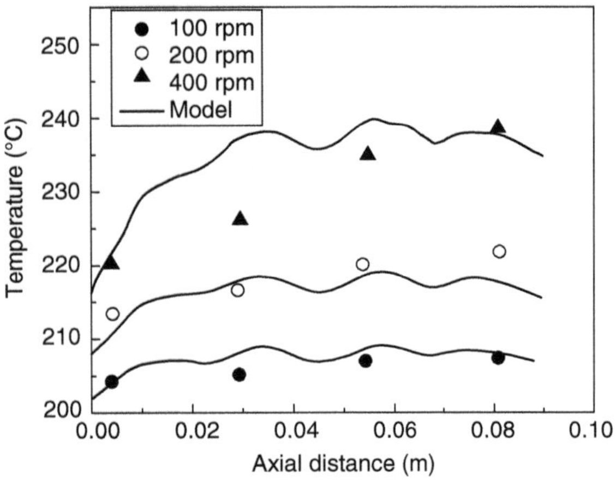

Figure 3.64 Experimentally and simulated results illustrating temperature values during half a turn of the screws.
Source: Ishikawa, Kihara & Funatsu 2000. Reproduced with permission of John Wiley & Sons.

of the rheological behavior of those materials (Mu & Zhao, 2008). Marchal (2005) critically reviewed different modeling techniques used in extrusion technology and included the study of viscoelastic effects on the extrusion operation as one of the challenges of extrusion research.

One of the major uses of numerical methods applied to TSE technology is to calculate parameters such as residence time distributions, strain distributions, local temperatures and pressures, degree of mixing, etc, which are important processing parameters to estimate thermomechanical treatments applied to the extruded product. Typical calculations of these parameters were described in previous sections for cases in which analytical solutions exist. The complexity of TSE geometry makes it impossible to find analytical solutions of the mathematical models involving TSE systems so all the determinations must be done either numerically or experimentally. Numerical calculations use the same approaches as those described in previous sections but parameters are numerically determined. Ganjyal and Hanna (2002) reviewed methods for the determination of RTD in extruders. From their review, it can be concluded that most studies focused on TSE and experimental methods. Ganjyal and Hanna (2002) proposed the use of a neural network model which is a statistically based method requiring significant and accurate experimental data to generate sound models. Thus, there is a need to develop new model-based methods to determine parameters that are elusive to experimental determination such as local pressures and temperatures, local strains and strain rates, as well as other parameters related to the degree of mixing in the process.

Poulesquen and Vergnes (2003) developed a method that used the commercial software LUDOVIC® to determine RTD in twin screw extruders without using adjustable parameters. The method is based on the continuum mechanics approach discussed in previous sections coupled to the theory of reactors. It combined the behavior of a group of ideal reactors having flows whose fluid mechanics were determined by LUDOVIC®. Effects of feed rate, screw speed, and melt viscosity on RTD were studied. RTD in twin extruders was also determined using combinations of ideal steady-state flow reactors (Kumar et al, 2008). In addition to these theoretical studies, there is an abundance of experimental research to determine RTD using die tracer systems combined with theoretical approaches (Kao & Allison, 1984; Puaux et al, 2000; Todd, 1975; Unlu & Faller, 2002; Wolf et al, 1986).

3.3.3 Commercial software

In previous sections mathematical models describing the non-isothermal flow of rheologically complex materials have been presented. In some cases, there exist closed analytical solutions of those models but they represent rough approximations to the real systems. 2D and 3D models require long and complex numerical solutions. For instance, the solution of a 3D non-isothermal flow in either a single

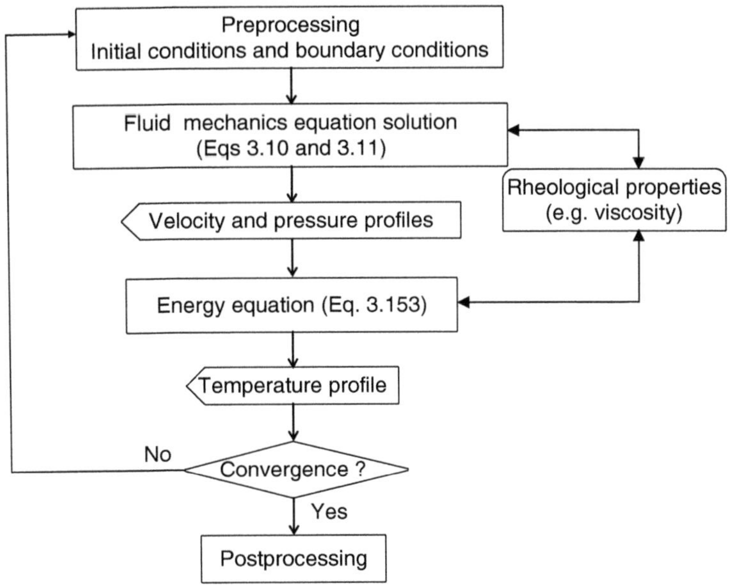

Figure 3.65 Flowchart illustrating the development of a numerical algorithm based on the finite element method (FEM) that can solve mathematical models describing extrusion processes, both SSE and TSE.

screw or a twin screw extruder represents a formidable task that involves the simultaneous solution of five partial differential equations. The solution consists of the velocity vector, pressure and temperature fields in the material inside the extruder channel. If only temperature and flow are considered, there are five partial differential equations: three equations of motion (one for each of the directions that form the 3D space), one energy equation and the continuity equation, with the local three velocity components in addition to local temperatures and pressures. These partial differential equations are coupled. Temperature affects the viscosity of the fluid and thus the velocity field. Velocity is coupled to the convection ($\underline{v}\cdot\nabla T$) and viscous dissipation ($\underline{\underline{\tau}}:\nabla\underline{v}$) terms in the energy equation. Hence these equations have to be solved simultaneously.

Figure 3.65 shows a general flow sheet commonly used to solve these equations. Setting the problem indicated in the chart as "initial condition" is the more time-consuming task. The flow domain needs to be discretized for the solution of the equations. For the discretization, it is necessary to create a 3D solid geometry representing the domain being studied. Programs as AutoCAD® or ProE® are used to create the 3D solid geometry. Once the solid 3D geometry is created, the discretization (meshing) of the flow domain is conducted. Although some programs provide an automatic meshing method, manual meshing, its fine tuning and preparation of the domain to set the numerical algorithm represents a challenge. It is important to utilize finer meshes in zones where there are large gradients of velocity, temperature and pressure which sometimes do not coincide. In the case of twin screw extruders, the flow domain changes with screw rotation which represents a serious challenge to the numerical analysis (Ishikawa et al, 2000). Figure 3.66 shows the mesh of an intermeshing co-rotating twin screw extruder. Next, the boundaries of the problem domain are fixed followed by the setting of the material properties. Once the setting of those initial conditions and fluid properties is complete, the simulation is started. Although the primary variables are the direct output of the problem solution, postprocessing of those data enables calculation of relevant parameters.

Two programs are often used for research and development in this area, Polyflow and LUDOVIC®, which are specifically dedicated to the solution of extrusion models. There exist other general programs that can be adapted to solve extrusion problems, such as Comsol®. In general, all the programs perform similar tasks and the convenience of one over the other is difficult to assess. Useful characteristics of these commercial packages include their ability to solve truly multiphysics models, which as discussed are a good representation of most extrusion processes.

As a conclusion to this chapter, it is worth emphasizing LUDOVIC® computer software which is a recognized

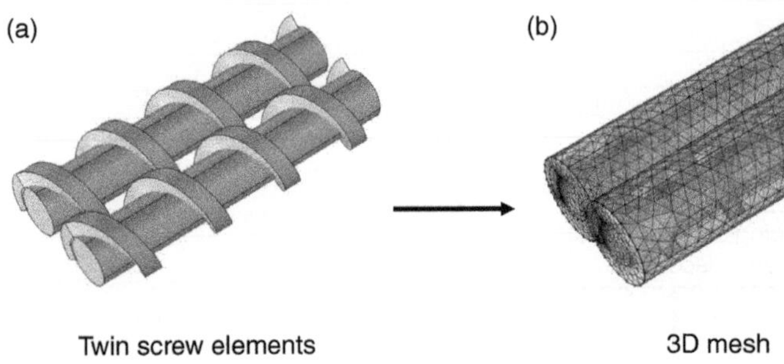

Figure 3.66 Illustration showing the discretization (meshing) of a section of an intermeshing co-rotating twin screw extruder. Source: Adapted from Ishikawa et al. 2000. Reproduced with permission of John Wiley & Sons.

package for the simulation of intermeshing co-rotating twin screw extrusion processes (Sciences Computers Consultants, 2013). The software is targeted to the processing of synthetic polymers, bio-based polymers, cosmetics, pharmaceuticals, etc. in which the material is mixed and processed by twin screw extrusion. The software environment consists of pre- and postprocessing user-friendly interfaces to set initial conditions, start calculations and visualize results. Preprocessing deals with aspects related to the screw design, description of the materials being processed, and the setting of the processing conditions. The description of materials is based on a database containing information on melting temperature, melting enthalpy, density, thermal conductivity, specific heat, and a number of rheological models for a variety of materials. The software also includes routines and procedures with application to reactive extrusion, a topic described in Chapter 5. In addition to the primary variables obtained from the mathematical models, namely the velocity, temperature and pressure fields, postprocessing enables the calculation of other variables, including the filling ratio, SME, residence time distribution, local strain rates and strains, cumulated strain, and local dissipated energy.

The approach of LUDOVIC® software aims at considering a local description of the flow field of polymeric melts in each screw element of the screw-barrel assembly of a twin screw extruder, including the die. Hence, the flow field is resolved by elementary models dedicated to each type of screw element. These models are 1D for right- and left-handed pitch screw elements, and the die whereas the models are 2D for kneading blocks. In each elementary model, results are obtained according to the material and process parameters (flow rate, screw speed, and temperature profile). Results of 2D models are averaged to provide 1D descriptions of processing variables. Heat balance is also considered in each screw element. The global description of the flow field in the extruder is given by combining the contribution of all elementary models.

A review of software specific to simulate twin screw extrusion processes describes the LUDOVIC® program and other similar programs (Anonymous, 2002).

References

Anonymous (2002) Twin screw extruder simulation programs – what can they offer? Plastics, Additives and Compounding 4(2): 22–26.

Berzin F, Tara A, Tighzert L, Vergnes B (2010) Importance of coupling between specific energy and viscosity in the modeling of twin screw extrusion of starchy products. Polymer Engineering and Science 50(9): 1758–1766.

Bigg D, Middleman S (1974) Mixing in a screw extruder. A model for residence time distribution and strain. Industrial and Engineering Chemistry Fundamentals 13(1): 66–71.

Bird RB, Armstrong RC, Hassager O (1987) *Dynamic of Polymeric Liquids*, vol 1, 2nd edn. New York: John Wiley.

Bird RB, Stewart WE, Lightfoot EN (2007) *Transport Phenomena*, rev. 2nd edn. New York: John Wiley.

Bouvier JM (2008) *Twin screw versus single screw in feed extrusion processing.* Petfood Forum, Chicago, USA, 14–16 April.

Bravo VL, Hrymak AN, Wright JD (2004) Study of particle trajectories, residence times, and flow behavior in kneading discs of intermeshing co-rotating twin-screw extruders. Polymer Engineering Science 44(4): 779–793.

Brent Strong A (1996) *Plastics: Materials and Processing.* New Jersey: Prentice-Hall.

Bruin S, Van Zuilichem DJ, Stolp W (1978) A review of fundamental and engineering aspects of extrusion of biopolymers in a single-screw extruder. Journal of Food Process Engineering 2(1): 1–37.

Brümmer T, Meuser F, Van Lengerich B, Niemann C (2002) Effect of extrusion cooking on molecular parameters of corn starch. Starch/Stärke 54(1): 1–8.

Capone C, Di Landro L, Inzoli F, Penco M, Sartore L (2007) Thermal and mechanical degradation during polymer extrusion processing. Polymer Engineering and Science 47(11): 1813–1819.

Chiruvella RV, Jaluria Y, Karwe MV (1996a) Numerical simulation of the extrusion process for food materials in a single-screw extruder. Journal of Food Engineering 30(3–4): 449–467.

Chiruvella RV, Jaluria Y, Karwe MV, Sernas V (1996b) Transport in a twin-screw extruder for the processing of polymers. Polymer Engineering and Science 36(11): 1531–1540.

Choo KP, Neelakantan NR, Pittman JFT (1980) Experimental deep-channel velocity profiles and operating characteristics for a single-screw extruder. Polymer Engineering and Science 20: 349–356.

Danckwerts PV (1957) Measurement of molecular homogeneity in a mixture. Chemical Engineering Science 7(1–2): 116–117.

Danckwerts PV (1958a) The effect of incomplete mixing on homogeneous reactions. Chemical Engineering Science 8 (1–2): 93–102.

Danckwerts PV (1958b) Local residence-times in continuous-flow systems. Chemical Engineering Science 9(1): 78–79.

Darnell WH, Mol EAJ (1956) Solids conveying in screw extruders. Society of Plastics Engineering Journal 12: 20–29.

Dhanasekharan M, Kokini JL (2000) Viscoelastic flow modeling in the extrusion of a dough-like fluid. Journal of Food Process Engineering 23(3): 237–247.

Dhanasekharan M, Kokini JL (2003) Design and scaling of wheat dough extrusion by numerical simulation of flow and heat transfer. Journal of Food Engineering 60(4): 421–430.

Della Valle G, Buleon A, Carreau PJ, Lavoie PA, Vergnes B (1998) Relationship between structure and viscoelastic behavior of plasticized starch. Journal of Rheology 42(3): 507–525.

Exton H (1978) *Handbook of Hypergeometric Integrals: theory, applications, tables, computer programs*. Chichester: Ellis Horwood p. 27.

Fenner RT (1977) Development in the analysis of steady screw extrusion of polymers. Polymer 18: 617–635.

Ficarella A, Milanese M, Laforgia D (2006) Numerical study of the extrusion process in cereals production: Part I. Fluid-dynamic analysis of the extrusion system. Journal of Food Engineering 73(2): 103–111.

Froment GF, Bischoff KB, De Wilde J (2011) *Chemical Reactor Analysis and Design*, 3rd edn. New York: John Wiley.

Ganjyal G, Hanna M (2002) A review on residence time distribution (RTD) in food extruders and study on the potential of neural networks in RTD modeling. Journal of Food Science 67(6): 1996–2002.

Gopalakrishna S, Jaluria Y, Karwe MV (1992) Heat and mass transfer in a single screw extruder for non-Newtonian materials. International Journal of Heat and Mass Transfer 35: 221–237.

Griffith RM (1962) Fully developed flow in screw extruders. Theoretical and experimental study. Industrial and Engineering Chemistry Fundamentals 1(3): 180–187.

Harper JM (1989) Food extruders and their applications. In: Mercier C, Linko P, Harper JM (eds) *Extrusion Cooking*. St Paul, Minnesota: American Association of Cereal Chemists, pp. 1–15.

Hyun KS, Spalding MA, Hinton CE (1996) Theoretical and experimental analysis of solids conveying in single-screw extruders. In: Proceedings of 54th SPE/ANTEC Conference, vol I: 199–206.

Ishikawa T, Kihara SI, Funatsu K (2000) 3-D Numerical simulations of non-isothermal flow in co-rotating twin screw extruders. Polymer Engineering and Science 40(2): 357–364.

Janssen L (1989) Engineering aspects. In: Mercier C, Linko P, Harper JM (eds) *Extrusion Cooking*. St Paul, Minnesota: American Association of Cereal Chemists, pp. 17–37.

Kao SV, Allison GR (1984) Residence time distribution in a twin screw extruder. Polymer Engineering and Science 24(9): 645–651.

Karapetsas G, Tsamopoulos J (2008) Steady extrusion of viscoelastic materials from an annular die. Journal of Non-Newtonian Fluid Mechanics 154: 136–152.

Kohlgrüber K (2008) *Co-Rotating Twin-Screw Extruders – fundamentals, technology, and applications*. Munich: Carl Hanser Verlag.

Kokini JL (1993) The effect of processing history on chemical changes in single- and twin-screw extruders. Trends in Food Science and Technology 4(10): 324–329.

Kroesser FW, Middleman S (1965) The calculation of screw characteristics for the extrusion of non-Newtonian melts. Polymer Engineering and Science 5: 230–235.

Kumar A, Ganjyal GM, Jones DD, Hanna MA (2008) Modeling residence time distribution in a twin-screw extruder as a series of ideal steady-state flow reactors. Journal of Food Engineering 84(3): 441–448.

Levenspiel O (1984) *The Chemical Reactor Omnibook*. Corvallis: Oregon State University.

Li Y, Hsieh F (1996) Modeling of flow in a single screw extruder. Journal of Food Engineering 27(4): 353–375.

Mackey KL, Ofoli RY (1990) Rheology of low- to intermediate-moisture whole wheat flour doughs. Cereal Chemistry 67(3): 221–226.

Marchal T (2005) Challenges of modeling the extrusion process. Plastics, Rubber and Composites 34: 265–270.

Meuser F, Van Lengerich B (1984) System analytical model for the extrusion of starches. In: Zeuthen P, Cheftel J, Eriksson C, et al. (eds) *Thermal Processing and Quality of Foods*. London: Elsevier, pp. 175–179.

Michaeli W (2003) *Extrusion Dies for Plastics and Rubber. Design and engineering computations*, 3rd edn. Munich: Carl Hanser Verlag.

Morgan RG, Steffe JF, Ofoli RY (1989) A generalized viscosity model for extrusion of protein doughs. Journal of Food Process Engineering 11: 55–78.

Morrison FA (2001) *Understanding Rheology*. New York: Oxford University Press.

Morton KW, Mayers DF (2005) *Numerical Solution of Partial Differential Equations: an introduction*. Cambridge: Cambridge University Press.

Mu MY, Zhao G (2008) Numerical study of non-isothermal polymer extrusion flow with a differential viscoelastic model. Polymer Engineering and Science 48(2): 316–328.

Poulesquen A, Vergnes B (2003) A study of residence time distribution in co-rotating twin-screw extruders. Part I: Theoretical modeling. Polymer Engineering and Science 43(12): 1841–1848.

Puaux JP, Bozga G, Ainser A (2000) Residence time distribution in a corotating twin-screw extruder. Chemical Engineering Science 55(9): 1641–1651.

Rauwendaal C (2001) *Polymer Extrusion*, 4th edn. Munich: Carl Hanser Verlag.

Reddy JN, Gartling DK (2010) *The Finite Element Method in Heat Transfer and Fluid Dynamics*, 3rd edn. Boca Raton, Florida: CRC Press.

Sastrohartono T, Jaluria Y, Karwe MV (1995) Numerical simulation of fluid flow and heat transfer in twin-screw extruders for non-Newtonian materials. Polymer Engineering and Science 35(15): 1213–1221.

Sciences Computers Consultants (2013) LUDOVIC: Logiciel d'Utilisation de DOubles VIs Corotatives, version 5.2. Information available through: www.scconsultants.com/.

Slattery JC (1999) *Advanced Transport Phenomena*. Cambridge: Cambridge University Press.

Steller RT (1990) Theoretical model for flow of polymer melts in the screw channel. Polymer Engineering and Science 30: 400–407.

Stevens MJ, Covas JA (1995) *Extruder Principles and Operations*, 2nd edn. New York: Chapman & Hall.

Syrjälä S (1999) On the analysis of fluid flow and heat transfer in the melt conveying section of a single screw-extruder. Numerical Heat Transfer, Part A: Applications 35(1): 25–47.

Tadmor Z, Gogos CG (2006) *Principles of Polymer Processing*, 2nd edn. New York: John Wiley.

Tadmor Z, Klein I (1970) *Engineering Principles of Plasticating Extrusion*. Polymer Science and Engineering Series. Florida: Robert E. Krieger.

Todd DB (1975) Residence time distribution in twin-screw extruders. Polymer Engineering and Science 15(6): 437–443.

Unlu E, Faller JF (2002) RTD in twin-screw food extrusion. Journal of Food Engineering 53(2): 115–131.

Vergnes B, Berzin F (2010) Predicting starch transformation in twin screw extrusion. Society of Plastic Engineers. Plastic Research Online. www.4spepro.org/pdf/002986/002986.pdf.

White JL, Szydlowski W, Min K, Kim MH (1987) Twin screw extruders; development of technology and analysis of flow. Advances in Polymer Technology 7(3): 295–332.

Wolf D, White DH (1976) Experimental study of the residence time distribution in plasticating screw extruders. AIChE Journal 22(1): 122–131.

Wolf D, Holin N, White DH (1986) Residence time distribution in a commercial twin-screw extruder. Polymer Engineering and Science 26(9): 640–646.

Yamsaengsung R, Noomuang C (2010) Finite element modeling for the design of a single-screw extruder for starch-based snack products. Proceedings of the World Congress on Engineering, vol. III, 30 June–2 July, London.

Yue M, Zhao G (2008) Numerical study of nonisothermal polymer extrusion flow with a differential viscoelastic model. Polymer Engineering and Science 48: 316–328.

4 The Generic Extrusion Process I: Thermomechanical Plasticating of Polymers and Polymer Melt Forming

The Generic Extrusion Process I (GEP I) is fundamental as it allowed the scientific and technical culture of extrusion processing to be created and formalized. GEP I involves thermomechanical plasticating of polymers in the screw-barrel assembly of a screw extruder, and polymer melt forming in a die assembly in line with the extruder barrel. As introduced in Chapter 1, GEP I is generically named as follows: Thermomechanical plasticating of polymers and polymer melt forming.

The first part of this chapter presents bio-based polymers and related bio-based plastics. As extrusion processing of synthetic polymers has been described extensively in numerous books, this chapter deals exclusively with some polymeric biomaterials which are intended for bio-based plastics. These polymers are destined to spread throughout the plastics world in the future, due to public concern about the effects of non-degradable plastics on the environment, climate change and limited fossil fuel resources; hence the focus of this chapter is on these polymers rather than on conventional synthetic polymers. Moreover, in contrast with synthetic thermoplastics, emerging bio-based thermoplastics have been treated rather discreetly in the scientific and trade literature.

Thermomechanical plasticating of polymers usually consists of feeding a solid polymer in the form of pellets or powder to a screw extruder through a hopper, compacting the solid polymer in the feed section of the screw-barrel assembly, melting polymeric materials in the compression section, possibly adequate mixing of polymeric melts with various additives in the mixing section, and finally pumping plasticized melts into the metering section, in order to deliver a steady stream of uniform molten polymer to the die assembly. As for energy inputs, two mechanisms are involved to ensure thermomechanical plasticating: conduction of heat from hot surfaces such as the barrel wall, and dissipation of heat caused by the shearing of polymeric material in the screw channel; hence the term "thermomechanical" is used to qualify both thermal and mechanical energy inputs in polymer plasticating. The engineering analysis of the material flow in the metering section of screw extruders, where the material is in liquid state, is described extensively in Chapter 3. The results of that analysis can be applied efficiently for extrusion processing of bio-based polymers once they are fully molten. The mechanism of polymer melting in the compression section of screw extruders, where polymeric materials are converted from solid to liquid state, has not yet been described in this book so we present and analyze the mechanism of polymer melting in the second part of this chapter.

The commercial objective of thermoplastics extrusion processing is to produce products with acceptable shapes at acceptable rates. This is called polymer melt forming, meaning that polymeric melts pass through a die assembly which is connected to the delivery end of the extruder barrel. The flow channel in the die changes its cross-section smoothly from that of the end of the extruder barrel to that setting the shape of the product. As seen in Chapter 2 (see section 2.1.3), various designs of die assembly are used in thermoplastics extrusion processing which have been developed for forming plastics materials based on synthetic polymers. Existing die designs and related

Extrusion Processing Technology: Food and Non-Food Biomaterials, First Edition. Jean-Marie Bouvier and Osvaldo H. Campanella.

engineering analysis can also be used advantageously for forming bio-based plastics through appropriate adjustments of specific geometrical dimensions and forming conditions. Such adjustments should be co-ordinated with a comprehensive understanding of the physical and rheological characteristics of bio-based polymeric melts.

In the plasticating process, polymeric materials undergo major physical modifications (such as glass transition, which is the transition from an amorphous or a partially crystalline solid state to a rubbery state, and melting transition, which is the transition from a rubbery state to a liquid state), and possibly chemical modifications (such as partial thermal depolymerization and/or covalently bonded three-dimensional cross-linking). In the polymer melt-forming process, polymeric biomaterials are converted back from liquid state to rubbery state, and possibly to solid state depending upon the forming conditions, the time-temperature history in particular; they may also undergo structural reorganization which affects the functional properties of the formed products. Glass and melting transitions are determinant phenomena in the plasticating and forming processes. In addition, understanding the rheological behavior of bio-based polymeric melts in relation to physicochemical changes due to extrusion processing is of paramount importance in the appropriate design of the metering section and the forming dies of extrusion processes. The third and fourth sections of this chapter are dedicated to examination of the transition phenomena as well as the rheological behavior of bio-based polymeric melts which are intended for bio-based plastics.

Finally, the fifth section presents case studies related to the thermomechanical plasticating process of bio-based polymers and forming bio-based thermoplastics.

4.1 The bio-based polymers and bio-based plastics

As suggested by Shen et al. (2009), in this chapter, the term "bioplastics" will be avoided due to its ambiguity, as it is sometimes used for plastics that are bio-based and sometimes for plastics that are biodegradable (including those that are made from fossil instead of renewable resources). Using the term "biopolymers" in contrast to the term "synthetic polymers" would also be inadequate as it could refer to any polymer derived from living organisms, and subjected to various processing routes for food and non-food uses, which may not necessarily be related to thermoplastics processing.

It is thus necessary to introduce clear definitions of technical terms to qualify the complex world of polymeric materials derived from renewable resources as opposed to those made from fossil resources.

4.1.1 Definitions

4.1.1.1 Biomass and biomaterials

"Biomass" is a carbon-based sustainable resource, which refers to the total mass of biological material produced by living organisms (Figure 4.1). "Biomaterials" is a generic term bringing together all materials derived from biomass, as well as related processed products, which are used by mankind.

"Bio-based polymers" may derive from various types of biomass such as wood, agricultural resources, animals, micro-organisms, etc. Processed products derived from bio-based polymers refer to polymers used for plastic applications (e.g. bio-based plastics, whether they are fully or partially produced from renewable raw materials), for non-food non-plastic applications (e.g. starch and cellulose derivatives), and for food uses.

Nature-derived polymers are abundant in the vegetal (wood, leaves, stems, fruits, roots, seeds, etc.) and animal kingdoms (muscles, shells, furs, etc.). The world biomass production is estimated to about 170 billion tonnes per year, of which only 3.5% (6 billion tonnes) are utilized by humans. As shown in Figure 4.1, most of these 6 billion tonnes are used for food (62%), about one-third for energy, paper and board, furniture and construction (mostly from wood), and only 5% (300 million tonnes) for other non-food uses such as clothing, chemicals, etc., of which bio-based plastics currently represent a very small part (0.8–0.9 million tonnes) (Shen et al., 2009).

It is worth noting that extrusion processing technology plays a significant role in the different conversion routes of biomaterials, such as food materials, high added-value papers, green chemicals, and emerging bio-based plastics, in particular.

Today, worldwide production of bio-based plastics is close to 1 million tonnes per year. This may seem to be of marginal importance compared with world plastics consumption (including thermoplastics, thermosets, and synthetic fibers) which amounted to approximately to 268 million tonnes per year in 2007. But accounting for the worldwide growth of plastics in the last 35 years (average yearly growth of 5.9% between 1971 and 2006;

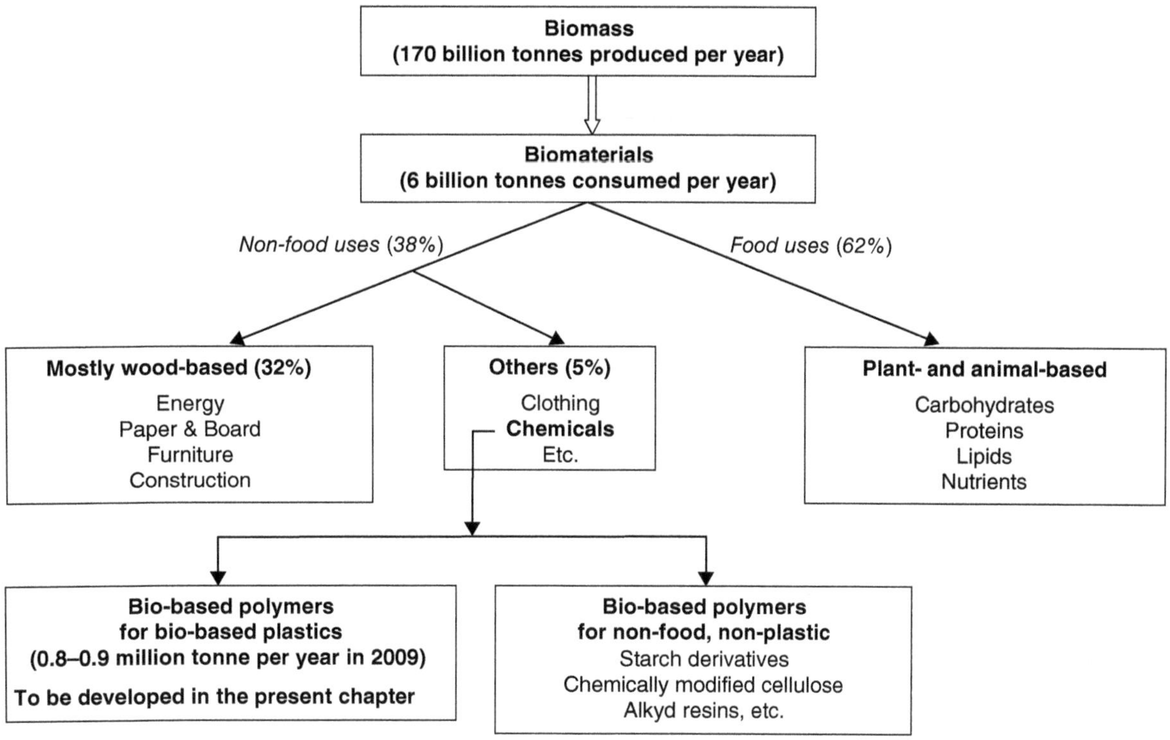

Figure 4.1 The segmentation of biomass and biomaterials. Role of bio-based plastics.

Simon & Schnieders, 2007) and the technical substitution potential of bio-based plastics (estimated at 85–90% for the coming years; Shen et al., 2009), experts in plastics macroeconomics and the plastics industry claim that worldwide production of bio-based plastics will range between 2.5 and 3.5 million tonnes per year in 2020 (Crank et al., 2005; Shen et al., 2009). Such projection is supported by the fact that prices of agricultural commodities will increase far less than the price of crude oil, as well as by the need to substantially reduce environmental impacts.

4.1.1.2 Bio-based plastics

Bio-based plastics are defined as plastics that are derived fully or partially from biological resources (renewable raw materials) and used for non-food purposes. Bio-based plastics can be, but are not necessarily, biodegradable. Natural fibers which are combined with polymeric matrices (natural fiber composites) are also included in the category of plastics, whether they are combined with bio-based polymers or with petrochemical polymers. In fact, the industrial use of natural fibers is growing and they have a low environmental footprint compared to their synthetic counterparts.

There are three main processing routes for producing bio-based polymers (Figure 4.2).

- *Processing route I*: use of bio-based polymers obtained from biomass by conventional fractionation (e.g. starch, cellulose, proteins, and natural fibers). In this category is included natural rubber, which is produced from the rubber tree, *Hevea brasiliensis*.
- *Processing route II*: use of bio-based polymers directly produced by bacteria (e.g. polyhydroxyalkanoates or PHAs).
- *Processing route III*: use of bio-based monomers obtained by fermentation or conventional chemistry, which are then polymerized through classic polymerization techniques; typical examples are polylactic acid (PLA) that is biodegradable, polyamides and also bio-based polyolefins (such as bio-based polyethylene, bio-based polyvinyl chloride, etc.) that are non-biodegradable.

The schematic illustrated in Figure 4.2 is an attempt to classify bio-based polymers which are already used,

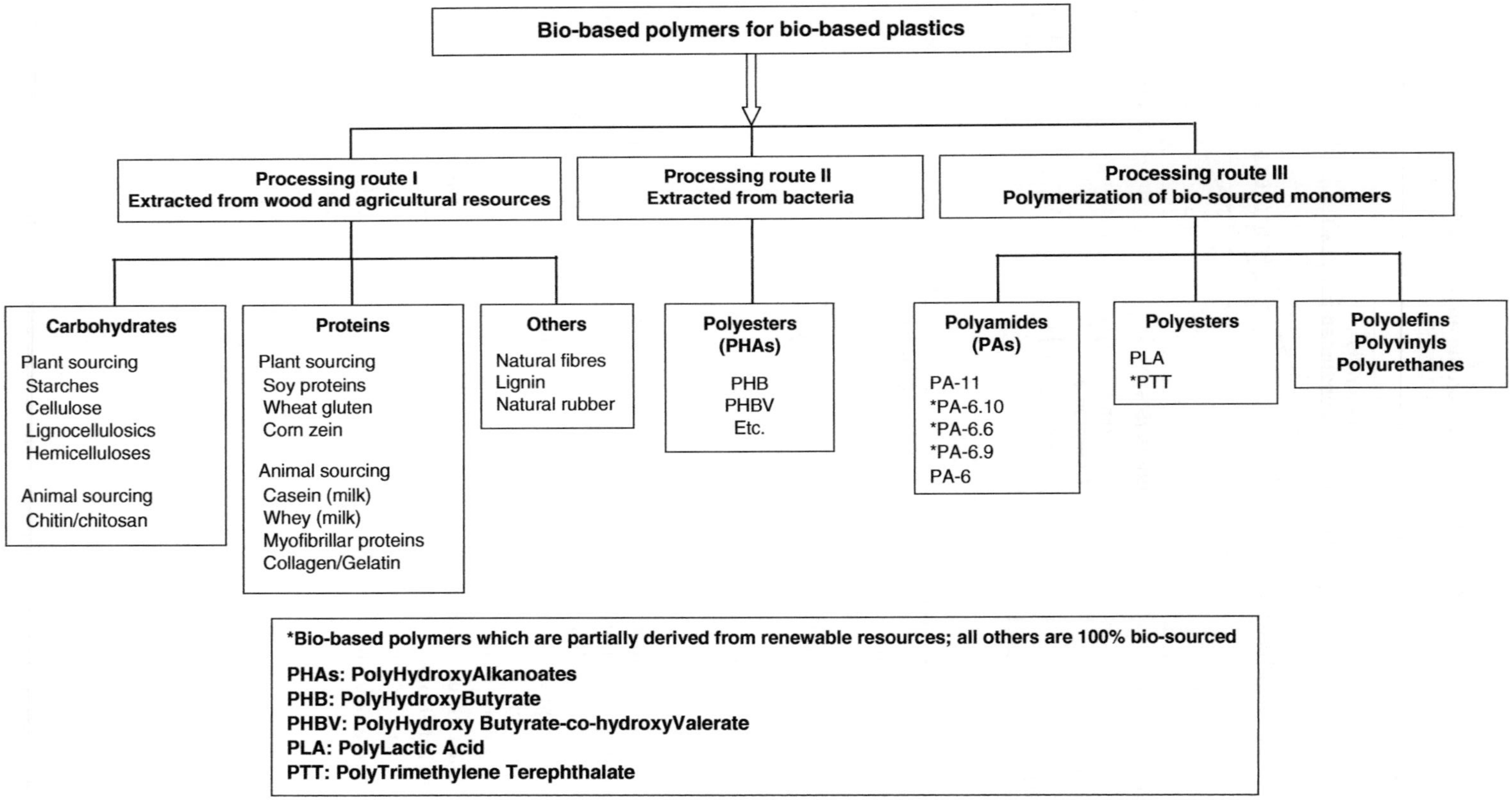

Figure 4.2 Classification of bio-based polymers for bio-based plastics.

or might be used in the future, to produce bio-based plastics. It must be mentioned that processing route I represents the largest market volume of today's bio-based plastics. This relatively high volume mainly includes thermoplastic starches, which are developing very rapidly with an average annual growth of nearly 50% since the early 2000s. Thermoplastic starches are still an emerging sector which should continue to grow in the coming decade as they can be an attractive alternative to petrochemical thermoplastics. By blending them with other co-polymers, a wide range of material grades with diverse properties is obtained.

Though polymers obtained by microbial production (processing route II) may offer interesting alternatives to some special applications, this category appears to be relevant to PHAs for which bulk volume applications appear to be a very long-term objective. In contrast, bio-based polymers related to processing route III should play an important role in the coming years for bulk volume applications and low-volume, high added-value applications. This category already comprises relevant bio-based polymers such as polyamide-11 and polylactic acid, which cover a wide range of technical applications; they are both produced in large-scale units and their market share should grow considerably in the coming years.

The purpose of the following sections of this chapter is to focus only on existing and emerging bio-based polymers and related bio-based plastics (whether they are fully produced from renewable resources or not). Natural fibers are included. Bio-based polyolefins, polyvinyls and polyurethanes, with similar characteristics to their petrochemical-based polymer counterparts, are not covered because these polymers as well as their process behavior in extrusion have been extensively presented in numerous books dealing with polymer chemistry and engineering.

4.1.2 Macromolecular characteristics of bio-based polymers

Here we present some of the emerging bio-based polymers which are produced via the processing routes introduced in section 4.1.1.2. Based on technical and scientific developments relative to bio-based polymers and derived plastics, the selection criterion for the polymers covered in this section is the technical substitution potential for 100% oil-sourced plastics in the next 10 years. Proximity to commercialization and to pilot scaling up, as well as the potential market volume, have been considered in making the selection. Only bio-based thermoplastics are included, as today the category of thermoplastics represents approximately 85% of the consumption of petrochemical polymers. Natural fibers are also included in this section, due to their technical substitution potential in polymer reinforcement composites.

4.1.2.1 Starch polymers

Starch is contained in a large variety of plant crops, such as cereals (50–80% dry basis), tubers (60–95% dry basis), and legumes (25–50% dry basis). This is the major storage polysaccharide in plants. The structure and macromolecular characteristics of starch polymers have been described extensively in many books dealing with biochemistry or food chemistry (e.g. Bemiller & Whistler, 2009; Cui, 2005). Thus, the purpose of this section is not to give a complete description of starch polymers but to recall their characteristics which must be kept in mind when dealing with either thermoplastic starches for non-food uses (in the present chapter) or extrusion-textured food products (Chapters 6 and 7).

Starch is stored in the form of granules with characteristics such as morphology, composition, and crystallinity that depend on their botanical origin. The size of starch granules generally ranges from 1 μm to 100 μm for the major sources of starch (maize, wheat, potato, rice, cassava). Starch is composed of a mixture of two polymers: amylose and amylopectin. The most common types of starch granules contain 20–30% amylose and 70–80% amylopectin, as well as limited amounts of lipids and proteins (Buleon et al., 1998). Some mutant genotypes contain less than 1% amylose (waxy maizes) whereas other genotypes may contain as much as 70% amylose (amylomaizes). Amylose is defined as an essentially linear macromolecule of (1–4) linked α-D-glucopyranosyl units (Figure 4.3a) with a molecular weight of 10^5–10^6 g/mol, which corresponds to 500–5000 glucose units per molecule. The amylose macromolecular chains can form single or double helices. In contrast, amylopectin is a highly branched molecule with a molecular weight of 10^7–10^9 g/mol. The macromolecule is based on short chains of (1–4) linked α-D-glucopyranosyl units which are joined together through α (1–6) branch points (Figure 4.3b).

The exact arrangement of chains within the amylopectin is not fully understood but the "cluster model" proposed by Robin et al. (1974) based on the model of French (1972) must be mentioned as it provides a useful conceptual basis for the understanding of the structure of the molecule (Hizukuri, 1986; Thompson, 2000).

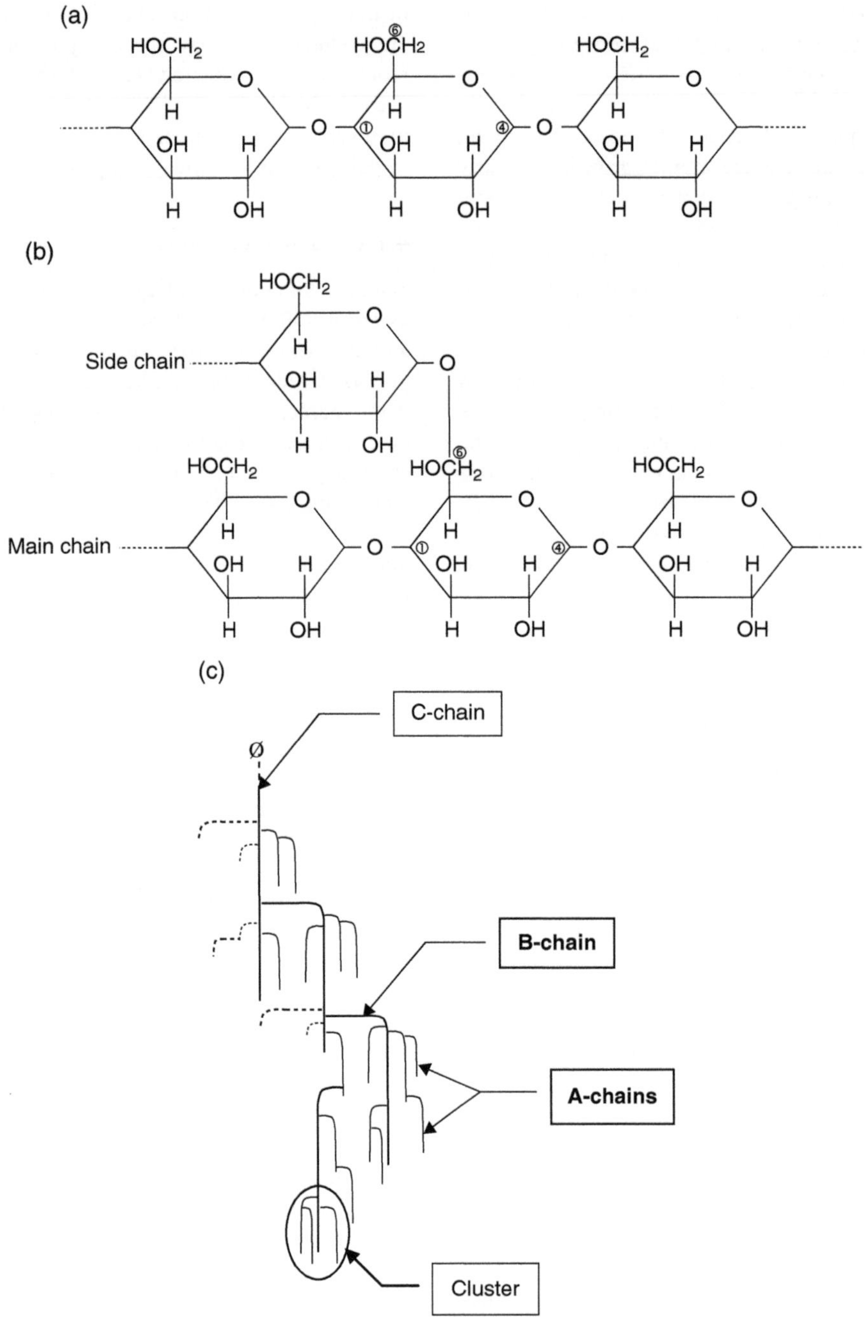

Figure 4.3 Macromolecular characteristics of starch polymers. (a) Amylose. (b) Amylopectin. (c) Cluster model proposed by Robin et al. (1974) for the structure of amylopectin. Source: (c) Adapted from Robin et al. 1974.

The cluster model is presented in Figure 4.3c where the amylopectin chains are classified into A-, B- and C-chains. The non-branched outer A-chains are linked at their potentially reducing group to an inner B-chain. Hence, the B-chains carry multiple other chains as branches while they themselves are linked to the one C-chain of the molecule. Therefore, the C-chain is the only one carrying a reducing group. The A-chains form double helices, which are arranged in clusters at regular intervals through the molecule (Robin et al., 1974).

Starch granules show a high degree of radial organization from a central hilum. This organization depends strongly on the botanical origin. Through the use of novel enzymes involved in starch biosynthesis and degradation, the fine structure of amylopectin has been further elucidated (Bertoft et al., 2012; Seetharaman & Bertoft, 2013). Amylopectin is predominantly responsible for the starch granule crystallinity. Zobel et al. (1988) reported that granule crystallinity varies between 15% and 45%. A-chains of amylopectin form left-handed double helices, which are grouped into crystalline zones (clusters according to the model of Robin et al., 1974). The ordered double helical amylopectin side chain clusters form crystalline lamellae, which are altered with amorphous lamellae consisting of the amylopectin branching regions. The crystalline lamellae (approx. 6 nm) and the amorphous lamellae (approx. 4 nm) form the semi-crystalline growth rings of 120–140 nm with approximately 16 repeat units. The semi-crystalline growth rings themselves are alternated with amorphous growth rings of similar size as illustrated in Figure 4.4. However, it must be noted that there is no clear limit between crystalline areas and amorphous areas. The exact location of amylose, proteins and lipids, and their interaction with amylopectin are still uncertain. It is strongly believed that a large portion of amylose is found in amorphous growth rings, with only small amounts associated with the semi-crystalline growth ring. The amylose molecules present in the semi-crystalline regions could be involved in forming double helices with the amylopectin short chains.

Starch granules show a complex three-dimensional network which consists of microcrystallites cross-linked with amorphous phases where water is located. Several types of crystallinity are observed in the native starch granules, called A-, B-, C-, or V-types (Buleon et al., 1998). A-type crystallites are associated mainly with cereal starches, while B-type crystallites are present generally in tuber starches. C-type crystallites are believed to be a combination of A- and B-types; they are usually associated with pea and various bean starches. A-type crystallites are denser and less hydrated than B-type crystallites. V-type crystallites result from complexes that amylose can form with other components such as the aliphatic tail of lipids. Amylose V-complexes are insoluble in water, even at extreme pressure and temperature conditions. Starch granules exhibit hydrophilic properties and strong intermolecular association via hydrogen bonding due to the hydroxyl groups on the granule surface.

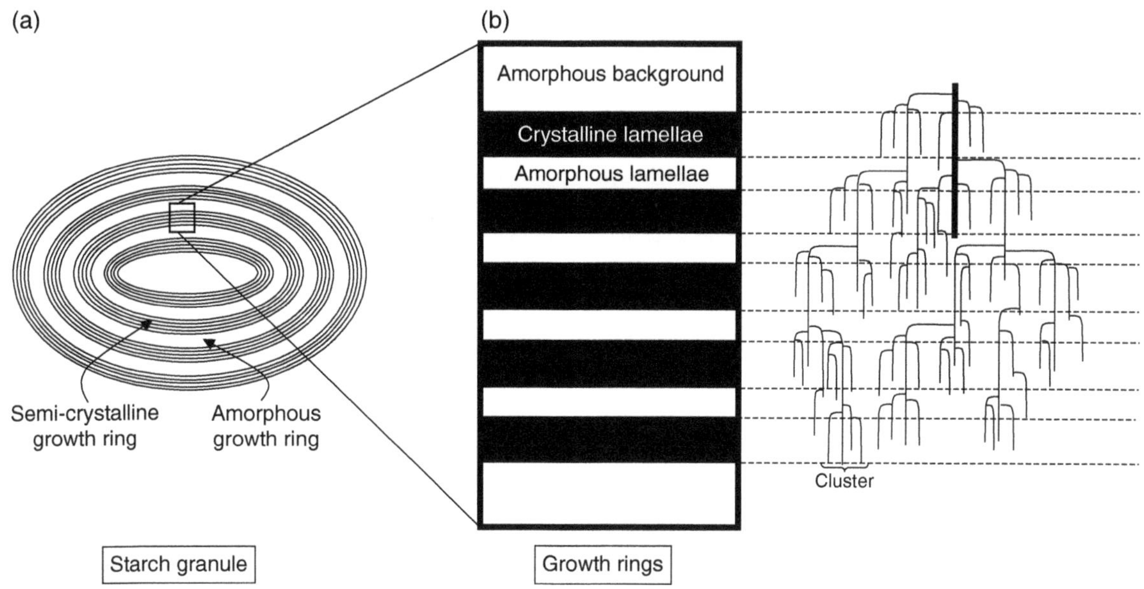

Figure 4.4 Structure of starch granule as proposed by Jenkins (1994). (a) The starch granule composed of concentric semi-crystalline and amorphous growth rings. (b) The alternating crystalline and amorphous lamellae within the semi-crystalline growth rings. Source: Adapted from Jenkins 1994.

The substitution potential of thermoplastic starches for polyolefins is very high as blends of thermoplastic starches with synthetic polyolefins are close to achieving the mechanical properties of pure synthetic polymers. Other applications making use of these materials include biodegradable packaging, fillers, etc. Thus, it is worth examining the behavior of thermoplastic starches in the following sections of the present chapter (see section 4.3.3.2.1).

4.1.2.2 Proteins

Proteins are complex macromolecules that may constitute 50% or more of the dry weight of living cells. They play a fundamental role in the structure and function of cells. Proteins possess an extraordinary diversity of functions, which arbitrarily may be classified in two main categories.

- Structural proteins (keratin, collagen, elastin, etc.) which are present in all tissues such as muscles, bone, skin, cellular membranes, etc. Their functionality is largely related to their fibrous structure.
- Proteins with biological activity which fulfill an active role in all biological processes, including enzymes, hormones, contractile proteins (myosin, actin, and tubulin), transfer proteins (hemoglobin, myoglobin, and transferrin), storage proteins (such as ovalbumin, casein, wheat proteins, soy proteins, zein, etc.).

Unlike polysaccharides which contain only the glucose monomer, different amino acids are the monomers of the protein molecules. For most proteins, the amino acid components belong to a restricted group of 20 different amino acids. These amino acids contain in their molecular structure at least one primary amino group ($-NH_2$) and one carboxyl group ($-COOH$); the primary amino group occupies a α-position with respect to the carboxyl group (Figure 4.5). R_x is a side chain of variable structure. Every amino acid has a characteristic side chain R, which influences its physicochemical properties and, as a result, the properties of the protein to which it belongs. According to the polarity of this side chain, four classes of amino acids can be distinguished.

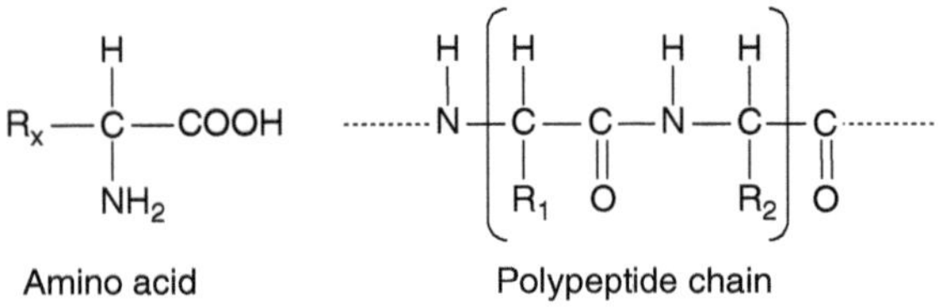

Figure 4.5 Molecular structure of amino acids and polypeptide chains.

- Amino acids with non-polar or hydrophobic side chains. Their hydrophobicity increases with the length of the aliphatic side chain.
- Amino acids with polar, uncharged (hydrophilic) side chains, which possess functional groups such as hydroxyl ($-OH$), amide ($-CO-NH_2$), and ($-SH$) groups able to establish hydrogen bonds.
- Amino acids with positively charged side chains (at pH close to 7).
- Amino acids with negatively charged side chains (at pH close to 7).

Side chains of the different amino acids which constitute protein molecules determine the structure of proteins.

The component amino acids are linked together by amide bonds called peptide bonds, forming polypeptide chains containing up to several hundred units (see Figure 4.5). Each protein is characterized by its conformation, that is, by its three-dimensional organization. Primary structure refers to the sequential order of amino acids in a protein. Secondary and tertiary structures relate to the three-dimensional organization of the polypeptide chain. Quaternary structure refers to the geometric arrangement among various polypeptide chains, which are linked together by non-covalent bonds in most cases. The complex three-dimensional organization of proteins creates:

- *fibrous proteins*, composed of polypeptide chains assembled along a common straight axis, which leads to the formation of fibers (collagen, keratin, elastin)
- *globular proteins*, composed of one or several polypeptide chains folded upon themselves to form a three-dimensional network (spherical and globular forms).

It is worth reviewing briefly the main sources of proteins which are used in film or sheet formation through thermoplastic processing.

- *Soy proteins.* The main constituents of soy proteins are the 7S and 11S fractions which have a molecular weight of 140–180 kDa and 300–400 kDa, respectively. The structure of the 11S fraction is based on ionic bonds and covalent disulfide cross-linking. Soy proteins have a relatively high solubility in water at pH value above or below their isoelectric point.
- *Corn proteins*, mainly composed of low molecular weight (zeins) and high molecular weight (glutelins) components. Corn gluten meal consists of approximately 65% zein and 30% glutelin. The zein fraction can be extracted with aqueous aliphatic alcohols of low molecular weight (soluble in 60–70% ethanol, for instance), and has a molecular weight of 18–45 kDa. Its solubility in water is

poor due to its high content of hydrophobic amino acids. The zein fraction is not constrained by disulfide bonds because the concentration of sulfur-containing amino acids is low (i.e. the number of cysteine amino acid molecule varies from 0 to 3). The glutelin fraction contains all proteins remaining after the extraction of the salt- and alcohol-soluble proteins. Unlike zein, corn glutelins contain significant amounts of cysteine amino acid molecules (i.e. 15–20 potential disulfide bonds per 1000 amino acid molecules), which form extensive cross-linked, three-dimensional networks of high molecular weight.

- *Wheat gluten*. In wheat, the principal storage proteins are the gliadin and glutenin fractions (80–85% of the endosperm protein content). Gliadins are soluble in 70–90% alcohol. Their molecular weight is approximately 36 kDa; the amino acid sequences are homologous and they appear to exist as single polypeptide chains rather than as associated units. Glutenins are insoluble in neutral aqueous solutions, saline solutions, or alcohol. Associated forms of glutenins have an average molecular weight of 150–3000 kDa. Subunits associate by intra- and intermolecular disulfide bonds.
- *Milk proteins*. Milk proteins are classified as either caseins or whey proteins. Casein represents 80% of the total milk protein and consists of α-, β-, and κ-casein with molecular weights ranging between 19 and 25 kDa. Casein contains relatively high levels of proline which leads to its emulsifying properties. Whey proteins (20% of the milk proteins) are water soluble and mainly consist of β-lactoglobulin (molecular weight 18 kDa), α-lactalbumin (molecular weight 14 kDa), and bovine serum albumin (molecular weight 66 kDa). β-Lactoglobulin and α-lactalbumin are globular proteins, with a rather uniform sequence distribution of non-polar, polar, and charged amino acids. Hence, these proteins fold intramolecularly.
- *Myofibrillar proteins*. The main proteins of muscle are myosin (470 kDa) and actin. Myosin contains two identical polypeptide chains, each with a high degree of α-helical structure. The molecule has two globular heads, which are responsible for its ability to interact with actin. Actin probably exists in muscle as double helical structure called fibrous actin or F-actin. Globular actin, or G-actin, is the monomeric form of the protein with a molecular weight of 43–48 kDa. Myosin and F-actin allow the formation of edible film with good mechanical properties.
- *Collagen/gelatin*. Collagen is abundant in tendons, skin, bone, the vascular system of animals, and the connective tissue sheaths surrounding muscle. Collagen is a long cylindrical protein which consists of three polypeptide chains wound around each other in a suprahelical fashion. Each chain has a molecular weight of about 100 kDa, yielding a total molecular weight of 300 kDa for collagen. Collagen molecules are linked end to end and adjacently to form collagen fibrils. When collagen is heated in solution, the fibrils shrink which involves disassembly of molecules in the fibrils and collapse of the collagen structure; non-covalent bonds are affected, and some intermolecular and intramolecular bonds and a few main-chain peptide bonds are hydrolyzed. This results in conversion of the collagen structure to an amorphous form, known as gelatin.

In general, proteins are considered amorphous. They are renewable heteropolymers which are considered as potential raw materials for making plastics, and can be used to produce edible bio-based packaging and materials. Thus, it is worth examining the thermoplastic behavior of proteins in the following sections of the present chapter (section 4.3.3.2.2).

4.1.2.3 Natural fibers

This section covers plant fibers (stem-based, in particular) which are suitable as environmentally friendly reinforcement for composites (Bledzki & Gassan, 1999; Mohanty et al., 2000; Nabi Saheb & Jog, 1999; Wambua et al., 2003). Our attention is mainly focused on non-wood bast fibers (flax, hemp, jute, ramie, etc.) because of their potential to replace glass fibers in reinforced composite materials.

As shown in Figure 4.6, non-wood bast fibers are extracted from the stems of plants. In fact, the stem consists generally of two very different types of fibers: bast and woody core fibers. The woody core fibers are thin-walled and short and have a chemical composition which resembles hard wood – approximately 40% cellulose, 20% hemicellulose, and 20% lignin. Bast fibers are located in the outer layer of the stem, which consists of fiber bundles. Non-wood bast fibers are composed of approximately 60–80% highly crystalline cellulose and other components such as hemicellulose (10–20%), lignin (3–4%), pectin (1–3%), mineral matter, fats, and waxes. The actual composition depends on the plant variety, growth conditions, and method of production; the degree of crystallinity of cellulose ranges from 60% to 97%.

Fiber bundles are composed of overlapping bast fibers and a cross-section consists of 10–40 elementary fibers which represent the plant cell. The length of elementary fibers closely depends on the plant and variety: for flax and hemp, the length is in the range of 4–77 mm and

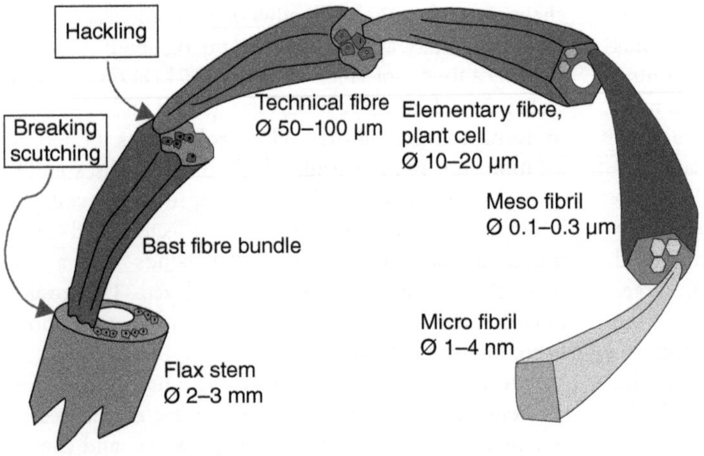

Figure 4.6 Schematic representation of a flax (*Linum usitatissimum* L.) fiber from stem to microfibril. Source: Bos, Müssig & Van den Oever 2006. Reproduced by permission of Elsevier.

5–55 mm, respectively while the equivalent diameter is in the range of 5–76 μm and 10–51 μm, respectively (Baley, 2005). Elementary fibers are composed of mesofibrils (equivalent diameter of 0.1–0.5 μm approximately) and mesofibrils are composed of microfibrils (equivalent diameter of 1–10 nm approximately). In bast fiber bundles, elementary fibers can be considered as a multilamellar structure based on highly crystalline cellulose merged into a polymeric matrix composed of amorphous hemicellulose and lignin, this natural composite being reinforced by covalent bonds, as well as by hydrogen bonds. In this matrix, cellulose microfibrils show a helical, supramolecular structure in which helix angle, the microfibril angle, confers a high elastic modulus to the fiber.

Natural fibers used in composites reinforcement correspond mainly to bast fiber bundles, currently called "technical fibers." For that purpose, they are separated from the stem via an appropriate process which comprises the following three important steps.

- *Retting*, aims at degrading the matrix materials to separate non-fibrous tissues. The retting method is generally based on an enzymatic, pectinolytic treatment.
- *Scutching*, consists of removing the woody part of the retted flax stem.
- *Heckling*, consists of separating the technical fibers, mainly composed of bast fiber bundles.

Are the mechanical properties of natural fibers attractive? According to their structure, they are already composite materials reinforced by cellulose microfibrils and these reinforcement characteristics are mainly determined by the cellulose content, the microfibril angle, and the length-to-diameter ratio. Table 4.1 gives those properties for the following natural fibers: flax, hemp, jute, and ramie. For a given cellulose content, the lower the microfibril angle, the higher the rigidity and strength of the fiber, and the lower the elongation at break. Moreover, the higher the length-to-diameter ratio, the higher the reinforcement characteristics; when the ratio gets close to 1, then fibers act as fillers in the polymeric matrix. Table 4.1 gives the mechanical properties under tensile loading (specific Young modulus, specific tensile strength, and elongation at break) of the four natural fibers, as well as those of the classic reinforcement fibers such as glass and carbon. It appears that the mechanical properties of natural fibers are rather dispersed. Actually, these properties result from various experimental data and for a given type of fiber, they may depend on the variety of the plant, the growth and harvesting conditions, and the postharvesting treatments (Baley, 2005). Nevertheless, Table 4.1 indicates that natural fibers have real potential for composites reinforcement, when comparing them with glass fiber in particular.

Baley (2002) and Charlet et al. (2007) have thoroughly studied the flax plant to investigate the effect of plant variety and the location of fibers in the stem. Their results show that the mechanical properties of flax fibers are increased when fibers are extracted from the middle of the plant. Specific Young modulus and specific tensile strength of fibers extracted from the *Hermès* variety of flax (density of 1.5 $kg.m^{-3}$) are 45 $GPa.m^3/kg^{-1}$ (+/− 24 GPa. m^3/kg^{-1}) and 0.97 $GPa.m^3/kg^{-1}$ (+/− 0.56 $GPa.m^3/kg^{-1}$) respectively; the elongation at break is 2.3 % (+/− 0.6%). Such mechanical properties are very encouraging, and they could even be improved through a dedicated genetic research program.

It is worth mentioning that the thermal stability of natural fibers may constitute a limitation for using them as reinforcement fibers for composites. Table 4.1 shows that the maximum processing temperature is around

Table 4.1 Comparison of reinforcement properties of natural and synthetic fibers.

Fibers	Cellulose content (%)	Microfibril angle (°)	*L/d	Density ($kg.m^{-3}$)	**Specific Young modulus (GPa. m^3/kg^{-1})	**Specific tensile strength (GPa.m^3/kg^{-1})	Elongation at break (%)	***Maximum processing temperature (°C)
Flax	64–71	10	1687	1.53	8–55	0.4–1.3	1–4	300–310
Hemp	78	6.2	960	1.07	32.7	0.36	1.6	160–200
Jute	61–71	8	25–200	1.44	18.4	0.27–0.54	1.5–1.8	160–200
Ramie	83	7.5	16–126	1.56	39–82	0.26–0.60	1.2–3.8	–
E glass (industrial fiber)	–	–	–	2.54	28–29	0.8–0.9	3	350–400
Carbon (Toray t300)	–	–	–	1.7–1.9	121–135	1.8–2.1	1.5	300–310

* Length-to-diameter ratio of the fiber.
** Specific values are given by the ratio of mechanical properties divided by density.
*** Values estimated by Leblanc (2010).
Source: Adapted from Baley 2005.

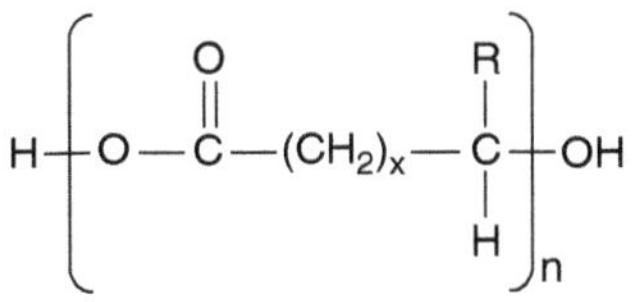

Figure 4.7 Molecular structure of polyhydroxyalkanoates.

180°C for most natural fibers. This estimation is based on the fact that natural fibers generally start losing their rigidity from 160°C, and that lignin polymer degrades at approximately 200°C. However, thermal degradation of crystalline cellulose occurs at temperatures higher than 320°C (Leblanc, 2010). Recently, Bledzki et al. (2008) observed that the degradation temperature of native flax fibers is 319°C, which is surprisingly high.

4.1.2.4 Polyhydroxyalkanoates

Polyhydroxyalkanoates (PHAs) are a class of aliphatic polyesters produced directly by bacterial fermentation of carbon substrates. Figure 4.7 shows the generic formula of PHAs where R can be either hydrogen or hydrocarbon chains of different lengths (methyl, ethyl, propyl, etc.). It must be noted that relevant polymers correspond to $x = 1$. A wide range of PHA homopolymers have been produced at the R&D scale but only a few have been commercialized in recent years; specifically, polyhydroxybutyrate homopolymer, or PHB ($R = -CH_3$), can be mentioned. PHA co-polymers have also been commercialized in the last decade. This includes the following co-polyesters: polyhydroxybutyrate-co-hydroxyalkanoates, such as polyhydroxybutyrate-co-hydroxyvalerate, or PHBV ($x = 1$; $R = -CH_3$ and $-CH_2CH_3$); polyhydroxybutyrate-co-hydroxyhexanoate or PHBHx ($x = 1$; $R = -CH_3$ and $-CH_2CH_2CH_3$).

The molecular weight of PHAs depends on the bacteria and conditions of synthesis, as well as on the separation and purification processes. PHAs are partially crystalline polymers with crystallinity within a range typical of thermoplastics. PHB and PHBV show a crystallinity of 60% and 51%, respectively (Chodak, 2008).

The production of PHAs consists of a three-step processing route: fermentation, extraction, and purification. In the fermentation step, the bacteria can be fed with a range of different carbon sources such as corn sugar, cane sugar, vegetable oil, fatty acids, etc. However, feedstocks currently used for PHA production are high-value substrates, and the challenge is to use lower value feedstock such as lignocellulosics which would require thermomechanical or enzymatic pretreatment. It must be noted that PHAs are still in an early stage of commercialization, and successful development of this class of polyesters would definitely require further improvements in fermentation yields and downstream extraction methods, to be economically viable. The various types of

PHAs are suitable for thermoplastic conversion technologies (such as injection molding, melt blowing, melt-spun fibering, thermoforming, etc.; Shen et al., 2009). They show potential technical substitution for polyvinyls (PVC) and polyolefins (high- and low-density polyethylene, polypropylene). They also show promising potential applications in the medical field in particular, due to their good biocompatibility and absorbability in human tissue.

Although bulk volume applications of PHAs will take time to develop, it is worth examining their thermoplastic behavior in the following sections (see section 4.3.3.2.3).

4.1.2.5 Bio-based polyamides

Polyamides are thermoplastics that have recurring amide groups (–CONH-) as an integral part of the main macromolecular chain. Polyamides are synthesized either from the polycondensation between amine and carboxylic groups (diamines and dicarboxylic acids, amino acids) or from the ring opening polymerization of lactams (Figure 4.8).

Table 4.2 shows the bio-based or partially bio-based polyamides commercially available today (PA-11 and PA-6.10), and the potential polyamides that may be produced from bio-based raw materials in the coming years (PA-6.6, PA-6.9 and PA-6) (Shen et al., 2009). PA-11 is the first 100% bio-based polyamide which results from the polycondensation of 11-aminoundecanoic acid monomer, which is produced from castor oil through a multistep chemical synthesis. PA-6.10 is a 62% partially bio-based polyamide (62% of biomass carbon) which results from the polycondensation of hexamethylenediamine (petrochemically produced from butadiene or propylene) and sebacic acid (bio-sourced from castor oil). PA-6.6 and PA-6.9 result from the polycondensation of hexamethylenediamine and adipic acid (PA-6.6) or azelaic acid (PA-6.9), respectively. Partially bio-based PA-6.6 and PA-6.9 will be available when adipic acid and azelaic acid can be produced via bio-based routes: adipic acid via fermentation of sugar (Li & Frost, 1999; Niu et al., 2002) or via chemical synthesis from oleic acid which is abundant in vegetable oils (Zahardis & Petrucci, 2007). PA-6 results from conventional ring opening polymerization of ε-caprolactam, which may be produced by fermentation of glucose having lysine as the precursor (Frost, 2005). Bio-based processing routes to produce adipic acid, azelaic acid, and ε-caprolactam are still at the R&D level, and when these routes will be economically feasible and successful at an industrial scale is unpredictable. However, these examples show the real potential of bio-based routes as future alternatives to conventional petrochemical routes.

An important specificity of polyamides is the presence of intermolecular hydrogen bonds due to the amide groups. The occurrence of hydrogen bonds depends upon the repeat units in the macromolecular structure: the higher the repeat units, the higher the density of hydrogen bonds. Of course, this affects the degree of

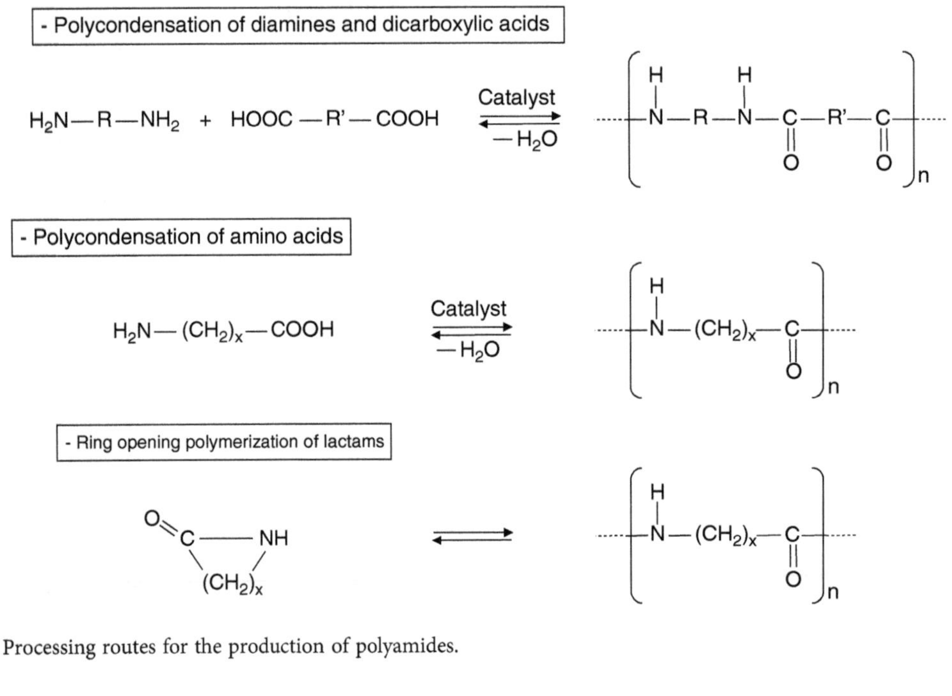

Figure 4.8 Processing routes for the production of polyamides.

Table 4.2 Potential bio-based or partially bio-based polyamides.

Polyamide	Monomers	Raw materials
*PA-11	11-aminoundecanoic acid $[H_2N-(CH_2)_{10}-COOH]$	Castor oil
*PA-6.10	**a.** Hexamethylenediamine $[H_2N-(CH_2)_6-NH_2]$	Butadiene, propylene
	b. Sebacic acid $[HOOC-(CH_2)_8-COOH]$	Castor oil
**PA-6.6	**a.** Hexamethylenediamine $[H_2N-(CH_2)_6-NH_2]$	Butadiene, propylene
	b. Adipic acid $[HOOC-(CH_2)_4-COOH]$	Glucose
**PA-6.9	**a.** Hexamethylenediamine $[H_2N-(CH_2)_6-NH_2]$	Butadiene, propylene
	b. Azelaic acid $[HOOC-(CH_2)_7-COOH]$	Oleic acid
**PA-6	ε-Caprolactam $[(CH_2)_5-NH-CO]$	Glucose

* Commercially available.
** At the R&D level, to be developed.

crystallinity and the properties of polyamides. The degree of crystallinity of PA-6.6, PA-6.10 and PA-6 is between 40% and 50% while it is in the range of 20–25% for PA-11. Key properties of polyamides are resistance to oils and solvents, toughness, fatigue and abrasion resistance, stability at high temperatures, and mechanical characteristics. Increasing crystallinity would result in higher density, stiffness, tensile and yield stress, chemical and abrasion resistance, dimensional stability but lower elongation, impact resistance, thermal expansion, and permeability.

Bio-based polyamides have a very high substitution potential for their petrochemical equivalents in several application areas, including engineering plastics (automotive industry, electrical and electronics, packaging, construction) and fibers.

4.1.2.6 Polylactic acid

Lactic acid, or 2-hydroxypropionic acid, can be produced by anaerobic fermentation of sugar solution (hexoses) and further purification. As lactic acid has an asymmetrical carbon, enantiomeric monomers, either D- or L-lactic acid, can be obtained by using an appropriate lactobacillus.

Two main processing routes are available to convert lactic acid to polylactic acid (Figure 4.9).

- Direct polycondensation of lactic acid; this route is not industrialized.
- Indirect processing route via the cyclic dimer (lactide) of lactic acid, and further ring opening polymerization to polylactic acid.

The second route is a continuous process which consists of oligocondensation of lactic acid to produce low molecular weight oligomers with *n*' ranging between 30 and 70. Oligomers are then depolymerized by changing operating conditions (temperature increase and pressure decrease) to give a mixture of lactide stereoisomers. After purification, the lactide is finally polymerized via catalyzed ring opening polymerization. By controlling the process chemistry, it is possible to select the stereoisomer of the lactide intermediates and thereby the stereoisomeric forms of polylactic acid polymer, hence leading to poly(L-lactic acid) or PLLA, and to poly(D-lactic acid) or PDLA. PLLA and PDLA are partially crystalline (C-PLA; approximately 40–60% crystallinity) while poly(D,L-lactic acid) is rather amorphous: it is called A-PLA. Figure 4.9 shows the processing routes for PLLA. Stereocomplexation between PLLA and PDLA (sc-PLA) may also be formed as long as L-lactide unit sequences and D-lactide sequences co-exist in a polymeric system.

The possibility to play with the two mesoforms, D and L, allows polymer producers to achieve the desired properties of PLA and therefore to produce value-added products. This reinforces the substitution potential of PLA for petrochemical thermoplastics. PLA producers agree on the potential for PLA to partially replace low- and high-density polyethylene, polypropylene, and polyethylene terephthalate, in particular. PLA has already gained real market importance and has been used for a wide range of applications, such as packaging, textiles, non-wovens, electronics, and agriculture. Examples of major end-products include extruded sheet for thermoformed products, biaxially oriented films, blow molded bottles, injection molded products, and fibers (Shen et al., 2009). PLA is considered in section 4.3.3.2.3.

4.1.2.7 Polytrimethylene terephthalate

Polytrimethylene terephthalate, or PTT, is a linear aromatic polyester produced by polycondensation of 1,3-propanediol with either purified terephthalic acid

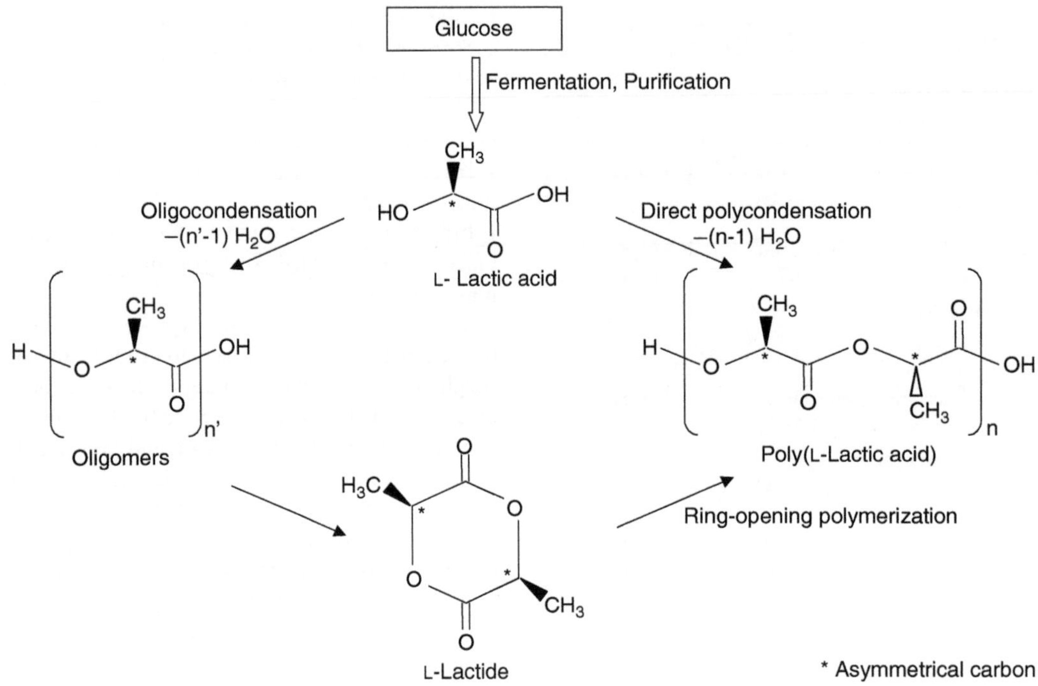

Figure 4.9 Processing routes for the production of poly(L-lactic acid).

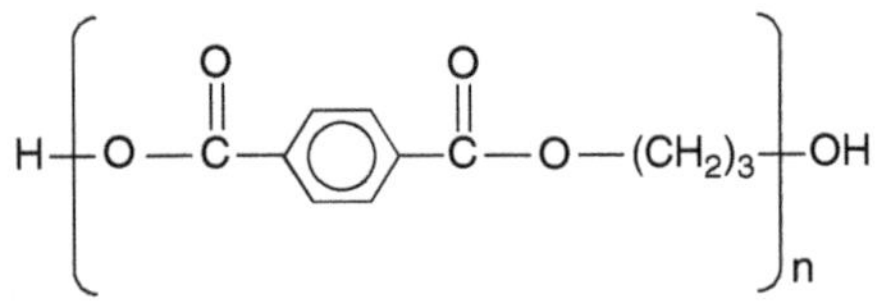

Figure 4.10 Molecular structure of polytrimethylene terephthalate.

(via esterification) or dimethyl terephthalate (via transesterification) (Brown et al., 2000). Figure 4.10 shows the formula of PTT.

Polytrimethylene terephthalate can be partially bio-based polyester when starting with bio-based 1,3-propanediol. In fact, 1,3-propanediol can be produced by aerobic fermentation of glucose, leading to high yield and productivity, hence giving real potential for bulk production of bio-based 1,3-propanediol (DuPont, 2007). In recent years, it has been proven that production of 1,3-propanediol from glucose is commercially viable.

In general, PTT is similar to other polyesters in average molecular weight and molecular weight distribution. It has a partially crystalline structure, with a crystallinity of approximately 25–30% (Wu et al., 2001). As PTT combines properties of polyethylene terephthalate (PET) and polybutylene terephthalate (PBT), its substitution potential is very high for nylon and PBT, and moderately high for PET. Moreover, since PTT production is similar to that of PET, it is possible to convert existing PET facilities to PTT production, which boosts the substitution potential. Thermoplastic behavior of PTT is included in following sections of this chapter.

4.2 Melting mechanism of polymer materials in screw extruders

During thermomechanical plasticating of polymers in screw extruders, between the feed hopper and the die a solid polymer is converted into a viscous melt. The melting zone of the screw extruder is defined as that portion of the screw (s) in which solid polymer and polymer melt co-exist. Polymer melting has a profound effect on both extruder performance and extrudate quality. Thus, a good understanding of the melting mechanism in screw extruders is of paramount importance, as this will allow the screw design and operating parameters to be controlled and optimized, in order to obtain good melt characteristics (free from unmelted material, homogeneous, no or limited polymer degradation).

The melting mechanism in screw extruders depends upon the type of extruder used, either the single screw extruder or the intermeshing co-rotating twin screw extruder. Melting mechanisms in these extruders are presented in this section. The sensitivity of the melting mechanism to process parameters is then analyzed and discussed.

4.2.1 Melting mechanism in single screw extruders: qualitative description

The melting mechanism in single screw extruders was first investigated by Maddock (1959); relevant experimental observations of the progress of polymer melting based on a visualization technique of cross-sections of the screw channel, at various axial distances, were reported. A few years later, Tadmor et al. (1967) performed extensive experimental investigations on melting in single screw extruders.

Experimental studies allow the progress of melting in single screw extruders to be illustrated, from the entry to the melting zone (Figure 4.11a) to complete melting of polymer pellets (Figure 4.11d). As soon as the polymer pellets touch the hot barrel wall, they melt and form a thin film of melted polymer between the barrel and the compacted bed of polymer pellets (Figure 4.11b). The melt film then grows and when its thickness δ_f becomes larger than the flight clearance δ, the melt flows towards the active flight flank of the screw channel (advancing flight), hence forming a melt pool in the rear of the channel in front of the advancing flight (Figure 4.11c). Since viscous heat generation occurs mostly in the upper melt film, most of the melting occurs at the barrel surface which feeds the melt pool. Therefore, the melt pool exerts a high pressure which deforms and forces the solid bed of polymer pellets towards the trailing flight flank; the solid bed of polymer is then continuously rearranged, while maintaining a constant height. A circulating flow is set up in the melt pool as a result of the relative velocity of the barrel and the root of the screw. As melting proceeds, the melt pool gradually grows at the expense of the solid bed of polymer until eventually the melted material fills the entire cross-section of the screw channel. The total length from the beginning of melting to the point where the width of solid bed of polymer drops to zero is called the "length of melting."

There are two sources of heat used for melting in the screw extruder. The first source of heat comes from the external barrel heaters. The second source comes from the mechanical energy supplied by the screw which is transformed into heat due to the high shear developed in the melt film. The extent of both heat sources depends on the operating conditions and the polymer properties. At high screw speed and low barrel temperature, most of the heat comes from viscous dissipation (autogenous

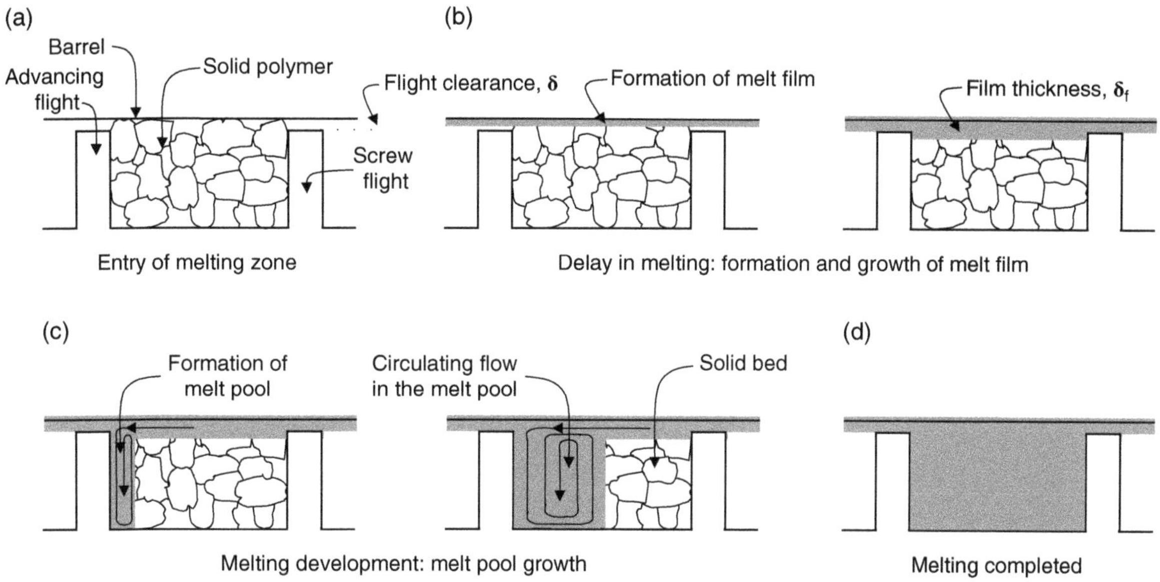

Figure 4.11 Melting mechanism of polymers in a single screw extruder.

operating). Conversely, at low screw speed and high barrel temperature, the heat required for melting may come from heat conduction through the entire thickness of the barrel and the melt film. In practice, viscous heat generation is much more efficient. In fact, due to the melting process which provides very high temperature gradients over a very short distance (the thickness of the melt film), a large amount of heat is generated by viscous dissipation. Furthermore, viscous heat generation is relatively uniform throughout the material, which considerably limits polymer degradation compared to heat conduction sourcing. Finally, the rate of melting would be far higher with viscous heat generation than without.

4.2.2 Engineering analysis of polymer melting in single screw extruders

In the 1960s, Tadmor performed a theoretical analysis of polymer melting in single screw extruders (Tadmor & Klein, 1970), and developed the Tadmor melting model which is classically used to give a quantitative description of polymer melting in single screw extruders.

In Tadmor's model, as already presented in Chapter 3 (section 3.2.3.1), the screw channel is well approximated by a rectangular prismatic channel. The channel is assumed to be stationary and the barrel rotating. Hence, the screw-barrel assembly is approximated by a flat plate (the unwound barrel) which is moving at constant velocity V_b over a rectangular channel (the unwound screw channel) at an angle θ (the helix angle) with respect to the down-channel direction (Figure 4.12).

This model is based on a differential mass and heat balance at the solid–melt interface in the unwound screw channel. For that purpose, a differential prismatic volume with a cross-section perpendicular to the solid–melt interface is considered. In the following description, the main results of the Newtonian melting model are presented. Readers interested in a full description of Tadmor's model should refer to pages 107–183 of the book published by Tadmor and Klein (1970).

4.2.2.1 Melting rate

The quantity of material which is converted from the solid state to the melt state over time can be estimated by the rate of melting per unit down-channel distance, $\dot{q}_m$, as:

$$\dot{q}_m = \Phi\sqrt{X} \tag{4.1}$$

where X is the width of the solid bed (in m) and the parameter Φ expressed as:

$$\Phi = \sqrt{\frac{V_{bx}\rho_m\left[\lambda_m(T_b - T_m) + \frac{\mu_a}{2}\Delta V^2\right]}{2\left[c_s(T_m - T_s) + \Delta H_m\right]}} \tag{4.2}$$

where:

V_{bx} is the velocity component of barrel in the x-direction (in m.s^{-1})

ΔV is a vector difference given by: $\Delta V = \left(V_b^2 + V_{sz}^2 - 2|V_b||V_{sz}|cos\,\theta\right)^{1/2}$ (Tadmor and Klein, 1970); with V_{sz}, the velocity of the solid bed in the down-channel direction z:

$$V_{sz} = \frac{Q_s}{\rho_s HX} = \frac{Q}{\rho_s H_o W} \tag{4.3}$$

V_{sz} is assumed to be constant; Q_s, Q, ρ_s, H_o and W are the mass flow rate of solid polymer, the mass flow rate fed into the extruder, the density of solid polymer, the height of the solid bed at $z = 0$ and the width of the screw channel, respectively. H is assumed to be constant and equal to the flight depth H_o.

T_b is the barrel temperature (in °C).

T_m is the polymer melting temperature (in °C).

T_s is the initial temperature of the solid polymer (in °C).

ΔH_m is the enthalpy of melting of the polymer (in J.kg^{-1}).

μ_a is the apparent viscosity of polymer melt (in Pa.s).

λ_m is the thermal conductivity of polymer melt (in W.m^{-1}.°C^{-1}).

ρ_m is the density of polymer melt (in kg.m^{-3}).

c_s is the specific heat of solid polymer (in J.kg^{-1}.°C^{-1}).

Φ is expressed in kg.m$^{-1.5}$ s^{-1}; it is a measure of the melting efficiency at given thermomechanical processing conditions. The numerator of Eq. 4.2 corresponds to the rate at which heat is being supplied for melting through conduction and viscous dissipation while the denominator corresponds to the heat required for converting the solid polymer at T_s to a melt at temperature T_m. For a given bio-based polymer, when referring to the group of variables of the numerator, there are potentially two possible ways to increase the rate of melting.

- Increase the barrel temperature, T_b, up to a value compatible with the thermal stability of the polymer. When the barrel temperature is increased, the heat conduction term increases. However, due to the temperature dependence of polymer melt viscosity, the viscous dissipation term consecutively decreases. Thus, the decision to act

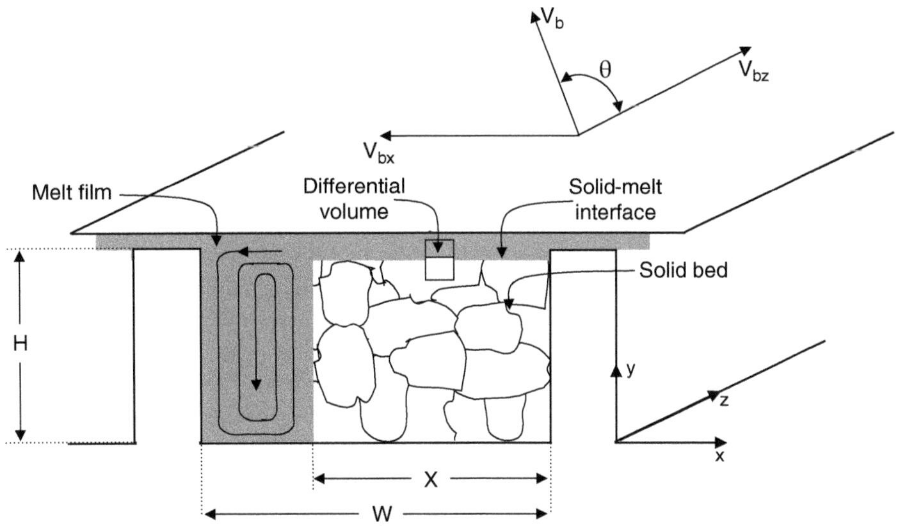

Figure 4.12 Cross-section of unwound screw channel in the melting zone of a single screw extruder. Source: Adapted from Tadmor & Klein 1970.

on the barrel temperature increase should account for the temperature dependence of the polymer melt viscosity to ensure an increased melting rate. There is an optimum barrel temperature for a maximum rate of melting.

- Increase the screw speed, N. When the screw speed is increased, the viscous dissipation term increases proportionally to N^2 but the effect is somewhat diminished by the rheological shear-thinning behavior of most polymer melts. Increasing screw speed also increases the velocity component of the barrel in the x-direction, which in turn increases the rate at which the melt is removed from the melt film and fed to the melt pool; this effect is proportional to N. Consequently, an increase of screw speed should increase the rate of melting. However, it must be noted that in normal operation of a single screw extruder, increased screw speed results in an increased flow rate in the extruder, thus invariably increasing the melting section of the screw extruder. Thus, increasing the screw speed would require care to ensure an increased rate of melting.

4.2.2.2 Solid bed profile

According to the melting mechanism, the solid bed of polymer occupies the forward portion of the channel, and has a rectangular cross-section. Eq. 4.1 indicates that the local melting rate is directly determined by the width of the solid bed. As melting proceeds, its width X decreases as a function of the down-channel distance, z, as follows:

$$\frac{X}{W} = \left(1 - \frac{\Phi z}{2\rho_s V_{sz} H W^{1/2}}\right)^2 \tag{4.4}$$

For a constant channel depth all along the melting zone (no compression), it is shown that the width of the solid bed decreases parabolically with the down-channel distance, z. As a consequence, the local melting rate will decrease accordingly. The highest melting rate is achieved at the start of melting, and then it drops with the down-channel distance in such a way that very long channel length is needed for complete melting, which means long barrel extruders. But, in practice, screw extruders have limited length while full melting of polymers is still required, free from unmelted material in the melt. In fact, the presence of unmelted material in polymeric melts would lead to products with depressed characteristics, for instance, bio-based plastics having poor physical and mechanical properties and food products with poor textural properties. Consequently, the length of melting zone for a given flow rate must be minimized while ensuring full melting of polymers. This is an important consideration in screw design and screw configuration which necessitates technical improvements and modifications to obtain a higher melting rate.

The key objective in improving melting performance is to maintain a relatively wide solid bed, over a substantial length of the screw channel. One way of reaching this objective consists of designing a screw which has a linearly decreasing channel depth such as:

$$H = H_o - Az \tag{4.5}$$

where A is the degree of tapering in the down-channel direction. Such screw design corresponds to monobloc plasticating single screw extruders described in section 2.1.2.1. In this case, the solid bed profile is given by the following equation:

$$\frac{X}{W} = \left[\frac{\Phi}{AV_{sz}\rho_s W^{1/2}} - \left(\frac{\Phi}{AV_{sz}\rho_s W^{1/2}} - 1\right)\left(\frac{H_o}{H_o - Az}\right)^{1/2}\right]^2 \tag{4.6}$$

The solid bed profile is a function not only of z, but also of A. Figure 4.13 shows the solid bed profile as a function of down-channel distance, for different values of A; A_l is defined in section 4.2.2.3. It is shown that melting proceeds more quickly with decreasing channel depth ($0 < A < A_l$), compared with a constant channel depth ($A = 0$).

From Eq. 4.6, it can be observed that an idealized condition is reached when the term $\Phi/(AV_{sz}\rho_s W^{1/2})$ becomes unity. Under this condition, the solid bed width is independent of down-channel distance and the melting rate is optimal. It remains constant and equal to unity until the melting is completed, and at this point it becomes zero, hence giving a solid bed profile equivalent to a step function (Figure 4.13; $A = A_l$). However, this condition cannot be achieved in practice because the channel is completely filled with solid polymer, leaving no room for the melt.

With a modular design of screw-barrel assembly which has constant channel depth (see the description in section 2.1.2.2), the screw pitch may be decreased in the compression section, in order to increase the width of the solid bed as melting proceeds. In practice, this consists of setting positive conveying screw elements with decreased pitches in the down-channel direction. In the melting zone, each time the mix of melt and solid polymer enters into a screw element with decreased pitch, a sharp increase of the local melting rate should be observed which globally should lead to a higher melting rate compared to that of a melting zone having screw elements with constant pitch and channel depth.

The use of multiple flighted screws may also be considered to increase the width of the solid bed in the melting process. Of course, this can be considered for monobloc and modular designs of screw-barrel assembly. The effect of the number of parallel flights on melting has been analyzed by Tadmor and Klein (1970). Actually, based on computer simulations, these authors have shown that the optimum is two channels in parallel, which resulted in a slight increase of the melting rate.

4.2.2.3 Length of melting

For a constant channel depth in the melting zone, $H = H_o$, complete melting is obtained when $X = 0$. Then, Eq. 4.4 gives the total length of melting, $Z_t(A = 0)$, or $Z_t(0)$:

$$Z_t(0) = \frac{2\rho_s V_{sz} H_o W^{1/2}}{\Phi} = \frac{2Q}{\Phi W^{1/2}} \tag{4.7}$$

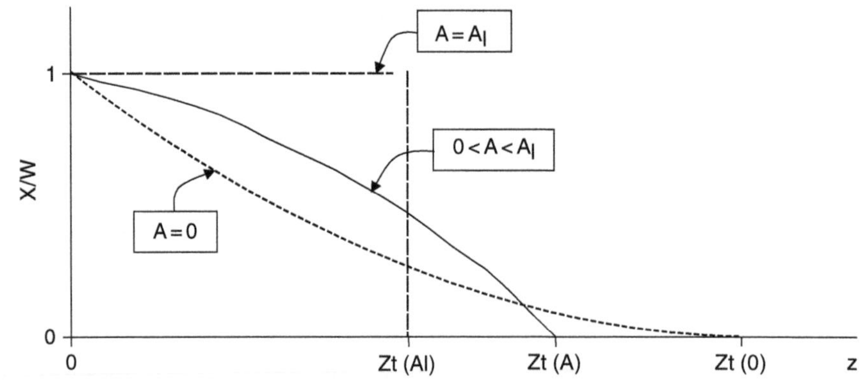

Figure 4.13 Solid bed profile as a function of down-channel distance (z) and degree of tapering (A).

The length of melting is proportional to the mass flow rate of the extruder and inversely proportional to the rate of melting. Thus, an increase of flow rate would need concomitant adjustments of other processing variables, to maintain full polymer melting and product characteristics. Figure 4.13 shows the solid bed profile for a constant channel depth ($A = 0$), and the total length of melting $Z_t(0)$ at $\frac{X}{W} = 0$. For a decreasing channel depth, the total length of melting $Z_t(A)$ is given by setting $X = 0$ in Eq. 4.6:

$$Z_t(A) = \frac{Q}{\Phi W^{1/2} H_o}\left(2 - \frac{AQ}{\Phi W^{1/2} H_o}\right) \quad (4.8)$$

Substituting Eq. 4.7 into Eq. 4.8, the latter equation becomes:

$$Z_t(A) = Z_t(0)\left(1 - \frac{A}{4H_o} Z_t(0)\right) \quad (4.9)$$

It is shown that the melting rate reaches a limit value for the degree of tapering $A_l = 2H_o/Z_t(0)$, for which $Z_t(A_l) = Z_t(0)/2$ (see Figure 4.13).

4.2.2.4 Temperature profile in the melt film

The temperature profile in the melt film is only a function of y. It is obtained from a differential heat balance in the melt film and given by the following equation:

$$T = -\frac{Br(T_b - T_m)}{2\delta_f^2} y^2 + \left(1 + \frac{Br}{2}\right)\frac{(T_b - T_m)}{\delta_f} y + T_m \quad (4.10)$$

where Br is the Brinkman number which represents the ratio between the viscous dissipation to the heat conduction resulting from the temperature difference $(T_b - T_m)$ imposed between the barrel wall and the solid–melt interface:

$$Br = \mu_a \Delta V^2 / \lambda_m (T_b - T_m) \quad (4.11)$$

Based on the temperature profile in the melt film, two situations can be distinguished:

- $Br < 2$: the temperature in the melt film is below the barrel temperature for all positions
- $Br > 2$: the temperature in the melt film can reach a maximum temperature at a position intermediate between the barrel wall and the solid–melt interface. Depending upon the value of Brinkman number, this temperature can be significantly higher than the barrel temperature.

As melting proceeds, one way to control the film temperature is to control the input of viscous dissipation through the relative velocity of the barrel, through the screw speed. In melting processes occurring in screw extruders, the Brinkman number, which is developed in the melt film, is usually higher than 2.

4.2.2.5 Effect of flight clearance

It is of paramount importance to examine the effect of this geometrical parameter on melting. In fact, in the engineering analysis developed in section 4.2.2.1, a zero flight clearance is assumed, which can be acceptable for a new screw and barrel but not for used equipment showing some wear. In practice, as discussed in section 2.2.3, some wear is shown along the screw-barrel assembly of single screw extruders, particularly in those sections where pressure build-up is developed such as in the melting and pumping sections. For instance, in the melting zone, wear is mainly due to abrasion. Thus, due to extruder operation, wear may appear and extend, and flight clearance δ increases.

The flight clearance affects the melting performance significantly. First, it increases the delay in the start of the melting which increases proportionally to flight clearance increases, or to the extent of wear; the larger the flight clearance, the longer the delay. Second, flight clearance decreases the melting rate, as it allows melt exchanges across the flight, thus changing the mass balance in the film to the detriment of the production of melt from the solid bed. It is shown that the melting rate decreases monotonically as the flight clearance increases. The melting model predicts that doubling and tripling the standard clearance cause a reduction in melting rate of about 25% and 35%, respectively (Rauwendaal, 2001). Thus, screw wear negatively affects the rate of melting and consequently ineluctably reduces overall performance of single screw extruders, via reduction of extruder productivity as well as decrease of product characteristics.

4.2.3 Melting mechanism in intermeshing co-rotating twin screw extruders

The flow of molten polymers in intermeshing co-rotating twin screw extruders has been studied more recently than that of single screw extruders. This was discussed in

Chapter 3 (section 3.2.3). However, the melting mechanism in such extruders has been investigated more recently, and though recent studies provide relevant data, more investigations are required to provide a complete description and modeling of polymer melting in these extruders.

In intermeshing co-rotating twin screw extruders, polymer melting usually occurs in the first pressure build-up section or the first processing section of the screw configuration. In this section, the screw configuration is generally either a set composed of right-handed screw elements followed by a left-handed screw element, or a set composed of right-handed screw elements followed by a kneading block and occasionally a short left-handed screw element. The length of the screw elements and the design of the kneading block (total length, thickness of kneading disks, staggering angle, in particular) are adjusted depending upon the characteristics of the polymeric material and the flow rate of the extruder.

Experimental results of polymer melting in intermeshing co-rotating twin screw extruders appeared in the 1990s. Todd (1993) presented experimental observations relative to the melting of PVC in an extruder configured with a kneading block, and equipped with a clam-shell barrel which allowed the evolution of polymer melting to be observed after stopping the process and cooling the barrel. In following years, several authors reported experimental observations by use of Todd's technique, by screw extractions, or by direct observations through glass windows. These studies mainly focused on melting sections equipped with a screw profile composed of right-handed screw elements followed by kneading blocks with right-handed staggering angles. All these studies showed that polymer melting in twin screw extruders occurred over very short lengths. Melting started in right-handed screw elements and ended in kneading blocks. The experimental studies have shown that the melting mechanism in twin screw extruders was totally different from that in single screw extruders. Instead of continuous formation of a melt film on the barrel wall and a clear partition between a melt pool and a solid bed of polymer pellets in the channel, in twin screw extruders a mix of melt and solid polymer pellets soon appeared in the channel, each solid polymer pellet being well surrounded by the molten material. As melting proceeds, the studies reported that the spaces between the solid pellets were progressively filled with molten material to give a solid-rich suspension at the beginning of the process which evolved to a melt-rich suspension, and finally to a totally molten phase.

More recently, Vergnes et al. (2001) carried out an extensive experimental study of the melting process in an intermeshing co-rotating twin screw extruder, including dead-stop experiments and sampling. The authors developed an experimental procedure which consisted of introducing small amount of black pigments with solid pellets after reaching steady-state extrusion processing conditions. Then, screw rotation and feeding were suddenly stopped, and the extruder barrel was quickly cooled by water circulation. After a few minutes, the barrel was removed by use of a hydraulic system, allowing direct access to the screws. Product samples were collected from the screws and cross-sections of these samples were obtained at different positions of the down-channel distance and analyzed by image analysis. The molten material appeared as black and the solid polymer as white. This type of experiment allowed the authors to investigate the effects of the screw profile in the melting zone (two sets of right-handed and left-handed screw elements, one set of right-handed screw element and kneading block followed by a short left-handed screw element), and of the main process parameters (screw speed and flow rate).

Figure 4.14 shows the melting zone of an intermeshing co-rotating twin screw extruder after barrel opening. From the inlet to the outlet, the melting zone was equipped with a right-handed screw element, then a kneading block and finally a left-handed screw element (total L/D of the melting zone: 3.6). Figure 4.14 shows clearly that polymer melting proceeds over a very short barrel length, from 3 to 4D. The qualitative observations made on the different cross-sections of samples collected from the screws allow establishment of a hypothesis for the melting mechanism which is illustrated in Figure 4.15, from the entry of the melting zone (Figure 4.15a) up to complete melting of polymer pellets (Figure 4.15f). Figure 4.15 refers to a screw profile in the melting zone composed of a right-handed screw element followed by a left-handed screw element. With this profile, the melting process proceeds as follows.

- *Figure* 4.15*b*: a thin film of molten polymer appears when contacting the barrel wall that has a temperature higher than the melting temperature of the polymer melt. This film starts to wet solid pellets which begin to form clusters, each pellet keeping its initial form. In this step, the channel is only partially filled and the pellets are not yet under pressure.
- *Figure* 4.15*c*: total filling of the channel appears in the right-handed screw element, one or two screw turns upstream from the left-handed screw element. This generates a pressure build-up and compaction of the solid

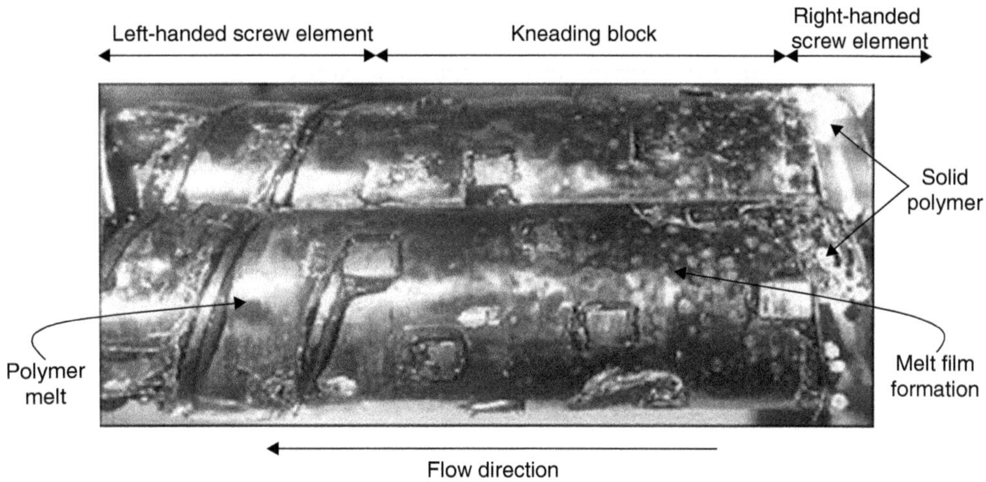

Figure 4.14 View of the melting zone of an intermeshing co-rotating twin screw extruder after barrel opening. Source: Vergnes et al. 2001. Reproduced with permission of Carl Hanser Verlag.

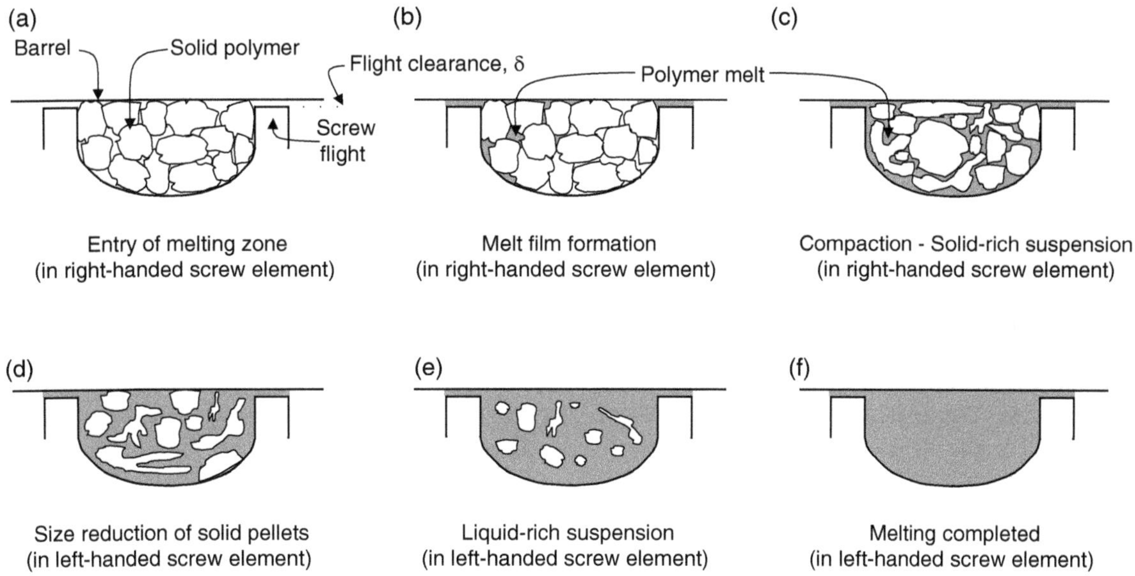

Figure 4.15 Melting mechanism of polymers in an intermeshing co-rotating twin screw extruder.

pellets, accompanied by important plastic deformation. The space between the deformed pellets is totally filled with molten polymer. A solid-rich suspension is obtained; it has reached the transition between the right-handed and left-handed screw elements.

- *Figure* 4.15*d*: the size of solid pellets surrounded by molten polymer decreases progressively.
- *Figure* 4.15*e*: a liquid-rich suspension is obtained.

Replacing the left-handed screw element by a kneading block does not significantly modify the melting mechanism. With a screw profile composed of a right-handed screw element followed by a kneading block, progressive reduction of the solid phase occurs in the kneading block.

Vergnes et al. (2001) quantified the evolution of the melting process, through the evolution of the solid fraction as a function of the down-channel distance; the solid fraction was calculated by image analysis of the different cross-sections of product samples collected from the screws. These authors have shown that for all processing conditions, the solid fraction decreased quasi-linearly as the down-channel distance increased, which leads to a constant melting rate; this result explains an important difference with the melting rate observed in single screw extruders. The solid fraction at the transition between the right-handed screw element and the restrictive element (left-handed screw element or kneading block) approximately ranged between 65% and 55%, which means that one-third to almost half of the solid pellets were molten in the right-handed screw element upstream from the restrictive element. Furthermore, the experiments have shown that decreasing the flow rate or increasing the screw speed, while all other processing variables remained unchanged, led to quicker melting, as the slope of the solid fraction versus down-channel distance was higher at lower flow rate and higher screw speed. The authors have attributed the increase of melting rate to higher viscous dissipation, or higher specific mechanical energy (SME) input when decreasing flow rate or increasing screw speed. Actually, the melting rate is very much dependent on the SME; the higher the SME, the quicker the melting.

The staggering angle of kneading disks did not have a significant influence on melting rate. This geometrical parameter affected only the extent of solid fraction at the transition position between the right-handed screw element and the kneading block; a neutral 90° staggering angle led to lower solid fraction (or higher extent of melting) at the transition position than that observed with a right-handed 55° staggering angle. This result is due to the higher filling ratio of the upstream right-handed screw element when using a neutral 90° staggering angle, hence leading to a higher extent of melting in the right-handed screw element with such profile. As for the melting zone, based on that observation it can be deduced that kneading blocks designed with neutral 90° staggering angle disks would be more efficient than kneading blocks designed with right-handed staggering angle disks.

4.2.4 Polymer melting: single screw extrusion versus twin screw extrusion

As a conclusion of the analysis of polymer melting in screw extruders, it is worth pointing out those important differences which characterize polymer melting in single screw extruders compared with intermeshing co-rotating twin screw extruders.

4.2.4.1 Melting rate and length of melting

It has been observed that polymer melting occurs over very short lengths of barrel in intermeshing co-rotating twin screw extruders, approximately over 4D, while it requires at least twice that length in single screw extruders. This is due to the intrinsic mechanism of polymer melting in each extruder. In the single screw extruder, polymer melting takes place only at the melt film–solid bed interface, where polymer melt is produced at the expense of solid polymer. As shown by Eq. 4.1, the local melting rate is proportional to $\sqrt{X}$. Thus, the melting rate is highest at the start of melting and then it drops as the width of the solid bed reduces. Furthermore, heat transfer from the circulating flow in the melt pool to the solid bed is negligible. Although the degree of tapering A is finely adjusted to optimize melting rate, full melting of polymer pellets still occurs over a relatively high length of barrel.

In intermeshing co-rotating twin screw extruders, the molten polymer that appears at the contact of barrel wall flows into all spaces between polymer pellets in the early stage of the melting process. The resulting mix of melt and solid pellets, which is highly viscous, generates an important amount of heat through viscous dissipation, leading to a constant melting rate almost from the start of melting up to completion of melting.

4.2.4.2 Flight clearance

Flight clearance is of great interest in thermomechanical plasticating extrusion. The reason for this is that wear of screw flights due to extruder operation results in a corresponding increase in flight clearance. The effect of flight clearance in single screw extruders was examined in section 4.2.2.5. In that case, an increase of flight clearance led to an increase in the delay of melting (early stage of melting mechanism) and an important decrease in the melting rate. In single screw extruders, the melting zone is exposed to screw wear through abrasion in particular, and as screw wear increases then the risk of generating unmelted polymer regions increases dramatically. To compensate for this phenomenon, an increase in screw speed can be applied if the extruder does not already operate at maximum screw speed. However, the increase of screw speed can be used in a very limited range. As wear increases, a significant reduction of the production rate soon occurs.

In intermeshing twin screw extruders, the effect of flight clearance on polymer melting has not been investigated systematically. In practice, it is well known that from a certain extent of wear in the melting zone, unmelted polymer may appear in the extruded melt. To compensate for this, an increase of screw speed is usually applied. In intermeshing co-rotating twin screw extruders, as flow rate and screw speed are independent processing variables, the increase in screw speed can generally be used over a significant time of operation to compensate for the wear and still preserve both the production rate and product quality. However, it should be noted that performance of intermeshing co-rotating twin screw extruders is less affected by flight clearance than that of single screw extruders.

4.2.4.3 Degree of process flexibility

Extrusion processing units often need modifications and adaptations of process conditions to maintain the performance of extrusion equipment (high production rate and product quality). Process engineers are rightly very attentive to process flexibility, and hence to the parameters they can manipulate to run the units consistently at optimum conditions. This is particularly important when dealing with new bio-based polymers and bio-based plastics with physical and chemical characteristics that differ significantly from those of conventional petrochemical polymers.

With a single screw extruder, process flexibility basically refers to the degree of tapering A, the barrel wall temperature, and to a lesser extent the screw speed. This is a rather limited degree of process flexibility, which generally leads to depression of the production rate of the extruder, in order to preserve product quality. However, the degree of flexibility of single screw extruders can be advantageously increased when using a modular design of the screw-barrel assembly. As described in section 2.1.2.2, this design enables process engineers to diversify the screw profile of the melting zone more extensively, hence leading to improved process performances.

With intermeshing co-rotating twin screw extruders, process flexibility is far higher since this equipment offers many more possibilities to vary the processing conditions and to design the screw profile of the melting zone appropriately. Furthermore, as flow rate and screw speed can be varied independently, process engineers can play with an extensive range of screw speeds in relation to the screw profile, while constant flow rate is maintained. Finally, the barrel wall temperature in the melting zone can be adapted according to the thermal characteristics of the polymer.

Obviously, intermeshing co-rotating twin screw extruders have a higher degree of process flexibility than single screw extruders, due to the ability to vary the mechanical energy input in the melting zone through various designs of screw profile and adjustments of screw speed in particular.

4.3 Physical transitions of bio-based polymers

During thermomechanical plasticating operation in screw extruders, polymers are generally heated from room temperature to their appropriate melt processing temperature. At room temperature, they are in solid state, either amorphous or partially crystalline, depending upon their microstructure and composition. Upon heating, polymers are converted from solid to liquid state. During that conversion, they undergo physical transitions such as glass transition and possibly melting transition for partially crystalline polymers. In die forming, the polymeric melts are further shaped and cooled. Upon cooling, they are converted back to solid state through physical transitions which lead to microstructure and hence physical characteristics which may differ from those of the original solid state, depending upon the processing conditions (time-temperature-shear history during thermomechanical plasticating, and the cooling rate during die forming, in particular).

Thorough understanding of the glass and melting transitions of bio-based polymers is of paramount importance in the appropriate design of processing conditions of GEP I, in order to obtain final materials with the expected properties.

4.3.1 Physical transitions of polymeric materials: generalities

The picture so far is that bio-based polymers consist of randomly coiled and entangled molecular chains showing no structural order, leading to the structure of an amorphous material with no widespread molecular order, and so their structure is non-crystalline. If an amorphous polymer is at very low temperature, its conforming chains can hardly move: it is frozen into a rigid glassy state. Upon uniform heating of the material, segments of the polymer chains become free to start moving (possibly 10–20 atoms, depending on the molecular structure) at the glass-rubber transition temperature, *Tg*. At that temperature, the elastic modulus of the material drops by several orders of magnitude (Figure 4.16a) while its specific volume, or thermal expansion coefficient, increases. On slow heating, amorphous polymers which have side groups in their molecular structure may show sub-*Tg* transition, an event detected by a slight elastic modulus drop. Amorphous bio-based polymer melts are viscous liquids when they are held at temperatures above their glass transition temperature. Further heating above *Tg* lowers the viscosity of the polymer melts; the rheological behavior of those melts is viscoelastic. Examples of amorphous bio-based polymers include proteins and amorphous PLA.

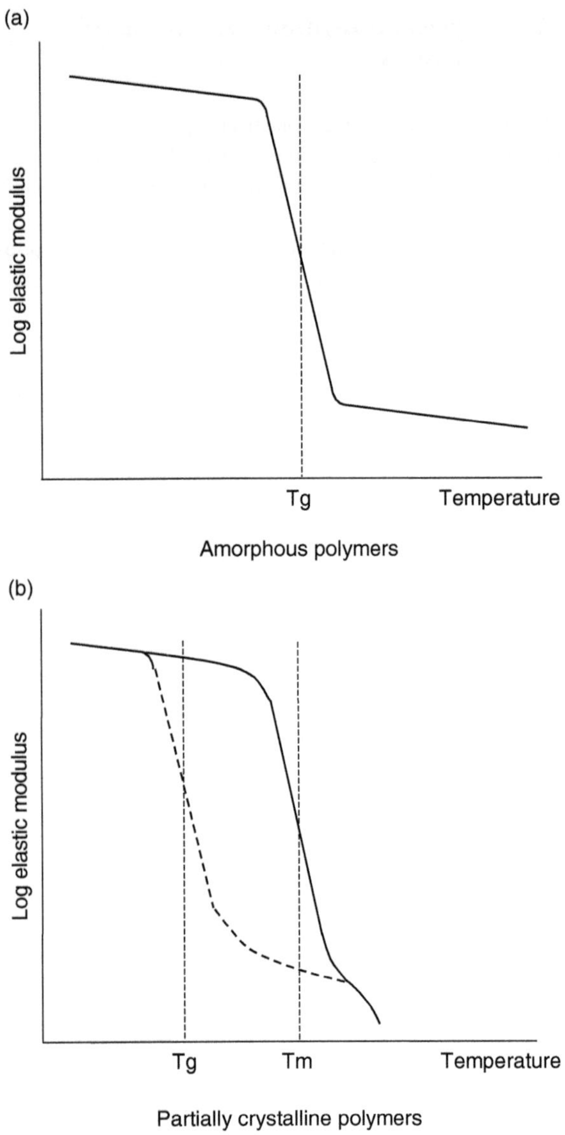

Figure 4.16 Physical transitions of amorphous (a) and partially crystalline (b) polymers.

Several thermoplastic polymers show regular polymer chains which are clustered together in local axial alignment to form crystallites; this is called the "fringed-micelle" model. Figure 4.17 is a schematic of randomly oriented crystallites suspended in and reinforcing a matrix of amorphous material of the same polymer type, thus leading to a partially crystalline polymer at room temperature. The traditional model shown in Figure 4.17 explains material behavior quite well if the degree of crystallinity is less than about 50%. Polymer crystallinity results either from plant growth in the case of native starches or from the chemistry of polymer chains in the case of polyamides and polylactic acid, for instance. A very cold, partially crystalline polymer is rigid and glassy. On heating, as the glass transition temperature is reached, the amorphous material becomes rubbery, but the crystals remain and can still confer substantial stiffness. Above *Tg*, the material softens with increases in temperature, as the matrix becomes more and more rubbery. The mobile matrix confers ductility and extra load will induce creep because of the uncoiling of the amorphous material. On further heating above *Tg*, the crystallites melt at the melting temperature *Tm*. Above *Tm*, the material is an amorphous, highly viscous and viscoelastic melt: further heating lowers its viscosity. Examples of partially crystalline bio-based polymers include native starches, PHAs, polyamides, C-PLA, sc-PLA, and PTT. On heating partially crystalline

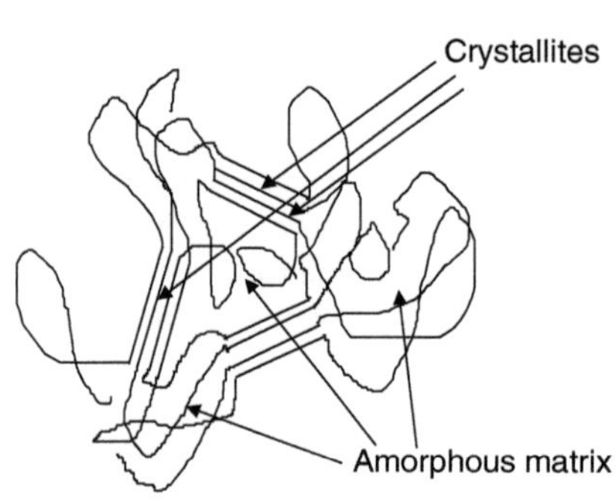

Figure 4.17 Model of partially crystalline polymers (or "fringed-micelle" model).

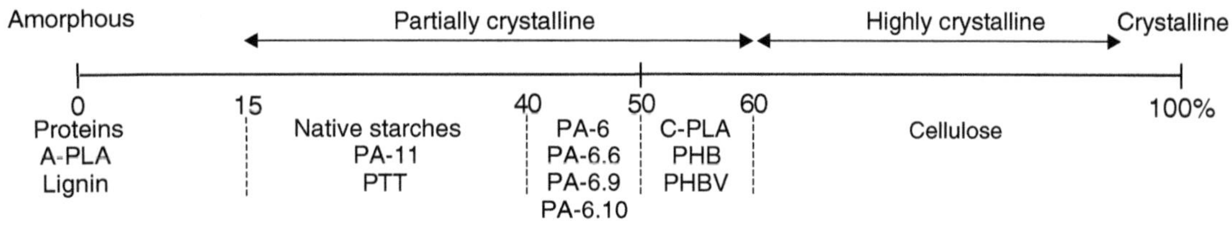

Figure 4.18 Degree of crystallinity of bio-based polymers.

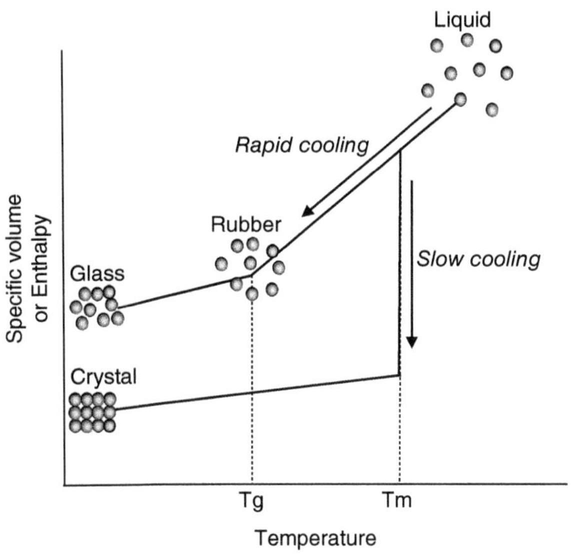

Figure 4.19 Relevant phase and state transitions of materials. Source: Aviad, Carvajal & Campanella 2009. Reproduced with permission of Springer Science+Business Media.

polymers, two physical transitions are observed: the glass transition and the melting transition which concerns the amorphous matrix and the crystallites, respectively (see Figure 4.16b).

Figure 4.18 is a summary of the bio-based polymers described in section 4.1 based on the degree of crystallinity. It is shown that the majority of bio-based polymers are either amorphous or partially crystalline, having a degree of crystallinity ranging from 0% to about 60%. For those polymers, estimation of the polymer conversion from the solid state to the liquid state requires good understanding of both glass and melting transitions. The conditions driving those transitions upon heating or cooling determine closely the microstructure and properties of bio-based polymer materials. Thus, we will examine in depth glass and melting phenomena in the following sections.

4.3.2 Glass and melting transitions: basics

The glass transition phenomenon was discovered in the 1930s; it has been studied since the early 1950s but still remains a debatable issue. Unlike first-order transitions characterizing a change of phase in a material, glass transition is a second-order transition that characterizes changes in the physical state of the material. At the glass transition region, continuous and significant changes in thermophysical and rheological properties such as the heat capacity and the material elastic modulus begin whereas no changes in entropy are observed. Conversely, melting of crystallites present in partially crystalline polymers is a first-order transition which characterizes a change of phase in the material, and shows discontinuous changes in enthalpy, entropy, and volume at the melting temperature. Both glass and melting transition phenomena are present and play an important role in polymer extrusion processing.

The glassy state of materials corresponds to a non-equilibrium solid state, in which the molecules forming the material are randomly arranged to occupy a volume larger than that of the crystalline state having a similar composition (Abiad et al., 2009). These glassy materials are referred to as amorphous solids. A glassy material is formed when a melt or a liquid having a disordered molecular structure is cooled below its crystalline melting temperature, *Tm*, at a rate sufficiently fast to avoid rearrangement of molecules that may result in crystallization. Figure 4.19 schematically illustrates the effect that the cooling rates have on the formation of crystalline and glassy solids. Moreover, different cooling rates may result in glasses with the same chemical composition but different structures and properties, due to the different thermal history. Thus, prediction of glass transition properties is not based on the assumption of an equilibrium state in which the properties are uniquely determined by the material itself and its state. The change in material structure with different heating or cooling rates, and probably with other treatments, results in the formation of glasses with different levels of energy (Wunderlich, 2005).

This energy can be estimated in terms of the material entropy, enthalpy, or free energy.

Since glassy solids are considered to be in a state of non-equilibrium, the stability of the material is therefore dependent on several factors, which include temperature, composition, plasticizer amount (water for most bio-based polymers), molecular weight and, as discussed, the thermal history undergone by the material before reaching the specific glassy state. Many of these factors affect the glass transition temperature range. The glass transition temperature range of an amorphous material affects its thermomechanical properties. At temperatures below their glass transition range, amorphous solids are stiff and glassy. Upon heating, these solids soften as they exceed their characteristic *Tg* range. At the glass transition range, other physical properties associated with increases in molecular mobility are also significantly affected and consequently vary accordingly.

For a given bio-based polymer, glass transition is a function of temperature, time, composition, molecular weight, and possibly water content. As the glass transition is a change in the physical state of the material, it can be considered as a kinetic phenomenon instead of one described by thermodynamic relationships that are only valid under equilibrium conditions. Many researchers have considered the glass transition phenomenon as a kinetic event as opposed to a thermodynamic transition such as melting. During the glass-rubber transition of bio-based polymers in the glassy state, shift in thermal properties such as heat capacity and coefficient of expansion can be described by smooth changes in these properties, which is clearly opposed to the observed discontinuities in properties such as enthalpy and volume observed during the melting of crystallites. Despite the intense debate on the type of transformation driving the conversion from a liquid to a glass upon cooling (and vice versa upon heating) and whether that transition can be described by equilibrium thermodynamics relationships like those used to describe condensation or crystallization, there is some agreement on the non-equilibrium condition of glass transition events.

Several theories describing the behavior of polymers and other materials near the glass transition temperature range have been stated, the major theories being free-volume, kinetic, and thermodynamic. These theories were reviewed extensively in a paper published by Abiad et al. (2009) and have been applied to predict glass transition of bio-based polymers with some success, the free-volume and the thermodynamic theories in particular.

The space within the polymer domain that is available for rotation and translational movements is considered as a free volume that will favor the mobility of macromolecules. Above the glass transition temperature range, the free volume increases linearly with temperature and so does the mobility of the polymeric molecules. Above the glass temperature range, when a polymer is cooled, the decreasing free volume reaches a point where there is insufficient space for long-range molecular motions. The free-volume theory explains the dependence of the glass transition temperature on cross-link density and molecular weight of the system. An increase in the glass transition temperature can be observed as the cross-link density of polymers increases, mainly due to an effective decrease in the free volume. As for the effect of molecular weight, as molecular weight of the polymer increases, the free volume decreases and thus the glass transition temperature must increase. Fox and Flory (1950) studied the relationship between glass transition and free volume for selected polymers as a function of molecular weight and the polymer relaxation time, which is a parameter strongly affected by the nature and structure of the polymeric material. This led to a model (Williams–Landel–Ferry or WLF model) which has been used to predict the glass transition temperature of starch materials (Roos & Karel, 1991).

The thermodynamic theory has been successful in predicting various phenomena that may be associated with the behavior of bio-based polymers. These include variation of the glass transition temperature as a function of molecular weight, cross-link density, and plasticizer content. Its applications also extend to predicting the glass transition temperature of binary polymer blends as a function of mass fractions and individual glass transition temperatures. Couchman and Karasz (1978) proposed a model to predict the glass transition temperature of mixtures based on the assumption that such a transition is a thermodynamic event rather than a second-order transition. In other words, they assumed that the entropy of mixing is a continuous function at the glass transition region, which yields the following equation:

$$\ln\left(T_g\right) = \frac{x_1 \cdot \Delta C_{p1} \cdot \ln\left(T_{g1}\right) + x_2 \cdot \Delta C_{p2} \cdot \ln\left(T_{g2}\right)}{x_1 \cdot \Delta C_{p1} + x_2 \cdot \Delta C_{p2}} \tag{4.12}$$

T_{gi} is the glass transition of the component i in the mixture whereas x_i is its mole fraction; $\Delta C_{pi} = \left(C_{pi}^{R} - C_{pi}^{G}\right)$ is the change in heat capacity where C_{pi}^{R} and C_{pi}^{G} are the heat capacities of the component i at the rubbery and glassy states, respectively. Equation 4.12 was also expanded for ternary polymer systems including a third term. Equation 4.12 has been used to predict the glass transition temperature of blends of bio-based polymers.

The glass transition of amorphous and partially crystalline polymers can be determined by continuously measuring various thermophysical and mechanical properties as a function of temperature, while heating or cooling the polymeric material. These measurements, which may include the material-specific volume, heat capacity, strain, conductivity, and viscoelastic properties, can be used to determine the glass transition range by identifying the temperatures at which these properties change significantly. Owing to the property measured, methods to characterize glass transition can be classified as calorimetric, thermomechanical, volumetric, and spectroscopic. As for bio-based polymers, these include mainly differential scanning calorimetry (DSC), temperature-modulated differential scanning calorimetry (MDSC), dynamic mechanical analysis (DMA), and dynamic mechanical thermal analysis (DMTA). Methods used to characterize glass transition have been reviewed extensively in a paper published by Abiad et al. (2009).

It has been shown that glass transition occurs over a range of temperatures rather than at one fixed temperature so it makes more sense to define a range of temperatures in which the transition occurs rather than a single temperature. Given the fact that glass transition is a non-equilibrium event, it is important to report the heating or cooling rates under which the analyses are conducted as this significantly affects the position of the endotherm.

The glass transition of polymeric systems is affected by three main factors.

- *Polymer-dependent factors* such as chain flexibility, molecular weight, size and polarity of lateral groups, etc. These factors concern bio-based polymers after extraction or polymerization, with no additives.
- *Process-related factors* such as time-temperature-shear history in thermomechanical plasticating, time-temperature history in material forming, in particular.
- *Formulation-related factors* such as type and content of plasticizers, water content, co-polymers and blends.

These factors are discussed in section 4.3.3, when referring particularly to bio-based polymers.

4.3.3 Glass and melting transitions of bio-based polymers

4.3.3.1 Anhydrous bio-based polymers in native or virgin states

Table 4.3 gives the glass transition and melting temperatures observed for the bio-based polymers which were presented in section 4.1; these concern bio-based polymers which are in either native state (starch, proteins) or virgin state (polyhydroxybutyrates, polyamides, polylactic acid, and polytrimethylene terephthalate). The table indicates also the method of measurement (DSC or DMA). It is worth commenting these values for each category of polymer.

4.3.3.1.1 Starch

For anhydrous native starches, Tg is experimentally accessible with difficulty. In fact, during heating, thermal degradation of starch polymers could take place before reaching Tg. A common practice is to add water to the starch and bring down its Tg so that it can be measured before thermal degradation takes place. Tg of anhydrous starch can then be determined on the basis of extrapolation or mathematical models. For instance, Roos and Karel (1991) reported a value of 243°C based on the Fox and Flory model while Chung et al. (2002) have estimated a value of 296°C based on the Couchman–Karasz equation (Eq. 4.12). The impossibility of directly measuring the glass transition of anhydrous starches, associated with the type of model used to estimate it, has generated an important debate and rather large dispersion of Tg values published in the literature. But, as seen in Table 4.3, Tg of anhydrous native starches is high, generally above 200°C, because of the high molecular weight of starch polymers (amylopectin, in particular), the high density of branching and the stiffening action of crystallites which decrease the mobility of macromolecular chains in amorphous domains; hence all these structural characteristics lead to increase the glass transition temperature of starch materials. It appears that Tg values of anhydrous native starches are higher than that of amylopectin (Orford et al., 1989), probably because of the effective contribution of amylose in the structural organization of starch system and thus in the reinforcement of the rigidity of the polymeric structure.

Native starch by itself is unsuitable for most non-food applications due to various disadvantages such as its brittleness and hydrophilicity. Moreover, its melting temperature is experimentally difficult to measure because it is close to the thermal decomposition temperature, but it can be estimated by use of the Flory–Huggins equation which results from the thermodynamic relation between the melting point of a crystalline polymer and diluent concentration (see Chapter 6, section 6.2.2.2). It must be pointed out that native starches show poor thermal processing ability. In order to overcome these

Table 4.3 Glass transition and melting temperatures of anhydrous bio-based polymers (in native or virgin states).

Bio-based polymer	Tg, °C	Tm, °C	Method of measurement	Reference
Starch:				
- Wheat-based	243	**	DSC	Roos and Karel (1991)
- Rice-based	250	**	DSC	Biliaderis et al. (1986)
- Rice-based	296	**	DSC	Chung et al. (2002)
- Waxy maize*	227	**	DSC	Orford et al. (1989)
Proteins:				
- Soy (7S fraction)†	120	-	DSC	Wang et al. (1996)
- Soy (11S fraction)†	150	-	DSC	Wang et al. (1996)
- Gluten (corn)	188	-	DMTA	Di Gioia et al. (1999)
- Zein	168	-	DMTA	Di Gioia et al. (1999)
- Glutelin	198	-	DMTA	Di Gioia et al. (1999)
- Gluten (wheat)	160	-	DMTA	Hoseney et al. (1986)
- Gliadin	121	-	DMTA	Kokini et al. (1995)
- Glutenin	160	-	DMTA	Cocero and Kokini (1991)
- Casein	144	-	DMTA	Kalichevski et al. (1993a)
- Myofibrillar proteins	215–250	-	DMTA	Cuq et al. (1997)
- Gelatin	210	-	DSC	Cuq et al. (1997)
- Collagen	200	-	DSC	Batzer and Kreibich (1981)
Polyhydroxyalkanoates (PHAs):				
- PHB	4	175	DSC	Chodak (2008)
- PHBV$	5	153	-	Chodak (2008)
- PHBV‡	−1	145	-	Chodak (2008)
Polyamides (PAs):				
- PA-6	65–75	228	DSC	Kohan et al. (2003)
- PA-6.6	46–66	260	DSC	Herrera Ramirez (2004)
- PA-11 (Arkema)	50	184–188	DSC	Boisot (2009)
Polylactic acid (PLA):				
- A-PLA	58	-	DSC	Averous (2004)
- C-PLA	55–60	160–170	-	Shen et al. (2009)
- sc-PLA	60–70	210–240	-	Shen et al. (2009)
Polytrimethylene terephthalate	45–75	225	-	Hwo et al. (1998)

*99% amylopectin.
† Soy fraction with 10% water content.
$ Poly(3-hydroxybutyrate-*co*-7 mol% hydroxyvalerate); Biopol D400G, HV 7% (Monsanto).
‡ Poly(3-hydroxybutyrate-*co*-20 mol% hydroxyvalerate).
** Melting temperatures of native starches vary roughly between 150°C and 230°C, depending upon the type of starch and the method of measurement (refer to Chapter 6, section 6.2.2.2).

disadvantages, native starch requires further conversion, leading to suitable processing ability. This includes mixing, extrusion and/or blending to obtain a pure or blended starch plastic or starch composites; this will be examined in section 4.3.3.2.1.

4.3.3.1.2 Proteins

The glass transition behavior of proteins depends on their structure, molecular weight and organization, as well as type and density of intermolecular interactions. Table 4.3 shows relatively high *Tg* values for proteins, and large differences in the glass transition temperatures. High density of interactions due to the presence of polar groups and hydrogen bonding leads to decrease in the mobility of amorphous polymer chains, resulting in high *Tg* values. Molecular organization and structural characteristics of the various proteins are probably responsible for the differences in the glass transition temperatures. *Tg* values of proteins range from 120°C for low molecular weight proteins (7S soy protein fraction, gliadin) to 215–250°C for high molecular weight proteins (such as myofibrillar proteins). Specific polydispersity of myofibrillar proteins, heterogeneity of intermolecular interactions within proteins, and the probable presence of some covalent cross-links or chain entanglements in myofibrillar proteins could be responsible for the high glass transition temperature and breadth ($\Delta T = 35°C$). Thermal degradation of myofibrillar proteins was observed to occur up to 250°C; that is the upper limit of the glass transition temperature range (Cuq et al., 1997) so the processing temperature of native myofibrillar proteins is rather restricted.

The addition of a plasticizer, and even cross-linking agents, is essential during the processing of proteins. This allows the extrusion processing ability of proteins to be optimized through modification of their three-dimensional organization and a decrease of their attractive intermolecular forces, resulting in decreased cohesion, elasticity, mechanical properties, and rigidity.

4.3.3.1.3 Polyamides and polyesters (PHAs, PLAs, PTT)

As seen in Table 4.3, polyamides and polyesters have relatively low *Tg* values. In fact, these polymers have very flexible backbone chains which result in better chain mobility, leading to low glass transition temperatures.

The melting temperature of bio-based polyamides ranges from approximately 180°C to 260°C, depending upon the number of carbon in the repeat units of the polyamides: the lower the number of carbon in the repeat units, the higher the melting temperature. For instance, the melting temperature of PA-11 is in the range of 184–188°C (Boisot, 2009) whereas PA-6.6 shows a melting temperature of 260°C (Herrera Ramirez, 2004). Actually, PA-11 has a low amide to methylene ratio (1:10) which makes it a polymer like polyethylene that has low melting temperature. In general, polyamides show good stability at elevated temperatures which lead to good extrusion processing properties.

Polyhydroxybutyrate and its co-polymers have low *Tg* values. *Tg* of the PHB homopolymer is only a little below room temperature. PHB co-polymers with rising hydroxyvalerate contents usually exhibit lower *Tg* values. A similar trend is observed for the melting temperatures. Polyhydroxyalkanoates show a very high susceptibility to thermal degradation which represents a real difficulty when processing these polymers. For instance, considering its melting temperature is 175°C, the extrusion processing temperature of the PHB homopolymer should be at least 190–200°C. At this temperature, although its onset decomposition temperature has been reported to be approximately 246°C (Rudnik, 2008), thermal degradation of PHB proceeds quickly which leads to substantial decrease in its molecular weight. Thus, lowering the extrusion processing temperature can be achieved by the addition of plasticizers, or by co-polymerization with higher alkanoates.

Typically, the glass transition temperature of PLA polymers ranges from 50° to 80°C. The melting temperature is related to the enantiomeric purity of the lactic acid stereo co-polymers. The stereocomplex PLA, which results from melt-blending PLLA and PDLA with a D/L ratio of 1:1, shows a melting temperature of 210–240°C, which is about 40–80°C higher than the enantiomerically pure, homocrystalline PLA; amorphous PLA shows no melting temperature. It should be noted that the major difficulty in processing PLA-based biomaterials is their limited thermal stability during melt processing. In fact, the onset decomposition temperature of typical PLA is about 254°C whereas it ranges from 287° to 297°C for homocrystalline PLA and stereocomplex PLA (Rudnik, 2008). The properties of the polymer therefore depend on the polymer formulation and the conditions of extrusion processing.

Polytrimethylene terephthalate (PTT) has a glass transition temperature of 45–65°C and a melting temperature that is roughly 200–240°C, equivalent to that of PA-6. Thus, PTT can be processed at relatively low temperatures, between 250° and 260°C. Accounting for its relatively good thermal stability (temperature at maximum degradation is approximately 361–370°C; Rudnik, 2008), PTT shows quite good processing properties.

4.3.3.2 Plasticized bio-based polymers

In general, bio-based polymers in native state (starch and proteins) and in virgin state (PHAs, PLA, in particular) show unsuitable processing ability and mechanical properties. To overcome this difficulty, bio-based polymers need to be plasticized and sometimes processed chemically. Thus additives combined with appropriate extrusion processing conditions allow bio-based polymers to be converted to valuable bio-based thermoplastics which can fulfill various applications.

Additives differentiate a plastic material from the virgin polymer and are used to fine-tune ultimate properties such as processing ability (decreased glass and melting transition temperatures, improved flow characteristics and melt forming, resistance against degradation during melt processing, etc.), mechanical and end-use characteristics (resistance to fire, UV stability, strength, flexibility, etc.). As for processing ability, plasticizers are very important additives as they allow transition temperatures to be decreased and flow characteristics to be improved. Plasticizers aim to lower the *Tg* of bio-based polymers. Plasticizers consist of miscible low molecular weight and low-volatility components that interact with polymer chains. Plasticizers have much lower glass transition temperatures than polymeric materials so polymer/plasticizer mixtures have *Tg* values significantly lower than pure polymers. The effect of a plasticizer on the glass transition can be explained by two mechanisms: (i) the plasticizer molecules screen off the attractive forces between the material molecules; and (ii) the plasticizer molecules increase the space between the material molecules. Both mechanisms provide a greater free volume and freedom for the molecules to move. In other words, when smaller molecules penetrate the interchain spaces of a given network, the average distance between these chains increases and as a result more free volume is created. Plasticizers generally contribute to decrease the melt viscosity, which makes further processing of the plasticized material at lower temperatures easier.

Plasticizers and other additives are usually incorporated into screw extruders during the thermomechanical plasticating process. It should be noted that bio-based polymers may show significant physicochemical changes such as decrease of molecular weight and increase of cross-link density, depending upon the macromolecular structure of the polymeric material and the time-temperature-shear history in the screw extruder. These physicochemical changes affect the glass transition temperature. In fact, polymers with shorter chains have larger free volumes and thus lower *Tg*. Similarly, an increase of cross-link density is expected to cause elevation of the glass transition temperature.

The purpose of this section is to examine the plasticization phenomenon of the bio-based polymers presented in section 4.1, focusing particularly on those which are emerging and have high technical substitution potential in the future, such as starch, proteins, PHAs, and PLA. Although polyamides have high technical substitution potential, plasticization of these polymers is not considered in this section because it has been already described extensively in the scientific and technical literature.

4.3.3.2.1 Thermoplastic starch

Except for the purpose of plastics reinforcement where native starch can be used as filler, native starch by itself is unsuitable for most plastic applications due to various disadvantages, the most important being its brittleness and hydrophilicity. Brittleness of starch is caused by the relatively high glass transition temperature. Moreover, when stored, this brittleness increases due to irreversible and fast recrystallization of amylose (ability of amylose macromolecules to form hydrogen bonds), and the retrogradation of amylopectin (reversible and slow process of crystallization of amylopectin macromolecules). Native starch must be further processed by mixing, extrusion, and/or blending to obtain a pure or blended starch plastic. Extrusion processing technology is particularly suitable for converting starch from native to thermoplastic states.

Native starch is destructured by use of low water contents (<20% wet weight basis) and plasticizers, leading to thermoplastic starch (commonly called TPS). TPS is generally produced by thermomechanically processing starch/water/plasticizer mixtures in a screw extruder at temperatures between 140° and 160°C, at high pressure and shear. If the mixture contains only water as the plasticizer ($Tg = -139$°C), an important decrease in *Tg* is observed for various starches when water content increases from 0% to 30%. For instance, in that range of water content, the glass transition temperature of amylopectin decreases from 225–230°C (estimated value; see section 4.3.3.1) at 0% water content down to approximately 8–10°C at 22.5% water content (experimental DSC data; Kalichevski et al., 1993b). Besides, it has been shown that *Tg* of starch remains constant above 30% water content, which has been attributed to the formation of a separate water phase outside the starch granules (Biliaderis et al., 1986). When thermomechanically processing starch/water mixtures at relatively low levels of water content (about 15–25%) starch polymer keeps its thermoplastic properties but as the processing temperature is higher than 100°C, the starch melt expands when exiting the die due to water flashing off, which can be a desired effect when producing loose-fill packaging materials (shock absorbers). If the production of non-expanded thermoplastic starch melt is desired, then non-volatile, high-boiling and low molecular weight plasticizers must be used.

Numerous plasticizers have been investigated such as glycerol, sugars, xylitol, sorbitol, maltitol, glycols, urea, ethanolamine, and formamide (Carvalho, 2008). The effect of glycerol as a plasticizer ($Tg = -93$°C) in starch is summarized in Table 4.4. As shown in the table, plasticization of wheat starch with glycerol and water leads to a marked decrease of the glass transition temperature when glycerol content increases (Averous, 2004). De Graaf et al. (2003) studied the effects of glycerol content in concentrations

Table 4.4 Tg values of thermoplastic wheat starch as a function of glycerol content.

Starch, %*	Water, %*	Glycerol, %*	Tg, °C**	Reference
74	10	16	43	Averous, 2004
70	12	18	8	Averous, 2004
67	9	24	−7	Averous, 2004
65	0	35	−20	Averous, 2004
74	6	20	41	De Graaf et al., 2003

* Percentages expressed in wwb (wet weight basis).
** Method of measurement: differential scanning calorimetry.

of 15%, 20%, and 25% (w/w) on the glass transition temperature of starches. The glycerol was mixed with four different starches having different amylopectin/amylose ratios (Am/AM): waxy maize (Am/AM = 1/99), wheat (Am/AM = 1/4), potato (Am/AM = 1/4) and pea (Am/AM = 2/1) and initial water contents of 9% (w/w), 8% (w/w), 13% (w/w), and 9% (w/w), respectively. The authors observed experimentally that the glass transition temperature linearly decreased with increases of the glycerol content. Moreover, it was shown that starches containing higher amount of amylopectin (such as waxy maize) had higher *Tg* than starches having less amylopectin (such as pea). In fact, due to its relatively low molecular weight with lack of side chains, amylose results in a greater free volume and mobility. Hence, amylose-rich starches show lower *Tg* than amylopectin-rich starches. From Table 4.4, it must be noted that the *Tg* value published by De Graaf et al. (2003) at glycerol content of 20% (wet weight basis) is comparable to the *Tg* value published by Averous (2004) for a similar composition (wheat starch 74%; water 10%; glycerol 16%).

According to the type of plasticizer and plasticizer/starch ratio, thermoplastic starches present a large range of *Tg* values and properties, which makes them particularly suitable to replace petrochemical plastics in many applications. When converting native starches into TPS in a screw extruder, the starch granules are fragmented. Under pressure and shearing, the crystalline order observed in native starch is completely destructured; the starch is melted, plasticized, and partially depolymerized. Compared to native starch, because of the mobility of the starch chains, recrystallization of plasticized starch may occur, leading generally to some crystallinity which depends upon the extrusion processing conditions (residence time, screw speed, temperature profile, and screw configuration). The crystallinity of TPS results from residual crystallinity of native starch (B-type, in particular) and from process-induced crystallinity (V- and E-type crystallinity), leading generally to the formation of B-, V- and E-type crystalline structures. As a consequence of their semi-crystalline character, the mechanical properties of TPS are characteristics of partially crystalline polymers with B-type crystallinity being the major factor influencing the mechanical behavior of TPS.

Extrusion processing technology plays an important role in the thermomechanical destructuring and plasticating process for native starches in order to produce homogeneous thermoplastic starch. This technology also allows starch blends (association of TPS and petrochemical polymers, with starch contents varying approximately from 30% to 80%) and starch composites (TPS reinforced with natural fibers) to be produced through mixing and reactive blending, in order to enlarge the scope of end applications.

4.3.3.2.2 Thermoplastic proteins

The thermoplastic behavior of proteins has been investigated to prepare edible and non-edible films, as well as sheets, through thermomechanical processing under low water content, similar to that for thermoplastic starch-based materials. Plasticizers are generally added to the protein matrices to improve their processability and to modify the properties of the final structures. As for starches, water is the natural and most effective plasticizer of proteins, enabling them to undergo the glass transition, facilitating deformation and processability of the polymeric matrix (Hernandez-Izquierdo & Krochta, 2008). Besides water, glycerol has also been used in thethermoplastic processing of proteins, as well as sorbitoland mono-, di- and oligosaccharides (Hernandez-Izquierdo & Krochta, 2008). All these studies have shown that the addition of plasticizers results in a decrease of the glass transition temperature.

The importance of *Tg* decrease depends on the physicochemical characteristics of both the plasticizer and proteins. For instance, Pommet et al. (2005) have investigated the influence of five selected plasticizers (water, glycerol, 1,4-butanediol, lactic acid, and octanoic acid) on the plasticization of wheat gluten, water having the lowest molecular weight, glycerol, 1,4-butanediol and lactic acid being similar in molecular weight and hydrophilicity, and octanoic acid being the largest and most hydrophobic molecule. The plasticizing effect, at the same molar content, was found to be identical for water, glycerol, and 1,4-butanediol. Lactic and octanoic acids had higher and lower plasticizing effects, respectively. When studying the plasticization of corn gluten meal, Di Gioia and Guilbert (1999) concluded that the plasticizing efficiency (e.g. decrease of *Tg*) at equal molar content was proportional to the molecular weight and inversely proportional to the percentage of hydrophilic groups of the plasticizer. Moreover, despite the strong plasticizing ability of water, its effect on corn gluten was limited because water cannot interact with the non-polar protein components responsible for the glass transition. It is assumed that polar plasticizers interact with readily accessible polar amino acids of proteins, whereas amphiphilic plasticizers interact with non-polar regions which are accessible with difficulty, and thus they have a larger effect on the net *Tg* of the system. It must be noted that at low water contents (<10% wet weight basis), water may act as an antiplasticizer by increasing the structural order of polymer chains (Hernandez-Izquierdo & Krochta, 2008).

Thermomechanical plasticating of proteins has been found to be very sensitive to the time-temperature-shear history applied in the extrusion process, as observed by Redl et al. (1999a) and Pommet et al. (2003) when studying the behavior of wheat gluten plasticized with glycerol. Depending on processing conditions, extrudates showed very different surface aspects, ranging from smooth surface aspects with high swell to disrupted surface appearances. The authors related this behavior to important macromolecular transformations that occur in the proteins, such as protein aggregation caused by cross-linking reactions (such as disulfide bonding). Formation of a strongly cross-linked structure increased the network density and decreased the mobility of polymeric chains, hence inhibiting their elastic recovery, and thus depressing the ability of the protein melt to support the strain applied during its extrusion through the die. The authors observed that increase of the network structure was induced by the thermomechanical processing conditions, the temperature and mechanical energy input, in particular.

As proteins are very complex and reactive systems, the choice and concentration of the used plasticizer must account not only for its molecular weight but also for its physicochemical characteristics and their relation to protein composition (the amino acid composition, in particular), to be able to optimize the processability of proteins. In the thermomechanical plasticating process, depolymerization and plasticization of proteins occur as a result of the mechanical shear and hydrophilic interactions, which results in a gel-like behavior. As temperature and time increase, covalent cross-linking is enhanced in a time-temperature dependent reaction. However, mechanical energy also plays a key role, as shear may promote an important decrease of the activation energy of the reaction. According to the functional and end-use characteristics of the plasticized and formed proteins, an optimum time-temperature-shear history must be applied in the thermomechanical extrusion process. Thus, the temperature profile and specific mechanical energy must be fine-tuned, putting particular emphasis on the reactivity of proteins, in order to ultimately control the network structure of the plasticized protein melt and thus its elastic response in the forming process. In that respect, twin screw extrusion technology would certainly be more appropriate for fine-tuning and applying such optimization.

4.3.3.2.3 PLA- and PHA-based materials

Plasticization of PLA melts is required to improve processability of PLA polymer and properties of PLA-based products, to satisfy the conditions of thermal and thermomechanical polymer processing and to compete with low-cost commodity polymers. Additives such as lactic acid, glycerol, polyethylene glycol, citrate esters, glucosemonoesters, and partial fatty acid esters have been investigated as potential plasticizers. Martin and Averous (2001) studied the effects of plasticizers (at two different plasticizer contents, 10 and 20 in wt%) on the thermal properties of amorphous PLA. Table 4.5 presents the glass and melting transition temperatures, as well as the degree of crystallinity of plasticized PLA systems (PLA/glycerol, PLA/polyethylene glycol, PLA/lactic acid). Plasticized PLA has *Tg* values lower than that of pure PLA and *Tg* decreases with increasing plasticizer content. The extent of *Tg* depression depended on the plasticizer. The lower *Tg* values of 12° and 18°C (compared with 58°C for pure PLA) were obtained with polyethylene glycols and lactic acid, respectively. A significant decrease in the melting temperature was also observed (10–15°C decrease). The degree of crystallinity ranged between 20% and 30%, and it tended to increase when plasticizer content increased, due to enhanced chain mobility. Polyethylene glycols (PEG with a molecular weight of 400 in particular) and lactic acid had a strong effect on the thermal transitions of PLA, and they can be considered as the most efficient plasticizers of PLA. In contrast, glycerol had a limited effect on the thermal transitions of PLA, and glycerol is a non-plasticizer of PLA.

As seen in section 4.3.3.1, due to its high susceptibility to thermal degradation, the homopolymer PHB shows

Table 4.5 Thermal properties of plasticized PLA.

PLA system Plasticizer content	*Tg, °C	*Tm, °C	Degree of crystallinity, %
Pure PLA	58	–	1
**PLA/Polyethylene glycol 400			
– 10%	30	147	26
– 20%	12	143	29
**PLA/Polyethylene glycol 1500			
– 10%	41	152	17
– 20%	30	148	25
PLA/Lactic acid			
– 10%	37	144	21
– 20%	18	132	24
PLA/Glycerol			
– 10%	54	142	24.3
– 20%	53	141	25.4

* Method of measurement: differential scanning calorimetry.
** Polyethylene glycols 400 and 1500 have molecular weights of 400 and 1500, respectively.
Source: Martin & Averous 2001. Reproduced by permission of Elsevier.

a narrow window for the processing conditions. Thermomechanical plasticating of PHB polymer would require special attention to the temperature profile and screw configuration: polymer melting over a rather short melting section and a progressive temperature decrease downstream, to minimize polymer degradation. The co-polymer PHBV shows easier plasticating conditions. Both polymers can be plasticized by citrate ester (Averous, 2004).

4.4 Flow properties of bio-based polymer melts

Knowledge of the flow properties of bio-based polymer melts is very important in analysis of extrusion processing. In this process, the first traces of melt appear in the melting section, and all downstream sections of the extruder, including die forming, are dedicated to converting bio-based polymer melts. Flow properties of melts determine to a large extent the characteristics of the extrusion process and downstream processing operations. Thus, knowledge of the flow properties (or rheological properties) of bio-based polymer melts allows appropriate design of the screw configuration and selection of process operating conditions.

4.4.1 Flow behavior: basics

Let us assume an idealized shear flow, shown in Figure 4.20, with two plates separated by a distance Y and the space between the two plates filled with a fluid. The upper plate of area A can move, while the lower plate is stationary. A shear stress, τ, is applied to the fluid through a shear force, F, which causes the upper plate to move at a constant velocity, V. A linear velocity gradient in the fluid results and assuming that there is no slip at the wall, the shear rate is calculated as:

$$\dot{\gamma} = V/Y \tag{4.13}$$

and the shear stress, or the stress required to achieve the shear flow, is:

$$\tau = F/A \tag{4.14}$$

Units of shear rate and shear stress are s^{-1} and Pa, respectively.

The shear viscosity, μ, or the internal resistance created by the fluid to the shear flow, is determined from the ratio of shear stress and shear rate:

$$\mu = \tau/\dot{\gamma} \tag{4.15}$$

The viscosity is usually expressed in Pa.s (or N.s/m^2), as well as in Poise (1 Pa.s = 10 Poise).

Equation 4.15 states that shear stress is directly proportional to shear rate. Fluids which behave in this manner are called Newtonian fluids. In a plot of shear stress versus shear rate, which is referred to as a flow curve, a Newtonian fluid will exhibit a linear relationship (Figure 4.21). Its viscosity is given by the slope of the line, which is independent of the shear rate. Fluids whose viscosity is dependent on the shear rate are called non-Newtonian (see Figure 4.21). The shear stress–shear rate relationship of such fluids is non-linear. Non-Newtonian fluids whose

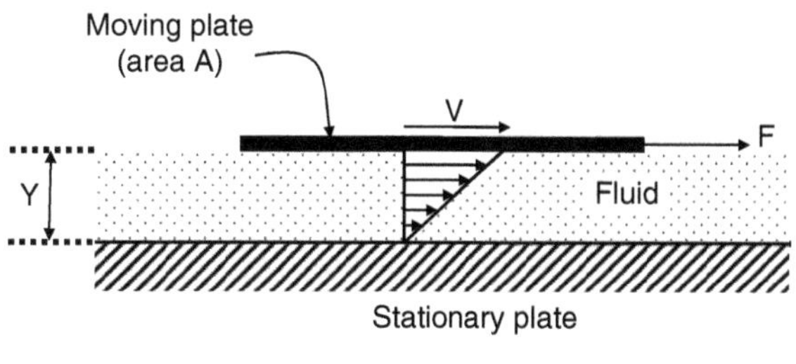

Figure 4.20 Shear deformation of an ideal fluid.

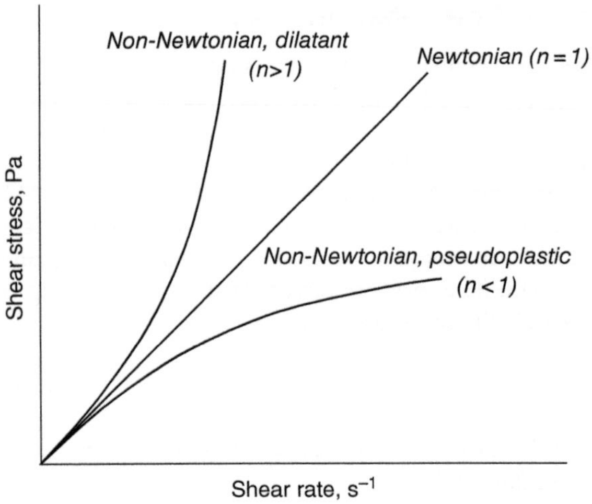

Figure 4.21 Flow curves of Newtonian and non-Newtonian fluids.

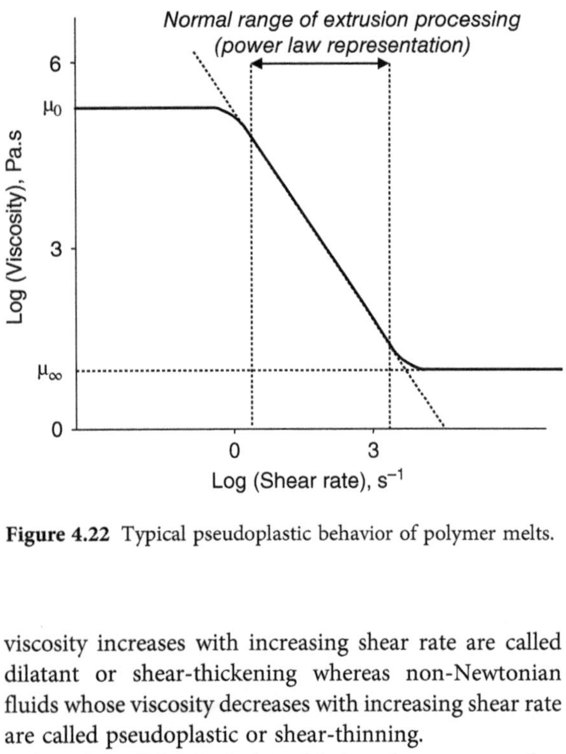

Figure 4.22 Typical pseudoplastic behavior of polymer melts.

viscosity increases with increasing shear rate are called dilatant or shear-thickening whereas non-Newtonian fluids whose viscosity decreases with increasing shear rate are called pseudoplastic or shear-thinning.

A variety of rheological models have been proposed to describe non-Newtonian behaviors. Because of its simplicity and ability to correlate observed rheological behaviors over the shear rate range of approximately 10–10^4 s^{-1}, the power law model has found extensive use as a constitutive equation describing the rheological behavior of polymeric melts. The power law model can be written as:

$$\tau = m\dot{\gamma}^n \tag{4.16}$$

where m and n are the consistency index (expressed in Pa. s^n or N.s^n/m^2) and the flow behavior index (dimensionless), respectively. This law is often referred to as the power law of the Ostwald and De Waele model. For pseudoplastic fluids $n < 1$ and for dilatant fluids $n > 1$. The degree of non-Newtonian behavior exhibited by a fluid can be measured by the extent to which the flow behavior index, n, deviates from 1.

Once the flow curve has been considered, it is necessary to discuss apparent viscosity. Apparent viscosity, μ_a, is the viscosity which exists at a particular value of $\dot{\gamma}$ and changes with differing values of $\dot{\gamma}$ for a non-Newtonian fluid. Apparent viscosity is defined as:

$$\mu_a = \tau/\dot{\gamma} = m\dot{\gamma}^{n-1} \tag{4.17}$$

From Eq. 4.17 it can be seen that the plotting of log μ_a versus log $\dot{\gamma}$ results in a straight line, which allows m and n indices to be determined from the intercept and the slope respectively. Similar information can be obtained from the rheological log τ versus log $\dot{\gamma}$ data.

As shown in Figure 4.22, the typical viscosity-shear rate curve for a pseudoplastic polymer melt over a wide range of shear rate illustrates the limitation of the power law model at both very low and very high shear rates. At very low and very high shear rates, the fluid behaves as a Newtonian fluid with viscosities μ_o (low shear rate value) and μ_∞ (high shear rate value), respectively. The level of shear rate encountered in polymer extrusion processing is generally comprised between roughly 10 and 5.10^3 s^{-1}. In that range, the viscosity-shear rate curve can then be described by the power law model reasonably well.

Besides the effect of shear rate, there are other variables that also affect the viscosity of polymer melts, such as temperature, pressure, the presence of plasticizers, and possible processing conditions which cause structural modifications and are time-temperature-shear dependent. It is important to understand the effect of these variables on the apparent viscosity. An Arrhenius equation relationship form has been extensively used to relate changes in temperature for relatively high temperature polymer melts, as follows:

$$\mu_a = (\mu_a)_T \exp\left(\Delta E_n/RT\right) \tag{4.18}$$

where μ_a is the apparent viscosity at temperature *T*, $(\mu_a)_T$ the apparent viscosity at a reference temperature, (ΔE_n) the activation energy for flow (in J/g.mol), *R* the gas constant ($R = 8.314$ J/g.mol), and *T* the absolute temperature (°K). As discussed by Peleg et al. (2002), it is worth pointing out that the use of thc Arrhenius model to account for the temperature effect on the apparent viscosity of polymeric melts may appear inadequate. In fact, when plotting ln μ_a as a function of *1/T* and calculating the activation energy (ΔE_n) by use of the gas constant *R*, the meaning of the activation energy is clearly questionable when dealing with bio-based polymeric systems. Hence eliminating the activation energy and gas constant from the equation and replacing $(\Delta E_n/R)$ by any parameter (i.e. *B* expressed in °K) is an alternative. The magnitude of *B*, the slope of the ln μ_a versus *1/T* relationship, remains a measure of the temperature effect which can be used to compare different systems or to calculate intermediate values of μ_a at a characteristic temperature range.

The effect of pressure on viscosity is relatively insignificant as long as the pressure does not exceed 35 MPa, which is generally the case in polymer melt extrusion processing.

A similar relationship has been employed to account for the effect of plasticizer content, as follows:

$$\mu_a = (\mu_a)_{Pl} \exp(K.Pl) \tag{4.19}$$

where μ_a is the apparent viscosity at plasticizer content *Pl*, $(\mu_a)_{Pl}$ the apparent viscosity at reference plasticizer content, *K* an empirical constant (non-dimensional), and *Pl* the plasticizer content (non-dimensional). Equation 4.19 assumes no interaction between the plasticizer and the other molecular species in the polymer melt. Therefore, it is better suited to situations where concentrations of plasticizer are small.

Structural modifications that affect the apparent viscosity of polymer melts may result from thermal alterations, and possibly chemical reactions, generated by the thermomechanical history of the extrusion process. The extent of structural modifications is reflected differently depending on the type of polymer. This will be discussed later in section 4.4.3, when examining the flow behavior of specific bio-based polymers.

So far in this section, polymer melts have been considered as purely viscous fluids and that the energy needed to shear the fluids is totally dissipated and so is not recoverable. Polymer melts exhibit viscoelastic behavior, which means that they may exhibit both viscous and elastic behaviors to some extent, depending upon the macromolecular characteristics of the melt (molecular weight dispersion, chain entanglements, etc.) and the time in which the deformation or flows occurs. Thus, the effects of the viscoelastic response are time dependent and the melt properties are a function of time and depend on the deformation history of the material. For a proper description of the flow behavior of a polymeric melt, its viscoelastic properties have to be taken into account, including its dependence on the deformation history. As an example, the swelling of an extrudate as it emerges from a die is caused by elastic recovery of the strain imparted to the polymer in the die assembly. The swelling phenomenon takes a finite time to develop fully and depends on the flow of the material through the die and the viscoelastic properties of the material. Hence, the viscoelastic properties of polymer melts significantly affect the final shape of formed extruded products.

Unfortunately, the viscoelastic models that include the dependence on deformation history are quite complex and difficult to apply to actual extrusion processes. Thus, when analyzing extrusion processes polymeric melts are generally considered to be viscous, time-independent fluids. This is particularly the case with emerging bio-based polymers whose rheological properties have not been extensively investigated.

4.4.2 Measurement of flow properties of polymer melts

Instruments for determining the rheological properties of fluids are generally referred to as rheometers. The most common rheometers based on the application of steady shear rate to the fluid are as follows.

- *Capillary rheometer*. This is basically a ram extruder with a piston that forces the polymer melt to pass through a capillary die. The measurement of the pressure drop along the capillary allows the calculation of the wall shear stress, while the wall shear rate is determined from the measured flow rate of melt through the capillary die, by measuring the displacement of the piston and the cross-sectional area of the capillary.
- *Melt index tester*. The principle is similar to that of the capillary rheometer but the piston operates at constant load (ASTM D1238) rather than a constant piston displacement.
- *Cone and plate rheometer*. The polymer melt is sheared in the gap between a stationary flat plate and a rotating conical plate. The cone plate rotates at constant angular velocity, which permits calculation of the shear rate. The torque necessary to rotate the cone allows the shear stress to be determined.

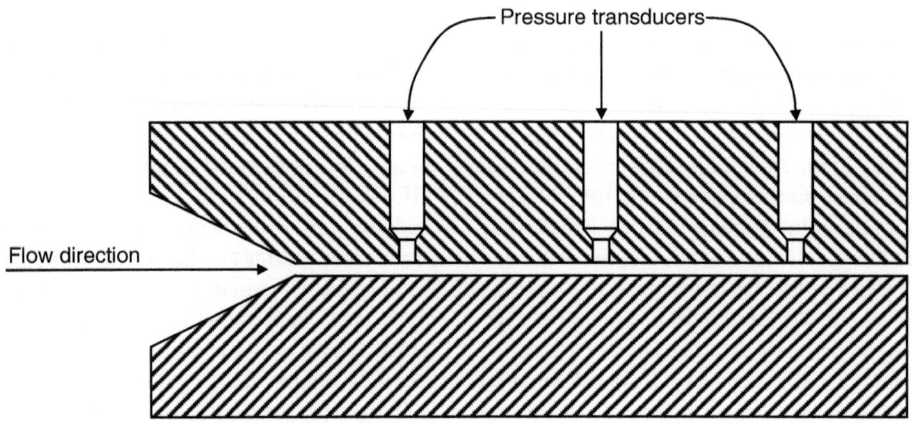

Figure 4.23 Schematic view of an extruder-fed die rheometer.

For a detailed description of these rheometers, the reader should refer to the book edited by Dealy and Wissbrun (1990).

The aforementioned rheometers are employed for off-line measurements of the material's rheological properties. They are not suitable for polymer melts which contain water as plasticizer, due to water flashing off as soon as the temperature reaches 100°C. Besides, a thermomechanical treatment is often needed to obtain a homogeneous melt that can flow, and such treatment induces significant structural changes and chemical modifications which affect the rheological properties of the melt. Consequently, classic off-line rheometry is not always appropriate for the study of the rheological behavior of bio-based polymer melts. For such materials, in-line rheometry is better adapted.

In that respect, it is worth presenting the principle of an extruder-fed die rheometer for in-line measurement of the rheological properties of time-temperature-shear dependent polymer melts. The die rheometer consists of a rectangular flow channel (slit die) or cylindrical flow channel (capillary die). The flow channel is equipped with pressure transducers at various axial locations (Figure 4.23). The die rheometer is connected to the barrel of the screw extruder which feeds the polymer melt under controlled conditions. By knowing the mass flow rate of polymer melt, Q, through the flow channel and the geometry of the flow channel (width, W_r, and height, H_r, for a rectangular flow channel; diameter, D_r, and length, L_r, for a cylindrical flow channel), the apparent shear rate, $\dot{\gamma}_{aw}$, at the wall can be derived. For a rectangular flow channel:

$$\dot{\gamma}_{aw} = \frac{6Q}{\rho_m W_r H_r^2} \tag{4.20a}$$

where ρ_m is the melt density. For a capillary flow channel:

$$\dot{\gamma}_{aw} = \frac{32Q}{\pi D_r^3} \tag{4.20b}$$

For a rectangular flow channel, the wall shear stress, τ_w, is given by:

$$\tau_w = \frac{H_r}{2}\left[\frac{1}{1+\frac{H_r}{W_r}}\right]\frac{\Delta P}{\Delta l} \tag{4.21}$$

where $\frac{\Delta P}{\Delta l}$ is the pressure gradient along the flow channel. In practice, W_r is much larger than H_r ($W_r >> H_r$); thus, Eq. 4.21 becomes:

$$\tau_w = \frac{H_r}{2}\frac{\Delta P}{\Delta l} \tag{4.22a}$$

For a capillary flow channel, the wall shear stress is given by:

$$\tau_w = \frac{D\Delta P}{4L} \tag{4.22b}$$

For a power law fluid, the correction of the shear rate (often referred to as Rabinowitsch correction) is needed. Then, for a rectangular flow channel, the actual shear rate at the wall becomes:

$$\dot{\gamma}_w = \frac{2n+1}{3n}\dot{\gamma}_{aw} \tag{4.23a}$$

and for a capillary flow channel, it becomes:

$$\dot{\gamma}_w = \frac{3n+1}{4n}\dot{\gamma}_{aw} \tag{4.23b}$$

with $n = d(\log \dot{\gamma}_{aw})/d(\log \tau_w)$. Finally, the apparent viscosity is calculated by:

$$\mu_a = \tau_w / \dot{\gamma}_w \tag{4.24}$$

Exit flow data can be used to calculate the first normal stress difference N_1, provided the velocity profile remains fully developed right up to the die exit. If the pressure gradient in the channel of the die rheometer is constant, the pressure can be extrapolated to the exit of the die. The exit pressure, P_{ex}, can be related to N_1 by the following equation (Han, 1974):

$$N_1 = P_{ex} + \tau_w\left(\frac{dP_{ex}}{d\tau_w}\right) \tag{4.25}$$

Although extrapolation of pressure data is practically difficult, exit pressure can be used to qualitatively evaluate the elasticity of polymer melts.

4.4.3 Rheological characteristics of bio-based polymer melts

The concepts of shear stress, shear rate, and apparent viscosity are important in analyzing and designing polymer extrusion processes, as well as downstream polymer processing operations (such as film blowing, injection molding, sheet forming, and fiber spinning). Melts from bio-based polymers are non-Newtonian, pseudoplastic fluids. The range of shear rates encountered in polymer processing is approximately 1–5000 s^{-1}. When including downstream processing operations, it is approximately 1–10,000 s^{-1}. As can be seen in Figure 4.22, within those ranges of shear rates, bio-based polymer viscosity is essentially dependent on shear rate.

For bio-based polymer melts, the flow behavior index, n, indicates the degree of non-Newtonian behavior. If n ranges from 0.75 to 1, the fluid is almost Newtonian but it is strongly non-Newtonian if n is less than 0.5. It turns out that thermoplastic starch (Martin et al., 2003), plasticized proteins (Redl et al., 1999b), and polylactic acid (Fang & Hanna, 1999) fall into this latter category. Polyamides (Rauwendaal, 2001) and polyhydroxyalkanoates (PHB and PHBV) (Yamaguchi & Arakawa, 2006) show relatively high flow behavior indices (over 0.6). The melt rheology of polytrimethylene terephthalate (PTT) is very similar to that of polybutylene terephthalate (PBT). The viscosity–shear rate relationship goes from moderately non-Newtonian at 10–15°C above melting temperature ($n \cong 0.5$) to quite Newtonian at temperatures of 40–50°C above melting temperature (Hwo et al., 1998). The latter observation shows that viscosity data and rheological behavior of bio-based polymer melts depend on processing conditions (shear and thermal histories, and possibly hydrolytic effect of water, etc.) which cause significant structural modifications of polymers. Bio-based polymer melts become more Newtonian as the severity of processing conditions increases, with an increase of the flow behavior index and a decrease of the consistency index. The influence of processing conditions must then be accounted for when investigating and modeling the rheological behavior of bio-based polymer melts.

The extrusion and viscous behavior of bio-based polymer melts depend on temperature, plasticizer content, and processing history undergone by polymers prior to viscosity measurements. In that respect, it is worth presenting relevant investigations related to bio-based polymers which are strongly non-Newtonian, in particular. Fang and Hanna (1999) used an extruder-fed capillary die rheometer to determine experimentally the shear viscosity of PLA polymers (amorphous and semi-crystalline) at two different barrel wall temperatures (150°C and 170°C) and at various shear rates ranging from 200 s^{-1} to about 1000 s^{-1}; the change of shear rate resulted from varying the screw speed of the extruder (from 30 to 150 rpm). Apparent viscosities of semi-crystalline PLA were slightly higher than those of amorphous PLA at both temperatures. The power law equation adequately described the behavior of PLA melts; a non-regression analysis allowed the flow behavior and consistency indices to be determined. Relatively low flow behavior indices have been found: n values were 0.17–0.19 at 150°C and 0.29–0.3 at 170°C. The consistency indices decreased considerably when increasing the barrel wall temperature: m values decreased from 6.0 10^5–6.5 10^5 Pa.s^n at 150°C down to 2.4 10^5 Pa.s^n at 170°C. But the barrel wall temperature did not reflect the melt temperature in the rheometer; therefore the effect of temperature on apparent viscosity could not be accounted for in the power law equation. Though the processing history in the extruder varied significantly due to the variation of screw speed in particular, its specific effects on the viscosity data were not included either.

Martin et al. (2003) have investigated the rheological behavior of thermoplastic starch. They carried out in-line rheometry measurements by use of an extruder-fed slit rheometer, as well as off-line measurements by use of a pre-shearing rheometer called Rheoplast® which aims to simulate extrusion of polymer melts. The apparent shear viscosity of plasticized molten starch was measured under conditions reflecting real processing conditions in terms of shear rates, temperature, composition (moisture and glycerol content), and thermomechanical processing. Thermomechanical processing was quantified based on measurement of the specific mechanical energy, SME. In the range of extrusion conditions, these authors have shown that viscosity data of starch melts followed the classic power law behavior. A general rheological model was derived from experimental results based on Eq. 4.17, with:

$$m = m_o \exp\left[\frac{E}{R}\frac{1}{T} - \alpha.MC - \alpha'.GC - \beta.SME\right] \qquad (4.26)$$

where m_o is the consistency index (Pa.s^n), E/R the reduced flow activation energy (°K), T the temperature (°K), MC the moisture content (w/w), and GC the glycerol content (w/w); α, α' and β are empirical parameters. The analysis of experimental data yielded coefficients of 10.9 and 4.7 for α and α' respectively, which confirms the plasticizing effects of moisture and glycerol on starch. Coefficient β was 5.9 10^{-3} kg/W.h, which confirms the effect of SME on the macromolecular degradation of thermoplastic starch. The flow behavior index was approximately 0.34. In the same study, through the analysis of exit pressures data, Martin et al. (2003) showed that thermoplastic modified starch exhibits elastic properties.

Little research has been done using extrusion technology for non-food uses of proteins and the rheological behavior of thermoplastic processed proteins has not been investigated extensively. Nevertheless, it is worth mentioning the excellent work done by Redl et al. (1999a, 1999b), who thoroughly studied gluten plasticized with glycerol through oscillatory shear. They proposed a general expression based on Eq. 4.26, to describe the viscous behavior of thermoplastic-treated gluten, and observed that the dependence of viscosity on temperature was comparable to that of molten starch. The dependence of viscosity on plasticizer content led to a coefficient α of 6.1. As seen earlier for thermoplastic starch, this value is lower than the dependence on moisture content for molten starch ($\alpha = 10.9$), but rather close to that for glycerol content ($\alpha' = 4.7$). An important difference for plasticized gluten comes from the energy term which was globally positive (β = 1.55 10^{-3} kg.W^{-1}.h^{-1}), whereas for thermoplastic-treated starch it was globally negative ($\beta = -5.9$ 10^{-3} kg.W^{-1}.h^{-1}). This means that shear viscosity of plasticized gluten increased with an increase in SME. This is likely due to covalent cross-linking, probably caused by disulfide bonds through free radical reaction mechanisms, leading to an increase of molecular weight. In contrast to thermoplastic starch, which is depolymerized by mechanical energy input, thermoplastic gluten shows an increasing network structure induced by the increase of SME.

As bio-based thermoplastics are expected to play an important role in the substitution of conventional petrochemical-based thermoplastics, flow properties of bio-based polymer melts should be further investigated, in order to better design and control appropriate polymer processing operations. In that respect, the development of rheological models based on Eqs 4.17 and 4.26 is strongly encouraged.

4.5 Case studies: emerging applications

4.5.1 Melting of polyamide-11 in a single screw extruder: exercise

Polyamide-11 is extruded in a single screw extruder, with the following screw geometry and operating conditions.

- Single flight, square pitch screw
- Screw diameter: Ds = 80 mm
- Channel depth of the feeding section: 16 mm
- Channel depth of the metering section: 4 mm
- Degree of tapering of the compression section (linear decrease of channel depth): A = 5.10^{-3}
- Flight width: e = 5 mm
- Flow rate of solid polymer: 100 kg/h
- Screw speed: 60 rpm
- Barrel wall temperature in compression section: 220°C
- Initial temperature of the solid polymer: 30°C

The physical properties of polyamide-11 were assumed constant in the compression section, as follows.

- Density of polymer melt: 1030 kg.m^{-3}
- Melting temperature: 185°C
- Thermal conductivity of polymer melt: 0.19 W.m^{-1}.°K^{-1}
- Enthalpy of melting: 39 kJ.kg^{-1}
- Specific heat of solid polymer: 1.75 kJ.kg^{-1}.°K^{-1}
- Apparent viscosity of polymer melt: 1000 Pa.s

a. Determine the quantity Φ, the melting efficiency.

b. Assuming constant channel depth of 16 mm in the melting zone, determine the solid bed profile $\frac{X}{W}(0)$ and the length of melting $Z_t(0)$. Plot the solid bed profile as a function of the down-channel distance z.

c. Considering the tapered channel in the melting zone, determine the solid bed profile $\frac{X}{W}(A)$ and the length of melting $Z_t(A)$. Plot the solid bed profile as a function of the down-channel distance z.

4.5.1.1 Solution

a. The quantity Φ is given by Eq. 4.2, where:

- $V_{bx} = V_b sin\,\theta = \pi.D_s N sin\,\theta = 0.2512 sin\,17.66° = 7.62\ 10^{-2} m.s^{-1}$
- $V_{sz} = 2.710^{-2} m.s^{-1}$ (given by Eq. 4.3)
- $\Delta V = (V_b^2 + V_{sz}{}^2 - 2|V_b||V_{sz}|cos\theta)^{1/2} = 0.226 m.s^{-1}$

$\Phi = 6.3810^{-2} kg.m^{-1.5} s^{-1}$

b. The solid bed profile $\frac{X}{W}(0)$ is given by Eq. 4.4, which leads to the following relationship: $\frac{X}{W}(0) = (1 - 0.3074z)^2$. Results are reported in Table 4.6 while Figure 4.24 shows the evolution of the solid bed profile as a function of the down-channel distance z. The length of melting is given by Eq. 4.7: $Z_t(0) = 3.26\ m$, which corresponds to approximately 16 turns long (0.20 m/turn).

c. The solid bed profile $\frac{X}{W}(A)$ is given by Eq. 4.6, which leads to the following relationship:

$$\frac{X}{W}(A) = \left[1.969 - (0.969)\left(\frac{H_o}{H_o - Az}\right)^{1/2}\right]^2.$$

Results are reported in Table 4.6 while Figure 4.24 shows the evolution of the solid bed profile as a function of the down channel distance z. The length of melting is given by

Table 4.6 Exercise, section 4.5.1: solid bed profile as a function of down-channel distance.

Z, m	$\frac{X}{W}(0)$	$\frac{X}{W}(A)$
0	1	1
0.2	0.88	0.937
0.6	0.665	0.80
1.2	0.40	0.55
1.8	0.20	0.25
2.4		0
2.5	0.054	
3.25	0	

Eq. 4.9: $Z_t(A) = 2.43m$, which corresponds to approximately 10 turns long (0.23 m/turn).

4.5.2 Extrusion processing of biodegradable starch-based loose-fill packaging foams

Loose-fill cushioning materials provide cost-effective protection for relatively light-weight products against shock or vibration during shipping and handling (Torok, 1990). Traditionally, loose-fill packaging materials have been prepared from petrochemical products such as polystyrene. Expanded polystyrene is the polymer most commonly used due to its low density, high resilience, and water resistance. Polystyrene-based loose-fill packaging foams are very light, with a bulk density of around 3 g/L, but disposal of used plastic packaging materials has become a public concern. In fact, such materials are experiencing growing pressure from existing and proposed environmental and disposal regulations, and from market-oriented sustainability initiatives. As they are

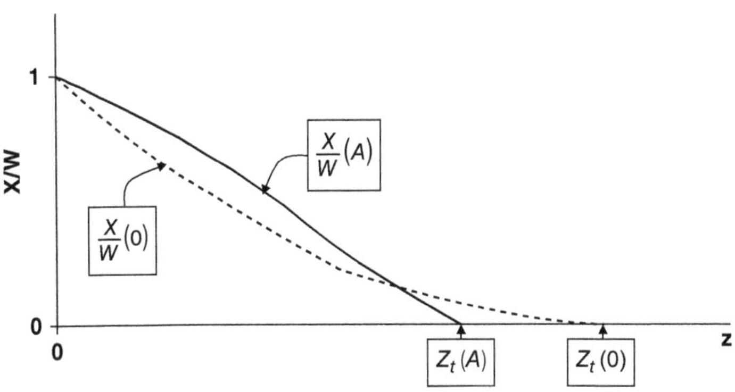

Figure 4.24 Solid bed profile as a function of down-channel distance (exercise, section 4.5.1).

light-weight and bulky, and not biodegradable, these materials create major disposal problems. Also, the manufacturing of petroleum-based loose-fill packaging materials increases oil imports in countries that do not have enough oil reserves. Thus, there is a growing market for bio-based, biodegradable loose-fill packaging foams that can be safely and effectively disposed of in soil or composting operations, while retaining current packaging foam performance requirements.

In the early 1990s, biodegradable starch-based loose-fill packaging foams were developed as a more environmentally friendly alternative. The starch comes from cereal-based sources (regular corn, waxy corn, sorghum, in particular). Several patents and papers have been published over the last 20 years relative to the formulation, processing, and properties of biodegradable starch-based loose-fill packaging foams (Nabar et al., 2006). As for the formulation, in addition to the starch component, various additives have been studied and proposed to fine-tune the end-use properties of the processed foams (such as bulk density, flowability, electrostatic adhesion, compressibility, resilience, in particular), and to impart more reproducible properties to starch foam. In general, additives included other biodegradable polymers (such as polyvinyl alcohol, polycaprolactone, cellulose acetate, starch esters, polylactic acid), plasticizers (such as glycerol, polyalkylene glycols), blowing agents (such as sodium carbonate, urea, ethanol, propanol, in addition to water), and nucleating agents (such as talc). However, starch is still the major component in the formulation in the range of 40–80% (wet basis).

Some authors have proposed combining starch with non-biodegradable polymers such as polystyrene or polymethyl methacrylate but the use of these polymers drastically lowers the biodegradability of the foams.

Starch-based foams are processed in both single and twin screw extruders but intermeshing co-rotating twin screw extruders are preferred when processing multicomponent formulations.

The properties of formulated starch-based loose-fill packaging foams (such as flowability, electrostatic adhesion, compressibility) compare favorably with commercial polystyrene-based foams. Their main drawbacks are higher bulk density (8–15 g/L), lower resilience, and dust creation. The determining advantage is biodegradability but they are water-sensitive (sometimes fully soluble in water), which is an issue for some applications. It must be underlined that the major issue in promoting biodegradable starch-based foams is their cost which is higher than that of polystyrene-based foams, due to two main factors: higher cost of the formulation and higher bulk density of the foam. In fact, the use of cost-effective additives is required when fine-tuning end-use properties of starch-based foams, including blowing and nucleating agents to minimize bulk density and non-starch thermoplastic polymers and plasticizers to improve extrusion processability of the blends, and foam compressibility and resiliency.

In the early 2000s, French inventors (Durand et al., 2000; Morillon & Morillon, 1999; Souveton et al., 2000) developed an alternative cost-competitive solution which uses a simple and economical formulation (based on 100% regular corn meal). Instead of using additives in the formulation, the inventors exploited the extrusion variables to thermomechanically convert the starch in such a way that the resulting foam develops the required physical characteristics. An industrial extrusion processing line has been developed by the Clextral Company, with a capacity of approximately 450 kg/h (Clextral, 2013). The Agripack Company (France) has operated such a processing line since 2000.

Figure 4.25 presents a flowsheet of the extrusion processing line for the production of biodegradable starch-based loose-fill packaging foams. The raw material (regular corn meal with an average particle size of 180–220 mμ; initial moisture content 13–14%; starch 75–78%; proteins 7.5–9.5%; fat 1–1.7%; minerals 04–05%) is stored in big-bags which are placed in a big-bag emptying station. This station is connected to a pneumatic conveyor which delivers the corn meal to a twin screw volumetric feeder. A second volumetric feeder may be included for feeding solid additives such as dyes, bactericides, etc. Both feeders feed the screw extruder. A pump for liquids allows water, and possibly water-soluble additives (dyes, for instance), to be added to the feeding section of the screw extruder. The main unit operation of the production line is the extruder. This is an intermeshing co-rotating twin screw extruder: EVOLUM® range; 53 mm screw diameter; L/D = 20. The screw configuration is classically composed of a feeding section, a compression section, and a shearing section. The latter section is particularly important as it determines the degree of starch conversion. In this particular process, where high levels of die melt expansion and a foam with suitable resiliency are required without use of functional additives, a high but controllable degree of starch dextrinization must be achieved. In fact, starch dextrinization generates low molecular weight components which contribute to starch plasticization, and so to die melt expansion and foam resilience; this can be considered as an autogenous dextrinization-induced plasticization of starch.

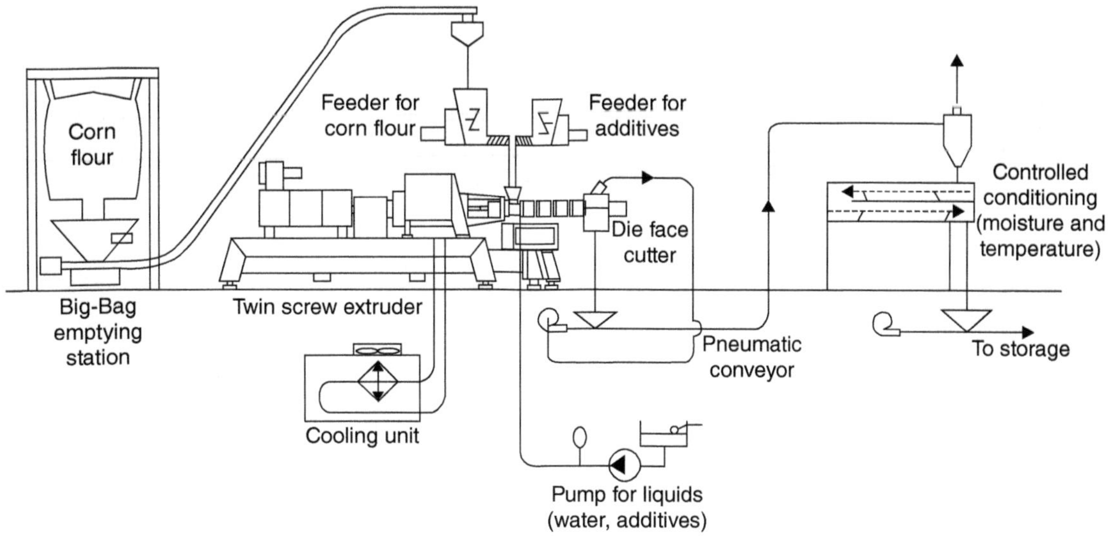

Figure 4.25 Flowsheet of the extrusion processing line for biodegradable starch-based loose-fill packaging foams. Source: Reproduced with permission of Clextral, France.

The targeted degree of dextrinization is achieved with a screw profile which favors relatively low residence time, high temperature and shear rate in the shearing section.

The extruder operates at high screw speeds (800 rpm and over). In these conditions, the specific mechanical energy (SME) ranges from 140 to 180 W.h/kg. The screw-barrel assembly of the extruder holds a die assembly which has inserts to define the shape of the resulting extrudates. Various shapes can be created but the most common shapes are similar to a cylindrical chip or to S-shape chip (for efficient locking under compression). The converted starch melt expands when it exits the die, due to water flashing off. A die-face cutter cuts the expanded ropes to give the final loose-fill packaging foams. After die-face cutting, the foams are transferred to a conditioner, or a conditioning room, which allows the starch-based foams to equilibrate in moisture and temperature. Figure 4.26 shows biodegradable starch-based loose-fill packaging foams (cylindrical and S-shape chips) produced by the extrusion processing line and Table 4.7 gives their standard process and physical characteristics.

4.5.3 Extrusion compounding of flax fiber-reinforced thermoplastics

Environmental and sustainability concerns have recently generated important interest in the development of fiber-reinforced composite materials based on renewable resources such as natural fibers and bio-based polymers, as pertinent alternatives (from technical and economical standpoints) to glass fibers and petrochemical polymers, respectively.

Engineered composite materials are made from the association of at least two constituent materials showing different mechanical properties, which remain separate and distinct within the finished structure. There are two categories of constituent materials: the matrix, also called resin, and the reinforcement. A polymeric matrix is used in many commercial composites while the reinforcements are often fibrous materials. The properties of the composite materials result from the large differences in mechanical properties of the polymeric matrix and fibers. For instance, thermoplastic polymers present tensile Young modulus of 1–2 GPa and tensile strength of 30–80 MPa whereas fibers show tensile Young modulus and tensile strength up to 500 GPa and 5 GPa, respectively. In other words, the mechanical properties of fibers and thermoplastic polymers differ considerably, from a factor of 10^2 to 10^4. This explains why adding fibers to a thermoplastic polymeric matrix leads to remarkable mechanical properties of the resulting composites; provided the matrix-fiber binding is effective, the fibers are oriented according to the direction of mechanical stresses, and the composite processing operates in optimal conditions. In fact, the major factors that govern the properties of fiber-reinforced thermoplastic composites are

Cylindrical chips S-Shape chips

Figure 4.26 Biodegradable starch-based loose-fill packaging foams. Source: Reproduced with permission of Clextral, France. For color detail, please see color plate section.

Table 4.7 Process and physical characteristics of biodegradable starch-based loose-fill packaging foams.

Type of loose-fill foam (100% starch content)	SME (W. h/kg)	Bulk density (g/liter)	Resiliency (%)*	WSI (%)**	Moisture content (%)***
Cylindrical chip	177	13	85	100	8.5
S-shape chip	140	8	81	100	9.2

* Elastic recovery after 50% uniaxial compression and release.
** Solubility in water after 30 seconds.
*** Moisture content after equilibrium at ambient temperature.
Source: Reproduced with permission of Clextral, France.

fiber-matrix adhesion, fiber orientation, and process-related factors (such as fiber volume fraction, fiber mixing, fiber-to-diameter ratio, fiber length distribution, extent of fiber degradation, in particular).

Engineered composite materials can be divided into two main categories, normally referred to as short fiber-reinforced materials and continuous fiber-reinforced materials. Natural fibers such as stem-based, technical fibers (flax, hemp, jute, ramie, etc.), described in section 4.1.2.3 (see Table 4.1), are suitable to reinforce polymer matrices, both thermoplastics and thermosets. And undoubtedly, there is growing interest in the use of natural fibers as reinforcing constituents in polymeric matrices, owing to their relevant advantages such as low density, low carbon footprint, renewability and biodegradability, relatively high specific strength and stiffness. Hence, new processing methods are presently under development, which aim to adapt existing technologies to be able to substitute conventional reinforcing fibers (such as glass fibers) and produce consistently natural fiber-reinforced thermoplastics or composites.

As an example, it is worth mentioning the ambitious development project initiated and led by FiMaLin® (2011). FiMaLin® was created in 2009 by French companies to build and promote a new flax industrial field in order to provide a range of high-performance flax fibers as sustainable reinforcement materials for thermoplastics and thermosets. Among others, FiMaLin® is developing an innovative, versatile extrusion processing method able to compound flax fiber-reinforced thermoplastics from various types of flax fibers and thermoplastic polymers.

Figure 4.27 shows the projected process diagram of an emerging extrusion compounding process, which starts from flax straw bales (resulting from field harvesting which comprises the following operations: plant pulling, plant turning, retting and flax straw rolling up; harvesting yield amounts approximately to 7 tonnes per hectare) and ends with fiber-reinforced compounds (semi-products

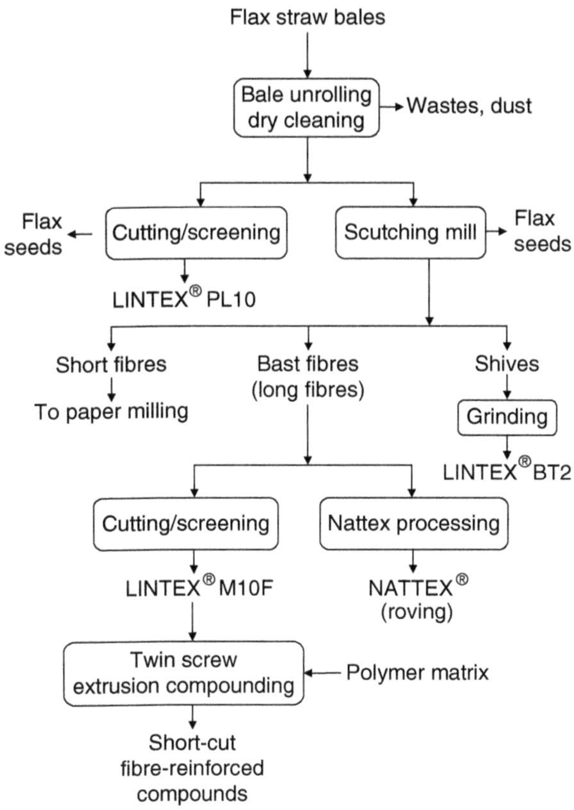

Figure 4.27 Process diagram of flax fiber-reinforced compounding. Source: FiMaLin®, 2011. Reproduced with permission of Dehondt Technologies.

which are further processed to make finished products through injection molding, for instance). The processing line is based on a fully continuous process which is composed of two main unit operations.

- Flax fiber handling and refining, to convert field-harvested flax plant into technical flax fibers.
- Twin screw extrusion compounding of flax fiber-reinforced thermoplastics, to produce highly homogeneous fiber-reinforced compounds.

4.5.3.1 Flax fiber handling and refining

Raw flax straw bales are unrolled and subjected to physical fractionation to separate flax seeds (approximately 29 kg per tonne of flax straw bale), flax straw (approximately 885 kg per tonne of flax straw bale), and wastes (rocks, soil; approximately 86 kg per tonne of flax straw bale). Flax seeds are directed to oil milling whereas flax straw is processed through different steps to provide various grades of technical flax fibers. Hence, flax straw is conveyed to a beater-cleaner in order to separate and remove wastes and dust. Dry-cleaned flax straw can either be ground and screened to provide Lintex® PL10 (whole plant fiber) or conveyed to a scutching mill to produce higher grades of technical fibers. A scutching mill consists of separating wooden core particles (that are shives which provide Lintex® BT2 after grinding; approximately 500 kg per tonne of flax straw bale) and bast fibers that are long fibers (approximately 214 kg per tonne of flax straw bale) and short fibers (approximately 171 kg per tonne of flax straw bale). Long fibers can be calibrated and screened to provide Lintex® M10F or upgraded through Nattex processing to give Nattex® (roving grade) whereas short fibers are directed to paper milling. Value addition of technical flax fibers increases in the following order: Lintex® BT2 < Lintex® PL10 < Lintex® M10F < Nattex®. Figure 4.28 shows the different grades of technical flax fibers.

4.5.3.2 Twin screw extrusion compounding

Lintex® grade fibers can be used to produce short fiber-reinforced compounds through extrusion processing. Extrusion compounding is generally carried out in a high L/D ratio (in the range of 40–48), intermeshing co-rotating twin screw extruder which consists of several processing sections in series.

- Polymer melting section where the polymer matrix is melted. Polymer granulates are fed into the transport subsection, and then compressed and melted by use of a typical screw configuration for polymer melting, as described in Chapter 2 (section 2.2.2; see Figure 2.20).
- Fiber feeding section where flax fibers are fed into the extruder barrel. Consistent feeding of fibers is required to ensure a constant mass flow rate of the fibers. With low-grade Lintex® fibers (such as Lintex® BT2), this can be done successfully with classic loss-in-weight screw-aided feeders. But with intermediate and high-grade Lintex® fibers (such as Lintex® PL10 and Lintex® M10F) which show complex flow behavior (entangled, low-density fibers), a special feeder design is required to ensure regular and stable flow of the fibers into the extruder barrel. In this section, the screw configuration is composed of right-handed pitch screw elements with large pitch, as fiber-to-polymer weight ratio may range from 10% to 40%.
- Laminar mixing section where polymer melt and flax fibers are mixed. In this section, the screw configuration

Figure 4.28 Different grades of technical flax fibers. Source: FiMaLin®, 2011. Reproduced with permission of Dehondt Technologies. For color detail, please see color plate section.

is composed of right-handed pitch screw elements combined with kneading blocks the design of which needs special care to obtain an intimate mixing of flax fibers and polymer melt, while preventing fiber degradation due to viscous dissipation.

• Necessary degassing section where residual water is flashed off, to prevent steam formation in final compounds.

• Compounding section where the fiber-polymer mix is conveyed to a shaping die to produce strands which are cooled and cut downstream, to produce short fiber-reinforced compounds (semi-products).

Twin screw extrusion compounding, which is in development, should be able to process any type of natural fiber (flax, hemp, etc.) with any source of polymer matrix (petrochemical polymers, bio-sourced polymers, biodegradable polymers) as long as the processing temperature of thermoplastic polymers is compatible with the susceptibility of natural fibers to thermal degradation. Hence, processing temperatures higher than 250°C cannot be envisaged. Undoubtedly, the demand for natural fibers (in particular, flax fibers owing to their physical and mechanical properties) in plastic composites is growing in many industrial sectors, owing to increasing pressure to use more environmentally friendly materials in manufactured products. However, it appears that natural fibers (including flax fibers) do not yet match the performance and consistency offered by existing reinforcement materials (such as glass fibers), due to the lack of material and process optimization and at the present time, natural fibers are able to substitute only lower grade fiber glass.

Natural fibers are used only to a limited extent in industrial practice due to relatively inadequate raw materials and fiber processing when dealing with polymer reinforcement. The primary drawback of natural fibers is associated with the lack of dedicated varieties and fiber extraction, which leads to insufficient physical and mechanical properties of resulting fibers. Increasing physical and mechanical properties of fibers through variety selection and appropriate fiber extraction should allow their reinforcement capability to be increased. Another drawback of natural fibers is their polar nature which is not compatible with the non-polar characteristics of most polymer matrices. Improving surface interactions between natural fibers and polymer matrices through appropriate surface modifications of fibers should favor dispersion levels in compounding and effective physico-chemical binding between both components, hence leading to efficient composites. Another drawback may be associated with conventional extrusion compounding processes which require technical adaptation (such as fiber feeding and laminar mixing) to account for the physical and chemical characteristics of natural fibers and allow their reinforcement potential to be revealed optimally, hence leading to consistent fiber-reinforced compounds. In that context, FiMaLin's initiative is relevant, as its development project for flax fibers is dedicated to minimizing the above drawbacks.

References

Abiad MG, Carvajal MT, Campanella OH (2009) A review on methods and theories to describe the glass transition phenomenon: applications in food and pharmaceutical products. Food Engineering Reviews 1(2): 105–132.

Averous L (2004) Biodegradable multiphase systems based on plasticized starch: a review. Journal of Macromolecular Science Part C: Polymer Reviews C44(3): 231–274.

Baley C (2002) Analysis of the flax fibres tensile behaviour and analysis of the tensile stiffness increase. Composites Part A: Applied Science and Manufacturing 33(7): 939–948.

Baley C (2005) Fibres naturelles de renfort pour matériaux composites. Techniques de l'Ingénieur, AM 5 130: 1–12.

Batzer H, Kreibich UT (1981) Influence of water on thermal transitions in material polymers and synthetic polyamides. Polymer Bulletin 5(11–12): 585–590.

BeMiller JN, Whistler RL (2009) *Starch – chemistry and technology*, 3rd edn. Food Science and Technology, International Series. Boston: Academic Press.

Bertoft E, Koch K, Åman P (2012) Building block organisation of clusters in amylopectin from different structural types. International Journal of Biological Macromolecules 50(5): 1212–1223.

Biliaderis CG, Page CM, Maurice TJ, Juliano BO (1986) Thermal characterization of rice starches: a polymeric approach to phase transition of granular starch. Journal of Agricultural and Food Chemistry 34: 6–14.

Bledzki AK, Gassan J (1999) Composites reinforced with cellulose based fibres. Progress in Polymer Science 24: 221–274.

Bledzki AK, Mamun AA, Lucka-Gabor M, Gutowski VS (2008) The effects of acetylation on properties of flax fibre and its polypropylene composites. eXPRESS Polymer Letters 2(6): 413–442.

Boisot G (2009) Mécanismes et modélisation mécanique de la deformation, de l'endommagement et de la rupture du PolyAmide 11 pur et renforcé choc. PhD dissertation, Mines ParisTech, Paris, France.

Bos HL, Müssig J, Van den Oever MJA (2006) Mechanical properties of short-flax-fibre reinforced compounds. Composites Part A: Applied Science and Manufacturing 37(10): 1591–1604.

Brown HS, Casey PK, Donahue JM (2000) Poly(Trimethylene Terephthalate): Polymer for Fibers. Shell Chemical Company, Westhollow Technology Center, Houston, Texas, USA. Information available through: www.technica.net/NF/NF1/eptt.htm.

Buleon A, Colonna P, Planchot V, Ball S (1998) Starch granules: structure and biosynthesis. International Journal of Biological Macromolecules 23: 85–112.

Carvalho AJF (2008) Starch: major sources, properties and applications as thermoplastic materials. In: Belgacem M, Gandini A (eds) *Monomers, Polymers and Composites from Renewable Resources*. Amsterdam: Elsevier, pp. 321–342.

Charlet K, Baley C, Morvan C, Jernot J.P, Gomina M, Bréard J (2007) Characteristics of *Hermès* flax fibres as a function of their location in the stem and properties of the derived unidirectional composites. Composites Part A: Applied Science and Manufacturing 38(8): 1912–1921.

Chodak I (2008) Polyhydroxyalkanoates: origin, properties and applications. In: Belgacem M, Gandini A (eds) *Monomers, Polymers and Composites from Renewable Resources*. Amsterdam: Elsevier, pp. 451–477.

Chung HJ, Lee EJ, Lim ST (2002) Comparison in glass transition and enthalpy relaxation between native and gelatinized rice starches. Carbohydrate Polymers 48: 287–298.

Clextral (2013) Private communication: Extrusion processing line for the production of biodegradable starch-based loose-fill packaging foams. Information available through: www.clextral.com/.

Cocero AM, Kokini JL (1991) The study of the glass transition of glutenin using small amplitude oscillatory rheological measurements and differential scanning calorimetry. Journal of Rheology 35: 257–270.

Couchman PR, Karasz FE (1978) A classical thermodynamic discussion of the effect of composition on glass-transition temperatures. Macromolecules 11: 117–119.

Crank M, Patel MK, Marscheider-Weidemann F, Schleich F, Hüsing B, Angerer G (2005) *Techno-economic feasibility of large-scale production of bio-based polymers in Europe*. Technical Report Series, European Commission, EUR 22103 EN. Information available through: www.jrc.es/.

Cui SW (2005) *Food Carbohydrates: chemistry, physical properties, and applications*. Boca Raton, Florida: CRC Press.

Cuq B, Gontard N, Guilbert S (1997) Thermoplastic properties of fish myofibrillar proteins: application to biopackaging fabrication. Polymer 38(16): 4071–4078.

De Graaf RA, Karman AP, Janssen L (2003) Material glass transition temperatures of different thermoplastic starches after extrusion processing. Starch/Stärke 55: 80–86.

Dealy JM, Wissbrun KF (1990) *Melt Rheology and Its Role in Plastics Processing. Theory and applications*. New York: Van Nostrand Reinhold.

Di Gioia L, Guilbert S (1999) Corn protein-based thermoplastic resins: effect of some polar and amphiphilic plasticizers. Journal of Agricultural and Food Chemistry 47(3): 1254–1261.

Di Gioia L, Cuq B, Guilbert S (1999) Thermal properties of corn gluten meal and its proteic components. International Journal of Biological Macromolecules 24: 341–350.

DuPont (2007) Fermentation: A New Take on Old Technology. Information available through: www2.dupont.com/Renewably_Sourced_Materials/en_US/proc-buildingblocks.html.

Durand D, Souveton G, Tabouillot P, Reynes P (2000) Procédé et installation de fabrication en continu d'un produit de calage expansé en matière biodégradable. French Patent 2817501.

Fang Q, Hanna MA (1999) Rheological properties of amorphous and semicrystalline polylactic acid polymers. Industrial Crops and Products 10: 47–53.

FiMaLin (2011) Private communication: Flax technical fibres for the reinforcement of thermoplastics and thermosets. Information available through: www.fimalin.com/.

Fox TG, Flory PJ (1950) Second-order transition temperatures and related properties of polystyrene. I. Influence of molecular weight. Journal of Applied Physics 21: 581–591.

French DJ (1972) Fine structure of starch and its relationship to the organization of starch granules. Journal of the Japanese Society for Starch Science 19: 8–12.

Frost JW (2005) Synthesis of caprolactam from lysine. PCT application, WO 2005123669.

Han CD (1974) On slit- and capillary-die rheometry. Journal of Rheology 18: 163–190.

Hernandez-Izquierdo VM, Krochta JM (2008) Thermoplastic processing of proteins for film formation – a review. Journal of Food Science 73(2): 30–39.

Herrera Ramirez JM (2004) *Les mécanismes de fatigue dans les fibres thermoplastiques*. PhD dissertation, Ecole Nationale Supérieure des Mines, Paris, France.

Hizukuri S (1986) Polymodal distribution of the chain lengths of amylopectins, and its significance. Carbohydrate Research 147: 342–347.

Hoseney RC, Zeleznak K, Lai CS (1986) Wheat gluten: a glassy polymer. Cereal Chemistry 63: 285–286.

Hwo C, Forschner T, Lowtan R, Gwyn D, Cristea B (1998) Poly (trimethylene phthalates or naphthalate) and copolymers: new opportunities in film and packaging applications. Future-Pak® 98 Conference, November 10–12, Chicago. Information available through: www.shellchemicals.com/.

Jenkins PJ (1994) X-ray and neutron scattering studies of starch granule structure. PhD dissertation, Robinson College, Cambridge, UK.

Kalichevski MT, Blanshard JM, Tokarczuk PF (1993a) Effect of water content and sugars on the glass transition of casein and sodium caseinate. International Journal of Food Science and Technology 28: 139–151.

Kalichevski MT, Jaroszkiewicz EM, Blanshard JM (1993b) A study of the glass transition of amylopectin–sugar mixtures. Polymer 34(2): 346–358.

Kohan MI, Mestemacher SA, Pagilagan RU, Redmond K (2003) Polyamides. In: *Ullmann's Encyclopaedia of Industrial Chemistry*, 7th edn. Weinheim: Wiley-VCH Verlag.

Kokini JL, Cocero AM, Makeda H (1995) State diagrams help predict rheology of cereal proteins. Food Technology 368: 74–82.

Leblanc J (2010) Les fibres naturelles, entre économie et écologie. Plastiques & Caoutchoucs Magazine 875: 47–49.

Li K, Frost JW (1999) Microbial synthesis of 3-dehydroshikimic acid: a comparative analysis of D-xylose, L-arabinose, and D-glucose carbon sources. Biotechnology Progress 15(5): 876–883.

Maddock BH (1959) A visual analysis of flow and mixing in extruder screws. Society of Plastics Engineers Journal 15: 383–389.

Martin O, Averous L (2001) Poly(lactic acid): plasticization and properties of biodegradable multiphase systems. Polymer 42: 6209–6219.

Martin O, Averous L, Della Valle G (2003) In-line determination of plasticized wheat starch viscoelastic behaviour: impact of processing. Carbohydrate Polymers 53: 169–182.

Mohanty AK, Misra M, Hinrichsen G (2000) Biofibres, biodegradable polymers and biocomposites: an overview. Journal of Applied Polymer Science 276/277: 1–24.

Morillon D, Morillon P (1999) Matériau de calage compressible biodégradable et procédé de fabrication. French Patent 2801041.

Nabar Y, Naravan R, Schindler M (2006) Twin-screw extrusion production and characterization of starch foam products for use in cushioning and insulation applications. Polymer Engineering and Science 46(4): 438–451.

Nabi Saheb D, Jog JP (1999) Natural fiber polymer composites: a review. Advances in Polymer Technology 18(4): 351–363.

Niu W, Draths KM, Frost JW (2002) Benzene-free synthesis of adipic acid. Biotechnology Progress 18(2): 201–211.

Orford PD, Parker R, Ring SG, Smith AC (1989) Effect of water as a diluent on the glass transition behaviour of malto-oligosaccharides, amylose and amylopectin. International Journal of Biological Macromolecules 11: 91–96.

Peleg M, Engel R, Gonzalez-Martinez C, Corradini MG (2002) Non-Arrhenius and non-WLF kinetics in food systems. Journal of the Science of Food and Agriculture 82: 1346–1355.

Pommet M, Redl A, Morel MH, Domenek S, Guilbert S (2003) Thermoplastic processing of protein-based bioplastics: chemical engineering aspects of mixing, extrusion and hot molding. Macromolecules Symposium 197: 207–217.

Pommet M, Redl A, Guilbert S, Morel MH (2005) Intrinsic influence of various plasticizers on functional properties and reactivity of wheat gluten thermoplastic materials. Journal of Cereal Science 42: 81–91.

Rauwendaal C (2001) *Polymer Extrusion*, 4th edn. Munich: Carl Hanser Verlag.

Redl A, Morel MH, Bonicel J, Vergnes B, Guilbert S (1999a) Extrusion of wheat gluten plasticized with glycerol: influence of process conditions on flow behaviour, rheological properties and molecular size distribution. Cereal Chemistry 76(3): 361–370.

Redl A, Morel MH, Bonicel J, Guilbert S, Vergnes B (1999b) Rheological properties of gluten plasticized with glycerol: dependence on temperature, glycerol content and mixing conditions. Rheologica Acta 38: 311–320.

Robin PJ, Mercier C, Charbonnière R, Guilbot A (1974) Litnerized starches. Gel filtration and enzymatic studies of insoluble residues from prolonged acid treatment of potato starch. Cereal Chemistry 51: 389–406.

Roos Y, Karel M (1991) Water and molecular weight effects on glass transitions in amorphous carbohydrates and carbohydrate solutions. Journal of Food Science 56(6): 1676–1681.

Rudnik E (2008) *Compostable Polymer Materials*. Amsterdam: Elsevier.

Seetharaman K, Bertoft E (2013) Review: Perspectives on the history of research on starch. Part V: On the conceptualization of amylopectin structure. Starch/Stärke 65(1–2): 1–7.

Shen L, Haufe J, Patel MK (2009) Product overview and market projection of emerging bio-based plastics. PRO-BIP 2009, final report commissioned by European Polysaccharide Network of Excellence and European Bioplastics, Utrecht University, The Netherlands.

Simon CJ, Schnieders F (2007) Business data and charts 2006. PlasticsEurope Market Research Group. Information available through: www.plasticseurope.org/.

Souveton G, Durand D, Tabouillot P, Reynes P (2000) Produit expansé biodégradable et son procédé de fabrication. French Patent 2817554.

Tadmor Z, Klein I (1970) *Engineering Principles of Plasticating Extrusion*. Polymer Science and Engineering Series. Florida: Robert E. Krieger.

Tadmor Z, Duvdevani I, Klein I (1967) Melting in plasticating extruders: theory and experiments. Polymer Engineering Science 7: 198.

Thompson DB (2000) On the non-random nature of amylopectin branching. Carbohydrate Polymers 43: 223–239.

Todd DB (1993) Melting of plastics in kneading blocks. International Polymer Processing VIII(2): 113–118.

Torok DB (1990) Cushioning: protecting the product and environment. Material Handling Engineering 45: 76–81.

Vergnes B, Souveton G, Delacour ML, Ainser A (2001) Experimental and theoretical study of polymer melting in a co-rotating twin screw extruder. International Polymer Processing XVI(4): 351–362.

Wambua P, Ivens J, Verpoest I (2003) Natural fibers: can they replace glass in reinforced plastics? Composites Science and Technology 63: 1259–1264.

Wang S, Sue HJ, Jane J (1996) Effects of polyhydric alcohols on the mechanical properties of soy protein plastics. Journal of Macromolecular Science Part A: Pure and Applied Chemistry 33(5): 557–569.

Wu J, Schultz JM, Samon JM, Pangelinan AB, Chuah HH (2001) In situ study of structure development in poly(trimethylene therephthalate) fibers during stretching by simultaneous synchrotron small- and wide-angle X-ray scattering. Polymer 42: 7141–7151.

Wunderlich B (2005) *Thermal Analysis of Polymeric Materials*. Berlin: Springer-Verlag.

Yamaguchi M, Arakawa K (2006) Effect of thermal degradation on rheological properties for poly(3-hydroxybutyrate). European Polymer Journal 42: 1479–1486.

Zahardis J, Petrucci GA (2007) The oleic acid–ozone heterogeneous reaction system: products, kinetics, secondary chemistry, and atmospheric implications of a model system – a review. Atmospheric Chemistry and Physics 7(5): 1237–1274.

Zobel HF, Young SN, Rocca LA (1988) Starch gelatinisation: an X-ray diffraction study. Cereal Chemistry 65: 443–446.

5 The Generic Extrusion Process II: Thermomechanical Micromixing and Reactive Extrusion

Chemical reactions are conducted mainly in batch or continuous reactors. Specifically for reactions involving chemical modifications of polymers or polymerization of monomers, they have been historically carried out under diluted conditions, thus avoiding problems caused by the mixing of high viscous reactant or products to achieve a homogeneous mixture. However, the use of solvents or diluents, which usually comprise 5–20 times the weight of the desired polymer product, requires costly facilities for recovery, vaporization, condensation, purification, storage, metering, and reuse. As extrusion technology has proven its efficiency and reliability in the polymer processing industry through the use of the Generic Extrusion Process I (Chapter 4), it has been recognized that the application of screw extruders could be extended into chemical reactions involving polymers. This has led to the development of the Generic Extrusion Process II (GEP II) which is aimed at resolving fluid mechanics/mixing and heat and mass transfer problems in high-viscosity, solvent-free reactive polymeric systems.

Since 1965, more than 1500 different patents have been granted to approximately 200 companies, with intense activity over the period 1965–1990. It must be noted that during this period, more activity in reactive extrusion has taken place in industrial laboratories (petrochemical and plastics companies, in particular) and equipment companies (extruder manufacturers) rather than in academic laboratories. The early interest of processing companies and extruder manufacturers in reactive extrusion is closely related to the ability of screw extruders to handle chemical and biochemical reactions involving high-viscosity polymeric systems. Hence, the large majority of patent applications concern the field of polymer reactive processing. A survey on the database Web of Science shows an increasing interest in the academic community, as more than 1500 reactive extrusion papers were published in the literature from the early 1980s to 2010, reaching approximately 220 papers per year since 2005 (Figure 5.1). Approximately 1000 publications relate to polymer science while almost 300 deal with chemical and process engineering.

Since energy and environmental concerns have become increasingly important in the processing industry, screw extruders are now being considered as very promising and efficient chemical reactors for continuous bulk reactive processing, and reactive extrusion is being considered as a sustainable processing technology. Hence, extensive applications of screw extruder-reactors are expected due to new developments in sustainability in the processing industry (e.g. innovative polymeric materials from renewable resources, green chemistry, cost-competitiveness of processes and products, sustainable processing).

The GEP II is qualitatively described in the first part of this chapter. Reactive extrusion is carried out continuously in screw extruder-reactors (single screw and twin screw) which have been found suitable for handling chemical reactions in complex media including both highly viscous homogeneous and heterogeneous systems, and others with varying viscosity. The qualitative description of GEP II is based on case studies taken from patents, open literature, and industrial applications, which cover many of the different types of reactive extrusion processes. The second part of this chapter gives an engineering analysis

Extrusion Processing Technology: Food and Non-Food Biomaterials, First Edition. Jean-Marie Bouvier and Osvaldo H. Campanella.

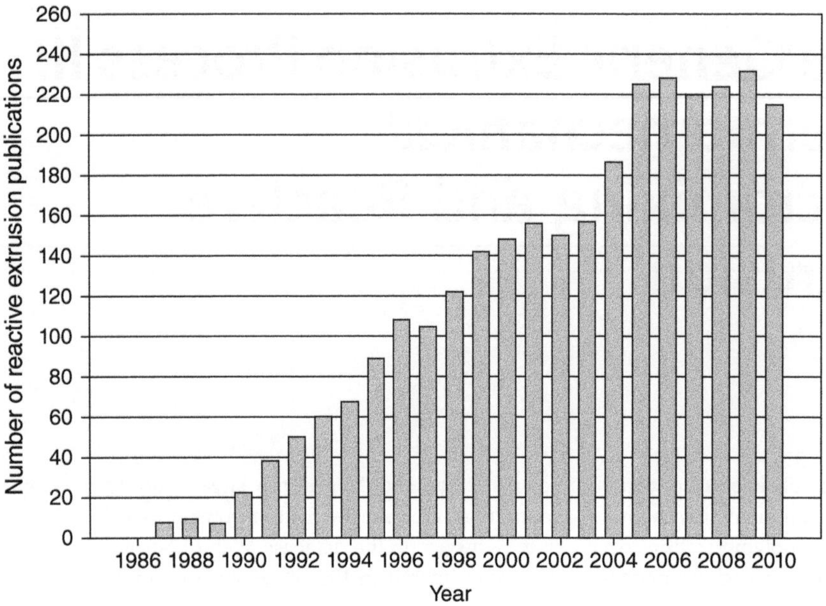

Figure 5.1 Number of reactive extrusion publications versus year of publication.

and perspectives of reactive extrusion processing with a focus on mixing capabilities of screw extruder-reactors to handle highly viscous materials. The third part of the present chapter describes the operation of well-recognized applications which illustrate the use of reactive extrusion in different processing industries. A focus on three case studies is proposed:

- casein-to-caseinate extrusion processing
- extrusion pulping of non-wood fibers
- enzymatic hydrolysis of starch.

5.1 Reactive extrusion: qualitative description

Vergnes and Berzin (2004) compare classic batch reaction processes involving relative diluted solutions with the process of reactive extrusion and list the following advantages of the latter.

- The reaction is conducted in the melt, in the absence of any kind of solvent.
- The extruder is able to handle very viscous materials, which is not the case for batch reactors.
- The range of processing conditions is much wider and also more flexible in an extruder-reactor (mixing conditions, high temperatures, modular geometry, etc.).

Drawbacks of the reactive extrusion process include:

- short residence times, so only fast reactions are possible in reactive extrusion
- the cooling capacity of an extruder-reactor is limited and reactions with high exothermicity could be difficult to manage in reactive extrusion processes unless good mixing and efficient heat exchange are achieved.

With regard to this subsection, it is worth reviewing the major process functions of extruder-reactors while examining the process configurations of typical reactive extrusion applications in the polymer processing industry as well as innovative approaches which have emerged from non-conventional fields of extrusion technology such as the speciality chemical and paper-milling industries. This will enable us to highlight processing characteristics of this important generic extrusion process.

5.1.1 Bulk polymerization

Reactive extrusion has been of great industrial interest since its beginnings, as shown by the hundreds of patents published since 1965, particularly on continuously polymerizing monomers in bulk reactions in order to avoid the use of solvents and enhance process productivity and flexibility (Brown, 1991). In bulk polymerization, a monomer or a mixture of monomers is converted to a high molecular weight polymer with little or no solvent dilution. Extruder-reactors have been designed to handle a pure monomer as feed or low-viscosity prepolymer from a continuous stirred tank reactor for final polymerization in the extruder. Both addition and condensation polymerization processes have been performed in extruder-reactors.

In the polymerization of monomers, such as in the case of polyamide synthesis by ring opening polymerization of lactams, ε-caprolactam has been polymerized in an intermeshing, co-rotating twin screw extruder-reactor (Bartilla et al., 1986; Menges & Bartilla, 1987). In this process, a mix of monomer and catalyst (or initiator) is fed to the extruder-reactor, while the temperature profile across the extruder is gradually increased from 70°C to 270°C. The modularity of the screw-barrel assembly of twin screw extruders enables several operations to be carried out in the same unit. The mix enters into a mixing section for intense mixing of the monomer and the catalyst, and then into the polymerization section to reach the desired degree of conversion. Residual monomer, which can cause undesirable foaming in the final extrudate, is removed in the venting section (where vacuum is applied), held at a temperature above the boiling point of the caprolactam monomer (that is, 220°C). Figure 5.2a shows the typical process profile of the twin screw extruder-reactor used for the polymerization of caprolactam.

In step-growth polymerization, such as in the case of polyurethanes, exact stoichiometry matching is critical for the production of high molecular weight polymers and often reactants are fed to the extruder-reactor as melts or liquids. In a process described by Ullrich et al. (1976), an intermeshing co-rotating twin screw extruder-reactor was used to polymerize polyurethane from a premix composed of low molecular weight polyester (prepolymer) and butane-1,4 diol (chain extender), and a di-isocyanate (monomer). The premix and di-isocyanate were pumped separately as melts into the feed section of the extruder-reactor at 90–120°C. In the polymerization section, intense mixing was carried out by means of appropriate screw configuration while a well-controlled temperature profile was applied aided by the modularity of the screw-barrel assembly of this type of extruder. The extruder-reactor was able to produce polyurethane sheet at a screw speed in the range 70–300 rpm and a range of residence time 0.8–2.5 min. Step-growth polymerization can be carried out successfully in twin screw extruder-reactors with process profiles similar to that shown in Figure 5.2a. One of the problems in these polymeric reactions is the formation of low molecular weight by-products such as water or alcohol during condensation of the monomers. In this case, efficient removal of these by-products by devolatilization in the venting section (or multistage venting sections) is required to optimize the equilibrium of the reaction and ensure high conversion of the polymerization reaction.

It is worth noting that addition polymerization processes using a peroxide initiator (such as processes to produce polyacrylates and related co-polymers) have also been performed successfully in intermeshing, co-rotating twin screw extruder-reactors (Brown, 1991).

In all cases of bulk polymerization, the viscosity of the reaction mixture may increase sharply as the polymerization proceeds. In addition to moving a higher viscosity material, the heat transfer process becomes more difficult with increasing viscosity. Extruder-reactors can be designed to control the temperature within the limits required by the process.

5.1.2 Reactive processing of polymers. Reactive plastics reprocessing

Reactive extrusion is currently used in industry to chemically modify virgin polymers as well as polymer alloys in bulk reactions, combining two traditionally separate operations such as the chemical reaction for the modification of molecules and macromolecules and polymer processing for the purpose of structuring and shaping the final products. Reactive processing of polymers is well known and used in the polymer processing industry. The use of reactive extrusion for producing packaging materials is well documented in a number of publications that describe both conventional and reactive extrusion processes for production of polyethylene terephthalate (PET) and PET/poyethylene blends for the production of recyclable materials (Awaja & Pavel, 2005). PET is widely used in the production of containers for liquids due to its high tensile and impact strength, chemical resistance, clarity, processability, color, and reasonable thermal stability. The use of PET in the beverage industry has been extensive but has increased further recently due to environmental pressures to reduce conventional waste management such as landfill and incineration (Pracella et al., 2002). Thus, new approaches have focused on technologies that can allow efficient and economic PET recycling to reduce wastes (Awaja & Pavel, 2005). It is thus opportune to choose one example in the area of reprocessing packaging wastes, such as postconsumer PET materials, to discuss the use of reactive extrusion in plastics recycling.

Two main processes have been applied to recycle postconsumer PET: chemical recycling (or chemolysis), which

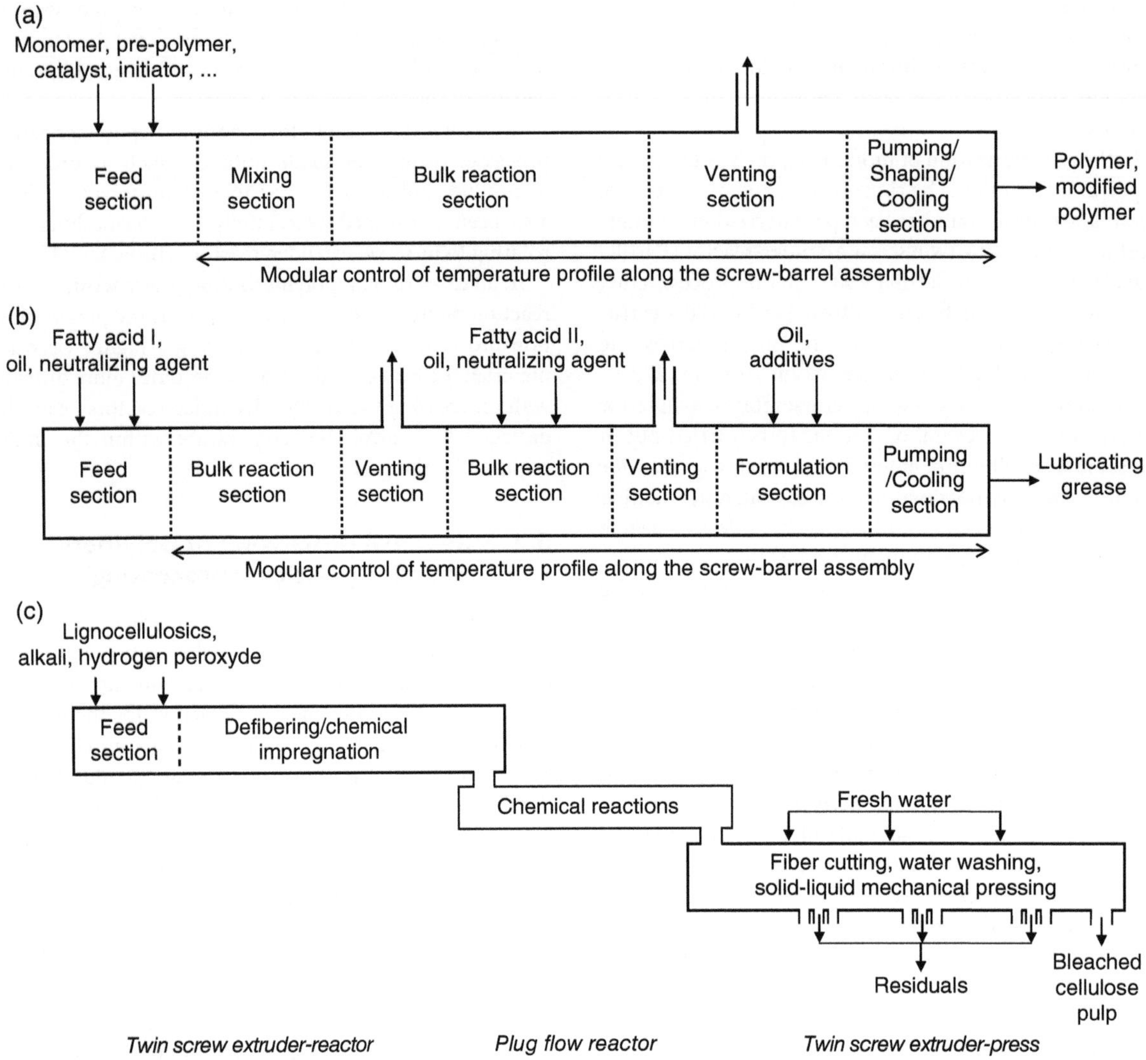

Figure 5.2 Simplified schematic representation of typical process profiles for reactive extrusion in intermeshing, co-rotating twin screw extruder-reactors. (a) Process profile for bulk polymerization and reactive plastics reprocessing. (b) Process profile for the synthesis of complex soap greases. (c) Process profile for reactive solid-liquid extrusion-pressing.

consists of the total depolymerization of the polymer into monomers or its partial depolymerization into oligomers, and mechanical recycling which aims at reprocessing washed and dried PET flakes in a normal extrusion process. Mechanical recycling of postconsumer PET is relatively simple, environmentally friendly and requires low investment. However, a significant reduction of molecular weight during melt reprocessing is observed, due to thermal and hydrolytic degradation reactions. Different methods have been reported over the past two decades to maintain the molecular weight of PET during reprocessing, such as operating under vacuum, adding stabilizers, applying solid-state polymerization (that is, heating PET above its glass transition temperature but below the melting temperature), or by use of chain extension processing (Awaja & Pavel, 2005).

Chain extension by reactive extrusion process has been the favorite method because its application is easy and economic (Awaja et al., 2004a, 2004b; Cardi et al., 1993; Incarnato et al., 2000). Extrusion chain extension is a process in which a di- or polyfunctional low molecular weight component reacts with the PET end groups to couple chains of the polymer and thus increase the polymer molecular weight. Chain extenders can be added in the reactive melting zone of the extruder-reactor where polycondensation is produced. Successfully used chain extenders include multiepoxides, di-isocyanates, dianhydrides, bis(oxazolines), etc. which react with the end groups -OH and -COOH of the PET polymer.

The effect of several process variables, such as shear rate, residence time, and chain extender concentration, was investigated in the extrusion chain extension process. It was found that short reaction times and a decrease in moisture content resulted in limited thermal and hydrolytic degradation and effective chain extension. Thus, PET extrusion recycling is carried out at short residence times with venting sections to extract volatiles. The process profile of the twin screw extruder-reactor illustrated in Figure 5.2a can be used for postconsumer PET recycling with chain extension processing.

5.1.3 Reactive extrusion in classic organic chemistry

Thousands of chemicals and novel materials are the result of reactive systems which historically have been performed in double-jacketed, batch or continuous stirred tank reactors. However, when carrying out chemical reactions in such reactors, it is necessary to operate with a highly diluted reactive system which involves high energy consumption, large processing volumes, and large processing equipment. It must be noted that the use of stirred tank reactors is far from being considered environmentally friendly. Hence, there is a highly technical potential substitution of classical organic reaction systems by reactive extrusion. Among others, lubricating grease manufacturing provides a relevant example which illustrates this substitution potential.

Lubricating grease is defined as a solid or semi-solid product consisting of a dispersion of a thickening agent in a liquid lubricant. Lubricating greases typically contain the following three main components: lubricating base oil, thickening agent, and additives. A wide variety of greases have been developed over the years with varying formulations and end-use properties. The main component in formulated greases is the thickening agent which is considered to influence their properties. Most lubricating greases use metal soaps as thickeners. These soaps are either simple molecules that are metal salts of long chain fatty acids (such as lithium 12-hydroxystearate), or complex molecules that generally comprise metal salts of a mixture of organic acids (such as complex lithium soap prepared from 12-hydroxystearic acid and azelaic acid). Lubricating greases have been generally produced by batch processes which used stirred tank reactors with processing capacities ranging from a few m^3 to 15–20 m^3. During processing of viscous materials, heat transfer and mixing capabilities of stirred tank reactors are generally limited, leading to low productivity processing; the whole production cycle of one batch may require from 8 to 24 hours depending upon the type of metal soap and grease formulation being produced. Thus, there is a need for a continuous lubricating grease manufacturing method able to provide significant process and economic advantages, and to offer operational flexibility which would enable end-users to produce various types of greases in the same equipment without significant changes.

Alexander (1983) patented an innovative method based on reactive extrusion for continuous grease formulation and manufacturing, with significant technical and economic advantages over conventional batch processing methods. The continuous process comprises the following steps:

- feeding the reactants (fatty acid such as 12-hydroxystearic acid solubilized in lubricating oil) and the neutralizing agent such as lithium hydroxide monohydrate (dissolved in water) into the feed section of a extruder-reactor
- efficiently mixing the materials while controlling the temperature to achieve a homogeneous mixture and saponification in the bulk reaction section
- flashing off water resulting from the dehydration of the reacting mixture in the venting section
- formulating the resulting mixture by adding extra lubricating oil and additives followed by complete homogenization in the formulation section
- removal of the finished lubricating grease in the pumping section.

Though this process is typically directed at simple soap greases, it involves multiple steps in series which are carried out at various temperatures and requires the addition and removal of different components at the different steps. The formation of complex soap greases is even more complicated since it involves additional ingredients and generally a multistage manufacturing technique such as that shown in Figure 5.2b. It must be noted that intermeshing, co-rotating twin screw extruder-reactors are able to operate continuously and successfully handle multiple process steps in one unit, with total residence time of a few minutes. Alexander (1983) underlined significant process advantages of reactive extrusion in grease manufacturing, such as better quality control, improved economics derived from lower capital costs, smaller space requirements, reduced energy consumption, less waste material, increased process productivity and flexibility with regard to the ability to handle different grease types.

5.1.4 Reactive solid-liquid extrusion-pressing

Although reactive solid-liquid extrusion-pressing was invented and developed 30 years ago, it is not yet well known by the reactive extrusion community. This innovative process is extensively described in this chapter (section 5.5.3.2). Reactive solid-liquid extrusion-pressing, which combines reactive extrusion and solid-liquid extrusion-pressing techniques, is used and well recognized in the paper-milling industry as an environmentally friendly technology for reactive extrusion pulping of non-wood fibers.

The process is continuous and uses two intermeshing, co-rotating twin screw extruders in series separated by a screw-type plug flow reactor. The process comprises the following functions (Figure 5.2c):

- feeding the raw materials (such as non-wood lignocellulosics) into the feed section of a twin screw extruder-reactor
- defibering of lignocellulosics in order to increase the specific surface area of fibrous solids
- impregnating chemicals (such as alkali and hydrogen peroxide) into the matrix of the fibrous material
- digesting lignin and bleaching cellulose pulp in the intermediate screw-type plug flow reactor
- cutting cellulose fibers to adjust fiber length, in a twin screw extruder-press
- washing cellulose pulp with water to remove residuals through solid-liquid mechanical pressing.

Hence, reactive solid-liquid extrusion-pressing operates continuously and allows several process functions to take place in the twin screw extruder which decrease the total residence time by a factor of 10 when compared to the conventional, semi-continuous pulping process (see Chapter 10, Table 10.5).

As discussed in section 5.5.3.3 of this chapter, reactive solid-liquid extrusion-pressing has also brought drastic improvements such as higher yield of fibers, energy and consumables saving, reduction of water consumption, and higher productivity. Hence, it has highlighted process functions which are relatively unknown in the conventional reactive extrusion world which is dealing mostly with polymer extrusion processing. New developments like reactive solid-liquid extrusion-pressing open up new processing avenues for reactive extrusion technology.

5.1.5 Processing characteristics of reactive extrusion

First, it is worth noting that reactive extrusion is a fully continuous process which combines not only thermomechanical treatment but also chemical reactions in screw extruders, converting them into chemical reactors. Reactive extrusion is characterized by bulk reactions with little or no solvent dilution, which brings tremendous process advantages. In the case of reactive processing of polymers, when comparing a process carried out in a continuous stirred tank reactor, with polymer concentration ranging from 5% to 25% (weight basis) and high consumption of energy, with a process performed in a screw extruder-reactor, where a bulk reaction is applied, a 25–75% energy saving (including thermal and electrical energies used for solvent recycling) is obtained, depending upon the reactive system. In reactive extrusion, the loss of solvent is insignificant, whereas the loss of solvent in a diluted solution process in a reactor may reach values ranging from 0.05% to 0.2% per year; reduction of depreciation and maintenance costs may range from 10% to 25% on average.

Unlike bulk reaction processing, processes which are carried out in a dilute solution require specific processing equipment (such as a dissolver, steam stripping, solvent purification, polymer drying, etc.) to recycle processing solvents, and recycling equipment is most likely to be emitting pollution (emission of volatiles to the atmosphere from the seals in equipment flanges, instrument ports, pumps and storage tank air displacement, water contamination). Average residence time in screw extruder-reactors is short, a few seconds or minutes, which leads to low processing volumes and thus important process benefits such as relatively higher equipment productivity, lower investment cost, and reduced footprint, in particular.

Typical equipment for reactive extrusion includes screw extruders, both single screw and twin screw technologies. Single screw extruder-reactors have the advantage of lower capital costs but they show relatively poor mixing capabilities and process flexibility. Intermeshing, co-rotating twin screw extruder-reactors are preferred for reactive extrusion due to their better mixing capability (efficient mixing of reactants in viscous homogeneous and heterogeneous media) and higher process flexibility (large range of screw configurations, multistage processing). Based on the different applications reviewed in sections 5.1.1 to 5.1.4, intermeshing co-rotating twin screw extruder-reactors allow several process functions to be combined in dedicated processing sections in series along the screw-barrel assembly, such as feeding, material defibering (in the case of fibrous materials), intense mixing, bulk reaction, venting, product formulation, fiber cutting (case of cellulose fibers), cooling and shaping, solid-liquid washing and pressing, etc. In addition, the modular design of screw-barrel assemblies with multiple injection ports for liquids offers a broad range of process flexibility, including sequential reaction processes and appropriate temperature profiles.

Many of the technological advantages of screw extruder-reactors can be summarized by their ability to transport large volume of viscous materials. In Chapter 3 an equation used to describe the flow of a viscous melt inside the extruder was derived and is given below:

$$Q = \alpha N - \frac{\beta \Delta p}{\mu} \tag{5.1}$$

The equation estimates the volumetric throughput Q in the extruder as a result of a balance between the drag flow, given by the first term in the equation, and a pressure flow moving on the opposite direction given by the second term in the right side of the equation. It is important to note that for viscous materials, the viscosity will reduce the pressure flow although the pressure gradient increases, resulting in a higher net volumetric throughput. Another advantage of reactions conducted at low solvent content is that the small quantity of solvent in the reacting mass going through the extruder-reactor improves the adherence of this mass on the equipment surfaces, favoring drag flow.

The kinetics of a reaction is a function of the reaction conditions and the conversion of reactants to products is highly dependent on the time under which those conditions are maintained. Thus it is essential to know the residence time distribution in extruder-reactors.

Of importance are the relations between viscosity and molecular size of the polymer and also between the conversion and the viscosity which would indicate that the conversion of the reaction could be followed by monitoring changes of viscosity (Figure 5.3).

5.2 Reactive extrusion: chemical reaction engineering approach

As seen in the introduction to this chapter, many reactions are considered in the reactive extrusion process, of which the most important include addition polymerization, polycondensation, and reactive processing of polymers. These reactions and processes are extensively described in dedicated books such as those edited by Xanthos (1992) and Janssen (2004), as well as pertinent reviews in specialized journals such as that published by

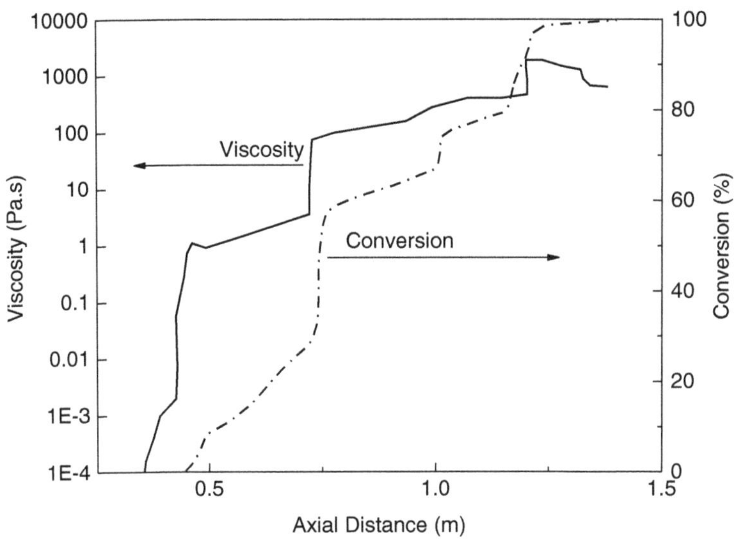

Figure 5.3 Schematic showing the changes of viscosity and reaction conversion in a typical reactive extrusion process.

Cassagnau et al. (2007), and readers interested in these specific applications should refer to these publications. Most of the applications concern the field of purely synthetic polymers, but innovative and relevant applications are being developed in emerging fields such as reactive processing of bio-sourced, biodegradable and non-biodegradable polymers (Kalambur & Rizvi, 2005, 2006; Kaseem et al., 2012; Moad, 2011; Raquez et al., 2008; Xie et al., 2006; Zhao et al., 2008), plastics recycling and reprocessing (Awaja & Pavel, 2005; Pracella et al., 2002), bioreactive extrusion (Govindasamy et al., 1997a–d), and reactive solid-liquid extrusion-pressing (section 5.1.4).

Regardless of the application, it is noteworthy to consider two aspects of importance in the area of reactive extrusion. One is related to the absence of relevant reaction kinetics data in the literature since most of the reported kinetics rely on dilute conditions that are not suitable for extrusion processes. The other aspect is related to the extremely viscous nature of the melts, which is exacerbated when the reactions involved in the process are polymerization reactions. As a consequence of the high viscosity of the melt, the mobility of polymer chains is limited, thus limiting chain growth polymerization reactions. This limitation on the mobility of polymeric chains, of a much larger extent than mobility limitations of monomers, causes an increase in the active monomer concentration in the molten mass and thus a momentary abrupt increase in the polymerization of the system, resulting in products with molecular weight and conversion higher than those expected from the reaction kinetics. This phenomenon is known as the gel effect (Janssen, 2004).

Another complication found in reactive extrusion processes is due to the concentrated nature of the reactive system. The removal of heat generated by both the reaction (for exothermic reactions) and the viscous dissipation during the process is severely limited, producing high temperatures that may cause degradation and limited conversion due to shifting of the reaction equilibrium toward the increase of monomer concentration. This phenomenon is known as a ceiling temperature (Janssen, 2004). Adequate mixing conditions can alleviate these problems and in that respect intermeshing, co-rotating twin screw extruders offer significant advantages, some of them already discussed in Chapter 3.

Since the data on polymerization and other reactions used in reactive extrusion processes are mainly based on dilute conditions and kinetics of reactions at conditions of high concentration are lacking, Janssen (2004) proposes simple and practical approaches for characterizing these kinetics which include a general kinetics approach and associated equations to describe the gel effect and the ceiling temperature mentioned before. Other kinetics includes co-polymerization and step-growth polymerization reactions. Regarding kinetics, in addition to the standards first, second and nth-order kinetics, practical methods to experimentally determine kinetics of reactive extrusion processes are described by Janssen (2004). These include thermoanalysis such as isothermal calorimetry (ITC) and differential scanning calorimetry (DSC); the DSC method is presented in Chapter 7 (section 7.1.4). Rheological methods, known as rheokinetics, are also used to evaluate reaction conversion (Biesenberger et al., 2004; Cioffi et al., 2002) in the extruder-reactor.

As discussed in Chapter 4, viscosity and viscoelastic properties of materials are highly sensitive to temperature, concentration, and molecular size of the reactive species so an objective has been to combine rheological data with chemical kinetics and express the conversion of the reaction in terms of the measured rheology. Polymerization of ε-caprolactone, a biodegradable and biocompatible synthetic polymer used in medical applications such as artificial skin and controlled drug release, has been performed using twin screw reactive extrusion. Owing to the large variation of viscosity of the reactive mixture as a consequence of the polymerization reactions, the conversion of the reaction was studied by monitoring changes in the viscosity of the melt during its transit through the extruder. In addition, rheological and physicochemical methods were used to characterize the viscosity and the molecular characteristics of the product to follow the reaction conversion in the process, as illustrated in Figure 5.3. This information can be integrated with details of the process to develop relationships between processing conditions and reaction conversion using specialized software such as LUDOVIC®, which is described in Chapter 3 (section 3.3.3).

Pressure drops based on online techniques to characterize the rheology of extruded materials have been discussed in the literature (Laun, 1983; Li et al., 2004; Padmanabhan & Bhattacharya, 1991; Vergnes et al., 1993); some are commercially available from companies such as Thermo Scientific Haake PolyLab OS System (www.rheologysolutions.com/thermo-scientific-haake-polylab-os-system-measuring-extruder-single-screw).

Chapter 3 describes the flow and thermomechanical treatment of a material in single and twin extruders, considering possible geometries and configurations. An important feature of single screw and to a much larger extent twin screw extruders is their ability to produce very efficient micromixing. Many reactive extrusion processes described in the literature have involved single and twin screw extruders or a combination of them (Janssen, 1998) but the better mixing and control of the twin screw extruder make this the preferred technology to operate reactive processes.

With the first known application of reactive extrusion presented in a patent assigned to the Dow Chemical Company (Stober & Amos, 1950), subsequent applications of this process included polymerization of nylon (Illing, 1969) and several polycondensation reactions (Mack, 1972). More references on reactive extrusion processes are listed in the review published by Cassagnau et al. (2007) to which readers should refer for specific reactive processes. From information and articles describing the general process, it becomes clear that a continuous plug flow reactor is a very close approximation to screw extruder-reactors (Meyuhas et al., 1973). Thus, the mathematical model describing the behavior and performance of a continuous plug flow reactor, which is presented in this section, gives the benchmark of an ideal system that can be used as a point of reference for the study of reactive extrusion processes.

5.2.1 The continuous plug flow reactor

5.2.1.1 Reaction kinetics

The approach to modeling any type of reactor, and specifically the continuous plug flow reactor (CPFR), is not based on the external form of the equipment or on the reaction taking place inside the reactor. Instead, models that describe the transfer of mass, heat and momentum are used, and thus the modeling and design of reactors are based on those transport equations (Froment et al., 2011). The transport equations applied to plug flow model reactors are the continuity and energy equations. Concerning the momentum equation, a simplified form assuming that only friction and pressure forces are present is considered (Froment et al., 2011). A similar analysis where drag and pressure flows were considered was presented in Chapter 3 to describe the flow of the melt in an extruder channel.

For a generalized chemical reaction of the type:

$$\nu_A A + \nu_B B \rightarrow \nu_C C + \nu_D D \tag{5.2}$$

the extent of the reaction ζ can be defined by the following equation:

$$d\zeta = \frac{dn_A}{\nu_A} = \frac{dn_B}{\nu_B} = \frac{dn_C}{\nu_C} = \frac{dn_D}{\nu_D} \tag{5.3}$$

where dn_A, dn_B, dn_C and dn_D are the differential changes in the number of moles of the reactants and products of the reaction and ν_A, ν_B, ν_C and ν_D are the corresponding stoichiometric coefficients, with the convention that the stoichiometric coefficients of the reactants are taken as negative and those of the products are taken as positive. By assuming that the initial number of moles of the compounds A, B, C and D are n_{A0}, n_{B0}, n_{C0} and n_{D0}, respectively, the integration of Eq. 5.3 results in:

$$\begin{cases} n_A = n_{A0} + \nu_A \zeta \\ n_B = n_{B0} + \nu_B \zeta \\ n_C = n_{C0} + \nu_C \zeta \\ n_D = n_{D0} + \nu_D \zeta \end{cases} \tag{5.4}$$

A fractional conversion with reference to one of the reaction species, e.g. the reactant A, is often used to follow the progress of the reaction:

$$x_A = \frac{n_{A0} - n_A}{n_{A0}} \tag{5.5}$$

Thus, moles of species A and the other species can be calculated in terms of the fractional conversion for species A as:

$$\begin{cases} n_A = n_{A0} - n_{A0} x_A \\ n_B = n_{B0} - \nu_B \dfrac{n_{A0}}{\nu_A} x_A \\ n_C = n_{C0} - \nu_C \dfrac{n_{A0}}{\nu_A} x_A \\ n_D = n_{D0} - \nu_D \dfrac{n_{A0}}{\nu_A} x_A \end{cases} \tag{5.6}$$

Although both approaches are measuring the progress of the reaction, it is important to define what kind of conversion is being used when solving design problems. For industrial applications, the conversion of the feed is of interest while for thermodynamic analysis the extent of the reaction is preferred (Froment et al., 2011). Regardless of the definition used, by combining Eqs 5.4 and 5.6, a relationship between the extent of the reaction and the fractional conversion for species A can be obtained as:

$$x_A = -\frac{\nu_A \xi}{n_{A0}} \tag{5.7}$$

If species A is the limiting reactant of the reaction, when n_A in Eq. 5.4 is zero, the corresponding extent of the reaction is the maximum that can be achieved from that reaction, calculated by the following equation:

$$0 = n_{A0} + \nu_A \xi_{max} \rightarrow \xi_{max} = -\frac{n_{A0}}{\nu_A} \quad (5.8)$$

Combining Eqs 5.8 and 5.7, the following is obtained:

$$x_A = \frac{\xi}{\xi_{max}} \quad (5.9)$$

To model a CPFR, also known as a tube reactor, a plug of the reacting mixture is considered as moving through the reactor and the reaction time is considered as the time the plug stays inside the reacting volume. Essentially, the ideal CPFR has poor mixing in the axial direction and excellent mixing in the radial direction, the latter to the point that the plug can be considered as a perfectly mixed reactor, where the concentration of the reactant and products as well as the temperature are considered uniform in that section. During the moving of the plug through the reactor, the extent of the reaction is changing from zero at the reactor entrance to the final extent at the exit of the reactor. As illustrated in Figure 5.4, the mass balance for one of the species of the reaction (in this case the limiting reactant *A*), assuming steady-state conditions and uniformity of chemical species concentration and temperature in a cross-section of the reactor, can be simply written by the following equation:

$$\frac{dF_A}{dV} = -r_A \quad (5.10)$$

where r_A represents the consumption (or generation) of the species A due to the chemical reaction.

Upon integration of the above equation, an expression to calculate the volume of the reactor to achieve the desired conversion of component *A* can be obtained by:

$$V = -\int_{F_{A0}}^{F_A} \frac{dF_A}{r_A} \quad (5.11)$$

For reactions involving liquids, it can be safely assumed that the volume of the reaction mixture does not change significantly so from Eq. 5.5 following equation results:

$$F_A = F_{A0}(1 - x_A) \quad and \quad dF_A = -F_{A0} dx_A \quad (5.12)$$

F_A is the molar flow rate of component *A* in a given position of the reactor and F_{AO} is the molar flow rate at the entrance of the reactor. Substituting of Eq. 5.12 in Eq. 5.11 yields:

$$\frac{V}{F_{A0}} = \int_{x_{Ao}}^{x_A} \frac{dx_A}{r_A} \quad (5.13)$$

The ratio V/F_{A0}, often called space time, has units of m^3.h/mol of species *A* and represents a measurement of the residence time of that reactant in the reactor. Since it is associated with the number of moles of *A* which change during the course of the reaction, a true residence time having units of time can be defined in the differential volume (indicated in Figure 5.4) as:

$$dt = \frac{dV}{F_A / C_A} \quad (5.14)$$

The residence time in the ideal CPFR to achieve the desired conversion of component A is obtained by integration of Eq. 5.14, which results in:

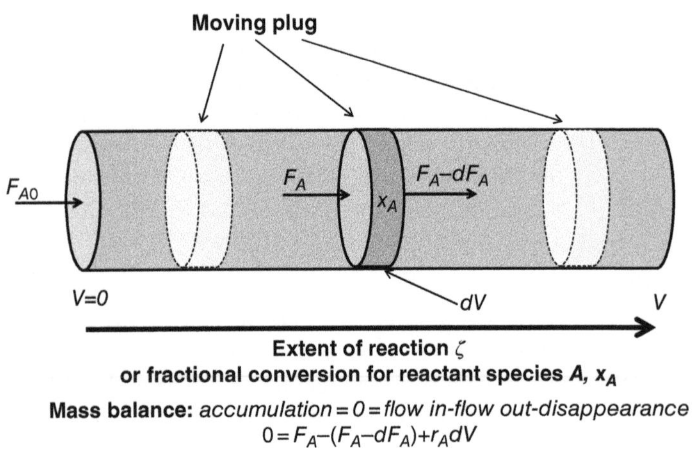

Figure 5.4 Balance of mass in a typical plug flow reactor.

$$t = \int_0^V \frac{dV}{F_A/C_A} \quad (5.15)$$

It must be noted that the volume of the reaction mixture can change due to changes in its density resulting from the conversion of reactants to products (Froment et al., 2011). Incorporation of reactants or other additives may also change the concentration of the reactants and their molar flow rate (e. g. for component A in Eq. 5.15). For liquids. the change of density of the reaction mixture is small and for processing conditions under which no reactants are added to the reaction mixture, the local concentration of the reactant (e.g. reactant A in the mixture) can be calculated as:

$$C_A = \frac{n_A}{V} = \frac{n_{Ao}(1-x_A)}{V} = C_{A0}(1-x_A) \quad (5.16)$$

Thus, by substituting Eqs 5.12, 5.13, and 5.16 into Eq. 5.15, the following equation to calculate the residence time in an ideal CPFR results:

$$t = \int_{x_{A0}}^{x_A} \frac{dx_A}{r_A} \quad (5.17)$$

Equation 5.17 simply expresses that if the kinetics of the reaction, written in terms of the fractional conversion of the limiting reactant (e.g. A using Eq. 5.6), is known, the residence time in the plug flow reactor can be calculated for the desired conversion of the reaction. This equation not only assumes ideal behavior and isothermal conditions but also that the reactants are in contact immediately after they are introduced into the reactor, i.e. they are contained in miscible phases and that the mixing of these phases to get a uniform concentration of the reactants in the reactive media is immediate. However, in real reactors, including those of the plug flow type, that is not the case. The flow pattern in reactors is complex not only due to complicated geometries, presence of flow stagnant regions, and baffle corners but also due to the fact that in many instances the viscosity of the reactive media is low and the fluid flow is turbulent. Thus the residence time of all fluid particles is not uniform, resulting in uneven conversion of reactant and products. In consequence, the residence time distribution (RTD) of fluid elements needs to be determined for proper reactor design.

5.2.1.2 Mixing in reactors and effects on the reaction conversion

Froment et al. (2011) describe methods to characterize mixing in chemical reactors, establishing clearly their pros and cons for application in research and production areas. The distinction between macro- and micromixing is established on scale lengths that go from the reactor size to the size of eddies formed in the turbulent flow of the reactive mixture. The type of flow is a notable difference between mixing in reactors consisting of diluted reactive phases and mixing in screw extruder-reactors used for reactive extrusion processes, due mainly to the high viscosity of the reactive media with resulting flows that are often laminar rather than turbulent in the latter. Despite differences in their flow characteristics, a number of approaches and methods to characterize mixing in plug flow reactors and its influence on the kinetics of the reaction bearing common scientific features with the flow in reactive extrusion processes are discussed in this section.

5.2.1.2.1 Time scales in reaction engineering

As in reactive extrusion, the evaluation of mixing limitations in reactors is based on comparisons between reaction rates and mixing time scales and needs to be defined. Many time scales are considered for the study of reaction rates in reactors which are associated not only with velocity profiles and the type of flow of the reacting media, notably turbulent and laminar flows, but also the reactive and product species profiles. For example, for regular reactors operating at diluted conditions and flows where the turbulent regime prevails, the ratio between the time scale given by the turbulent flow in the reactor, τ_K, and the time scale associated to the diffusivity of the reactive species and the energy of turbulent eddies, known as the Batchelor time scale, τ_B, is related to the Schmidt dimensionless number defined as $Sc = \frac{\nu}{D}$, by the following equation:

$$\frac{\tau_B}{\tau_K} = \frac{1}{S_c^{1/2}} \quad (5.18)$$

where ν is the kinematic viscosity of the reactive medium and D the diffusion coefficient of the reactive species in that medium. For high-viscosity liquids, the Schmidt number is larger than 1 and time scales associated with the concentration field of the reaction species are smaller than those associated with the flow field. Thus, the distribution of reactive species in the volume of the reactor created by the liquid flow plays a fundamental role in the overall reaction rate. The influence of mixing limitations of this rate is measured in terms of the Damköhler number, which for a specific reaction is defined as (Froment et al., 2011):

$$Da_{I,j} = \frac{t}{t^j_{reaction}} \tag{5.19}$$

where the subscript j refers to the reaction j and t is the residence time in the reactor estimated from the ratio between the volume of the reactor and its incoming flow. Based on the diffusion rates of the reactive species in a mixture, two Damköhler numbers can be distinguished. The first, named Da_I, is the dimensionless number defined above whereas a second Damköhler number, Da_{II}, is used to establish a comparison between the reaction time with a characteristic time associated with the diffusion of the reactive species. Since this diffusion of reactive species is directly related to the mixing characteristics of the reactor, a description of the mixing pattern is essential to fully characterize the process and define that specific Damköhler number. Thus, although different models and methods are used to quantify the degree of mixing in chemical reactors, it is always stressed that both macro- and microscale phenomena affect the reactor behavior and all mixing details should be considered in the analysis (Froment et al., 2011).

Models used to incorporate mixing in reactors include the continuity equation for each of the species present in the reactive mixture, which for species A is given below:

$$\frac{\partial C_A}{\partial t} + \underline{v} \cdot \nabla C_A = \nabla D_{Am} \nabla C_A + r_A(\underline{C}) \tag{5.20}$$

In the above equation the first term on the left side is the accumulation or loss of species A in the reactor over the reaction time, which for steady-state processes can be considered zero. The second term on the left side is the transport of species A due to convection, which as noted is dependent on the flow in the reactor described by the velocity profile $\underline{v}$. On the right side, the first term represents the transport, by molecular diffusion, of species A in the medium m.

The case contemplated in the above equation is that of a diffusivity which is not constant. However, the approximation of constant diffusivity, although at times far from reality, is often used to obtain closed and simpler solutions to the above complex mathematical equation. Solutions to the above equation serve to provide information on many aspects of the mixing process. The second term on the right side of the equation, as discussed, represents the consumption or generation of species A due to the chemical reaction. It must be noted that the equation describing the kinetics of the reaction may involve one or more species, so in general terms, the reaction kinetics is expressed as a function of a vector concentration $\underline{C}$ that includes all or the species participating in the reaction kinetics model. Other equations to model the process include the conservation of mass or continuity equation, the momentum and the energy equations. These equations were given in Chapter 3 but applied to conditions pertinent to screw extrusion processes which include steady-state conditions and negligible inertial forces owing to the high viscosity of the melts. The mass balance or continuity equation, the momentum and the energy equations are repeated in this chapter and given below as Eqs 5.21, 5.22 and 5.23 respectively:

$$\frac{\partial(\rho \underline{v})}{\partial t} + \nabla \rho \underline{v} = 0 \tag{5.21}$$

$$\frac{\partial(\rho \underline{v})}{\partial t} + \nabla . \rho \underline{v v} = -\nabla p + \rho \underline{g} + \nabla \cdot \underline{\underline{\tau}} \tag{5.22}$$

$$\rho c_p \left(\frac{\partial T}{\partial t} + \underline{v} \cdot \nabla T \right) = \nabla \cdot k \nabla T + \left(\underline{\underline{\tau}} : \nabla \underline{v} \right) + \sum_j (\Delta h_k) r_k \tag{5.23}$$

Equation 5.23 assumes that changes in the kinetic and potential energies are small compared to enthalpy changes, and those enthalpies involved in the chemical reactions are included in the last term of the right side of Eq. 5.23. In addition, the flow in the reactor can be turbulent, which affects the fluid velocity profile $\underline{v}$ included in the above equations.

The solution of this set of partial differential equations along with the rheological behavior of the reactive fluid and the appropriate boundary conditions represents a formidable mathematical task that provides information on species concentrations, velocities, pressure, and temperature profiles in the domain of the reactor. These profiles, in turn, provide the necessary information to estimate the progress of the reaction and the distribution of unreacted reactants and products in the reactor.

Since macro- and microscale phenomena affect the reactor behavior, mixing characteristics of reactors have been studied by a variety of methods. Froment et al. (2011) list most of those methods which include complex numerical approaches. However, the difficulty in obtaining solutions to these complex models that could be suitable for the design of industrial reactors has prompted researchers to use experimental approaches to measure mixing and RTD in reactors. Some of those approaches have been extended to the extrusion and reactive extrusion areas.

5.2.1.2.2 Concept of residence time distribution

Specifically, RTD methods are based on the concept of age distribution functions and experimentally based on the response of the reactor to a tracer impulse injected at the reactor inlet (Froment et al., 2011). Monitoring the concentration of tracer at the reactor outlet enables an empirical characterization of its mixing and flow behavior. Two ideal extreme phenomena can be visualized: for a perfectly mixed reactor, the outlet concentration of the tracer will decay continuously due to the wash-out effect that the incoming flow imparts to the concentration of the tracer. Conversely, for an ideal continuous plug flow tubular reactor with perfect radial and no axial mixing, the tracer will not exit until a time equal to the mean residence time, calculated by the ratio between the reactor volume V and the feed flow rate F, has elapsed.

Experimentally, determination of the RTD function is performed by following the concentration, at the exit of the reactor, of a tracer injected in the reactor inlet. If the concentration of the tracer at the outlet is denoted by $C(t)$, the RTD or exit age distribution is determined as:

$$f(t) = \frac{C(t)}{\int_0^{\infty} C(t)dt} \tag{5.24}$$

With this RTD function, the mean residence time in the reactor is calculated by the following equation:

$$\bar{t} = \frac{V}{F} = \int_0^{\infty} tf(t)dt \tag{5.25}$$

The RTD function can also be used to estimate the variance σ^2 that characterizes the spread around the average as:

$$\sigma^2 = \int_0^{\infty} (t-\bar{t})^2 f(t)dt \tag{5.26}$$

Figure 5.5 illustrates a plot of a typical RTD function as a function of the transit time in the reactor. It has been established that the experimentally determined RTD function is a reflection of the spatial characteristics of the macro flow and consequently the extent of mixing in the reactor (Froment et al., 2011); thus it is commonly used to characterize mixing and its influence on the conversion of the reactions. The link of the mixing characteristics in the reactor with the chemical reaction is based on

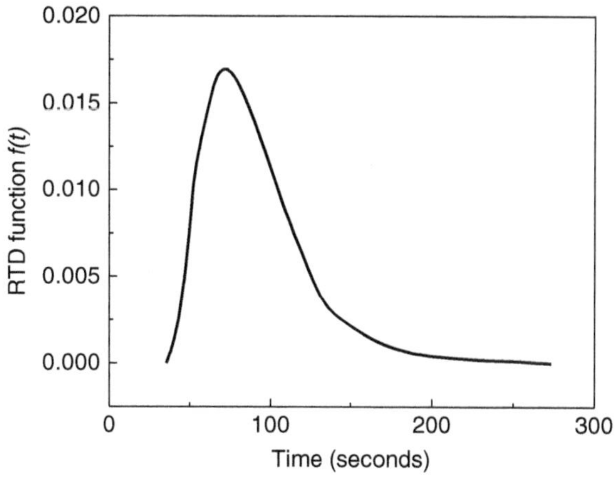

Figure 5.5 Typical RTD function determined in a standard reactor.

the reaction extent of a reacting species (e.g. reactant A) present in a large number of fluid elements. Thus, the average concentration of the reactive species over the fluid elements is calculated as:

$$\bar{C}_A = \int_0^{\infty} C_A(t)f(t)dt \tag{5.27}$$

The concentration of the reactant A, C_A, is associated with the reaction kinetics r_A occurring in the fluid elements. For a given kinetics, the disappearance of species A in a fluid element can be obtained by solving the following equation along with its initial condition:

$$\begin{cases} \frac{dC_A(t)}{dt} = -r_A(C_A(t),T) \\ C_A = C_{A0} \quad at\ t=0 \end{cases} \tag{5.28}$$

where T is the temperature of the reaction. Froment et al. (2011) provide an interpretation of this model by considering that Eq. 5.28 can be applied to small batch reactors associated with the fluid elements that move independently in the reactor. Thus, for isothermal reactions that follow first-order kinetics, i.e. only depending on $C_A(t)$, it can be assumed that the conversion in the fluid element only depends on the residence time of the element and not on its trajectory through the reactor. For higher order reactions and non-isothermal conditions, the rate of reaction will depend on the encounter of two reactants and the local temperature. For this case, in addition to the residence time, the trajectories of these fluid elements must be known along with local temperatures and concentrations which are determined by the simultaneous solution of the system of partial differential equations given by Eqs 5.20 to 5.23.

By assuming complete micromixing in a fluid element, Eq. 5.28 can be replaced by:

$$\begin{cases} \frac{d\bar{C}_A}{dt} = -\bar{r}_A(C_A(t),T) \\ \bar{C}_A = \bar{C}_{A0} \ at \ t=0 \end{cases} \tag{5.29}$$

where mean values of the concentration and the reaction kinetics are considered. For fast non-linear reactions and non-isothermal conditions, $\bar{r}_A(C_A,T) \neq r_A(\bar{C}_A,\bar{T})$ and thus interactions between non-uniformities in concentration and temperatures within a fluid element should be evaluated. In other words, when complete micromixing is assumed in a fluid element, the effect of micromixing on the macroscale averaged reaction is not rigorously accounted for, especially for situations in which non-isothermal conditions and higher order reactions prevail. Conversely, for first-order kinetics and isothermal conditions Eqs 5.27 and 5.28 can be used to calculate the mean concentration of reactant A at the reactor outlet. The value of the average concentration at the reactor outlet can be easily demonstrated for a first-order reaction expressed as:

$$\begin{cases} \frac{dC_A}{dt} = -kC_A \\ C_A = C_{A0} \ at \ t=0 \end{cases} \tag{5.30}$$

with the solution:

$$C_A(t) = C_{A0}e^{-kt} \tag{5.31}$$

By substituting Eq. 5.31 into Eq. 5.27, the following equation is obtained:

$$\bar{C}_A = \int_0^\infty C_{A0}e^{-kt}f(t)\,dt \tag{5.32}$$

As explained above, in an ideal plug flow reactor, a fluid element will not exit the reactor until a time equal to the mean residence time $\bar{t}$ is elapsed. Mathematically, this is described by the Dirac function expressed as:

$$f(t) = \delta(t-\bar{t}) = \begin{cases} 1 \ for \ t=\bar{t} \\ 0 \ for \ t \neq \bar{t} \end{cases} \tag{5.33}$$

By substituting Eq. 5.33 into Eq. 5.32, the concentration of component A in a fluid element at the exit of an ideal plug flow reactor can be calculated as:

$$\frac{\bar{C}_A}{C_{A0}} = e^{-k\bar{t}} \tag{5.34}$$

Likewise, for a perfectly mixed reactor, the RTD function can be expressed as an exponential decay (Froment et al., 2011):

$$f(t) = \frac{1}{\bar{t}}e^{-t/\bar{t}} \tag{5.35}$$

This upon substitution in Eq. 5.32 yields an equation to calculate the concentration of component A at the exit of a perfectly mixed reactor:

$$\frac{\bar{C}_A}{C_{A0}} = \int_0^\infty \frac{1}{\bar{t}}e^{-(k+1/\bar{t})t}dt = \frac{1}{(1+k\bar{t})} \tag{5.36}$$

The mixing behavior of reactors has also been studied by other empirical models known as the axial dispersion and tanks-in-series models. The axial dispersion model is based on Eq. 5.20 in which, by assuming steady-state conditions, one-dimensional convection and diffusion flows, takes the form (Froment et al., 2011):

$$u\frac{\partial C_A}{\partial z} = \frac{\partial}{\partial z}\left(D_a\frac{\partial C_A}{\partial z}\right) + r_A(\underline{C}) \tag{5.37}$$

where u is the z-component of the mean (plug flow) velocity through the reactor, and D_a is an axial dispersion coefficient obtained from experiments. The axial dispersion coefficient is related to the Peclet number (Pe) by the following equation:

$$Pe = \frac{uL}{D_a} \tag{5.38}$$

where L is a characteristic length related to the reactor. The magnitude of the axial dispersion coefficient (or Pe number) distinguishes the mixing behavior in the reactor. It ranges from perfect mixing in the axial direction, i.e. $D_a \to \infty$ (or $Pe \to 0$), describing a perfectly mixed reactor, to the condition $D_a \to 0$ (or $Pe \to \infty$) distinguishing an ideal plug flow reactor. These two extreme behaviors as well as the behavior of real reactors with low axial and high axial dispersions (defined by the Pe number) are shown in Figure 5.6. The axial dispersion coefficient can be

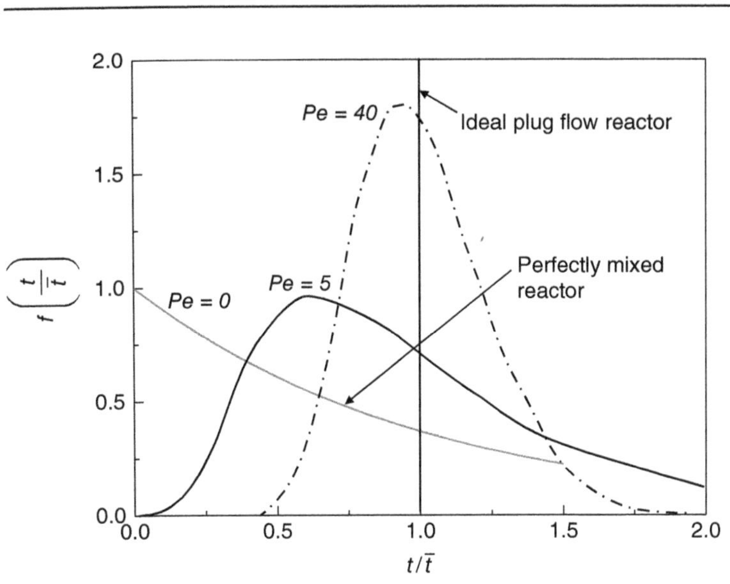

Figure 5.6 Typical RTD functions for ideal (perfectly mixed and ideal plug flow) reactors and real reactors having different axial dispersions (defined by the Peclet number).

determined experimentally by injecting a tracer in the reactor and fitting the value of D_a so that the model solution and the experimental curve agree. Since the injection of the tracer and its concentration recording at the reactor outlet is a dynamic phenomenon, the transient term $\partial C_A/\partial t$ must be included in Eq. 5.37 for the fitting procedure that includes the simultaneous solution of Eq. 5.37 linked to a non-linear fitting routine.

For situations in which the mixing behavior cannot be represented by the models described above, multizone models have been used to describe the mixing characteristics of the reactor. To develop these models, the reactor is divided into zones where idealized mixing or flow patterns are assumed to obtain the mean concentration of the species at the reactor outlet. These models were popular in the years 1950 to approximately 1980 but lately their use has been dropped because they are sensitive to ill-defined parameters such as, for example, the feed conditions. Froment et al. (2011) gives a comprehensive review of these methods to characterize mixing characteristics of reactors. Multizone models include tanks-in-series which are schematically illustrated in Figure 5.7. A simplified tanks-in-series model assumes n perfectly mixed reactors of the same volume V_t/n with constant flow rate Q entering and leaving each tank. A mass balance for a specific species in the tank i yields:

$$\frac{V_t}{n}\frac{dC_i}{dt} = QC_{i-1} - QC_i \tag{5.39}$$

The injection of a quantity M of a tracer in the first tank, mathematically expressed through the Dirac function as:

$$at \quad t=0 \quad \begin{cases} C_0 = \dfrac{M}{Q}\delta(t) \\ C_{i\neq 0} = 0 \end{cases} \tag{5.40}$$

gives the following RTD function (Froment et al., 2011):

$$f(t/\bar{t}) = \frac{C(t/\bar{t})}{\int_0^{\infty} C(t/\bar{t})dt} = \frac{n^n}{(n-1)!}\left(\frac{t}{\bar{t}}\right)^{n-1} e^{-n\frac{t}{\bar{t}}} \tag{5.41}$$

The variance of this RTD function can be calculated by substituting Eq. 5.41 into Eq. 5.26 to yield:

$$\sigma^2 = \frac{1}{n} \tag{5.42}$$

The above equation shows that the larger is the number of tanks fitting the RTD curve, the narrower is the distribution of the RTD function, eventually approaching the plug flow reactor behavior.

Residence time distribution functions as a function of time for different numbers of tanks-in-series (different n values) are illustrated in Figure 5.8. For $n = 1$, Eq. 5.41 simplifies to the RTD function of a perfectly mixed reactor, and when the value of n increases, RTD functions approach behaviors characterized by reactors with some degree of axial dispersion. The larger the value of n, the less is the axial dispersion, and for $n \to \infty$ the RTD function will be that for an ideal plug flow reactor.

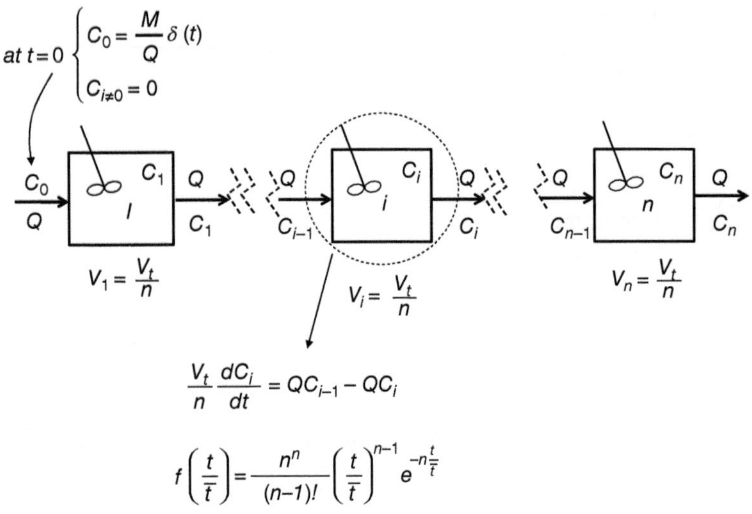

Figure 5.7 Multizone models used to characterize mixing in standard reactors and including main variables utilized to develop the models.

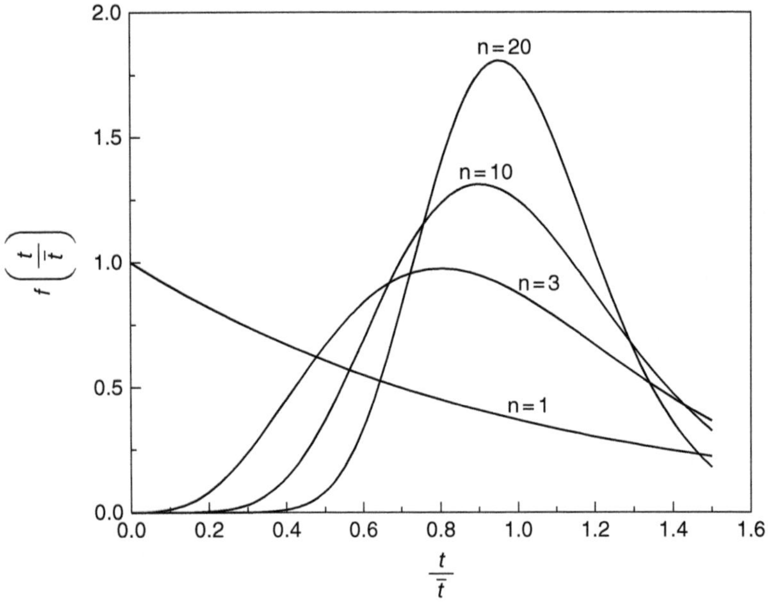

Figure 5.8 Simulated RTD functions estimated from multizone models consisting of one to several perfectly mixed reactors in series.

Multizone models do not account for effects of micromixing in the macromixing characteristics of the reactor.

5.2.1.2.3 Use of the RTD concept to describe mixing in screw extruder-reactors

Although mixing is a unit operation of relevance to many processing areas, it is still viewed in a fragmentary way (Ottino, 1994). Two approaches have been used for the study of mixing. One has been applied in the area of reactors and has been developed mainly to study the behavior of non-ideal reactors. Some aspects of this approach are discussed in this section. Diluted conditions in typical reaction systems mean that in many cases a turbulence flow will be the prevailing flow regime in these reactors. Macro- and micromixing characteristics of flows in reactors have been described using theoretical models that apply to turbulent flows. Froment et al. (2011) describe many of the methods used in the reactor area. Experimental methods to characterize macromixing have been based on the RTD concepts described in this section.

The other approach used to the study of mixing is associated with polymer processing and due to the high viscosity of polymers, the mixing processes are designed as laminar mixing. Both approaches appear to be unrelated, but the work of Ottino and collaborators has focused on development of a more general theory unifying these approaches, which is described in this section.

Given the complex flows existing in extruders, experimental approaches based on RTD concepts have been used to characterize mixing in extrusion processing and in particular in the area of reactive extrusion. The different unit operations that an extruder system may perform in different zones of the extruder have prompted researchers to apply the multizone method described above. For instance, Poulesquen and Vergnes (2003) used combinations of ideal reactors associated with specific sections of the extruder and pulse trace experiments were used to determine the RTD functions that were fitted using a combination of ideal reactors. A similar approach was used by Puaux et al. (2000) by fitting experimental RTD curves with a combination of ideal reactors. This approach may be criticized because the fitted parameters do not bear a physical meaning that could be associated with the process and the research becomes a non-linear fitting exercise. Caution should also be exercised in these analyses because often the lack of physical meaning of the fitted parameters may affect some of the conclusions drawn, in particular regarding implications that the obtained parameters may have on the degree of mixing, which invariably result in an incorrect analysis of the data (McGuire & Blackburn, 2008). Considerable research has been conducted using RTD methods to measure RTD in extruders. Two are mentioned here but reactive extrusion is more concerned with the type of mixing and the efficiency of mixing and that is the focus given to this chapter.

5.2.2 Mixing in screw extruder-reactors

Concepts of mixing have been discussed in Chapter 3 (see section 3.2.3.1.6) where, owing to the laminar flow characteristics of most of extrusion operations, the importance of the melt velocity profiles and fluid element trajectories was noted. From the analysis done in Chapter 3, it is noteworthy to observe the velocity profile existing in the extruder channel, which makes the trajectories of specific fluid elements different. Thus, these trajectories become key parameters to be quantitatively evaluated, especially given their relevance to the reaction conversion during a reactive extrusion process.

In reactive extrusion operations, effective mixing is essential to achieve the desired conversion of reactants in products. Tucker (1991) describes methods to quantitatively evaluate the degree of mixing and notes that in order to measure the degree of mixing, it is important to define the scale used for the measurements. The concept is exemplified with a simple analysis involving mixing of two materials having different colors. The degree of mixing varies if the examination is done with the naked eye or with a powerful microscope. Concerning reactive extrusion, reactants and products are conveyed through the extruder-reactor and in that flow, different fluid elements move with different trajectories. The spatial variation of the composition of these reactants and products becomes very relevant when defining the kinetics of the reaction.

Thermodynamically, if sufficient time has elapsed it would be expected that the concentration of the reactive species would distribute fairly uniformly within the extruder-reactor. However, times involved in reactive extrusion are short and the distribution of these species will depend on both the bulk flow of the melt in the extruder and the molecular diffusion of these species in the mixture. In that sense, these two phenomena are represented by the Damköhler numbers. Tucker (1991) provides a pictorial description of the phenomena and effects on the mixing efficiency by assuming the mixing of round black blobs surrounded by a white material. As a consequence of the deformation produced by the bulk flow, the black blobs smear into streaks, increasing the interfacial area between these two materials, but the white and black colors are still distinct. Conversely, the diffusion of these colors does not change the shape of the blobs; instead, due to the molecular diffusion of the white blob phase into the black blob phase, they become grey. Indeed, this type of mixing is relevant to extrusion operations but it must be noted that for reactive extrusion, the diffusion of the reactive species is associated with chemical reactions whose conversion will depend on the concentration of these reactants in specific locations of the extruder.

Ottino (1994) clearly distinguishes types of mixing and approaches used for the study of reactive mixing. One type of mixing is associated with the mixing of reactive species in reactors, which is mathematically described by models that characterize non-ideal behaviors observed in different types of reactors. These models were discussed in the previous section. The other "type" of mixing is that found during the processing of highly viscous polymers, known as laminar mixing. Although it appears that these types of mixing are different and unrelated, it is argued that they could be described by a general theory based on first principles that relate the transport of species by both bulk flow and molecular diffusion with the fluid mechanics of these flows. In fact, laminar mixing can also be described by Eqs 5.20 to 5.23.

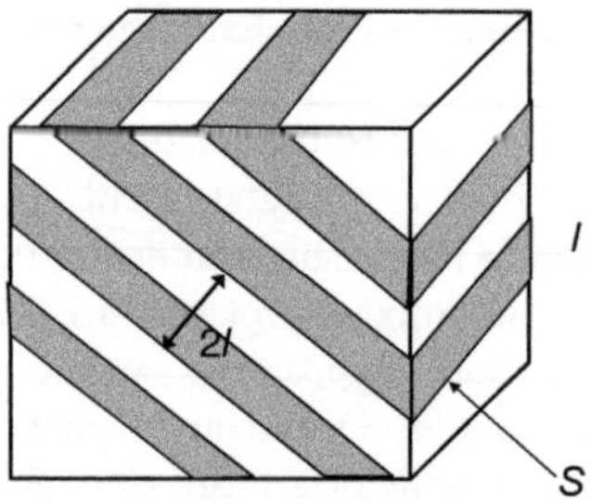

$$l = \frac{1}{A_v} = \frac{V}{S/2}$$

Figure 5.9 Schematic showing typical layered or lamellar structures indicating the definition of the striation thickness.

5.2.2.1 Basics of laminar mixing

A notable characteristic of the mixing of highly viscous polymers is that they are mixed by stretching and folding actions, which results in lamellar structures that have a wide distribution of length scales (Ottino, 1994). The formation of these structures is associated with the laminar mixing in the processing unit, in which fluid mechanics and establishment of velocity profiles and particles trajectories play a key role. However, interdiffusion of the mixture components depends on the interfacial area separating those components which is affected by the mixing characteristics. Diffusion of those components affects local compositions which in turn significantly affect the rate of reactions. Thus, it is essential to quantitatively assess the degree of mixing in terms of the interfacial area between components. One of those variables is the interfacial surface area per unit volume, designated by A_v and known simply as the interfacial area. A_v is one of the few measured variables associated with mixing which become larger when mixing proceeds (Tucker, 1991). The interfacial surface per unit volume is related to the interfacial area S and its corresponding volume V as:

$$A_v = \frac{S}{V} \tag{5.43}$$

As mentioned, a characteristic of the laminar flow is the production of layered or lamellar mixtures that can be conveniently represented by a variable known as the striation thickness, l, which is defined as one half the thickness of the repeating layered and lamellar unit (Tucker, 1991). The definition of the striation thickness for a lamellar mixture of two immiscible components is schematically illustrated in Figure 5.9 (Tucker, 1991). Thus, for a layered mixture, the interfacial area is related to the striation thickness by the following equation:

$$l = \frac{1}{A_v} = \frac{V}{S/2} \tag{5.44}$$

Both striation thickness and interfacial area per unit volume are variables which can be simply measured. However, they may fail to accurately describe the mixing quality, especially when the mixture is characterized by striations that have a distribution of thickness. In addition, lamellar mixtures consisting of regular layers may appear to the eye to be better mixed than mixtures having layers that do not follow a regular pattern, although the values of striation thickness and the interfacial area would be indicating the opposite (Tucker, 1991).

Mixing materials in standard reactors has been discussed in the previous section, where some of the approaches used to assess mixing behavior in reactors were described. For low-viscosity liquids, turbulence and diffusion phenomena are used as key factors to assess the mixing performance of well-mixed reactors. However, when the materials mixed are very viscous, neither diffusion nor turbulence can really assist the mixing action (Spencer & Wiley, 1951). Rather, mixing is effected by the complex, continuous flow existing in the mixer unit, specifically in this book the laminar flow in extruders.

To gain an insight into the effects of laminar flow on the deformation and degree of mixing of lamellar structures, it is worth analyzing the deformation of a cube element in a simple shear flow, as illustrated in Figure 5.10. As noted in the figure, the orientation of the cubical fluid region at the start of the shear deformation is established by the normal to three of its surfaces indicated by the unit vectors $\underline{N}_1$, $\underline{N}_2$ and $\underline{N}_3$. As a consequence of the velocity profile and the laminar conditions of the flow, the line AF moves with a higher velocity than the line BE; hence

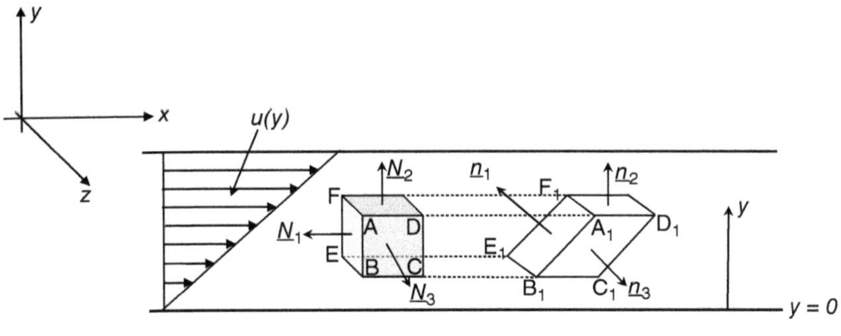

Figure 5.10 The deformation of a cube element in a typical shear flow created by a unidimensional flow $v(y)$.

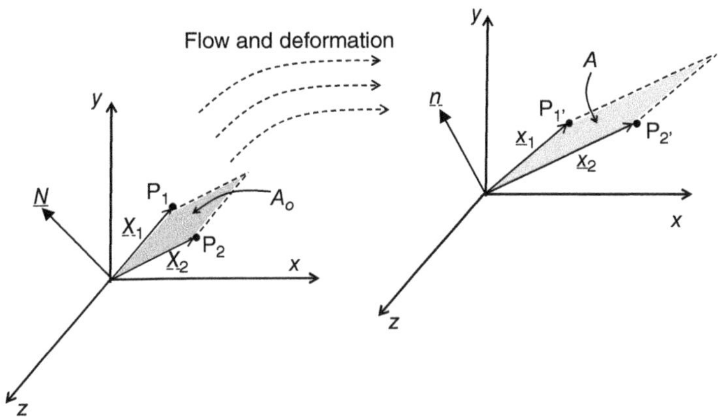

Figure 5.11 Deformation of a surface in a specific flow, characterization and estimation of the changes in the surface as a consequence of the flow.

the corresponding unit normal to that surface will change from $\underline{N}_1$ to $\underline{n}_1$; similar changes are also observed on the unit vector $\underline{N}_3$, which as a consequence of the shear flow changes to $\underline{n}_3$. However, for the assumed unidirectional shear flow the orientation of the surface characterized by the normal vector $\underline{N}_2$ does not change because the particles located in line AD have the same velocity. In other words, for these lines located at a distance y from $y = 0$, the distances between A and D, and A_1 and D_1, do not change as a consequence of the shear flow.

Consider a fluid element with an interfacial area ABEF with normal given by the unit vector $\underline{N}_1$, which as a consequence of the shear strain γ changes to the interfacial area $A_1B_1E_1F_1$ with a normal given by the unit vector $\underline{n}_1$. The relationship between the areas of these two surfaces, denoted by A_o and $A(\gamma)$, can be calculated by the following equation (Rauwendaal, 1991):

$$\frac{A(\gamma)}{A_o} = \left(1 + \gamma^2\right)^{1/2} \tag{5.45}$$

Spencer and Wiley (1951) and later Tadmor and Gogos (2006) presented a more generalized approach to assessing the deformation of a material in a pure shear flow. By defining a selected area A_o, associated with an interfacial region of the fluid under study, the deformation of the area due to the application of a flow or strain can be followed by a Lagrangian approach (Figure 5.11). The initial area A_o limited by the vectors $\underline{X}_1$ and $\underline{X}_2$ giving the location of two points on the surface, e.g. P_1 and P_2, can be estimated by the vector or cross-product between the two vectors $\underline{X}_1 \times \underline{X}_2$, which results in a vector with a direction normal to the area (indicated by $\underline{N}$ in Figure 5.11) and a magnitude equal to the area of the parallelogram formed by the two vectors. As a consequence of the flow or deformation, the locations of the initially selected points change, being now identified by the vectors $\underline{x}_1$ and $\underline{x}_2$, whose vector or cross-product $\underline{x}_1 \times \underline{x}_2$ defines a new normal, $\underline{n}$, and a new magnitude for the area, A (see Figure 5.11). Using a simple shear flow, Tadmor and Gogos (2006) determined the direction and magnitude of the normal vectors $\underline{N}$ and $\underline{n}$, thus extending the use of Eq. 5.45 for an initial arbitrary orientation of the fluid element:

$$\frac{A(\gamma)}{A_o} = \left(1 - 2\gamma cos\alpha_x cos\alpha_y + \gamma^2 cos^2\alpha_x\right)^{1/2} \tag{5.46}$$

where α_x and α_y are the angles formed between the unit vector to the surface in the initial orientation and the co-ordinates x and y, respectively. For the specific case represented in Figure 5.10 and the interfacial area ABEF $\alpha_x = 180^{\circ}$ and $\alpha_y = 90^{\circ}$, which upon substitution in Eq. 5.46 simplifies to Eq. 5.45. Conversely, if the surface ABCD is considered, the unit normal to that face is given by $\underline{N}_3$, so $\alpha_x = 90^{\circ}$ and $\alpha_y = 90^{\circ}$, and the change of area as a result of the simple shear flow of magnitude γ for that specific surface becomes $\frac{A(\gamma)}{A_0} = 1$. Both Eqs 5.45 and 5.46 predict that when the interfacial area per unit volume increases by the action of mixing, the striation thickness is decreased as indicated below:

$$l(\gamma) = \frac{2}{A_{v0}\left(1 - 2\gamma cos\alpha_x\, cos\alpha_y + \gamma^2 cos^2\alpha_x\right)^{1/2}} \quad (5.47)$$

where A_{v0} is the initial interfacial area per unit volume of the element of fluid. Under the action of a unidirectional shear flow, it can be demonstrated that the striation thickness is related to the applied shear strain by the following equation (Mohr et al., 1957a, 1957b):

$$l(\gamma) = \frac{l_o}{\gamma x_i} \quad (5.48)$$

where l_o is the initial striation thickness and x_i the volume fraction of the minor component i. Eq. 5.48 is not entirely applicable to extrusion processing because the shear flow existing in screw extruders is not unidirectional. Hence an analysis similar to the one done in Chapter 3 (section 3.2.3.1.6) is necessary to evaluate changes in striation thickness due to flow existing in extrusion processes. As already shown in Chapter 3 (section 3.2.3.1.5), even for simple situations involving ideal isothermal conditions and Newtonian melts, shear rates and shear strains are not uniform; in fact, they are strongly dependent on the location of the element of fluid in the extruder channel so striation thickness will vary accordingly. Furthermore, the flow of melt in screw extruders can exhibit extensional flow combined with shear flow and the relationship between the surface of an initially selected interfacial fluid region and the surface of the same region after flow or deformation is far from being simple.

The approach commonly used to find the deformation of a selected region of fluid as a consequence of an existing flow in a mixer or a screw extruder is the use of Lagrangian co-ordinates that follow the positions that an element of fluid takes during its flow or deformation. The use of Lagrangian co-ordinates is schematically illustrated in Figure 5.12 where at a previous time t' the particles designed by P and Q in an element of fluid are separated by a length defined by the vector $d\underline{r}'$. The location of the element of fluid is defined by the vector $\underline{r}'$. At the current time t, the element of fluid is deformed as a consequence of the flow or deformation and the distance between the particles P and Q changes to $d\underline{r}$. The location of the element is now specified by the vector $\underline{r}$. The deformations illustrated in Figure 5.12 can be expressed mathematically in a Cartesian co-ordinate system as:

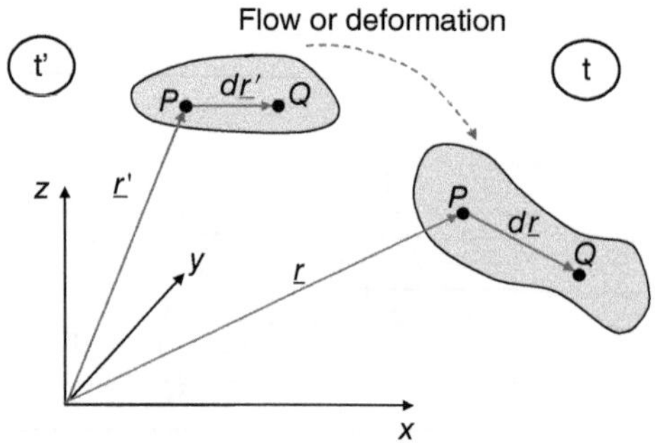

Figure 5.12 Lagrangian approach used for the study of deformation/strain of a fluid element as a consequence of a flow or a deformation.

$$d\underline{r}' = \begin{bmatrix} dX \\ dY \\ dZ \end{bmatrix} \quad and \quad d\underline{r} = \begin{bmatrix} dx \\ dy \\ dz \end{bmatrix} \quad (5.49)$$

Due to the flow or deformation, the element of fluid moves and is deformed, so the Lagrangian approach provides the position of the fluid element at the past time t' (which is defined by the vector $\underline{r}'$) whereas at the current time t its position is defined by the vector $\underline{r}$. In other words, the position of the fluid element at the past time t' depends on the fluid element being "labeled" with its position at the current time t defined by the vector $\underline{r}$ (Morrison, 2001). Mathematically, it is expressed as $\underline{r}' = \underline{r}'(\underline{r})$, which allows one to relate the distances between the two points at the past time t' and the current time t as:

$$dX = \frac{\partial X}{\partial x}dx + \frac{\partial X}{\partial y}dy + \frac{\partial X}{\partial z}dz = d\underline{r} \cdot \frac{\partial X}{\partial \underline{r}} \quad (5.50a)$$

$$dY = \frac{\partial Y}{\partial x}dx + \frac{\partial Y}{\partial y}dy + \frac{\partial Y}{\partial z}dz = d\underline{r} \cdot \frac{\partial Y}{d\underline{r}} \quad (5.50b)$$

$$dZ = \frac{\partial Z}{\partial x}dx + \frac{\partial Z}{\partial y}dy + \frac{\partial Z}{\partial z}dz = d\underline{r}\,\frac{\partial Z}{d\underline{r}} \tag{5.50c}$$

The above equations can be written in tensor form as (Morrison, 2001):

$$d\underline{r}' = d\underline{r}\cdot\frac{\partial \underline{r}'}{\partial \underline{r}} \tag{5.51}$$

The term $\partial\underline{r}'/\partial\underline{r}$ is a tensor known as the deformation gradient $\underline{\underline{F}}(t',t)$ and it is determined from the kinematics of the fluid flow. If the deformation gradient tensor is substituted in Eq. 5.51 the following results:

$$d\underline{r}' = d\underline{r}.\underline{\underline{F}}(t',t) \tag{5.52}$$

Equation 5.52 indicates that if the deformation gradient tensor associated with the flow kinematics is known, the deformation of an element fluid can be determined. Following a similar procedure, it can be shown that:

$$d\underline{r} = \underline{\underline{F}}^{-1}(t',t)\cdot d\underline{r}' \tag{5.53}$$

where $\underline{\underline{F}}^{-1}(t',t)$ is known as the inverse deformation gradient tensor, defined as:

$$\underline{\underline{F}}^{-1}(t',t) = \frac{d\underline{r}}{d\underline{r}'} \tag{5.54}$$

The deformation gradient and the inverse deformation gradient are tensors that describe the deformation history of each fluid element and have key relevance in evaluating mixing operations, in particular laminar mixing such as that existing in screw extruders. They are also used to estimate other tensors such as the Cauchy and Finger tensors which are important measurements when large strains are applied to the materials under different types of flow, notably shear and extensional flows, both of relevance in extrusion. Morrison (2001) gives equations to calculate not only both deformation gradients but also the Cauchy, $\underline{\underline{C}}(t',t)$, and the Finger, $\underline{\underline{C}}^{-1}(t',t)$, strain tensors for shear and extensional flows. For simplicity, the dependence of these tensors on past *t'* and current *t* times is not indicated in the equations.

Ottino et al. (1981) and Ottino (1989) give equations that utilize these tensors as well as the rate of deformation tensor $\underline{\underline{\dot{\gamma}}}$ defined in Chapter 3 (see Eq. 3.95) to characterize deformations of line, areas, and volumes of element of fluids (Figure 5.13):

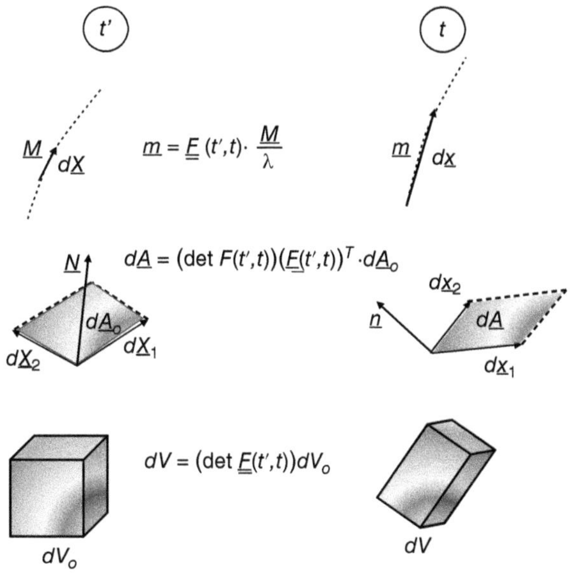

Figure 5.13 Definitions of parameters associated with a specific surface and used to characterize their changes as a consequence of flows or deformations.

$$\underline{m} = \underline{\underline{F}}(t',t)\cdot\frac{\underline{M}}{\lambda} \tag{5.55}$$

$$d\underline{A} = (\det F(t',t))\left(\underline{\underline{F}}(t',t)\right)^{T}\cdot d\underline{A}_o \tag{5.56}$$

$$dV = \left(\det\underline{\underline{F}}(t',t)\right)dV_o \tag{5.57}$$

Parameters associated with interfacial surfaces such as the length stretch λ, the area stretch η, the specific rate of stretching of the material element $\alpha(\underline{X},t)$, and the mixing efficiency $e(\underline{X},\underline{N},t)$ can also be calculated from these tensors (Chella & Ottino, 1985; Ottino, 1989):

$$\lambda = \left(\underline{\underline{C}}:\underline{M}\underline{M}\right)^{1/2} \tag{5.58a}$$

$$\eta = \left(\det\underline{\underline{F}}(t',t)\right)\left(\underline{\underline{C}}^{-1}:\underline{N}\underline{N}\right)^{1/2} \tag{5.58b}$$

$$\alpha(\underline{X},t) = \frac{\dot{\eta}}{\eta} = -\underline{\underline{\dot{\gamma}}}:\underline{n}\underline{n} \tag{5.58c}$$

$$e(\underline{X},\underline{N},t) = \frac{\textit{actual rate of stretching}}{\textit{upperbound}} = \frac{\dot{\eta}/\eta}{\left(\underline{\underline{\dot{\gamma}}}:\underline{\underline{\dot{\gamma}}}\right)^{1/2}} = \frac{-\underline{\underline{\dot{\gamma}}}:\underline{n}\underline{n}}{\left(\underline{\underline{\dot{\gamma}}}:\underline{\underline{\dot{\gamma}}}\right)^{1/2}} \tag{5.58d}$$

Of relevance to the mixing efficiency in extruders and other mixing processes, Ottino et al. (1981) defined average efficiency indexes as:

$$< eff(z) > |_0^z = \frac{\overline{e(\underline{X},t)\left(\dot{\underline{\underline{\gamma}}} : \dot{\underline{\underline{\gamma}}}\right)^{1/2}}}{\overline{\left(\dot{\underline{\underline{\gamma}}} : \dot{\underline{\underline{\gamma}}}\right)^{1/2}}} \quad (5.59)$$

$$< eff(t) > |_0^t = \frac{\overline{\overline{e(\underline{X},t)\left(\dot{\underline{\underline{\gamma}}} : \dot{\underline{\underline{\gamma}}}\right)^{1/2}}}}{\overline{\overline{\left(\dot{\underline{\underline{\gamma}}} : \dot{\underline{\underline{\gamma}}}\right)^{1/2}}}} \quad (5.60)$$

One line over the variable or groups of variables indicates an area-averaged mean value whereas two lines over variables or groups of variables indicate volume-averaged mean values. The angle brackets in the above two equations indicate the mean value of the variable included.

The mathematical foundations to quantitatively link mixing characteristic parameters with the pertinent fluid mechanics of the process are given by Ottino et al. (1981). The list of parameters presented here is not exhaustive; instead, it represents some key parameters to quantitatively evaluate a somewhat elusive term defining mixedness of the material in screw extruders. For more details on mathematical derivations, further definitions and other relevant parameters, refer to Ottino (1989).

Second-order tensors such as the deformation gradients and the Cauchy and Finger tensors can be represented by 3 × 3 matrices; thus in the above equations the term "det" denotes the determinant of the matrix, the superscript "T" denotes the transpose of the matrix, and the colon ":" a matrix product whereas a dot "." represents a product between a vector and a matrix and vice versa. The dyadic product between two vectors, for example the unit vectors normal to specific surfaces at the current and initial times, are designated respectively by $\underline{n}\,\underline{n}$ or $\underline{N}\,\underline{N}$ (defined below). The dyadic product results in a second-order tensor. The normal vectors are associated with mixed surfaces because the surface of a fluid element region is characterized by a vector whose magnitude is the area of the parallelogram formed by the vectors forming its sides, e.g. $d\underline{X}_1$ and $d\underline{X}_2$ for past time t', and $d\underline{x}_1$ and $d\underline{x}_2$ for current time t, with directions given by the normal to the surfaces indicated by $\underline{N}$ and $\underline{n}$ respectively (see Figure 5.13). In addition, the orientation of lines such as $\underline{M}$ and $\underline{m}$ can be defined from information of the flow kinematics (see Figure 5.13). Thus, the deformation of lines and surfaces of fluid elements located in the interface of two immiscible materials upon mixing is characterized not only by the kinematics of the flow but also the orientations of these lines or areas which are defined as (Ottino, 1989):

$$\underline{M} = \frac{d\underline{X}}{|d\underline{X}|} \quad (5.61a)$$

$$\underline{N} = \frac{d\underline{A}_o}{|d\underline{A}_o|} \quad (5.61b)$$

$$\underline{m} = \frac{d\underline{x}}{|d\underline{x}|} \quad (5.61c)$$

$$\underline{n} = \frac{d\underline{A}}{|d\underline{A}|} \quad (5.61d)$$

Real mixtures that are characterized for lamellar structures do not have a regular pattern after mixing; hence a distribution function of striation thickness must be defined and measured. Ottino (1994) gives relevant equations that can provide relationships between the stretching and deformation of lamellar structures with diffusion of the reactants in a mixture and the chemical reaction.

Chella and Ottino (1985) have recognized that there are two conceptual problems associated with laminar mixing. One deals with how mixing is measured whereas the other focuses on establishing relationships between mixing measurements and the fluid mechanics of the material flow in the mixer. In regard to this chapter, the latter is related to the complex flows existing in screw extruders.

The laminar mixing process in a mixer, for example a screw extruder, can be visualized starting as a two-phase system separated by a relatively small interfacial surface. For instance, Figure 5.14 schematically illustrates the mixing of two materials identified by white and black colors and initially separated by a horizontal interfacial surface. As a consequence of mixing, the increase in interfacial area is accompanied by a significant decrease in the thickness of the striations (Chella & Ottino, 1985). The stretching and break-up of the interfacial area as well as the thinning of the striation thickness present in the mixed phases are related to the fluid mechanics of the flow through complex relationships. The transcendence of the fluid mechanics pertinent to the flow in the extruder thus becomes evident. In that respect, the following two sections include a brief description of mixing behavior in single and twin screw extruder-reactors. Some of these concepts were discussed in Chapter 3 but new aspects are described in the following two sections, in particular how deformations of the material can be used to evaluate characteristics of the mixed material

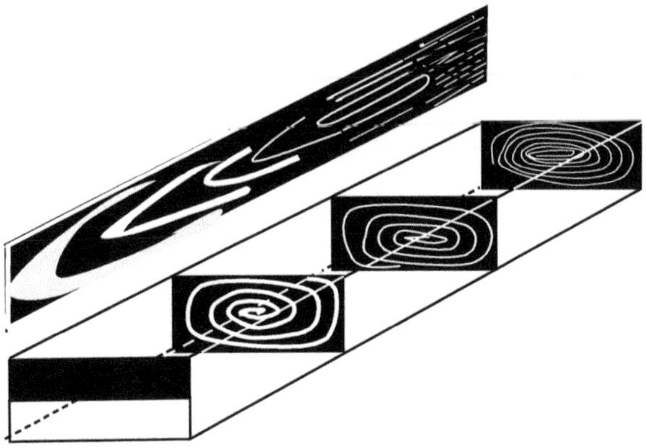

Figure 5.14 Schematic illustrating the formation of a lamellar structure generated by the flow in a screw extruder when two adjacent fluids are fed into the process. Source: Adapted from Chella & Ottino 1985. Reproduced with permission of the American Chemical Society.

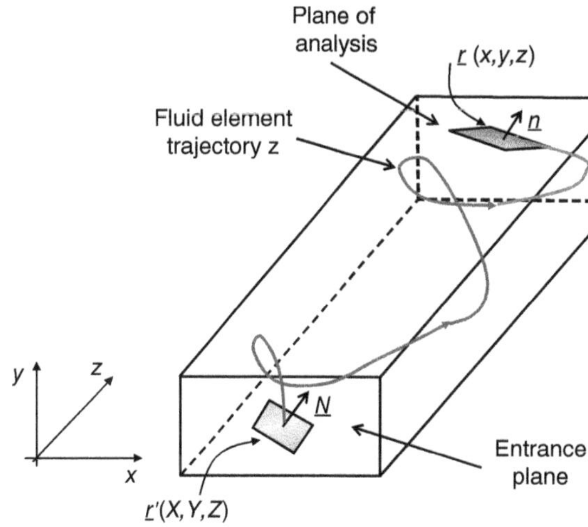

Figure 5.15 Trajectories of a fluid element in the channel of a single screw extruder. Source: Adapted from Chella & Ottino 1985. Reproduced with permission of the American Chemical Society.

such as area stretching, mixing efficiency and striation thickness, to name a few.

5.2.2.2 Mixing behavior in single screw extruder-reactors

For simplicity, only a model describing the extrusion of Newtonian fluids under isothermal conditions will be considered in this section. However, although extension to processes involving non-isothermal conditions and non-Newtonian melts requires numerical solutions of the mathematical models representing them, the approach followed is similar to the one presented in this section.

It was demonstrated in Chapter 3 that for isothermal conditions and Newtonian melts, the kinematics of the flow in single screw extruders can be represented by dimensionless velocity profiles in the down-channel $u_z(\xi)$ and the cross-channel $u_x(\xi)$, given by Eqs 3.193 and 3.63, and reproduced below as Eqs 5.62 and 5.63 respectively:

$$u_z(\xi) = \xi + 3\frac{Q_p}{Q_d}\xi(1-\xi) \tag{5.62}$$

$$u_x(\xi) = 2\xi - 3\xi^2 \tag{5.63}$$

The ratio Q_p/Q_d is defined in Chapter 3 by Eq. 3.72 as:

$$\frac{Q_p}{Q_d} = -\frac{H^2}{6\mu V_{bz}}\left(\frac{\partial p}{\partial z}\right)\frac{F_p}{F_d} \tag{5.64}$$

F_p and F_d are shape factors associated with the drag and pressure flows defined by Eqs 3.68 and 3.69 respectively. For shallow screws, i.e. $H/W << 1$, Figure 3.22 shows that those shape factors are close to 1. Down-channel velocity profiles were calculated using Eq. 3.193 in Chapter 3 (Eq. 5.62 in this chapter) and plotted in Figure 3.24 for different back pressure conditions. Additionally, trajectories of different fluid particles were obtained by combining the down-channel velocity component with the cross-channel velocity component given by Eq. 5.63 (Eq. 3.63 in Chapter 3), resulting in a trajectory schematically illustrated in Figure 5.15. The circular trajectory of a fluid element in the cross-section area of the channel was obtained from the condition of zero net flow in that cross-section resulting in Eqs 3.196 and 3.197, given below as Eqs 5.65 and 5.66 respectively:

$$\text{For } 0 \le \xi_c \le \frac{2}{3} \qquad \xi = \frac{1}{2}\left(1 - \xi_c + \sqrt{1 + 2\xi_c - 3\xi_c^2}\right) \tag{5.65}$$

$$\text{For } \frac{2}{3} \le \xi \le 1 \qquad \xi_c = \frac{1}{2}\left(1 - \xi + \sqrt{1 + 2\xi - 3\xi^2}\right) \tag{5.66}$$

These equations indicate that the position of a fluid element in the upper portion of the channel ξ determines the corresponding and complementary position in the lower portion of the channel ξ_c. The arrangement of these two trajectories determines the cycling flow of a specific fluid element in the channel. As discussed in Chapter 3, these trajectories allow estimation of the residence time of a fluid element in a cycle. For the upper part, $t_u(\xi)$, and the lower part, $t_l(\xi)$, of the channel of width W, these residence times can be respectively estimated as (Rauwendaal, 1991):

$$t_u(\xi) = \frac{W}{|u_x(\xi)|} \quad (5.67)$$

$$t_l(\xi) = \frac{W}{|u_x(\xi_c(\xi))|} \quad (5.68)$$

And the residence time of a fluid element in a cycle t_{cycle} is calculated as:

$$t_{cycle}(\xi) = t_u(\xi) + t_l(\xi) \quad (5.69)$$

The fraction of the cycle time that the element of fluid spends on the upper portion of the channel is calculated as:

$$f_u(\xi) = \frac{t_u(\xi)}{t_{cycle}(\xi)} \quad (5.70)$$

By substituting Eqs 5.67, 5.68, and 5.69 into Eq. 5.70, the following expression is obtained (the same equation is also given as Eq. 3.198 in Chapter 3):

$$f_u(\xi) = \frac{1}{1 + \frac{|u_x(\xi)|}{|u_x(\xi_c)|}} = \frac{1}{1 - \frac{\xi(2-3\xi)}{\xi_c(2-3\xi_c)}} \quad (5.71)$$

In Chapter 3, values of this fraction were plotted as a function of the position of a fluid element trajectory in the lower region of the channel ξ_c (Figure 3.50). Figure 5.16 shows values of this fraction as a function of the location of the fluid element trajectory in the upper region of the channel; in the same plot is also shown the corresponding position of the fluid element trajectory in the lower region of the channel, i.e. ξ_c. At regions close to the barrel surface ($\xi \rightarrow 1$), both the position of the fluid element trajectory in the lower region ξ_c and the fraction of time that the fluid element spends on the upper region $f_u(\xi)$ converges to zero. Conversely, for $\xi_c = 2/3$, the fraction of time spent by that specific element of fluid in the upper region of the channel reaches its maximum value, i.e. $f_u(\xi) = 0.5$. This, along with results shown in Figure 5.16, clearly illustrates that fluid elements spend more time in the lower part of the channel than in the upper region. These flow characteristics have a profound effect on the deformation of the melt and the mixing of different components during the extrusion process.

A simple but more detailed analysis of the effects that these different residence times have on the deformation of fluid elements was done by Rauwendaal (1991), who hypothetically and schematically visualized the deformation of fluid elements for pure shear flows in different regions of

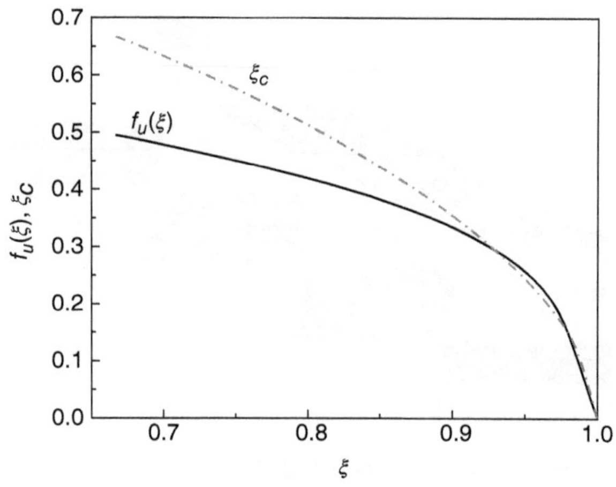

Figure 5.16 Trajectory of a fluid element in a cross-section of the extruder channel in terms of the location of the fluid element in the lower region of the channel ξ_c as a function of the location that the same fluid element has in the upper region during a cycle, and fraction of time that the fluid element spends in the upper region $f_u(\xi)$.

the channel cross-section. As shown in Figure 5.17, the different areas are denoted by A_u, A_l, B_u and B_l. The subscripts u and l indicate upper and lower regions respectively whereas A and B indicate proximity to the barrel/screw surfaces and center of the channel, respectively. With the assumption of negligible velocity component in the direction normal to the channel, the figure shows very distinctive deformation patterns depending on the location of the fluid elements. In the shearing A region, close to the surfaces of the screw and the barrel, shear rates have a different sign. In the upper region (A_u), shear rates are positive whereas in the lower region (A_l), they are negative which may cause demixing of the material (Rauwendaal, 1991). Whereas the circulation in the inner region B the shear in the upper region B_u is complemented by the shear in the lower region B_l. As discussed the residence time of a fluid element in a cycling trajectory of the cross-section in region A is longer than in the central region (region B), resulting in different deformations of the fluid elements clearly evidenced by comparing Figures 5.17a and 5.17b. An expression to estimate the residence time of a fluid element over the length of the extruder was obtained in Chapter 3 (Eq. 3.199), given below as Eq. 5.72:

$$t(\xi) = \left(\frac{L}{3V_b\left(1 + \frac{Q_p}{Q_d}\right) sin\theta cos\theta} \right) \cdot \frac{3\xi - 1 + 3\sqrt{1 + 2\xi - 3\xi^2}}{\xi\left(1 - \xi + \sqrt{1 + 2\xi - 3\xi^2}\right)} \quad (5.72)$$

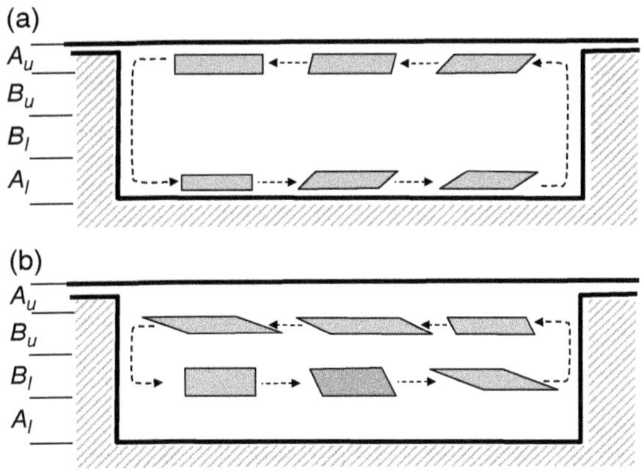

Figure 5.17 Deformation of fluid elements in pure shear flows at different regions of the channel cross-section of a single screw extruder. (a) Deformation in cycles whose trajectories are in the proximity of the barrel and screw surfaces. (b) Deformation in cycles with trajectories closer to the center of the channel.

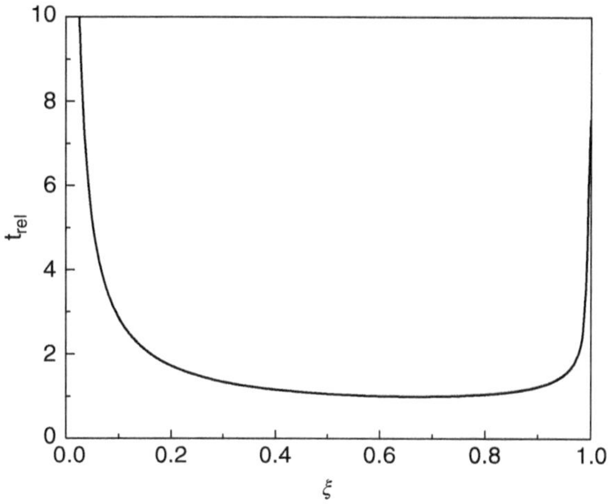

Figure 5.18 Relative residence time (defined by Eq. 5.75) as a function of the fluid element location at the upper region of the cross-section of the screw channel.

where L is the axial length of extruder (or the length of the shaft). At position $\xi = 2/3$, as indicated by Eq. 5.63, the cross-section velocity component $u_x(\xi)$ is zero. Thus, in the absence of that velocity component the residence time becomes a minimum, t_{min}, which is determined by substituting $\xi = 2/3$ in Eq. 5.72 to give:

$$t_{min} = \frac{3L}{2V_b\left(1+\dfrac{Q_p}{Q_d}\right) sin\theta cos\theta} \tag{5.73}$$

In Chapter 3, the above equation was further simplified by using the relationships $V_{bz} = V_b\cos\,\theta$ and $L = z\,\sin\,\theta$ to yield the following equation (see Eq. 3.200 in Chapter 3):

$$t_{min} = \frac{3z}{2V_{bz}\left(1+\dfrac{Q_p}{Q_d}\right)} \tag{5.74}$$

where z is the down-channel distance. The relative residence time can be estimated by the ratio of Eqs 5.72 and 5.73 as:

$$t_{rel} = \frac{t(\xi)}{t_{min}} = \frac{2}{9}\cdot\frac{3\xi - 1 + 3\sqrt{1+2\xi-3\xi^2}}{\xi\left(1-\xi+\sqrt{1+2\xi-3\xi^2}\right)} \tag{5.75}$$

Equation 5.75 shows that over a broad region at the center of the channel, the relative residence time is close to 1 (i.e. close to the minimum value given by Eq. 5.74) but increases significantly close to the surface of the screw ($\xi \to 0$) and the surface of the barrel ($\xi \to 1$) (Figure 5.18). Of more interest is to determine the effect of the ratio Q_p/Q_d (by Eq. 5.64 the ratio defines the pressure gradient) on the extrusion residence time, which can be estimated by Eq. 5.72. This is illustrated in Figure 5.19 for positive and negative pressure gradients and also for the pure drag flow ($Q_p/Q_d = 0$). The parameters used to generate the curves shown in Figure 5.19 include a channel helix angle (θ) of 20°, an axial length (L) of 2.5 m, a barrel with an internal diameter (D) of 10 cm and a screw speed (N) of 200 rpm. The latter translates on a peripheral speed of the screw $V_b = \dfrac{\pi ND}{60}$ of approximately 1.0 m/s. As noted in the figure, the residence times of elements of fluid in the upper region of the channel $\left(\text{i.e. } \xi \geq \dfrac{2}{3}\right)$ increase when the location of the trajectory is closer to the barrel surface. A similar plot of residence time as a function of the location of the trajectory of a fluid element moving in the lower region of the channel $\left(\text{i.e. for } \xi_c \leq \dfrac{2}{3}\right)$ would show residence times increasing when the trajectory of the fluid element becomes closer to the screw surface.

The effect of pressure gradient (Q_p/Q_d ratio) is clearly evidenced in Figure 5.19 where increases in pressure gradients (more negative Q_p/Q_d ratios) result in longer residence times. Based on the transit time of elements of fluids in different locations of the channel and the corresponding shear rates on those locations, an expression

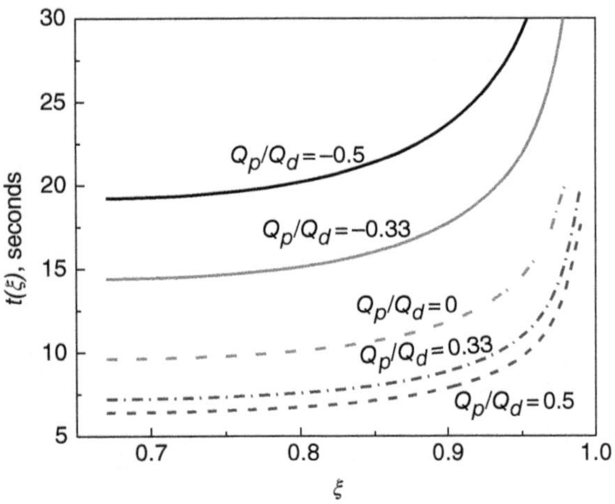

Figure 5.19 Residence time of different fluid elements defined by the location of their trajectories in the upper region (ξ) for different pressure gradients (defined by the Q_p/Q_d values).

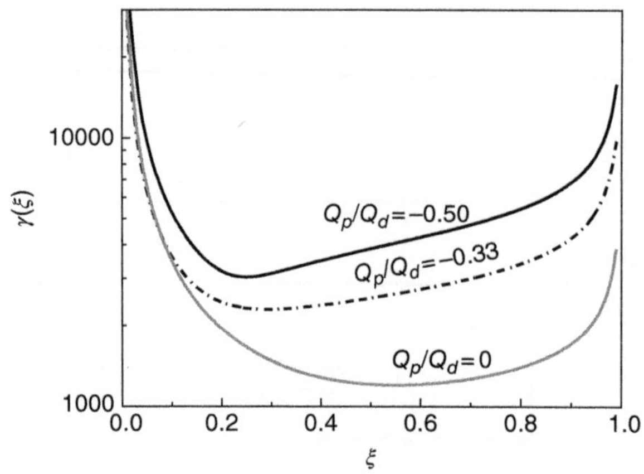

Figure 5.20 Strain applied to fluid elements defined by the location of their trajectories in the upper region (ξ) for different pressure gradients (defined by the Q_p/Q_d values).

was derived in Chapter 3 (Eq. 3.126) to evaluate the full strain applied to elements of fluids that move along the extruder and follow the helical path schematically shown in Figure 5.15. The equations to calculate the strain of a fluid element in a channel location are given below as Eqs 5.76, 5.77, and 5.78:

$$\gamma(\xi) = \frac{1}{3H}\left(\frac{1}{1+Q_p/Q_d}\right)\cdot\left[E_1\left(\xi,\theta,Q_p/Q_d\right)+E_2\left(\xi,\theta,Q_p/Q_d\right)\right] \tag{5.76}$$

The functions $E_1(\xi,\theta,Q_p/Q_d)$ and $E_2(\xi,\theta,Q_p/Q_d)$ are given below:

$$E_1\left(\xi,\theta,Q_p/Q_d\right) = \frac{2f_u(\xi)}{cos\theta}\frac{\left[(1-3\xi)^2+\frac{cotg^2\theta}{4}\left(1+3\frac{Q_p}{Q_d}-6\xi\frac{Q_p}{Q_d}\right)^2\right]^{1/2}}{\xi_c(1-\xi_c)+f_u(\xi)\cdot(\xi-\xi_c)\cdot(1-\xi-\xi_c)} \tag{5.77}$$

$$E_2\left(\xi,\theta,Q_p/Q_d\right) = \frac{(1-f_u(\xi))}{sin\theta}\frac{\left[4(1-3\xi_c)^2tan^2\theta+\left(1+3\frac{Q_p}{Q_d}-6\xi_c\frac{Q_p}{Q_d}\right)^2\right]^{1/2}}{\xi_c(1-\xi_c)+f_u(\xi)\cdot(\xi-\xi_c)\cdot(1-\xi-\xi_c)} \tag{5.78}$$

Figure 5.20 illustrates total strain of fluid elements as a function of their vertical location for pure drag flow ($Q_p/Q_d = 0$) and for different pressure gradients. It clearly shows that the total strain depends on the location and the pressure gradient. Fluid elements located close to the center experience lower strains whereas fluid elements located close to the barrel and the screw surfaces are subjected to larger shear strains.

When the pressure gradient is defined by the ratio Q_p/Q_d equal to −0.33 or smaller, the shear strain reaches a minimum in a region near $\xi = 1/3$. After that region, the shear strain does not vary significantly until $\xi = 0.9$. The larger the pressure gradient, the larger the variation of the shear strain, which causes non-uniform mixing and a higher strain is applied to the melt (see Figure 5.20). Knowing the shear strain, the increase in the interfacial area can be evaluated by Eqs 5.45 or 5.46. Furthermore, the striation thickness can be evaluated by Eqs 5.44 or 5.47. However, as explained above, the equations that estimate increases in interfacial areas and striation thickness should be used with caution because they are derived only for a unidirectional shear flow which is not the type of flow commonly encountered in screw extruder-reactors. Instead, the flow in screw extruder-reactors is characterized by complex shear deformations combined with extensional flow.

Despite the simplicity of the fluid mechanics model used in this section, the analytical equations obtained serve not only to quantitatively examine the effect of many important processing variables such as pressure gradient and screw speed on the mixing characteristics of extruder-reactors but also to explore the effect of geometric characteristics of the extruder-reactor such as helix angle, height of screw flight, and extruder length on the mixing performance. Rauwendaal (1991) illustrates the effects of pressure gradient and helix angle on the total strain, which can be estimated by this simplified model using Eqs 5.76, 5.77, and 5.78.

An aspect that has not been considered in this approach to assessing mixing in single screw extruders is the initial orientation of the interfaces or their reorientation due to interruption of the shear flow by mixing elements. Rauwendaal (1991) discusses the effect of interface orientation, considering simple equations that take into account the application of shear flows until they are interrupted by a short mixing section. For instance, the following equation predicts the increase of the interfacial area as a consequence of an applied shear strain until a mixing region is reached in the flow:

$$\frac{A_1(\gamma_1)}{A_o} = \frac{1}{2}\gamma_1 \tag{5.79}$$

where γ_1 is the shear strain at which a element of fluid is exposed before the mixing section. By assuming the same amount of shear strain applied between n mixing sections, the change in the interfacial area can be estimated by the following equation:

$$\frac{A_n(\gamma_1)}{A_o} = \left(\frac{\gamma_1}{2}\right)^n \tag{5.80}$$

Although derived from a simplified model, Eq. 5.80 provides a qualitative description showing the efficacy of mixing zones in extruder-reactors. Rauwendal (1991) assesses the efficacy of mixing zones in the extruder-reactor profile by comparing the increase of the interfacial area produced by the presence of n mixing zones with n shear strain exposures (given by Eq. 5.80) with the increased interfacial area after the material is subjected to a shear strain of magnitude $n\gamma_1$, which is calculated by the following equation:

$$\frac{A_1(n\gamma_1)}{A_o} = \frac{1}{2}n\gamma_1 \tag{5.81}$$

Equation 5.82 gives the ratio between the interfacial areas $A_n(\gamma_1)$ and $A_1(n\gamma_1)$, which is plotted in Figure 5.21 as a function of the magnitude of the shear strain γ_1 applied among the mixing regions and the number of mixing regions n in the extruder-reactor profile:

$$\frac{A_n(\gamma_1)}{A_1(n\gamma_1)} = \frac{(\gamma_1/2)^n}{n\gamma_1/2} \tag{5.82}$$

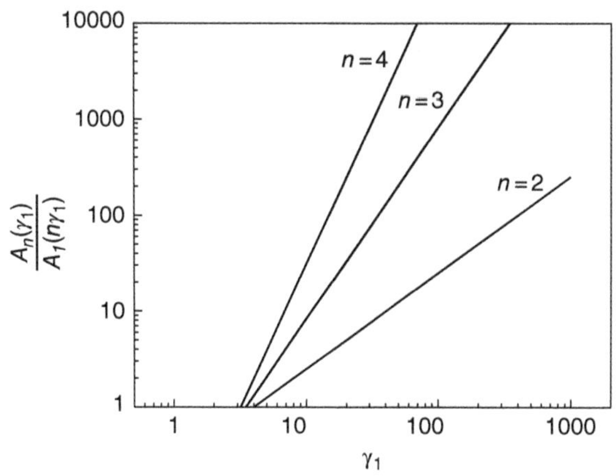

Figure 5.21 Increase of the surface area due to the introduction of mixing regions.

The figure clearly shows that the interfacial area among components can be substantially increased through the incorporation of mixing zones.

Estimation of the total strain undergone by an element of fluid and the RTD function of the extruder-reactor enables the calculation of a parameter that is used to measure the mixing performance of extruder-reactors. It is defined as the weighted average total strain (WATS) and given as (Bigg & Middleman, 1974; Rauwendaal, 1991):

$$WATS = \int_0^\infty \gamma(t) f(t) dt \tag{5.83}$$

Calculation of the WATS is relatively simple but it is a quantity that cannot be measured experimentally. It relies only in the trajectory of the element of fluid and the RTD function, and so the parameter does not explicitly account for effects of micromixing, which are of key importance in reactive extrusion to estimate local reaction rates. Chella and Ottino (1985) list disadvantages in using this parameter to assess mixing and its effects on associated chemical reactions. They include:

- the difficulty of quantitatively relating strains to experimentally accessible measurement of mixing
- the assumption of only unidirectional shear flow in an extruder may result in an oversimplification that can hide mixing mechanisms
- no explicit consideration of interface orientation precludes a systematic analysis of the influence of the mode of feed
- only mean values are obtained, but information about the distribution of mixedness is important when simultaneous

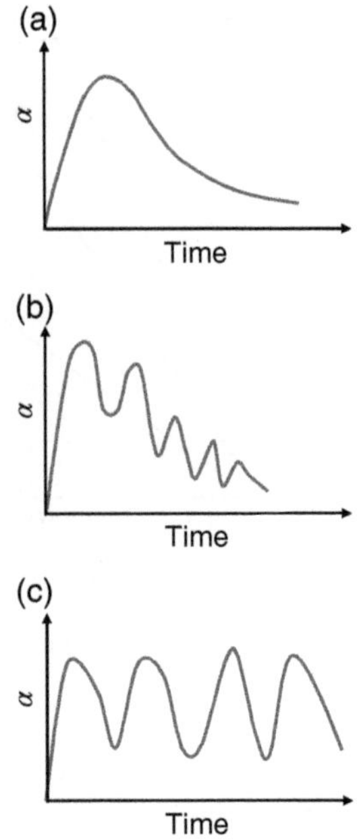

Figure 5.22 Schematic illustrating characteristic features of the specific rate of stretching parameter (α) during mixing in a reactor or extruder-reactor. Source: Adapted from Chella & Ottino 1985. Reproduced with permission of the American Chemical Society.

diffusion and reaction occurs. The latter is extremely important when assessing reactive extrusion processes.

The importance of characterizing the properties of the interfacial surfaces using parameters such as area stretch, surface orientation, specific rate of stretching of an element of fluid (Lagrangian approach), which were defined in section 5.2.1.3 for arbitrary kinematics, has been stressed out by Chella and Ottino (1985). For instance, characteristic features of the specific rate of stretching parameter (α) could be associated with micromixing phenomena in the reactor or extruder-reactor.

Figure 5.22 schematically illustrates different types of stretching of interfacial surfaces. Figure 5.22a shows the rate of stretching of the interfacial area of an element of fluid for stationary bound flows, in which over long periods α decays as t^{-1}. Figure 5.22b illustrates the behavior of the parameter α in flows that promote moderate orientation, where a succession of t^{-1} decaying patterns are observed with the average value of the parameter α going to zero. Flows that promote strong orientation of the interfacial surfaces are characterized with a rate of stretching parameter that oscillates around an average value that does not decay over long periods (Figure 5.22c). Moderate and strong orientation of interfacial surfaces is sought in extrusion processing and has been the practical reason for introducing screw geometries that promote mixing zones in the screw profile. This behavior may be beneficial to reactive extrusion processes to decrease the diffusion path of slowly diffusing reactants.

Chella and Ottino (1985) analyzed the mixing of a single screw extruder based on the kinematics described by Eqs 5.62 and 5.63 and also by a more refined kinematics that considers that an element of fluid may also have a vertical velocity component, especially in regions close to the screw flights. The latter refinement was used only under conditions that require it, whereas most of the analysis was conducted using the simplified 2D flow given by Eqs 5.62 and 5.63. To estimate properties inherent to the interfacial areas of elements of fluid and their deformations created by the mixing flow, a number of differential planes/surfaces (about 200–300) were created (Lagrangian approach) in the feed plane. The location and orientation of each element were identified in correspondence with the position of the interfacial area in the feed. The number of elements of fluid selected was sufficient and more sensitive than standard RTD measurements. The kinematics of the flow was used in conjunction with the definition of the relevant parameters given by Eqs 5.58, 5.59, and 5.60. The parameters were averaged over the number of selected element of fluids. Of particular interest was the calculation of the interfacial area density A_v defined by Eq. 5.43. The averaged value of this interfacial area over the number of elements of fluid was calculated as:

$$< \ln A_v > |_0^z = < eff(z) > |_0^z \frac{\overline{\left(\underline{\underline{\dot{\gamma}}} : \underline{\underline{\dot{\gamma}}}\right)^{1/2}}}{u_z} z \tag{5.84}$$

If the kinematics of the flow is known, associated parameters (e.g. strain rate and deformation tensors) can be found from tables for the respective flows (Morrison, 2001; Ottino, 1989). Equations such as Eq. 5.84 can be used to establish upper mixing bounds. For instance, by assuming that $eff(z) = 1$, Eq. 5.84 simplifies to the following:

$$< \ln A_v > |_0^L \le \frac{\left[4-3\left(1-\left(\frac{Q_p}{Q_d}\right)^2\right)\right] \cos^2\theta}{\left(1+\frac{Q_p}{Q_d}\right) \sin(2\theta)} \cdot \left(\frac{L}{H}\right) \tag{5.85}$$

The proportionality constant in the above equation required for the equality of the equation is a function of processing parameters (e.g. Q_p/Q_d), geometric parameters (e.g. θ and L/H), and the feed orientation given by the vector normal to the interface $\underline{N}$ (Chella & Ottino, 1985). On the basis of this mixing bound, several conclusions can be drawn concerning optimal mixing conditions. An analysis of Eq. 5.85 shows that mixing can be improved by: (i) increasing pressure gradients (more negative Q_p/Q_d values) at constant L/H and θ; (ii) increasing L/H at constant Q_p/Q_d and θ; and (iii) increasing θ for $\theta > \theta_{min}$ or decreasing θ for $\theta < \theta_{min}$ where θ_{min} is defined as:

$$\theta_{min} = tan^{-1}\left(\frac{\left(1+3\left(\frac{Q_p}{Q_d}\right)^2\right)^{1/4}}{\sqrt{2}}\right) \tag{5.86}$$

where:

$$\frac{\pi}{6} \leq \theta_{min} \leq \frac{\pi}{4} \tag{5.87}$$

The other condition for optimizing mixing that can be inferred from the upper mixing bound is to decrease H at constant L/H and θ.

Results of this approach in terms of the interfacial stretching area ratio are illustrated in Figures 5.23 to 5.26. Figure 5.23 shows changes in the averaged parameter over the extruder length at difference pressure gradients. It can be noted that the interfacial area stretch ratio changes almost linearly with the axial distance with some moderate oscillations due to the reorientation of these surfaces in the flow. The effect of the pressure gradient is also fairly evident and clearly shows that high values of back pressure (more negative Q_p/Q_d values) promote the stretching of the interfacial areas. Figure 5.24 illustrates an important phenomenon that has been experimentally observed when the extruder is fed at different locations to promote better mixing. These processing benefits to improve mixing efficiency are easily achievable in intermeshing, co-rotating twin screw technologies and are of significant relevance to the area of reactive extrusion. Figure 5.25 shows the effect of the screw geometry (specifically helix angle θ) and the back pressure on the average area stretch measured at the exit of the extruder. Minimum mixing is achieved when the helix angles are between 30° and 60° ($\pi/6$ to $\pi/3$ radians) which is in general agreement with the values obtained from the analysis of the upper bounds (Eq. 5.87). The figure also shows that for all geometries, higher pressure gradients favor mixing. Figure 5.26 shows the general trends observed in the previous figure in terms of the pressure gradient and the helix angles. It also illustrates that the average stretching of the interfacial areas estimated by the WATS approach (calculated and calibrated for the case of pure drag flow

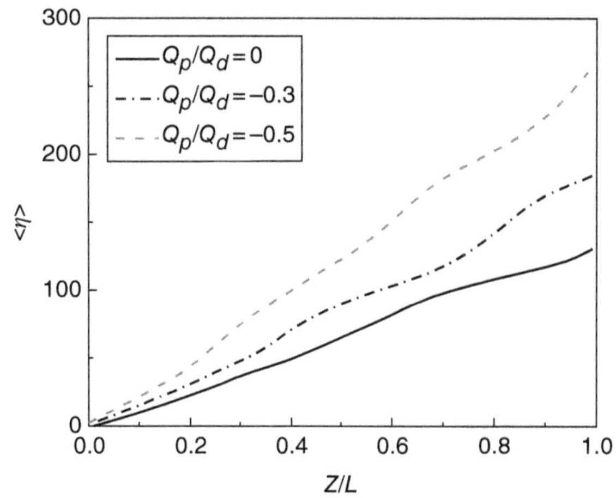

Figure 5.23 Average area stretch $<\eta>$ evolution during the flow of a Newtonian melt in a single screw extruder as a function of the pressure gradient (defined by Q_p/Q_d). Simulations were performed using a single screw extruder with the following characteristics: $L/H = 50$, $W/H = 15$, $\theta = 15°$, and fed with a vertical feed interface. Source: Adapted from Chella & Ottino 1985. Reproduced with permission of the American Chemical Society.

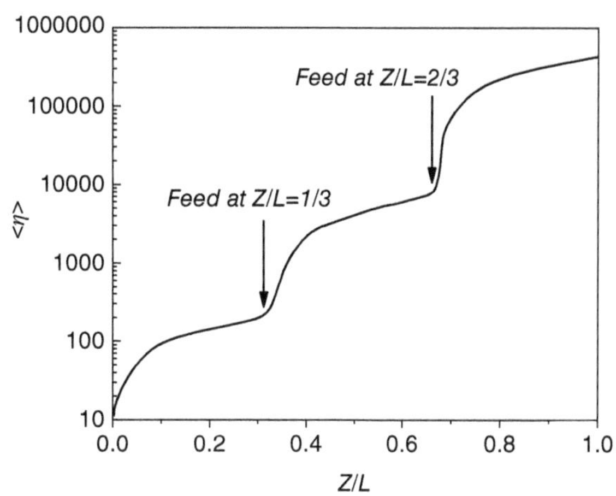

Figure 5.24 Effect of multiple feedings on the average area stretch $<\eta>$ evolution of the flow of a Newtonian fluid in a single screw extruder. Simulations were performed using a single screw extruder with the following characteristics: $L/H = 50$, $W/H = 15$, $\theta = 15°$, $Q_p/Q_d = -0.3$, and fed with a vertical feed interface. Source: Adapted from Chella & Ottino 1985. Reproduced with permission of the American Chemical Society.

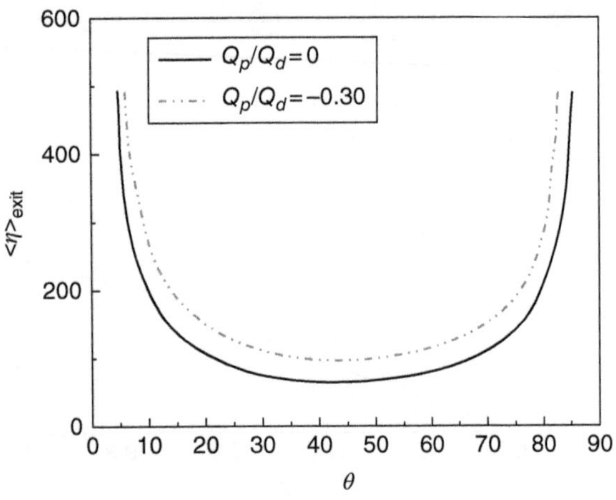

Figure 5.25 Effect of the flight angle and the pressure gradient on the average area stretch measured at the exit of the extruder $<\eta>_{exit}$ during the flow of a Newtonian fluid in a single screw extruder. Simulations were performed using a single screw extruder with the following characteristics: $L/H = 50$, $W/H = 15$, fed with a vertical feed interface. Source: Adapted from Chella & Ottino 1985. Reproduced with permission of the American Chemical Society.

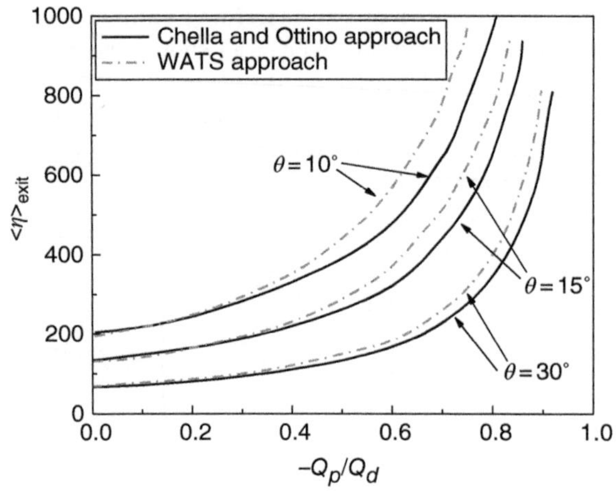

Figure 5.26 Effect of the pressure gradient and the flight angle on the average area stretch measured at the exit of the extruder $<\eta>_{exit}$ during the flow of a Newtonian fluid in a single screw extruder. Simulations were performed using a single screw extruder with the following characteristics: $L/H = 50$, $W/H = 15$, fed with a vertical feed interface. Source: Adapted from Chella & Ottino 1985. Reproduced with permission of the American Chemical Society.

$Q_p/Q_d = 0$) overestimated the stretching of the interfacial areas, especially at high pressure gradients. For the calculations, values of $H/L = 50$ and $W/H = 15$, which represent a middle size extruder, were used. Other variables are included in the figure legends (Figures 5.23–5.26).

5.2.2.3 Mixing behavior in intermeshing, co-rotating twin screw extruder-reactors

Due to the more complicated geometry of intermeshing, co-rotating twin screw extruders, much less analytical work can be found regarding their flows and mixing characteristics. Thus, the analysis of the flow and mixing will be centered on Newtonian melt fluids under isothermal conditions. The flow in a twin screw extruder is assumed to be steady and laminar, and the usual assumption of a rotating barrel and a stationary screw is still considered. Similarly to the analysis of single screw extruders, the barrel is unwound. Although this simplification facilitates the fluid mechanic analysis, it is necessary to bear in mind that it is only applicable to shallow channels (Maheshri & Wyman, 1980). As before, the flow of the melt is created by the barrel moving with a velocity $V_b = N\pi D$ with components $V_{bx} = V_b \sin\phi$ and $V_{bz} = V_b \cos\phi$ in the cross-channel and down-channel respectively whereas ϕ is the helix angle. The down-channel and cross-channel velocity components are analyzed in a similar manner as for single screw extruders, including the boundary no-slip conditions at the surfaces of the barrel and the screw, and also the assumption that both velocity components are a function of the vertical direction y. With those assumptions, the cross-channel velocity component can be expressed as:

$$\frac{v_x}{V_b sin\phi} = u_x(\xi) = \xi(2-3\xi) \tag{5.88}$$

Figure 5.27 illustrates the flow in the channel where, as in single screw extruders, the trajectory of an element of fluid in the lower and upper region of the channel are identified by the locations ξ_c and ξ, respectively. By assuming no leakage in the cross-channel direction, the figure clearly shows that the particles travel following a close loop between the locations ξ_c and ξ, which can be calculated by the same relationship used for single screw extruders (given by Eqs 5.65 and 5.66). It is important to note that the path of an element of fluid is not a closed loop due to the down-channel velocity component that can be calculated by the following equation:

$$\frac{v_z}{V_b cos\phi} = u_z(\xi) = \xi - 3\frac{H^2}{6\mu V_b cos\phi}\left(\frac{\partial p}{\partial z}\right)\xi(1-\xi) \tag{5.89}$$

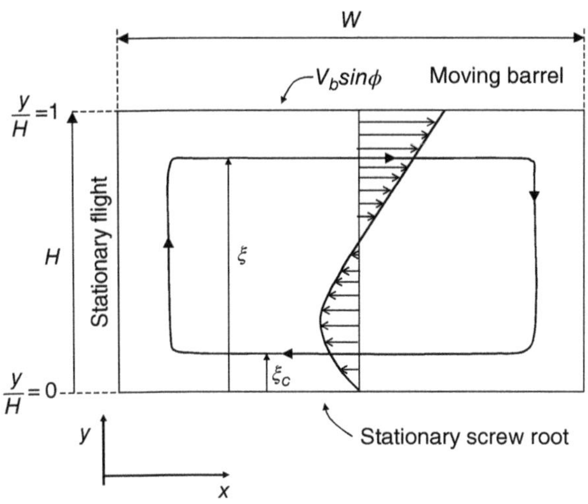

Figure 5.27 Velocity profiles in the middle part of the screw chamber in a twin screw extruder and corresponding trajectories of elements of fluid. Moving barrel in the x-direction and the screw root and flights are stationary. Source: Maheshri & Wyman 1980. Reproduced with permission of John Wiley & Sons.

The down-channel flow Q is then estimated by integrating the down-channel velocity component u_z as done in Chapter 3 (see Eq. 3.66) by assuming that it is only a function of the vertical direction, to give:

$$Q = WH\,V_b cos\phi \int_0^1 u_z(\xi)d\xi = \frac{WHV_b cos\phi}{2} - \frac{WH^3}{12\mu}\left(\frac{\partial p}{\partial z}\right) \tag{5.90}$$

In intermeshing co-rotating twin screw extruders, the extruder chamber is assumed to provide perfect positive displacement action with no down-channel leakage past the advancing second screw lands. Thus, the flow rate is assumed to be equal to the volumetric displacement rate of the second screw land down the first screw channel (Maheshri & Wyman, 1980). The velocity of advance by the second screw land depends on the channel depth so a mean velocity can be used and defined as:

$$\bar{V}_z = \frac{(V_b + V_s)}{2cos\phi} \tag{5.91}$$

where V_s is the linear velocity of the screw root. The net down-channel positive displacement flow rate Q can be calculated as:

$$Q = WH\bar{V}_z = WH\frac{(V_b + V_s)}{2cos\phi} \tag{5.92}$$

By equating Eqs 5.90 and 5.92, an equation for the pressure profile can be obtained:

$$\frac{\partial p}{\partial z} = -\frac{6\mu}{H^2}\left(V_b sin\phi\, tan\phi + \frac{V_s}{cos\phi}\right) \tag{5.93}$$

By substituting Eq. 5.93 into Eq. 5.89, an expression for the down-channel velocity can be obtained:

$$u_z(\xi) = \xi V_b cos\phi + 3\xi(1-\xi)\left(V_b sin\phi\, tan\phi + \frac{V_s}{cos\phi}\right) \tag{5.94}$$

If the reference frame is changed from a stationary screw to a stationary second screw land, the velocity of the second screw land $\pi ND/cos\ \phi = V_b/cos\ \phi$ must be subtracted from the above equation to give the down-channel velocity referred to a stationary screw land:

$$u_{z,s}(\xi) = \xi(2-3\xi)V_b sin\phi\, tan\phi + (1-\xi)(3\xi-1)\frac{V_s}{cos\phi} \tag{5.95}$$

where $u_{z,s}$ indicates that the down-channel velocity component is measured with respect to a stationary screw land. With respect to the stationary screw lands, if no leakage in the down-channel direction is assumed, the net flow in the down-channel direction is zero so the element of fluid moves between complementary channel depths in the down-channel direction in a manner similar to the cross-channel direction (Maheshri & Wyman, 1980). A schematic representation of those down-channel flows is depicted in Figure 5.28 which shows two separate flow patterns. One, with clockwise fluid motion, is in the region from the screw root to point F whereas the other, with counter-clockwise circulation, is in the region from point F to the barrel surface. Maheshri and Wyman (1980) estimated the trajectory of elements of fluid in any channel depth ξ_1 and its complementary location ξ_{1c} by the following equation:

$$\xi_{1c} = \frac{(1+C-\xi_1) \pm \left[(1-C)^2 + 2\xi_1(1+C) - 3\xi_1^2\right]^{1/2}}{2} \tag{5.96}$$

where the constant C is the channel depth that separates the lower circulation region from the upper circulation

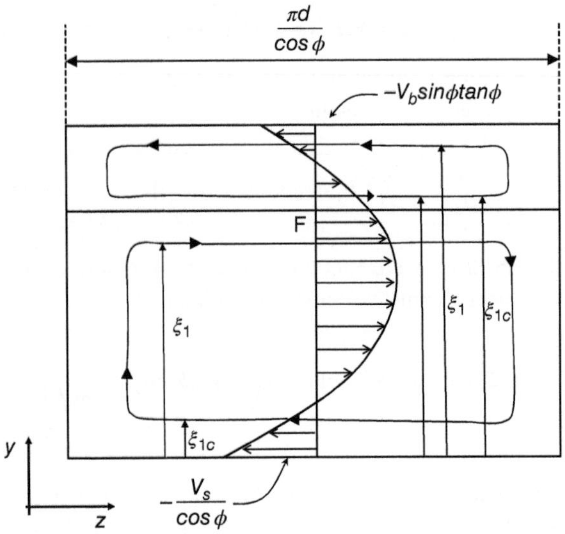

Figure 5.28 Down-channel velocity profile, the screw root and the barrel moving in the negative z-direction and the second screw lands stationary. Source: Maheshri & Wyman 1980. Reproduced with permission of John Wiley & Sons.

region (indicated by point F in Figure 5.28). The constant C can be calculated by the following equation:

$$C = \frac{1}{1 + \frac{D}{d} sin^2\phi} \tag{5.97}$$

The positive sign is used in Eq. 5.96 when $\xi_1 > C$, whereas the negative is used when $\xi_1 < C$.

In an intermeshing, co-rotating twin screw extruder-reactor the trajectory of an element of fluid is given by a pair of complementary channel locations due to the cross-channel velocity component, just as for a single screw extruder. However, in the single screw extruder that element of fluid will also advance toward the die moving between these locations following a helical path as schematically illustrated in Figure 5.15. Conversely, in an intermeshing, co-rotating twin screw extruder-reactor, the element of fluid advances in the down-channel until it approaches the leading intermeshing second screw land which seals the forward path. At that point the element of fluid moves to a complementary channel and returns toward the trailing second screw land. During the backward displacement of the element of fluid (with respect to the second screw land) along the channel length it moves again through a set of complementary depths due to the cross-channel circulation. Since the complementary channel depths in the down- (denoted by ξ_1 and ξ_{1c}) and cross-channel (denoted by ξ and ξ_c) directions are different, the element of fluid will move through a different set of complementary channel depths to the cross-channel circulation on its return trip (Maheshri & Wyman, 1980). Based on these trajectories and velocities, shear rates in the x- and z-directions as well as the magnitude of the total shear rate in different locations of the channel can be calculated as:

$$\dot{\gamma}_x = \frac{du_x}{d\xi} = \frac{V_b}{H} sin\phi(2-6\xi) \tag{5.98}$$

$$\dot{\gamma}_z = \frac{du_{z,s}}{d\xi} = \frac{V_b}{H} sin\phi\, tan\phi(2-6\xi) + \frac{V_s}{H}\frac{(4-6\xi)}{cos\phi} \tag{5.99}$$

And the magnitude of the shear rate is calculated as:

$$\dot{\gamma} = \left(\dot{\gamma}_x^2 + \dot{\gamma}_z^2\right)^{1/2} = \frac{V_s}{H}\cdot\left[\left(\frac{V_b}{V_s}\right)^2 tan^2\phi(6\xi-2)^2 + \frac{(4-6\xi)^2}{cos^2\phi} + 2\left(\frac{V_b}{V_s}\right) tan^2\phi(2-6\xi)(4-6\xi)\right]^{1/2} \tag{5.100}$$

As discussed in Chapter 3, evaluation of the shear strain requires knowledge of the time during which the shear rate is acting. The concept was used in Chapter 3 to evaluate the total shear strain applied to an element of fluid during the element trajectory in a single screw extruder. Trajectories on twin screw extruders are more complex because elements of fluid move in cross-channel and down-channel trajectories determined by Eq. 5.65 or Eq. 5.66 and Eq. 5.96. It must be noted that the total shear rate expressed by Eq. 5.100 is a function of the channel depth location ξ and will be acting on a particular specific element of fluid until that element moves to either the cross-channel or the down-channel complementary channel depth position indicated by either ξ_c or ξ_{1c}. That time from an initial location of the fluid element given by co-ordinates (x_0, y_0, z_0) to a final location given by co-ordinates (x_f, y_0, z_f) can be calculated as:

$$t_0 = \frac{\left[\left(x_f - x_0\right)^2 + \left(z_f - z_0\right)^2\right]^{1/2}}{\left[v_x^2 + v_{z,s}^2\right]^{1/2}} \tag{5.101}$$

and the shear strain applied during that time is calculated as:

$$\gamma_0 = \dot{\gamma}_0 t_0 \tag{5.102}$$

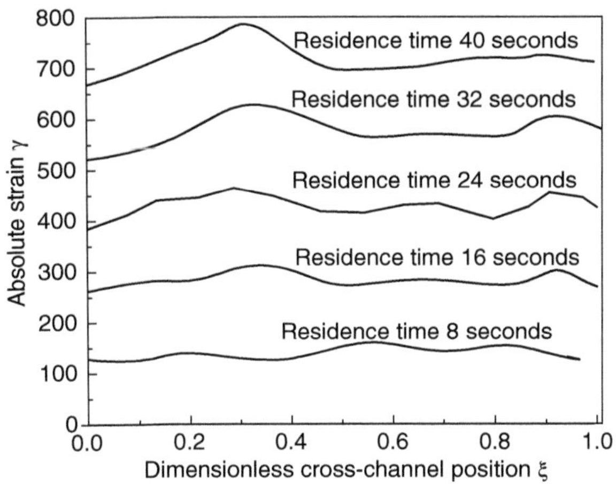

Figure 5.29 Total absolute shear strains applied to different elements of fluid as a function of the dimensionless direction ξ and different residence times. Calculations were done assuming an extruder with a root diameter $d = 76.2$ mm ($V_s = \pi Nd$), channel depth $H = 19$ mm ($D = d + H$), channel width $W = 25.4$ mm, $N = 30$ rpm and $\phi = 30^{\circ}$. Source: Maheshri & Wyman 1980. Reproduced with permission of John Wiley & Sons.

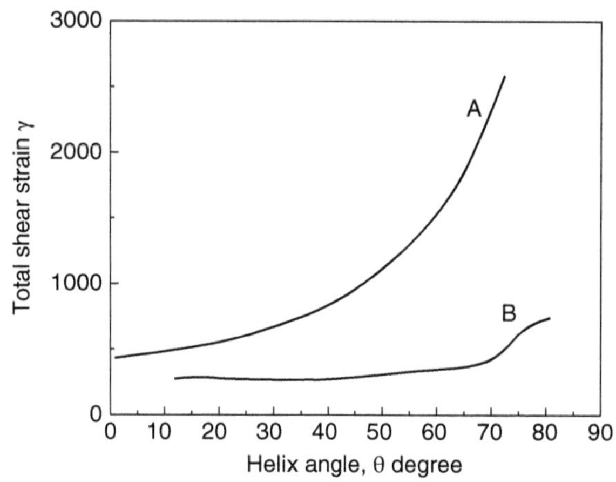

Figure 5.30 Total shear strain as a function of helix angle. Elements of fluid were considered in the center of the channel. Curve (A): absolute shear strain estimated from Eqs 5.100–5.102. Curve (B): shear strain calculated taking into account shear rate direction. Calculations were done assuming an extruder with a root diameter $d = 76.2$ mm ($V_s = \pi Nd$), channel depth $H = 19$ mm ($D = d + H$), channel width $W = 25.4$ mm, $N = 30$ rpm and residence time 40 seconds. Source: Maheshri & Wyman 1980. Reproduced with permission of John Wiley & Sons.

The subscript used in the total shear rate $\dot{\gamma}_0$ indicates that this value is calculated on the element of fluid located at the selected initial channel depth y_0. When the element of fluid moves from the initial location, due to the cross-channel flow, the complementary location is calculated by Eqs. 5.65 o 5.66. But if the element of fluid moves to a new location due to down-channel movement, the complementary location is calculated by Eq. 5.96.

The procedure for estimating the total strain in the new location is similar to that described above. This calculation can be repeated for all the complementary locations that an element of fluid occupies before exiting the extruder. Figure 5.29 illustrates values of the total strain as a function of the vertical position ξ for a central plane x-z and different residence times. Results indicate that the total strain does not vary much across channel depth but is highly dependent on the residence time. It was shown in this chapter that in a single screw extruder, the RTD across the channel depth varies from a low value at $\xi = 2/3$ to almost infinity close to the screw root and barrel surface (see Figure 5.19). It was also demonstrated that the total strain is highly dependent on the location ξ (see Figure 5.20); in other words, in a single screw extruder, some of the elements of fluid are strained considerably while others not. Thus, products converted in a single screw extruder-reactor have fractions of the material that are poorly mixed while others are well mixed. Conversely, in an intermeshing, co-rotating twin screw extruder-reactor, all the material experiences a similar strain history (see Figure 5.29). Although results were found from relationships obtained after applying a number of simplifying assumptions, general trends appear to indicate a larger uniformity of the strain distribution when twin screw extruders are used. Uniformity of applied strains across the channel depth is an indication of the superior mixing capabilities of intermeshing, co-rotating twin screw extruder-reactors.

Figure 5.30 shows the total strain applied to the material as a function of the helix angle for an element initially located at the center of the channel (curve A). It is observed that total strain increases with the helix angle. The same results are observed for elements of fluid initially located in other positions. In contrast, the mixing capabilities of single screw extruders decrease when the helix angle increases from 10° to approximately 40° (see Figure 5.25).

Since there are some concerns about the simplified approach used in this section which was proposed by Maheshri and Wyman (1980) to calculate total strain, a different approach was presented by the same authors.

The concern was that for the total strain to be related to the laminar mixing, the elements of fluid must be continually sheared in one direction. For single screw extruders, the total strain is a good indication of mixing because given the nature of the cross- and down-channel flows the strains applied tend to reinforce each other despite the fact that the flow can "unmix" the product (see Figure 5.17), which is not the case for the flow in intermeshing, co-rotating twin screw extruder-reactors. Maheshri and Wyman (1980) proposed an alternative calculation of the shear strain which is based on the shear rate direction. Results showed a significant reduction in the magnitude of the applied strain but some of the trends and conclusions extracted were similar in terms of the uniformity of the applied shear strain in the ξ direction. Other results obtained with the alternative calculation of strain were different; for instance, in Figure 5.30 the total strain calculated using the shear rate direction is weakly affected by the helix angle (curve B).

Extension to more complex rheology and non-isothermal conditions, considering a leakage flow, etc., can be done using numerical approaches, some of which have been discussed in Chapter 3.

Other theoretical, numerical, and experimental studies concerning mixing in twin screw extruders include works by Todd (1975), Wyman (1975), Altomare and Ghossi (1986), Kalyon et al. (1988), Kalyon and Sangani (1989), Ganzeveld and Janssen (1994), Vergnes et al. (1998), and Shearer and Tzoganakis (2000). In principle, the approach developed by Chella and Ottino (1985) to study the mixing characteristics of single screw extruders is applicable to mixing in intermeshing, co-rotating twin screw extruder-reactors.

5.2.3 Heat transfer mechanisms in extruder-reactors

5.2.3.1 Heat transfer models

Chemical reactions are strongly dependent on temperature so efficient control of reactions in a reactive extrusion process requires suitable control of temperature of the reactive materials and products involved in the extruder operation. The rate of many reactions increases with temperature and that dependence is well described by the Arrhenius relationship. As a rule of thumb and for some values of energy of activation, it can be estimated that for each 10°C change in temperature, the rate of reaction approximately doubles. Davis (1992) lists important considerations to take into account concerning heat transfer in the reactive extrusion process.

- The kinetics of the reaction, in terms of selectivity and yield, is very sensitive to temperature.
- It is important to prevent localized overheating which results in a high reaction rate and possible product degradation.
- The extrusion operation encompasses many stages having different functions along the axial direction, including chemical and biochemical reactions, so knowledge of temperature distribution in that direction is essential to monitor and control processes involved in the extruder, notably reactions in reactive extrusion.
- Heat transfer is directly related to the extruder and process scale-up.
- Precise temperature control is essential to maintain products of uniform consistency.

In order to describe the heat transfer phenomena in an extrusion process, is important to recognize how heat energy is incorporated into the product. This is further discussed in Chapter 6 where is indicated that the heat energy input in the extruded material is due to mechanical energy and the resulting heating by friction plus external heating through the barrels. To describe those phenomena, two different approaches are used. The first considers the extruder as a whole and a balance of energy is considered including all involved energies in the process. The second approach describes the heat transfer through the barrel; this is a more local approach in which each section of the screw-barrel assembly has to be considered (Mottaz & Bruyas, 2001).

A balance of energy, including all energy components (in terms of power) involved in the standard extrusion process, was proposed by Mottaz and Bruyas (2001). However, for a reactive extrusion process the thermal energy generated (for exothermic reactions) or consumed (for endothermic reactions) in the process must be included. Thus the balance of energy would be:

$$\dot{P}_{material} = \dot{P}_{mech} + \dot{P}_{heating} + \dot{P}_{reaction} - \dot{P}_{cooling} - \dot{P}_{losses} \tag{5.103}$$

where $\dot{P}_{material}$ is the power absorbed by the material whereas $\dot{P}_{mech}$, $\dot{P}_{heating}$, $\dot{P}_{reaction}$, $\dot{P}_{cooling}$, and $\dot{P}_{losses}$ are the powers associated with the mechanical energy generated by friction, the heating power supplied through the barrel, the power associated with the energy (enthalpy) involved in the reactions of the reactive extrusion process, the cooling power exchanged through the barrel, and the heating power lost in the process, respectively. In Chapter 3 (section 3.2.1.1), a similar global balance for standard extrusion processes was considered but it included terms involved in the evaporation of water and the melting of extrudate which implicitly do not appear in Eq. 5.103 but they are included in the power energy absorbed by the material $\dot{P}_{material}$. The relationship between the global balance of energy and the transformations undergone by the material during the process have been useful to quantify the extent of these transformations, specifically for starch (Della Valle et al., 1989).

The second approach focuses on the analysis of the heat transfer from a local standpoint where the balance of energy is expressed in terms of local temperatures. One way of describing the balance of energy at a microscopic or local level has been discussed in section 5.2.1.2.1 (Eq. 5.23). Eq. 5.23 provides the instant change in the enthalpy of the reactive mixture, given by the term $\rho c_p \frac{\partial T}{\partial t}$, which can be rewritten as:

$$\rho c_p \frac{\partial T}{\partial t} = \nabla \cdot k \nabla T + \left(\underline{\underline{\tau}} : \nabla \underline{v}\right) + \sum_j (\Delta h_k) r_k - \rho c_p \underline{v} \cdot \nabla T \qquad (5.104)$$

Thus, the rate of enthalpy change in the extruded product is estimated from the conduction of heat through the material, the generation of heat due to mechanical energy dissipation, the heat involved in the chemical reaction, and the heat convection due to the flow of material in the extruder. Solution of this partial equation, along with the fluid mechanics and rheology of the melt, enables the estimation of the material temperature evolution over time at different locations of the extruder channel, i.e. $T(\underline{x}, t)$. In addition, appropriate initial and boundary conditions are required. Specifically for the extrusion process, the boundary conditions are given by the heat transferred to the material through the barrel (involving both heating and cooling).

The solution of Eq. 5.104 provides a comprehensive description of the temperature field in the extruder that can be used to estimate many other parameters related to the process. However, in most cases solution of the equation requires numerical approaches, some of them described in Chapter 3. An integrated version of this model provides a simpler and more practical tool to analyze the heat transfer process in extruders as well as a way to identify and characterize parameters that contribute to its optimization. The integration is performed in the cross-channel direction so that temperature in one section is considered uniform and varying along the extruder shaft. The equation was used by Mottaz and Bruyas (2001) which, adapted to the case of a reactive extrusion process, is given below:

$$\Delta H_{material} = q + h_{mb} \cdot A_b \cdot (T_b - T_m) + h_{ms} \cdot A_s \cdot (T_s - T_m) + \sum_j (\Delta h_k) r_k \qquad (5.105)$$

It provides the change in the enthalpy of the material in different zones of the extruder. That change in enthalpy depends on the type and state of material (solid, semi-solid, melt), and consequently its thermal properties, notably heat capacity and the latent heat of melting. As indicated in the equation, the convective heat transfer between the material and the barrel, and the convective heat transfer between the material and the screw given by its respective heat transfer coefficients h_{mb} and h_{ms}, must be known to estimate the energy input to the material. The heat transfer between the screws and the material, however, can be ignored as there are not many commercial extruder equipment that incorporates cooling through the screws.

It has been demonstrated that in general, most of the control of heat transfer to the extrudate is internal, i.e. controlled by the extrudate itself and not the heat transfer through the barrel wall. Davis (1992) proposed analyzing heat transfer mechanisms in extruders by estimating the overall heat transfer coefficient applied to a geometry formed by concentric layers (Figure 5.31). The figure illustrates a two-piece barrel characterized by their respective thermal conductivities k_1 and k_2 and the convection coefficients in the material, characterized by h_i, and in the barrel, characterized by h_o. The cylindrical geometry model is a good fit to single screw extruders but its application to twin screw extruders may be questionable. However, Davis (1992) estimates that results of this simple model, when compared with estimations based on comprehensive numerical models, are within 10% and can be used with some confidence. In addition, the use of the simpler cylindrical model could help to understand heat transfer mechanisms and establish qualitative trends.

The overall heat transfer based on the barrel area can be estimated by general heat transfer calculations described by Datta (2002), which applied to an extruder yield:

$$q = U_i A_{bi} (T_b - T_{ext}) \qquad (5.106)$$

where U_i is the overall heat transfer coefficient based on the internal area of the barrel, A_{bi} is the area of the barrel which is wetted by the extrudate, and T_b and T_{ext} are the barrel and extrudate temperatures in an axial position of the extruder, respectively. The overall heat transfer coefficient U_i can be evaluated by considering resistances in series to the heat transfer process. These are given by

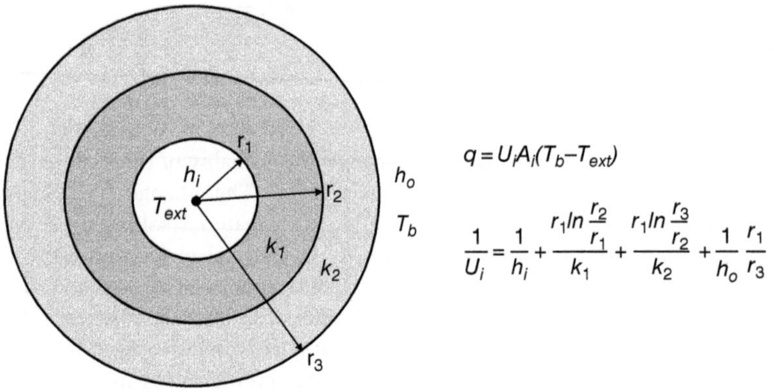

Figure 5.31 Schematic illustrating heat transfer though a composite cylinder considering both conduction and convection (internal and external).

the internal resistance due to internal convection characterized by h_i, resistances due to conduction through the two-piece barrel of thermal conductivities k_1 and k_2, and the resistance due to convection in the barrel given by h_o. The overall resistance can then be calculated as:

$$\frac{1}{U_i} = \frac{1}{h_i} + \frac{r_1 ln \frac{r_2}{r_1}}{k_1} + \frac{r_1 ln \frac{r_3}{r_2}}{k_2} + \frac{1}{h_o}\frac{r_1}{r_3} \qquad (5.107)$$

A similar calculation could be made if the overall heat transfer coefficient is based on the external surface defined by the radius r_3; in this case the overall coefficient should be multiplied by the external area of the barrel $A_{b,ex}$. Concerning the area considered, it should be noted that the value of the heat exchanged q does not change. In addition, for calculation of the enthalpy gained by the material, the coefficient U_i and the internal area A_{bi} should replace the coefficient h_{mb} and A_b in Eq. 5.105. Regardless of how external heating is transferred to the material (e.g. via heated jackets, electrical or induction heatings), values of h_o are much larger (and heat resistance much lower) than the other terms and can be neglected. Thus Eq. 5.107 is simplified to:

$$\frac{1}{U_i} = \frac{1}{h_i} + \frac{r_1 ln \frac{r_2}{r_1}}{k_1} + \frac{r_1 ln \frac{r_3}{r_2}}{k_2} \qquad (5.108)$$

Values of h_i representing the internal convection coefficient of the extrudate, which depend not only on its properties but also on the barrel geometry, are reported by Mottaz and Bruyas (2001) for a variety of situations including the nature of the extrudate and the type of extruder. In general, these values are significantly lower than 1000 W/m^2.K and larger for intermeshing, co-rotating twin screw extruders than for single screw extruders. The other aspect considered in estimating these heat exchange mechanisms is the area of heat exchange. This has a direct relationship with the degree of filling (*DF*) in the extruder section, of which a dimensionless form can be estimated by the following equation:

$$DF = \frac{Q}{\rho \cdot N \cdot W \cdot A_{bi}} \qquad (5.109)$$

Thus, when DF is less than 1 the value of A_{bi} in Eq. 5.106 must be replaced by $DF \cdot A_{bi}$. Q is the mass flow rate, ρ the density of the melt, N the screw speed, and W the width of the channel. Davis (1992) estimated the resistances involved in Eq. 5.108 using different design parameters, such as a liner inserted into a larger outer barrel and two materials such as stainless steel and carbon steel. Calculations also included the presence of an air gap between the materials and its effect on the overall heat transfer coefficient.

A simpler situation showing the effects of the barrel material and the internal heat transfer coefficient on the overall heat transfer coefficient can be estimated for a single layered cylindrical model with radius $r_1 = 25$ mm and $r_2 = 75$ mm (and $r_3 = 0$) by Eq. 5.107. Results are illustrated in Figure 5.32 showing the overall heat transfer coefficient as a function of the thermal conductivity of different materials used in the construction of the barrel and different values of the barrel extrudate convection coefficient h_i. It is clear from the figure that when the heat transfer convection for the interface barrel-extrudate is low, and controlling the heat transfer, the effect of the thermal conductivity of the material composing the barrel is not important. Conversely, for higher values of h_i the type of material used to construct the barrel may have a larger influence on the overall heat transfer. Thus, the

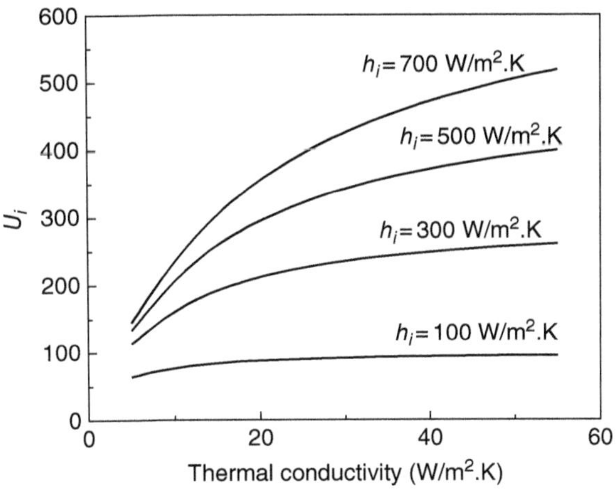

Figure 5.32 Values of overall heat transfer in a barrel assembly formed by a single layer as a function of the thermal conductivity of the barrel material and convection coefficients between the barrel and the extrudate.

coefficient h_i is of extreme importance in the overall heat transfer.

It has been found that the heat transfer coefficient h_i, also known as the film coefficient, increases with increases in the screw speed and with decreases in the clearance between the barrel and the screw. The clearance may act as a boundary layer over which the temperature varies linearly between the barrel and the extrudate bulk temperatures. For larger clearances, a design condition producing low film coefficients, the effect of the barrel design is less important (Davis, 1992). In general, tight clearances contribute positively to increase the overall heat transfer coefficient. From that standpoint, the self-wiping action of intermeshing, co-rotating twin screw extruders may also contribute positively to increase the film coefficient h_i and the overall heat transfer coefficient.

5.2.3.2 Estimation of heat transfer coefficients

For estimation of the film coefficient, adapted relationships between the Nusselt, Prandtl and Reynolds dimensionless numbers such as those described in Chapter 9 could be used. A general equation model is given by Mottaz and Bruyas (2001) as:

$$Nu = \frac{h_i D_{ext}}{k_{melt}} = aPr^b Re^c \left(\frac{\mu_{melt}}{\mu_{melt,w}}\right)^d \qquad (5.110)$$

where D_{ext} is the external screw diameter, ρ_{melt} the melt density, and μ_{melt} and $\mu_{melt,w}$ the melt viscosity calculated at the bulk and the barrel wall temperatures respectively; a, b, c and d are empirical constants. The Prandtl and Reynolds numbers are respectively given as:

$$Pr = \frac{c_{p,melt}\mu_{melt}}{k_{melt}} \qquad (5.111)$$

$$Re = \frac{D_{ext} u_{melt} \rho_{melt}}{\mu_{melt}} \qquad (5.112)$$

Despite the successful use of correlations such as those given above for simpler problems, the complexity of the flow and geometry in extruders makes the validity of these relationships less reliable and determining them experimentally is more convenient. The experimental procedure is simple provided that accurate measurements of temperature at different locations and heat flow can be obtained. If the heat flux q_{barrel} is measured for a given temperature difference between the barrel and the bulk extrudate and assuming a barrel made of a single material, the overall heat transfer coefficient can be calculated as:

$$U_i = \frac{q_{barrel}}{\pi D_i L (T_{barrel} - T_i)} \qquad (5.113)$$

and the film heat transfer coefficient calculated as:

$$h_i = \frac{1}{\frac{1}{U_1} - \frac{r_1 ln \frac{r_2}{r_1}}{k_{barrel}}} \qquad (5.114)$$

Accurate temperature measurement is essential not only to experimentally estimate the film heat transfer coefficient but also for control and analysis of the extruder-reactor performance. Davis (1992) described methods to accurately measure temperatures in different locations of the extruder channel, whereas Mottaz and Bruyas (2001) took a very comprehensive experimental approach including data from 100 thermocouples located in different barrel and product positions to determine the temperature distributions in these zones; a pilot co-rotating twin screw extruder (Clextral BC 45) with different screw profiles and processing conditions was used for the study. Results showed considerable non-uniformity of the temperatures in the barrel and the product in axial, radial and azimuthal directions. Measurements of both temperature and pressure for different extrusion conditions indicated very close coupling between temperature and pressure during conversion of the extrudate material. In addition to an experimental approach, to study heat transfer in twin screw extrusion a heat transfer model was developed for predicting thermal changes in the barrel and the material, particularly when the operating conditions changed.

5.2.4 Coupling of transport phenomena and chemical reactions

As mentioned in section 5.1, work in the area of reactive extrusion has maintained a stable output in terms of patents and potential commercial applications complemented with a continuous increasing trend in research output. It is not the objective of this section to present an exhaustive review of all reactive extrusion processes that have been conducted over the years, which would be an impossible task to do in a systematic and comprehensive manner given the amount of existing information about types of reactions and the large variety of technologies and conditions used. Instead, the challenges faced in the area and approaches utilized for solving them with special consideration given to fundamental aspects are described.

Kowalski (1992) identified the mixing capability of the extruder as one of the challenges in reactive extrusion and listed three technological developments in the mixing area that have contributed to improved reactive extrusion processes: i) technologies to efficiently move a viscous melt consisting of partially miscible or immiscible phases through the reaction zone of the extruder; ii) technologies that promote an increase and renewal of the interfacial area between the phases; and iii) developments using green chemistry and energy savings. The impact of mixing on the production of high-quality products is exemplified by Kowalski using a case study which describes the halogenation of butyl rubber. Butyl rubber is a co-polymer of isobutylene with isoprene used in tubes for tires. For some applications, the polymer is modified throughout halogenation with chlorine or bromine (Kowalski, 1992). For better product quality, it is desirable to halogenate most of the isoprene, which provides only 2% of the overall co-polymer. In addition to this selectivity problem, the reactive extrusion process faces other challenges. There is a great difference between the viscosity of the polymer and the halogen gas (about 10^{10}), and the diffusion of the gas through the polymer is extremely slow. The other problem is related to the selectivity of the reaction because there are two competing reactions in the halogenation of the co-polymer, and one of them, the halogenation of the butyl group, has to be avoided in favor of the halogenation of the isoprene co-polymer.

Attempts to develop this reaction in an extruder-reactor clearly showed that mixing was a key requirement to successfully develop a high quality product. For instance, it was observed that mixing in single screw extruder-reactors or extruders running at conditions that promote circulation in the cross-section of the channel did not result in a good-quality product. Experimental studies showed that fluid circulation was also associated with a condition of a channel full of polymer (Bigio et al., 1985; Gailus & Erwin, 1981). Kinetics and selectivity of the reaction were also negatively affected when halogen bubbles were moving with the material in a circulatory flow in the cross-channel area. A major step in obtaining an acceptable product concurred with using a reaction zone that was partially filled in where contact between the polymer phase and the halogen gas was achieved through a good contact between the two phases. Thus, further optimization of the reaction was obtained by changing screw configuration, screw speed, feed rate and other conditions, all with the general objective of increasing the interfacial area but recognizing that adequate mixing was promoting those beneficial changes.

A second book on reactive extrusion (Janssen, 2004) states that it is difficult to find a polymer process in which mixing is not involved. With applications using screw extruders as chemical reactors, it is important to recognize that mixing is a key operation in ensuring that the distance between components is small enough to overcome problems caused by the normally very slow diffusion of reactants in a reactive polymeric mixture. Janssen (2004) pointed out that good homogenization is essential to assure good stoichiometry within the normally short time for the reaction in the extruder-reactor, and that the size of homogeneities, defined by parameters such as striation thickness can be efficiently reduced by diffusion despite its extremely slow rate. The relationship between diffusive effects and size or path of inhomogeneities or striation thickness can be captured by the dimensionless mass Fourier number, F_{oM}, given by the following equation:

$$F_{oM} = \frac{Dt}{r_s^2} \tag{5.115}$$

where D is the diffusion coefficient of a particular reactant in the reactive melt, t is the reaction time closely associated with the extruder residence time, and r_s is the characteristic length of the inhomogeneity closely associated with striation thickness l. It is easy to imagine that processes operating under conditions of large Fourier numbers will promote high conversions in the reaction. Large Fourier numbers can be achieved by increasing the reaction time by modification of either the screw profiles or the extrusion conditions. However, there are limitations to this approach that are imposed by the continuous nature of the extrusion process in contrast with long reaction times and batch processing using standard stirred tank reactors. The option of increasing the diffusion rate is limited to the types of components involved in the specific reaction but given the polymeric nature of many of the systems that use reactive extrusion, not much improvement can be achieved in terms of increasing molecular diffusion. Thus, if a large Fourier number represents a criterion for recognizing an efficient reaction, a practical option that could improve the reactive extrusion process is the reduction of the characteristic length r_s through efficient mixing.

Concerning the nature of the components and phases making up the reactive mixture, three types of mixing modes have been recognized: disperse mixing; distributive mixing; and longitudinal mixing (Janssen, 2004). Disperse mixing occurs when the components of the reactive mixture are thermodynamically incompatible. For instance, where two phases exist, disperse mixing will cause the dispersion of bubbles/droplets of the minor component in a matrix of the major component. The use of mixing in this case is to promote the reduction in size of the disperse phase to increase the interfacial area. The presence of interfaces between the phases implies the presence of capillary forces which in general oppose the shear forces applied during the process. Thus, in many instances, the effect of capillary forces cannot be ignored. A dimensionless number known as the capillary number is defined as $Ca = \tau R/\sigma_{12}$ where τ is the local stress at the interface of phases 1 and 2, R is the radius of the interface and σ_{12} the interfacial tension between the two phases.

Distributive mixing occurs when the phases of components are miscible. Hence, the purpose of mixing is to distribute one component on the other as homogeneously as possible. No interfacial surfaces are involved. Local shear rates and shear strains applied to the element of fluids are of importance and need to be described, which was done in previous sections.

Longitudinal mixing is associated with the damping of disturbances that can occur at the feeding sections of the extruder-reactor which may have a large impact on the extrusion operation.

In addition to enhancing mass transfer, mixing is also associated positively with the heat transfer phenomenon. Temperature is a key variable in any reaction kinetics, so temperature inhomogeneities in the reacting mass will also have a major effect on the efficiency of the reaction. Similar to the mass transfer of reactant, a Fourier dimensionless number applied to thermal homogenization, F_{oT}, can be defined as:

$$F_{oT} = \frac{\alpha t}{r_s^2} \tag{5.116}$$

where α is the thermal diffusivity of the material, which in general is not greatly affected by the processing conditions. Thus, efforts to increase the thermal Fourier number should be focused on reduction of the characteristic length that can be done through mixing. Some concepts related to the heat transfer phenomenon and its influence to reactive extrusion are presented in section 5.2.5.

Since compatibility between components is a material-based criterion, in addition to knowledge of the type of mixing in the process it is also important to have a preliminary indication between the miscibility of the reacting phases. Janssen (2004) introduces concepts based on traditional thermodynamics to describe mixtures of polymers, polymers, and monomers. Miscibility of phases will depend on changes in the free energy, defined as $\Delta G = \Delta H - T\Delta S$, which will be affected by enthalpic ΔH and entropic ΔS changes of the system as well as the temperature of the process. From an entropic standpoint, compatibility can be reached when the molecules are small (large positive entropy changes during the process) or when the reacting molecules are alike (small enthalpy changes). Miscibility or compatibility between monomers and the polymer forming non-crystalline amorphous phases can be estimated with the Flory–Huggins equation expressed as:

$$\frac{\Delta G_m}{RT} = \phi_1 \ln \phi_1 + \frac{\phi_2}{x_n} \ln \phi_2 + \chi \phi_1 \phi_2 \tag{5.117}$$

where ΔG_m is the free energy of the reactive mixture, ϕ_1 and ϕ_2 are the volume fractions of the monomer and the polymer respectively, x_n is the degree of polymerization of the polymer and χ is an interaction parameter related to solubility parameters (also known as δ parameters) by the following equation:

$$\chi = \frac{V_0}{RT}(\delta_1 - \delta_2) \tag{5.118}$$

The subscripts for the solubility parameters (δ parameters) correspond respectively to the monomer and the polymer. The closer are the solubility parameters of both components, the more compatible are the components (Janssen, 2004). And the following criterion is used to characterize the miscibility of a polymer and a monomer or a polymer and any small molecular component:

$$|\delta_1 - \delta_2| < 3.5x10^3 \left(J/m^3\right)^{1/2} \tag{5.119}$$

whereas for two polymers the criterion is given as:

$$|\delta_1 - \delta_2| < 1.3x10^3 \left(J/m^3\right)^{1/2} \quad (5.120)$$

It has been described that mixing under the laminar conditions found in reactive extrusion processing proceeds by stretching and folding the extruded material, producing lamellar structures. Ottino and collaborators have attributed a key role to the fluid mechanics of the flow in the mixer and other pieces of equipment such as screw extruders not only to improve the efficacy of the mixing process but also in the interplay between these lamellar structures with diffusion and reaction phenomena.

To analyze the interplay between mixing, diffusion and reaction, it is necessary to consider the balance of mass for any component of the reaction (reactant or product), which was applied to the component A and expressed by Eq. 5.20. By assuming a constant diffusivity of species i in the media, diffusion and fluid velocity only in direction x, Eq. 5.20 simplifies to:

$$\frac{\partial C_i}{\partial t} + u_x \frac{\partial C_i}{\partial x} = D_i \frac{\partial^2 C_i}{\partial x^2} + r_i \quad (5.121)$$

A similar equation representing the balance of energy can be written in terms of the temperature change simply by substituting the diffusion coefficient D_i by the thermal diffusivity α and the concentration of the species i by the temperature.

Equation 5.121 involves several phenomena associated with the diffusion and reaction of reactants with their characteristic times and lengths. If a local volume of an element of fluid in the extruder-reactor is considered, there are three characteristic times that are associated with the local process: the characteristic diffusional time, the time scale of the reaction, and a characteristic time associated with the stretching and deformation of the surface forming that element of fluid, the latter related to the fluid mechanics of the flow. Ranz (1979) and Ottino (1994, 1989) proposed transformations of the space x and time t into a dimensionless reduced time Γ and a "warped" time τ:

Reduced time, Γ:

$$\Gamma = \frac{t}{t_c} \quad (5.122)$$

Warped time, τ:

$$\tau = \frac{1}{t_c}\int_0^t \eta^2(t')dt' \quad (5.123)$$

where η is the area of stretch associated with the element of fluid and t_c is a characteristic time.

The space co-ordinate is made dimensionless using the striation thickness $l(t)$:

$$\zeta = \frac{x}{l(t)} \quad (5.124)$$

Parameters associated with the deformation of the element of fluid that depend on the fluid mechanics of the flow are defined as (Ottino, 1989):

Stretching function:

$$\alpha(t) = -\frac{dln\, l(t)}{dt} = -\frac{1}{l(t)}\frac{dl(t)}{dt} \quad (5.125)$$

Stretch at time t (obtained by integration of Eq. 5.58c):

$$\eta(t) = e^{\int_0^t \alpha(t')dt'} dt' \quad (5.126)$$

Time-averaged stretching function:

$$< \alpha(t) > = \frac{1}{t}\int_0^t \alpha(t')dt' \quad (5.127)$$

Based on these definitions, the following characteristic times are defined (Ottino, 1989, 1994):

Characteristic fluid mechanics time $t_F(t)$:

$$t_F(t) = \frac{1}{< \alpha(t) >} \quad (5.128)$$

Instantaneous diffusion time t_D:

$$t_D = \frac{l^2(t)}{D_i} \quad (5.129)$$

Characteristic diffusion time:

$$t_{D0} = \frac{l_0^2}{D_i} \quad (5.130)$$

Characteristic reaction time t_R:

$$t_R = \frac{1}{k} \quad (5.131)$$

where l_0 is the initial striation thickness and k a constant that defines the chemical reaction. It must be noted that, for consistency, the units of k have to be a reciprocal time (e.g. seconds). Thus, the reaction term in Eq. 5.121 is defined as an apparent first-order reaction (Ranz, 1979). Based on these characteristic times, similar to the approach discussed for reactors (see section 5.2.1.2.1), two Damköhler numbers are defined. The first Damköhler number relates the mixing characteristics or fluid mechanics of the flow with the reaction characteristic time and is defined as:

$$Da_I^{(L)} = \frac{t_F}{t_R} \tag{5.132}$$

whereas the second Damköhler number, as defined in section 5.2.1.2.1, provides a ratio between the diffusion and reaction characteristics times, defined as:

$$Da_{II} = \frac{t_D}{t_R} \tag{5.133}$$

The superscript (L) in the first Damköhler number indicates the use of Lagrangian co-ordinates, which are described in section 5.2.1.3.1. From the definitions given above, it can be noted that Da_I can be interpreted as the ratio between a characteristic length over a velocity (see Eq. 5.125) and a reaction characteristic time. Thus, the ratio between the second and the first Damköhler numbers equals the Peclet number defined by Eq. 5.38 and associated with the axial dispersion of reactive species in the reactor:

$$\frac{Da_{II}}{Da_I} = \frac{t_D/t_R}{t_F/t_R} \approx \frac{\dfrac{l^2(t)}{D_i}}{\left(\dfrac{l(t)}{dl(t)/dt}\right)_{avg}} \approx \frac{dl(t)/dt \cdot l(t)}{D_i} = Pe \tag{5.134}$$

The incorporation of the fluid mechanics of the flow bringing parameters associated with the stretching of the elements of fluid, which are defined above, would enable the establishment of relationships between the fluid deformation in the mixer or extruder-reactor with the diffusion and reaction phenomena.

5.2.5 Basic principles of process engineering in reactive extrusion

In order to better understand the influence of the above defined parameters on mixing, diffusion and reaction phenomena, it is important to relate them to the conservation of mass of the reactive components (Eq. 5.121). In terms of these variables and parameters, the equation of balance of mass for the species i can be expressed in two different ways (Ottino, 1989):

$$\frac{\partial c_i}{\partial \Gamma} = \eta^2(\Gamma)\Delta_i\left(\frac{t_c}{t_{D0}}\right)\frac{\partial^2 c_i}{\partial \zeta^2} + \beta_i r_i \tag{5.135}$$

where c_i is a dimensionless concentration for the species i, whereas Δ_i and β_i are the diffusion coefficient ratio and the initial concentration for the species i respectively, and r_i is the consumption or production of species i due to the reaction. As noted in Eq. 5.135, the convection term of the balance of mass (second term on the left side of Eq. 5.121) has been associated with the diffusion term through the square stretching parameter $\eta(\Gamma)$. Thus, it has a double contribution to the transport of the species i; by increasing the stretching of the surface, the diffusion process increases and also promotes the mass flux. Eq. 5.135 also shows that the diffusion term is augmented by a factor $\eta^2(\Gamma)$. By using the warped time τ an equation for the mass balance of a reactive species based on this time is obtained:

$$\frac{\partial c_i}{\partial \tau} = \Delta_i\left(\frac{t_c}{t_{D0}}\right)\frac{\partial^2 c_i}{\partial \zeta^2} + \frac{1}{\eta^2(\tau)}\beta_i r_i \tag{5.136}$$

As noted in this equation, the reaction rate is affected by the factor $1/\eta^2(\tau)$. In practical terms, this indicates that increases in the surface stretching may change the process from being diffusion controlled to be reaction controlled. It also establishes the importance of mixing in the overall kinetics of the reactive extrusion process.

Equations 5.135 and 5.136 can be solved using appropriate initial and boundary conditions. Initial conditions are assumed as unmixed or segregated components and the boundary conditions by considering symmetry conditions (Chella & Ottino, 1984). Using the above equations and the defined parameters, two extreme cases can be studied: slow reactions and fast reactions.

5.2.5.1 Slow reactions

In this case the characteristic reaction time is much longer than the characteristic diffusion time. Although these are not typical conditions for a reactive extrusion process, the analysis could be useful to establish bounds. For these conditions:

$$Da_{II} = \frac{t_D}{t_R} = \frac{\dfrac{l^2(t)}{D_i}}{t_R} = \frac{l^2(t)}{D_i t_R} \ll 1 \tag{5.137}$$

However, it must be considered that at the start of the process, the initial striation thickness could be large enough such that $l_0^2/D_i > t_R$ and the control of the reaction is due to diffusion. During the reactive extrusion process the mixing action makes the striation thickness to decrease, as $l(t) = l_0\, e^{-\alpha(t)t}$ so $t_D \approx t_R$ and the reaction is controlled by both diffusion and reaction. Moving into the extruder reaction zone, the striation thickness decreases further due to the mixing action and $t_D << t_R$. Beyond that zone, the reaction is solely controlled by the chemical reaction and no longer affected by mixing in the extruder-reactor. Under those conditions the mass balance for species i is given by the equation:

$$\frac{dc_i}{dt} = \beta_i r_i \tag{5.138}$$

5.2.5.2 Fast reactions

For a fast reaction the characteristic reaction time is very short and the second Damköhler number is much higher than one; mathematically this is expressed as:

$$Da_{II} = \frac{t_D}{t_R} = \frac{l^2(t)}{D_{Am}t_R} >> 1 \tag{5.139}$$

Given the high reaction rate, once reactants are in close proximity they react immediately and the reaction zone reduces to a plane (Ottino, 1989, 1994). The control, however, is due to the rate of stretching of the surfaces and the decrease in striation thickness. Chella and Ottino (1984) demonstrated that for $t_R << t_D$, the mass balance for species i given by Eq. 5.136 reduces to:

$$\frac{\partial c_i}{\partial \tau} = \Delta_i \frac{\partial^2 c_i}{\partial \zeta^2} \tag{5.140}$$

Equation 5.140 indicates that the overall reaction rate is controlled by diffusion of the species to the reaction plane. In addition, the diffusion of reactants to the reaction plane must satisfy the overall reaction stoichiometry (Ottino, 1989). A schematic of changes in the characteristic times as a function of the residence time (or age) of an element of fluid through a reactive extrusion process is shown in Figure 5.33, where is clearly shown that the overall reaction rate can change from diffusion to reaction control while the material moves in the extruder.

The use of these concepts and equations was better illustrated in an example given by Ottino (1989) that could be easily applied to a reactive extrusion process if

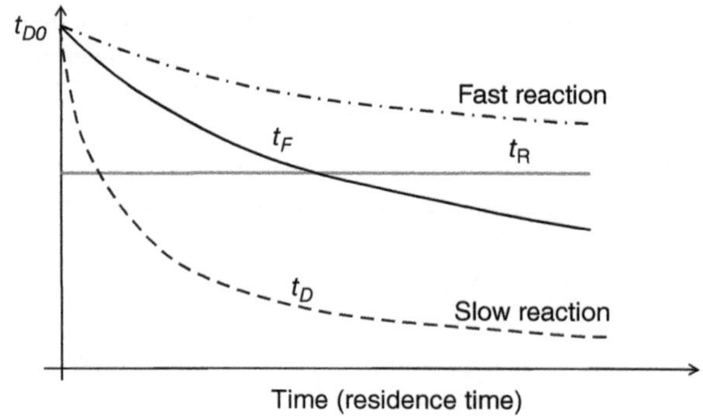

Figure 5.33 Schematic of the changes in the characteristic times as a function of the residence time (or age) of an element of fluid through a reactive extrusion process for slow and fast reactions.

the fluid mechanics of the flow were known. The reaction mechanism is illustrated below:

$$A + B \rightarrow P \ (\textit{specific rate of reaction } k_2)$$
$$R + B \rightarrow S \ (\textit{specific rate of reaction } k_3)$$
$$\varepsilon = \frac{k_2}{k_3}$$

Using the mass balance, based on the reduced time, for each component (Eq. 5.135) and assuming first order in each component reactions, Chella and Ottino (1984) obtained the following equations:

$$\frac{\partial c_A}{\partial \Gamma} = \eta^2(\Gamma)\Delta_A\left(\frac{t_C}{t_{D0}}\right)\frac{\partial^2 c_A}{\partial \zeta^2} - \left(\frac{t_C}{t_R}\right)c_A c_B \tag{5.141a}$$

$$\frac{\partial c_B}{\partial \Gamma} = \eta^2(\Gamma)\Delta_B\left(\frac{t_C}{t_{D0}}\right)\frac{\partial^2 c_B}{\partial \zeta^2} - \left(\frac{t_C}{t_R}\right)\beta_B c_B(c_A + \varepsilon c_R) \tag{5.141b}$$

$$\frac{\partial c_R}{\partial \Gamma} = \eta^2(\Gamma)\left(\frac{t_C}{t_{D0}}\right)\frac{\partial^2 c_R}{\partial \zeta^2} + \left(\frac{t_C}{t_R}\right)c_B(c_A + \varepsilon c_R) \tag{5.141c}$$

These equations were solved assuming that the two components A and B in the reactor were completely segregated (i.e. unpremixed). Other parameters used were: $\varepsilon = 0.1$, $\Delta_A = \Delta_B = 1$, $\beta_B = 0.5$ initial diffusion time $t_{D0} = 2.25 \times 10^5$ seconds, final diffusion time $t_{Df} = 2.25 \times 10^{-1}$ seconds (i.e. a reduction in striation thickness $l_0/l(t) = \sqrt{t_{D0}/t_{DF}} = 10^3$) and a characteristic reaction time $t_R = 2.25 \times 10^{-1}$ seconds. Two flows were considered, a shear flow given by the kinematics $v_x = \dot{\gamma}y$, $v_y = 0$, and an extensional flow with the kinematics given by $v_x = \dot{\varepsilon}x$; $v_y = -\dot{\varepsilon}y$ (Morrison, 2001). In both cases the same number of surfaces was considered, all having the same orientation

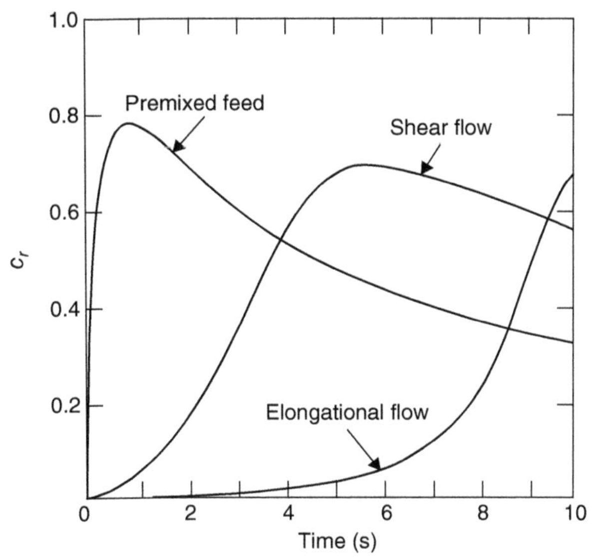

Figure 5.34 Average concentration of a component R as a function of the time that the elements of fluid spend in a mixer reactor, for unpremixed mixtures and prevalent shear and extensional flows in the reactor. For comparison, a premixed feed reacting in a shear flow-based reactor is presented.

defined by a normal to the surfaces in the *y*-direction. Magnitude of the shear $\dot{\gamma}$ and strain $\dot{\varepsilon}$ rates was selected to produce the same striation thickness reduction in the total residence time of the elements of fluid in the mixer that was 10 seconds. Results in terms of the average concentration of the component *R* as a function of the time that the elements of fluid spent in the mixer are shown in Figure 5.34. It is clearly shown in the figure that a mixer with a shear flow is more efficient in terms of product conversion than a mixer in which the flow is dominated by an extensional flow. For comparison, the conversion of a reacting system that is premixed (i.e. the initial condition corresponds to the reacting species A and B molecularly mixed) run in a mixer governed by a shear flow is illustrated. Other applications of co-polymerization reactions of interest to reactive extrusion are described by Ottino (1994).

Analysis of the effects of mixing on the chemical conversion of reactants in reactive extrusion processes presents many challenges due to the high viscosity of the media and the laminar characteristic of the flow, which makes it difficult to determine the phenomena controlling the process. Two different approaches have been used to study the role of mixing on reactive extrusion. One incorporates both theoretical and experimental aspects using tools previously developed for application studies in the chemical reactor area. Many of those studies have been reviewed in a number of journal articles (e.g. Cassagnau et al., 2007; Ganzeveld & Janssen, 1994) and books (e.g. Janssen, 2004; Xanthos, 1992). Readers interested in specific applications of reactive extrusion can obtain information from the many reviews and publications in the area.

The other approach is the one described in this chapter which presents a generalized theory in which the emphasis is on the fluid mechanics of the specific flow in the reactive extrusion equipment. The flow characterization serves to quantitatively evaluate many useful material properties associated with the lamellar structure of the surfaces that have an important effect on the diffusion and reaction of reactive species in a high-viscosity mixture. Examples given in this chapter have dealt with simple flows, most of them even having analytical solutions. Extension to more complicated flows follows a similar approach to the one presented in this chapter. Provided the kinematics of the flow, either analytical or numerical, is obtained, the characterization of mixing and its effects on the lamellar structure of the extruded material and the conversion of the reactions in a reactive extrusion process can be performed in the same way as above. Analytical and numerical methods of studying the kinematics of Newtonian and non-Newtonian flows under isothermal and non-isothermal conditions were provided in Chapter 3.

5.3 Reactive extrusion applications and processing lines

5.3.1 The classes of chemical reactions in reactive extrusion

Reactive extrusion has historically referred to combining chemical reactions and polymer melt extrusion into a single process carried out continuously in either a single screw or a twin screw extruder. Basically, features that make an extruder-reactor attractive as a chemical reactor include the ability to operate continuously, to add reactants at any barrel module, to mix them intensively with appropriate heating or cooling, to remove volatiles through devolatilization when required, to combine various process functions in series such as reactive extrusion combined with product shaping and structuring, reactive extrusion combined with extrusion-pressing, etc. Discrete chemical processes can be carried out in specific barrel modules, which are isolated from surrounding barrel modules by including restrictive screw elements (such as reverse screw elements or kneading blocks) in the screw profile. Extruder-reactors are characterized by laminar mixing which has remarkable ability to intensively mix reactants in viscous media or to effectively mix reactants with differing viscosities and volume ratios. The main limitations of an extruder-reactor as a chemical reactor are the limited residence time (most frequently 1–5 min) and heat transfer capability. Hence, this imposes strict requirements on the kinetics of the targeted chemical reaction and on the energy involved.

The processing characteristics of reactive extrusion have been extensively described and studied in the literature, and have received full attention from industry and academia since the 1960s with intense activity in classic extrusion polymer processing, and emerging activity in new fields such as, among others, biomaterials extrusion processing, bioreactive extrusion processing, and reactive solid-liquid extraction.

Based on the numerous investigations that have been published since the emergence of reactive extrusion, it is worth reviewing the typical classes of chemical reactions, most of them related to polymer extrusion processing.

• *Bulk polymerization* (Brown, 1991). Bulk polymerization consists of converting a monomer, a mixture of monomers, or a pre-polymer to a high molecular weight polymer with little or no solvent dilution. In principle, all types of polymerizations can be carried out in a screw extruder-reactor, particularly in intermeshing co-rotating twin screw extruder-reactors: the free-radical addition polymerizations and step-growth polycondensations. But, in practice, though extruder-reactors show multiple process advantages, industrial applications of bulk polymerization remain limited, due to either technical difficulties or economic constraints.

• *Chemical modification of polymers* (Brown, 1991). Chemical modification of polymers aims at modifying the chemical and/or physicochemical properties of polymers, while preserving their physical and mechanical properties. Typical chemical modification of polymers in a screw extruder-reactor includes: (i) grafting reactions which consist of grafting one or more monomers onto a polymer backbone (such as free radical grafting of relatively chemically inert polyolefins with maleic anhydride); (ii) functionalizing reactions, or introduction of functional groups into a polymer backbone (such as halogenation of polyolefins and butyl rubbers); (iii) modifying existing functional groups (such as capping carboxylic acid end-groups of polyethylene terephthalate with epoxides to improve thermal stability of PET).

• *Reactions for the control of molecular weight of polymers* (Brown, 1991). This includes a broad range of chemical modifications such as: (i) interchain co-polymer formation which is defined as reaction of two homopolymers to form co-polymers with improved morphology and processing capability (particularly useful for compatibilizing immiscible polymers such as polypropylene and PET); (ii) coupling/cross-linking reactions which involve the reaction of a homopolymer with a bi- or polyfunctional coupling agent, or a cross-linking agent to modify molecular weight of polymeric materials by chain extension or branching (chain extension of polyamides such as nylon 6 and 6/6 through a coupling reaction with polyepoxides, or chain extension of polyesters such as PET through reaction with 2,2′-bis(2-oxazoline)); (iii) upgrading rheological properties and material morphology by cross-linking (such as dynamic vulcanization of thermoplastic elastomers); (iv) controlling the molecular weight and hence the rheological properties of polymers (such as free radical degradation of polypropylene and other polyolefins).

• *Reactive solid-liquid extraction (or reactive solid-liquid extrusion-pressing).* This application emerged in the 1980s with the development of extrusion pulping of non-wood cellulose fibers in the paper-milling industry (for a full description, refer to section 5.3.3). The process consists of controlling the alkaline degradation of lignocellulosic structure as well as the peroxidic bleaching of cellulose fibers through reactive extrusion in intermeshing, co-rotating twin screw extruders. In reactive solid-liquid extraction, reactive media include two types of reactants that are structured solid such as lignocellulosics (complex 3D structures of lignin and cellulose) and low molecular weight liquid reagents such as alkali and hydrogen peroxide. Intermeshing, co-rotating twin screw extruder-reactors proved their efficacy in the required mixing of such reactants in a short residence time.

Besides reviewing classes of reactions, due to sustainability and environmental concerns, it is worth highlighting some emerging applications in which development can be envisaged owing to the determinant role of reactive extrusion.

• Chemical modification of starch, for food and non-food uses in particular (Govindasamy et al., 1997a-c; Kalambur & Rizvi, 2006; Kaseem et al., 2012; Moad, 2011; Xie et al., 2006). Starch has received great attention as a possible base material for the manufacture of sustainable polymeric materials, as an alternative to petrochemical polymers, owing to its renewability and biodegradability. However, native starch cannot be used directly in most applications. It must be physically and chemically modified or blended with other materials to meet the required properties. Conventionally, starch modification has been performed in batch or continuous stirred tank reactors in aqueous diluted conditions. Aqueous starch suspensions are then considered (typically 35–45 wt% starch and 10–30 wt% sodium chloride or sodium sulfate), while the reaction requires long residence times (typically in the range of 2–24 h). Conventional processes show high operating costs (due to salt and water removal) and low productivity. Reactive extrusion offers realistic solutions to these issues, which are discussed and promoted in the literature. The chemical modifications covered in the literature are: grafting of monomers from starch such as ring opening of epoxides, esterification, phosphorylation and silylation; graft polymerization from starch by radical-induced grafting or the ring opening polymerization of lactones; reactive compatibilization with polyesters and polyolefins; cross-linking of starch; and the degradation of starch thermally or catalyzed by acid or enzymes.

- Biodegradable polymers (Averous, 2004; Kalambur & Rizvi, 2005; Martin & Averous, 2001; Mohanty et al., 2000; Raquez et al., 2008; Zhao et al., 2008). Biodegradable plastics (such as poly(ε-caprolactone), polylactides, and aliphatic-aromatic polycondensates) can be converted into environmentally friendly products. However, the use of such bulk materials is still restricted to a few applications because of their insufficient thermomechanical properties and relatively high cost. Hence, recent research work proposes combining these biodegradable polymers with inexpensive organic and inorganic fillers (such as starch granules or silicate-type particles) through reactive compatibilization by reactive extrusion, in order to optimize the properties of resulting blends/composites and reduce their cost. Reactive extrusion can then be used to produce biodegradable aliphatic polyesters through catalyzed ring opening polymerization, to chemically modify biodegradable polymers (through grafting, for instance), to chemically melt-blend starch with polyesters, and to reinforce biodegradable polymers using (nano) fillers.
- Reactive solid-liquid extrusion-pressing, through the combination of reactive extrusion and solid-liquid extrusion-pressing. As described in section 5.3.3, reactive extrusion pulping of non-wood fibers brings drastic improvements in sustainable processing (higher yield of fibers, energy and consumables saving, reduction of water consumption and higher productivity). It is worth promoting the process principle in green chemistry for those applications which aim at extracting renewable chemicals or functional ingredients from agro feedstock. Nowadays, the technological status of reactive solid-liquid extraction allows process strategists and technologists to consider it as a serious candidate to thermochemically modify biomass resources (for the production/extraction of intermediate chemicals from wood or non-wood materials, or of food ingredients such as natural polysaccharides from seaweeds), or to chemically compatibilize natural fibers (for the production of natural fiber-reinforced composites). Hence, this innovative reactive extrusion technology should be considered seriously by the processing industry to boost the production of chemicals from renewable resources instead of petrochemical resources on one hand, and to promote sustainable processing in green chemistry on the other hand. Concrete developments are currently emerging in this direction.

It must be noted that emerging applications relative to these three domains deal mainly with intermeshing, co-rotating twin screw extruder-reactors. In addition to the aforementioned domains, reactive extrusion technology also provides beneficial advantages in plastics recycling and reprocessing to upgrade plastics wastes (Awaja & Pavel, 2005; Pracella et al., 2002) as well as in organic chemistry (speciality chemistry and fine chemistry) as an alternative to conventional batch stirred tank reactors, where the processing conditions can likely be implemented in extruder-reactors, thus taking advantage of the operational benefits and sustainability of reactive extrusion in these important industrial sectors.

5.3.2 Case study 1: casein-to-caseinate extrusion processing

Casein represents about 82% of milk proteins and can be isolated by acidifying skim milk to the protein isoelectric point (pH = 4.6). The casein protein can be precipitated by use of either mineral acid (such as hydrochloric or sulphuric acid, in particular), or lactic acid (through lactic fermentation), leading to acid casein, which is the typical starting material for the production of caseinates. After acid precipitation, acid casein is insoluble and nonfunctional. Thus, its use as a functional ingredient (in the food industry such as bakery, biscuit manufacturing, dairy, meat processing and also in non-food industries such as those producing speciality paper, and speciality fine chemicals) requires its solubilization through neutralization to pH 7. Acid casein has carboxyl groups (-COOH) which react with alkalis to provide salts, or caseinates. For instance, neutralization of acid casein with sodium hydroxide (NaOH) proceeds following the reaction:

$$R-COO^{-}\,{}^{+}H + Na^{+}\,{}^{-}OH \rightarrow R-COO-Na + H_2O$$

The reaction rate of neutralization is very fast as long as reacting species are well mixed and distributed in the reaction medium. This can be done either under diluted conditions (such as in batch stirred tank reactors where water facilitates the bulk diffusion of the reacting species) or under concentrated conditions (such as in continuous extruder-reactors where efficient laminar mixing occurs). Thus, the conversion of acid casein into caseinates by use of alkalis is currently carried out through two different classes of processes:

- semi-continuous processing under diluted conditions
- continuous processing under concentrated conditions.

5.3.2.1 Semi-continuous processing under diluted conditions

This is the conventional process. It consists of carrying out the reaction in batch, double-jacketed stirred tank reactors (typical reactor volume of 10 m^3) in which acid casein is dispersed in water (solids content of 18–25%, wwb). The resulting colloidal suspension is continuously stirred and heated to 70°C, while an alkali solution (e.g. sodium hydroxide, potassium hydroxide) is added to maintain a pH of 6.6–7. As the reaction between acid casein and alkali proceeds, the colloidal casein suspension becomes a homogeneous caseinate solution. Though the solids content is low, the resulting solution is relatively viscous. The whole processing time for batch neutralization may take from 30 minutes to 2 hours. During this relatively long residence time, a Maillard reaction may take place between residual lactose and free amine groups, which confers an undesired light yellow color and off-flavors to the final product, hence lowering its quality. After completion of neutralization, the caseinate solution is then dried through continuous processing to provide caseinate in powder form (96–97% dry solids); this can be done either by roller drying combined with attrition drying or by spray drying. Roller drying enables processing at solids content significantly higher than those used in spray drying.

However, the drying stage of the conventional, semi-continuous process is costly (high capital and operating costs) owing to high diluted conditions of the process, which require high processing volumes and energy consumption. The specific energy consumption of the whole process amounts approximately to 2400–2500 kW.h/tonne finished product (97% dry solids) for the spray drying option and 1100–1200 kW.h/tonne for the roller drying option. Thus, process developments have been conducted to reduce the production cost and thus carry out the neutralization at higher protein concentration.

5.3.2.2 Continuous extrusion processing under concentrated conditions

The idea of using twin screw extrusion processing technology to produce caseinates from acid casein emerged in the 1980s (Tossavainen et al., 1986). At the same time, two French companies, Clextral and Pillet, joined their competences to develop a continuous, high-concentration extrusion process, as an alternative to the semi-continuous, low concentration process. Relevant studies carried out in the 1990s, such as those of Fichtali and Van De Voort (1991), Barraquio and Van De Voort (1991) and Fichtali et al. (1995), must also be mentioned as significant contributions to the development and understanding of casein-to-caseinate extrusion processing. The use of continuous extrusion processing technology to convert acid casein into caseinates relevantly illustrates the potential of reactive extrusion to improve existing batch or semi-continuous costly processes. In fact, alternative extrusion processing technology relies on performing casein neutralization under concentrated conditions, in order to reduce processing volumes and energy consumption and hence increase process productivity and cost competitiveness. Figure 5.35 presents a flowsheet of a typical casein-to-caseinate twin screw extrusion processing unit (Clextral, 2013). The unit is composed of three main unit operations:

- raw material handling, acid casein in particular
- twin screw reactive extrusion where acid casein is converted into caseinate
- postprocessing operation, to produce dry powder caseinate.

Raw material handling: acid casein

Acid casein (9–10% moisture content, wwb) is homogenized and stored in silos. It is then conveyed to the extrusion unit, where it is metered into the first barrel module of the extruder by use of a loss-in-weight feeder, as accurate feeding is required. The alkali is usually fed separately into the extruder either in powder form (for instance, calcium hydroxide) or in liquid state (for instance, sodium hydroxide solubilized in water). Liquid water is also added, to obtain the solids concentration required for the chemical reaction.

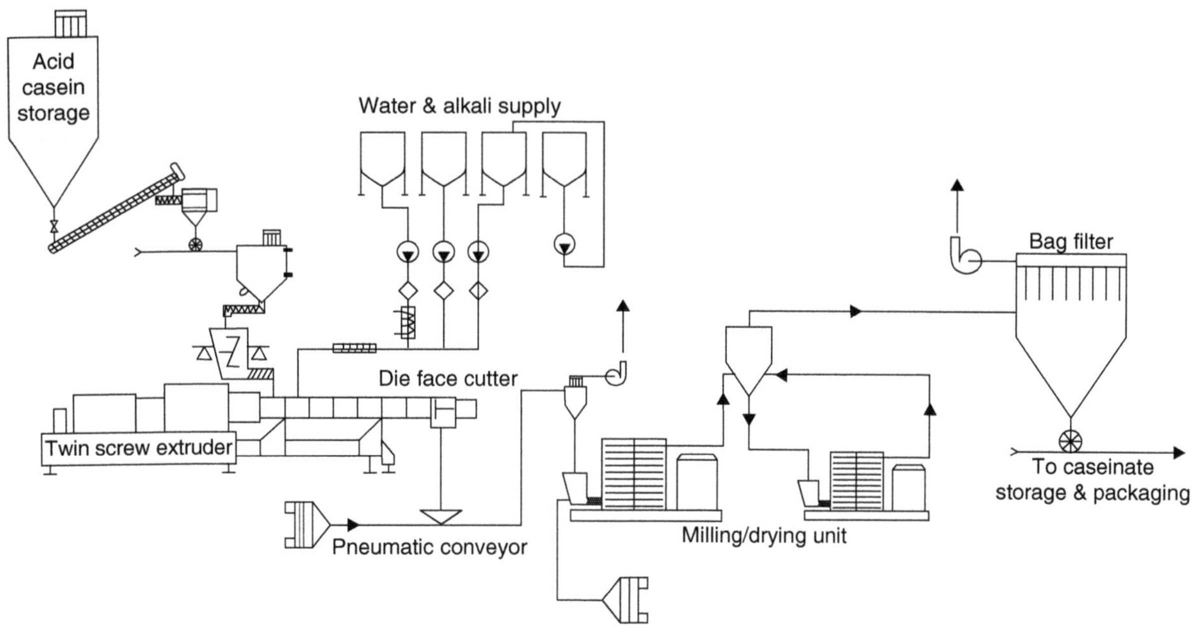

Figure 5.35 Casein-to-caseinate twin screw extrusion processing line. Source: Reproduced with permission of Clextral, France.

Twin screw reactive extrusion

The chemical conversion of acid casein into caseinates is carried out in a co-rotating twin screw extruder-reactor; the L/D ratio and screw speed typically range from 16 to 20 and from 200 to 250 rpm, respectively. Casein-to-caseinate extrusion processing usually operates at very high solids concentration (typically in the range of 75–82% wwb). In general, the screw-barrel configuration of the extruder consists of a transport and compression section, a melting and reaction section, and finally a cooling section. The reacting species are fed to the transport section of the extruder which is composed of relatively large-pitch, right-handed screw elements; the acid casein (typically in powder state) is fed to the first barrel module, while the alkali is usually fed at the beginning of the second barrel module. The reaction medium is compressed and conveyed to the mixing and reaction section which generally consists of a combination of relatively low-pitch, right-handed and left-handed screw elements whose designs enable quick melting of the acid casein and intense mixing of the protein melt with the alkali. In this section, casein neutralization then occurs quasi-instantly once the reaction medium is well mixed and homogenized: the average residence time in the mixing and reaction section does not exceed 6–8 seconds, while the reaction temperature is in the range of 120–130°C.

The resulting caseinate is cooled efficiently in the last cooling section of the screw-barrel assembly of the extruder-reactor, to prevent the development of secondary reactions (such as the Maillard reaction and protein denaturation) and reduce product stickiness at the die exit. The molten caseinate is finally directed to the shaping die where caseinate strands are extruded and die-face cut to produce caseinate flakes. It is worth using an air-cooled die-face cutter to cool the product efficiently at the die exit and to handle the stickiness of the product consistently.

The specific mechanical energy (SME) of the casein-to-caseinate extrusion process varies from 240 to 270 W.h/kg, depending upon the dry solids content of the reacting medium in the extruder-reactor. Compared with the semi-continuous low-concentration process, the high-concentration extrusion process is characterized by a very different temperature-time processing history in the extruder-reactor, as the processing temperature is increased from 70°C to 120–130°C. In addition, the overall extrusion processing time is reduced from approximately 0.5–2 hours down to 15–25 seconds.

The extrusion processing conditions must be controlled cautiously (the screw configuration, screw speed, and temperature profile, in particular) to ensure complete casein neutralization and good product quality (functional and nutritional properties).

Postprocessing operation
The caseinate flakes are pneumatically conveyed to a two-stage milling-drying unit which mills and dries the product in the same processing unit, and converts it into dried powder. As hot air is used in the process, a bag filter in a cyclone allows air and dry caseinate to be separated. Dry caseinate is then stored and packaged.

The specific energy consumption in the extrusion process amounts approximately to 400–500 kW.h/tonne finished product (97% dry solids), hence leading to very significant energy savings when compared with semi-continuous, low-concentration processes.

5.3.2.3 Performances of casein-to-caseinate twin screw extrusion processing

The production of caseinates from acid casein through high-concentration reactive extrusion shows improved performance when compared to conventional, semi-continuous low-concentration processes (such as spray drying or roller drying). Several process advantages can be noted.

- A very important increase of process productivity owing to the reduction of processing time (one to two orders of magnitude reduction). Besides, casein-to-caseinate extrusion processing is a fully continuous process.
- A very significant energy saving, which may vary from at least 200% (comparison with the roller drying option) up to 500% (comparison with the spray drying option).
- Approximately 100% reduction of footprint, due to the integration of several process functions in the twin screw extruder-reactor such as casein melting, intense laminar mixing, casein neutralization, melt cooling and flaking, which makes the reactive extrusion process very compact, hence allowing important savings of capital costs.
- The ability to process dry casein as well as casein curds and to produce all types of caseinates (sodium, potassium, calcium, etc.) with the same unit, which makes the reactive extrusion process very flexible and versatile.

According to these process advantages, twin screw reactive extrusion represents an energy-saving, flexible, compact and versatile process for converting continuously acid casein to caseinates. The extrusion-processed caseinates (sodium and potassium, in particular) are very soluble and extremely heat stable over a wide range of conditions. When used in products above pH 6, those proteins exhibit excellent emulsification, water binding, thickening, whipping-foaming, and gelling characteristics. Thus, they have found wide acceptance as emulsifiers and water binders in a wide variety of formulated food products (such as meat and cheese products, margarine, toppings, cream substitutes, coffee whiteners, etc.), and as foam stabilizers. Furthermore, due to the flexibility of reactive extrusion, different levels of product quality (functional and end-use properties) can be reached by appropriately varying processing variables such as dry solids concentration, screw speed, screw profile, and temperature profile.

5.3.3 Case study 2: extrusion pulping of non-wood fibers

Although the information technology revolution is substituting electronic information for paper, paper is still an essential material with a growing consumption. It is estimated that the worldwide demand for paper and cardboard will increase by an average of about 2.5% in the coming decade. The annual consumption of paper per capita depends very much on the standard of living, varying from 1 kg in the poorest countries to 310 kg in the US, hence leading to an average consumption of approximately 56 kg per capita. This means that today's worldwide consumption reaches 380 million tonnes and will probably exceed 470–480 million tonnes in 2020.

Wood pulps from softwood and hardwood are essential raw materials for paper-making purposes, due to the low price and availability of wood, particularly in industrialized countries. However, for emerging and poor countries where wood resources are scarce, non-wood cellulosic resources (such as cotton, hemp, flax, etc.) may constitute an appropriate class of raw materials for paper making and contribute significantly to the paper supply of these countries. Also, in industrialized countries, non-wood pulps cover effectively high added value speciality paper applications such as filter papers, cigarette and other similar papers, technical papers and speciality printing and writing papers, due to the requirement of these applications for high-quality pulps with exceptional properties.

Recently, the pulp and paper industry has shown a real concern for protection of the environment. During recent decades, considerable progress has been made in tackling environmental issues, such as increased recycling of paper, sustainable management of tree plantations and forests, and progressive implementation of clean pulp and paper technologies. However, technical progress and improvements in sustainable processing have particularly impacted the pulp and paper industry in industrialized countries (Orzechowska, 2004). Nevertheless, questions remain about its important growth in emerging countries and poorer regions of the world. The use of non-wood fibers as well as the promotion of clean pulp and paper technologies (such as twin screw extrusion pulping) may contribute to creating a sustainable pulp and paper industry in developing countries which have abundant cellulosic annual plants and fiber crops resources.

5.3.3.1 Semi-continuous chemical pulping of non-wood fibers

Chemical pulping was traditionally used for non-wood fiber crops. This involves:

- kraft or alkaline cooking in batch digesters for several hours (approximately 8 hours) under diluted conditions (water-to-fiber weight ratio of 5–10) together with chemical reagents (20–30% dry matter weight) and at elevated temperatures (140–160°C)
- washing and bleaching with excess water, often in Hollander beaters, to separate chemicals and extracted binding components (such as lignin and pectin) from the pulp fibers, hence producing clean bleached fibers. Typical processing time may be up to 12 hours in conventional batch processing.

The whole chemical pulping process requires large quantities of water (approximately 30 m^3 per tonne of fiber pulp in the most modern milling plants where water recycling is optimized), and produces a liquid effluent containing 1 tonne of dry matter (that is, for a typical process yield of 50%) and about 0.4 tonne of residual chemicals per tonne of fiber pulp. In large-scale pulping plants (500,000+ tonnes per year, as for chemical pulping of wood fibers) the effluent is processed in such a way that almost all dissolved chemicals are recovered and reused, and the dissolved fiber matter is incinerated, hence generating more energy than needed for both fiber pulping and effluent recovery. However, small-scale pulping plants (around 10,000 tonnes per year) are not able to run a recovery process efficiently, so the plant faces the costly problem of discarding the effluent, using new chemicals, and lacking the energy generated through energy recovery. Besides, the whole processing time of chemical pulping is more than 20 hours, which makes the productivity of this process very low and its operating cost very high. However, it must be noted that the traditional chemical pulping process produces a very high-quality pulp (long fiber length) with exceptional end-use properties. Effluent problems of chemical pulping process combined with its low productivity tend to condemn small-scale chemical pulping plants, unless alternative technology allows non-wood fibers to be processed in an environmentally friendly way at reasonable costs.

Extrusion pulping has established itself as a technically and economically viable processing technology for the mechanical or chemomechanical pulping of non-wood fibers. It provides a much higher yield and uses much less water and chemicals. Moreover, extrusion pulping allows fiber cutting to desired length, so that the pulps can be handled by regular paper-making systems. However, extrusion-processed pulps are characterized by lower strengths than chemical pulps, and some lignin is still linked to the fiber. Twin screw extrusion pulping technology has contributed successfully to the processing of non-wood fibers (such as cotton fibers, hemp and flax bast fibers, in particular) for the pulp and paper industry for almost 25 years. The process uses twin screw extrusion processing technology as a continuous reactor in which fibers are mechanically or chemomechanically converted at high shear and short time processing. Hence, it is worth presenting twin screw extrusion pulping process as an original and relevant example of reactive extrusion.

5.3.3.2 Continuous twin screw extrusion pulping process

The idea of using twin screw extrusion processing technology to produce cellulosic pulps from non-wood annual plants and from wood emerged in the second part of the 1970s from a fruitful partnership between Clextral and the Technical Center for Paper in France. The basic principles of extrusion pulping were developed in the 1980s with successful results with non-wood fibers such as cotton combers, cotton linters, hemp and flax bast fibers (De Choudens & Angelier, 1990; De Choudens et al., 1984, 1987; Kurdin & Bohn, 1984). Based on these results, a generic twin screw extrusion pulping process (also called BiVis by the inventors) has been designed and industrialized to continuously digest, cut, bleach, and wash cellulosic fibers (De Choudens et al., 1989).

Figure 5.36 presents the flowsheet of a typical twin screw extrusion pulping process for the production of paper pulp from cotton linters (Combette, 2010). Cotton linters are agricultural by-products generated from the cotton crop. Basically, cotton flowers are composed of combers (long cellulosic fibers, a high value-added part of the flower) which represent approximately 40% (wwb) of the flower, and two main by-products (seeds and linters) which are processed in oil-milling plants. In such plants, linters are separated from the seeds through a delinting operation, and oil is fractionated from the protein-based meal.

Cotton linters constitute an interesting source of raw material to paper-making purposes; it is particularly relevant to cotton-producing countries, where wood resources are lacking, hence allowing them to develop a domestic pulp and paper industry through the valorization of renewable sources of raw material.

On average, dry cotton linters are composed of cellulosic fibers (approximately 88% wwb) and wastes (approximately 12% wwb). Cellulosic fibers are mainly composed of cellulose (approximately 73% of total linters) together with bound components (such as water, waxes, fats, pectins, proteins, minerals; approximately 15% of total linters). Of course, the composition and physical characteristics of cotton linters depend closely on the botanical origin of the cotton, harvesting conditions, storage conditions of raw cotton seeds, and operating variables of delinting. Hence, rigorous identification of the composition and physicochemical characteristics of cotton linters is of paramount importance to adapt the design of the upstream unit in the pulping line and the operating conditions of the extrusion pulping process.

The extrusion pulping processing line separates wastes from cellulosic fibers and pulps/washes cellulose fibers. Hence, the line is based on a fully continuous process which is composed of three main unit operations (see Figure 5.36).

- Dry-cleaning operation, to clean raw cotton linters or separate cellulosic fibers.
- Twin screw extrusion pulping of cellulosic fibers, to remove the bound components through chemomechanical treatment and bleaching.
- Postprocessing operation, to wash and press the resulting cellulose fibers.

Dry-cleaning unit operation

Raw cotton linters consist of compressed bales of about 250 kg. The bales are opened and conveyed to a cyclone separator to remove the heaviest wastes. Cotton linters are then conveyed to a beater-cleaner to separate and remove dust while a metal detector allows possible metal pieces to be detected and eliminated. Dry-cleaned cellulosic fibers are then conveyed to a densifier-feeder which ensures a constant mass flow rate of the fibers. The design and operating of the densifier-feeder are very important, since consistent feeding closely determines the efficiency and stability of the pulping process downstream, and the operation of the first BiVis extruder in particular. Thus, the densifier-feeder is specially adapted to the complex flow behavior of entangled fibers like cotton linters; it ensures regular and stable plug flow of the fibers with the aid of several sensors which drive operating variables of the feeder.

Twin screw extrusion pulping

The original BiVis pulping process is designed with two 240 mm screw diameter, co-rotating twin screw pulping extruders (or BiVis machines; refer to Figure 1.5) in series.

The first twin screw extruder (or BiVis machine 1) ensures pre-cutting of linters, as well as alkaline impregnation (using sodium hydroxide) and precooking. Cotton linters are fed into the transport section of the extruder which is composed of right-handed pitch screw elements (relatively large pitch). Then, cellulosic fibers enter into several processing sections in series which are basically composed of a right-handed pitch screw element (relatively small pitch) combined with a restrictive screw element (typically specially designed left-handed screw elements) where high compression and shear forces are applied to cause defibration, fibrillation, and shortening of the fibers. Recycled effluents as well as sodium hydroxide are added in the processing sections while excess effluents may be pressed out of the fiber mass and extracted through barrel filters placed upstream from the restrictive screw elements. Intense shear mixing occurs in the processing sections when the fiber mass flows through the restrictive screw element, and provides rapid temperature increase and impregnation of chemicals into the fiber mass. Under such conditions, lignin softens and fibers are easily defibrated whereas sodium hydroxide partly decomposes the lignin, which also facilitates defibration.

Cooking of cellulosic fibers takes place at relatively high fiber concentration (about 35%) with a temperature at the extruder outlet of over 90°C. Shear-induced heating fulfills the conditions required to dissolve waxes and fats, and concomitant shear mixing enables the appropriate mixing of sodium hydroxide in the reactive mixture. As the residence time in the extruder is relatively low (in the range of 25–35 s)

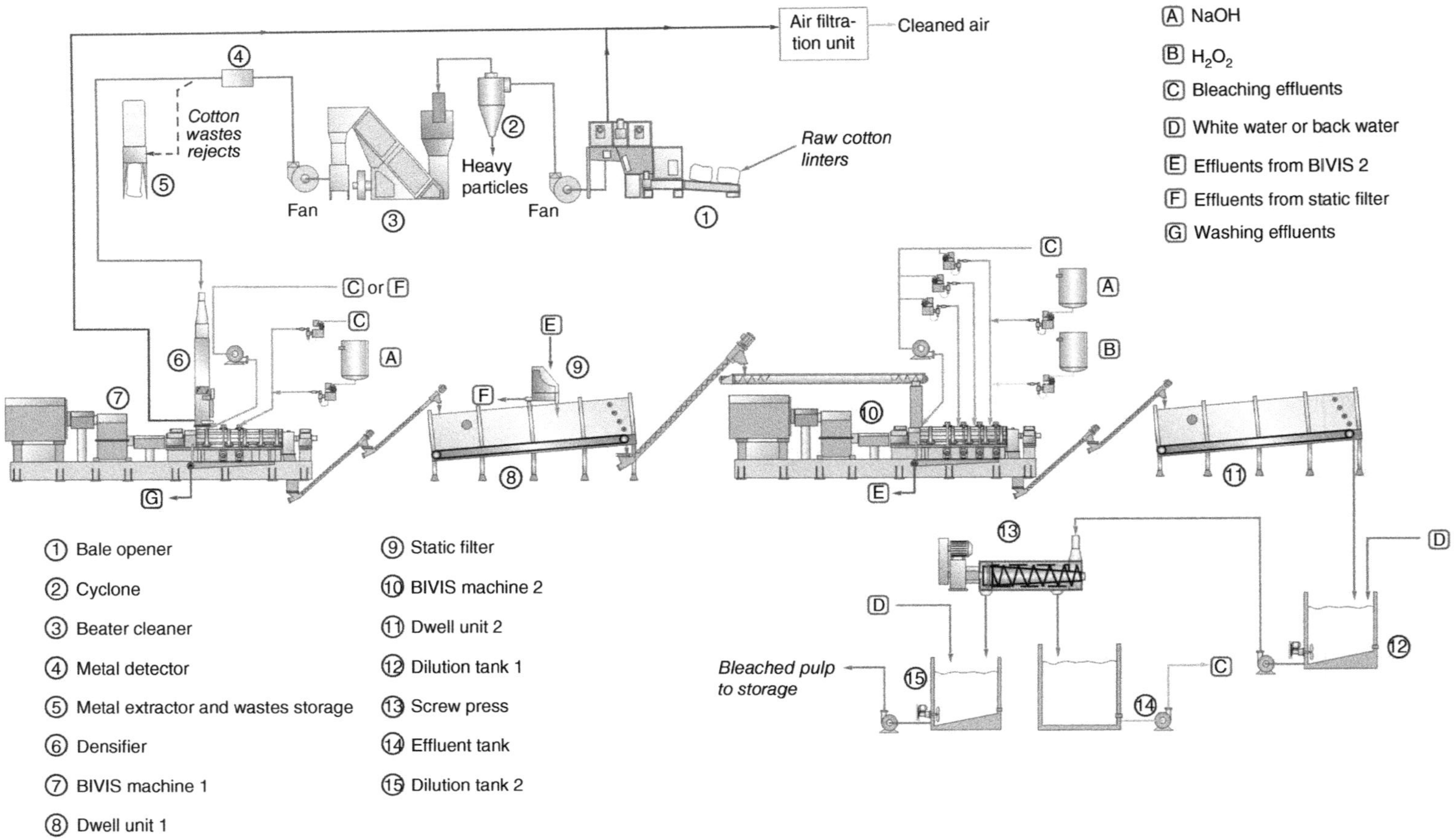

Figure 5.36 Twin screw extrusion pulping processing line for cotton linters (Combette, 2010). Source: Reproduced with permission of Clextral, France.

and complete treatment of the cellulosic fibers cannot be achieved, chemical conversion is then pursued and completed in dwell unit 1, which consists of a continuous, temperature-controlled plug flow reactor where the fibers stand for approximately 60 minutes to complete the chemical reactions needed for the required fiber cooking.

The cooked pulp that emerges from the dwell unit 1 is conveyed to the second twin screw extruder (or BiVis machine 2) which ensures pulp washing, fiber cutting, and pulp bleaching. The cooked pulp is first washed in three contacting stages in series where recycled bleaching effluents are added. The screw profile of a contacting stage is based on a right-handed screw element combined with a left-handed screw element (same principle as for BiVis machine 1, but with adapted screw designs to press and cut the pulp fibers appropriately), while the barrel holds filters upstream from the left-handed screw element in order to extract the expressed effluents from each contacting stage. In the last processing section of the extruder, bleaching chemicals (a solution of hydrogen peroxide and a solution of sodium hydroxide) are added for physical impregnation. Pulp solid content at the extruder exit ranges from 35% to 40%. The bleaching reaction is completed in dwell unit 2 where the fibers stand for 90 minutes.

Hydrogen peroxide can be used as a bleaching agent. It allows environmentally friendly (or totally chlorine-free) pulp bleaching to be carried out successfully with high-yield pulps (such as extrusion-processed chemomechanical pulps) to a brightness greater than 80% ISO. During H_2O_2 bleaching, it is well recognized that the HOO^- anion is the most probable active species in alkaline media ($H_2O_2 + {}^-OH \rightarrow H_2O + HOO^-$). It allows carbonyl groups of lignin to be oxidized, and hence lignin chromophores to be destroyed, as shown in Figure 5.37.

Postprocessing operation

It consists of washing the bleached pulp. For that purpose, the bleached pulp is conveyed to a stirred dilution tank where it is diluted to 3.5% solids (wwb) by use of recycled white water. Then, the diluted bleached pulp is pumped to a screw press where water is continuously expressed in order to eliminate dissolved matter resulting from the bleaching reaction (or bleaching effluents). After screw pressing, a 35% solids pulp is obtained, which is diluted again in a stirred tank where the dilution is adjusted to 3–3.5% solid content (wwb) while the pH of the pulp is decreased to the appropriate value required by the process. Finally, the diluted suspension of the bleached pulp is transferred to a storage chest before refining.

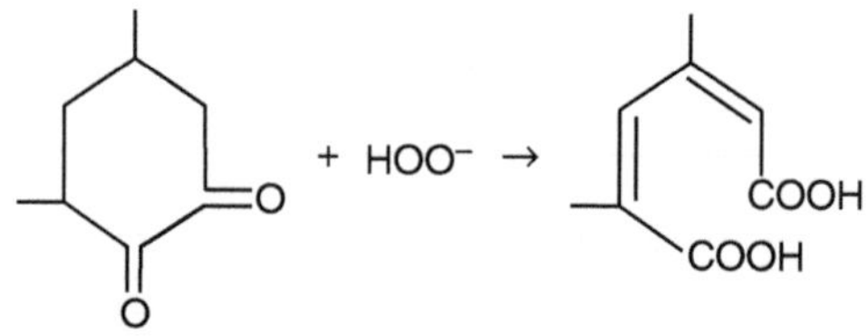

Figure 5.37 Oxidation of carbonyl groups of the lignin molecule by hydrogen peroxide.

5.3.3.3 Performance of twin screw extrusion pulping process

The production of chemomechanical pulps from cotton linters by the fully continuous twin screw extrusion pulping process shows improved performance compared with conventional semi-continuous chemical pulping processes. The extrusion pulping process presents several advantages.

- Approximately 50% increase of fiber pulp yield. Depending on cotton linters quality, typical fiber pulp yields of 76–82% are obtained, which enables a significant reduction in consumption of raw materials and production of wastes per tonne of fiber pulp.
- Energy saving of 20–40%. It must be noted that extrusion pulping does not consume steam, while conventional chemical pulping consumes approximately 1 tonne of steam per tonne of fiber pulp.
- Reduction of 20–30% in the consumption of chemical reagents. The consumption of sodium hydroxide and hydrogen peroxide per tonne of fiber pulp is in the range of 60–65 kg and 45–55 kg, respectively.
- Approximately 100% saving of fresh water, owing to the processing of concentrated media and the optimization of water recycling (bleaching effluents, white water or back water).
- Reduction of 100% in footprint, due to the integration of several process functions in intermeshing, co-rotating twin screw extruders such as fiber cutting, chemical impregnation and fiber washing of concentrated fibrous media, which makes the extrusion pulping process very compact.
- The ability to process cotton linters of varying quality, with an extensive range of wastes content (in the range of 10–30% wwb), which makes the extrusion pulping process very flexible.
- The ability to produce totally chlorine-free fiber pulps, owing to the use of hydrogen peroxide for bleaching, which makes extrusion pulping a clean process.

With these advantages, twin screw extrusion pulping represents a clean, flexible, compact, and versatile process for chemomechanical pulping of lignocellulosics, particularly raw materials from non-wood annual plants. The production of fiber pulps from cotton by-products (combers and linters), and from hemp and flax bast fibers is now a successful reality in the pulp and paper industry. Speciality papers such as banknote paper (from cotton), cigarette paper and filter paper (from hemp and flax) are currently made with extrusion-processed fiber pulps. The extrusion-processed pulp (from cotton linters, for instance) can also be used for making writing and printing papers that meet the highest international standards. In the future, with its high level of technical and economical performance, twin screw extrusion pulping should undoubtedly extend its application to other raw materials such as straws, bagasse, sisal, kenaf, abaca, etc.

5.3.4 Case study 3: enzymatic hydrolysis of starch

5.3.4.1 Conventional processing

Glucose syrups are used extensively not only in food formulations but also they are important chemical intermediates to be converted into high-value products (e.g. penicillin). They are products of the hydrolysis of starch and consist of mixtures of polymers of D-glucose. This includes any purified and concentrated aqueous solution of sweeteners derived from the hydrolysis of starch (Mulvihill, 1992).

Industrial manufacture of these hydrolyzates from starches includes two processing steps: liquefaction and saccharification (Slominska, 1993). The liquefaction step combines two starch modification mechanisms. The first is the gelatinization of starch, which includes its previous hydration, that is used to open the starch structure to facilitate enzyme attack. The second mechanism is the dextrinization of the starch to a level that prevents its crystallization or retrogradation upon further processing (Underkofler et al., 1965).

The liquefied stage consists of a mixture of oligo- and polysaccharides known as maltodextrins; during starch saccharification the liquefied starch is converted from a mixture of maltodextrins into a mixture of low molecular weight sugars. When the hydrolysis is taken to full completion and the hydrolyzate contains the maximum amount of low molecular weight sugars, the product is known as a dextrose hydrolyzate. Otherwise, after partial hydrolysis the mixture contains products with a range of molecular weight sugars and it is known as glucose syrup.

Conventional processes to produce glucose syrups include starch hydrolysis by acid, enzymes or their combination (Fulbrook, 1984; Guzmán-Maldonado et al., 1995). The liquefaction processing time, depending on the starch and hydrolysis conditions, can be around 30–90 minutes. Conversely, saccharification is a lengthy process of about 12–96 hours (Pontoh & Low, 1995) and hence requires the use of large reactors operating in batch mode. Other processes for starch hydrolysis have been investigated, for instance for conditions of low moisture content (Van Der Veen et al., 2006) or by the incorporation of new enzymes (Richardson et al., 2002).

The conversion of starch in lower molecular components can be followed by a parameter known as the dextrose equivalent (DE). The DE is a measure of the relative reducing power of these starch derivatives with anhydrous glucose, which is given the value of 100 whereas for unhydrolyzed starch the value assigned is 0. Thus, this parameter is associated with the molecular structure of the modified starch and gives a good indication of the extent of hydrolysis. The DE is related to the molecular structure of the hydrolyzed product or maltodextrin and specifically is inversely proportional to the degree of polymerization and the number average molecular weight, M_n of the maltodextrin molecule (Rong et al., 2009). It is also related to the rheological properties of concentrated and diluted solutions of these maltodextrins or glucose syrups (Dokic et al., 1998). The degree of polymerization (DP) denotes the number of glucose units in individual components of starch hydrolyzates. For example, DP1 is a dextrose, DP5 is a maltopentaose, and so on. The DE has been measured by several methodologies which include colorimetric and osmometric methods (Fitton, 1979; Lane & Eynon, 1923; Rong et al., 2009).

Acid hydrolysis of starch is produced when a starch slurry is heated with hydrochloric or sulphuric acid (pH adjusted to 1.8–2) at a temperature of 120–140°C. Improvement of the hydrolysis and higher glucose yield are obtained by using a combined acid/enzyme process that also has the benefit of reducing retrogradation (Fullbrook, 1984). In some cases the use of enzymes allows the production of hydrolyzates with a DE in the range 30–55. Other processes include the acidification of the starch slurry to produce hydrolyzates with a DE in the range 5–7, which after pH adjustment to 6–6.5, heating at 85°C and further treatment with α -amylase for about 30 minutes produces a hydrolyzate of DE about 12.

The discovery of thermostable α-amylases has enabled the use of higher processing temperatures at around

100–110°C for the liquefaction process. In a typical process (Richardson et al., 2002), a starch slurry of concentration 30–40% solids content containing an enzyme (e.g. a thermostable α-amylase) and calcium ions is heated in a jet cooker at about 110°C for 5–6 minutes to produce a maltodextrin hydrolyzate of low DE. The hydrolyzate is then flashed to 100°C and held for 1–2 hours to achieve DE values of about 12. The hydrolyzate is further cooled to 60°C and treated with a second enzyme (glucoamylase) for saccharification of the starch.

The conventional process for saccharification of starch presents many problems. First, the non-specific acid hydrolysis is not able to yield glucose syrups with more than 80–85% dextrose under practical conditions (30–40% solids content). The glucose syrup product is in acid medium that needs to be neutralized with a base, which requires the use of expensive corrosion resistance materials. But it is the relatively low product yield and excessive formation of by-products that seriously limit the use of acid hydrolysis (Linko, 1992).

A combined acid-enzyme hydrolysis process, however, is not commercially used because the hydrolysis produces highly branched saccharides that are resistant to the action of glucoamylase for further saccharification to glucose. Thus, high-dextrose hydrolyzates are produced entirely by an enzymatic process using a combination of enzymes, e.g. α-amylases and glucoamylases (Reeve, 1992). The liquefaction step is used to prepare hydrolyzates of DE between 7 and 15, under conditions that prevent the formation of maltulose, a disaccharide formed by glucose and fructose, due to the inability of glucoamylase to hydrolyze this compound. Another drawback in the conventional process is related to the use of high concentrations of saccharifying enzymes as well as extended hydrolysis times (24–96 hours) at relatively high solid contents (30–40%) that cause reactions producing saccharides from glucose (Govindasamy et al., 1995; Labout, 1985).

Processing of starch in an extruder has been suggested as a novel approach to circumvent many of the drawbacks of the conventional hydrolysis processes discussed above. In addition to a reduced liquefaction process, the saccharification time can be reduced when the starch is pretreated using extrusion. However, generic differences between starches from different botanical sources necessitate the use of a particular liquefaction technique appropriate to individual requirements (Knudsen & Karkalas, 1969; Oates, 1997).

Most research on potential applications of extrusion technology for hydrolysis of starch has focused on starches derived from cereal sources such as corn, wheat, and potato (Tester et al., 2006). Very few reports are available on extrusion processing of starches derived from crops indigenous to the tropical regions, including sago starch (Govindasamy et al., 1995). Sago starch is an agronomically important indigenous crop of South East Asia and its utilization may lead to conservation of agriculturally marginal areas of land (Wang et al., 1996). Use of sago starch for hydrolysis is currently limited by the resistance of the starch to commercial enzymes as well as the high paste viscosity formed during gelatinization of the starch (Wang et al., 1996). The ability of extruders to handle high-viscosity products offers a unique opportunity to process this type of starch. A comprehensive study on the liquefaction and saccharification of sago starch using both single screw and twin screw extrusion technology is described in this section.

Thermomechanical gelatinization and liquefaction of starch has been commercially achieved to pretreat starches prior to enzymatic saccharification either by conventional batch stirred tank technology using soluble glucoamylase or using an immobilized enzyme packed bed reactor to obtain products with DE values as high as 98 (Linko, 1992). In addition, both single and twin screw extruders were used to treat starches in the production of glucose syrups (Ben-Gera et al., 1983; Darnoko & Artz, 1988; Korn & Harper, 1982; Wenger et al., 1981). Recently, extrusion-enzyme liquefaction processes have been used as complementary processes for successful concentration of proteins from sorghum flour (De Mesa-Stonestreet et al., 2010, 2012). The mechanical action of the extrusion process simultaneously degrades the starch and disrupts protein bodies of the sorghum flour that limit functionality and digestibility of sorghum proteins. However, the enzyme-assisted extrusion processing step makes starch liquefaction and its removal easier and faster, which results in a very pure sorghum protein concentrate that has good digestibility and improved functional properties for food and beverage applications (De Mesa-Stonestreet et al., 2012).

5.3.4.2 Enzyme-assisted reactive extrusion in single screw extruders

In preceding sections of this chapter and also in Chapter 3, it has been demonstrated that single screw extruders are less efficient in terms of mixing and product uniformity than twin screw extruders. However, they have been considered as potential candidates for liquefaction reactions where the overall objective is the pretreatment of the starch to facilitate an enzymatic saccharification reaction. Govindasamy et al. (1997a) explored the feasibility of using a single screw extruder as a bioreactor to pretreat sago starch for subsequent saccharification. A single screw laboratory extruder (19 mm screw diameter; L/D 20:1; temperature controlled) was used for the study.

The study was conducted in two parts, the first part aimed at the optimization of the process and conditions for the production of liquefied starch that could be used as a suitable substrate for subsequent saccharification and the second part focused on the saccharification of resulting products. The effects of processing variables such as feed moisture content, enzyme concentration, and product temperature on starch degradation during the reactive extrusion process were studied, specifically by determination of the following parameters: dextrose equivalent (DE), water solubility index (WSI), water absorption index (WAI), and starch degree of degradation (DGR). Determination of WSI and WAI as well as their physical interpretation are described in Chapter 6, whereas measurement of DE and its physical interpretation were explained in the preceding section. DGR was determined using high pressure size exclusion (HPSEC) chromatography analysis (Govindasamy et al., 1997a).

To assess the effects of extrusion conditions on the physicochemical properties of the extrudates to be used as a substrate of a saccharification process, an experimental design based on five levels and three variables listed above was used. Moisture content varied from approximately 20% to 40% (80–60% solids content), enzyme concentration (thermostable α-amylase from *Bacillus licheniformis*, Termamyl 120 L) was in the range 1.5–6.5%, and product temperature was in the range 70–100°C. Results showed that the thermomechanical degradation of the starch (characterized by DE values) in the extruder was promoted by the presence of enzyme and was also dependent on feed moisture (Figure 5.38). The dependence on feed moisture content is related to its level in the feed. As illustrated in the figure, at low enzyme concentration and below 30% moisture content, starch degradation was characterized by extremely low DE values (in the range 0.3–3), which indicate that starch degradation was reduced. Conversely, at higher feed moisture and high enzyme concentrations, starch degradation was high. The high moisture content required for enzymatic hydrolysis of sago starch in an extruder has also been reported for other type of starches in a reactive extrusion process measured in terms of increases of DE values and the release of oligosaccharides (Hakulin et al., 1983; Reinikainen et al., 1986).

Higher processing temperatures did not favor the degradation of starch and extrudates with lower values

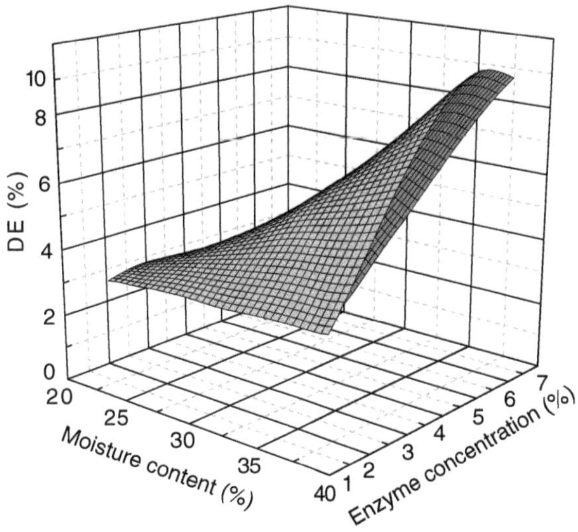

Figure 5.38 Influence of feed moisture and enzyme concentration on DE of sago starch extrudates produced in a single screw extruder. Source: Adapted from Govindasamy, Campanella & Oates 1997a.

of DE were formed. Although this effect was observed only at moisture levels above 26.5%, the phenomenon could not be attributed to the thermal inactivation of the enzyme used (Termamyl 120 L) as the product temperatures were in the range 70–100°C, which is below the reported deactivation temperature of the enzyme. Inactivation of the Termamyl 120 L enzyme has been reported for barrel temperatures above 105°C (Likimani et al., 1991; Roussel et al., 1991) and feed moisture contents in the range 16–19%. Based on extrusion trials conducted in a twin screw extruder, Reinikainen et al. (1986) concluded that relatively high extrusion moisture levels are needed for rapid conversion of wheat starch even at relatively low enzyme levels. The general recommendation for these enzymatic reactive extrusion processes is high feed moisture and relatively low product temperatures.

The WSI for liquefied products varied in a range of 6–35%. Enzyme concentration had a strong positive effect on WSI (Figure 5.39). The WSI was also dependent on product temperature, as greater amounts of soluble material were produced at the higher extrusion temperature used (97.5°C). This result supports the assumption that little or no enzyme inactivation occurred in the temperature range investigated. As gelatinization and liquefaction proceed simultaneously in the reactive extrusion process, a low thermal energy input at high moisture content may prevent proper gelatinization, thereby impeding

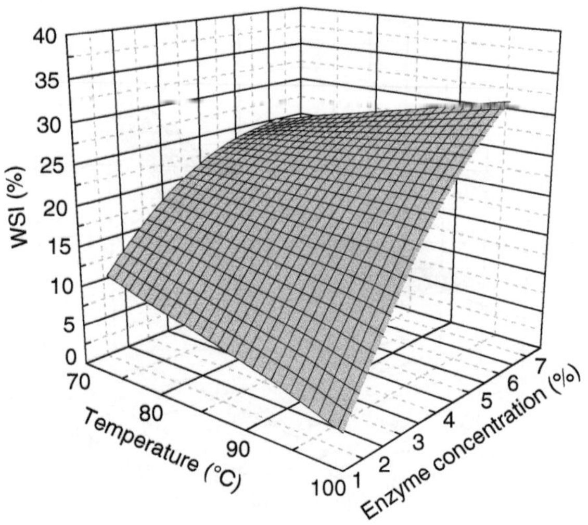

Figure 5.39 Influence of product temperature and enzyme concentration on WSI of sago starch extruded with a moisture content of 30% in a single screw extruder. Source: Adapted from Govindasamy, Campanella & Oates 1997a.

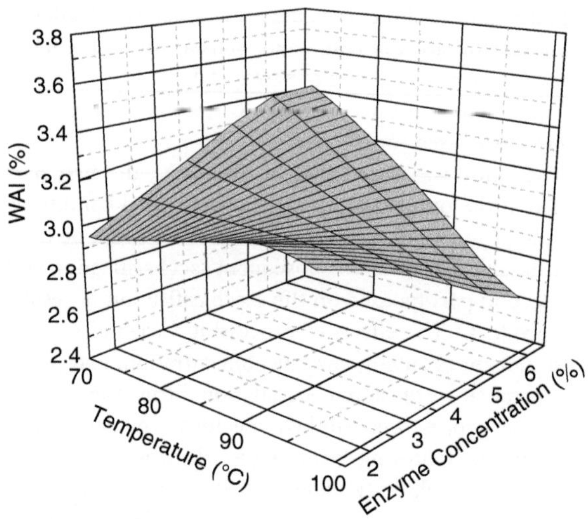

Figure 5.40 Influence of product temperature and enzyme concentration on WAI of sago starch extrudates processed in a single screw extruder at moisture content 30%. Source: Adapted from Govindasamy, Campanella & Oates 1997a.

accessibility of the enzyme during the liquefaction and saccharification processes (Govindasamy et al., 1997a).

The WAI of the extruded samples ranged from 2.8 to 3.6 g/g and decreased with increases in feed moisture. Dextrinization of the starch occurred at higher enzyme concentrations and was dependent on the barrel temperature. The influence of temperature was more evident at low enzyme concentration (1.5%), showing that increasing barrel temperature resulted in increased WAI values. These results suggested that gelatinization of starch was the fundamental mechanism under these processing conditions in a single screw extruder. The compact granular starch structure was loosened to some extent as a consequence of the increase in the thermal energy, which facilitated the water absorption capabilities of the liquefied starch with the consequent increase in WAI values for low enzyme concentrations (Figure 5.40). The figure also shows that at higher enzyme concentration (6.5%), there is a significant decrease in WAI as a result of dextrinization of the starch at higher temperatures.

By use of liquid chromatography techniques, Govindasamy et al. (1997a) showed that the amylopectin fraction of starch was more susceptible to α-amylase activity than the amylose fraction. A similar trend was demonstrated when sago starch was extruded under high-moisture, low-temperature conditions without enzyme (Lim, 1993). The DGR of the starch was influenced by moisture content and enzyme concentration. There was a moisture threshold at around 26.5% below which little degradation was evident. Increasing the feed moisture content above this value resulted in a sharp increase in degradation.

The application of single screw extrusion as a pretreatment of the substrate for saccharification was conducted in a different study in which the samples obtained from the enzyme-assisted single screw extrusion were subjected to batch saccharification reaction using the enzyme glucoamylase. The saccharification reaction previously optimized in terms of temperature, glucoamylase and substrate concentrations, and reaction time was conducted on liquefied samples produced by the enzyme-assisted single screw extrusion described above (Govindasamy et al., 1995). The saccharification reaction was conducted over a 24-hour period in a stirred reactor reaching a saccharification yield in the range 47.6–73.3%. Feed moisture and enzyme concentration were the important extrusion variables that affected the subsequent saccharification. This may reflect the fact that during the time available for saccharification, the greater availability of shorter chain polymeric material, resulting from extensive breakdown in the extruder at the expense of the thermomechanical and enzymatic treatment, enhanced saccharification.

Although it had been previously demonstrated for other starches that DE of the pretreated starch (substrate) before saccharification affects both the rate and final glucose yield (Hakulin et al., 1983; Linko et al., 1983; Nebesny, 1992; Slominska, 1993), water solubility, measured by the WSI, of this pretreated substrate appeared to be the greater contributor to the successful saccharification in the study performed by Govindasamy et al. (1995) on sago starch. The WSI relates to the solubility of some starch components still having a polymeric structure but with lower molecular weight (dextrins) resulting from the enzyme-assisted extrusion treatment. As the preferred substrate for enzymatic hydrolysis of starch with glucoamylase is the one having large molecular size (Guzmán-Maldonado et al., 1995), the ease with which those molecules go into solution would affect saccharification. In that sense, the enzyme-assisted reactive extrusion was able to improve the yield of the saccharification reaction. In the study on sago starch, the optimum DE in the pretreated sample was around 6. Pretreated substrates with higher values of DE exhibited a decline in the saccharification test. This value is not unique for all starches; for instance, Nebesny (1992) found an optimum DE value of 11 for potato starch hydrolysis, probably due to its different starch architecture and different mechanism of starch dextrinization in the enzyme-assisted extrusion process.

In the conventional process for the production of glucose syrup, the formation of maltulose is an impurity that remains in the saccharification process due to the inability of glucoamylase to hydrolyze this compound and can affect yield. The use of an extrusion process for the gelatinization and liquefaction of the starch showed that maltulose was not detected in the pretreated liquefied starch used as a substrate for the saccharification reaction.

Although it was demonstrated that the enzyme-assisted single screw reactive extrusion process could be used as a bioreactor to pretreat sago starch for subsequent saccharification, the concentration of the thermostable α-amylase required (1.5–6.5%) is rather high so the efficacy of the process in terms of cost and economic viability could be challenged. To alleviate this possible cost, a similar study was conducted using a twin screw extruder as the bioreactor.

5.3.4.3 Enzyme-assisted reactive extrusion in intermeshing, co-rotating twin screw extruder

Extruded product properties depend mostly on the molecular transformations of the biopolymers comprising the extrudate formulation. Specifically for starch, there are transformations at the mesoscale level such as disruption of starch granules, whereas at the molecular level transformations are related to disruption of the amylose and amylopectin fractions of the starch molecules, including their partial depolymerization. These transformations are generated in the different sections of the extruder by effects of variables such as temperature, pressure, shear, and residence time. A complete understanding of the interrelationship between the extrusion process, the thermomechanical history and the transformation of these biomolecules involves measurements of those processing variables as well as good characterization of the transformations undergone by the molecules participating in the extrusion process at all scale levels (the subject is fully covered in Chapter 6).

It has been shown for several starch sources that twin screw extrusion can be utilized in the pretreatment of starch for subsequent enzymatic saccharification of the pre-extruded starch (Darnoko & Artz, 1988; Govindasamy et al., 1997b, 1997d). Pretreatment of the starch (sago starch) in the twin screw extruder was studied by Govindasamy et al. (1997b, 1997d) without enzyme and in the presence of a thermostable α-amylase (Thermamyl 120 L). The success of extrusion-cooking in the simultaneous gelatinization and liquefaction of starch for glucose syrup production has prompted other alternatives to the overall process such as the initiation of the saccharification process by incorporation of an enzyme (glucoamylase) in the extruder just before the die. The good mixing and heat transfer capabilities of twin screw extruders have allowed the incorporation of saccharification enzymes, such as glucoamylases, in the extruder accompanied by a decrease in the product temperature near the die plate from 120°C to 65°C to avoid denaturation of the enzyme. This approach resulted in increases of glucose yield for corn (Chouvel et al., 1983) and barley (Hakulin et al., 1983) starches. Thus, for the sago starch enzymatic study described in this section, a pretreatment including the incorporation of amyloglucosidase was performed.

5.3.4.3.1 Thermomechanical treatment of starch without enzymes

The main objective in this study was to examine how the thermomechanical history affects the properties of the extruded product in terms of its suitability for subsequent saccharification. Thus, the effects of the material thermomechanical history, characterized by processing variables such as die pressure, product temperature and specific mechanical energy (SME), on the product transformations were determined. For the study Govindasamy et al. (1996) used an intermeshing, co-rotating laboratory twin screw extruder with a 25 mm screw diameter; L/D 16:1, a mixing/shearing section equipped with bilobal kneading disks; and temperature controlled (Clextral BC21 from Clextral SAS, Firminy, France). Extrusion processing conditions were moisture content that varied in a range 34–47% (wet basis), screw speed that varied in a range 315–485 rpm and barrel temperature that ranged from 81°C to 149°C. Other parameters were kept constant. Structural changes induced during extrusion processing of sago starch were characterized in terms of WSI, WAI, degree of gelatinization (DG), DE and the extrudate chromatographic profiles. DG was determined by a method developed by Wootton et al. (1971) and Owusu-Ansah et al. (1983), which is based on the complexation of iodine with the amylose fraction of the sample.

Examination of results showed that all three variables (moisture content, screw speed, and barrel temperature) affected thermophysical and physicochemical properties of the extrudates. Previous studies on extrusion of sago starch under low-moisture and high-temperature conditions showed significant fracture of the granules and formation of stable starch fragments (Lim, 1993). As a comparison, the study of Govindasamy et al. (1996), which was intended to produce a substrate for the saccharification process, was conducted under lower temperatures and higher moisture conditions and showed fewer changes in the starch molecular structure than those reported by Lim (1993).

Scanning electron microscopy examining the starch at a mesoscale level showed that as a result of the thermomechanical treatment under relatively mild conditions (115°C temperature and 41% moisture content), starch granules were fused in a coarse mass due to gelatinization. By examining the starch at the molecular and structural level, Figure 5.41 illustrates high pressure size exclusion liquid chromatography (HPSEC) chromatograms for (a) native sago starch and (b) thermomechanically modified sago starch. The chromatogram of the native starch gives a fraction of approximately 70% for amylopectin (Ap peak), 27% for amylose (Am peak), and a small peak (Im peak) of approximately 3% of the total starch eluting between the amylopectin and the amylose peaks. In contrast, the extruded sample is showing three peaks but slightly changed. Peak A corresponds to a fraction very close to the amylopectin fraction, but it should be noted that the area of the intermediate peak B corresponding to the intermediate fractions has increased at the expense of the area corresponding to the amylopectin fraction (peak A) whereas the area of the peak corresponding to the amylose fraction (peak C) was not affected.

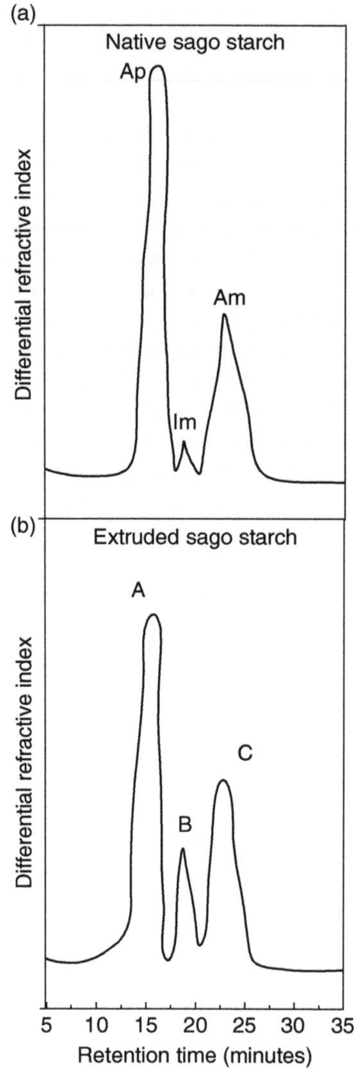

Figure 5.41 HPSEC chromatograms of (a) solubilized native sago starch and (b) solubilized extruded sago starch. For solubilization 1 N NaOH solution was used, and the starch concentration was 0.5%. Extrusion conditions were 47% moisture content, 135°C barrel temperature and 300 rpm screw speed. Source: Govindasamy, Campanella & Oates 1996. Reproduced with permission of Elsevier.

This chromatogram shows that the extrusion process under these mild conditions generates some intermediate molecular size fractions coming from the amylopectin molecules and did not affect the amylose fraction (Govindasamy et al., 1996). Furthermore, since most of the changes were associated with the amylopectin fraction, an index for DGR of starch in the extrudates was arbitrarily assigned to a decrease in percentage of material present in the range of molecular weights corresponding to amylopectin; hence degree of degradation was measured in the range of 0–10%. Limited degradation, as indicated in Figure 5.41 by increasing amounts of intermediate material at the expense of the amylopectin fraction, is typical of shear-induced degradation. Under these mild conditions it was found that variations in starch degradation with processing conditions were insignificant, suggesting that changes in the processing conditions (barrel temperature and screw speed) at high moisture extrusion were not enough to cause significant starch degradation. Related to this low modification of the starch structure are the low DE values obtained for the extrudates. It was noted, however, that DE increased with decrease in moisture content and increases in barrel temperature, indicating that although mild conditions were used, to some extent, they still favor some degradation of the starch molecules. A statistically based model showed that DE was significantly affected by the square of the screw speed.

The effects of those processing variables on DE and starch degradation can be explained on a thermodynamic basis. Rauwendaal (2001) proposed an equation that describes the volumetric rate of entropy generation ($\dot{S}_{gen}$) during the non-isothermal flow of an incompressible fluid, which could be applied to the flow of a melt inside the extruder:

$$\dot{S}_{gen} = \left[\frac{\nabla k \cdot \nabla T}{T} + \frac{\underline{\underline{\tau}} : \nabla \underline{v}}{T}\right] \tag{5.142}$$

For the extrusion process k would be the thermal conductivity of the melt, T its absolute temperature whereas $\underline{\underline{\tau}}$ and $\underline{v}$ are the local stress tensor and the fluid velocity in the extruder channel, respectively. The first term on the right-hand side represents the entropy increase of the product due to the external heating supplied to the product by conduction through the heating elements. The second term on the right side of the equation represents the entropy increase due to friction, where two effects are distinguishable: the effect of the screw speed and the effect of the fluid stress. The equation states that the rate of increase of the entropy of the extruded material is given by the contribution of two energy sources: a heat flow term and a mass flow term. Increases in the temperature of the product either by external heating or friction, or both contributions increase the internal energy of the product which can be regarded as a requisite for initiating a reaction in the thermophysically transformed material. In addition, entropy generation is associated with irreversible changes in the material that can occur inside the extruder, caused by both the heat flow and shear/friction (Wang et al., 1992). At constant moisture content, as the screw speed increases the heat generated by friction increases. On the other hand, at constant screw speed, decreasing the moisture content of the product would result in an increase in the product viscosity and so does the local stress. For non-Newtonian fluids, the shear stress is proportional to the apparent viscosity of the product μ_a and the magnitude of the shear stress is estimated as:

$$\tau = \mu_a \dot{\gamma} \tag{5.143}$$

The shear rate $\dot{\gamma}$ is related to the screw speed as follows:

$$\dot{\gamma} = cN \tag{5.144}$$

where c is a constant that depends on the geometry of the screw and N is the screw speed. By substituting Eq. 5.144 into Eq. 5.143, the following equation is obtained:

$$\tau = c\mu_a N \tag{5.145}$$

The above equation shows that the effect of screw speed is more important for degradation of the starch at low moisture content, i.e. when the viscosity is high. Conversely, at high moisture content the product viscosity is low and probably the effect of screw speed on the shear stress becomes less important. Thus, the effect of screw speed on the degradation of starch depends on moisture content. At low moisture content, the molecular degradation (or generated entropy) is affected by both the screw speed and the shear stress. By substituting Eqs 5.144 and 5.145 into Eq. 5.142, it can be shown that the term $\underline{\underline{\tau}} : \nabla \underline{v}$ becomes $\mu_a N^2$. Results from Govindasamy et al. (1996) showed a strong correlation between starch degradation (measured by DE, WSI, WAI, and degree of gelatinization) and the square of the screw speed.

Changes in the microstructure of the starch were also observed in the measured values of WSI and WAI. Specifically, WSI values provide an indication of molecular degradation of starch and it was found that sago starch increased the water solubility from 0.2% for native sago starch to values ranging from 4.5% to 18.1%, depending on the extrusion conditions. The volume of the swollen granules forming a gel when the extrudate is mixed with water, which is indicated by the WAI values, is associated with the thermophysical treatment of the extrudate as well. Native starches have low WAI values, i.e. they do not have the capacity to absorbe large amounts of water. Instead, thermophysically processed starches have enhanced capacity for absorbing water and consequently they are characterized by high WAI values. Govindasamy et al. (1996) found that native sago starch has WAI values of the order of 1.6 g/g dry matter but they changed to values in the range 2.7–7.0 g/g dry matter after twin screw extrusion processing under mild conditions of high moisture and low temperatures. In contrast, sago starch extruded in a single screw extruder was characterized by WAI values significantly lower (in the range 1.9–5.2 g/g dry matter) even though the range of moisture used (15–38%) in the single screw extrusion was significantly lower (Lim, 1993). Extruded samples produced using the twin screw extruder gave parameters comparable with those of products obtained from other starches (Govindasamy et al., 1996). Low WAI and WSI values are related to the inaccessibility of some compact structure to water and possibly enzyme attack. The relationship between measured values of WAI and WSI indicates the type of mechanism responsible for the increased water solubility of extruded starches. In this study, dextrinization does not seem to be the predominant mechanism occurring in the thermomechanical process. Results presented by Govindasamy et al. (1997d), which are illustrated in Figure 5.42, show that under high moisture conditions and relatively low temperatures used in this starch pretreatment, WAI increases exponentially with increasing WSI. In other words, solubility does not increase at the expense of the water absorption capacity and dextrinization of starch molecules is not significant.

Figure 5.43 shows changes in WSI with feed moisture and barrel temperature at a screw speed of 400 rpm. The WSI goes through a minimum for barrel temperature at approximately 101°C for all the moisture levels tested. For temperatures below 101°C the generated entropy (see Eq. 5.142) is mainly due to friction. For temperatures above approximately 101°C the melt viscosity decreases so the generated entropy is lowered, and the contribution of heat conduction through the heating elements becomes an important component for thermal degradation of starch (higher entropy generation). Govindasamy et al. (1996) extended the study to the characterization of starch gelatinization and how it was affected by the extrusion conditions, and results showed that gelatinization of the starch was the primary mode of starch modification under these conditions. The main conclusion of the study was that thermomechanical processing of sago

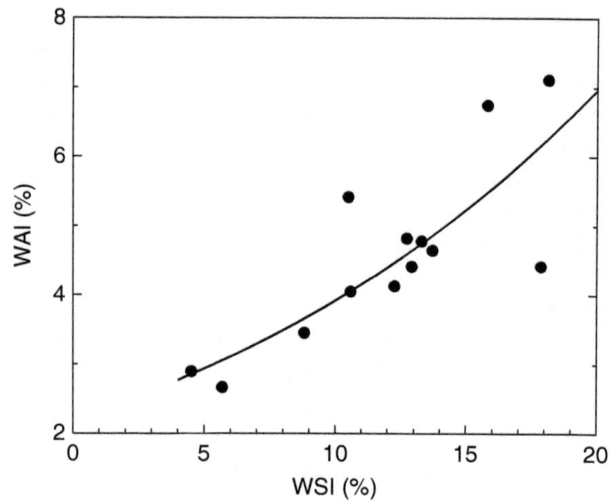

Figure 5.42 Relationship between WAI and WSI for sago starch extrudate in a twin screw extruder. Source: Adapted from Govindasamy, Campanella & Oates 1997d.

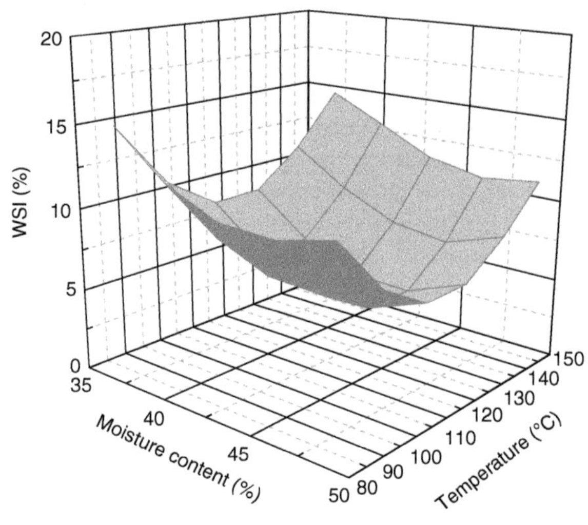

Figure 5.43 Influence of feed moisture and barrel temperature on WSI of sago starch extruded in a twin screw extruder at a screw speed of 400 rpm. Extrusion conditions are given in the text. Source: Adapted from Govindasamy, Campanella & Oates 1997d.

starch at high moisture contents resulted in limited starch degradation, a phenomenon seen at the meso and molecular scales, and only starch gelatinization was observed. The mild extrusion conditions used for the sago starch pretreatment resulted in a lower yield for enzymatic saccharification (using glucoamylase) of the pretreated starch carried out in a stirred tank reactor.

The other limitation observed in the saccharification reaction was the high viscosity of the pretreated starch suspension that precluded the use of pretreated starch as substrate at concentrations higher than 10% making the process economically unfeasible. The study showed some processing indicators that related to the suitability of the pretreated starch to be successfully saccharified. Die pressure (an indicator of melt viscosity) was a useful indicator to predict the success of a subsequent saccharification. Extruded products that showed high die pressure (large melt viscosity) were poorly saccharified (Govindasamy, 1995). A global trend between saccharification and specific mechanical energy (SME) was obtained and is illustrated in Figure 5.44. The figure shows that higher input of mechanical energy leads to the formation of pretreated starch that is not easily saccharified. The general trend is that saccharification is poor probably due to some structural characteristics of the pretreated starch substrate that limit the action of the glucoamylase enzyme.

A global trend is also observed in Figure 5.45 when WSI is plotted as a function of SME. It appears that solubility of the starch is related, although weakly, to the starch transformation even at high moisture extrusion conditions. In general, saccharification and WSI were less influenced by SME and this could be related to the fact that both parameters are not based on unique structural parameters but instead it may result from a complex combination of different phenomena (Govindasamy, 1995). Given the limiting condition to saccharification imposed by the high viscosity of the suspensions prepared with the pretreated starch, a follow-up study was conducted focusing on combining the thermomechanical treatment in the intermeshing, co-rotating twin screw extruder in the presence of a thermostable α-amylase.

5.3.4.3.2 *Thermomechanical treatment of starch with an enzyme*

Govindasamy et al. (1997b) used an intermeshing, Clextral co-rotating laboratory twin screw extruder (25 mm screw diameter; L/D 16:1, three mixing/shearing sections in series, two with kneading disks and one with reverse screw element; temperature controlled) for the study. The screw configuration was modified to provide a longer residence time of the starch in the extruder. The processing variables utilized were barrel temperature

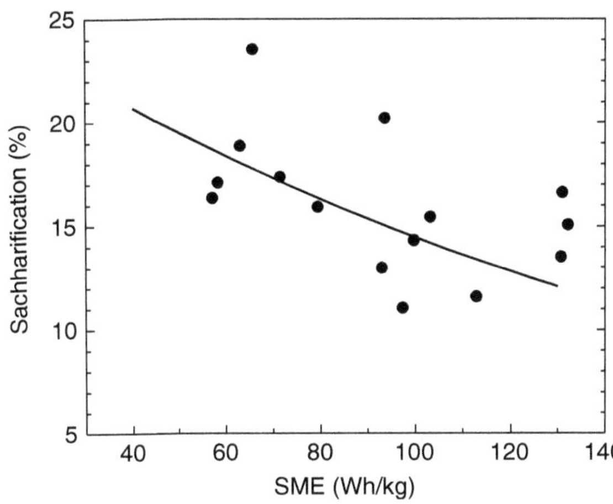

Figure 5.44 Global trend showing the effects of SME on saccharification of extruded sago starch. Extrudates were prepared in a twin screw extruder without the use of enzymes. Source: Adapted from Govindasamy 1995.

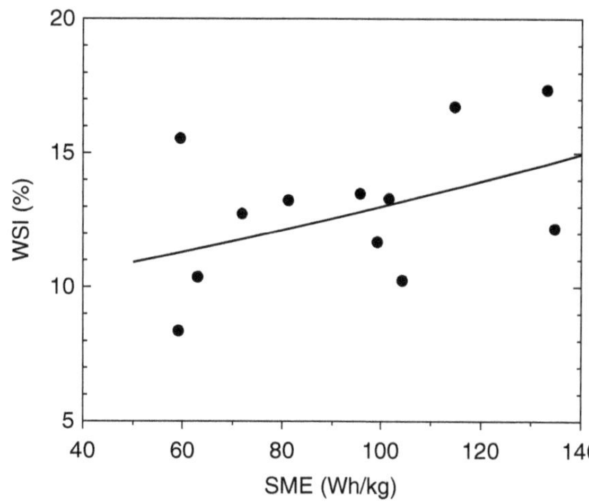

Figure 5.45 Global trend showing the effects of SME on WSI of extruded sago starch processed in a twin screw extruder. Source: Adapted from Govindasamy 1995.

(*T*), screw speed (*N*), enzyme concentration (*E*), and feed moisture (*M*) which were varied in the ranges 70–130°C, 70–190 rpm, 0–1%, and 28.5–50.5% (wet basis) respectively. Other parameters were kept constant.

The study explored the feasibility of utilizing a co-rotating twin screw extruder as a bioreactor for the thermomechanical gelatinization and liquefaction of sago starch using an enzyme concentration lower than the one used in the single screw system described in section 5.3.4.2. As before, the extent of starch hydrolysis was evaluated using HPSEC chromatography to measure DGR and oligosaccharide content, as well as DE, WSI, and WAI. Product temperature, die pressure, and SME were also determined. The thermostable enzyme used was *Bacillus licheniformis* α-amylase (Thermamyl 120 L). Unlike in the preceding study in which no enzyme was used, the incorporation of the thermostable enzyme was able to promote gelatinization and enzymatic liquefaction in the twin screw extruder even at relatively low enzyme concentrations (range 0–1%). Feed moisture and enzyme concentration were found to be the most critical variables affecting the extrusion process. In addition, processing temperature was another important parameter to be considered due to the possible enzyme deactivation. Increased breakdown of starch molecules revealed by increases in DE and water solubility showed that hydrolysis of the starch granules was the fundamental reaction occurring during the process (Govindasamy et al., 1997b). The presence of the enzyme during the enzyme-assisted extrusion process ensured a higher breakdown of the starch with lower SME inputs.

These results clearly showed that the enzymatic reactive extrusion process could be used either for the production of maltodextrins of low DE values (DE 0–10) or as a pretreatment of the starch to be used as the substrate in the saccharification process for the production of glucose syrups with a relatively low energy consumption. The more striking differences with the results obtained in the preceding study, i.e. when the starch was extruded without the presence of the α-amylase enzyme, can be observed in the chromatogram illustrated in Figure 5.46 (compare with chromatograms shown in Figure 5.41). For the starch extruded with 39.5% moisture content, a screw speed of 130 rpm, a barrel temperature of 100°C and 0.5% enzyme concentration, there was appreciable degradation of the amylopectin and amylose fractions. Extrudates were characterized by different types of fractions, peaks A_1 and A_2 corresponding to fractions with molecular weights smaller than the amylopectin molecule resulted from the enzyme-assisted extrusion, peak B with molecular weight in the range 10^3–10^5 daltons, and peaks C_1 and C_2 corresponding to oligosaccharides species (from G_1 to G_{10}). In this case, both amylopectin (Ap) and amylose (Am) were susceptible to the attack of the α-amylase to produce a range of smaller

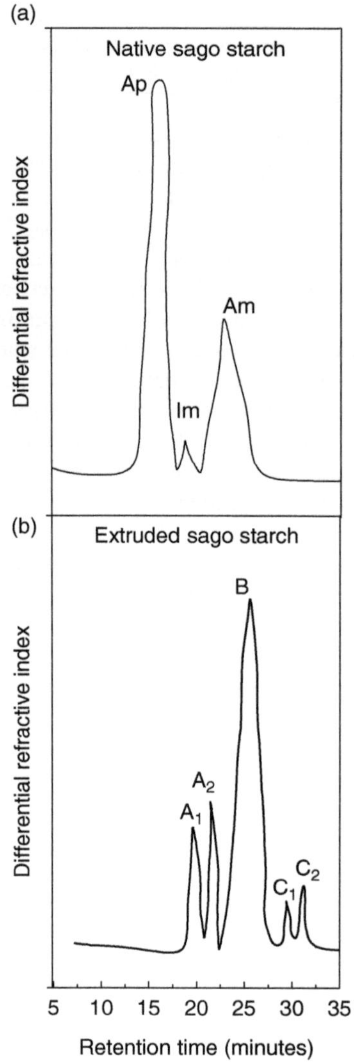

Figure 5.46 HPSEC chromatograms of (a) solubilized native sago starch and (b) solubilized extruded sago starch. For solubilization 1 N NaOH solution was used, and the starch concentration was 0.5%. Extrusion conditions were 50.5% moisture content, 100°C barrel temperature and 130 rpm screw speed. Source: Adapted from Govindasamy, Campanella & Oates 1997b.

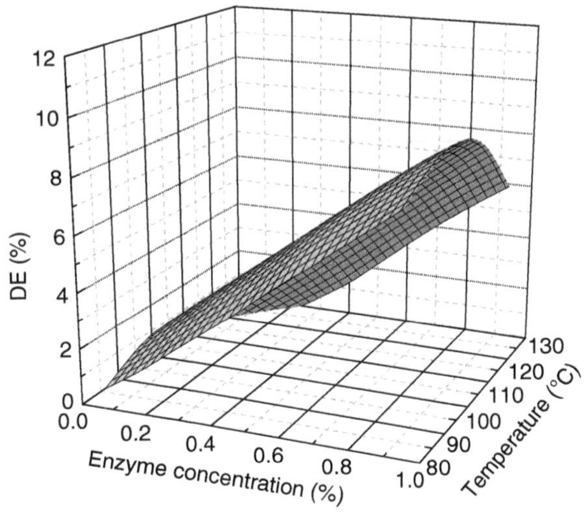

Figure 5.47 Influence of enzyme concentration and barrel temperature on DE of extrudate produced by twin screw extrusion at 39.5% moisture content at 130 rpm in the presence of enzymes. Source: Adapted from Govindasamy, Campanella & Oates 1997b.

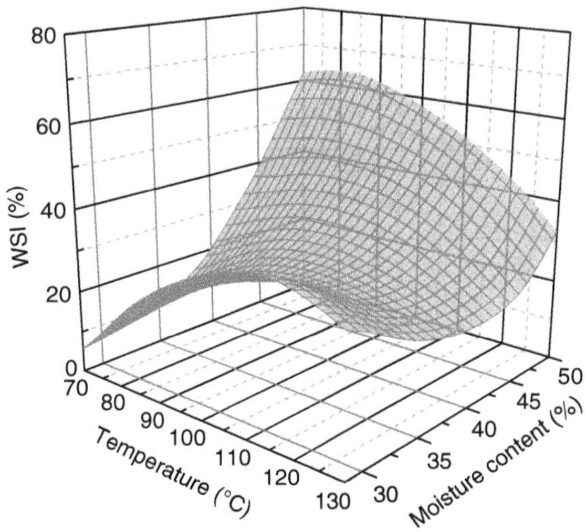

Figure 5.48 Influence of barrel temperature and moisture content on WSI of extruded sago starch processed in a twin screw extruder at 130 rpm with an enzyme concentration of 0.5% (Thermamyl 120 L). Source: Adapted from Govindasamy, Campanella & Oates 1997b.

molecular weight macromolecules and oligosaccharides (Govindasamy et al., 1997b). In contrast, in the single screw extruder and in the presence of α-amylase, the amylopectin fraction was preferentially and slightly degraded whereas the effect on amylose was not significant.

Figure 5.47 illustrates the effect of barrel temperature and enzyme concentration on DE values which reached values up to 10%. It is clear from the figure that the effect of the enzyme concentration on the DE values at any given barrel temperature appears to change linearly with enzyme concentration. For all enzyme concentrations, effect of the barrel temperature on the DE values is small for all temperatures studied, which clearly demonstrates the thermal stability of the α-amylase. A maximum of DE is observed at barrel temperatures of approximately 105°C; higher temperatures showed a decline in the DE values (see Figure 5.47), probably due to inactivation of the enzyme. Lower values of DE for temperatures below 105°C could be associated with insufficient gelatinization and liquefaction of the starch at low temperatures. That is in agreement with results shown in Figure 5.48. Specifically, at a low moisture content of 28.5%, WSI values show a maximum at a barrel temperature close to 105°C. The figure also shows that the influence of the barrel temperature is dependent on the feed moisture. The highest WSI (around 60%) is attained following extrusion under conditions of highest feed moisture level (50.5%) and lowest barrel temperature (70°C).

It is important to note how the enzyme attacks the starch granules so scanning electronic microscopy studies were conducted to determine differences in starch granules incubated with 0.25% α-amylase (Thermamyl 120 L) for 4 hours in a solution of pH 6.0 and 30 ppm Ca^{2+} (Figure 5.49a), starch extruded at 70°C, 130 rpm, 0.50% Thermamyl 120 L, and 50.5% moisture content (Figure 5.49b), and starch extruded at 100°C, 0.50% Thermamyl 120 L, and 50.5% moisture content (Figure 5.49c). In the micrographs, I indicates intact granules, H honeycomb structure and P pores. It is clear that in the absence of shear, the granular structure is maintained despite the hydrolytic action of the enzyme was able to develop internal pores (Figure 5.49a). At temperatures below 100°C, some of the integrity of the starch granules is maintained but some breaking of the granules is also noted as well as the presence of more pores due to the combined action of the enzyme and shear (Figure 5.49b). For temperatures of 100°C or higher, there was loss of granular integrity and the action of the enzyme, shear and temperature (below the inactivation temperature of the enzyme) was able to create a honeycomb structure (Figure 5.49c). It is hypothesized that the presence of a honeycomb structure in addition to pores and cracks can facilitate the penetration of the glucoamylase enzyme and thus favor the saccharification reaction. Studies on the saccharification reaction using pretreated starch as a substrate showed a significant improvement with respect to the process when the starch was extruded in the absence of enzymes (Govindasamy et al., 1997c). Measurements of die pressure (an indicator of melt viscosity) were able to predict the susceptibility of the starch to saccharifying enzymes. In general, measured higher melt viscosities manifested as high die pressures resulted in extrudates with poor glucoamylase susceptibility (Figure 5.50). As in the preceding study, the effect of SME on product properties such as DE, DGR, WSI, and WAI and oligosaccharide formation exhibited a global trend but unlike the study on the starch treatment without enzyme, starch hydrolysis was improved at lower SME (Figure 5.51). In general, very high values of saccharification (98%) were achieved after 10 hours of saccharification reaction using a susbtrate processed with a barrel temperature of 100°C, a screw speed of 130 rpm, an enzyme concentration of 0.5%, and a moisture content of 39.5%. An important fact was that of achieving a suitable starch pretreatment at SME around 27 W.h/kg.

Thus, this enzyme-assisted reactive twin screw extrusion process could be used to pretreat sago starch for successful saccharification with relatively lower mechanical energy input. In contrast, pretreatment without enzyme requires a SME in the range 57–131 W.h/kg, and with a considerably lower yield of the saccharification reaction (50–70%).

Through these case studies extrusion has been shown as a promising technology to perform reactive extrusion processing. Specifically, this has the potential to be used in the simultaneous gelatinization and liquefaction of starch, a pretreatment used to obtain a substrate for the subsequent saccharification reaction used in the

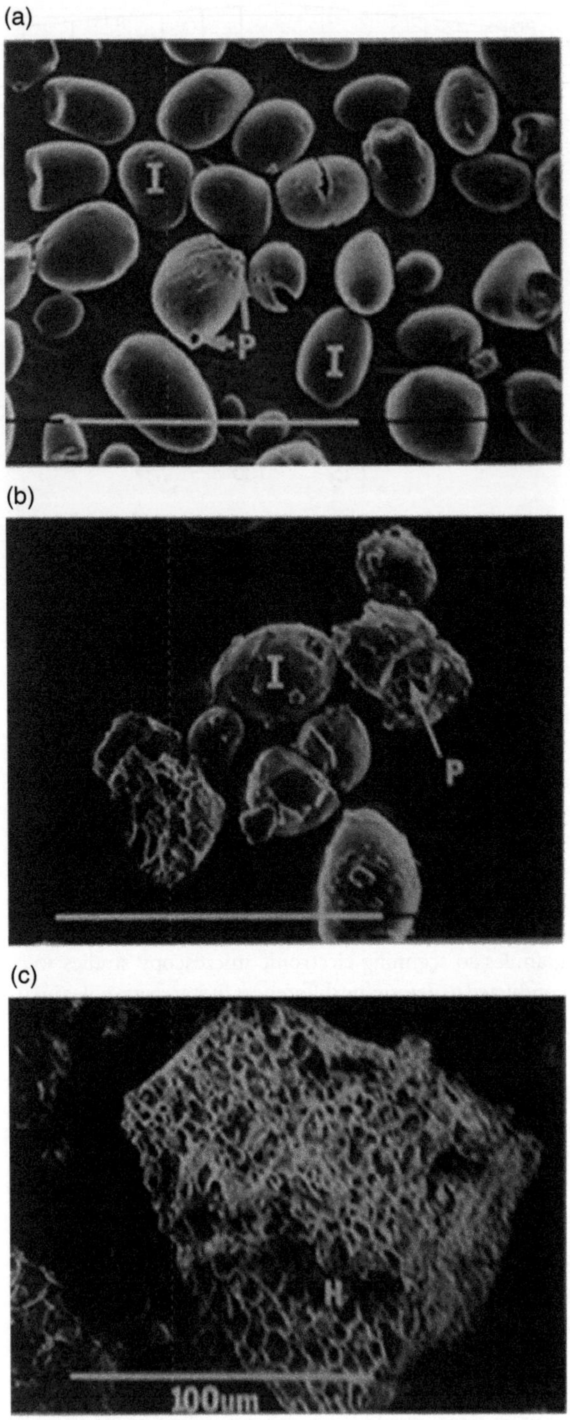

Figure 5.49 Scanning electron micrographs of the enzyme hydrolyzed extrudates prepared by an enzyme-assisted extrusion in a twin screw extruder. (a) Sago starch granules degraded after 4 h incubation with a 0.25% Thermamyl 120 L enzyme at 40°C, pH 6.0 and 30 ppm Ca^{2+}. (b) Extrudate processed in a twin screw extruder at 70°C, 39.5% moisture, 130 rpm, and an enzyme concentration of 0.50% (Thermamyl 120 L). (c) Extrudate processed in a twin screw extruder at 100°C, 50.5% moisture, 130 rpm, and an enzyme concentration of 0.50% (Thermamyl 120 L). (I) indicates intact granules, (H) honeycomb structure and (P) pores.
Source: Adapted from Govindasamy, Campanella & Oates 1997b.

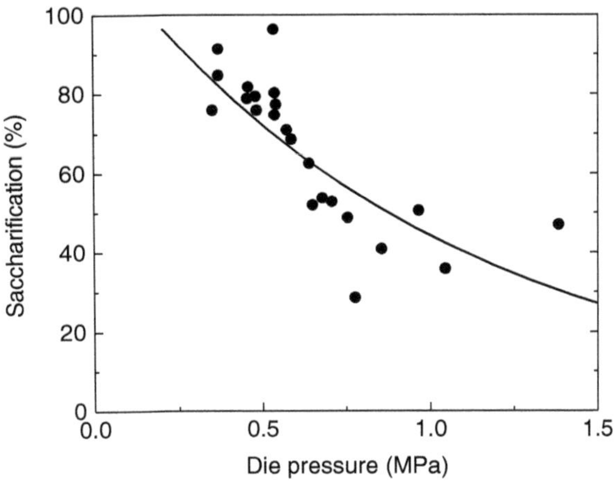

Figure 5.50 Effect of die pressure on the saccharification of extrudates processed in a twin screw extruder. Source: Adapted from Govindasamy, Campanella & Oates 1997c.

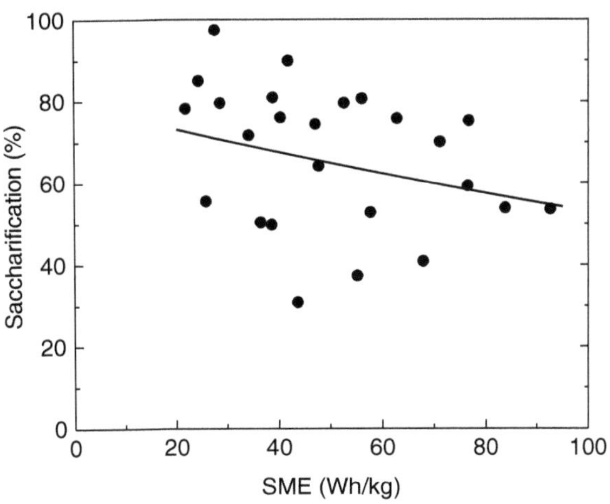

Figure 5.51 Effect of SME on the saccharification of extrudates processed in a twin screw extruder. Source: Adapted from Govindasamy, Campanella & Oates 1997c.

production of glucose syrup. Initiation of the saccharification process during extrusion processing is another innovative alternative to improve the rate of the reaction and this is discussed in section 5.3.4.3.3.

5.3.4.3.3 Continuous liquefaction and saccharification of starch in twin screw extruder

Initiation of saccharification in the extruder operation by incorporating the enzyme glucoamylase just before the die element has been successfully used with different starch sources such corn starch (Chouvel et al., 1983) and barley starch (Hakulin et al., 1983). Furthermore, Linko (1992) suggested that to obtain the highest glucose yield, sequential gelatinization, liquefaction and saccharification, each at its optimum processing conditions, in one continuous extrusion process is necessary. The work on cereal starches showed that the highest glucose yield was obtained when most of the water and the liquefying α-amylase enzyme were added into the first port of a long barrel co-rotating twin screw extruder and the rest of the enzyme in the fourth section (Linko, 1992). The saccharifying enzyme glucoamylase suspended in a buffer solution of pH 4.5 was added in the last section with a dosing rate to provide a total water level of about 70%. The approach of introducing the different enzymes was able to eliminate retrogradation of the starch before saccharification which is an important benefit of twin screw extrusion because it would allow the saccharification reaction to start with low DE substrates. Results from Linko (1992) also showed that the subsequent batch saccharification time could be significantly reduced from about 2 days to about 10 hours.

The study conducted by Govindasamy (1995) used the same extrusion equipment and screw configuration as the preceding study. In order to examine the enhancement of this simultaneous liquefaction/saccharification approach with respect to the enzyme-assisted liquefaction described above, conditions for the enzyme-assisted thermomechanical gelatinization/liquefaction that gave the lowest glucose yield were selected. The study focused on the effects of feed total moisture content, glucoamylase concentration and reaction time for the glucoamylase hydrolysis of sago starch (reaction time was varied by introducing the enzyme at two different ports of the screw-barrel assembly) on the amount of glucose released in the extruder and after 10 hours of saccharification reaction (Govindasamy, 1995). Total feed moisture content varied in the range 39.5–50% and glucoamylase concentration in the range 0–0.75% was introduced at different sections of the screw-barrel assembly to vary the reaction time. Results showed that both the initial and subsequent saccharification were dependent on the feed moisture, enzyme concentration and the stage where the glucoamylase was introduced in the extruder. Specifically, glucose

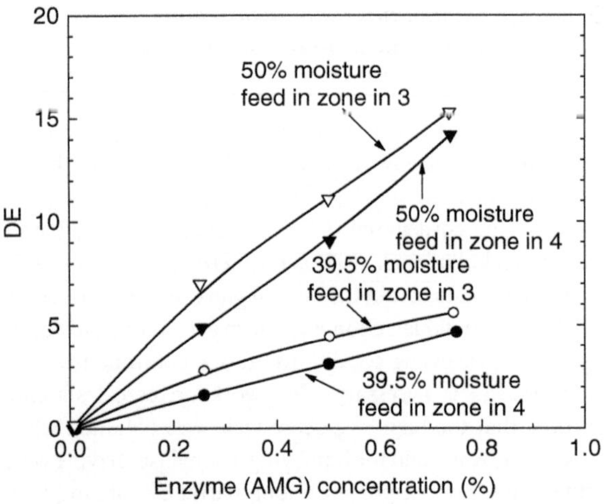

Figure 5.52 Effect of enzyme concentration (AMG), moisture content and stage of addition of the enzyme in a twin screw extruder on DE. Zone 4 is a section close to the die and zone 3 is the section previous to the die. Source: Adapted from Govindasamy, Campanella & Oates 1995.

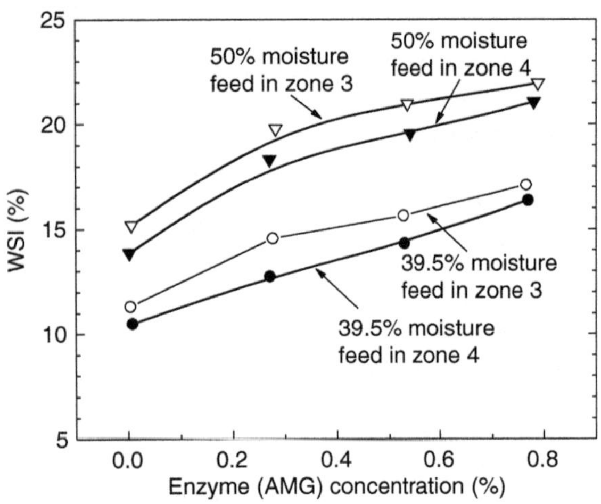

Figure 5.53 Effect of enzyme concentration (AMG), moisture content and stage of addition of the enzyme in a twin screw extruder on WSI. Zone 4 is a section close to the die and zone 3 is the section previous to the die. Source: Adapted from Govindasamy, Campanella & Oates 1995.

yield in the dual enzyme-assisted thermomechanical process was improved by using the highest concentration of glucoamylase (0.75%) and a relatively long reaction time; the total feed moisture content was 50%.

It appears that during extrusion processing, the presence of α-amylase promotes the activity of the glucoamylase. The increased conversion of sago starch was also evidenced by increases in DE and WSI after the extrusion process (Figures 5.52 and 5.53). Although glucose yield was high after the dual enzyme-assisted thermomechanical extrusion, further saccharification of that starch did not result in an increased saccharification yield. An explanation for that could not be found during the study (Govindasamy, 1995). Liquefied samples used in the subsequent saccharification studies were collected immediately at the extruder outlet in liquid nitrogen, freeze dried, milled to powder and stored in air-tight plastic containers. A hypothesis considered was that changes occurring during storage affected the susceptibility of those treated substrates to further saccharification.

References

Alexander AG (1983) Process for continuously manufacturing lubricating grease. US Patent 4,392,967.

Altomare RE, Ghossi P (1986) An analysis of residence time distribution patterns in a twin screw cooking extruder. Biotechnology Progress 2: 157–163.

Averous L (2004) Biodegradable multiphase systems based on plasticized starch: a review. Journal of Macromolecular Science, Part C: Polymer Reviews C44(3): 231–274.

Awaja F, Pavel D (2005) Review: recycling of PET. European Polymer Journal 41: 1453–1477.

Awaja F, Daver F, Kosior E (2004a) Recycled poly(ethylene terephthalate) chain extension by a reactive extrusion process. Polymer Engineering and Science 44(8): 1579–1587.

Awaja F, Daver F, Kosior E, Cser F (2004b) The effect of chain extension on the thermal behaviour and crystallinity of reactive extruded recycled PET. Journal of Applied Analysis and Calorimetry 78: 865–884.

Barraquio VL, Van De Voort FR (1991) Sodium caseinate from skim milk powder by extrusion processing: physicochemical and functional properties. Journal of Food Science 56(6): 1552–1561.

Bartilla T, Kirch D, Nordmeier J, Prömper E, Strauch Th (1986) Physical and chemical changes during the extrusion process. Advances in Polymer Technology 6(3): 339–387.

Ben-Gera I, Rokey GJ, Smith OB (1983) Extrusion cooking of grains for ethanol production. Journal of Food Engineering 2(3): 177–187.

Biesenberger JA, Kumar R, Garritano R, Starita JM (2004) Rheokinetics for reactive polymer processing. Polymer Engineering and Science 25: 301–304.

Bigg D, Middleman S (1974) Mixing in a screw extruder. A model for residence time distribution and strain. Industrial and Engineering Chemistry Fundamentals 13: 66–71.

Bigio DI, Boyd JD, Erwin L, Gailus DW (1985) Mixing studies in the single screw extruder. Polymer Engineering and Science 25(5): 305–310.

Brown SB (1991) Chemical processes applied to reactive extrusion of polymers. Annual Review of Materials Science 21: 409–435.

Cardi N, Po R, Giannotto G, Occhiello E, Garbassi F, Messina G (1993) Chain extension of recycled poly(ethylene terephthalate) with 2,2′-Bis(2-oxazoline). Journal of Applied Polymer Science 50(9): 1501–1509.

Cassagnau P, Bounor-Legaré V, Fenouillot F (2007) Reactive processing of thermoplastic polymers: a review of the fundamental aspects. International Polymer Processing 22(3): 218–258.

Chella R, Ottino JM (1984) Conversion and selectivity modifications due to mixing in unpremixed reactors. Chemical Engineering Science 39(3): 551–567.

Chella R, Ottino JM (1985) Fluid mechanics of mixing in a single-screw extruder. Industrial and Engineering Chemistry Fundamentals 24(2): 170–180.

Chouvel H, Chay PB, Cheftel JC (1983) Enzymatic hydrolysis of starch and cereal flours at intermediate moisture contents in a continuous extrusion-reactor. Lebensmittel-Wissenschaft & Technologie 16: 346–353.

Cioffi M, Ganzeveld KJ, Hoffmann AC, Janssen LPBM (2002) Rheokinetics of linear polymerization. A literature review. Polymer Engineering and Science 42: 2383–2392.

Clextral (2013) Private communication: Twin screw extrusion processing line for the production of caseinates from acid casein. More information available through: www.clextral.com/.

Combette P (2010) Cotton linters for paper making. Clextrusion, 19: 8–9. More information available through: www.clextral.com/.

Darnoko D, Artz WE (1988) Twin screw extrusion as a continuous pretreatment process for enzymatic hydrolysis of cassava. Journal of Food Science 53: 1792–1794, 1799.

Datta AK (2002) *Biological and Bioenvironmental Heat and Mass Transfer*. New York: Marcel Dekker.

Davis WM (1992) Heat transfer in extruder reactors. In: Xanthos M (ed) *Reactive Extrusion. Principles and Practice*. New York: Hanser Publishers, pp. 257–282.

De Choudens C, Angelier R (1990) Les pâtes chimico-thermo-mécaniques blanchies obtenues par le procédé BiVis. Revue ATIP 44(3): 137–146.

De Choudens C, Angelier R, Combette P (1984) Pâtes mécaniques de résineux, pâtes chimicomécaniques de feuillus. Revue ATIP 38(8): 405–416.

De Choudens C, Angelier R, Lombardo G (1987) Pâtes chimico-mécaniques blanchies obtenues par le procédé « BiVis ». Revue ATIP 41(2): 63–68.

De Choudens C, Angelier R, Combette P, Lesas C (1989) Procédé de fabrication de pâtes chimicomécaniques ou chimi-cothermo-mécaniques blanchies. French Patent 8711082.

Della Valle G, Kozlowski A, Colonna P, Tayeb J (1989) Starch transformation estimated by the energy balance on a twin-screw extruder. Lebensmittel-Wissenschaft & Technologie 22(5): 279–286.

De Mesa-Stonestreet NJ, Alavi S, Bean SR (2010) Sorghum proteins: the concentration, isolation, modification and food applications of kafirins. Journal of Food Science 75(5): R90–R104.

De Mesa-Stonestreet NJ, Alavi S, Gwirtz J (2012) Extrusion-enzyme liquefaction as a method for producing sorghum protein concentrates. Journal of Food Engineering 108(2): 365–375.

Dokic P, Jakovljevic J, Dokic-Baucal L (1998) Molecular characteristics of maltodextrins and rheological behaviour of diluted and concentrated solutions. Colloids and Surfaces A: Physicochemical and Engineering Aspects 141(3): 435–440.

Fichtali J, Van De Voort FR (1991) Pilot plant production of caseins using extrusion processing. II Sodium caseinate production. Milchwissenschaft 46(8): 479–483.

Fichtali J, Van De Voort FR, Diosady L (1995) Performance evaluation of acid casein neutralization process by twin-screw extrusion process. Journal of Food Engineering 26(3): 301–318.

Fitton MG (1979) Rapid determination of dextrose equivalent by cryoscopy. Starch/Stärke 31(11): 381–384.

Froment GF, Bischoff KB, De Wilde J (2011) *Chemical Reactor Analysis and Design*, 3rd edn. New York: John Wiley.

Fullbrook PD (1984) The enzymatic production of glucose syrups. In: Dziedzic SZ, Kearsley MW (eds) *Glucose Syrups, Science and Technology*. London: Elsevier Applied Science, pp. 65–115.

Gailus DW, Erwin L (1981) Experiments in extensive mixing in laminar flow: II. The single screw extruder with mixing section. Paper presented at the 39th Annual Meeting of the Society of Plastics Engineers, ANTEC Technical Papers, 27: 639–642.

Ganzeveld KJ, Janssen LPBM (1994) Role of mixing and rheology in reactive extrusion. Industrial and Engineering Chemistry Research 33: 2398–2403.

Govindasamy S (1995) Extrusion processing of sago starch for glucose syrup production. PhD thesis, National University of Singapore.

Govindasamy S, Campanella OH, Oates CG (1995) Influence of extrusion variables on subsequent saccharification behavior of sago starch. Food Chemistry 54: 289–296.

Govindasamy S, Campanella OH, Oates CG (1996) High moisture twin-screw extrusion of sago starch: 1. Influence on granule morphology and structure. Carbohydrate Polymers 30(4): 275–286.

Govindasamy S, Campanella OH, Oates CG (1997a) The single screw extruder as a bioreactor for sago starch hydrolysis. Food Chemistry 60: 1–11.

Govindasamy S, Campanella OH, Oates CG (1997b) Enzymatic hydrolysis of sago starch in a twin-screw extruder. Journal of Food Engineering 32: 403–426.

Govindasamy S, Campanella OH, Oates CG (1997c) Enzymatic hydrolysis and saccharification optimisation of sago starch in twin-screw extruder. Journal of Food Engineering 32: 427–446.

Govindasamy S, Campanella OH, Oates CG (1997d) High moisture twin screw extrusion of sago starch. II. Saccharification as influenced by thermomechanical history. Carbohydrate Polymers 32(3–4): 261–274.

Guzmán-Maldonado H, Paredes-López O, Biliaderis CG (1995) Amylolytic enzymes and products derived from starch: A review. Critical Reviews in Food Science and Nutrition 35: 373–403.

Hakulin S, Linko YY, Linko P, Seiler K, Seibel W (1983) Enzymatic conversion of starch in twin-screw extruder. Starch/Stärke 35: 411–414.

Illing G (1969) Direct extrusion of nylon products from lactams. Modern Plastics 46: 70.

Incarnato L, Scarfato P, Di Maio L, Acierno D (2000) Structure and rheology of recycled PET modified by reactive extrusion. Polymer 41(18): 6825–6831.

Janssen LPBM (1998) On the stability of reactive extrusion. Polymer Engineering and Science 38: 2010–2019.

Janssen LPBM (2004) *Reactive Extrusion Systems*. New York: Marcel Dekker.

Kalambur S, Rizvi SSH (2005) Biodegradable and functionally superior starch-polyester nanocomposites from reactive extrusion. Journal of Applied Polymer Science 96: 1072–1082.

Kalambur S, Rizvi SSH (2006) An overview of starch-based plastic blends from reactive extrusion. Journal of Plastic Film and Sheeting 22(1): 39–58.

Kalyon DM, Sangani HN (1989) An experimental study of distributive mixing in fully intermeshing, co-rotating twin screw extruders. Polymer Engineering and Science 29(15): 1018–1026.

Kalyon DM, Gotsis AD, Yilmaze U, et al. (1988) Development of experimental techniques and simulation methods to analyze mixing in co-rotating twin screw extrusion. Advances in Polymer Technology 8(4): 337–353.

Kaseem M, Hamad K, Deri F (2012) Thermoplastic starch blends: a review of recent works. Polymer Science Series A 54: 165–176.

Knudsen FE, Karkalas J (1969) Enzymatic hydrolysis of starch hydrolyzates. Starch/Stärke 21: 284–291.

Korn SR, Harper JM (1982) Extrusion of corn for ethanol fermentation. Biotechnology Letters 4: 417–422.

Kowalski RC (1992) Fit the reactor to the chemistry. Case histories of industrial studies of extruder reactions. In: Xanthos M (ed) *Reactive Extrusion. Principles and Practice*. New York: Hanser Publishers, pp. 7–32.

Kurdin JA, Bohn WL (1984) *Mechanical pulping by extrusion*. Tappi Pulping Conference, San Francisco, USA, pp. 265–274.

Labout JJM (1985) Conversion of liquefied starch into glucose using a novel glucoamylase system. Starch/Stärke 37: 157–161.

Lane JH, Eynon L (1923) Determination of reducing sugars by means of Fehling's solution with methylene blue as internal indicator. Journal of the Society of Chemical Industry 42(2): T32–T37.

Laun HM (1983) Polymer melt rheology with a slit die. Rheologica Acta 22: 171–185.

Li PX, Campanella OH, Hardacre AK (2004) Using an in-line slit-die viscometer to study the effects of extrusion parameters on corn melt rheology. Cereal Chemistry 81: 70–76.

Likimani TA, Sofos JN, Maga JA, Harper JM (1991) Extrusion cooking of corn/soybean mix in presence of thermostable α-amylase. Journal of Food Science 65: 99–108.

Lim LYY (1993) *Physical modification of sago starch for industrial uses*. Masters thesis, National University of Singapore.

Linko P (1992) Twin-screw extrusion cooker as a bioreactor for starch processing. In: Kokini JL, Ho CT, Karwe MV (eds) *Food Extrusion Science and Technology*. New York: Marcel Dekker, pp. 335–344.

Linko P, Hakulin S, Linko YY (1983) Extrusion cooking of barley starch for the production of glucose syrup and ethanol. Journal of Cereal Science 1(4): 275–284.

Mack WA (1972) Bulk polymerization in screw-conveyor reactors. Chemical Engineering 79: 99–102.

Maheshri J, Wyman C (1980) Mixing in an intermeshing twin screw extruder chamber: combined cross and down channel flow. Polymer Engineering and Science 20: 601–606.

Martin O, Averous L (2001) Poly(lactic acid) : plasticization and properties of biodegradable multiphase systems. Polymer 42: 6209–6219.

McGuire PA, Blackburn S (2008) Comments on "Residence time distribution in a corotating twin screw extruder" by J.P. Puaux, G. Bozga and A. Ainser [Chemical Engineering Science 55, 1641–1651], 63(15): 4033–4034.

Menges G, Bartilla T (1987) Polymerization of ε-caprolactam in an extruder: process analysis and aspects of industrial application. Polymer Engineering and Science 27(16): 1216–1220.

Meyuhas GS, Moses A, Reibenba Y, Tadmor Z (1973) Continuous polymerization in extruder reactor. Journal of Polymer Science, Part C: Polymer Letters 11: 103–111.

Moad G (2011) Chemical modification of starch by reactive extrusion. Progress in Polymer Science 36: 218–237.

Mohanty AK, Misra M, Hinrichsen G (2000) Biofibres, biodegradable polymers and biocomposites: an overview. Journal of Applied Polymer Science 276/277: 1–24.

Mohr WD, Saxton RL, Jepson CH (1957a) Mixing in laminar-flow systems. Industrial and Engineering Chemistry 49: 1855–1856.

Mohr WD, Saxton RL, Jepson CH (1957b) Theory of mixing in the single-screw extruder. Industrial and Engineering Chemistry 49: 1857–1862.

Morrison FA (2001) *Understanding Rheology*. New York: Oxford University Press.

Mottaz J, Bruyas L (2001) Optimized thermal performance in extrusion. In: Guy R (ed) *Extrusion Cooking – Technologies and Applications*. Cambridge, UK: Woodhead Publishing, pp. 51–82.

Mulvihill PJ (1992) Crystalline and liquid dextrose products: production, properties and applications. In: Schenck FW, Hebeda RE (eds) *Starch Hydrolysis Products – Worldwide Technology, Production and Applications*. New York: VCH Publishers, pp. 121–176.

Nebesny E (1992) Changes of carbohydrates and molecular structure of dextrins during enzymatic liquefaction of starch. Starch/Stärke 44: 398–401.

Oates CG (1997) Towards an understanding of starch granule structure and hydrolysis. Trends in Food Science and Technology 8: 375–382.

Orzechowska A (2004) The pulp and paper industry… this overview shows that times have changed. Pulp and Paper Canada 105(6): 19–21.

Ottino JM (1989) *The Kinematics of Mixing: Stretching, Chaos, and Transport*. New York: Cambridge University Press.

Ottino JM (1994) Mixing and chemical reactions, a tutorial. Chemical Engineering Science 49(24), Part A: 4005–4027.

Ottino JM, Ranz WE, Macosko CW (1981) A framework for description of mechanical mixing of fluids. AIChE Journal 27: 565–577.

Owusu-Ansah J, Van De Voort FR, Stanley DW (1983) Physicochemical changes in cornstarch as a function of extrusion variables. Cereal Chemistry 60(4): 310–324.

Padmanabhan M, Bhattacharya M (1991) Flow behaviour and exit pressures of corn meal under high-shear-high-temperature extrusion conditions using a slit die. Journal of Rheology 35: 315–343.

Pontoh J, Low NH (1995) Glucose syrup production from Indonesian palm and cassava starch. Food Research International 28: 379–385.

Poulesquen A, Vergnes B (2003) A study of residence time distribution in co-rotating twin screw extruders. Part I: Theoretical modeling. Polymer Engineering and Science 43(12): 1841–1848.

Pracella M, Rolla L, Chionna D, Galeski A (2002) Compatibilization and properties of poly(ethylene terephthalate)/polyethylene blends based on recycled materials. Macromolecular Chemistry and Physics 203: 1473–1485.

Puaux JP, Bozga G, Ainser A (2000) Residence time distribution in a corotating twin-screw extruder. Chemical Engineering Science 55: 1641–1651.

Ranz WE (1979) Applications of a stretch model to mixing, diffusion, and reaction in laminar and turbulent flows. AIChE Journal 25(1): 41–47.

Raquez JM, Narayan R, Dubois P (2008) Review: Recent advances in reactive extrusion processing of biodegradable polymer-based compositions. Macromolecular Materials and Engineering 293(6): 447–470.

Rauwendaal C (1991) Mixing in single extruders. In: Rauwendaal C (ed) *Mixing in Polymer Processing*. New York: Marcel Dekker, pp. 129–240.

Rauwendaal C (2001) *Polymer Extrusion*, 4th edn. Munich: Hanser Publishers.

Reeve A (1992) Starch hydrolysis: processes and equipment. In: Schenck FW, Hebeda RE (eds) *Starch Hydrolysis Products – Worldwide Technology, Production and Applications*. New York: VCH Publishers, pp. 79–120.

Reinikainen P, Suorti T, Olkku J, Mälkki Y, Linko P (1986) Extrusion cooking in enzymatic liquefaction of wheat starch. Starch/Stärke 38: 20–26.

Richardson TH, Tan X, Frey G, et al. (2002) A novel, high performance enzyme for starch liquefaction. Journal of Biological Chemistry 277: 26501–26507.

Rong Y, Sillick M, Gregson CM (2009) Determination of dextrose equivalent value and number average molecular weight of maltodextrin by osmometry. Journal of Food Science 74(1): C33–C40.

Roussel L, Vieille A, Billet I, Cheftel JC (1991) Sequential heat gelatinization and enzymatic hydrolysis of corn starch in an extrusion reactor. Optimization for a maximum dextrose. Lebensmittel-Wissenschaft & Technologie 24: 449–457.

Shearer G, Tzoganakis C (2000) The effects of kneading block design and operating conditions on distributive mixing in twin screw extruders. Polymer Engineering and Science 40(5): 1095–1106.

Slominska L (1993) Studies in the modification of enzymatic saccharification process. Starch/Stärke 45: 88–90.

Spencer RS, Wiley RM (1951) The mixing of very viscous liquids. Journal of Colloid Science 6(2): 133–145.

Stober KE, Amos JL (1950) Method for polymerizing styrene. US Patent 2,540,790.

Tadmor Z, Gogos CG (2006) *Principles of Polymer Processing*, 2nd edn. New York: John Wiley.

Tester RF, Qi X, Karkalas J (2006) Hydrolysis of native starches with amylases. Animal Feed Science and Technology 130: 39–54.

Todd DB (1975) Residence time distribution in twin-screw extruders. Polymer Engineering and Science 15: 437–443.

Tossavainen O, Hakulin S, Kervinen R, Myllymäki O, Linko P (1986) Neutralization of acid casein in a twin screw cooking extruder. Lebensmittel-Wissenschaft & Technologie 19: 443–447.

Tucker CL III (1991) Principles of mixing measurement. In: Rauwendaal C (ed) *Mixing in Polymer Processing*. New York: Marcel Dekker, pp. 101–127.

Ullrich M, Meisert E, Eitel A (1976) Process for the production of polyurethane elastomers. US Patent 3,963,679.

Underkofler LA, Denault LJ, Houe EF (1965) Enzymes in the starch industry. Starch/Stärke 6: 179–184.

Van Der Veen ME, Veelaert S, Van Der Goot AJ, Boom RM (2006) Starch hydrolysis under low water conditions: a conceptual process design. Journal of Food Engineering 75(2): 178–186.

Vergnes B, Berzin F (2004) Modelling of flow and chemistry in twin screw extruders. Plastics, Rubber and Composites 33: 409–415.

Vergnes B, Della Valle G, Tayeb J (1993) A specific slit die rheometer for extruded starchy products. Design, validation, and application to maize starch. Rheologica Acta 32: 465–476.

Vergnes B, Della Valle G, Delamare L (1998) A global computer software for polymer flows in corotating twin screw extruders. Polymer Engineering and Science 38(11): 1781–1792.

Wang SS, Chiang WC, Zheng XG, Zhao B, Cho MH, Yeh AI (1992) Application of an energy equivalent concept to the study of the kinetics of starch conversion during extrusion. In: Kokini JL, Ho CT, Karwe MV (eds) *Food Extrusion Science and Technology*. New York: Marcel Dekker, pp. 165–176.

Wang WJ, Powell AD, Oates CG (1996) Sago starch as a biomass source: raw sago starch hydrolysis by commercial enzymes. Bioresource Technology 55: 55–61.

Wenger LG, Rokey GJ, Ben-Gera I, Smith OM (1981) Enzymatic conversion of high moisture shear extruded and gelatinized grain material. US Patent 4,2806,058.

Wootton M, Weeden D, Munk N (1971) A rapid method for the estimation of starch gelatinization in processed foods. Food Technology Australia 23: 612–615.

Wyman CE (1975) Theoretical model for intermeshing twin screw extruders: axial velocity profile for shallow channels. Polymer Engineering and Science 15: 606–611.

Xanthos M (1992) *Reactive Extrusion. Principles and Practice.* New York: Hansen Publishers.

Xie F, Yu L, Liu H, Chen L (2006) Starch modification using reactive extrusion. Starch/Stärke 58: 131–139.

Zhao RX, Torley P, Halley PJ (2008) Emerging biodegradable materials: starch- and protein-based bio-nanocomposites. Journal of Materials Science 43: 3058–3071.

6 The Generic Extrusion Process III: Thermomechanical Cooking and Food Product Texturization

In the last three to four decades, extrusion-textured products have increased tremendously in the food and feed market, and food market analysts are predicting widespread development of new products in the coming years. Extrusion-textured food products result from the Generic Extrusion Process (GEP) III, which arose from a remarkable technology transfer of GEP I, which is described in Chapter 4. When applying GEP I to nature-derived polymers (starch, proteins) with water as the plasticizer, a new method of continuous food cooking emerged, called the Generic Extrusion Process III (or GEP III). GEP III cooks starch- and protein-based materials in the screw-barrel assembly of the extruder, and texturizes the resulting cooked melt in a die assembly in line with the extruder barrel, or in a separate processing unit downstream from the extruder barrel. As introduced in Chapter 1, GEP III is generically named as follows: thermomechanical cooking and food product texturization.

Extrusion processing of nature-derived polymers (from now on called "biopolymers") for the purpose of food cooking, commonly called "extrusion-cooking," has initiated the development of many new cereal-based food and feed products. Extrusion-cooking of cereal-based raw materials uses the two main process functions of GEP III: thermomechanical cooking of biopolymeric materials and texturization of the resulting extrusion-cooked melts. Extrusion-cooking processes convert formulated mixes composed of biopolymers with functional ingredients (such as texture- and flavor-aided ingredients) and take advantage of ancillary processing units (such as preconditioning, direct steam heating, vent stuffing, for instance) in order to fine-tune the end-use characteristics of the numerous extrusion-textured food products.

The two main process functions of GEP III are described qualitatively in the first part of this chapter. Unlike conventional hydrothermal cooking (such as dough baking or steam cooking), which runs under mild processing conditions (high water content, long residence time and moderate temperatures), extrusion-cooking typically operates under severe processing conditions (low water content and high temperature) somewhat alleviated by short residence times. Under these extreme processing conditions, biopolymers behave like polymeric melts, but the material transformation does not follow classic mechanisms such as gelatinization of starches and denaturation of proteins under moderate pressure, temperature and shear conditions and at high moisture content as those used in hydrothermal cooking. In addition to the thermal energy input which is defined by the time-temperature history, extrusion-cooking also incorporates mechanical energy to the material that results from the shearing and stretching of the material by the action of the screw-barrel assembly, leading to a thermomechanical conversion of biopolymers which is defined by the time-temperature-shear history in the extruder.

The mechanism of thermomechanical transformation of biopolymers is thus far more complex than transformations that occur during hydrothermal cooking processes. At the exit of the screw extruder, the thermomechanically cooked biopolymeric melt is shaped and cooled, and possibly textured in a die assembly. The mechanism of product texturization is specific to the extrusion-cooking process and results also from the time-temperature-shear history in the texturization die. The second part of this chapter gives an engineering analysis of the main process functions of GEP III: the mechanism of thermomechanical

Extrusion Processing Technology: Food and Non-Food Biomaterials, First Edition. Jean-Marie Bouvier and Osvaldo H. Campanella.

conversion of biopolymers and the mechanism of product texturization. The interrelation between both process functions is also covered, mainly focusing on the link between the process history and the product characteristics.

The productivity, versatility, and reliability of extrusion-cooking have made it one of the favorites and currently it is widely used in the cereal processing industry, both in food and feed areas, particularly when viscous biopolymeric materials, mixing of solids or liquid ingredients, or special texturization are involved. Major commercial applications have been introduced over the years which have required significant engineering development to design extrusion processing lines adapted to the specificities of extrusion-cooking technology. The third part of this chapter presents relevant industrial applications of extrusion-cooking technology and typical extrusion processing lines. A description of four case studies is provided.

- Extrusion processing of directly expanded cereals.
- Extrusion processing of pellet-to-flake breakfast cereals.
- Extrusion processing of aquafeeds.
- High moisture extrusion processing of protein-based materials.

6.1 Food extrusion-cooking: qualitative description

As introduced previously, extrusion-cooking is a unique form of food cooking and texturization because it converts biopolymeric raw materials under unusually severe process conditions with processing and product characteristics that depend on the application. In general, processing conditions are characterized by low water contents (in the range of 14–32% wwb), high temperatures (100–160°C) and short residence times (8–40 seconds). Due to these extreme processing conditions, extrusion-cooking is recognized as a high-temperature, short-time (HTST) cooking process.

In the screw-barrel assembly of a screw extruder, the raw materials are thermomechanically converted into a viscous melt and subjected to various processing operations, in order to obtain a cooked melt with the required composition and rheological properties. The cooked melt is further shaped and textured in the die and possibly in other downstream processing units. In general, extruded food products show cellular structures due to the steam-induced expansion of the cooked melt, either at the die exit for directly expanded products or in downstream puffing units for deferred expanded products. The extrudate structure can be optimized and is a function of the molecular and macromolecular characteristics of the biopolymer melt. The use of structure-modifying ingredients as well as the rheological behavior of the cooked melt resulting from the right choice of extrusion processing conditions, and possible additions of rheology-modifying ingredients, all affect the cellular characteristics of the extruded product.

Consequently, under thermomechanical processing conditions such as extrusion-cooking, it becomes obvious that the physical features of raw materials, as well as the composition of the formulation to be extrusion-cooked, are particularly important. Therefore, it is essential to present and discuss the functional role of each component in the formulation as well as the processing characteristics of extrusion-cooking, based on the long-term experience established by food extrusion users.

6.1.1 Thermomechanical cooking of biopolymer-based systems

6.1.1.1 Raw materials and formulation

Extrusion-textured food and feed products are made from a diverse range of raw materials. The components of the formulation have different functional roles which contribute to the quality profile of the final products, which is defined by their texture, appearance, flavor and taste, and nutritional profile. Extrusion-textured products have basic structured matrices that are formed by the biopolymers starch and protein in the raw materials, such as cereal and tuber flours, which are starch-rich feeds, or soy materials, wheat gluten, and fish meal which are protein-rich feeds. The basic structured matrix has functional ingredients with functions aimed at fine-tuning the product quality profile. Thus, the design of food extrusion formulations is rather complex due to the incorporation of ingredients which may be multifunctional.

It is worth reviewing the major ingredients of food formulations in extrusion-cooking. For this review, a classification of ingredients is based on their functional role, as shown in Table 6.1. Four different categories of ingredients can be distinguished.

6.1.1.1.1 Category I: matrix-forming materials

The structured matrix of extrusion-textured products is created by forming a viscous melt from the biopolymers, which is steam-expanded by water flashing off, thus

Table 6.1 Food extrusion formulation: classification of typical ingredients.

Ingredient	Category I	Category II	Category III	Category IV
Cereals (flours, grits)	X			X (flavor)
Potato (granules, flour)	X			X (flavor)
Proteins (flours, slurries)	X		X	X (flavor, color, nutrition)
Water		X		
Sugars (sucrose)		X		X (flavor)
Malt		X		X (flavor, color)
Milk powder				X (flavor, color)
Bran (or dietary fibers)			X	X (nutrition)
Vegetable oils and fats		X		X (nutrition)
Emulsifiers			X	
Sodium chloride				X (flavor)
$CaCO_3$, $NaHCO_3$			X	
Colorants				X (color)

Table 6.2 Typical starch and protein sources for the major extrusion-textured food products.

Extrusion-textured product	Starch source	Protein source
Human foods		
– Snacks (DX, pellets, multigrain chips[1])	Corn, potato, rice, wheat	
– Breakfast cereals (DX, flakes)	Corn, wheat, oat, rice	
– Crispy flat bread	Wheat, rice	
– Co-filled biscuits	Wheat, corn	
– Precooked flours (pregels, instant drinks, etc.)	Wheat, corn, rice	
– Textured vegetable protein		Soya, wheat gluten
– Wet fibrated protein (WFP)		Soya, wheat gluten, surimi, meat
Animal feeds		
– Dry pet foods	Wheat, corn	Meat, soya, gluten
– Pet treats (bone analogs, chew toys, etc.)		Meat, soya, gluten, casein
– Aquafeeds	Wheat, corn gluten	Fishmeal, gluten, soya

DX: directly expanded.
[1] Non-expanded intermediate products (low extent of gelatinization in the cooker-extruder).

promoting the formation of an alveolar structure leading to the production of a cellular solid after cooling. The characteristics of the process which enable the generation of the structured matrix to be generated are presented in sections 6.1.1.2 and 6.1.1.3. The biopolymeric melt forms the cell walls of steam bubbles and controls the growth and shrinkage of the cells depending upon the average polymer size as well as the rheological behavior and the physicochemical characteristics of the melt.

The matrix of extruded products may be formed from either starches or proteins. Table 6.2 presents starch and protein sources of the major types of extruded products. The main sources of starch are cereals and tuber crops like potato, whereas proteins are mainly provided by soybeans, cereals, and animal sources. The structure and functional properties of starch and proteins were described in Chapter 4 (refer to sections 4.1.2.1 and 4.1.2.2 for the description of their structural characteristics).

Starch materials are very good matrix-forming biopolymers because naturally they are mostly in an amorphous state at temperatures below their glass transition temperatures and change to a rubbery viscous melt for temperatures higher than their glass transition temperatures. As starch materials are composed of two major constituents, amylose (with a relatively low molecular weight of 10^5–10^6 g/mol) and amylopectin (with a relatively high molecular weight polymer of 10^7–10^9 g/mol) (see Chapter 4, section 4.1.2.1), it is worth discussing their respective potentials for melt expansion in extrusion-cooking. Chinnaswamy and Hanna (1988b) observed that the bulk density of extrusion-cooked starch-based materials (with different amylose/amylopectin ratios) decreased quasi-linearly from 300–350 kg.m^{-3} to approximately 150 kg.m^{-3} with increasing amylose content from 0% to 70%. In addition, Della Valle et al. (1997) found that at any temperature (in the range of 130–160°C) and moisture content (in the range of 20.5–27.5% wwb), the volumetric expansion of extrusion-cooked amylose/amylopectin mixes increased significantly when amylose content increased from 0% to 70%. The authors concluded that volumetric expansion was much higher for high amylose contents for melt viscosities of extrusion-cooked starch mixes in the approximate range of 200–500 Pa.s.

These two studies show clearly that volumetric expansion of extrudates is favored by amylose compared with its amylopectin counterpart. This shows that the average polymer size of amylose is more appropriate for large volumetric expansion, whereas the average polymer size of amylopectin is too large for optimum expansion. Amylopectin, which has a higher molecular weight and gives low expansion, requires reduction of its average polymer size to obtain an optimum expansion. This can be done during extrusion-cooking by use of mechanical shearing. It is worth noting that these results concern only the volumetric expansion of extrudates, with no consideration of the directions of expansion that are defined as radial and axial expansion. Thus, the extent of either radial or axial expansion is of paramount importance as they determine the final structure and texture of extrudates. In this respect, amylopectin and amylose both play a complementary role in the process to favor one direction of expansion or the other. This will be discussed later in this chapter (section 6.2.3).

Proteins at high concentrations are also used to form expanding matrices of extruded products. Amorphous globular proteins from soy and cereals can be dispersed from their native bodies into a free-flowing, expandable continuous melt, thanks to the shearing/mixing action of the extruder. Similarly, animal-sourced proteins such as meat slurries, poultry meal, and fishmeal can also be converted successfully into a continuous melt with a viscosity that allows some of the vaporizing steam to be retained, forming an alveolar-like structure.

It shsould be noted that starch-rich extruded products are highly expanded (snacks, breakfast cereals, flat bread, co-filled biscuits) whereas protein-rich extruded products show either intermediate expansion (textured vegetable protein) or low expansion (dry pet foods, aquafeeds). Pet treats and wet fibrated protein (WFP) are non-expanded, with well-defined chewy and fibrous textures, respectively.

6.1.1.1.2 Category II: rheology-modifying ingredients

Extrusion processability of biopolymers and successful melt expansion require dedicated ingredients with a functional role aimed at fine-tuning extrusion-cooking conditions (plasticizing effect) and rheological properties of the cooked biopolymeric melts (viscosity-modifying effect).

Table 6.1 shows four ingredients included in this category (water, sugars, malt, oils and fats). Water acts as the plasticizer in the extrusion-cooking process of biomaterials. As seen in Chapter 4 (section 4.3.3), an important decrease in the glass transition temperature (Tg) is observed for starch and proteins when water content increases from 0% to 30%. According to Flory's law (1953), the melting temperature of starch, Tm, also decreases when water content increases. Moreover, as seen in Chapter 4 (section 4.4.3; Eq. 4.26), the apparent viscosity of biopolymeric melts decreases when water content increases. Thus, water acts as a plasticizer of biopolymeric molecules, helping the formation of the melt, and can control the melt viscosity to levels that produce optimal expansion. Hence, water can be considered a key functional ingredient to fine-tune the processability of matrix-forming biopolymers.

Extrusion formulations may include sugars (in case of breakfast cereals, flat bread, co-filled biscuits, for instance) to confer a sweet taste to the extruded products. Sugars as well as malt also act as plasticizers and viscosity reducers. Sugars dissolve in the aqueous phase during the initial mixing stage of processing and their effect on the extrusion-cooking process depends on concentration. At low levels of addition (<5%), sugars have little effect on the biopolymeric melt. At high levels of addition (>5%), sugars usually replace part of the starch in the recipe, hence reducing the potential for expansion in the melt. In addition, the presence of sugars in the biopolymeric melt tends to increase the melting temperature of starch, Tm, and decrease the glass transition temperature, Tg, thus significantly favoring melt shrinkage during steam-induced expansion and consequently reducing melt expansion (refer to section 6.2.3.3 for a description of the shrinkage phenomenon). A solution to compensate for the effect of sugars on the melt expansion is to reduce added water, in order to restore the relative concentration of starch to water.

The presence of sugars in the formulation reduces the glass transition temperature of biopolymeric melts. Kalichevsky et al. (1993) showed that the degree of plasticization of amylopectin by sugars appears to increase in the order: sucrose < glucose < fructose. Sucrose is a disaccharide and has a higher molecular weight than monosaccharides, and a higher *Tg* (70°C) than glucose (38°C) and fructose (7°C). Hence, sucrose has a reduced plasticizing effect when compared with the monosaccharides and for that reason it is often used in extrusion formulations. Kalichevsky et al. (1993) also observed that at low water contents (in the range of 10–20%), the presence of 10% sucrose reduces the *Tg* of amylopectin by 10–43°C.

Oils and fats act as lubricants. They are rapidly dispersed into fine oil droplets (a few microns in size) in the biopolymeric melts during the extrusion-cooking process, and are trapped in the continuous phase. The structure of the system resembles that of an oil/water emulsion where the dispersed oil droplets weaken the structure of the continuous melt. Thus, the resulting oil-in-water viscous melt shows a lower viscosity compared to that of the same fat-free melt. Oils and fats significantly affect the extrusion-cooking process at low levels (<3–4%); at high levels (>5–6%) they tend to protect the biopolymers from the shearing effect to such an extent that low or no expansion is obtained because the undisturbed molecular structure is not favoring the expansion process. For instance, when oat flour is extrusion-cooked, this phenomenon may be observed if a standard screw profile is used, because oat flour has a relatively high level of lipid content (approximately 9%). This phenomenon is also observed with partially defatted soy meal when fat content exceeds 6%, to produce textured vegetable protein. A practical solution to this issue would be to increase the shearing action of the screw profile, in order to restore the mechanical energy input and the extent of thermomechanical cooking of the fat-rich biopolymeric raw material.

6.1.1.1.3 Category III: expansion-modifying ingredients

This category includes functional ingredients which affect melt expansion in the extrusion-cooking process: the overall expansion, or volumetric expansion, as well as the bubble size distribution of expanded products. Starch-based extruded products show several dispersed phases lying within the continuous phase of starch polymers. Dispersed phases are formed by proteins (derived from cereals, pulses or animal proteins), by fibrous materials (such as bran from grains), and by insoluble nucleating agents (such as calcium carbonate and minerals).

In any starch-based formulation, the different types of protein present form dispersed phases in the resulting biopolymeric melt and in extruded products: water-soluble proteins (such as albumins) which are denatured and coagulated, hydrated globular proteins, cereal proteins, meat proteins. Under low water content extrusion-cooking conditions (<30%), the formed particle proteins have reduced size (<20 μm) due to the shearing action of the screws, and are dispersed in the continuous phase of the melt by the mixing action of the screw configuration. Fibrous materials remain physically stable during extrusion-cooking. The presence of dispersed phases (particles of proteins, bran) in the continuous melt significantly affects the potential of the starch melt for expansion. In fact, dispersed phases tend to disrupt the cell walls of the expanding melt, reducing volumetric expansion and leading to low cell size structures. Nucleating agents such as insoluble salts, minerals and even bran particles also affect melt expansion, as they favor bubble nucleation at the die exit, and consequently increase the number of bubbles appearing in the biopolymeric melt exiting the forming die of the extruder.

Adding emulsifiers when extrusion-cooking a starch-based formulation leads to the formation of amylose-lipid complexes, which occurs during die melt cooling. Amylose-lipid complexes, which affect expansion, are insoluble in water and form crystallites in the continuous amorphous phase of molten starch (see section 6.2.3.2.2.2).

6.1.1.1.4 Category IV: quality-enhancing ingredients

As for many food products, the quality profile of extruded products is defined by sensory acceptability factors such as appearance, flavor, and texture (Bourne, 1982), and nutritional characteristics. The texture of extruded products is closely related to their cellular structures which are significantly affected by ingredients in categories I, II, and III. Appearance (as defined by the shape, color, surface aspect, etc.), flavor (or taste), and nutrition are also very important factors which add value to extruded products. Thus, besides texture, extrusion formulations must contain functional ingredients which optimize the quality profile in terms of the properties described below.

- *Taste* (food products for humans) or palatability (feed products for animals) can be enhanced by flavoring components which already play a functional role related to the previous categories; this concerns potato flour (or potato granules), protein-based materials (such as meat slurries in pet foods, fishmeal in aquafeeds), sugars, malt, milk powder, and sodium chloride. Specific flavoring ingredients are usually added to the extruded products downstream of the extrusion-cooking and drying processes in a coating unit. Flavoring ingredients may also be added directly to the extrusion formulation, particularly when the time-temperature history is compatible with the thermal stability of those ingredients. This processing approach is called "flavor-in extrusion-cooking" and particularly concerns extrusion processes which produce non-expanded products at the die exit, such as multigrain chips, wet fibrated protein, pet treats, and semi-moist pet foods. Readers interested in a detailed description of flavoring in extrusion-cooking should refer to the article by Maga (1989).

• *Color* can be handled by adding color precursors such as reducing sugars and proteins, that may consist of malt and milk powder, in order to favor non-enzymatic browning or Maillard reaction products. Colorants such as carotenoid pigments (i.e. β-carotene, annatto, etc.) and meat-like pigments can be added to the extrusion formulation to reinforce the color of end-products. For bicolored snacks or fruit loop cereals, natural food colors are added to the cooked biopolymeric melt at the die level before expansion. Readers interested in a detailed description of coloring in extrusion-cooking should refer to the article by Berset (1989).

• *Nutrition* can be improved by using nutritive matrix-forming raw materials (such as cereal flours, proteins) and adding specific ingredients (such as dietary fibers, vegetable oils, etc.), which bring the required nutritional properties to the extruded products. Readers interested in a detailed description of nutrition in extrusion-cooking should refer to the article by Asp and Björck (1989).

As seen in Table 6.1, most ingredients present in extrusion formulations are multifunctional. The multifunctionality and specificity of the extrusion-cooking process mean that the design of extrusion formulations is a complex subject whose understanding and control often result from long-term experience.

6.1.1.2 Process functions in the screw-barrel assembly of screw extruders

Formulations using only starch and proteins do not provide optimum expansion. Optimum extrusion processing conditions must be selected in accordance with the macromolecular structure and physicochemical properties of the biopolymers used in the formulation. In food extrusion-cooking, the screw-barrel assembly (single screw and intermeshing co-rotating twin screw extruders) is configured based on the principles presented in Chapter 2 (see section 2.1.2). The main streams of the food formulation and water are fed separately into the feed throat of the feeding section. The powdery raw mix is delivered by a feeder (either volumetric or gravimetric) through the opening of the feeding barrel while water is added by a metering pump through an injection port specific for liquids. The powdery raw mix and water are further mixed and transformed in the following sections of the screw-barrel assembly (compression, melting, mixing, cooking, cooling, etc.) where secondary streams (solids or liquids) may be added on-line depending upon process requirements.

Thermomechanical cooking occurs in the screw-barrel assembly, where the conversion of biopolymers is usually conducted at low moisture contents (below 30–32% wet basis) for the majority of food extrusion applications. In these conditions, biopolymer processing leads to a series of structural changes. In the case of starchy materials which are composed of two types of semi-crystalline macromolecules stored in granules, amylose and amylopectin, the structural changes include the loss of crystallinity followed by the loss of granular integrity, depolymerization of starch polymers (or dextrinization), and the formation of complexes between amylose and lipids. These changes result from the shear- and thermal-induced conversion of the starch, at relatively high temperatures (usually above 100°C) and short residence times (below 40 s), which leads to a cooked material with particular properties which are different from those of starch processed by conventional hydrothermal cooking.

Figure 6.1 shows the starch granule conversion in extrusion-cooking. This process may vary from compaction to complete disruption and degradation. A few formulations (such as precooked cereal flours, multigrain chips) may require water contents higher than 30–32% (in the range of 30–40% wet basis) in order to reduce the effects of mechanical energy input and starch dextrinization. By varying the relative contributions of the mechanical and thermal components of the thermomechanical action, it is possible to process numerous formulations and produce a large number of products. However, the extruder must be properly designed to obtain the required starch conversion through the setting of suitable process conditions (L/D ratio of the screw-barrel assembly, screw profile, screw speed, moisture content, temperature profile, in particular). Thermomechanical extrusion processing of starch-rich raw materials leads to a rather homogeneous, single-phase melt with the required rheological properties due to the relative uniformity of its polymeric

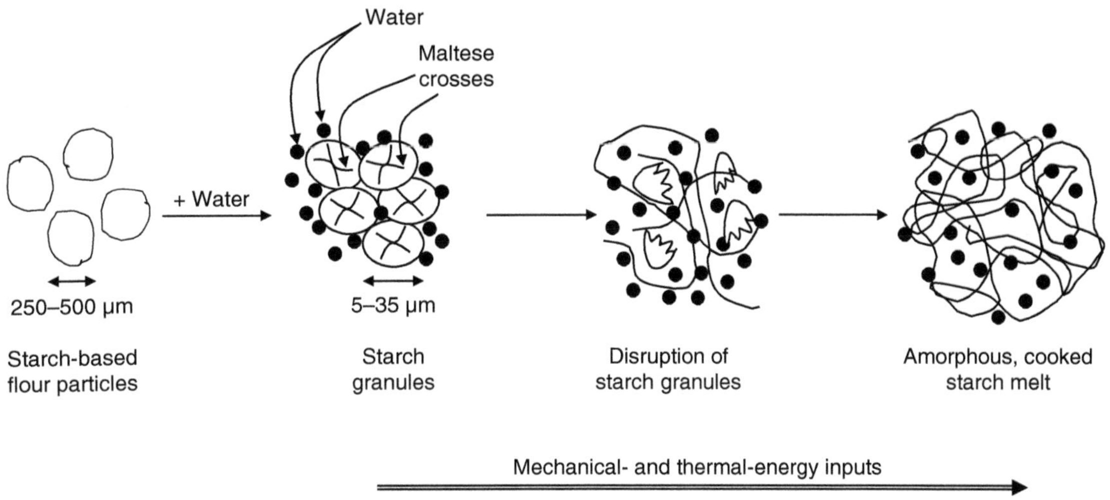

Figure 6.1 Thermomechanical cooking of starch-based materials (water content <30–32%; temperature >100°C; residence time <40 s).

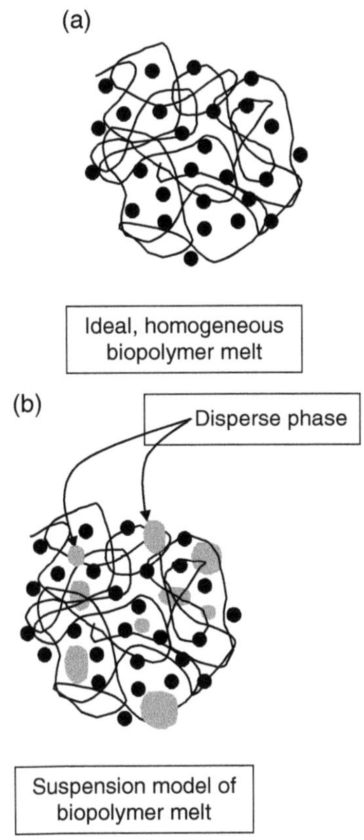

Figure 6.2 Schematic representation of structural characteristics of biopolymer melts: ideal, homogeneous biopolymer melt (a); suspension model of biopolymer melt (b).

macromolecular structure, mainly formed by one type of monomer and stabilized by intermolecular forces such as hydrogen bonds (Figure 6.2a).

Protein-rich materials are based on proteins which may be mixed with other polymers such as polysaccharides (soy-based raw materials, for instance). Proteins contain a wider range of functional chemical groups than polysaccharides and are therefore more reactive. The polymeric system shows a wide range of interaction energies for its stabilization (hydrophobic, cation-mediated electrostatic interactions and covalent bonds in particular). This leads to very complex polymeric systems in which the physicochemical changes occurring in the extrusion-cooking process, especially at high temperature (120–165°C) and low moisture content (approximately 22–30%), are not well described. It is generally agreed that proteins exhibit the following structural modifications in the screw-barrel assembly of a screw extruder (Mitchell & Areas, 1992).

- *Denaturation*: unfolding of globular proteins, with consequent loss of secondary and tertiary structures.
- *Association* (or aggregation), due to protein–protein interactions (such as disulfide bonds, hydrophobic and electrostatic interactions), which leads to protein insolubilization. Association occurs when heating the material, and the major interaction forces involved in this thermal aggregation are of hydrophobic and ionic nature.
- *Disruption* of some or all aggregates caused by heat and shear, to form concentrated solutions or a protein-rich melt phase. The amount of insoluble material in the

continuous phase depends on the extrusion temperature and water content. Obviously, insoluble material increases with increasing extrusion temperature and decreasing water content.

- Possible *formation of non-disulfide covalent bonds* at high temperatures.

The extent of these structural changes varies significantly depending upon the type of proteins. For instance, casein can be extruded at temperatures as low as 80°C and soy proteins are extruded successfully at temperatures of at least 120°C. But blood protein, in contrast, cannot be extruded even at temperatures as high as 180°C (Areas, 1992). It is worth noting that the extrusion process for wet fibrated protein (WFP), called high moisture extrusion-cooking (HMEC), does operate at high moisture contents (in the range of 65–75% wet basis), in order to obtain a free-flowing continuous protein-based melt.

The resulting molten material which flows through the extruder is rather heterogeneous as it may consist of mixed solutions of different polymers (polysaccharides and proteins), and/or mixed solutions having unequal conformations of the same protein (denatured and non-denatured), and/or a solution having insoluble material in particulate form which could be present initially in the raw material or arise during extrusion processing (e.g. process-generated protein aggregates which were described on p.249). Thus, the molten material can be assimilated to a water-in-water emulsion which no longer behaves as a single-phase melt, but more as a two-phase melt with characteristics and composition that depend upon the raw materials and the processing conditions (Mitchell & Areas, 1992).

Mitchell and Areas (1992) proposed a model, called the "suspension model," for the extrusion of biopolymers, considering the biopolymeric melt as a two-phase system. This model is an extended version of the mechanism of protein texturization of Tolstoguzov (1988). The suspension model considers that the extruded biopolymeric melt consists of a homogeneous continuous phase, with a particulate insoluble phase (Figure 6.2b). The insoluble material can be present initially in the feed mix (such as bran fibers, insoluble salts) or arise during extrusion processing (such as process-generated protein aggregates, crystallites, amylose–lipid complexes, possible crystalline ungelatinized aggregates). Insoluble particulate materials in biopolymeric melts act as fillers in the continuous phase and, consequently, there is a critical volume fraction of insoluble material above which the molten material cannot flow in the extruder and cannot form a final coherent structure.

As a first approximation, considering the two-phase system as a close packing of spheres, a critical volume fraction of 0.64 can be obtained. It turns out that the higher the volume fraction of insoluble material, the higher the apparent viscosity of the two-phase melt, and the lower the ability of the two-phase melt to flow in the screw-barrel assembly and provide a coherent product structure necessary for optimum processing and subsequent expansion. The suspension model predicts that the success of extrusion-cooking of biopolymeric systems depends closely on the amount of insoluble particulate material initially present in the feed or produced under the extrusion process. Extrusion processing of starch-rich formulations with low-volume fractions of the particulate phase results in homogeneous biopolymeric melts which are easier to process. Protein-rich formulations, whose volume fraction of the particulate phase is highly dependent on the type of protein and the extrusion processing conditions, require special care when formulating the feed mix and selecting the extrusion-cooking conditions. Soy-based formulations generate a negligible amount of particulate phase under standard extrusion conditions, and thus these formulations are processed easily as well as soy flours with approximately 50% protein content. Soy isolate, however, with approximately 90% protein content, may require special extrusion conditions that limit the generation of particulate phase. Blood plasma protein cannot be extrusion-cooked properly, probably because the volume fraction of insoluble material generated by extrusion conditions exceeds the critical value.

It is noteworthy that the presence of reducing agents would enhance protein dissociation in the screw-barrel assembly and thus facilitate the flow of the melt. The addition of lipids at low levels (<5%) may be beneficial to protein extrusion processing, as lipids may prevent the formation of insoluble material by binding with the hydrophobic sites of proteins. The use of polysaccharides in the feed mix enhances the processability of protein-rich formulations, decreasing the concentration of the insoluble particulate phase and reinforcing the continuous phase of the polymeric melt.

Figure 6.2a shows an ideal melt which results from homogeneously dispersed random coil biopolymers interacting through entanglements, and plasticized by water molecules. It is obtained through loss of order and subsequent loss of granule integrity (starch conversion), and loss of non-covalent (hydrogen and hydrophobic interactions) and covalent (disulfide) bonds between protein molecules. Figure 6.2b presents the suspension model of the melt which consists of a continuous phase and a disperse phase with insoluble particles having lower water content than the surrounding polymeric matrix.

6.1.1.3 Ancillary processing units

Thermomechanical cooking of biopolymers requires significant process assistance to extruders, provided by ancillary equipment which was described in Chapter 2 (see section 2.3). This section focuses on ancillary processing units which allow the thermal energy input to be increased, and the moisture and temperature of the melt to be controlled. Increases in thermal energy input can be provided mainly by two ancillary processing units, preconditioning of raw material upstream to the extruder and direct steam heating in the cooking section of the extruder, while the control of moisture and temperature of the melt can be enhanced by the assistance of degassing and vent stuffing in a dedicated processing section of the screw-barrel assembly, before die texturization. Figure 6.3 shows a typical extrusion processing unit equipped with its three ancillary processing units; they can be installed easily on modular designs of screw-barrel assemblies of single screw and intermeshing co-rotating twin screw extruders.

6.1.1.3.1 Preconditioning

Preconditioning of raw materials is used in thermomechanical cooking processes with the aim of reinforcing thermal energy input, in order to limit the effect of shearing and consequently the extent of starch dextrinization and protein aggregation. The purpose of preconditioning is to preheat and prehydrate biopolymer-based raw materials such as flours and grits by mixing them with steam and water. In practice, steam and liquid water are added in the feeding section of the preconditioner; steam is injected at the bottom whereas liquid water is added at the top. The slightly superheated steam heats up the particles of raw material and its condensation contributes to hydrating the particles. Complementary liquid water is necessary to obtain the required water content of the preconditioned mix needed for further extrusion processing. During preconditioning, steam, liquid water and particles of the raw materials must be well mixed. This is handled by double-shaft counter-rotating preconditioners, which offer better mixing and hence more homogeneous prehydrated mixes than single-shaft preconditioners.

Preconditioners are characterized primarily by their free volume, filling ratio, and specific capacity. The filling ratio expresses the ratio of the mix volume (based on mix bulk density) to the free volume of the preconditioner, the mix volume being measured when the preconditioner is stopped. The specific preconditioning capacity (*SCP*) expresses the ratio between the mass flow rate of mix and the volume of the preconditioner. In practice, *SCP* ranges from 1 to 6 kg/h.L. In food extrusion processing, the filling ratio of the preconditioner (F_r) usually ranges from 40% to 55%, and the average residence time ($\bar{t}$) may vary between 2 and 15 minutes depending upon the process requirements. $\bar{t}$ and *SCP* can be determined by the following equations:

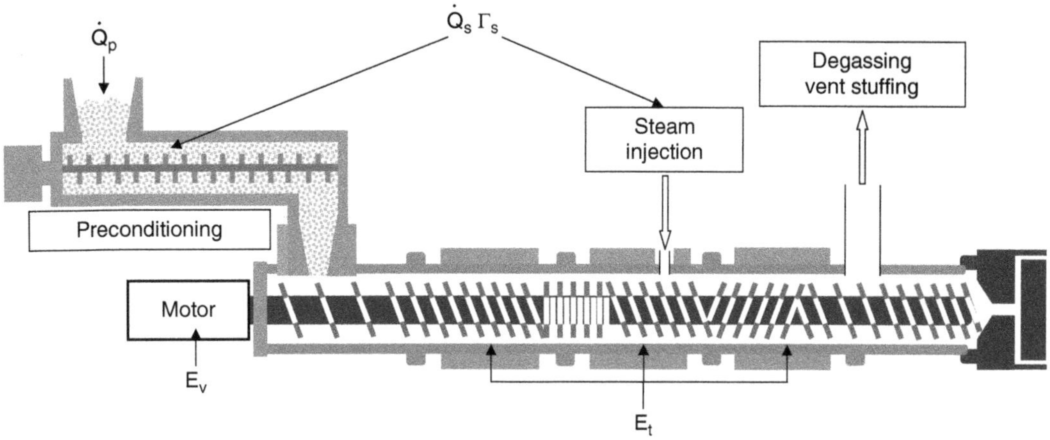

Figure 6.3 Typical food extrusion-cooker with three ancillary processing units (preconditioning, direct steam injection, degassing/vent stuffing).

$$\bar{t} = \frac{V_p F_r \rho_f}{\dot{Q}_p} \quad (6.1)$$

$$SCP = \frac{\dot{Q}_p}{V_p} = \frac{F_r \rho_f}{\bar{t}} \quad (6.2)$$

where V_p, ρ_f, $\dot{Q}_p$, are the volume of the preconditioner, the bulk density of the feed mix particles and the mass flow rate of feed mix in the preconditioner (including the water absorbed by the mix in the preconditioner), respectively.

The preconditioning conditions can be governed by the paddle configuration and operating variables (throughput of feed mix, shaft speed, temperature, and water flow rate). Varying paddle angle in any direction allows control of the intensity of axial flow of particles in the preconditioner (forward flow and reverse flow, when paddle angle is positive and negative, respectively). This in turn helps to combine axial and radial components of mixing. Paddle configuration results from the arrangement of forward and reverse sections to finally optimize the residence time, filling ratio, and mixing efficiency in the preconditioner. For a given throughput, the shaft speed and the temperature in the preconditioner are usually in the ranges of 150–250 rpm and 75–90°C, respectively while the water content of the hydrated feed mix exiting the preconditioner is generally between 18% to 24%, depending upon the associated extrusion-cooking operation. Depending on the water content of the preconditioned mix, preconditioning allows the glass transition temperature of biopolymers to be reached, thus decreasing the shearing stress in the melting zone of the screw-barrel assembly and preserving the macromolecular integrity of the biopolymers.

Preconditioning is generally used in extrusion-cooking processes which produce precooked flours, pellet-to-flake cereals, snack pellets, dry pet foods, and aquafeeds.

6.1.1.3.2 Direct steam heating

This consists of injecting slightly superheated steam in direct contact with the material flowing into the screw-barrel assembly. The steam is injected in a zone of relatively low pressure located between the melting and cooking sections. Steam brings thermal energy to the extrusion-cooking process, which enables a reduction in the processing temperature and controls the full conversion of biopolymers. Direct steam heating is more efficient when processing protein-rich formulations, as it contributes to controlling protein–protein association and consequently the volume fraction of insoluble particulate material in the biopolymeric melt. Direct steam heating is currently used in dry pet food extrusion processing and aquafeed extrusion processing, by incorporation of steam injection kits in the equipment.

6.1.1.3.3 Degassing and vent stuffing

During degassing and vent stuffing, some moisture flashes off and is extracted quickly from the biopolymeric melt. Simultaneously, the melt temperature decreases down to approximately 100°C, or even lower values if degassing and vent stuffing are carried out under vacuum. Volatile components such as off-flavors are also extracted during degassing and vent stuffing. It is worth pointing out that flash degassing results in a rapid increase of the apparent viscosity of the biopolymeric melt due to the reduction in moisture content and temperature. Thus, it is important to account for it when designing the pumping section of the screw-barrel assembly and the die assembly. Flash degassing is often applied in snack pellets extrusion processing which requires accurate control of the melt temperature and viscosity for the next processing operation, which is the die forming. Vent stuffing is currently used in pet food and aquafeed extrusion processing.

6.1.1.4 Energy inputs

The extent of thermomechanical cooking is defined by the net energy input, E_T, provided to the product in the extrusion system (i.e. in the preconditioner and the screw extruder). Schematically indicated in Figure 6.3, energy inputs are:

- the viscous dissipation of the net mechanical energy used to turn the screw(s), E_v. E_v is determined by the drive power, as follows: $E_v = M_d \cdot \omega$; where M_d and ω are the screw torque (N.m) and the screw speed (rads/s)
- the heat transferred through the extruder barrel, E_t
- the latent heat from steam, $\dot{Q}_s \Gamma_s$, injected directly into the mix in the preconditioner and/or through the extruder barrel.

Therefore:

$$E_T = \frac{E_v + E_t + \dot{Q}_s \Gamma_s}{\dot{Q}} \quad (6.3)$$

where $\dot{Q}_s$, Γ_s and $\dot{Q}$ are the total mass flow rate of steam, the latent heat of steam and the mass flow rate of feed mix (including the water absorbed by the mix in the preconditioner and the screw extruder), respectively.

Specific mechanical energy (SME) and specific thermal energy (STE) can be calculated as follows:

$$SME = \frac{E_V}{\dot{Q}} \quad (6.4)$$

$$STE = \frac{\dot{Q}_s \Gamma_s + E_t}{\dot{Q}} \quad (6.5)$$

SME and STE are expressed in $kJ.kg^{-1}$ or in $W.h.kg^{-1}$ ($1\ W.h.kg^{-1} = 3.6\ kJ.kg^{-1}$). SME values range from 200 to 650 kJ/kg for most extrusion-cooking applications. SME is a key process parameter as it reflects the extrusion-cooking conditions through screw speed, moisture content, screw configuration, etc., and it is commonly used to adjust the product conversion, to optimize and scale up the process appropriately. From the process standpoint, it is worth noting that process SME is proportional to three factors: the average apparent viscosity of the melt, $\bar{\mu}_a$, the square of the average shear rate, $\bar{\dot{\gamma}}^2$, and the average residence time in the screw-barrel assembly, $\bar{t}$, as follows:

$$SME \propto \bar{\mu}_a \bar{\dot{\gamma}}^2 \bar{t} \quad (6.6)$$

This means that, depending on the relative values of these three factors, in theory different thermomechanical conversion profiles can be obtained for the same SME. In fact, for a given food extrusion application, at constant SME and melt apparent viscosity, thermomechanical processing conditions may vary from relatively high shear rate and short time conditions to relatively low shear rate and long time conditions. In the first case, the process applies relatively high mechanical energy input whereas in the second case, the process uses relatively high thermal energy input. Consequently, though the process SME is constant, the functional properties of resulting extrudates may vary significantly, depending upon the ratio of mechanical to thermal energy input. For instance, this occurs when comparing physical and textural characteristics of directly expanded products produced in very short single screw extruders (such as Collet extruders) and those obtained in typical configurations of both single screw and intermeshing co-rotating twin screw extruders (with longer screw-barrel assemblies).

Table 6.3 gives an overview of typical food extrusion processing conditions (ranges of SME, L/D, water content, screw speed, average residence time and die melt temperature) for the major extrusion-textured food products. In general, high SME extrusion-cooking processes are carried out on short extruder lengths (low L/D), at relatively low water contents and short residence times. As the SME input of extrusion-cooking processes decreases with concomitant increase of the thermal energy input, both L/D and water content used for the process increase. Very high-capacity extrusion-cooking processes (such as dry pet foods and aquafeeds), which use high levels of thermal energy inputs and moderate or low SME inputs, show rather low residence times and high screw speeds. The WFP extrusion process, characterized by low levels

Table 6.3 Typical processing conditions for the major extrusion-textured food products.

	Typical ranges of extrusion processing conditions						
Range of SME ($W.h.kg^{-1}$)	L/D[1]	Water content (% wwb)	Screw speed (rpm)	Residence time (s)	Melt temperature (°C)[2]	Products
120 and over	8–16	14–20	350–550	8–20	130–160	DX snacks, DX breakfast cereals, crispy flat bread, textured vegetable protein, co-filled biscuits
80–120	24–32	22–30	150–250	25–30	85–120	Snack pellets, cereal flakes, precooked flours, pet treats
60–80	20–28	26–32	150–200	20–30	50–80	Multigrain chips[3]
	18–24	20–25	500–800	10–16	100–140	Dry pet foods
60 and below	20–28	22–26	500–800	8–12	100–140	Aquafeeds, semi-moist pet foods
	26–36	65–75	350–450	25–40	155–165	Wet fibrated protein

[1] Length-to-diameter ratio of the screw-barrel assembly.
[2] Melt temperature range in the die assembly.
[3] Non-expanded intermediate products (low extent of gelatinization in the cooker-extruder).
DX, directly expanded; SME, specific mechanical energy.

of SME input (below 60 W.h.kg^{-1}), but high thermal energy input, operates at very high water content (over 65% wet basis) and relatively high temperature (over 150°C). These conditions produce a protein-based melt that is homogeneous with a fibrous texture.

6.1.2 Texturization of extrusion-cooked melts

Continuous thermomechanical cooking of biopolymeric melts is generally combined with another unit operation to shape and texturize the product. Two major possibilities are considered.

- *On-line product shaping and texturization*, carried out by direct expansion through a die assembly that terminates the screw-barrel assembly of the screw extruder; this is called "direct expansion extrusion processing" (or DX extrusion processing). This process applies to numerous directly expanded extrusion-cooked products (DX snacks and ready-to-eat [RTE] cereals, crispy flat bread, co-filled biscuits, some precooked flours, textured vegetable protein, dry pet foods, aquafeed pellets).
- *Downstream product shaping and texturization*, carried out in different unit operations of the processing line. This process applies to a few products such as pellet-to-flake cereals, snack pellets, pet treats, multigrain chips, and fibrated vegetable protein.

6.1.2.1 On-line product shaping and texturization

During on-line shaping and texturization, the cooking section of the screw extruder brings the cooked melt to a rheological state compatible with satisfactory die shaping and texturization. In the pumping section of the screw-barrel assembly, the pressure and temperature of the biopolymeric melts are about 30–160 bar and 130–160°C, respectively. In such conditions, water is in liquid state and perfectly mixed with the melt. When it emerges from the die, the cooked melt suddenly goes from an ambient high pressure to atmospheric pressure. This pressure drop causes extensive flashing off of internal moisture and the resulting water vapor pressure causes the melt to expand.

As seen in Figure 6.4, the expansion phenomenon at the die exit comprises the nucleation and growth of water vapor bubbles in the extrudate matrix. The elasticity and viscosity of the matrix allow the melt to swell and expand very quickly. At the early stages of expansion, bubbles grow independently and do not collapse but bubble rupture soon occurs because the loss of cell wall resistance can no longer sustain the water vapor pressure inside the bubble, which results in water vapor loss and the formation of an open foam structure in the extrudate. The heat necessary to vaporize liquid water causes very fast evaporative cooling of the melt, down to approximately 100°C. Based on the energy balance, several researchers have calculated the volume of steam release during the water flashing off and

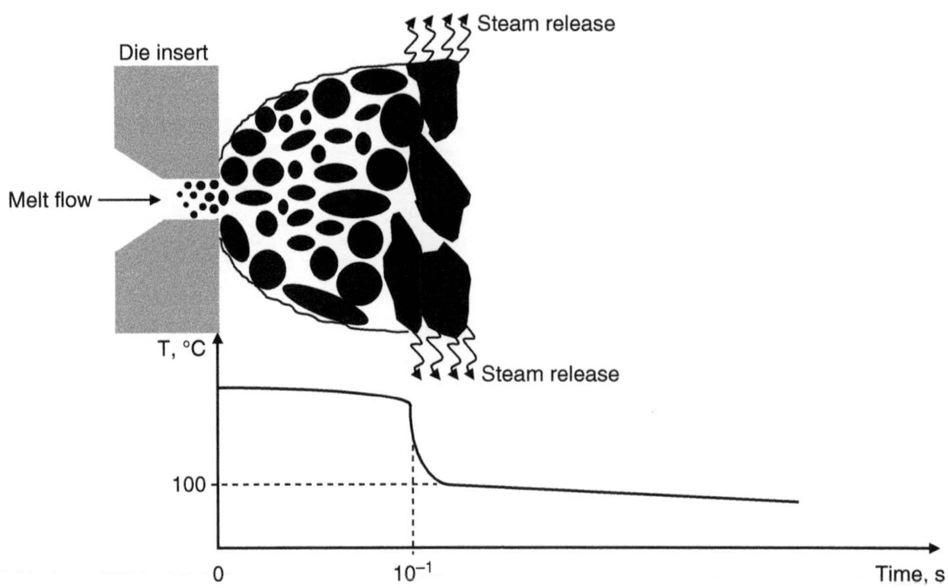

Figure 6.4 Schematic representation of steam-induced expansion of biopolymeric melt at the die exit.

compared it to the total cell volume of the final expanded product. They have shown that the total cell volume represents less than 60% of the volume of steam release (Tharrault, 1992). As melt expands at the die exit, the pressure and temperature drops occur quasi-instantaneously (that is, within about 0.1 s). After completion of steam release, the expanded product further cools down due to convective heat exchange with the ambient air.

Based on visual observations of melt expansion at the die exit, it must be noted that though water flash-off drives expansion and shaping, it does not determine the final volume, shape, and cell structure of directly expanded products. Thus, the expansion mechanism requires thorough investigation in order to better control the textural properties of extruded food products. This is presented in section 6.2.3.

6.1.2.2 Downstream product shaping and texturization

In this case, the screw-barrel assembly incorporates a forming die. The extruder cooks the raw mix and delivers the resulting melt to the die assembly. There, two main processing configurations can be distinguished.

- The die assembly continuously forms non-expanded, non-textured semi-products, and the final texture is defined in downstream processing units. For instance, in the production of pellet-to-flake cereals, snacks pellets and multigrain chips, the extruder-die system supplies preformed, cooked starch-based melts which are further processed to finalize the shape and develop the texture of the end-products. In these different cases, downstream processing units may be critical as they contribute importantly to the final quality attributes of extrusion-textured products.
- The die assembly continuously forms and texturizes non-expanded products whereas downstream processing units are used to finalize the shape and appearance of the end-products (such as pet treats and fibrated vegetable protein). In these cases, downstream processing units are less critical because their contribution to the final quality attributes of extrusion-textured products is less important.

6.2 Engineering analysis of process functions

The qualitative description of food extrusion-cooking, as presented in section 6.1, involves important process functions. The engineering analysis of these process functions plays a fundamental role in the process performance and product quality, specifically surface aspect, shape, and texture. For most of the food extruded products, performance and quality factors depend strongly on the thermomechanical cooking supplied in the process, including preconditioning and direct steam heating, as well as on die shaping and texturization. It is therefore worth analyzing in depth the unit operations involved in extrusion-cooking, and particularly the physicochemical mechanisms that govern their process functions and their influence on the design of related equipment. The purpose of this section is to focus on the following process functions.

- Preconditioning
- Extrusion-cooking
- Steam-induced die texturization

An expansion chart of directly expanded food extruded products is discussed, as a guide to handling extrusion-cooking conditions, in relation to product structural characteristics.

6.2.1 Preconditioning

The designs of preconditioners were presented in Chapter 2 (see section 2.3.1.2). The aim of this section is to analyze the operation of preconditioners, in terms of preconditioning of biopolymers. Preconditioning has long been associated with extrusion-cooking processes. It was first used in feed extrusion processing specifically for the processing of full-fat soy, dry pet foods and aquafeeds, to overcome process limitations of single screw extruders such as low flexibility and mixing efficiency. Currently, preconditioning is frequently introduced in many other applications relative to food and feed extrusion processing with both single screw and intermeshing co-rotating twin screw extruders. It is recognized as contributing positively to extrusion process performance and the quality of food extruded products. Benefits include:

- preheating, hydration, and possibly precooking of biopolymer-based raw materials, through significant thermal energy input
- reduction of shear stress in the melting section of the screw extruder, hence reducing macromolecular degradation of biopolymers and loss of nutrients due to the high shear environment, and enhancing product quality
- increase of extruder throughput, hence increasing productivity
- increasing the lifetime of screws and barrels through hydration of raw materials, which softens the feeds and reduces their abrasiveness, hence decreasing wear costs.

Although preconditioning is often used, it has been developed and applied using empirical approaches, with rather limited understanding of its governing process functions. Based on the qualitative description of preconditioning presented in section 6.1.1.3, three main process factors need thorough engineering investigation: thermal energy requirements, rate of heating and hydration time, and residence time distribution.

6.2.1.1 Thermal energy requirements

An important function of preconditioning is to heat up the raw materials (cereal-based flours or grits) to the temperature required for the subsequent extrusion-cooking operation. The preconditioning temperature is governed by the energy balance in the preconditioner where the temperature of the raw mix and the liquid water increases from the inlet temperature (usually ambient temperature) to the targeted exit temperature of the preconditioned mix (in the range of 80–90°C). The energy utilized to raise that temperature comes from the latent heat supplied by the steam (Figure 6.5). Assuming that the preconditioner operates adiabatically, that all steam is converted into liquid water, and that the outlet temperature of the preconditioned mix is around 85°C, an energy balance applied to the preconditioner allows estimation of the amount of steam required (as expressed in kg steam/100 kg dry mix), as a function of moisture content of the mix at the exit of the preconditioner. This estimation represents the minimum amount of steam supply and directly defines the outlet temperature of the preconditioned mix. In practice, an extra amount of steam (from 10% to 20%) is needed to account for heat energy losses in the preconditioner.

6.2.1.2 Rate of heating and hydration

In preconditioning, heat and liquid water must be uniformly distributed within the particles of the raw materials, to avoid temperature and moisture variations in the preconditioned biopolymer-based mixes, consequently avoiding different temperature-shear histories in the melting section of the extruder. In engineering terms, preconditioning can be defined as the process which transfers heat and mass between components of a three-phase gas (steam)/liquid (water)/solid (flours, grits) medium.

When feeding the steam and liquid water into the preconditioner, the steam heats the feed particles and liquid water through its latent heat released during condensation. Condensed steam, together with water, generates a thin film of water around the particles forming the raw materials (flours, grits). Thus, when the cold particles are surrounded by the hot saturated steam, the temperature and moisture content of the particles increase significantly, but not instantaneously. Two factors govern the rate of heating and hydration of the particles. The first

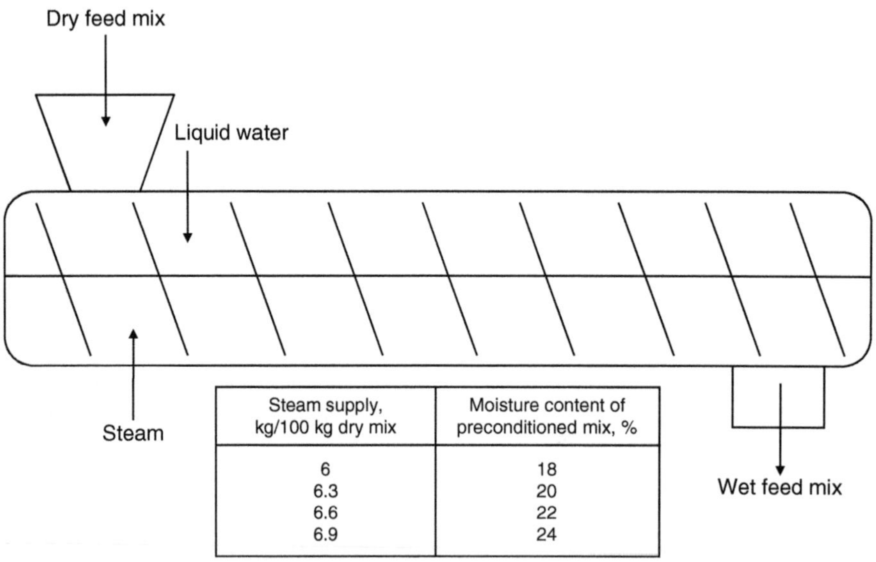

Steam supply, kg/100 kg dry mix	Moisture content of preconditioned mix, %
6	18
6.3	20
6.6	22
6.9	24

Figure 6.5 Preconditioning unit: steam supply versus moisture content.

factor is the film resistance at the particle surface which relates to the quality of the contact between steam or liquid water and the particles. The better the fluid/solid contact or the higher the degree of mixing in the preconditioner, the lower the film resistance. The second factor is the rate of heat and moisture flow inside the particles; this is the internal resistance governed by the heat and water diffusivities, which are characteristics inherent to the particles. The higher the heat and water diffusivities, the higher the rate of heat and moisture flows inside the particles, and the lower the internal resistance. In the case of starch- and protein-based materials at ambient temperature, the heat and water diffusivities are nearly 10^{-7} m^2/s and 10^{-9} m^2/s, respectively; diffusivity coefficients increase when temperature increases. The ratio between the heat diffusivity and water diffusivity is around 100, meaning that heat transfer is much faster than water transfer within the particles. Therefore, for a given temperature, water transfer from the liquid water to the particles controls the preconditioning conditions, reflected by the time-temperature history in the preconditioner.

The mixing efficiency of the preconditioner determines the film resistance while heat and water diffusivities govern the internal resistance. The relative importance of the film and internal resistance terms is usually measured by the dimensionless Biot number (B_i), which is classically used for modeling the heat and mass transfer phenomena. The Biot number is characterized by the ratio of the internal resistance to the film resistance in the heat/mass transfer. At time $t = 0$, suppose a spherical particle of dry flour is suddenly surrounded by a film of liquid water, moisture flows from the film to the particle at a rate which depends on the particle size, particle diffusivity and the Biot number for mass transfer (or B_{iM}). For a small Biot number ($B_{iM} < 0.1$), the main resistance is in the film around the particle; this would be the case in preconditioners with poor mixing leading to poor dispersion of liquid water and uneven distribution of water around the flour particle. For a high Biot number ($B_{iM} > 10$), the main resistance is on the diffusion of water inside the particle; this would be the case in preconditioners with perfect mixing leading to good dispersion of liquid water and homogeneous distribution of water around the flour particle. Preconditioners currently used in industrial processes would show intermediate efficiency of mixing, meaning that B_{iM} should be close to 1, i.e. both film and internal resistance terms affect the moisture transfer in the flour particles.

The rate of water flow inside the flour particles is governed by the following equation that represents the mass transfer through a spherical particle by assuming negligible convection, constant water diffusion coefficient and no chemical reaction:

$$\frac{\partial C}{\partial t} = D\left[\frac{1}{r^2}\frac{\partial}{\partial r}\left(r^2\frac{\partial C}{\partial r}\right) + \frac{1}{r^2 sin^2\theta}\left(\frac{\partial^2 C}{\partial \varphi^2}\right) + \frac{1}{r^2 sin^2\theta}\frac{\partial}{\partial \theta}\left(sin\theta\frac{\partial C}{\partial \theta}\right)\right] \tag{6.7}$$

where C is the water content (hydration process) at any point of the particle, t is the time, and D is the water diffusivity of the water in the particle. Knowing the water diffusivity coefficient, solutions of Eq. 6.7 can be used to predict the moisture concentration within the particle as well as the diffusion time (or hydration time) for different particle sizes. Techniques for solving Eq. 6.7 as well as the analytical solutions are discussed in several books such as Crank (1975), Carslaw and Jaeger (1959), and Datta (2002). Graphical solutions are expressed through charts which are a function of the Biot number, the location and the dimensionless mass Fourier number, F_{oM}, defined as follows (see Chapter 5, section 5.2.4):

$$F_{oM} = \frac{Dt}{(R/3)^2} \tag{6.8}$$

where t and R are the diffusion time and the radius of the particle, respectively.

The solutions of the mass balance equations that account for both film and internal resistance allow evaluation of moisture concentration at the center (C_o) and other locations of the particle as a function of these dimensionless numbers and the moisture concentration at the surface, C_s (Datta, 2002). By knowing the water diffusivity coefficient in biopolymer-based particles, the time necessary to hydrate the particles homogeneously can be predicted. Accounting for values of water diffusivity at 60°C, 80°C, and 90°C provided by Karathanos et al. (1991) and Vagenas and Karathanos (1991), and by assuming that $C_o = 0.99C_s$, i.e. a negligible moisture gradient within the particle, it is possible to estimate the hydration time for homogeneous hydration of a particle as a function of the particle diameter.

Figure 6.6 shows the calculated data that represent the minimum time required to hydrate starch-based flour particles at various temperatures (60, 80, and 90°C) for $B_{iM} = 1$, which would correspond to an intermediate efficiency of mixing in industrial preconditioners.

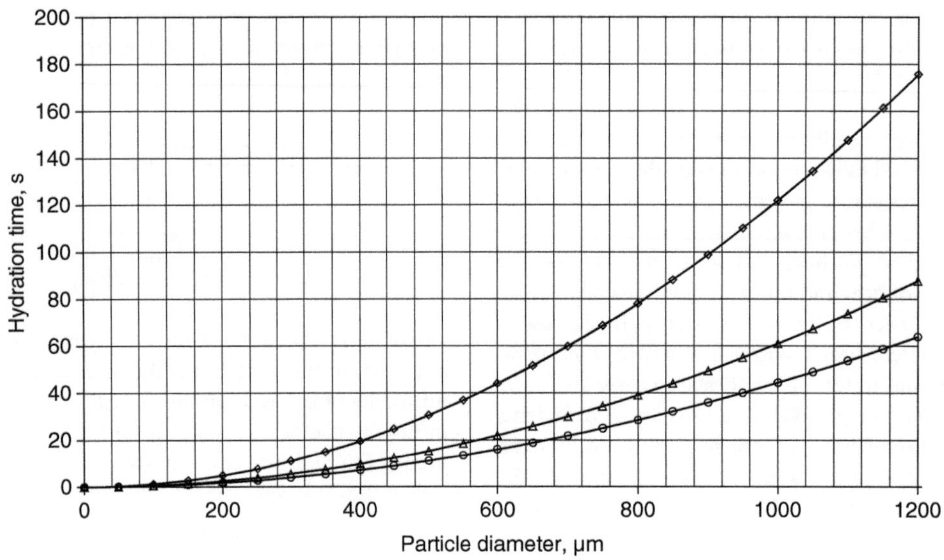

Figure 6.6 Hydration time of biopolymer-based particles as a function of particle size and temperature (◇ 60°C; △: 80°C; ○: 90°C).

Knowing the temperature and the particle size, Figure 6.6 can be used to estimate the minimum hydration time of flour particles in preconditioning. For instance, for a particle size of 400 μm, the minimum hydration time at 80°C would be approximately 10 s. But in practice, it must be kept in mind that flour mixes show a range of particle sizes. Thus, estimation of the minimum hydration time in preconditioning must account for the biggest particles rather than the average sizes in order to guarantee satisfactory preconditioning of all particles of the mix. For instance, in pellet-to-flake extrusion processing, the raw mix consists of corn grits with an average particle size ranging from 350 to 400 μm. In addition, the mix may contain a small percentage of particles as large as 800μm. In that case, the minimum hydration time to account for those feed characteristics would be approximately 40 s at 80°C.

It is worth noting that this minimum hydration time cannot be considered as the targeted average residence time in the preconditioning process. Obviously, the average residence time must account for not only the minimum hydration time of cereal-based particles (flours, grits) but also the time required for complete distribution of water within the macromolecular network at the molecular level. Hence, the average residence time required for preconditioning cereal-based particles must be significantly longer than the minimum hydration time defined above, in the range of 4–6 times the minimum hydration time.

6.2.1.3 Residence time distribution

Preconditioners offer various processing times and degrees of mixing according to their operating conditions (such as shaft speed and throughput) and paddle configuration. Describing the flow of flour particles in this equipment is difficult, as a theoretical analysis of the pertinent fluid mechanics is very complex. Experimental determination of the residence time distribution (RTD) of a preconditioner yields information about the conveying process. In particular, it provides the time history of particles in the preconditioner, which needs to be adequate for the required hydration time, as seen in section 6.2.1.2. RTD can also be used to analyze the mixing process in the preconditioner, which contributes to the effectiveness of the fluid/particle contact. Increasing axial mixing results in homogenizing any compositional variations of the raw mixes and improves the performance of the process. But increasing axial mixing at the same time generates an increased dispersion of the residence time, which means that different particles receive different preconditioning times, leading to uneven preconditioning and hydration.

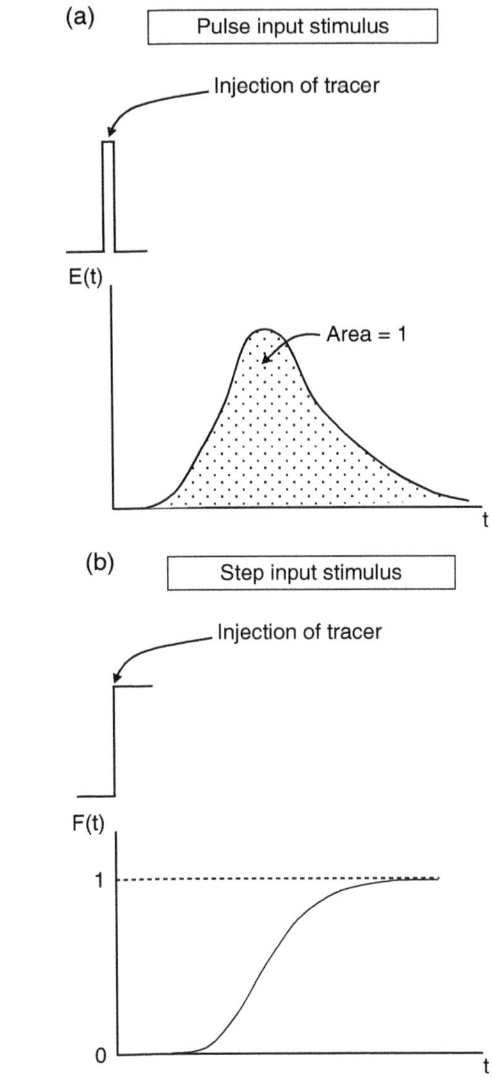

Figure 6.7 Classic stimulus response techniques for the determination of residence time distribution in processing devices.

The RTD is determined by measuring the output response of a change in the input. This is referred to as the stimulus response method, as discussed by Levenspiel (1984). The system is disturbed by a stimulus and the response of the system to the stimulus is measured. Two common stimuli can be used: the pulse input and the step input (Figure 6.7). The pulse input stimulus consists of injecting at time $t = 0$ a certain quantity of a tracer in the inlet of a preconditioner; while it operates under steady state. The tracer concentration is then measured at the outlet of the preconditioner versus time; the ideal pulse input is of infinitely short duration, theoretically described by a delta function or Dirac function. The response to the pulse input stimulus (tracer concentration versus time) is usually a bell-shaped curve. The experimental concentration versus time curve is usually normalized to give an $E(t)$ function so that the area under the response curve is unity (Figure 6.7a).

The step input stimulus consists of injecting continuously at time $t = 0$ a constant concentration of tracer in the inlet of a preconditioner, while the device operates at steady state. The tracer concentration is then measured at the outlet as a function of time. The step input, described by a Heaviside function, is easy to accomplish from an experimental standpoint. The response of a step input stimulus (tracer concentration versus time) in general is an S-shaped curve. The experimental concentration versus time curve can also be simply normalized to give an $F(t)$ function, so that the curve rises from 0 to 1 (Figure 6.7b), the $F(t)$ curve being the integral of the $E(t)$ curve:

$$F(t) = \int_0^t E(t)dt \qquad (6.9)$$

Bouvier (1996) investigated the RTD in a 500 L twin-shaft counter-rotating preconditioner which consisted of two different sections: a forward flow section (66% of the axial length) and a reverse flow section (34% of the axial length). A step input stimulus was applied using soy grits as a tracer. The experiment consisted of feeding the preconditioner with corn grits (low protein content) at steady state (temperature, water swelling and filling ratio equilibrated). Then, at $t = 0$, the feeding of corn grits was replaced by the feeding of soy grits (approximately 50% protein content). Soy grits had similar particle characteristics (particle size distribution) to corn grits so the fluid mechanics of the material in the preconditioner was not modified when switching from corn grits to soy grits. For $t > 0$, samples were collected at the preconditioner exit and the protein content was analyzed to determine the response curve through the protein concentration increase versus time. Figure 6.8 presents an experimental residence time distribution of material flowing into the preconditioner, through the normalized $F(t)$ curve. It shows a typical sigmoidal curve with an inflection point that represents nearly the average residence time. The curve reveals that the residence time of grits particles in the preconditioner ranged from a few seconds to approximately 250 s for a throughput and shaft speed of 750 kg/h and 150 rpm, respectively.

It is also worth investigating the effects of throughput and shaft speed on the RTD using the same 500 L twin-shaft counter-rotating preconditioner. Experimental data are presented in terms of the dimensionless time scale $\theta = t/\bar{t}$, where $\bar{t}$ is the average residence time; then, the average residence time corresponds to $\theta = 1$. Figure 6.9a presents the effect of throughput which varied from 500 kg/h to 1000 kg/h. F(θ) curves show similar profiles, meaning that increasing throughput does not significantly change the dispersion of residence time. Figure 6.9b presents the effect of shaft speed which varied from 90 rpm to 210 rpm. The F(θ) curves show that the distribution of residence times is significantly affected by shaft speed, and a change in the flow behavior is observed at shaft speeds above 120 rpm. When preconditioning at 90 rpm and 120 rpm, a lower dispersion of residence times is observed, compared with preconditioning at higher shaft speeds (150 rpm and over).

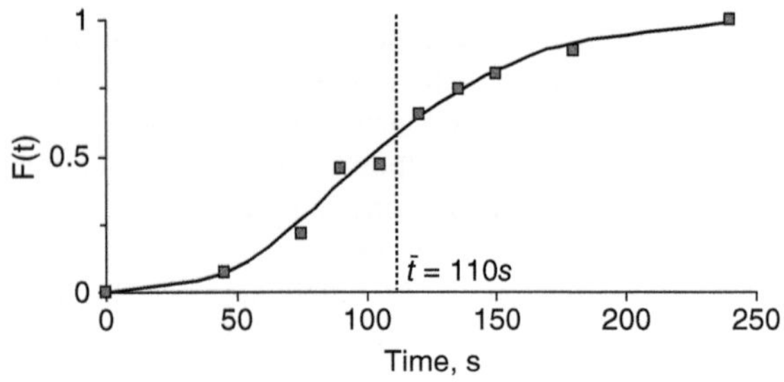

Figure 6.8 Typical *F(t)* curve as a function of time in a 500 L twin-shaft counter-rotating preconditioner (throughput: 750 kg/h; shaft speed: 150 rpm); ■: experimental data; ……: model (Eq. 6.11; *J* = 5).

The experimental RTD data can be modeled easily using the simple tanks-in-series model (Levenspiel, 1984). Application of the model enables visualization of the flow behavior in the preconditioner, which can be used to assess whether the actual flow is close to that observed in ideal plug reactors or in perfectly mixed tanks. The tanks-in-series model consists of considering the preconditioner as an arrangement of *J* stirred tanks in series, determining the value of the parameter *J* that fits experimental RTD curves. Depending on the RTD curve considered, *E(θ)* or *F(θ)*, the models are given by the following equations:

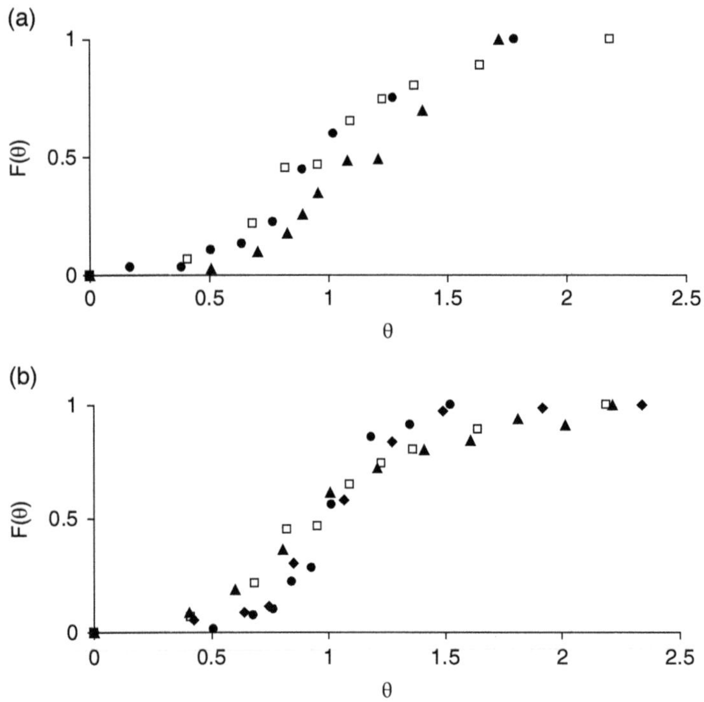

Figure 6.9 (a) Experimental *F(θ)* versus *θ*: effect of throughput (●: 500 kg/h; □: 750 kg/h; ▲: 1000 kg/h) at constant shaft speed (150 rpm). (b) Experimental *F(θ)* versus *θ*: effect of shaft speed (●: 90 rpm; ◆: 120 rpm; □: 150 rpm; ▲: 210 rpm) at constant throughput (750 kg/h).

$$E(\theta) = \frac{J(J\theta)^{J-1}}{(J-1)!} e^{-J\theta} \tag{6.10}$$

$$F(\theta) = 1 - e^{-J\theta}\left[1 + J\theta + \frac{(J\theta)^2}{2!} + \cdots + \frac{(J\theta)^{J-1}}{(J-1)!}\right] \tag{6.11}$$

where J would express the efficiency of axial mixing in the preconditioner. When J is close to 1 (that corresponds to a perfectly mixed device), the preconditioner is characterized by very high axial mixing with a very large distribution of residence times. In contrast, when $J \rightarrow \infty$ (that corresponds to a plug flow device), the preconditioner is characterized by perfect radial but no axial mixing and a single distribution of residence time. In practice, when $J > 10$, the preconditioner shows a flow behavior close to plug flow. As shown in Figure 6.8, the tanks-in-series model fits the experimental RDT in the preconditioner well and therefore the model can be used to describe flow behavior in the preconditioner when varying the two important operating conditions, throughput and shaft speed.

Figure 6.10 gives the evolution of the parameter J of the model as a function of shaft speed, for three different throughputs – 500, 750, and 1000 kg/h. An increase of shaft speed from 30 to 240 rpm at 750 kg/h shifts the flow behavior from a plug flow to perfectly mixed tank. Actually, at throughputs of 750 and 1000 kg/h, the shift occurs over a rather short range of shaft speeds, that is between 90 rpm and 150 rpm. As seen in Figure 6.10, at 500 kg/h throughput, though shaft speeds lower than 90 rpm were not investigated, the shift seems to occur over a range of lower shaft speeds.

Based on RTD observations and modeling, the flow behavior of twin-shaft counter-rotating preconditioners could be characterized as follows. For specific preconditioning capacities of 1.5 kg/h.L and over (for the definition of specific capacity, refer to section 6.1.1.3), the preconditioners show two different flow behaviors depending upon the shaft speed: plug flow behavior when shaft speed is lower than 90 rpm and perfectly mixed flow behavior over 150 rpm, while a flow transition occurs between 90 and 150 rpm. For specific capacities lower than 1.5 kg/h.L, the shift from plug flow to perfectly mixed flow tends to start at lower shaft speeds.

The practical implication of RTD is preconditioning efficiency, which depends closely on mixing efficiency defined by both radial and axial mixing. If the distribution of residence times in preconditioners is broad, some of the feed particles may spend only a short time in the preconditioner and receive insufficient treatment. This is a particular problem when the raw materials show a wide distribution of particle sizes. In fact, shorter times may generate moisture gradients within the particles, particularly the biggest ones, resulting in particles with various moisture contents at the preconditioner exit; this would introduce heterogeneities during cooking in the extruder, and possibly uncooked particles in the biopolymeric melt. Intermediate distribution of RTD must be present in preconditioning in order to avoid too short times and to obtain sufficient axial mixing. This means values of 6–7 for J, which corresponds to shaft speeds of 150–180 rpm. Besides, such conditions are appropriate for reaching the right level of radial mixing in the preconditioner.

6.2.2 Extrusion-cooking

During extrusion-cooking, food biopolymers undergo several significant structural changes such as starch gelatinization and melting, protein denaturation, and complexations between ingredients (such as amylose–lipid complexation). Besides, biopolymers are subjected to quite severe processing conditions (high temperature and pressure, mechanical shear stresses) which cause

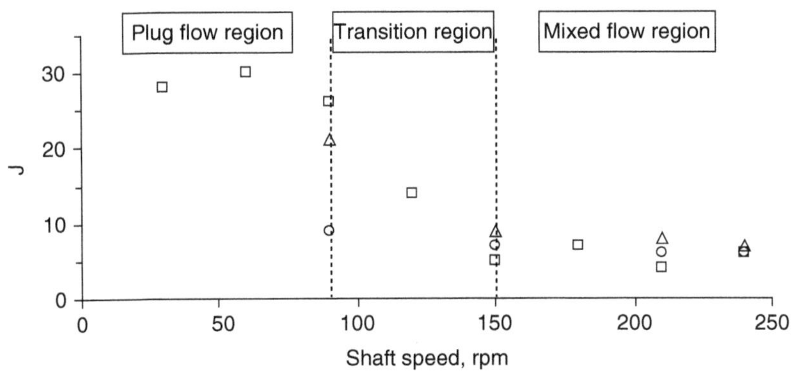

Figure 6.10 J versus shaft speed in preconditioning (○: 500 kg/h; □: 750 kg/h; △: 1000 kg/h).

various physicochemical reactions to occur such as starch degradation (or starch dextrinization) and protein aggregation. These molecular transformations aim to convert food polymers into a viscoelastic biopolymeric melt with rheological properties that determine the characteristics of the extruded products. The understanding of physicochemical changes occurring in biopolymer-based materials during extrusion-cooking is of paramount importance in controlling extruder design appropriately, notably screw-barrel configuration as well as processing variables such as temperature, screw speed, moisture content, and feed formulation to optimize quality and targeted texture profile of end-products.

Among all the biopolymer-based raw materials used in extrusion-cooking, starches play a determinant role. Therefore, this section will focus on the physicochemical changes of starch during extrusion-cooking.

6.2.2.1 Phase transitions of starch

Phase transitions of starch are influenced by the type and intensity of energy inputs. Thermal energy is an important component in the cooking process, either transferred from steam (through preconditioning and/or direct steam heating in the extruder barrel) and/or from the heaters through the barrel, and/or generated by viscous dissipation.

As seen in Chapter 4 (see section 4.1.2.1), starch is a partially crystalline polymer. The "fringed-micelle" model can be used to represent the structure of starch composed of crystalline regions that covalently cross-link amorphous regions of flexible chain segments. During heating, that partially crystalline polymer undergoes phase transitions that depend on the temperature. As schematically illustrated in Figure 6.11, these transitions include:

- the glass transition of amorphous phase, characterized by the glass transition temperature, *Tg*
- the conversion of crystalline phase, characterized by the melting temperature, *Tm*.

As described in Chapter 4 (see section 4.3.3.1), at *Tg* chain segments of the starch amorphous phase become more mobile, changing from a solid glassy state to a highly viscous rubbery state. However, some chain segments of the crystalline phase remain stable and start to move at a different temperature in a way that the overall glass transition is observed over a large range of temperatures. In contrast, petrochemical thermoplastic polymers have glass transition occurring over a small range of temperatures, whereas glass transition events of starch polymers occur over a large range of temperatures (over 10°C). The glass transition temperature is strongly affected by the water content: *Tg* decreases significantly when water content increases due to the plasticizing effect of water. Several studies have been carried out on the glass transition of starch materials (native starch, gelatinized starch, extrusion-cooked starch). The published data on the values of *Tg* of starch materials show significant variability depending upon the origin and thermal history of starch, and the experimental methods used to identify the glass transition (Biliaderis et al., 1986a; Kaletunc & Breslauer, 1993; Kalichevsky et al., 1992; Tomka, 1991; Zeleznak & Hoseney, 1987).

The water dependence of *Tg* can be described by the Couchman–Karasz equation presented in Chapter 4 (Eq. 4.12, section 4.3.2). When this equation was applied to the amylopectin-water system, Kalichevsky et al. (1993)

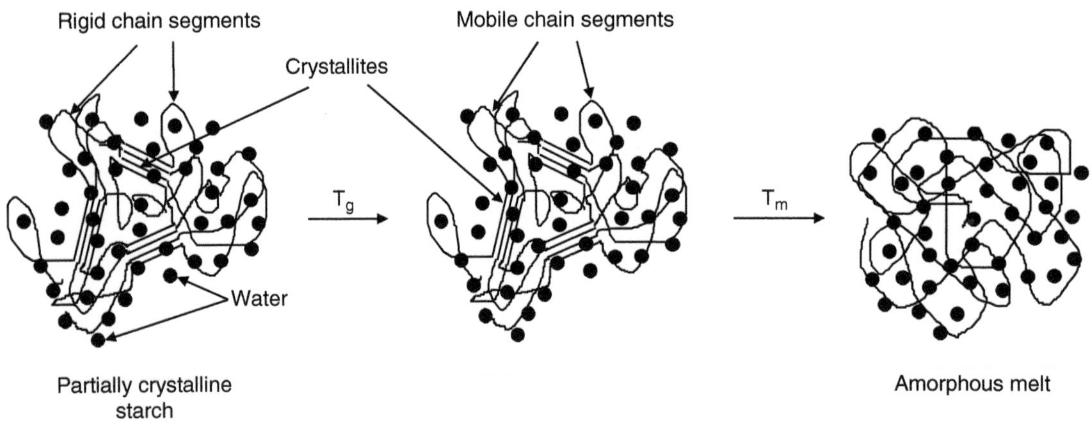

Figure 6.11 Phase transitions of partially crystalline starch.

observed that the predicted *Tg* fell well below the observed *Tg*, probably because the assumption that ΔC_p is independent of temperature is not valid. This is even more important in polymer-plasticizer systems where *Tg*s are far apart. Assuming that ΔC_p is inversely proportional to temperature, Kalichevsky et al. (1993) used the following modified equation:

$$T_g = \frac{w_1 \cdot \Delta C_{p1} \cdot T_{g1} + w_2 \cdot \Delta C_{p2} \cdot T_{g2}}{w_1 \cdot \Delta C_{p1} + w_2 \cdot \Delta C_{p2}} \tag{6.12}$$

When Eq. 6.12 is applied to the amylopectin–water system (with $\Delta C_{p1} = 0.425\, J.g^{-1}.°K^{-1}$ and $T_{g1} = 227°C$ for starch; $\Delta C_{p2} = 1.94\, J.g^{-1}.°K^{-1}$ and $T_{g2} = -139°C$ for water; w_1 and w_2 the weight fractions of starch and water respectively), a good fit to the experimental data is obtained. Thus, Eq. 6.12 can be used to predict *Tg* of starch as a function of water content in a limited water environment such as that existent in food extrusion processing; that is, in the range of approximately 2.5–35% (wwb).

Glass transition data as a function of water content can be used to develop a state diagram for starch materials. The state diagram consists of plotting the *Tg*–water content relationship associated with the phase transitions of starch. Figure 6.12 presents a state diagram for starch where the glass transition temperature has been estimated from Eq. 6.12 applied to the amylopectin–water system. It shows the separation of the glassy region below the glass transition line and the rubbery region above the glass transition line. The state diagram allows the prediction of the different starch phases that can be expected in food processes such as hydrothermal and extrusion-cooking; it describes the water content and temperature region at which starch polymers will undergo the appropriate phase

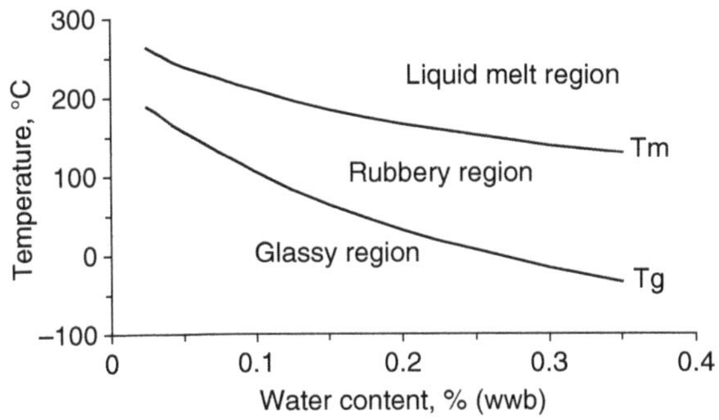

Figure 6.12 State diagram of starch materials.

changes according to the different unit operations involved in the process.

When reaching the melting temperature, *Tm*, crystalline regions disappear, leading to an amorphous biopolymeric melt. The melting mechanism of starch is discussed in section 6.2.2.2.

6.2.2.2 Hydrothermal conversion of starch

Starch conversion above the glass transition temperature has been extensively investigated. It takes many forms which depend on the main processing conditions such as water content, thermal energy input, and possible mechanical energy input. As a synthesis, it is worth exploiting two relevant well-known investigations carried out by Donovan (1979) and by Wang et al. (1991). These authors have experimentally studied the effect of water content on the conversion of two types of starches in a shearless environment by the use of differential scanning calorimetry (DSC). In these experiments, starches were exposed to different combinations of water plasticization and thermal energy, leading to a hydrothermal processing of starch.

Donovan (1979) first showed a diphasic transition of potato starch subjected to a heating rate of 10°C/min at different volume fractions of water ranging from 0.28 up to 0.81 (or weight fractions of water ranging from 0.200 up to 0.733). The volume fraction of water equalled the total volume of water divided by the total volume of starch (with a density of 1.55) plus water; whereas the weight fraction of water is expressed as the weight of water divided by the weight of starch plus water. Figure 6.13 presents DSC thermograms of potato starch obtained by Donovan (1979). At high water content (73.3% w/w), the thermogram shows a single endotherm at a temperature of approximately 66°C (peak temperature). This endotherm is attributed to starch gelatinization and the peak temperature gives the corresponding gelatinization temperature. The first endotherm gradually decreases with decreasing water content, while a second endotherm occurs at higher temperature when the content of water is decreased; it appears initially as a high temperature shoulder on the first endotherm when water content drops to 53.4% (w/w). The second endotherm shifts to higher temperature with decreasing water content until it becomes a single endotherm which is observed at very low water content (34.6% w/w). Donovan concluded that at low water contents, the transition at higher temperature corresponded to the melting of crystallites present in the starch granule. Similar observations were subsequently

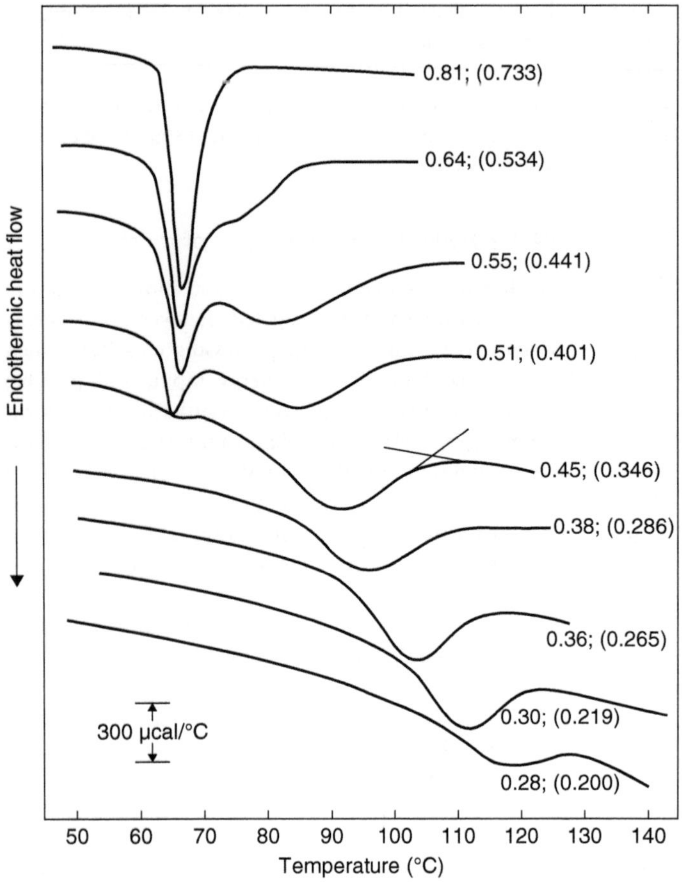

Figure 6.13 DSC thermograms of potato starch obtained at a heating rate of 10°C/min (labeled with volume fraction of water present; in brackets: weight fraction of water).
Source: Donovan 1979. Reproduced with permission of John Wiley & Sons.

reported by several authors including Biliaderis et al. (1986b) and Liu et al. (1991), confirming the occurrence of a melting endotherm at moisture contents below 60% (w/w).

Wang et al. (1991) used a DSC scan rate of 5°C/min to identify the DSC peak temperature of waxy corn–water mixtures with water ranging from zero to 99% (w/w). The authors made similar experimental observations to Donovan (1979). Figure 6.14 illustrates the relationship between the DSC peak temperature and the water content of the sample obtained by Wang et al. (1991). At water contents higher than 60% (w/w), a constant peak temperature of about 71°C is observed. As water content decreases from 60% (w/w), the peak temperature increases linearly from 71°C up to 230°C for anhydrous waxy corn.

Experimental observations made by Donovan (1979) and Wang et al. (1991) show fairly clearly that starch granules generally experience two different conversion regimes during DSC measurements that depend upon the amount of water available in the sample. Above 60% (w/w) water content where the peak temperature is constant and equal to 71°C, the amount of water available is sufficient to complete the gelatinization of starch; this is the "gelatinization regime" characterized by a single phase transition. Starch gelatinization includes the irreversible loss of molecular order due to water plasticization followed by granule swelling and amylose leaching without disruption of the granular structure. This occurs once the glass transition of the amorphous regions has been completed. The extent of amylose leaching increases when temperature increases. Some low molecular weight amylopectin may also be leached, depending on the type of starch and conditions. On the other hand, amylose that is complexed with lipids is inhibited from leaching and remains together with essentially all amylopectin inside the granules. A water dispersion of fully gelatinized starch granules is composed of shrunk starch granules (called "ghosts") dispersed in an amylose-rich (with traces of

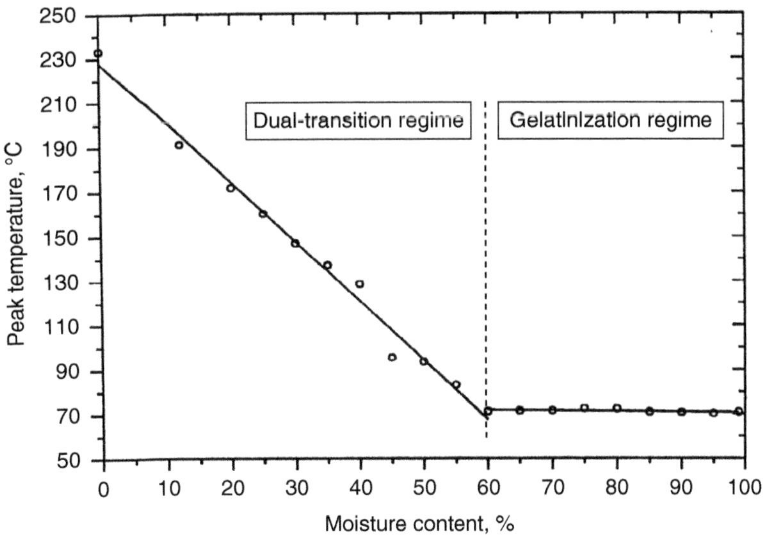

Figure 6.14 DSC-peak temperature (heating rate of 5°C/min) versus water content (w/w) for waxy corn.
Source: Wang, Chiang, Zhao, Zheng & Kim 1991. Reproduced with permission of John Wiley & Sons.

amylopectin molecules) continuous amorphous phase. Below 60% (w/w) water content, water becomes limiting for gelatinization and the peak temperature starts to rise to a higher value, meaning that another regime of starch conversion occurs.

Structural changes of starch that occur during conversion in limited amounts of water are somewhat different to those occurring during gelatinization with excess water. As the relative amount of water is reduced, some starch granules begin to gelatinize as in the phenomenon observed in the presence of excess water. However, once the available water is exhausted through physicochemical interactions, the co-operative plasticization process stops and further loss of crystalline structure would depend on another contributing factor such as thermal energy input, which would ensure the increase of molecular mobility. Starch materials must then be heated to higher temperatures than those observed under excess water conditions, leading to conversion of starch of a biphasic nature.

Wang et al. (1991) suggested that when water content decreases from 60% (w/w), a part of starch begins to undergo phase transition through melting. Below 60% water content, a "dual-transition regime" is observed which is characterized by the co-existence of both gelatinization and melting in the starch conversion process. In this regime, as water content decreases, the extent of melting gradually increases to the detriment of gelatinization, leading to a fully converted starch composed of a continuous amorphous phase over the dual-transition regime. Though both conversion regimes lead to fully converted starch with regard to the complete loss of crystallinity, it must be kept in mind that the resulting converted materials show significant differences in their physicochemical properties and textural attributes.

The occurrence of the two conversion regimes has been confirmed by several investigators who studied different types of starch. Wang et al. (1992) gathered published data in the same chart presented in Figure 6.15. This figure shows that the two conversion regimes, gelatinization and dual-transition, proceed similarly for three different kinds of starch (potato, rice, and corn) having different amylose-to-amylopectin ratios, different degrees of crystallinity, and different granule sizes. The gelatinization temperature (in the gelatinization regime) as well as the slope of the linear increase of the peak temperature when water content decreases (in the dual-transition regime) are specific to the type of starch: the gelatinization temperature varies approximately between 60°C and 75°C. Waxy corn starch and potato starch show the highest and the lowest slopes of peak temperature increase, respectively. The shift from complete gelatinization to the dual-transition regime occurs at constant water content (that is, around 60% w/w), regardless of the type of starch. This confirms that the minimum water content for complete gelatinization of starch is around 60% (w/w) which corresponds approximately to 14 water molecules per monomer unit (anhydroglucose) of starch. It is worth pointing out that in starch conversion, the occurrence of the two transition regimes with a demarcation point at 60% (w/w) water content is valid for all types

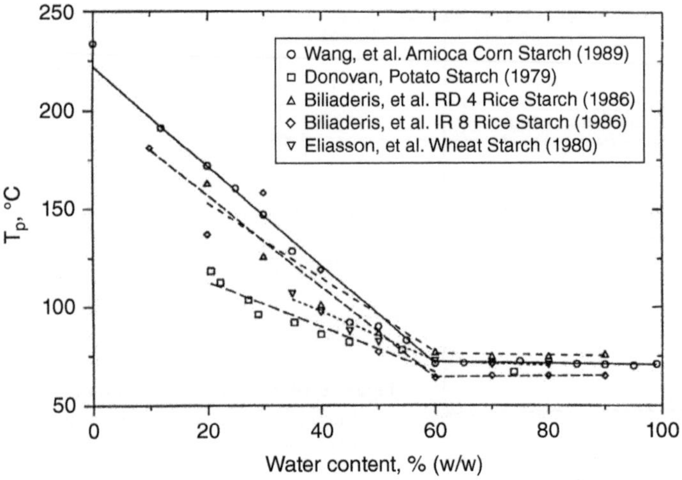

Figure 6.15 DSC-peak temperature versus water content for various starches. Source: Wang, Chiang, Zheng, Zhao, Cho & Yeh 1992. Reproduced with permission of Taylor & Francis.

of starch, thus leading to a unified concept which can be applied to study the behavior of starch–water systems in various processing techniques, including extrusion-cooking.

Based on the assumption that 14 is the minimum molecular stoichiometric ratio for complete gelatinization of starch (which corresponds exactly to a water content of 61% w/w), Wang et al. (1991) proposed a phase diagram which related water contents to gelatinization and melting of starch. For a binary starch–water system, the phase diagram involves two phases: a gel phase for gelatinized starch and a melt phase for melted starch. Above 61% water content, a 100% converted starch would consist of a single gel phase. Below 61%, a 100% converted starch would consist of two phases, and the phase diagram allows the ratio of gel phase to melt phase to be determined depending on initial water content.

The melting temperatures of anhydrous starches (T_m^o) range roughly between 150°C and 230°C. These values depend on the type of starch and the amylose-to-amylopectin ratio (Colonna et al., 1989; see Figure 6.15). In addition, published values show large dispersion due to the different measurement methods used. It is worth pointing out that for a given anhydrous starch, the melting temperature T_m^o is quite close to the glass transition temperature; moreover, for amylopectin-rich starch and cereal starches, T_m^o is close to the thermal decomposition temperature (200–220°C), which makes the melting operation of anhydrous starches critical. Nevertheless, the DSC investigation of the dual-transition conversion regime of starch–water systems should allow the melting temperatures of anhydrous starches (T_m^o) to be determined with an acceptable significance. For instance, as reported by Wang et al. (1991), dry bone waxy corn samples were melted in DSC pans around 230°C; at 10% (w/w) water content, waxy corn starch should have a peak temperature of approximately 200°C according to Figure 6.14. Though starch gelatinization and melting may not be at the thermodynamic equilibrium, the dependence of the endotherm temperature (T_m) upon water content can be represented by use of the Flory–Huggins equation:

$$\left(\frac{1}{T_m}-\frac{1}{T_m^o}\right)=\frac{R}{\Delta H_u}\cdot\frac{V_u}{V_1}\left(\upsilon_1-\chi_1\upsilon_1^2\right) \tag{6.13}$$

where T_m is the melting temperature measured with the water–starch mixture (K), T_m^o is the melting temperature of the anhydrous starch (°K), R is the gas constant (8.31 J.mol^{-1}.K^{-1}), ΔH_u is the enthalpy of melting per mole of monomer unit (or anhydroglucose), V_u/V_1 is the ratio of the molar volume of the monomer unit to the molar volume of water (dimensionless), υ_1 is the volume fraction of water (dimensionless), and χ_1 is the Flory–Huggins starch–water interaction parameter (dimensionless).

The Flory–Huggins equation can be used to reasonably estimate the melting temperatures of native starch–water systems, in a limited water environment such as that applied in food extrusion processing; that is, in the range of approximately 2.5% to 35% (wwb), hence completing the state diagram of starch presented in Figure 6.12. Calculation of the melting temperature versus water content requires the values of the melting temperature of anhydrous starch (T_m^o) and the enthalpy of melting per mole

Table 6.4 Melting temperature and enthalpy of melting of anhydrous starch materials.

Starch material	T_m^o, °C	ΔH_u, kJ/mol anhydroglucose	Reference
Waxy maize starch	277	25.4	Farhat et al. (2000)
	246	29.3	Shogren (1992)
	237	50.2	Wang et al. (1991)
Potato starch	168	56.5	Donovan (1979)
Maize starch	205	41.0	Colonna and Buléon (1994)
Wheat starch	233	28.5	Colonna and Buléon (1994)
	210	29.3	Lelievre (1974)

of monomer unit (ΔH_u). Table 6.4 gives some values of melting temperature and enthalpy of melting for typical starches, which were estimated from different experimental investigations by the Flory–Huggins equation. As seen in Table 6.4, the values are rather scattered depending upon the experimental conditions used by the authors. The melting transition line shown in Figure 6.12 is derived from the melting characteristics published by Farhat et al. (2000), with $\chi_1 = 0.5$. The melting transition line separates the rubbery region from the liquid melt region.

The state diagram of starch will be used in section 6.3.1 to describe the water content and temperature region at which starch polymers undergo the appropriate phase according to the unit operations of extrusion-cooking processes designed for the production of breakfast cereals.

6.2.2.3 Thermomechanical conversion of starch

In thermomechanical processing such as extrusion-cooking, starch materials are exposed to a combination of heat and mechanical energy inputs, in limited water contents ranging roughly from 14% to 32% (wet weight basis). Moreover, as seen in Chapter 4 (section 4.2), polymer melting in screw extruders occurs over very short residence times (less than 10 s, depending upon the extrusion process). Under these processing conditions, the water content is much below the amount of water necessary for gelatinization and the melting time is too short for transforming starch by thermal energy alone. Thus, another processing factor such as mechanical energy input allows the starch to be fully converted into an amorphous melt.

In thermomechanical processing of starch materials, structural changes include the loss of granular integrity, the loss of crystallinity, the depolymerization of polymeric molecules, and the formation of amylose–lipid complexes. The degree of starch conversion which results from these structural changes increases with increasing severity of processing conditions. The degree of starch conversion can be evaluated by use of various analysis techniques such as optical microscopy (loss of birefringence), DSC (loss of gelatinization endotherm), Water Solubility Index or WSI (solubilized polysaccharides), Water Absorption Index or WAI (gel phase of polysaccharides), alkaline viscosity (starch depolymerization), and Rapid ViscoAnalyzer or RVA (pasting properties). These techniques will be described in Chapter 7 (section 7.1). Other techniques must also be mentioned such as X-ray (loss of crystallinity and formation of amylose–lipid complexes), iodine complexation (amylose–lipid complexes), and intrinsic viscosity (starch depolymerization). Intrinsic viscosity is a measurement commonly used in the study of the chemistry of macromolecules and has units of inverse concentration, e.g. L/g or dL/g. Its value is related to the structure and size of the polymeric material in solution; in general, lower values of intrinsic viscosity correspond to polymeric materials having smaller molecular size in solution. The intensity of thermomechanical processing is typically estimated by the specific mechanical energy (SME) input.

Starch structural changes during thermomechanical processing have been investigated by several researchers since the 1980s. Published studies report relevant correlations between the extent of starch structural changes and the SME applied during the thermomechanical process. Thus, it is worth using some of these studies which allow starch conversion to be reviewed and discussed in relation to the intensity of the thermomechanical process.

Kirby et al. (1988) studied the relationship between the process SME and physical characteristics (microstructure and solubility) of extruded products, when achieving a range of SME values in extrusion processing of maize grits. In the study, the authors used an intermeshing co-rotating twin screw extruder with varying screw configurations (four screw configurations ranging from low to high shear configuration), moisture content (in the range of 18–24% wwb) and temperature (in the range of 90–150°C for the final two sections of the screw-barrel assembly), to obtain SME values ranging from about

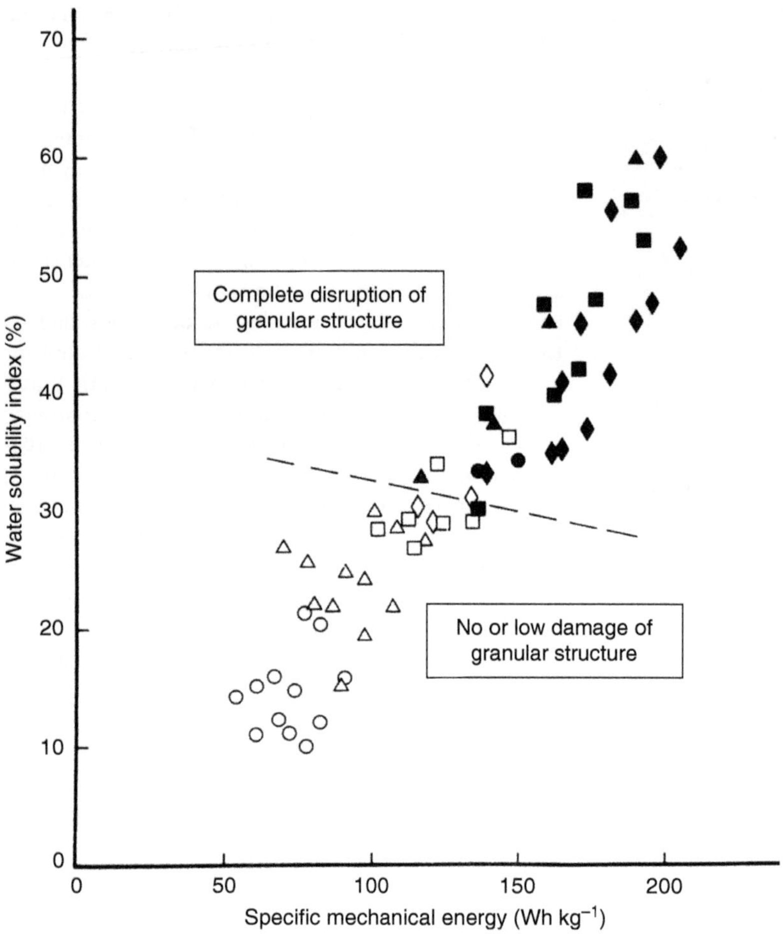

Figure 6.16 WSI as a function of SME for maize grits (the four symbols refer to four screw configurations; closed symbols correspond to melts with completely disrupted maize particles). Source: Kirby, Ollett, Parker & Smith 1988. Reproduced with permission of Elsevier.

50 to 200 W.h.kg^{-1}. Figure 6.16 shows the evolution of the WSI of extrudates as a function of process SME. A linear increase of the WSI with increasing SME is observed, meaning that an increasing amount of soluble polysaccharides is generated when increasing the severity of the thermomechanical process. The WSI-SME plot shows clearly three different domains characterized by the structural state of grit particles and starch granules. At low SME (below about 115 W.h.kg^{-1}), grit particles remained intact, comprising largely ungelatinized starch granules with only the outer surface of the grit containing gelatinized granules. At high SME (above about 140 W.h.kg^{-1}), no particle or granule structure remained. A transition domain occurred for the SME values of approximately 115–140 W.h.kg^{-1}, where the resulting extrudates were composed of a mix of gelatinized and ungelatinized starch granules.

These results are coherent with those reported by Guy and Horne (1988) who observed that starch granules were partly disrupted into a continuous starch phase at SME input of less than 70 W.h.kg^{-1} while at SME inputs higher than 140–165 W.h.kg^{-1}, complete disruption of the granular structure was observed. These studies clearly indicate that in thermomechanical extrusion, a minimum SME input (approximately 115–140 W.h/kg^{-1}) is necessary for complete starch conversion. This minimum SME depends on the screw configuration used and the associated processing variables. In general, high shear screw configurations will be characterized by relatively low minimum SMEs for complete starch conversion, and vice versa. In the SME range of 70–140 W.h.kg^{-1}, incomplete starch conversion might be obtained. In that range, an increase of the starch conversion is expected with increasing SME input. Kirby et al. (1988) have also determined

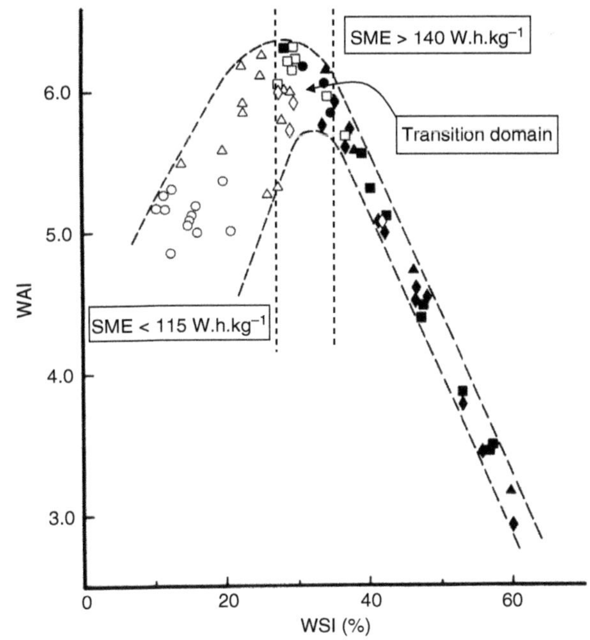

Figure 6.17 WAI as a function of WSI for maize grits (the four symbols refer to four screw configurations; closed symbols correspond to melts with completely disrupted maize particles). Source: Kirby, Ollett, Parker & Smith 1988. Reproduced with permission of Elsevier.

the WAI of the extruded products obtained in the SME range of 50–200 W.h.kg^{-1}.

Figure 6.17 illustrates the relationship between the WAI and WSI, which shows a maximum WAI. The increase of WAI with WSI that corresponds to low SME (below 115 W.h.kg^{-1}) is due to an increasing proportion of gelatinized starch granules reaching complete gelatinization at the maximum WAI. The decrease of WAI with WSI that corresponds to high SME (above 140 W.h.kg^{-1}) results from starch depolymerization, hence contributing to increase the amount of soluble polysaccharides to the detriment of gelatinized starch. At the maximum WAI, there is a transition domain characterized by SME ranging roughly from 115 to 140 W.h.kg^{-1}. Actually, a significant amount of extrusion-cooking applications proceed according to this transition domain (see Table 6.3), where complete starch conversion is required while minimizing WSI or the extent of starch degradation as much as possible. In food extrusion of starchy materials, the WAI–WSI curve can be considered as a "master" curve, guiding extrusion practitioners in the pertinent selection of processing conditions such as screw configuration, screw speed, and water content in particular. In fact, it is of paramount importance to handle the co-ordinates of the transition domain with high flexibility; that is, being able to obtain a large range of WAI/WSI ratios, from high WAI-low WSI extruded products (in the case of intermediate levels of SME: 80–120 W.h.kg^{-1}; see Table 6.3) to low WAI-high WSI extruded products (in the case of high levels of SME: 120 W.h.kg^{-1} and over; see Table 6.3), and anything in between, depending upon the functional properties of end-products. The art of food extrusion practitioners is to design a robust and appropriate home-made WAI-WSI master curve, and handle it wisely according to the different process and product challenges that have been fixed for the product.

As seen in Figure 6.16, an increase of SME leads to a linear increase of WSI, which results from starch depolymerization, or dextrinization, an aspect that was evidenced by studies measuring intrinsic viscosity and also through gel permeation chromatography (Colonna et al., 1984). Colonna et al. (1984) reported that only α-(1,4) linkages of the starch polymers were affected as the percentage of α-(1,6) linkages did not change during extrusion-cooking. The authors concluded that both amylose and amylopectin were susceptible to molecular degradation, which would occur as random chain scission. Starch depolymerization in extrusion processing is a result of mechanical energy input which, for a given screw configuration, depends on the operating conditions of the extruder such as screw speed, temperature, and water content as well as the type of processed starch. In general, in the water content range of 18–28% and temperature range of 120–170°C, the extent of starch dextrinization increases with decreasing temperature or water content and increasing screw speed. The effects of temperature and water content are viscosity dependent. In fact, the lower the temperature or water content, the higher the apparent viscosity of the molten starch and the higher the effect of the mechanical energy input on starch dextrinization.

When examining extruded samples in DSC, an endotherm of fusion in the temperature range of 90–110°C reveals the presence of amylose–lipid complexes. Meuser et al. (1987) studied the yield of amylose–lipid complexation during extrusion processing of wheat starch in the presence of glycerol monostearate. The authors showed that the yield of complexes increased gradually when increasing the extrusion temperature (from about 120°C to 200°C) and the SME from about 150 kJ/kg to 640 kJ/kg, leading to a maximum complexation of 27% at the highest temperature and SME. Crystallization of amylose–lipid complexes occurs during melt cooling, mainly in the die assembly or downstream.

6.2.2.4 Kinetics of thermomechanical conversion of starch

Thermomechanical conversion of starch is a complex phenomenon which consists of multiple reactions, so in this subsection a model for starch conversion is presented based on the main phase transition or reactions involved in the process such as gelatinization, melting, depolymerization, and amylose–lipid complexation. In the model, extruders are regarded as reactors in which different forms of energies are applied to convert raw food ingredients into functional extruded products. The main question which interests process engineers is the rate at which these transitions or reactions occur while accounting for the input of thermal and mechanical energies. Then, for better control of the extrusion operation, an understanding of process kinetics is of prime importance.

Assume the following reaction scheme for starch conversion:

$$RS + H_20 \rightarrow CS$$

where *RS* and *CS* are raw starch and converted starch, respectively. The extent of starch conversion is a function of the order of reaction, rate of conversion, and conversion time. In general terms, the form of the conversion rate, *r*, which expresses the rate of the disappearance of raw starch is represented as:

$$r = -\frac{d[RS]}{dt} = k_r[RS]^{n'}[H_2O]^{m'} \qquad (6.14)$$

where m' and n' are orders of reaction for H_2O and *RS*, respectively, k_r is a proportionality constant called conversion rate constant, $[H_2O]$ and $[RS]$ are the concentrations of water and native starch, respectively, and *t* is the time. For a starch conversion in a large excess of water, the water concentration is constant during the process and the reaction has order n'. When the order of reaction is 0, i.e. the starch conversion follows a zero-order kinetics, a linear relationship between the concentration of one of the reactants (raw starch in this case) and the time of the conversion is obtained. If order of reaction is 1, then a first-order kinetics conversion occurs, leading to a logarithmic relationship between the concentration of one of the reactants and the time of the conversion.

The kinetics of starch conversion has been investigated by researchers who generally took advantage of a global rate approach. Assuming that DSC endotherm area is directly proportional to the mass of unconverted starch at any time, investigators used DSC endotherms as the indicator of starch conversion and followed its evolution as a function of time. The degree of conversion, *X*, during starch processing due to thermal or mechanical energy was evaluated by the decrease of the DSC endotherm area (Wang et al., 1989). Assuming that the DSC endotherm area is nil when starch is fully converted, the degree of starch conversion can be defined as follows:

$$X = \frac{\textit{Amount of converted starch at time t}}{\textit{Initial amount of unconverted starch}} = \frac{\Delta H_o - \Delta H_t}{\Delta H_o} \qquad (6.15)$$

where ΔH_t and ΔH_o are the enthalpy changes of partially cooked starch at time *t* and raw starch at time $t = 0$, respectively. Based on this definition, the degree of starch conversion *X* is a non-dimensional variable that varies from 0 to 1 when time increases from $t = 0$ to the time at which starch is fully converted. The conversion rate of raw starch during processing can then be expressed as follows:

$$r = \frac{dX}{dt} = k_r X^{n'} \qquad (6.16)$$

From experimental data, Eq. 6.16 allows the order of reaction to be determined. In extrusion processing of raw starch, published data have reported both zero-order and first-order kinetics (Gogoi et al., 1995; Lai & Kokini, 1991). An interesting approach to modeling the rate constant of starch conversion has been proposed by Wang et al. (1992). Considering zero-order kinetics, the authors developed a concept which assumes that thermal and shear energies are thermodynamically additive and interactive, hence proposing that the degree of conversion and the conversion rate constant can be described as follows:

$$X = k_r.t \qquad (6.17)$$

and:

$$k_r = k_T + k_s + (k_T \cdot k_s)^{0.5} \qquad (6.18)$$

where k_T is the thermal-induced rate constant and k_s is the shear-induced rate constant. An Arrhenius type of relationship was used to express changes of these parameters as a function of temperature:

$$k_T = k_{T,o}\,exp(-E_T/RT) \qquad (6.19)$$

and:

$$k_s = k_{s,o}\, exp(-E_s/\tau v_u) \tag{6.20}$$

where E_T is the thermal activation energy, E_s is the shear activation energy, τ is the shear stress, v_u is the molar volume of monomer unit (or anhydroglucose), and $k_{T,o}$ and $k_{s,o}$ are pre-exponential coefficients, respectively.

The rate constant due to thermal activation, k_T, can be evaluated by using the following correlation given by Wang et al. (1989):

$$k_T = \frac{\dfrac{T}{Tp}}{74.97\left(\dfrac{T}{Tp}\right)^2 - 167.21\dfrac{T}{Tp} + 93.38} \tag{6.21}$$

where k_T is expressed in min^{-1}, T and Tp are the operating temperature and the DSC peak temperature, respectively. Tp is a constant for a specific starch at specified water content. This correlation was derived from different kinetics data of waxy corn starch, rice starch and potato starch, with the assumption of zero-order kinetics. Eq. 6.21 is valid for $0.63 < \frac{T}{Tp} < 1.06$. For example, with raw corn meal at 20% water content, the measured value for Tp was 151.6°C, and thus Eq. 6.21 is applicable for temperatures between 95°C and 160°C (Gogoi et al., 1995). In that example, the estimated k_T from Eq. 6.21 at 95°C was 0.035 min^{-1}. It means that in the case of an extrusion-cooking process operating at 95°C and 20% water content, the degree of starch conversion would reach 0.035 for a residence time of 1 min. Under the same conditions, in terms of water content and residence time, if the operating temperature is Tp, the process would give a degree of conversion of approximately 88%, which means that the contribution of thermal energy input to the conversion of starch materials in extrusion processing is highly predominant at extrusion temperatures close to Tp values. Eq. 6.21 is useful for estimating the rates of phase transitions of various starch materials at various water contents and temperatures, knowing that the source of starch and water content determine the value of Tp, which results from DSC measurements. It is worth noting that thermal activation energy, E_T, for the conversion of various starches at temperatures higher than 100°C varies from about 42 to 188 kJ/mol.

Conversion of waxy corn starch induced specifically by shear energy has been studied by Zheng and Wang (1994) at relatively high water contents in the range 30–35% (w/w), and low temperatures (from 21°C to 100°C), by use of a capillary rheometer and a single screw extruder. The authors also determined the shear rate constant, k_s, hence leading to the shear activation energy, E_s, and the pre-exponential coefficient, $k_{s,o}$. The shear activation energy for starch conversion was in the range of 2–290 cal/mol which is two or three orders of magnitude smaller than that of thermal activation energy. The pre-exponential coefficient was in the ranges of 6.6 10^6 to 20 min^{-1} at 35% water content, and 3.3 10^4 to 43 min^{-1} at 30% water content. The observed shear activation energy and pre-exponential coefficient varied importantly with processing temperature: the higher the operating temperature, the lower the activation energy and pre-exponential coefficient. Based on DSC endotherms, Wang et al. (1992) observed relatively high conversion during extrusion processing of waxy corn in a single screw extruder at relatively low temperatures (100°C and lower). For instance, at 80°C and 200 rpm, the total conversion degree was 0.864, with a shear energy contribution as high as 0.845, while the thermal contribution was low (0.019) as determined by Eq. 6.21. At 100°C and 200 rpm, the total conversion degree was significantly lower (0.708), with a shear energy contribution of 0.684 and a thermal contribution of 0.024.

These results show that high degrees of starch conversion (based on the evolution of DSC endotherms relative to phase transitions such as gelatinization and melting) can be obtained by use of shear energy contribution, at water contents of 30–35% (w/w) and at relatively low operating temperatures, i.e. at temperatures where thermal energy contribution is negligible. But these data cannot realistically be used to estimate the shear-induced rate constant in typical extrusion processing conditions (lower water contents and higher temperatures).

In extrusion processing of starch-based materials, the operating conditions in the screw-barrel assembly of screw extruders are characterized by water contents of approximately 14–26% (wwb) and temperatures of 130–180°C (product temperature). Under these conditions, the thermal energy contribution to starch conversion is usually predominant compared with the mechanical energy contribution. Eq. 6.21 is then useful to estimate the extent of starch phase transitions according to the time-temperature history in the screw-barrel assembly. Thermomechanical cooking of starch-based materials does not only include starch phase transitions, but also reactions such as depolymerization of starch polymers, or dextrinization. And as mentioned in section 6.2.2.3, shear energy does contribute importantly to the depolymerization of starch polymers.

During single screw extrusion of wheat starch, Diosady et al. (1985) studied the kinetics of molecular degradation, and developed the following reaction model derived from first-order reaction kinetics:

$$X' = f exp(k'\tau.t + b) + (1 - f) \qquad (6.22)$$

where X' is the extent of molecular degradation of starch polymers defined by the ratio of intrinsic viscosities as $[\eta_i]'/[\eta_{rs}]$, where $[\eta_i]'$ is the intrinsic viscosity of the final product and $[\eta_{rs}]$ is the intrinsic viscosity of raw starch, f is the fraction of the starch which is fully cooked, k' is the reaction rate constant (in $m^2.N^{-1}.s^{-1}$), τ is the mean shear stress (in $N.m^{-2}$), t is the time spent in the cooking section of the extruder (in seconds), and b is a constant. Plotting X' as a function of $\tau.t$ allowed the determination of the constants in Eq. 6.22 from extrusion experiments under the following operating conditions: temperature range of 79–121°C, water content range of 25–30%, and screw speed range of 50–100 rpm. The authors obtained values for k' and b of 0.39 $10^{-6}\,m^2.N^{-1}.s^{-1}$ and −0.08, respectively. The approach proposed by Diosady et al. (1985) is pertinent to estimating the kinetics of starch degradation in extrusion-cooking.

This brief review of the kinetics of thermomechanical conversion of starch indicates that there is not a simple and robust model to estimate the rate of thermomechanical conversion of starch materials, and consequently to design appropriately the screw profile and operating conditions to obtain a desired degree of conversion.

Accounting for important structural changes occurring during starch conversion, a two-stage reaction scheme for starch conversion in a limited amount of water such as that in thermomechanical cooking can be assumed to fit closely with real conditions during starch extrusion processing. This scheme consists of two irreversible reactions in series, with a first reaction that describes gelatinized/melted starch (GMS) from raw starch (RS) and a second reaction which corresponds to the depolymerization of gelatinized/melted starch (PDS), as follows:

$$RS \overset{k_1}{\rightarrow} GMS \overset{k_2}{\rightarrow} PDS \qquad (6.23)$$

where k_1 and k_2 are the rate constants for the first and second reactions describing starch conversion, respectively. Based on the literature review, a zero-order kinetics for the first reaction and first-order kinetics for the second reaction could be reasonably assumed.

By considering the first reaction illustrated in Eq. 6.23 as characterized by the rate constant k_1, an attempt to estimate the value of k_1 could be undertaken using Eq. 6.21. Suppose an extrusion-cooking process which converts raw corn meal at relatively high SME (120 $W.h.kg^{-1}$ and over), where the water content and operating temperature are 20% (wwb) and 160°C, respectively. Under these conditions, considering that Tp = 151.6°C, the estimated value of the rate constant would be 2.8 min^{-1}. Then, the minimum residence time to achieve full conversion of regular corn meal would be approximately 21 s. Actually, though the process operates thermomechanically and not entirely thermally, this value for the residence time compares well with the range of residence time observed in high SME applications to produce directly expanded extrudates (see Table 6.3). Now, suppose an extrusion-cooking process which converts raw corn meal at an intermediate level of SME (in the range of 80–120 $W.h.kg^{-1}$), where the water content and operating temperature are 20% (wwb) and 140°C, respectively. Under these conditions, the estimated value of the rate constant would be 0.32 min^{-1}. Then, the minimum residence time to achieve full conversion of regular corn meal would be about 188 s. This residence time is far longer than that for applications running with intermediate SME such as snacks pellets and cereal flakes which are currently characterized by residence times slightly lower than 1 min. Thus, the shear energy contribution to the conversion degree of starch must be important when the temperature is significantly lower than Tp.

For the second reaction included in Eq. 6.23, the approach developed by Diosady et al. (1985) could be used to estimate the rate constant k_2. But validated kinetic models are not yet available to estimate k_2 with a high level of confidence. The set of reactions schematically illustrated by Eq. 6.23 indicates that as soon as the first reaction stage proceeds, the gelatinized and melted starch polymers are subjected to depolymerization, or dextrinization. The rate of starch depolymerization corresponds to the rate at which WSI increases and the extent of depolymerization depends on the final WSI values. As seen in section 6.2.2.3, starch depolymerization is highly shear dependent or highly SME dependent. More specific studies are needed to produce a workable kinetic model for k_2, particularly in close coherence with the WAI–WSI master curve presented in Figure 6.17.

6.2.2.5 Rheological properties of starch-based melts

Knowledge of the viscous behavior of extrusion-cooked molten starch at low water content is required to determine optimal processing conditions and to better control the quality of end-products. As seen in Chapter 4 (section 4.4.3), the viscous behavior of glycerol plasticized thermoplastic starch for non-food applications depends on temperature, water and plasticizer contents, as well as thermomechanical treatment. Thus, factors affecting the viscous behavior of extrusion-cooked molten starch for food applications are mainly temperature, water content, and thermomechanical history in the screw-barrel assembly of the extruder.

Viscosity data and rheological models that take these factors into account have been reported in the literature. Martin et al. (2003) have summarized the main rheological studies performed on extrusion-cooked starch materials, including measurement techniques and models utilized for the description of the rheological behavior of these materials. A general pseudoplastic behavior is reported in a shear rate range 10^1–$10^3\ s^{-1}$, with an Arrhenius dependence of viscosity on temperature and an exponential dependence on water.

Structural modifications of starch according to the processing history have a strong influence on the viscosity of extrusion-cooked starch-based melts. Disruption of starch granules increases the melt viscosity whereas depolymerization decreases it. Senouci and Smith (1988a) proposed a model for the viscosity of molten potato granules, which contained a term depending on screw speed to account for structural modifications of starch. But this approach affected the viscosity differently according to the type of extruder and screw geometry used. Consequently, as proposed by Vergnes and Villemaire (1987), the effect of starch modification on the melt viscosity is better reflected by SME values specifically to describe the impact of macromolecular depolymerization. Vergnes and Villemaire (1987) proposed the following viscosity model for extrusion-cooked starch-based melts:

$$\mu_a = m\dot{\gamma}^{n-1} \tag{6.24}$$

where m and n are the consistency index (expressed in $Pa.s^n$, or $N.s^n/m^2$) and the flow behavior index (dimensionless), respectively:

$$m = m_o\ exp\left[\frac{E}{R}\frac{1}{T} - \alpha.MC - \beta.SME\right] \tag{6.25}$$

and:

$$n = n_o + \alpha_1.T + \alpha_2.MC + \alpha_3.SME + \alpha_{12}.T.MC + \alpha_{13}.T.SME + \alpha_{23}.MC.SME \tag{6.26}$$

where T is the temperature of the melt (in K), MC is the water content (g H_2O/g total wet weight), SME (in kW.h/metric ton), and m_o, α, β, E/R, n_o, α_i, and α_{ij} are coefficients which are calculated from multiple linear regressions of experimentally determined values of m and n.

Vergnes and Villemaire (1987) have carried out off-line measurements with maize starch, under controlled thermomechanical conditions using a preshearing rheometer called Rheoplast® which is able to simulate extrusion processing of polymeric melts. They have studied the effects of shear rate in the range 10–$10^3\ s^{-1}$, temperature in the range 110–170°C, water content in the range 0.21–0.33 and SME in the range 0.5–210 W.h/kg (assuming that the density of molten starch is constant and equal to $1330\ kg/m^3$ for an average water content of 0.27). The authors observed that molten maize starch exhibited a pseudoplastic, shear-thinning behavior ($n < 0.4$) and that the apparent melt viscosity, μ_a, decreased when temperature and water content increased. In addition, the apparent melt viscosity decreased significantly with increases of SME due to the starch depolymerization, which was confirmed by decrease of the intrinsic viscosity and increase of WSI measured on the material exiting the rheometer. The values of the coefficients of Eq. 6.25 and Eq. 6.26 for maize-based starch melt are reported in Table 6.5 (Vergnes & Villemaire, 1987).

Della Valle et al. (1995) studied the viscous behavior of potato starch processed in an intermeshing co-rotating twin screw extruder using the specially designed slit die rheometer described above for shear rates ranging from 10 and $600\ s^{-1}$, at temperature ranging from 120 to 190°C, water content ranging from 22 to 35% (wwb), and SMEs ranging between about 110 to 320 W.h/kg. Molten potato starch exhibited a pseudoplastic behavior, with a flow behavior index, n, and a consistency index, m, ranging from 0.34 to 0.65 and from about 1000 to 7600 $Pa.s^n$, respectively. The values of the coefficients of Eq. 6.25 and Eq. 6.26 for potato-based starch melt are reported in Table 6.5. It is observed that the dependency of m on temperature (coefficient E/R) and on depolymerization (coefficient β) is higher than that observed for maize starch (see Vergnes & Villemaire, 1987). This difference may be explained by the larger molecular weights of potato starch polymers, mainly amylose. Thus, compared with maize starch under identical conditions of extrusion, potato starch shows higher melt viscosity. But, if the comparison is made at similar degrees of polymerization or similar values of intrinsic viscosity, the difference in melt viscosity between the two starches is considerably reduced. This means that in thermomechanical extrusion, if the molecular weight of processed

Table 6.5 Values of coefficients of the rheological models (Eqs 6.24 – 6.26) of starch-based melts (ns, no significant effect in a 5% confidence interval).

Starch (Reference)	m_o (Pa.s^n)	E/R (°K)	α	β (kg/W.h)	n_o	α_1 (°C)$^{-1}$	α_2	α_3 (kg/W.h)	α_{12} (°C)$^{-1}$	α_{13} (kg/W.h.°C)	α_{23} (kg/W.h)
Maize (Vergnes & Villemaire, 1987)	7.36	4250	10.6	4.26×10^{-3}	0	6.59×10^{-4}	0.112	–	7.28×10^{-3}	–	–
[1]Maize (70% amylose) (Della Valle et al., 1996)	1.74×10^{-5}	9275	9.4	ns	−1.16	7.93×10^{-3}	1.31	ns	ns	ns	ns
[2]Maize (47% amylose) (Della Valle et al., 1996)	2.86×10^{-2}	7190	19.1	3.3×10^{-3}	1.81	7.36×10^{-3}	−1.93	-1.76×10^{-2}	ns	7.14×10^{-5}	3.31×10^{-2}
[2]Maize (23.5% amylose) (Della Valle et al., 1996)	0.163	6140	18.6	2.1×10^{-3}	3.54	1.54×10^{-2}	−3.19	-3.14×10^{-2}	ns	1.47×10^{-4}	4.06×10^{-2}
[3]Maize (0% amylose) (Della Valle et al., 1996)	1.36×10^{-4}	9700	26.1	5.5×10^{-3}	−1.02	7.20×10^{-3}	2.54	5.60×10^{-4}	ns	ns	ns
Potato (Della Valle et al., 1995)	0.34	5710	9.45	7.41×10^{-3}	14.33	−0.997	−60.6	0.044	0.418	-2.1×10^{-4}	−0.0341

[1] High-amylose maize Eurylon (70% amylose content).
[2] Starch blends obtained by mixing batches of high-amylose maize and waxy maize.
[3] Waxy maize (99% amylopectin content).

starch materials is reduced to the same level, identical melt viscosity can be found for conditions of water content and temperature, regardless of the botanical origin of the starch.

After deadstops and barrel opening, Della Valle et al. (1995) observed that potato starch exhibited early melting in the unfilled section of the screw extruder (conveying section), for values of SME higher than 110 W.h/kg, leading to a continuous molten phase before the material is compressed in the filled melting section. This melting mechanism is rather different from the melting mechanism typically observed during twin screw extrusion, which occurs very rapidly after compaction in the last right-handed screw element preceding a restrictive element (such as a left-handed screw element or kneading block). Such melting mechanism relies on numerous observations made on cereal starches (such as maize starch, wheat starch, maize grits, and wheat flour). The authors suggested that early melting of potato starch in extrusion processing might be due to its transition temperatures which are lower than values measured in cereal starches. For instance, DSC data on thermal transition of starch show that the melting temperature of potato starch is 15–30°C lower than that of cereal starches at similar water contents of about 15–30% (Donovan et al., 1983; Wang et al., 1992). Early melting in the conveying section of the extruder along with the high melt viscosity in the shearing sections leads to high energy requirements during extrusion processing of potato starch. As a consequence, macromolecular degradation was increased due to the more intense thermomechanical treatment. All extrusion practitioners know that potato starch exhibits a more difficult processing behavior than cereal starches. Ways to improve the processability of potato starch may include shortening the conveying section, reducing the barrel wall temperature in the conveying and melting sections, and reducing the screw speed slightly, in order to handle potato starch melting and depolymerization appropriately.

Della Valle et al. (1996) studied the viscous behavior of maize starches with various amylose contents (from 0% to 70% amylose content) processed in an intermeshing co-rotating twin screw extruder equipped with a specially designed slit die rheometer, over a wide range of shear rates (between about 10 and $2 \times 10^3\ s^{-1}$), at process SMEs higher than 115 W.h/kg, meaning that starch granules were totally disrupted and their crystalline structure had totally disappeared. As shown in Table 6.5, the coefficient α increases significantly when the amylose content decreases, meaning that the plasticizing action of water is more effective on the amylopectin polymer. The authors also observed a continuous increase of n and a decrease of m with increasing temperature. The values of the activation energy for pure starches (high amylose maize and waxy maize) are significantly larger than those of the two maize blends (47% and 23.5% amylose contents). This is probably due to a lower domain of temperature applied for pure starches (100–170°C), compared with those applied for the two maize blends (140–190°C). Below $10^3\ s^{-1}$ shear rate, though the molecular weight of amylopectin is much larger than that of amylose after processing, waxy starch shows lower melt viscosity values whereas amylose-rich starch has a more pronounced shear-thinning behavior.

Similar results are reported by Campanella et al. (2002) using an online rheometer developed by Li et al. (2004) to determine the rheological properties of extrudates. Della Valle et al. (1996) interpreted this result in terms of entanglements. In fact, as seen in Chapter 4 (section 4.1.2.1), the macromolecular structure of amylopectin shows clusters composed of short chain branches and despite its high molecular weight, it has been suggested that this feature reduces its ability to create entanglements, leading to lower viscosity when compared with that observed on extrudates prepared with amylose-rich starch. Above $10^3\ s^{-1}$ shear rate, a cross-over of the flow curves of amylomaize and waxy maize was observed.

Though the results shown above concern pure starches (maize starch and potato starch) without any added ingredient, the aforementioned rheological models proposed by Vergnes and Villemaire (1987) and Della Valle et al. (1995, 1996) can be used advantageously to estimate the apparent viscosity of starch-based melts in food extrusion-cooking. The effects of shear rate, water content, temperature, and SME on the apparent viscosity are particularly well described, as they vary in ranges which fit fairly well with extrusion-cooking conditions.

It is worth mentioning the report of Wang et al. (1990) who studied the viscous behavior of wheat flour processed in an intermeshing co-rotating twin screw extruder configured for the production of crispy flat bread. The extruder was equipped with a capillary-type rheometer which was placed between the screw-barrel assembly and a slit die plate (flat bread die forming). The geometry of the rheometer allowed the apparent wall shear rates to vary between 20 and $500\ s^{-1}$. Water content and process SME ranged from 18.4 to 23.8% (wwb) and from about 80 to 170 W.h/kg, respectively. In the experiments, the authors kept the melt temperature in the rheometer within the range of 170–185°C. Assuming isothermal conditions in the rheometer, a power law model was proposed to account for the effects of wall shear rate, $\dot{\gamma}$, water content, *MC*, and *SME*, as follows:

$$\mu_a = \mu^* \dot{\gamma}^{n-1}(1-\beta'.SME)exp(\alpha'.MC) \qquad (6.27)$$

where μ^* is a reference viscosity, n is the flow behavior index, β' and α' are constants. Based on the experimental data, the values of these four parameters were as follows: $\mu^* = 3.62 \times 10^5$ Pa.s; $n = 0.15$; $\alpha' = -8.71$; $\beta' = 9.8 \times 10^{-4}$ kg/kJ. This indicates that wheat flour in molten state would show a highly shear-thinning behavior, which was in good agreement with the study of Guy and Horne (1988). This model has been used to estimate the effects of various ingredients which may be added to the formulation of crispy flat bread such as bran, sodium chloride, sucrose, whey, lecithin, and palm oil (Wang et al., 1993). The model specifically developed for extrusion-cooked wheat-based melts can help extrusion process developers who deal with extrusion processing of crispy flat bread, as well as directly expanded breakfast cereals.

Starch-based melts also exhibit viscoelastic behavior. However, studies on the elastic properties of cereal-based melts are very limited, specially using on-line approaches. The measurement of the elastic component of molten starch associated with the first normal stress difference (N_1), is critical because conventional rheometers have difficulties in promoting the thermophysical treatment applied in an extrusion operation. The study of Senouci and Smith (1988b) related the entrance and exit pressure losses produced in a slit viscometer die with the elastic properties of potato starch. However, the validity of the exit pressure method is questionable as it may be subjected to flow disturbances, velocity rearrangements or melt expansion which affect the stress near the die exit, hence invalidating the fully developed flow assumption.

6.2.3 Steam-induced die texturization

The major applications of the extrusion-cooking process are for the production of expanded food products with a cellular structure that is generated by water flashing off either at the die exit (direct expansion) or in postprocessing units (downstream expansion). The cellular structure of extruded products defines their texture which represents an important quality attribute. Then, the understanding of the expansion phenomenon is of paramount importance, to allow product manufacturers to optimize textural characteristics according to process conditions. The purpose of this section is to study thoroughly the mechanisms of the steam-induced die expansion of directly expanded extrudates in relation to extrusion processing conditions and the characteristics of the extrudate.

6.2.3.1 Visualization of extrudate expansion

Fan et al. (1994) suggested that die expansion of extrusion-cooked starch-based melts consisted of an expansion phase followed by a shrinkage phase. Arhaliass et al. (2003) developed a visualization technique based on a CCD camera combined with an image-processing procedure to observe and qualitatively assess the dynamics of both the expansion and shrinkage phases at the die exit of an intermeshing co-rotating twin screw extruder. The injection of a tracer at the exit of the die insert and analysis of its trajectory enable the identification of the characteristic times of the expansion phenomenon. Figure 6.18 depicts an expanding starch-based melt at the die exit. This allows direct observation of the expansion phenomenon, the formation of steam-induced bubbles at the exit of a die insert, the evolution of the cell

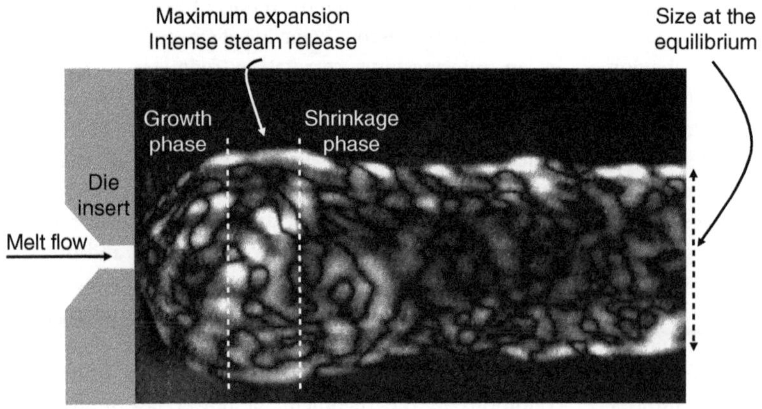

Figure 6.18 Visualization of steam-induced melt expansion at the die exit. Source: Arhaliass et al. 2003. Reproduced with permission of Elsevier.

network, and the partial recovery of the melt shape. Figure 6.18 also confirms the existence of two phases in the steam-induced die expansion which are described as follows.

6.2.3.1.1 Growth phase

The growth phase takes place quasi-adiabatically. It comprises the nucleation and growth of steam bubbles, up to a maximum expansion until steam is released to the environment. According to Arhaliass et al. (2003), the characteristic time of this growth phase varies between about 40 and 90 milliseconds which confirmed that the adiabatic growth phase occurs very quickly. At maximum expansion, the steam release also occurs very quickly, causing an open cellular structure and quick cooling of the melt down to approximately 100°C due to evaporative cooling caused by the flashing off of water. During steam release, the water content of the starch-based melt decreases significantly from 14–26% down to about 6–20% (wet basis), depending upon the characteristics of the melt at the die exit (specifically temperature and water content), and the melt state changes from a liquid to a rubbery state. When steam release is completed, based on the state diagram illustrated in Figure 6.12, the melt temperature may either reach the glass transition when the water content is low (below 8%) or be significantly over the glass transition if the water content is relatively high (above 10%). In both cases, the extrudate reaches a glassy state with high strength that somewhat freezes its deformation.

Expanding melt at the die insert can be considered as a network of steam bubbles surrounded by a viscous melt. Assuming that the melt is homogeneous and isothermal, each steam bubble follows the same expansion mechanism, and consequently behaves similarly to the expansion of the whole melt. Thus, the experimental data obtained from the expanding melt can be used to figure out the behavior of steam bubbles. Consider a symmetrical spherical vapor bubble surrounded concentrically by a viscous medium. The bubble growth is controlled by the pressure difference (ΔP) between the vapor pressure of water at the melt temperature inside the bubble and the atmospheric pressure, and by the viscosity of the melt as well as the surface tension (Amon & Denson, 1984). Taking into account the boundary conditions of the melt–vapor interface and neglecting the inertial effects (the melt is highly viscous) and surface tension effects, the dynamics of bubble growth in a viscous melt can be described by the following equation:

$$\frac{1}{R_b}\frac{dR_b}{dt} = \frac{V_m + R_b^3}{4(\mu_a)_d V_m}\Delta P \qquad (6.28)$$

where R_b is the bubble radius, V_m is the volume of viscous melt surrounding the bubble, $(\mu_a)_d$ is the melt apparent viscosity in the die insert, and $\Delta P = (P_v - P_a)$ is the pressure difference between the pressure inside the bubble, that is, the vapor pressure at melt temperature in the die insert, and atmospheric pressure. Kokini et al. (1992) have reduced Eq. 6.28 to Eq. 6.29 where only water vaporization and viscous forces control the volumetric melt expansion:

$$\frac{1}{R_b}\frac{dR_b}{dt} \propto \frac{\Delta P}{(\mu_a)_d} \qquad (6.29)$$

where $\Delta P/(\mu_a)_d$ is the driving force of melt expansion at the die exit. Eq. 6.29 means that volumetric expansion of extrusion-cooked melts at the die should be controlled by the driving force of expansion. Though important approximations are introduced in this simplified model, we will examine its ability to predict extrudate expansion later in the chapter.

6.2.3.1.2 Shrinkage phase

Shrinkage starts when steam release is completed. At this stage, the driving force of expansion equals zero, or the pressure inside the bubbles is equal to atmospheric pressure. At that condition, the liquid water of the melt is in equilibrium with its vapor phase at the corresponding temperature denoted by T_e. The temperature when shrinkage starts to occur is 100°C at high water content but is slightly higher than 100°C at relatively low water content since the vapor pressure of water in the extrudate may be lower than the vapor pressure of pure water, due to water absorbed tightly to the solid phase. According to Arhaliass et al. (2003), the characteristic time of that phase as measured from maximum expansion to the equilibrium recovery varied between about 25 and 115 milliseconds. Due to its quasi-instantaneity, melt shrinkage might be due to the elastic recovery of the rubbery melt after maximum expansion. During shrinkage, the structure of the expanded melt changes from an open to a closed cellular structure. At the equilibrium of shrinkage, the extrudate reaches a temperature at which bubble wall movement ceases because the apparent viscosity of the extrudate matrix becomes extremely high as it approaches the glass transition temperature. An extension of the work of Amon and Denson (1984) and further modifications described above were conducted by Hailemariam et al. (2007) who developed a model describing the dynamics of a bubble growing in a viscoelastic medium.

In a study to test the effects of different fibers on the expansion of fiber-enriched corn meal extrudates, Pai et al. (2009) found that viscoelastic properties of the melt are also important in controlling the expansion of extrudates. Determining apparent viscosity of melts by using a capillary viscometer under conditions similar to those used in the extrusion process as well as the squeezing flow technique to evaluate the extensional viscosity of fiber-enriched doughs, Pai et al. (2009) showed that both the apparent viscosity and the extensional viscosity of the melt may have an important effect on the expansion mechanism. Similar results were observed by Robin et al. (2012a) in a study to evaluate the effects of selected fibers on the extrusion and expansion of fiber-enriched formulations.

Mitchell et al. (1994) defined a shrinkage domain as the area comprised between two curves shown in Figure 6.19: the melt temperature curve, T_e, where the vapor pressure of the water in the bubble equals the atmospheric pressure and a temperature (associated to T_g) where the viscosity of the extrudate becomes too high to allow movement of bubble wall. This is the temperature when the product structure sets. Fan et al. (1994) stipulated that this temperature is about ($T_g + 30°C$). The extent of shrinkage is as high as the water content is high, or as the content of plasticizing medium (water and sugars for instance) is high when the melt starts to shrink.

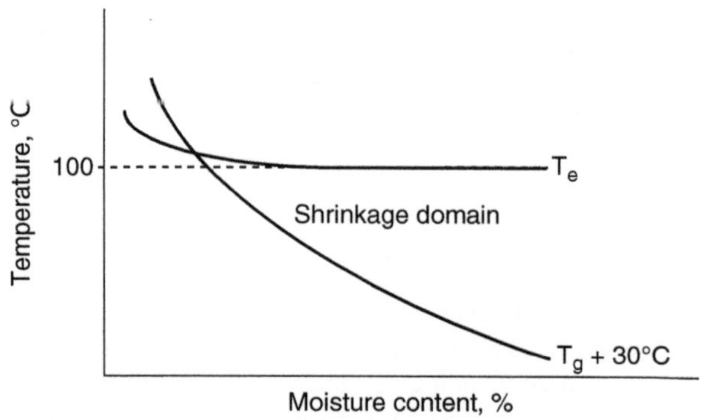

Figure 6.19 Schematic representation of the shrinkage domain of extrudate at the die exit. Source: Adapted from Mitchell et al. 1994.

6.2.3.2 Effect of process factors on extrudate expansion

6.2.3.2.1 Expansion indices and methodology

After completion of shrinkage at the die exit, the extent of melt expansion reached remains and becomes an important quality factor. Expansion of extrudates produces highly aerated products with relatively low bulk densities ranging roughly from 35–220 g/L for lighter products (such as directly expanded snacks and cereals) to 350–650 g/L for more dense products (such as directly expanded pet foods and aquafeed pellets). The large range of products that result from direct expansion extrusion requires a high level of experience and process control in order to handle the melt expansion appropriately, and also the product structure and structure-related properties such as texture and surface aspect of finished products. This is possible through investigation of the relationship between expansion criteria and extrusion-cooking factors.

In the 1980s, several researchers observed that melt expansion is not isotropic, and depends upon the extent of diametral and longitudinal expansion. Alvarez-Martinez et al. (1988) proposed introducing dimensionless expansion criteria to account for diametral and longitudinal expansion. These are the sectional expansion index (SEI) and the longitudinal expansion index (LEI) at the equilibrium and after the shrinkage phenomenon (Figure 6.20):

$$SEI = \frac{S_e}{S_d} \tag{6.30}$$

and:

$$LEI = \frac{v_e}{v_d} = \frac{\rho_d(1-MC_d)}{\rho_e(1-MC_e)}\left[\frac{1}{SEI}\right] \tag{6.31}$$

where S_e and S_d are the cross-sections of the extrudate and the die insert, respectively, v_e and v_d are the axial velocity of extrudate and the melt flow velocity in the die insert, ρ_e and ρ_d are the apparent density of extrudate and the melt density at the die, respectively, and MC_e and MC_d are the moisture content of the extrudate and the moisture content of the melt at the die, respectively. Finally, the volumetric expansion index, VEI, is defined as: VEI = SEI.LEI. VEI is inversely proportional to the density of extrudate.

Table 6.6 gives values of SEI and LEI indices obtained with extrusion-cooked pure starch-based materials such as corn grits, wheat flour, and a mix of wheat flour and

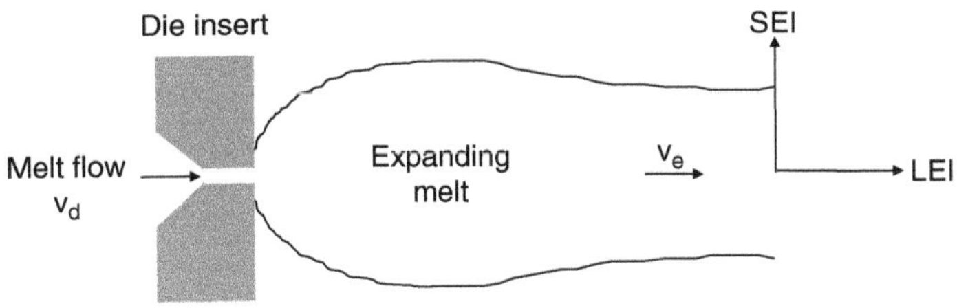

Figure 6.20 Dimensionless expansion indices of directly expanded extrudates.

Table 6.6 Examples of SEI and LEI values for extrusion-cooked starch-based materials.

Starch material	SEI	LEI
Corn grits	17.6	0.44
Wheat/potato (50/50)	8.5	0.77
Wheat flour	13.6	1.37

LEI, longitudinal expansion index; SEI, sectional expansion index.

potato flour (50/50). Corn grits show higher SEI and lower LEI. Note that LEI was less than 1 for corn grits and wheat/potato mix which means that the axial velocity of the extrudates was lower than the velocity of the melt in the die insert. Conversely, when expanding wheat-based melt, the axial velocity of extrudate increased in comparison with that of the melt in the die insert.

The relative importance of the SEI and LEI defines the extent of anisotropy of the product during expansion. Consider a cylindrical melt in the die with a radius R_d and a length l_d. If the melt expansion is isotropic, the radius and length of the expanded cylinder will become $R_e = \kappa . R_d$ and $l_e = \kappa . l_d$, respectively. Then, SEI and LEI will be equal to κ^2 and κ, respectively. Thus, an isotropic expansion should satisfy the following equation:

$$LEI = SEI^{0.5} \qquad (6.32)$$

Dimensionless expansion criteria are particularly suitable to describe the structure of the extrudate. An expansion chart illustrating values of LEI as a function of SEI can be used to present experimental observations when varying process factors. Figure 6.21 illustrates that expansion chart with iso-VEI curves ranging from VEI = 4 to VEI = 20, which corresponds to an expansion behavior of most directly expanded products. Figure 6.21 displays the isotropic expansion curve. The region above that curve corresponds to extrudates for which expansion occurs preferably in the axial direction (LEI-oriented expansion) whereas the region below the curve corresponds to extrudates for which expansion occurs preferably in the radial direction (SEI-oriented expansion). The expansion chart is commonly used to illustrate the effect of process factors on melt expansion. Basically, the expansion phenomenon depends on the rheological properties of the extrusion-cooked melt, and on die design which affects those properties. These rheological properties of the melt are affected by many process factors, which can be classified as follows.

- Factors relative to the mix formulation and composition fed into the extrusion-cooking process.
- Factors relative to thermomechanical cooking which occurs in the screw-barrel assembly of the extruder, including preconditioning if used.
- Factors relative to die shaping and texturization which occur in the die assembly.

6.2.3.2.2 Effect of mix formulation and composition

As seen in section 6.1.1.1, mix formulations for extrusion-cooking processes consist of components of four main categories. Specifically, components belonging to categories I (matrix-forming materials), II (rheology-modifying ingredients), and III (expansion-modifying ingredients) must be considered when dealing with die melt expansion. Thus, the effect of mix formulation (nature of components and composition) is reviewed in this section.

Matrix-forming biopolymeric materials

The maximum degree of melt expansion is related to starch content; the higher the starch content in the formulation, the higher the degree of expansion. For directly expanded products, a starch content of at least 60% (dry basis) is required for good expansion.

It is generally agreed that the decrease of particle size also leads to an increase of melt expansion (Chauhan & Bains, 1985; Mohamed, 1990). Desrumaux et al. (1998) studied the effects of corn grits particle size (in the range of about 100–520 microns) and the size distribution (the size distribution coefficient varied from 0.36 for a narrow distribution to 0.58 for a wide distribution)

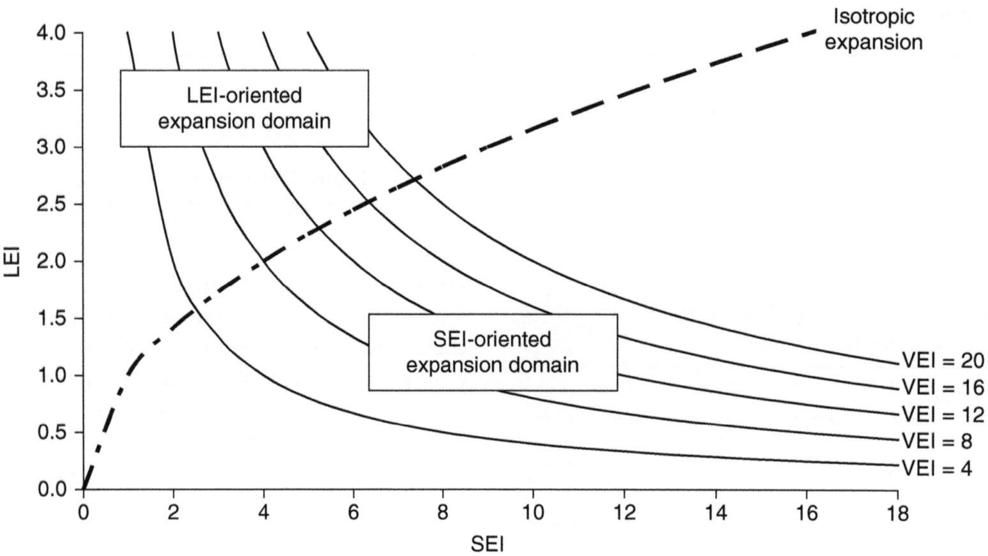

Figure 6.21 Expansion chart of directly expanded extrudates.

on the melt expansion criteria applied to direct expansion extrusion-cooking at constant SME (about 470 kJ/kg). These authors found that LEI and hence VEI decreased (about 35% decrease) when the average particle size of corn grits increased from about 200 to 500 microns, while SEI did not change significantly. Also, the authors observed that melt temperature at the die decreased (about a 10°C decrease) when the grits particle size increased. Thus, the decrease of LEI might be due to the decrease of melt temperature and the resulting increase of melt viscosity. They found that particle size distribution had no significant effect on melt expansion in the range investigated. It must be noted that such observations were obtained during twin screw extrusion processing. Results might have been different in single screw extrusion processing as single screw extruders are naturally more sensitive to particle size characteristics of the raw materials than twin screw extruders (see Chapter 4, section 4.2).

An important characteristic of the raw material is the amylose-to-amylopectin ratio in starches. Regardless of the approach used in exploring the role of amylose-to-amylopectin ratio in the expansion phenomenon, either by using reconstituted feeds by mixing pure amylose and amylopectin to reach a given ratio or using starches in which ratios change naturally, all studies agreed that decreasing the amylopectin content from normal (that is 75–70%) to about 30% as in high amylose corn leads to a decrease in SEI and very significant increase in values of LEI. For instance, Chinnaswamy and Hanna (1988b) studied the effect of amylose content on the expansion ratio of extrusion-cooked starch-based materials (with different amylose/amylopectin ratios); the expansion ratio is equivalent to SEI. The authors found a maximum expansion ratio at the optimum amylose level of 50%. Harper and Tribelhorn (1992) carried out extrusion-cooking experiments on a laboratory-scale single screw extruder through a factorial experimental plan in which three different native corns were used: waxy corn (0% amylose), regular cornmeal (28% amylose), and high-amylose corn (70% amylose). Extrusion processing conditions were barrel temperature, moisture content, and screw speed in the ranges of 150–170°C, 16–24% wwb and 80–120 rpm, respectively. The authors found that high-amylose corn produced extrudates with lower SEI and higher LEI, while regular cornmeal gave extrudates with the largest SEI and the smallest LEI. Generally, expansion indices of regular cornmeal were very close to those obtained from amylopectin meal. Conversely, the SEI of extrudates obtained from high-amylose corn was reduced by half compared to that of regular cornmeal while the LEI was three times larger than that of regular cornmeal. Harper and Tribelhorn (1992) also observed that the SEI was inversely related to the LEI. This general trend was further confirmed by several studies such as those of Tharrault (1992), Bouzaza et al. (1996), and Robin et al. (2010).

Adding protein-rich materials to starch-based formulations leads to significantly altered melt expansion criteria. Voisin (1993) measured a linear increase of LEI when adding soy flour (from 0% to 5%) or wheat flour (from 0% to 25%) to corn grits in direct expansion twin screw extrusion-cooking. At 5% soy flour or 25% wheat flour content, LEI increase was approximately 25% but SEI did not vary significantly. Voisin (1993) has observed similar results for LEI when extrusion-cooking a mix of wheat flour (50%) and potato starch granules (50%) in the same processing conditions. At 5% soy flour content, a LEI increase of about 25% and a SEI decrease of about 40% were obtained, leading to a notable decrease of VEI.

Rheology- and expansion-modifying ingredients

Though their role in mix formulations is important, the effects of functional ingredients such as fatty acids, emulsifiers, sugars, whey, fibers, salts, nucleating agents, etc. on melt expansion criteria have not been extensively investigated in the literature. This section reviews the data obtained by some authors, in order to enlighten extrusion practitioners who have to face the complexity of designing formulations for extrusion-cooking processes.

Effect of oils and emulsifiers

The presence of oil and fat in the extrusion formulation is able to form a dispersed phase that protects starch granules and has a lubricant effect during the compression and shearing action in the extruder (Guy, 2001). The protection of the starch granule by oil prevents the molecular dispersion of the starch and the presence of oil in the extrusion process tends to have an effect opposite to plasticization and increases the viscosity of the melt. This is due to the fact that starch molecules are less degraded and more intact. Emulsifiers are special forms of lipids, which have melting points higher than triglycerides but behave as oils that provide lubrication in the extrusion processes. Several types of emulsifiers such as fatty acids, monoglycerides and their esters form complexes with amylose, which provide a protective layer at the surface of starch granules during extrusion-cooking. The complexation of starch with lipids, specifically its linear fraction amylose, has an impact on the quality of the final extruded product in terms of its nutritional and textural properties.

Desrumaux et al. (1999) studied the role of fatty acids present in corn grits on melt expansion, in direct expansion twin screw extrusion-cooking at constant process conditions. The total content of free fatty acids varied from 0.16% (that is, the content of raw corn grits at normal storage conditions) to 0.8% (wwb) by adding free fatty acids to raw corn grits. Added free fatty acids had the same composition as that of the raw corn grits. Raw corn grits with relatively high contents of free fatty acids (higher than 0.16%) may result from enzymatic hydrolysis of triglycerides, depending upon storage conditions. As seen in Figure 6.22, melt expansion at the die was significantly affected by the presence of free fatty acids. Desrumaux et al. (1999) observed a reduction of about 60% in the SEI and an increase of about 48% in the LEI when free fatty acids increased from 0.16% to 0.8%. Both indices changed quasi-linearly, leading to a slight decrease of VEI. This result is particularly important as it reveals that melt expansion is highly sensitive to low changes in the content of free fatty acids of raw starch-based materials. As expansion behavior at the die drives the structural characteristics of directly expanded extrudates, it is worth checking the content of free fatty acids of the raw materials when facing structure-related quality problems in extrusion-cooking.

Voisin (1993) studied the effect of emulsifiers such as glycerol monostearate (GMS) and polyglycerol saturated ester (PGSE) on die melt expansion during twin screw extrusion of corn grits or wheat flour/potato starch granules (50/50). The content of emulsifier varied from 0% to 1.15% (dwb). Under these conditions die melt expansion was affected significantly by the presence of emulsifiers. As seen in Figure 6.22, a reduction of about 35% in the SEI and an increase of 32% in the LEI, at a GMS content of 1.15%, were observed. With corn grits, VEI remained constant when GMS content increased while VEI decreased slightly with GMS added to wheat flour/potato starch granules (50/50). In the presence of PGSE, the effect of emulsifier followed the same tendency as that observed when GMS was the emulsifier, but the changes of SEI and LEI were less than 10% for 1.15% emulsifier content (see Figure 6.22).

According to the physical chemistry of starch–lipid systems, the effect observed when free fatty acids and emulsifiers are present in starch-based formulations might be due to amylose–lipid complex formation, leading to a more rigid starch matrix that is less elastic. The effect is more significant as the complexation between the ingredients is higher. In fact, GMS exhibits a high complexation index whereas PGSE shows a low complexation index. As radial expansion is favored by the elastic properties of the melt, the loss of elasticity due to amylose–lipid complexation may explain the decrease of SEI and consequently the increase in LEI when the content of free fatty acids or emulsifiers is increased. It is worth noting

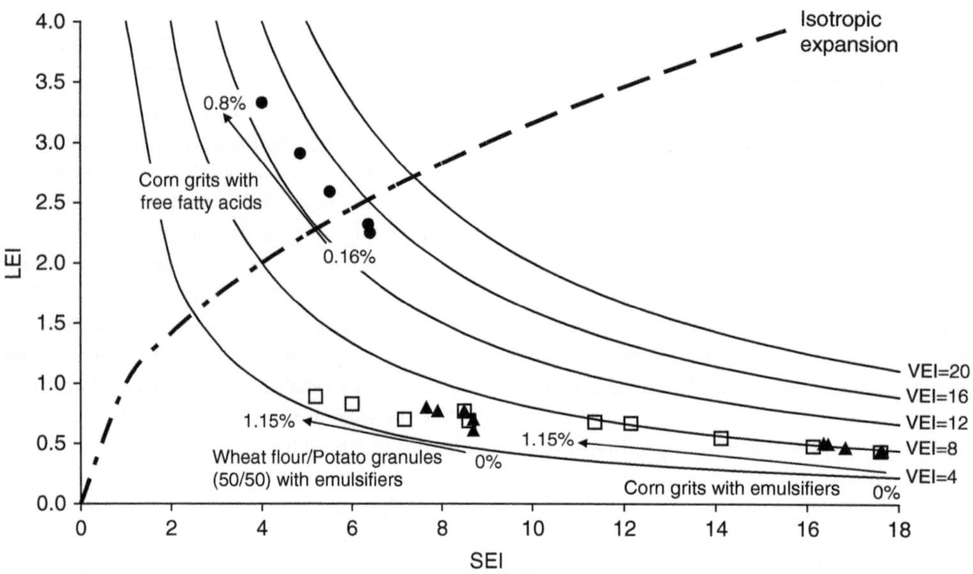

Figure 6.22 Effect of free fatty acids and emulsifiers on SEI and LEI (□: GMS; ▲: PGSE).

that adding free fatty acids or emulsifiers allows the direction of melt expansion to change drastically, with no or slight modification of the volumetric expansion. It means that it is possible to modify significantly the structure of extrudates at constant VEI, or at constant density, which is valuable from a practical standpoint.

Effect of sugars

Voisin (1993) studied the effect of sucrose (in the range of 0–11.5% dwb) on die melt expansion during twin screw extrusion of corn grits or wheat flour/potato starch granules (50/50). As illustrated in Figure 6.23, a very important decrease of SEI in the presence of corn grits (decrease of about 65%) and an increase of SEI in the presence of wheat flour/potato starch granules (increase of 34%), at a sucrose content of 11.5% (dwb) is observed. As for LEI, with corn grits a very slight increase of LEI was found (increase of about 15%) when sucrose content increased from 0% to about 3%. Above 3%, LEI then decreased to reach the original value obtained with pure corn grits. With wheat flour/potato starch granules, LEI showed a continuous decrease (decrease of about 25% at 11.5% dwb sucrose content).

Wang et al. (1993) investigated the effects of sucrose and whey (68% lactose) on die melt expansion of extrusion-cooked wheat flour in a twin screw extruder, at constant SME (600 kJ/kg). The authors observed an important decrease of SEI (50% decrease with both ingredients) when sucrose and whey increased from 0% to 15.8% (dwb) and from 0% to 7.8% (dwb), respectively. LEI showed a slight increase (about 10%) when sucrose content increased from 0% to about 7.5%, and then a slight decrease from 7.5% to 15.8%; at 15.8% sucrose content, the value was identical to that obtained with pure wheat flour.

In the aforementioned extrusion-cooking experiments, sugars replaced part of the starch in the mix. Then, adding sucrose or whey to starch–water mixes led to reduction of the apparent viscosity and elasticity of the resulting melt, and consequently the SEI index at all sugar contents as well as the LEI index at relatively high sugar content. As seen in section 6.1.1.1, sugars also contribute to the plasticization of starch during extrusion processing of starch/water/sugar systems, and consequently promote the decrease of glass transition temperature of sugar-rich starch-based formulations. As discussed in section 6.2.3.1, the decrease of the glass transition temperature increases the extent of the shrinkage phase, hence reducing melt expansion indices and particularly the LEI when sugar content is relatively high. However, this does not explain the effect of sucrose during extrusion of a mix composed of 50/50 wheat flour/potato starch granules, where SEI increased and LEI decreased with increase in the sucrose content. These results clearly show the complexity of the physical chemistry involved in the expansion process, and consequently the difficulty of making predictions when sugars are added to mix formulations.

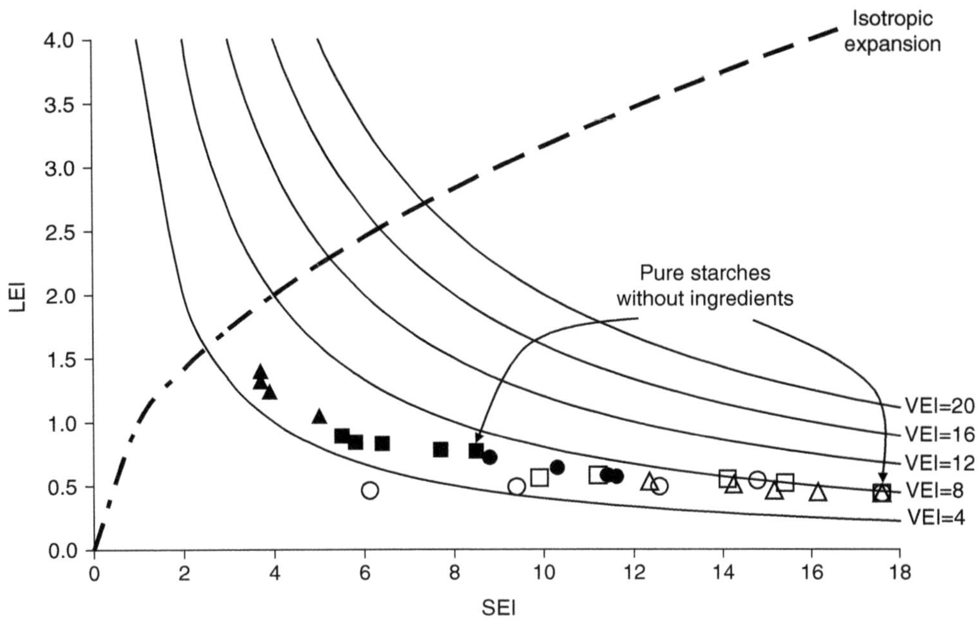

Figure 6.23 Effect of sucrose, sodium chloride and sodium bicarbonate on SEI and LEI (open symbols: corn grits; closed symbols: 50/50 wheat flour/potato granules; ●, ○: sucrose; □, ■: sodium chloride; △, ▲: sodium bicarbonate).

Effect of dietary fibers

Basically, dietary fibers are non-starch polysaccharides and are important ingredients as they have a large impact on the nutritional properties of extruded foods. The thermophysical treatment applied during extrusion can affect the structure of dietary fibers, in most cases enhancing the nutritional properties of the extruded products. Dietary fibers, however, may have drastic and negative effects on the expansion of extrudates when added in concentrations larger than 25%. It is hypothesized that the compact structure of dietary fibers is not compatible with the starch phase and a potential phase separation formed during the expansion of extrudates tends to increase the extensional viscosity of the melt, a property closely related to the formation of the cellular structure of extrudates, and negatively affects the expansion of the product (Pai et al., 2009).

The addition of dietary fibers to enrich foods is reviewed by Robin et al. (2012b), where it is claimed that increasing the dietary fiber content in extruded products is a challenging task because it often leads to reduced expansion volumes, higher densities, harder textures, less crispy and thus less preferred products. However, it is important to differentiate between soluble and insoluble dietary fibers because soluble fibers such as inulin may deliver higher expansion and more favorable textures compared to insoluble fibers such as cereal bran. The differences in the functional expansion characteristics between soluble and insoluble dietary fibers may be explained by their observed effects on product expansion. It has been demonstrated (Kale et al., 2010; Pai et al., 2009; Robin et al., 2012b) that the modification of dietary cereal fibers could contribute favorably to the expansion and quality of fiber-enriched extrudates. Dietary fibers which have shown physiological and nutritional benefits are classified as those that can be modified in an extrusion process and Robin et al. (2012b) review most of the work done up to date focusing on the changes of dietary fiber content as a consequence of the extrusion process. Among the effects observed can be included the significant increase of fiber solubility with the intensity of the extrusion treatment, measured by the SME, a process linked more to the mechanical rather than the thermal treatment (Robin et al., 2012b).

Insoluble dietary fibers contained in cereals such as wheat are composed of a xylose linear backbone and branched molecules consisting of arabinose and glucose, and also small amounts of mannose, galactose, and glucoronic acid. The compositions of other fiber sources are reviewed by Robin et al. (2012b) as well as the effects of the extrusion process on the structure of the molecules, where it was observed that mostly xylose and arabinose oligosaccharides are affected. It is worth mentioning the study of Wang et al. (1993) who investigated the effects of bran on die melt expansion of extrusion-cooked wheat flour in a twin screw extruder, at constant SME (600 kJ/kg). The authors observed an important decrease of SEI (40% decrease) and a slight increase of LEI (almost 40% increase) when bran content increased from 0% to 16.4% (dwb). However, VEI decreased slightly. Actually, bran particles tend to favor steam nucleation when melt expansion starts, and to rupture the cell walls of steam bubbles during expansion, hence limiting expansion phenomenon in the radial direction.

Further studies on fiber enrichment of extrudates have focused on the use of other flours and hydrocolloids that could improve the nutritional characteristics of the extrudates measured by their glycemic index (Brennan et al., 2008a, 2008b, 2012).

Effect of sodium chloride

In general, adding sodium chloride to pure starch in extrusion-cooking significantly affects die melt expansion. As seen in Figure 6.23, adding sodium chloride in the range of 0–4.5% (dwb), Voisin (1993) observed a decrease of SEI (decrease of about 45% and 35% with corn grits and wheat flour/potato starch granules, respectively, at salt content of 4.5% dwb) and an increase of LEI (increase of about 30% and 15% with corn grits and wheat flour/potato starch granules, respectively, at salt content of 4.5% dwb). With corn grits alone, the author observed a slight increase of VEI when the content of sodium chloride increased from 0% to 1.1%, and then a continuous decrease from 1.1% to 4.5% salt content. This is in agreement with the results of Chinnaswamy and Hanna (1988a).

Wang et al. (1993) studied the effect of sodium chloride during extrusion-cooking of wheat flour with a salt content varying from 0% to 3.9% (dwb). At low salt content (below 0.5%), SEI showed an increase of about 10%, while LEI decreased slightly (about 7% decrease), causing a slight increase of VEI. Above 0.5% salt content, SEI decreased continuously (SEI reduction of about 35% at 3.9% salt content, compared to the value obtained with pure wheat flour) whereas LEI showed an increase of about 25% at 3.9% salt content. Globally, VEI decreased significantly from 0.5% to 3.9% salt content.

Effect of sodium bicarbonate and sodium citrate

As seen in Figure 6.23, with corn grits, Voisin (1993) observed that sodium bicarbonate (in the range of 0–1.15% dwb) significantly affected expansion indices (about 30% decrease in SEI and 20% increase in LEI) while VEI decreased slightly. Sodium citrate (in the range of 0–1.15% dwb) had no significant effect.

The role of these two salts in the expansion indices of a mix composed of wheat flour/potato starch granules (50/50) was also investigated (Voisin, 1993). As seen in Figure 6.23, sodium bicarbonate (in the range of 0–1.15% dwb) led to an important decrease of SEI (about 55% decrease at 0.6% salt content) and a continuous increase of LEI (about 80% increase at 1.15% salt content), and VEI remained almost constant. When adding sodium citrate (in the range of 0–1.15% dwb), the effect was less important than that of sodium bicarbonate but the same tendency was observed: about 25% decrease of SEI and 25% increase of LEI. As for VEI, a significant decrease of about 25% was obtained.

The effect of the two salts on die melt expansion was definitely more important with sodium bicarbonate, meaning that the anion counterpart might predominantly drive the functional role of acidic salts. Besides, Voisin (1993) noticed that by adding sodium bicarbonate to two starch-based formulations, a brownish color developed, probably due to the fact that this salt could catalyze Maillard reactions during extrusion-cooking.

Figures 6.22 and 6.23 show clearly that die melt expansion is very sensitive to the biopolymer base and starch type, as well as to the ingredients which are added to food extrusion formulations. In the case of a corn grits base, the addition of ingredients modifies SEI to a large extent, and LEI to a lesser extent. With a 50/50 wheat flour/potato starch granules mix, both LEI and SEI varied extensively. The resulting structure of the extrudates was generally anisotropic. In fact, with pure starch materials, die melt expansion occurred preferably in the radial direction. However, the addition of ingredients such as free fatty acids and emulsifiers tended to promote melt expansion in the axial direction, and consequently to reduce the structural anisotropy of extrudates. Globally, an inverse relationship between SEI and LEI is observed. It must also be noted that the effects of ingredients are complex and difficult to predict, as they depend very much on the starch-based formulation.

6.2.3.2.3 Effect of thermomechanical cooking

Among the operating parameters which drive thermomechanical cooking, water content in the screw-barrel assembly affects melt expansion much more than any other parameter. With starch-based materials, a maximum expansion is generally obtained at low water contents (about 15–16%, wet basis). Beyond that level, as water content increases, SEI and VEI generally decrease. Alvarez-Martinez et al. (1988) studied the expansion of corn extrudates prepared in a single-screw extruder under various operating conditions. The results showed the highly significant effect of moisture, as SEI was reduced by half when water content was increased from 16% to 22%. However, moisture had a less significant effect on LEI. Interestingly, Harper and Tribelhorn (1992) showed that SEI values for high-amylose corn (70% amylose) were not as strongly influenced by the moisture content as regular cornmeal, while the effect of moisture on LEI values was notably greatest for high-amylose corn when compared to regular cornmeal.

Barrel wall temperature is another important operating parameter that directly influences die melt temperature. The effect of die melt temperature on LEI and SEI is almost identical for all starch-based materials. With increasing barrel wall temperature, VEI generally increased, particularly because the resulting melt temperature at the die was increased. Harper and Tribelhorn (1992) studied the effect of barrel wall temperature on expansion of corn extrudates prepared with a single screw extruder. With regular cornmeal, as melt temperature increased, LEI increased and SEI decreased very slightly. The authors observed that the decrease of SEI was significantly higher when high-amylose corn was used. However, it is worth discussing the effect of die melt temperature on melt expansion rather than the effect of barrel wall temperature, as die melt temperature results not only from heat exchange along the barrel but also from viscous dissipation in the die.

The effects of moisture content and temperature during extrusion-cooking are relevant because they allow the physicochemical mechanism which occurs during melt expansion to be interpreted. It is clearly observed that increasing the moisture content and temperature during extrusion-cooking shifts expansion from radial to longitudinal. This general trend has been further confirmed by Della Valle et al. (1997) and Robin et al. (2010). Higher moisture contents and melt temperatures lead to lower melt viscosities, which may favor disentanglement and alignment of the starch molecules as they flow through the screw-barrel assembly and die, leading to decreasing viscoelasticity of extrusion-cooked melts and a subsequent reduction in their ability to store energy during their flow through the die. Less stored energy then leads to decreased radial expansion or SEI, and consequently to increased axial expansion or LEI.

In extrusion-cooking, other operating parameters such as screw speed, throughput, and screw profile must be accounted for. However, it is inappropriate to assume that their effects on die melt expansion are independent. In fact, together with water content and barrel temperature, they are highly interactive, hence determining the time-temperature-shear history in the screw-barrel assembly of the extruder and consequently, the extent of conversion of the biopolymeric formulations. It is thus more relevant to discuss their influence by use of the parameter SME. Figure 6.24 presents the evolution of SEI and LEI as a function of SME, based on a set of various directly expanded extrudates produced from corn grits in a twin screw extruder (Clextral, 2013a). The extrudates were obtained by maintaining a constant throughput and barrel temperature, while varying the process conditions such as screw profile (different L/D ratios of the cooking

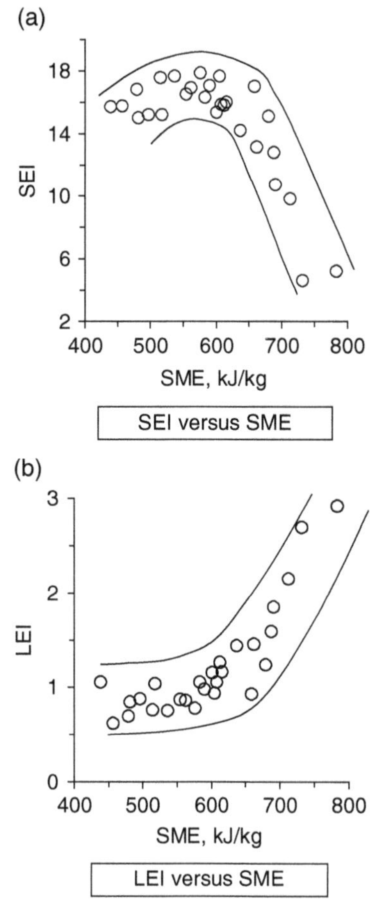

Figure 6.24 Expansion indices as a function of SME: (a) SEI; (b) LEI (Clextral, 2013a). Source: Reproduced with permission of Clextral, France.

section of the screw-barrel assembly), screw speed (in the range of 250–450 rpm), and water content (in the range of 14–19% wwb). These conditions allowed SME to vary from about 430 to 790 kJ/kg which ensured full gelatinization of the starch but at various extents of starch dextrinization. When SME increases, a slight increase of SEI is observed up to 500 kJ/kg followed by a maximum SEI over a rather large domain of SME (from about 500 to 640 kJ/kg). Above 640 kJ/kg, SEI decreases sharply due largely to starch dextrinization, hence causing an important reduction of starch elasticity and consequently an important decrease of SEI. In the SME range of 430–640 kJ/kg, as seen in section 6.2.2.3, the extent of starch dextrinization increases significantly and progressively depresses the ability of the starch melt to expand in the radial direction due to a loss of elasticity. However, LEI shows an inverse tendency compared with SEI; LEI remains almost stationary in the SME range of about 430–560 kJ/kg, then it increases continuously when SME increases from 560 to 790 kJ/kg. An inverse relationship between SEI and LEI is again observed.

It is worth illustrating the effects of screw configuration, screw speed, and water content independently by use of the SEI-LEI expansion chart. Figure 6.25a focuses on the effect of screw profile on expansion indices (Clextral, 2013a). The influence of six screw profiles has been investigated with pure corn grits (no ingredient added), at constant operating variables (throughput, screw speed, water content, temperature profile) and die assembly, using a twin screw extruder (total L/D of 18). The screw profiles differ only from the configuration in the cooking section (type and local L/D of shearing elements), as follows.

- *Profile A*: two separate stages of shearing, each composed of one conjugated, single-flight, left-handed pitch element (L/D ratio of 0.9 for each element) with standard half-moon openings (see Chapter 2, Table 2.2); the two stages are separated by one conjugated, single-flight, right-handed pitch element (L/D ratio of 0.9).
- *Profile B*: one single stage of shearing, composed of one conjugated, single-flight, left-handed pitch element (L/D ratio of 1.8) with large half-moon openings.
- *Profile C*: one single stage of shearing, composed of one bilobe kneading block (neutral 90° staggering; L/D ratio of 0.9), followed by one conjugated, single-flight, left-handed pitch element (L/D ratio of 0.9) with standard half-moon openings.
- *Profile D*: one single stage of shearing, composed of one conjugated, single-flight, left-handed pitch element (L/D ratio of 1.8) without openings but the opening area for melt flow across the shearing element is similar to that of profile B.
- *Profile E*: one single stage of shearing, composed of one trapezoidal, single-flight, left-handed pitch element (L/D ratio of 1.8) with rectangular shaped openings.
- *Profile F*: three separate stages of shearing each composed of one conjugated, single-flight, left-handed pitch element (L/D ratio of 0.9 for each element) with decreasing half-moon openings from large to small size openings; the three stages are separated by one conjugated, single-flight, right-handed pitch element (L/D ratio of 0.9).

The total length of shearing elements of profile F equals 2.7, as opposed to profiles A–E with a total length of shearing elements of 1.8D. Consequently, profile F allowed more mechanical energy to be supplied compared with the other profiles at constant processing variables. As seen in Figure 6.25a, extrudate obtained from profile F is well separated from the other extrudates. Profile F produces extrudates with the lowest SEI and the highest LEI and VEI values; its expansion is more isotropic than that of the other extrudates. Profiles A–E with less shearing intensity tended to produce extrudates with SEI and LEI ranging from about 15 to 17, and from 0.8 to 1.2, respectively, showing a highly anisotropic expansion.

Figure 6.25b gives some insight into the role of screw speed and water content, with various screw profiles (Clextral, 2013a). Profile G has one single stage of shearing, which consists of one bilobe kneading block (neutral 90° staggering; L/D ratio of 1.2). With profile G, a formulated mix of corn grits and a special design of die assembly have been used in the extrusion-cooking process, which explains the position of extrudates in the LEI-oriented expansion domain of the expansion chart. As shown in Figure 6.25b for profiles A, D and G, both SEI and LEI increase when water content decreases. When screw speed increases (profiles A and C) SEI decreases and LEI increases. With profile G, the effect of screw speed significantly increases LEI while SEI increases very slightly. In general, in the SEI-oriented expansion domain at high VEI, SEI decreases when screw speed increases. But when VEI and/or SEI are relatively low or when die melt expansion moves to the LEI-oriented expansion domain, the effect of screw speed on SEI may change at some point, such that SEI may increase slightly when screw speed increases while LEI always increases. It is worth discussing the extrudates of profile F in Figure 6.25b (cross symbols). Screw profile F allows the directions of melt expansion to be varied significantly at an almost constant volumetric expansion (constant VEI), by changing screw speed and water content. This means that it is possible to vary the

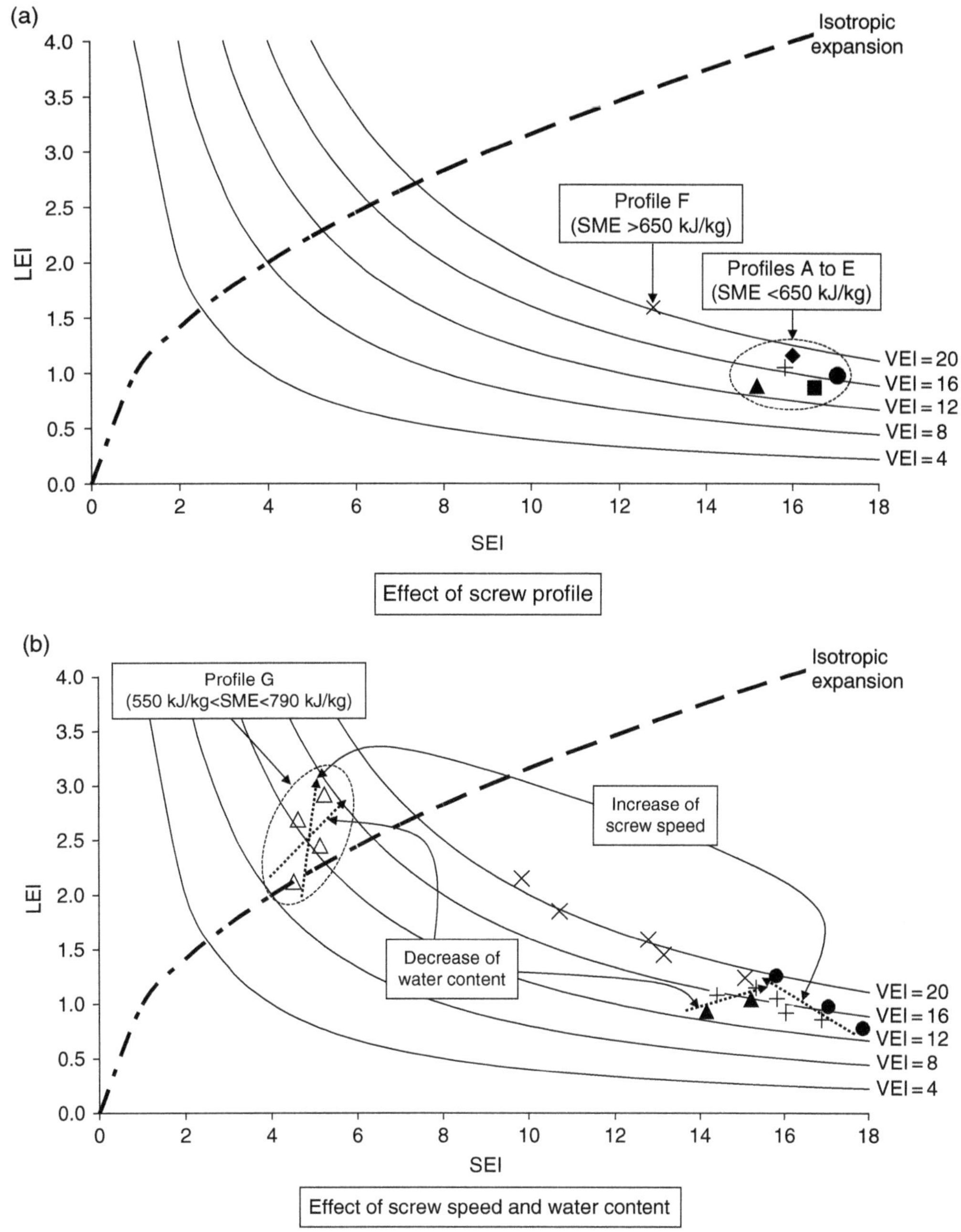

Figure 6.25 Effect of extrusion processing conditions on die melt expansion: (a) effect of screw profile; (b) effect of screw speed and water content (+: profile A; ■: profile B; ●: profile C; ▲: profile D; ◆: profile E; ×: profile F; △: profile G) (Clextral, 2013a). Source: Reproduced with permission of Clextral, France.

structure of extrudates to a large extent at constant bulk density, by use of processing variables such as screw speed and water content.

The different screw designs discussed above allow the die melt expansion to vary significantly in terms of VEI, SEI, and LEI. In practice, once a biopolymeric matrix is defined depending upon the targeted extruded product, the design of the screw profile constitutes the first important decision for extrusion operators when dealing with food extrusion processes and products. A large variety of screw designs is possible. Then, for a given formulation, throughput and die assembly, the manipulation of operating variables such as water content, screw speed, and barrel temperature allows die melt expansion and extrudate structure to be fine-tuned. In general, once process SME reaches the minimum value that guarantees full gelatinization of starch, further increase of SME leads to decreased SEI and increased LEI, and progressive movement from SEI-oriented domains to LEI-oriented domains in the expansion chart. This tendency may work successfully at constant VEI, if appropriate choices of screw profile and operating variables are made, and if the process SME remains within a reasonable range of values which depends on the targeted products. Over that range, VEI may then decrease dramatically due to excessive starch depolymerization. The present discussion clearly shows that the expansion chart may constitute an efficient database for process optimization.

6.2.3.2.4 Effect of die shaping and texturization

The last section of the screw-barrel assembly is often designed to adjust and control the temperature and viscosity of the cooked melt, and pump it into the die assembly. The die assembly then distributes the melt flow evenly into several unitary flows according to the final shape and size of the extruded product. As the melt flows into the die, it is subjected to velocity changes as well as shear and deformation that affect its thermorheological characteristics. In direct expansion extrusion processing, when the melt enters into the final die hole or die insert, it will reach its ultimate thermorheological process which would impose the expected physical characteristics such as shape, size, surface aspect, structure, and texture through steam-induced expansion. The control of die shaping and texturization is not simple as the expansion phenomenon is very sensitive to the interaction between the die insert design and the thermorheological characteristics of the melt. The purpose of this section is to discuss the influence of die insert geometry and melt characteristics on expansion indices.

Use of the visualization technique presented in section 6.2.3.1 allowed Arhaliass et al. (2003) to observe the effects of insert design on die melt expansion, at varying insert diameters (in the range of 5–8 mm) and insert land length (in the range of 12–80 mm), during extrusion-cooking of corn grits at constant process SME (500 kJ/kg). The throughput per insert varied from about 31.6 to 42.1 kg/h, hence applying mass fluxes (or feed rate of melt per unit area) from about $180.10^3\ g.s^{-1}.m^{-2}$ to $600.10^3\ g.s^{-1}.m^{-2}$. Figure 6.26 shows two pictures which illustrate the effect of insert diameter (diameters of 5 and 8 mm) on die melt expansion, at constant insert land length (29 mm) and throughput per insert. It shows clearly that insert diameter significantly affects melt

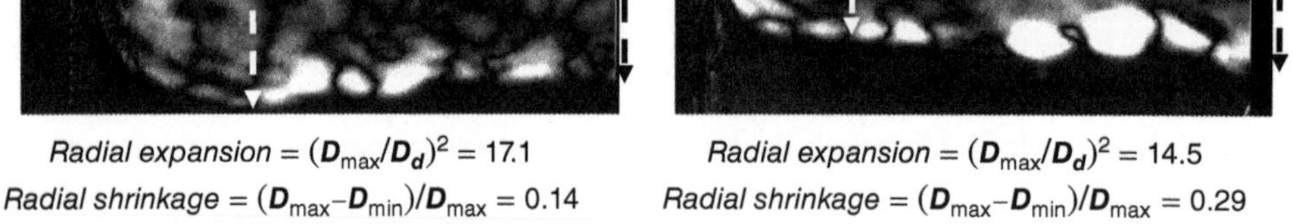

Figure 6.26 Effect of insert diameter on melt expansion. Source: Arhaliass et al. 2003. Reproduced with permission of Elsevier.

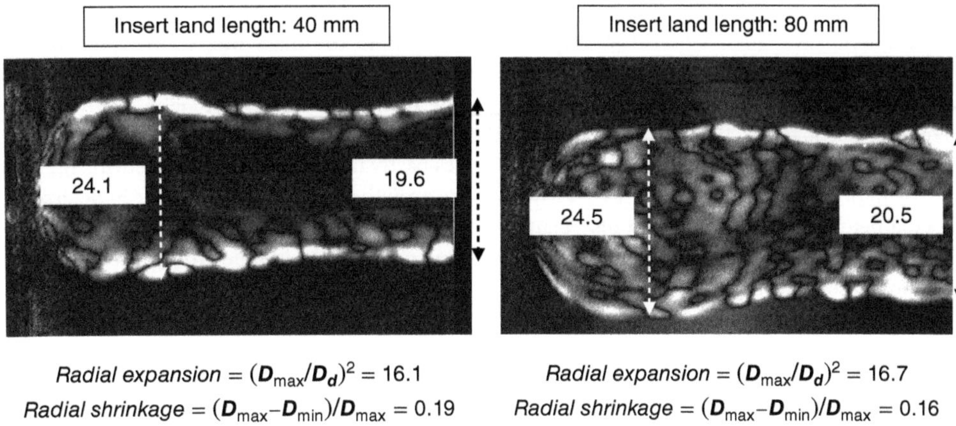

Figure 6.27 Effect of insert land length on melt expansion. Source: Arhaliass et al. 2003. Reproduced with permission of Elsevier.

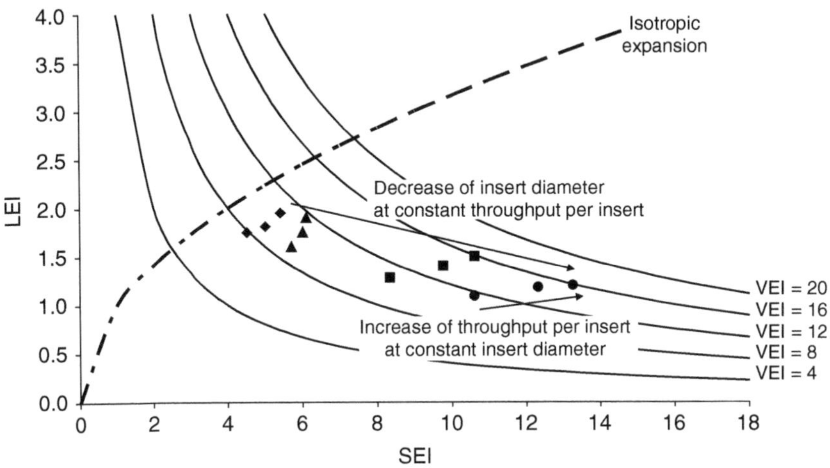

Figure 6.28 Effect of die insert geometry on die melt expansion (●: 5 mm; ■: 6 mm; ▲: 7 mm; ◆: 8 mm; throughputs per insert: 31.6 kg/h, 36.9 kg/h and 42.1 kg/h). Source: Arhaliass et al. 2003. Reproduced with permission of Elsevier.

expansion. Of course, the larger the insert diameter, the larger the maximum diameter of expanding melt. But the larger the insert diameter, the lower the radial expansion and the larger the radial shrinkage. This means that die inserts characterized by low mass fluxes would result in extrudates with relatively low SEI values. Figure 6.27 shows two pictures which illustrate the effect of insert land length (land length of 40 and 80 mm) on die melt expansion, at constant insert diameter (6 mm) and throughput per insert. As illustrated, the insert land length has very little effect on radial expansion and radial shrinkage.

Figure 6.28 presents the expansion chart of final extrudates (after completion of shrinkage and cooling) resulting from the investigation carried out by Arhaliass et al. (2003), while varying both the insert diameter (in the range of 5–8 mm) and the insert land length (in the range of 12–80 mm), at constant process SME (500 kJ/kg). The chart shows that the structure of directly expanded extrudates is strongly affected by the mass flux flowing into the final die insert. As the mass flux increases due to the decrease of insert cross-section (at constant throughput per insert), SEI and VEI increase significantly whereas LEI decreases. But, as the mass flux increases due to the increase of throughput per insert (at constant insert diameter), SEI and LEI increase slightly. As with die swell in plastics forming, the increase of SEI when insert diameter decreases might be associated with the elastic properties of starch-based melts. It is generally agreed among researchers that the swelling of extrudates emerging from a capillary is typical of non-Newtonian viscoelastic liquids and is believed to be related to their elastic properties. It is known that extrusion-cooked starch-based melts are non-Newtonian viscoelastic fluids, thus favoring die melt expansion in the radial direction as a result of the recovery of the elastic deformation imposed in the die insert. And, as the die insert diameter decreases at constant insert land length, the elastic energy stored by the melt while flowing in the die insert increases, causing larger SEI due to the recovery of the elastic deformation induced by flow restriction at the die outlet.

As seen in section 6.2.3.1, bubble growth in a high viscosity fluid can be described by Eq. 6.29 which predicts that the rate of bubble growth should be proportional to the driving force of melt expansion $\Delta P/(\mu_a)_d$. This tells us that the lower the apparent viscosity of the melt for a given pressure difference, the greater the rate of bubble growth. And the higher the pressure difference or the melt temperature at the exit of the die insert, the greater the rate of bubble growth. Thus, a linear relationship between the volumetric expansion index of the extrudates and the driving force of melt expansion should be experimentally observed.

Figure 6.29 shows the plot of VEI as a function of $\Delta P/(\mu_a)_d$ by use of the data obtained by Arhaliass et al. (2003). With the exception of VEI data derived from the smaller insert diameter (5 mm), the graph of VEI versus $\Delta P/(\mu_a)_d$ shows an acceptable linear correlation. This is in good agreement with the investigation of Kokini et al. (1992). With an insert diameter of 5 mm, VEI values are relatively low compared with those predicted by the bubble growth

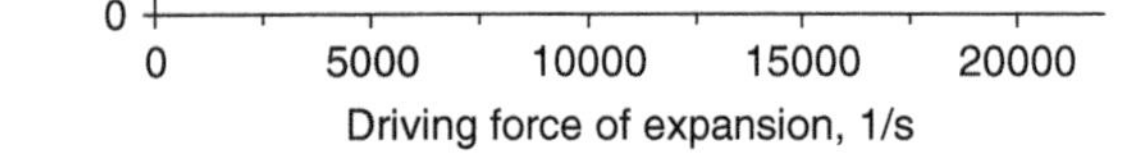

Figure 6.29 VEI as a function of driving force of expansion (◇: 5 mm, △: 6 mm, ○: 7 mm, □: 8 mm, at constant insert land length of 29 mm; ▲: insert land length varying from 12 to 80 mm, at constant insert diameter of 6 mm). Source: Arhaliass et al. 2003. Reproduced with permission of Elsevier.

model. Reducing the insert diameter will cause an increase in the mass flux of melt and therefore reduce the melt apparent viscosity due to the increase of shear rate in the insert. At first sight, this would be expected to increase and not reduce VEI. The answer to this issue lies partly in considering bubble shrinkage as well as bubble growth as an important factor in determining the VEI of final extrudates. Hence, for a given type of extruded food product, the bubble growth model could give a fairly good prediction of VEI within a certain range of mass flux, to keep the extent of melt shrinkage low. Finally, with the exception of data related to a 5 mm insert diameter, the linear relationship between VEI and $\Delta P/(\mu_a)_d$ is the following: $VEI = 9.10^{-4}\Delta P/(\mu_a)_d + 6.05$ (with $R^2 = 0.92$).

6.2.3.3 Extrudate shrinkage

As shown in section 6.2.3.1, melt shrinkage starts at temperature T_e, where the driving force of expansion equals zero, and proceeds as long as the extrudate temperature remains over a temperature of about $(T_g + 30°C)$ (see Figure 6.19). The extent of shrinkage domain depends on how these temperatures are affected by mix formulation and the process SME.

If the mix formulation contains significant quantities of low molecular weight ingredients which affect the physicochemistry of the water–starch system, both T_e and $(T_g + 30°C)$ lines will be changed. In fact, the addition of low molecular weight solutes such as sugars or salts may significantly lower the water vapor pressure at equilibrium and hence increase T_e values. Though there are no data available on the thermodynamics of formulated water–starch systems at high temperatures, it is well known that T_e should increase when low molecular weight solutes are added in significant quantities as in extrusion-cooking; the T_e line will switch to the T_e' line. Concerning the $(T_g + 30°C)$ line, as already seen in Chapter 4 (section 4.3.2), the addition of low molecular weight ingredients in water–starch systems such as plasticizers reduces the glass transition temperature and the melt viscosity. Similarly, at relatively high process SME which corresponds to a relatively high extent of starch dextrinization, a significant decrease of the average molecular weight of starch polymers would lead to a decrease of the glass transition temperature, and hence the $(T_g + 30°C)$ line of the resulting melt; the $(T_g + 30°C)$ line will switch to the $(T_g' + 30°C)$ line. As shown in Figure 6.30, adding low molecular weight solutes to water–starch systems or operating at high process SME would increase the shrinkage domain, and have a significant effect on the density and structure of resulting extrudates.

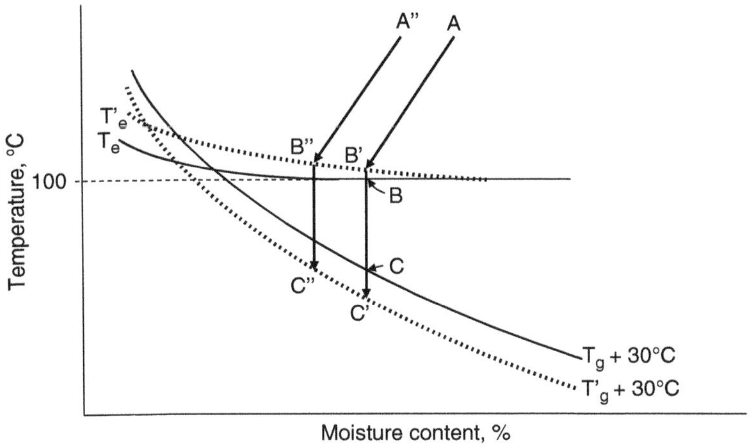

Figure 6.30 Schematic representation of the shrinkage domain with formulated mixes and high process SME in extrusion-cooking. Source: Adapted from Mitchell et al. 1994.

Let us consider a standard starch-based mix which is extrusion-cooked at constant moisture content and temperature conditions, and at a standard process SME (in the range of 450–600 kJ/kg). If the resulting melt exits the die insert at point A on the phase diagram, according to Figure 6.30 it will enter the shrinkage domain at point B, and will shrink proportionally to the segment BC. Suppose the same starch-based mix is enriched with significant quantities of low molecular weight soluble ingredients and is extrusion-cooked under the same process conditions defined by the moisture content, temperature, and SME. The resulting melt will exit the die insert at point A and enter the shrinkage domain at point B' and shrink according to the path from B' to C'. The magnitude of shrinkage for that formulated water–starch mix will then be proportional to the length of the segment B'C', and its shrinkage will be substantially greater than that of the standard starch–water mix (B'C' > BC). For the same water–starch mix processed at high SME (SME >650 kJ/kg), the magnitude of shrinkage will be even greater. In addition, the decrease of melt viscosity due to the addition of low molecular weight ingredients will reduce the melt temperature at the die exit. Consequently, a combination of ingredients that favor shrinkage, lower melt temperature and high process SME will cause large extrudate shrinkage and substantial increase in product density.

In practice, for a given mix formulation and throughput, the shrinkage phenomenon must be accounted for during extrusion-cooking of starch-based materials. And when the mix composition leads to significant extrudate shrinkage, appropriate processing conditions such as water content, screw profile, screw speed, and temperature profile must be selected in order to control the extent of shrinkage according to the required extrudate density and structure. For instance, by reducing the water content of the formulated water–starch mix, the resulting melt may exit the die insert at point A" on the phase diagram, shrinkage will start at point B" and follow the path from B" to C", resulting in a shrinkage magnitude proportional to the length of the segment B"C", with B"C" < B'C'. Hence, a lower water content extrudate will shrink less and have a lower bulk density while keeping the mix composition constant.

6.2.3.4 Expansion chart of directly expanded extrudates: main trends

In extrusion-cooking, steam-induced shaping and texturization of directly expanded extrudates is very important as it defines relevant properties such as shape, density, structure, and texture of end-products. These properties result from the expansion phenomenon at the die exit as characterized by expansion indices (SEI and LEI) and shrinkage. It is worth synthesizing the effects of the main process factors on extrudate expansion and structure by use of the expansion chart. The example of corn-based raw materials is illustrated in Figure 6.31.

6.2.3.4.1 Mix formulation and composition

In general, when typical extrusion processing conditions are applied (SME in the range of 450–600 kJ/kg and mass flux per insert in the range of 400–800.10^3 g.s^{-1}), pure starch-based materials (without functional ingredients) are characterized by relatively high SEI and low LEI values (SEI >10; LEI <1.5). Extrudate expansion is definitely

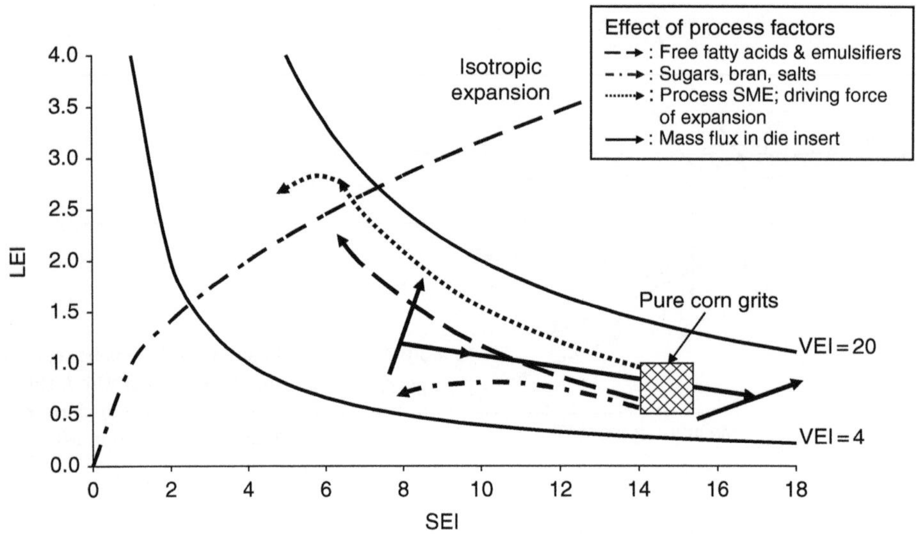

Figure 6.31 Expansion chart of directly expanded extrudates: main trends associated with the increasing effect of the major process factors (corn-based raw materials).

favored in the SEI direction because of the elastic properties of starch polymers. Adding amylose-complexing ingredients such as free fatty acids, emulsifiers, and oils will cause an increase of LEI and a decrease of SEI due to a significant loss of polymer elasticity. VEI may decrease slightly if the content of ingredients remains in practical ranges of classic industrial formulations. The changes promoted by these ingredients clearly show that they may induce an isotropic extrudate expansion. Adding plasticizing ingredients such as sugars, malt, and whey as well as ingredients that promote bubble nucleation such as bran and salts will notably decrease SEI and slightly modify LEI, with an important decrease of VEI. With these ingredients, extrudate expansion is often affected significantly by extrudate shrinkage which markedly depresses VEI and therefore increases extrudate density. If high extrudate expansion is required, then the content of such ingredients must be kept as low as possible and appropriate conditions of thermomechanical cooking must be adapted to target the expected extrudate density and structure. In practice, such ingredients may be successfully used to optimize and control extrudate expansion and structure.

6.2.3.4.2 Thermomechanical cooking

For a given formulation and required conditions for die shaping and texturization, thermomechanical cooking is controlled by four main variables: water content, screw profile, screw speed, and temperature profile, keeping in mind that the process throughput is generally maximized. By controlling these variables, the extent of thermomechanical cooking, or process SME, could be fine-tuned according to process and product requirements. In general, increasing process SME in the range of 500–650 kJ/kg decreases SEI and increases LEI, and therefore crosses the isotropic expansion line; this can be done at constant VEI, which allows the extrudate structure to be changed at constant density. Under severe extrusion-cooking conditions, when SME exceeds 650–700 kJ/kg, both SEI and LEI tend to decrease. Under these conditions, VEI decreases significantly due to excessive starch dextrinization, which causes extrudate shrinkage.

6.2.3.4.3 Die shaping and texturization

For a given formulation and a defined thermomechanical cooking imposed on the material, die shaping and texturization processes depend closely on the mass flux through the die insert, as well as on the thermorheological characteristics of the melt at the die exit. Mass flux in the die insert is determined by the insert diameter (or equivalent diameter for non-cylindrical inserts) and the mass throughput per insert. Increasing the mass flux at a constant throughput per insert (achieved by reduction of insert diameter) will increase SEI and VEI, and will decrease LEI to a greater extent, hence leading to increasing anisotropy in the extrudate expansion. An insert of low diameter then allows expansion of the extrudate to be developed in the radial direction, due to the elastic properties of a starch-based melt. Increasing the mass flux at constant insert diameter which increases throughput per insert will significantly increase both SEI and LEI. It must be noted that VEI increases almost linearly when the driving force of melt expansion increases.

In direct expansion extrusion processing, regardless of the variations in the process variables, when SEI increases LEI generally decreases, leading to an inverse relationship between these two expansion indices (see Figures 6.22, 6.23 and 6.25b: screw profile F). An inverse relationship between SEI and LEI has also been observed by other researchers (Alvarez-Martinez et al., 1988; Bouzaza et al., 1996; Harper & Tribelhorn, 1992; Robin et al., 2010; Tharrault, 1992). Thus, the relationship between SEI and LEI can be represented by a classic power law equation: $LEI = K(SEI)^a$, where K and a are constants. For example, using data from Voisin (1993) and Desrumaux et al. (1999), who specifically studied the effects of emulsifiers and free fatty acids on the expansion of corn-based melts, the values of the constants K and a were 7.2 and −1.1 for the corn/emulsifier systems, and 11.1 and −0.85 for the corn/free fatty acids systems, respectively.

Though the aforementioned trends are generally valid when adding ingredients one by one to pure starch-based materials, it turns out that quantification of the expansion behavior of formulations with multi-ingredients is still a difficult task. In fact, with such formulations the physical chemistry events that occur in the extrusion-cooking process are very complex, and therefore the physicochemical and rheological properties of the resulting melt are unpredictable. Consequently, the best approach to designing suitable extrusion-cooking processes lies in building relevant databases from multiple experimental runs, trying to utilize the aforementioned trends to minimize the number of experimental trials. The expansion chart presented in section 6.2.3.2.1 (see Figure 6.21) is an efficient and reliable tool for organizing experimental results, and optimizing food extrusion processes and extruded products. Extruders can be regarded as food cookers in which raw food ingredients are converted into structured products and the main question that interests product designers is how to control the structure of such products according to consumer acceptance. This aspect will be discussed in Chapter 7.

6.3 Examples of industrial applications: food extrusion processing lines

The cooking characteristics of extrusion technology enable cereal processing industries to process any starch- and protein-based formulations, at high productivity and product diversity. Compared with conventional batch hydrothermal cooking at temperatures of 90–110°C and 45–90 minutes processing time, continuous thermomechanical cooking operates at relatively high temperatures and short residence times. This is crucial in making extrusion-cooking processes more productive; in particular, in addition of thermal heating mechanical energy also contributes to the cooking of the material. Although both hydrothermal and thermomechanical cooking processes ensure full conversion of biopolymers, there are substantial differences between the two types of cooking processes. Due to the simultaneous inputs of both thermal and mechanical energies as well as the conversion of biopolymers in limited water environments, the extrusion-cooking process provides specific macromolecular structures and conformations and molecular weight distributions, which are significantly different from those obtained in hydrothermal cooking, hence leading to a wide diversity of functional properties of extruded biopolymeric melts.

Over the last few decades, the success of extrusion-cooking technology has led to important developments in extrusion-textured products and related food extrusion processes. Table 6.3 has given an overview of the major products and processes which are recognized worldwide. Extrusion-cooking belongs to the "second-stage conversion" process in the whole cereal and protein processing industry, from raw agricultural products to consumers. In the cereal and protein processing industry, "first-stage conversion" processes generally consist of grain or seed milling processing to physically fractionate the main constituents of these raw materials and produce basic raw ingredients such as flours, grits, bran, oils, etc. Extrusion-cooking is a second-stage conversion process aimed at converting starch- and protein-based flours and grits into functional products for either the business-to-business (B-to-B) market or the business-to-consumer (B-to-C) market.

In spite of the large diversity of products and processing conditions, industrial extrusion-cooking processes correspond to the same core process shown in Figure 6.32. This core process generally starts with starch- or protein-based raw materials (flours or grits) and processes them through four main unit operations:

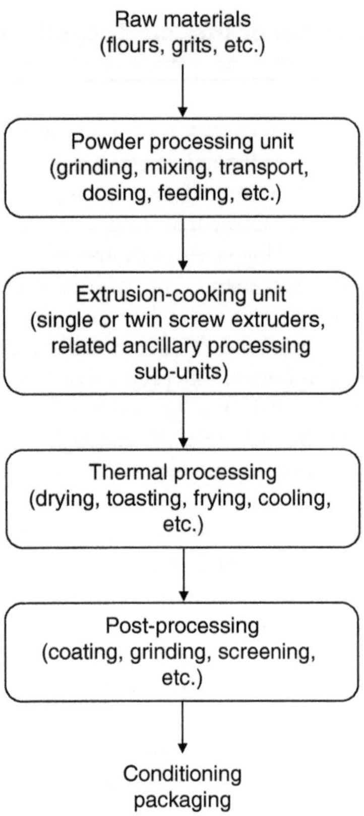

Figure 6.32 Core process diagram of industrial extrusion-cooking processes.

• powder processing (grinding to fine-tune particle size characteristics, mixing biopolymer-based raw materials with functional ingredients, dosing and feeding)
• extrusion-cooking in a single or twin screw extruder (including possible ancillary subunits such as preconditioning, direct steam heating, venting, cutting and forming, etc.)
• thermal processing (such as drying, toasting, oil frying, cooling, in particular)
• postprocessing treatments (such as flavored-oil or flavored-sugar coating, grinding and screening).

Some examples of relevant industrial extrusion-cooking processes are now presented to illustrate the diversity of food extrusion processing and extrusion-textured food products mentioned in Table 6.3.

6.3.1 Breakfast cereals extrusion processing

Cooked wheat, barley or oats in the form of porridge have been common foods for many hundreds of years, and they are still very popular in several countries. However, pre-processed, cereal-based products, such as ready-to-eat cereals to be eaten directly out of the package, are quite new. When one speaks of ready-to-eat cereals, or breakfast cereals, the first thought that comes to mind is that of corn flakes for breakfast. Since they were first produced more than a century ago by W.K. Kellogg in the United States, corn flakes have lead a vigorous development in the cereal industry, and stimulated creativity in the marketing and designing of breakfast cereal products.

The popularity of breakfast cereals stems from their nutritional content. Such products can simultaneously provide energy (350–400 kcal/100 g), nutrients (vitamins, minerals), and healthy components (e.g. dietary fiber, essential fatty acids) and well-being through organoleptic acceptability given by factors such as appearance, taste, and texture. Throughout the long history of breakfast cereals developments, advertising has always underlined the health potential of the products, and consumers do recognize such beneficial input. Today, the impact of nutrition on health and disease control is becoming a major concern of food designers, food manufacturers, and consumers. Thus, the healthy eating trend increasingly influences food choices and breakfast cereals do not escape these developments.

Over the last three decades, extrusion-cooking technology has played a decisive role in the innovation and development of breakfast cereal products. The extrusion-cooking process allows continuous cooking of a large range of recipes using various cereals to produce different shapes and textures at satisfactory costs. Two types of extrusion-cooked breakfast cereals can be found on the market.

• *Directly expanded breakfast cereals.* Cereal flours and/or grits are cooked with selected ingredients at very low moisture content (usually below 20%). The process may use single or twin screw extruders with configurations and operating conditions that generally lead to relatively high mechanical cooking.
• *Pellet-to-flake breakfast cereals.* Cereal flours and/or grits are cooked with ingredients at a moisture level in the range of 22–26%. They are usually processed in twin screw extruders and the configuration and operating conditions are selected to apply a low cooking mechanical component, as opposed to the production of directly expanded breakfast cereals.

This section describes the two main extrusion processes which allow the production of these two types of cereals.

6.3.1.1 Direct expansion extrusion-cooking process

In Table 6.3, this process relates to SME values of 120 W.h/kg and over. Figure 6.33 presents a typical flowsheet of a twin screw extrusion processing line for the production of directly expanded breakfast cereals. The process consists of four main unit operations (Clextral, 2013b).

- Powder processing unit: mixing of raw materials and functional ingredients.
- Extrusion-cooking unit: thermomechanical cooking in the screw extruder and die texturization.
- Drying unit: may include toasting, in the case of crisp rice, for instance.
- Flavored-sugar coating/drying and cooling unit.

The dry raw materials, usually a mixture of flours (corn, wheat, rice or oats) and functional ingredients (modified starches, bran, sugar, emulsifiers, sodium chloride, calcium phosphate, etc.), are gravimetrically metered in a batch blender and then premixed. A horizontal batch blender, whose agitator has two reverse-spiral ribbons to provide uniform mixing, is commonly used. The premixed recipe is fed into a circular buffer bin and transferred to the extruder feeder by a screw conveyor. The feeder delivers a uniform, continuous flow of premixed formulation into the extruder. Two types of feeder are normally used: a volumetric screw-type feeder, the more common and cheaper type, and a loss-in-weight feeder. The feeder has either a single screw or self-cleaning twin screws; the twin screw feeder delivers a more constant flow rate than the single screw feeder, and it handles fairly sticky materials better with greater accuracy. If more precision is required in the feed flow, a loss-in-weight feeder is used to provide a constant mass flow rate of dry premix to the extruder.

The extruder cooker in the processing line may be a single screw or an intermeshing co-rotating twin screw

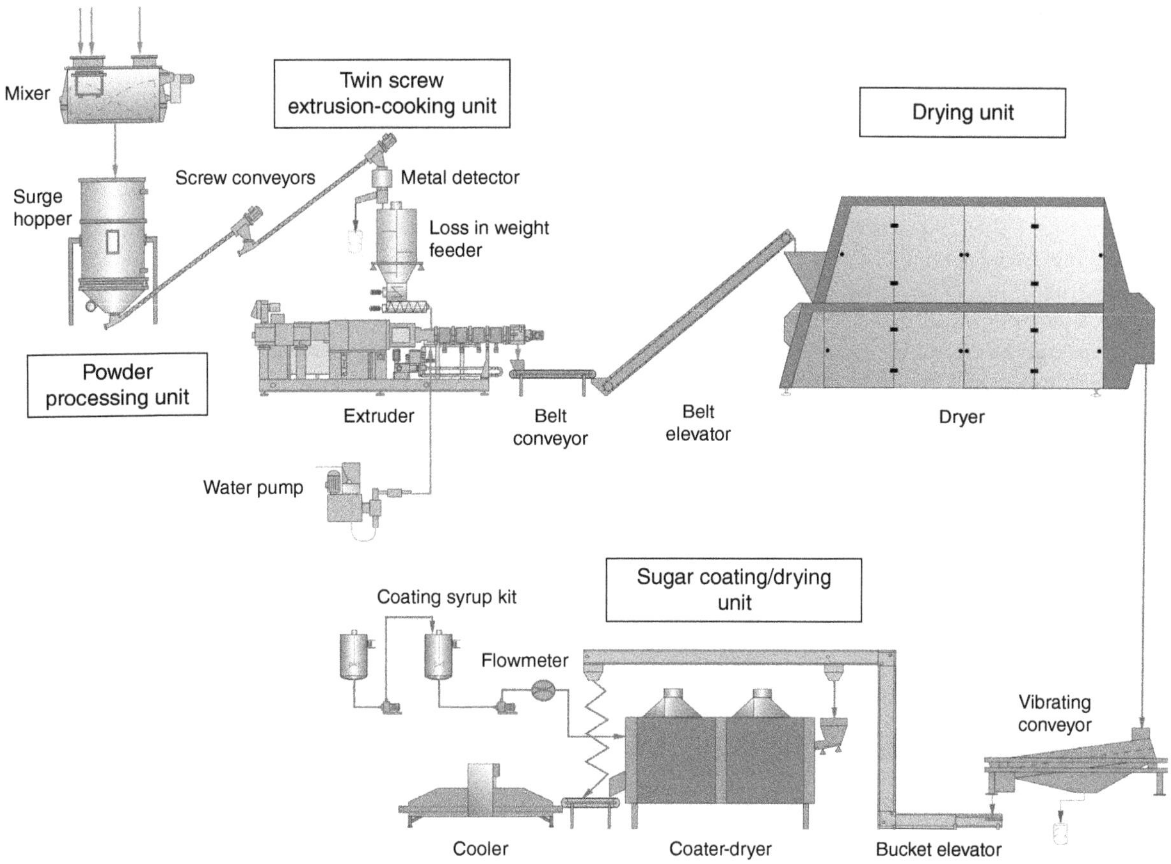

Figure 6.33 Twin screw extrusion processing line for the production of DX breakfast cereals. Source: Reproduced with permission of Clextral, France.

extruder. The mixture is processed in a relatively short extruder with a quite simple screw configuration which has a cooking section in the terminal position. In this section, the material is intensively sheared to produce a cooked melt. The water needed for the cooking process comes from the raw materials but some moisture adjustments are made by a volumetric pump connected directly into the feeding section of the extruder; the total moisture content in the extruder may range from 16% to 20%. The screw speed is usually set between 200 and 450 rpm which, combined with the screw profile and moisture content, makes it possible to adjust shear stresses and mechanical work in the cooking section. These processing conditions mean that the extruder cooks the material more mechanically than thermally. Malt syrup may be added directly to the melt as it enters the cooking section. It is worth recalling the use of the WAI-WSI curve presented earlier (section 6.2.2.3) in Figure 6.17, and noting the targeted domain of the curve for directly expanded breakfast cereals. Actually, WAI and WSI characteristics of such products correspond to the domain starting from the maximum of WAI and beyond. The use of the WAI-WSI curve then allows quick selection of the thermomechanical cooking conditions (through screw speed, screw configuration, and water content, as well as barrel temperature) according to the targeted physicochemical characteristics (WAI, WSI) and physical properties of the resulting extrudates.

The extruder has a die assembly at the end of the screw-barrel system, through which the product is extruded. The cooked melt at the die at high pressure (above the vapor pressure of water) and high temperature (above 100°C) undergoes considerable expansion on exiting the die, owing to the flash pressure reduction and the water evaporation: the melt expands directly at the die. Water evaporation causes a rapid fall in the temperature of the melt, which becomes more and more viscous. It hardens quickly into a very aerated structure that gives the final product a very pronounced crispy/crunchy texture. The rate of expansion depends on the rheological and thermal properties of the molten material and on the geometry of the die insert. The action of the die is supplemented by a cutter which directly cuts the expanding strands as they leave the die in order to give the desired shape to the final product. The die assembly is a very important part of the extruder as it determines the quality profile and product consistency as well as bulk density, texture, and shape. Table 6.7 summarizes the main processing conditions (thermomechanical cooking and die texturization) of the well-known generic DX extrusion-textured breakfast cereals.

After cutting, the extruded products are transferred to the drier either by a belt elevator or through a pneumatic conveyor. At the drier inlet, the extruded products have a moisture content of 7–10%; that is reduced to 2.5–3.5% to give the right crispness as they leave the drier. Drying takes place continuously on single pass or multipass conveyor belt driers, with heat transfered by convection. The drier may be fitted with electrical resistances, a gas burner or even steam, to heat the ventilating air. The drying temperature is usually between 140°C and 160°C; the residence time is optimized by varying the belt speed, and ranges between 3 and 8 minutes. The orientations of the product change from pass to pass and hence its different surfaces are exposed to the drying air, which results in a more uniform drying. The drier may have a toasting section for browning of breakfast cereals (crisp rice, for example); this occurs at temperatures ranging between 150°C and 200°C. Fluidized-bed driers may also be used to dry and toast expanded breakfast cereals.

Dried expanded breakfast cereals are often coated with flavored sugar syrup (about 80° Brix). The products are

Table 6.7 Typical processing conditions for generic DX extrusion-textured breakfast cereals.

	Thermomechanical cooking			Die texturization	
Product	Screw speed (rpm)	Barrel temperature (°C)	SME (kJ/kg)	Insert mass flux (kg/h.mm^2)	Bulk density (g/l)
Ball (corn-based)	300–450	130–150	400–450	4–5	40–60
Crisp rice (rice-based)	300–400	160–180	380–450	3–4	110–120
Loop (oat-based)	200–300	140–160	320–400	5–6	180–220
Cup (wheat-based)	250–350	110–130	620–700	3–4	120–140
Bran stick	200–300	115–135	550–620	2–3	180–200

then blended in a cylindrical rotating drum, and exposed to the spray of coating syrup. The drum is designed to create a folding action in the product bed, to facilitate contact between the product and syrup and ensure a uniform coating. The flavored sugar syrup is handled in two jacketed tanks-in-series: syrups are batch prepared in the first tank, while the second tank continuously feeds the spray system of the drum.

Finally, the sugar-coated breakfast cereals are dried and cooled in a conveyor belt drier before packaging. As seen in Figure 6.33, an optional coating technology which combines sugar coating and drying operations in one processing device can be used. In this technology, coating occurs within the first two-thirds of the length of the unit while drying occurs within the whole length of the unit. This allows each layer of sugar coating to be dried as soon as it is applied, which leads to more efficient drying and lower loss of syrup. This innovative design brings important advantages in terms of coating quality (homogeneous coating, no product agglomeration) and production efficiency (longer operating time before cleaning); also, less space is required.

In conclusion, the process diagram of direct expansion extrusion-cooking process may be designed by use of the state diagram presented in section 6.2.2.1. The process diagram is shown in Figure 6.34a, from the inlet of the screw extruder to the exit of the cooling unit.

6.3.1.2 Pellet-to-flake extrusion-cooking process

As indicated in Table 6.3, this process involves a SME range of 80–120 W.h/kg. Figure 6.35 presents a typical flowsheet of a twin screw extrusion processing line for the production of pellet-to-flake breakfast cereals. The process consists of six main unit operations (Clextral, 2013b).

- Powder processing unit.
- Extrusion-cooking unit.
- Pellet forming unit.
- Drying/toasting unit.
- Flavored-sugar coating unit.
- Drying/cooling unit.

The mixing of the dry raw materials (corn flour, corn grits, wheat flour, and oat flour) and the functional ingredients (bran, sugar, emulsifiers, etc.) is identical to that illustrated in Figure 6.33. The mix enters the cooking section of the process using either a volumetric or a loss-in-weight feeder. The cooking section consists first of a preconditioner which preheats and prehydrates the mix. The preconditioning process takes place continuously in either a single-shaft or a counter-rotating twin shaft system. The mixture is humidified by spraying water into the preconditioner through spray nozzles and preheated by steam injection through a row of injectors directed upwards. The mix then reaches a temperature of 80–85°C and a moisture content of 18–20%. Preconditioning is essential in the process to bring thermal energy to the whole thermomechanical cooking process.

The preconditioner feeds the extruder-cooker which is usually an intermeshing co-rotating twin screw extruder which can handle operating conditions with a high level of flexibility and provide a better consistency of cooking, as opposed to a single screw extruder. The mixture (starch-based materials and various ingredients) is processed in a relatively long extruder with an L/D ratio of between 25 and 30. It has a relatively complex screw configuration with several cooking sections in series, each composed of screw elements giving moderate shear; low-pitch, right-handed conveying screw elements are placed between cooking sections. The screw configuration ends with a transport section designed to facilitate heat transfer between the biopolymeric melt and the barrel in order to cool the molten material (below 100°C) before shaping in the die. This last section can include a degassing barrel, which is designed particularly to increase the cooling efficiency. The screw speed rarely exceeds 200 rpm; combined with a relatively high water content (22–26%) and a screw configuration that provides reduced shear, this has the overall effect of satisfactorily handling the mechanical and thermal energy inputs to the product. Malt syrup is usually added directly to the extruder through the feeding section or the early stages of the cooking section.

As for the optimization of thermomechanical cooking, the use of the WAI-WSI curve presented earlier (section 6.2.2.3) in Figure 6.17 must be mentioned. In fact, WAI and WSI characteristics of extrusion-cooked starch-based melts for the production of flakes products correspond to the domain close to the maximum of WAI or slightly before. For pellet-to-flake extrusion-cooking process, extrusion practitioners aim to design process conditions which allow WAI and WSI values to be the highest and the lowest, respectively. This process trend optimizes the textural characteristics of resulting flakes.

After thermomechanical cooking in the screw-barrel assembly, the molten material is then fed into a multihole die assembly in order to shape and cut vitreous, unexpanded or very slightly expanded pellets (5–6 mm diameter, 6–8 mm long). These can be produced either by die-face cutting or delayed cutting where the die produces ropes which are then cut. After cutting, the temperature and moisture content of the pellets are 80–95°C

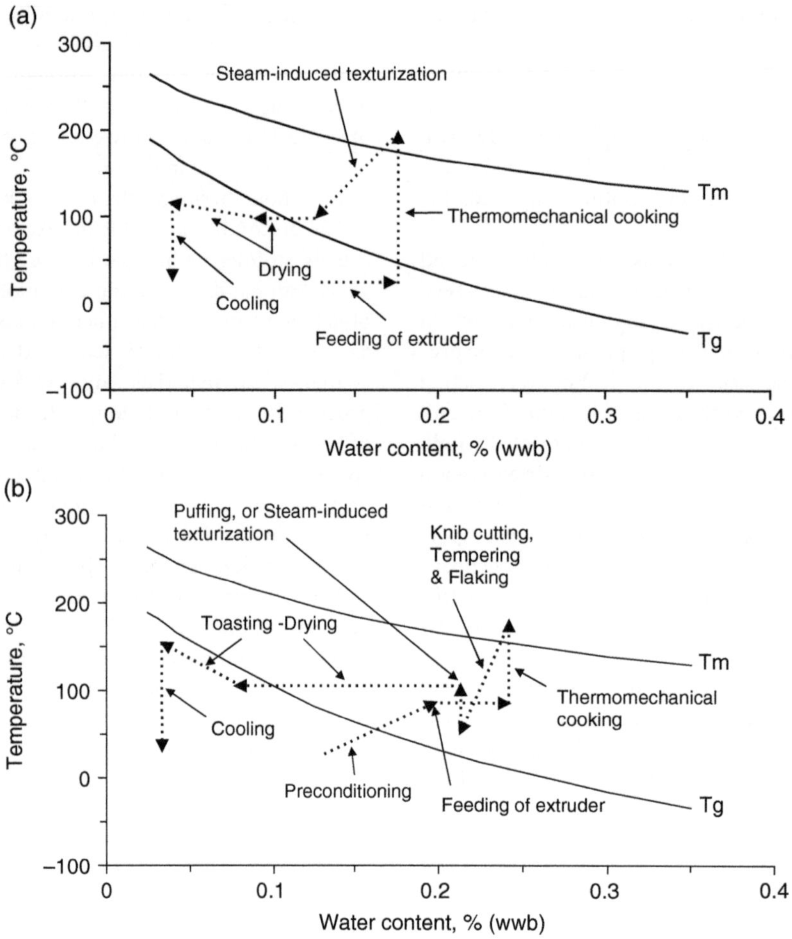

Figure 6.34 Process diagram of twin screw extrusion-cooking processes for the production of breakfast cereals: (a) directly expanded cereals; (b) pellet-to-flake cereals.

and 20–22% respectively. Next, pellets are transferred to a tempering drum for cooling (40–60°C), and also to prevent the sticking of pellets and to regularize the internal moisture gradient. The extent of cooling is controlled by a counter-current flow of ambient air in direct contact with the pellets. The drum rotates at a constant speed and is slightly tilted, so as to give a plug flow behavior and control the dwell time before flaking. As an option, the L/D ratio of the twin screw extruder-cooker may be limited and the extruder used just as a cooker. In this case, the hot cooked melt produced by the twin screw extruder is fed into a slowly turning single screw former. This gently kneads and cools the melt, and allows its temperature to be controlled for die-face cutting. This optional process configuration brings more flexibility as it decouples cooking and pellet forming.

After leaving the drum, the pellets are transferred to a vibratory feeder which supplies single pellets in a uniform and continuous flow, and disperses the pellets across the whole width of the flaker. The pellets then fall directly into the flaking rolls zone where they are flattened. The flaker consists of two rollers of the same size, one of them being movable so that the space between the rollers can be adjusted, while keeping them parallel, in order to regulate the thickness of the flakes. When the flaker is operating, the rollers tend to heat up due to friction; care must therefore be taken to ensure that they are constantly cooled by an internal cooling system, to keep the space between the

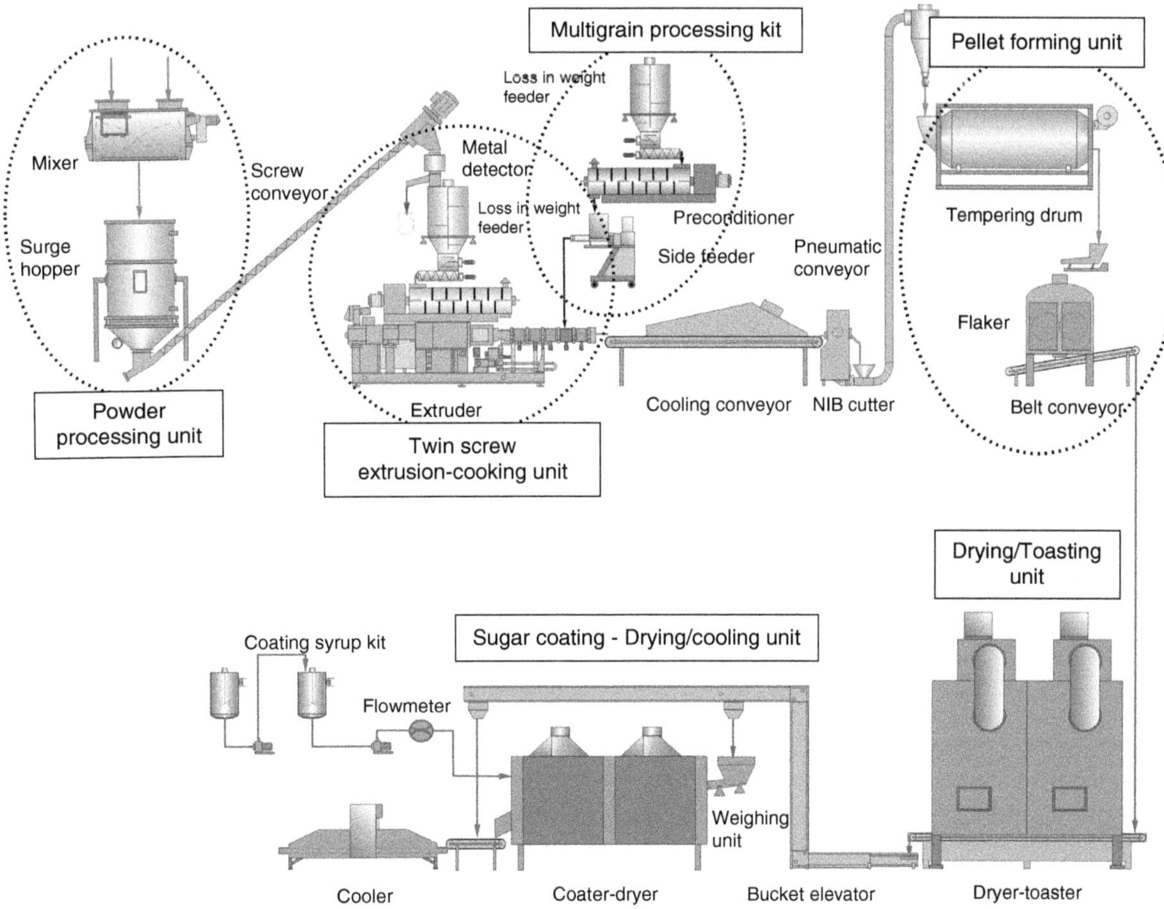

Figure 6.35 Twin screw extrusion processing line for the production of pellet-to-flake breakfast cereals. Source: Reproduced with permission of Clextral, France.

rollers constant. Scraper blades separate the rubbery flakes from the roll surface.

Rubbery flakes enter the drying-toasting unit at a temperature and moisture content of about 32–38°C and 18–20% moisture content, respectively. Because of the uniform heat transfer and flake treatment at relatively short dwell times, air impingement technology is much preferred to traditional rotary or conventional conveyor belt technologies. An air impingement drier generally has three subunits: the first uses high temperatures (220–270°C) to remove moisture and puff the flakes, while the second acts as a toaster at temperatures ranging from 160°C to 200°C, so as to give the specified expansion, crunchiness, and color to the flakes. The third subunit cools the puffed, toasted flakes to stop the toasting process and prepare the product for further processing or packaging. Additional processing may include flavored-sugar coating, as described in section 6.3.1.

As seen in Figure 6.35, the additional multigrain processing unit can be associated with the twin screw extruder to enable manufacturers of breakfast cereals to produce multigrain flakes. This unit is mobile and consists of a side feeder (see Chapter 2, section 2.3.2.4) which allows the direct addition of various types of cereal-based grits (such as precooked rice grits) into the last processing section of the screw-barrel assembly to produce dual-textured flakes. The side feeder may even be fed by a double-shaft preconditioner, which aims to preheat and prehydrate the cereal grits upstream according to the process and product specifications.

It must be noted that this pellet-to-flake extrusion processing line allows production of not only a large variety of cereal flakes but also all typical directly expanded cereals. For that purpose, additional equipment relative to direct expansion processing is supplied such as die-face cutting assembly and a dedicated screw profile. Line suppliers do propose relevant designs which allow quick and easy switching from one type of product to the other, leading to a highly flexible and productive extrusion processing line.

The typical process diagram of pellet-to-flake extrusion-cooking process is shown in Figure 6.34b, from the inlet of the preconditioner to the exit of the cooling unit.

6.3.2 Aquafeed extrusion-cooking process

6.3.2.1 Technical issues of aquafeed production

The annual worldwide growth of aquaculture was 8.9% per year on average during the last 20 years while in the same time the growth rate for wild captured fishes was only 1.2% per year, and 2.8% per year for the land-based meat production. Today, aquaculture amounts to about 50 million tonnes/year. Some forecasts (FAO, 2004) predict a production in 2030 of 83 millions tonnes for aquaculture grown fish, while capture fisheries will remain constant with a production of 93 millions tonnes.

Farmed aquatic animals, on land or marine based, are cultured in closed environments: basins, tanks, or cages built in the ocean. It is therefore necessary to control the required feed in order to optimize breeding conditions. The feed has to comply with specific physical and nutritional requests.

- *Pellet size.* It is necessary to provide the right size adapted to the mouth of the animal and with perfect calibration. Size of aquafeed pellets generally ranges from 0.5 mm up to 30 mm; some larvae need aquafeeds with sizes as low as 0.2–0.3 mm.
- *Texture.* This relates to hardness, brittleness, and cohesiveness. Depending upon the fish, aquafeed pellets must be water stable to minimize pollution and give high ingestion rates. They must be cohesive, while preserving nutrient composition.
- *Density.* Some fishes live at the bottom of water (demersal species) while others stay in the middle of the water (pelagic species). Some fish move quickly to the feed (salmonids), while others wait for slow sinking pellets to come to them. Therefore, the density of aquafeed pellets has to be adapted to the behavior of the animals in order to increase the chance for them to ingest the feed. Sinking and floating properties depend on product density.
- *Storage and handling.* After manufacturing, aquafeed pellets are put in sacks or big bags, transported, stored, and finally distributed manually or through mechanical systems such as pneumatic conveyors and automatic distribution units. The pellets must resist abrasion, attrition, and crushing. They should not create fines which are costly sources of pollution and uneaten feed.

Physical properties of aquafeed pellets depend on the processing technology which cooks and texturizes the required feed formulation. Aquafeed pellets can be manufactured by either pelletizing technology or extrusion-cooking technology. Extrusion-cooking technology has played a determining role in the rapid development of aquaculture, and is nowadays widely employed and recognized, as extruded pellets show much better quality compared to pellets made using pelletizing technology. The ability to control the floating and sinking properties of aquafeed pellets, along with the ability to increase the fat content, has led extrusion-cooking technology to take a strong position in the aquafeed producing market. Although it is a more expensive technology than pelletizing technology, its payback is definitely beneficial due to the positive impact on process productivity, process flexibility, environmental issues, and feed conversion ratios.

One technological challenge in aquafeed production relates to the production of various sizes of aquafeed pellets (including micro-aquatic feeds often used as starters) on the same extrusion processing line, with high levels of flexibility and productivity. This section will present such a flexible processing line.

6.3.2.2 Aquafeed extrusion processing line

In Table 6.3, the aquafeed extrusion-cooking process relates to SME values of 60 W.h/kg and below. Figure 6.36 presents a typical flowsheet of a twin screw extrusion processing line for the production of aquafeed pellets. The processing line consists of four main unit operations (Clextral, 2013b).

6.3.2.2.1 Feed mix preparation unit

This is the first step of the line, which selects the raw materials, reduces the particle size and mixes the ingredients. Raw materials selection and feed mix preparation have a significant impact on final product texture, uniformity,

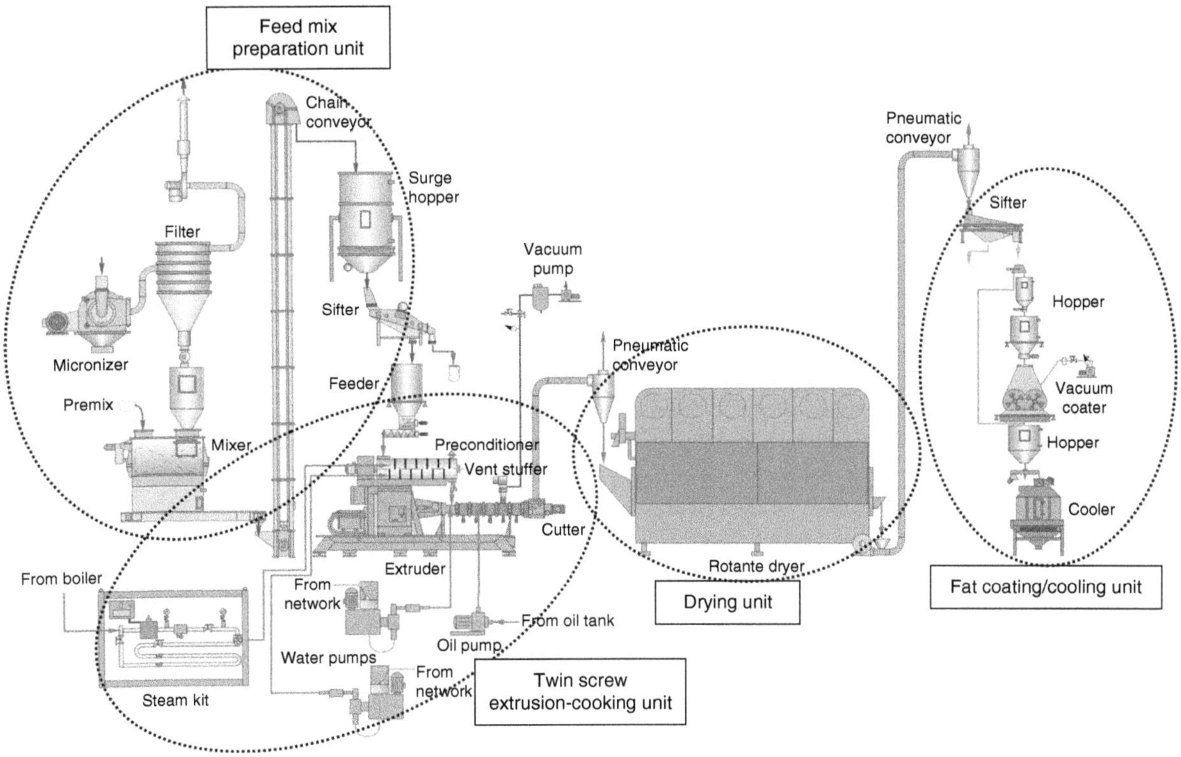

Figure 6.36 Twin screw extrusion processing line for the production of aquafeed pellets. Source: Reproduced with permission of Clextral, France.

nutrition, and economics. Once the ingredients are selected, an important operation in feed mix preparation includes particle size reduction and mixing of ingredients. Particle size reduction concerns mainly biopolymer-based raw materials (starchy and proteinaceous ingredients), and attention must be paid to it as it may bring many benefits such as:

- *wear reduction and capacity increase of the extruder*. The lower the particle size, the lower the wear through abrasion in the melting section of the screw configuration, and the higher the melting and cooking capacity of the extruder. Besides, particle size should account for the final size of the pellets. In general, grinding guidelines are the following: for die openings up to 3 mm, the largest particle size should not be larger than one-third of the die opening; for die openings larger than 3 mm, the largest particle size should not be larger than 1.5 mm
- *pellet quality improvement*. Smaller particles improve pellet durability and water stability, and decrease pellet friability. Obviously, during extrusion-cooking of smaller particles, more complete and homogeneous melting and cooking of the biopolymer-based ingredients occur, which tend to improve the strength and mechanical resistance of the pellet structure
- *positive nutritional effects*. Finely ground feed ingredients may bring beneficial nutritional effects, such as improving digestibility and bio-availability of nutrients, due again to the more complete and homogeneous melting and cooking of the ingredients. This is particularly important since most species of fish, crustacea and shellfish have a very short digestive tract.

However, particle size reduction needs to be optimized as ultra-fine grinding may produce damage in the biomacromolecules from a nutritional standpoint because the process of size reduction generates heat, due to the high frictional and attrition forces. Starch can be damaged or dextrinized and proteins denatured at high temperatures and high shear forces existing in the grinders. Added supplements can be destroyed or modified (less bio-availability). Care must then be taken to control the conditions of particle size reduction, the temperature increase in particular.

Hammer mills and air swept pulverizers are commonly used for grinding the raw materials in aquafeed processing. Hammer mills offer high efficiency, low heating, and reduced aspiration requirements but they are limited in the finished particle size that can be achieved. Typical finished ground ingredients are in the range of 95–99% less than 500 microns, with an average particle size of 200–300 microns. Specially equipped hammer mills are able to grind feed ingredients as fine as 95–99% less than 250 microns, with an average particle size of 100–175 microns. Air swept pulverizers can achieve finished ground ingredients in the range of 95–99% less than 150 microns, with an average particle size of 40–75 microns.

The milled feed ingredients are conveyed to a mixer in which the main flour-based stream is homogeneously mixed with additives. Then, it is fed to the extrusion-cooking unit.

6.3.2.2.2 Extrusion-cooking unit

The second step of the aquafeed processing line aims at cooking the dry feed mix, texturizing and shaping the aquafeed pellets. This unit operation consists of three subunits: preconditioning, thermomechanical cooking, and die texturization and shaping.

- *Preconditioning of dry feed mix* (section 6.2.1). As discussed, the advantages that preconditioning brings include increased extrusion capacity, reduced starch damage and nutrient loss, because part of the mechanical energy normally supplied in the extruder is replaced by thermal energy supplied in the preconditioner. Besides, preconditioning leads to a significant reduction of screw wear in the extruder, as a result of good hydration of the feed in the preconditioner which reduces the abrasiveness of the feed particles.
- *Thermomechanical cooking of preconditioned feed mix.* The mix is submitted to a controlled, thermomechanical cooking treatment which consists of converting biopolymer-based components into an amorphous state, and mixing amorphous biopolymers with all other feed ingredients, to obtain a homogeneous melt at the exit of the cooking section. In particular, starch needs to be well cooked and fully gelatinized to improve its binding effect when achieving pellet texturization. Thermomechanical cooking of feed mix in an extruder requires two energy inputs: mechanical energy input, defined mainly by screw speed and screw configuration, and thermal energy input, determined by direct steam heating and indirect barrel heating.

Two types of cooker extruder can be used in aquafeed production: single screw extruder and intermeshing co-rotating twin screw extruder. Single screw extruders have been extensively used, particularly for processing simple recipes and producing basic pellets. As discussed in Chapter 3 (section 3.2.4), the single screw extruder has major drawbacks including: low flexibility in terms of rheology of the mix, specifically fat-enriched feed formulations; dependency of the major operating variables screw speed and mix throughput; low cooling efficiency in the last barrel section, which leads to a lack of temperature control of the cooked melt; low mixing degree in the screw channel, which gives heterogeneity in the melt matrix. These factors limit the operational range of single screw extruders. Twin screw extruders are able to process consistently, with high levels of flexibility, a large range of raw materials because of their positive pumping action; it is possible to starve feed the machine which makes screw speed and throughput independent; cooling efficiency in the last part of the barrel is high, which allows good control of melt temperature; and the mixing degree in the screw channel is very high, which gives a homogeneous melt.

Although twin screw extruders are more expensive than single screw extruders, they have been extensively used in the aquafeed industry since the early 1980s, mainly due to their high level of process flexibility and productivity, and pellet consistency. In addition, new developments of twin screw extruders bring advanced functionalities as they are able to operate at higher screw speeds (from 500 to 800 rpm) and L/D ratios of 16–20, in order to apply the required time-temperature-shear history in the screw-barrel assembly that determines the cooking extent of the feed mix. As far as heat transfer is concerned, the design of barrel modules of modern twin screw extruders is optimized and they are able to exchange more heat than previous models, in the cooling section in particular. Vent stuffing is often used in the cooling section, when sinking pellets are produced. Improved heat transfer capability and vent stuffing are of course important in aquafeed extrusion processing, as they allow for accurate control of die melt temperature and expansion, and consequently the right pellet density can be obtained.

- *Texturization and shaping* of aquafeed pellets, through a die system equipped with a die-face cutter. Although the die-cutter assembly is the cheapest subunit in the aquafeed processing line, it is one of the most important as it affects the quality of the finished pellets. The design and process variables of the die-cutter assembly aim at converting the rheological properties of the cooked melt into a pellet with an expanded structure, and the resulting structure defines the end-use characteristics of the pellets such as bulk density, durability, water stability, etc. For micro-aquatic feeds that are smaller than 2 mm, specific pre-die and die design allows for directly expanded pellets to be extruded consistently down to 0.5 mm. Finally, but not least, the design of the die-cutter assembly needs to allow aquafeed producers to quickly change pellet sizes and reduce production downtimes to a minimum.

6.3.2.2.3 Drying unit

After die-face cutting, the wet aquafeed pellets are conveyed to a dryer in order to reduce the moisture content from 22–24% down to about 8%. Of course, moisture reduction is the main purpose of the dryer: ideally, 2325 kJ/kg is required for water removal in the drying process. Besides the implication of the drying process in production costs, the impact of drying on pellet quality can vary tremendously, depending upon the design of the dryer and drying parameters. The dryer aims at applying heat energy uniformly over the whole surface area of the pellets by the action of turbulent air flow which surrounds the pellets, this air being maintained at a low partial vapor pressure. The heat energy is transferred to the pellets by heat convection with a transfer rate that strongly depends upon the degree of turbulence around the pellets. However, that heat energy is transferred within the pellets by heat conduction from the surface through to the center which depends mainly upon the temperature gradient and porosity of the pellets. During heating, the temperature of the water contained in the pellets and its partial vapor pressure increase, thus generating a moisture gradient which induces water diffusion out of the pellets; the turbulent drying air then carries along the moisture leaving the surface of the pellets.

Several factors control water removal from the pellets, such as:

- *air-related factors*: air flow characteristics around the pellets, temperature and relative humidity of drying air
- *pellet-related factors*: initial moisture content and temperature, porosity, and size.

The design of the dryers should be optimized from aerodynamics and thermal standpoints; the principle of contacting drying air and pellets is also an important design factor, in order to remove moisture with respect to pellet quality (low fines generation, low damage to pigments and palatability ingredients) and drying economics (low energy losses, low moisture variance). In aquafeed processing lines, extruded pellets are generally dried on one of the following dryer designs: horizontal belt dryer (single or multiple passes); vertical counter flow dryer, where the pellets move downwards while the drying air goes up; fluidized bed dryer, where the pellets are fluidized by the drying air. In Figure 6.36, the Rotante drying technology is shown (Clextral, 2013b). Its design is based on a rotary multiscrew assembly which allows pellets to move as an approximate plug flow while the slow screw rotation ensures good contact between pellets and the drying air. Compared with the aforementioned drying technologies, the Rotante drying technology presents the following advantages:

- microclimate inside the screws, which leads to reduced drying time and accurate control of temperature and humidity
- high energy yield, higher than 75%
- low fines generation
- no retention zone, which avoids product cross-pollution and produces perfect homogeneity of product drying
- easy cleaning and low maintenance – small number of wear parts and easy access.

Though significant improvements in pellet drying have been achieved during the last decade, further developments are still required to operate more consistently with regard to pellet quality, hygienic operation, and operating cost.

6.3.2.2.4 Fat coating and cooling unit

Aquafeed pellets are lipid-rich feedstuffs, and fat must be added to reach the targeted composition, according to fish needs (energy and nutrition). Fat addition is commonly done after drying. As hot pellets come out of the dryer, oil is added through coating. This can be carried out in two ways.

- *Under atmospheric pressure*: oil is sprayed on to the pellets in an inclined rotating drum; oil temperatures around 40°C are common to ensure enough fluidity of the oil and thus distribute it easily on the pellets. Oil diffuses from the surface through to the center and is stored in cavities of the pellet structure. The design of the spraying system together with the rotating effect of the drum should ensure homogeneous exposure of the pellets to the oil spray. Continuous, atmospheric fat coating currently allows 30% total oil content to be achieved.
- *Under vacuum pressure*: air is withdrawn from the various cells and pores inside the pellets; when the desired vacuum pressure is reached, spraying of oil takes place. Then, the vacuum is released which presses the oil into the pellet structure. The capillary forces also help the oil

to penetrate some distance into the pellets, while the surface stays relatively dry. It is possible to achieve more than 40% total oil content (salmon feed, for example). Since its introduction in the late 1980s, vacuum coating has become well recognized and used as it brings significant advantages with regard to pellet quality (higher yield of fat addition, dry surface of pellets, ability to maintain different added ingredients inside the pellets). It operates via a batch process, with a typical cycle time of 4–7 minutes.

In Figure 6.36, a semi-continuous vacuum coating system is depicted.

The cooling step is required after fat coating to remove excess heat, and thus prevent any condensation which can occur in storage bins or final packaging. Condensation occurs when air–water vapor mixture is cooled to or below the corresponding dew point temperature. At high relative humidity approaching 100%, the dew point equals the ambient temperature so only a slight amount of cooling of the ambient air will generate moisture condensation. At low relative humidity, say 30%, larger temperature changes can occur before moisture condensation. Practically, the pellet temperature should be cooled to within about 5°C of the temperature to which the package will be exposed during storage or transportation. This means that in winter, chilled air may have to be used to lower the temperature of the pellets to safe levels while in summer, ambient air is usually satisfactory.

Cooling of pellets is carried out by flowing ambient or chilled air through a pellet bed via a centrifugal fan. The required residence time to cool the pellets depends on the three following variables.

- The *convective heat transfer* between the cooling air and the pellets, and the thermal conductivity of pellets which depends on pellet density and porosity; the higher the density, the lower the porosity, and the higher the thermal conductivity.
- The *size of pellets*: the higher the pellet diameter, the higher the residence time.
- The *extent of temperature decrease*, from pellet temperature at the inlet of the cooling unit to the required pellet temperature at the outlet.

Horizontal belt coolers and vertical counter flow coolers can be used to cool aquafeed pellets. However, vertical coolers are widely recognized as the standard cooling equipment in aquafeed processing lines, as they offer many advantages over horizontal coolers such as higher thermal efficiency, better sanitation, lower footprint, and low operating cost. In Figure 6.36, a vertical counter flow cooling system is depicted for this processing line.

Though humans have raised fish for thousands of years, the development of modern, intensive aquaculture is relatively young (about 30 years old). Fish meal and fish oil are still the preferred dietary protein and fat sources for feeding fishes. However, fish meal and fish oil are finite resources, and modern aquaculture cannot continue to grow without depleting world fish stocks. Besides, protein and fat sources from fish origins are relatively expensive. The uncertainties surrounding the price and sustainability of fishmeal in the coming years make the development of alternative protein and fat sources (plant proteins and vegetable oils) a promising alternative. The evolution of feed raw materials combined with other important factors such as increasing feed conversion ratios, reducing environmental impact, and improving feed economics are ongoing challenges for feed producers. Consequently, aquafeed production needs highly flexible and productive technology to fit with those trends. Extrusion technology has already satisfied the needs of aquaculture for about 30 years, and has contributed significantly to its growth. Extrusion technology, and particularly twin screw extrusion, is the right technology to help with the coming challenges of aquaculture. Due to its reliability, flexibility, and productivity, it will undoubtedly remain the key technology in aquafeed production to produce high-quality aquafeed pellets and contribute to the expansion of a sustainable aquaculture industry.

6.3.3 High-moisture extrusion-cooking process

In the early 1980s, a new extrusion-cooking process was developed consisting of plasticizing food mixes at high moisture content ranging from 50% to 80%. This process was applied to emulsify and prepare gel cheese analogs, to prepare fat substitutes or texturize meat analogs (Akdogan, 1999; Cheftel et al., 1994; Noguchi, 1989). Most of the studies published on the high-moisture extrusion-cooking (or HMEC) process and products have described the texturization of various sorts of proteins at high moisture content. Protein texturization is not really new in extrusion-cooking, as food manufacturers have adapted extrusion technology for the production of textured vegetable proteins, notably soy proteins, for many decades. In consumer food products, textured vegetable proteins are mostly used to extend meat products and improve their functionality with regard to fat and water retention. Textured vegetable proteins are produced either on single screw or twin screw extruders, at relatively low moisture content (<30% wwb), and present an expanded structure. HMEC process and products are conceptually different, as the process aims at fibrating protein-based formulations at high moisture content (>60% wwb), and thus texturizing wet meat-like products from either vegetable proteins (soya flour, soy protein concentrate and isolate, vital wheat gluten, in particular) or animal proteins (meat by-products obtained by mechanical separation, fish proteins, in particular). Hence, the resulting products from the HMEC process are called wet fibrated protein (or WFP). Applications of HMEC involve mainly twin screw extruders that have much better conveying capabilities than single screw extruders.

The HMEC process uses a complex screw configuration with a high length/diameter ratio (L/D between 26 and 36), followed by a long cooling die (Figure 6.37).

- *Conventional feeding and compression sections* that represent about 35% of the axial length of the screw. Raw flours are fed into the feed section, while meat slurries are injected into the compression section by use of a slurry pump.
- *Plasticating/cooking section* that represents about 45–55% of the axial length of the screw. This comprises several mixing sections in series ($3 \leq n \leq 6$) and raises the temperature inside the barrel which results mainly from the heat energy supplied through the barrel; viscous dissipation also contributes to temperature increase. The protein melt is plasticized at 140–170°C while screw speed ranges from 250 to 450 rpm. During plasticating, proteins exhibit important structural modifications, as described in section 6.1.1.2 (such as denaturation, aggregation, and disruption of aggregates). Starch-based ingredients are generally added to enhance the fiber formation. With soy-based formulations, HMEC processing produces a rather homogeneous protein-based melt.
- *Die-feeding section* that represents about 10–20% of the axial length of the screw. Together with the front die plate, this section is very important as it defines the initial conditions (temperature profile, velocity field) of the flow which enters into the fibrating die.
- *Fibrating die*, 1–5 meters long, equipped with independent cooling sections to control the temperature decrease of the melt while it flows into the die. The cross-section of the fibrating die can be either circular or rectangular. The average velocity of the melt in the die ranges from 2 to 5 m/min, hence developing a laminar flow. The shearing effect then allows flow lines or lamellae within the melt to be developed, and protein macromolecules to be aligned in the flow direction. On cooling in laminar flow, the protein-based melt exhibits a phase transition from a liquid to a gel state, which allows fiber formation to occur. Fiber characteristics depend strongly on the velocity profile of the liquid melt before gelling. For a fixed die opening, melt velocity and temperature gradient within the melt govern the development of fibers inside the protein matrix. Wet fibrated protein generally shows length-wise strength higher than cross-wise strength, when measuring tensile strengths of textured samples (Noguchi, 1989).

The HMEC processing unit produces intermediate products which are further postprocessed through conventional meat processing operations. WFP products present a smooth surface, an elastic and coherent structure. According to flow and heat transfer characteristics in the die channel, various types of surface aspect and fibrous structure can be obtained, which range from profile A to profile B in Figure 6.38.

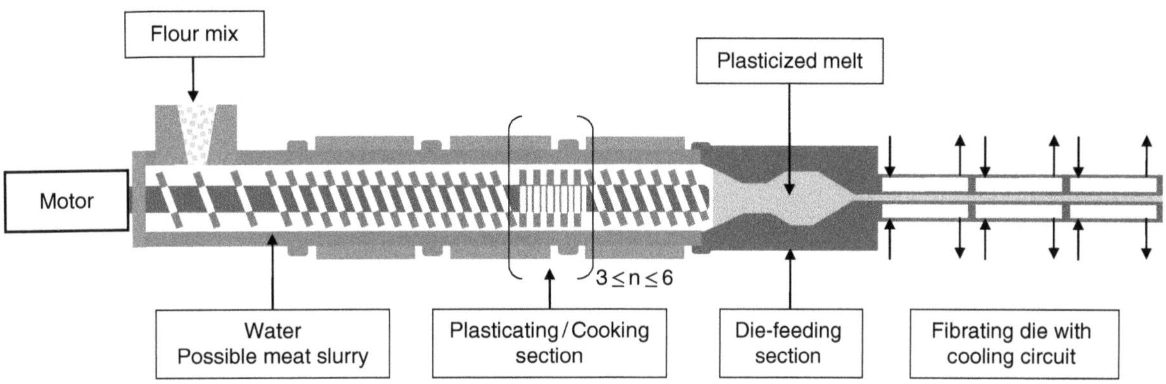

Figure 6.37 High-moisture extrusion-cooking unit for the production of wet fibrated protein (Clextral, 2013b). Source: Reproduced with permission of Clextral, France.

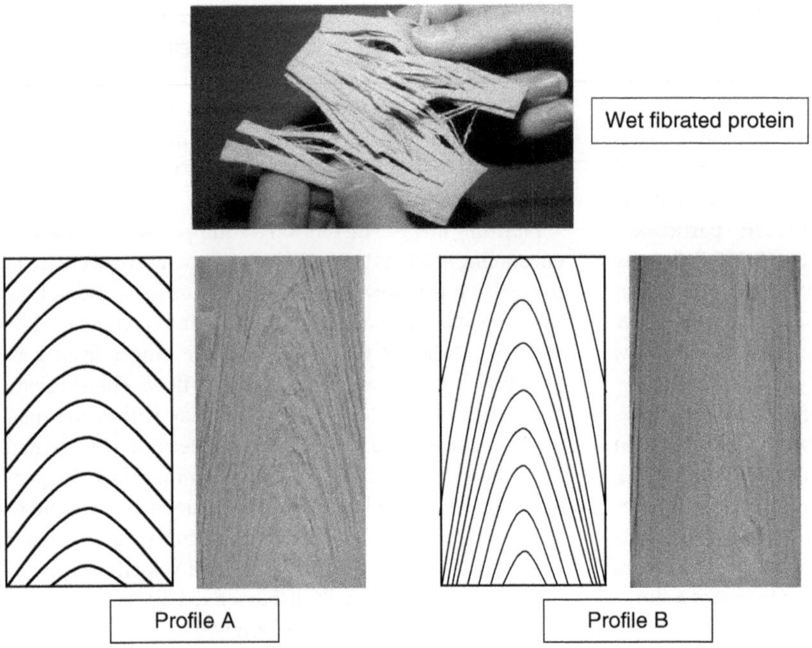

Figure 6.38 Texture profiles of wet fibrated protein. Source: Reproduced with permission of Clextral, France. For color detail, please see color plate section.

- *Profile A*: a rough surface, with relatively flat velocity profile, leading to short and big fibers; meat analog flakes can be easily produced from this profile.
- *Profile B*: a smooth surface, with a well-developed velocity profile, leading to long and numerous thin fibers; meat analog slices can be obtained from this profile.

Wet fibrated protein products do offer potential for innovation in human foods as well as for petfoods. In human foods, this comprises fat-controlled meat analog products such as vegetarian products from vegetable proteins, as well as fibrated meat by-products (mechanically deboned poultry meat, beef meat, pork meat, surimi, in particular). In petfoods, this might include specialities with specific functional attributes (such as healthy petfoods). The finished products can be conditioned through canning, freezing, and vacuum processing.

References

Akdogan H (1999) High moisture food extrusion. International Journal of Food Science and Technology 34: 195–207.

Alvarez-Martinez L, Kondury KP, Harper JM (1988) A general model for expansion of extruded products. Journal of Food Science 53(2): 609–615.

Amon M, Denson CD (1984) A study of the dynamics of foam growth: analysis of the growth of closely spaced spherical bubbles. Polymer Engineering Science 24(13): 1026–1034.

Areas JAG (1992) Extrusion of food proteins. Critical Reviews in Food Science and Nutrition 32(4): 365–392.

Arhaliass A, Bouvier JM, Legrand J (2003) Melt growth and shrinkage at the exit of the die in the extrusion-cooking process. Journal of Food Engineering 60: 185–192.

Asp NG, Björck I (1989) Nutritional properties of extruded foods. In: Mercier C, Linko P, Harper JM (eds) *Extrusion Cooking*. St Paul, Minnesota: American Association of Cereal Chemists, pp. 399–434.

Berset C (1989) Color. In: Mercier C, Linko P, Harper JM (eds) *Extrusion Cooking*. St Paul, Minnesota: American Association of Cereal Chemists, pp. 371–385.

Biliaderis CG, Page CM, Maurice TJ, Juliano BO (1986a) Thermal characterization of rice starches: a polymeric approach to phase transitions of granular starch. Journal of Agricultural and Food Chemistry 34(1): 6–14.

Biliaderis CG, Page CM, Maurice TJ (1986b) On the multiple melting transitions of starch/monoglyceride systems. Food Chemistry 22(4): 279–295.

Bourne MC (1982) *Food Texture and Viscosity. Concept and measurement*. New York: Academic Press.

Bouvier JM (1996) Engineering analysis of preconditioning in the extrusion-cooking process. Cereals Foods World 41(9): 738–740.

Bouzaza D, Arhaliass A, Bouvier JM (1996) Die design and dough expansion in low moisture extrusion-cooking process. Journal of Food Engineering 29: 139–152.

Brennan M, Merts I, Monro J, Woolnough J, Brennan CS (2008a) Impact of guar and wheat bran on the physical and nutritional quality of extruded breakfast cereals. Starch/Stärke 60(5): 248–256.

Brennan M, Monro JA, Brennan CS (2008b) Effect of inclusion of soluble and insoluble fibres into extruded breakfast cereal products made with reverse screw configuration. International Journal of Food Science and Technology 43(12): 2278–2288.

Brennan M, Menard C, Roudaut G, Brennan CS (2012) Amaranth, millet and buckwheat flours affect the physical properties of extruded breakfast cereals and modulates their potential glycaemic impact. Starch/Stärke 64(5): 392–398.

Campanella OH, Li PX, Ross KA, Okos MR (2002) The role of rheology in extrusion. In: Welti-Chanes J, Barbosa-Canovas GV, Aguilera JM (eds) *Engineering and Food for the 21st Century*. Food and Preservation Technology Series. Boca Raton, Florida: CRC Press, pp. 393–414.

Carslaw HS, Jaeger JC (1959) *Conduction of Heat in Solids*, 2nd edn. New York: Oxford University Press.

Chauhan GS, Bains GS (1985) Effect of granularity on the characteristics of extruded rice snack. Journal of Food Technology 20: 305–309.

Cheftel JC, Kitagawa M, Quéguiner C (1994) Nouveaux procédés de texturation protéique par cuisson-extrusion à teneur élevée en eau. In: Colonna P, Della Valle G (eds) *La Cuisson-Extrusion*. Paris: Lavoisier, pp. 45–84.

Chinnaswamy R, Hanna MA (1988a) Expansion, color and shear strength properties of corn starches extrusion-cooked with urea and salts. Starch/Stärke 40(5): 186–190.

Chinnaswamy R, Hanna MA (1988b) Relationship between amylose content and extrusion-expansion properties of corn starches. Cereal Chemistry 65(2): 138–143.

Clextral (2013a) Private communication: data base of extrusion-cooking conditions applied to starch-based directly expanded extrudates. Information available through: www.clextral.com/.

Clextral (2013b) Private communication: twin screw extrusion processing lines for the production of extruded food and feed products. Information available through: www.clextral.com/.

Colonna P, Buléon A (1994) Transformations structurales de l'amidon. In: Colonna P, Della Valle G (eds) *La Cuisson-Extrusion*. Paris: Lavoisier, pp. 18–43.

Colonna P, Doublier JL, Melcion JP, de Monredon F, Mercier C (1984) Extrusion cooking and drum drying of wheat starch. I. Physical and macromolecular modifications. Cereal Chemistry 61(6): 538–543.

Colonna P, Tayeb J, Mercier C (1989) Extrusion cooking of starch and starchy products. In: Mercier C, Linko P, Harper JM (eds) *Extrusion Cooking*. St Paul, Minnesota: American Association of Cereal Chemists, pp. 247–319.

Crank J (1975) *The Mathematics of Diffusion*, 2nd edn. New York: Oxford University Press.

Datta AK (2002) *Biological and Bioenvironmental Heat and Mass Transfer*. New York: Marcel Dekker.

Della Valle G, Boché Y, Colonna P, Patria A (1995) The extrusion behaviour of potato starch. Carbohydrate Polymers 28: 255–264.

Della Valle G, Colonna P, Patria A (1996) Influence of amylose content on the viscous behavior of low hydrated molten starches. Journal of Rheology 40(3): 347–362.

Della Valle G, Vergnes B, Colonna P, Patria A (1997) Relations between rheological properties of molten starches and their expansion behaviour in extrusion. Journal of Food Engineering 31: 277–296.

Desrumaux A, Bouvier JM, Burri J (1998) Corn grits particle size and distribution effects on the characteristics of expanded extrudates. Journal of Food Science 63(5): 857–863.

Desrumaux A, Bouvier JM, Burri J (1999) Effect of free fatty acids addition on corn grits extrusion cooking. Cereal Chemistry 76(5): 699–704.

Diosady LL, Paton D, Rosen N, Rubin LJ, Athanassoulias C (1985) Degradation of wheat starch in a single-screw extruder: mechano-kinetic breakdown of cooked starch. Journal of Food Science 50: 1697–1699, 1706.

Donovan JW (1979) Phase transitions of the starch-water system. Biopolymers 18: 263–275.

Donovan JW, Lorenz K, Kulp K (1983) Differential scanning calorimetry of heat-moisture treated wheat and potato starches. Cereal Chemistry 60: 381–387.

Fan J, Mitchell JR, Blanshard JMV (1994) A computer simulation of the dynamics of bubble growth and shrinkage during extrudate expansion. Journal of Food Engineering 23: 337–356.

FAO (2004) *State of World Fisheries and Aquaculture*. Rome: FAO Fisheries Department.

Farhat IA, Blanshard JMV, Mitchell JR (2000) The retrogradation of waxy maize starch extrudates: effect of storage temperature and water content. Biopolymers 53: 411–422.

Flory PJ (1953) *Principles of Polymer Chemistry*. New York: Cornell University Press.

Gogoi BK, Wang SS, Yam KL (1995) Applicability of a shear induced rate constant model to conversion of corn meal in a twin-screw extruder. Journal of Food Processing and Preservation 19: 53–63.

Guy RCE (2001) Raw materials for extrusion cooking. In: Guy R (ed) *Extrusion Cooking – Technologies and Applications*. Cambridge, UK: Woodhead Publishing, pp. 5–28.

Guy RCE, Horne AW (1988) Extrusion and co-extrusion of cereals. In: Blanshard JMV, Mitchell JR (eds) *Food Structure – Its Creation and Evaluation*. London: Butterworths, pp. 331–349.

Hailemariam L, Okos M, Campanella OH (2007) A mathematical model for the isothermal growth of bubbles in wheat dough. Journal of Food Engineering 82(4): 466–477.

Harper JM, Tribelhorn RE (1992) Expansion of native cereal starch extrudates. In: Kokini JL, Ho CT, Karwe MV (eds) *Food Extrusion Science and Technology*. New York: Marcel Dekker, pp. 653–667.

Kale MS, Pai DA, Hamaker BR, Campanella OH (2010) Structure-function relationships for corn bran arabinoxylans. Journal of Cereal Science 52: 368–372.

Kaletunc G, Breslauer KJ (1993) Glass transition of extrudates: relationship with processing-induced fragmentation and end-product attributes. Cereal Chemistry 70(5): 548–552.

Kalichevsky MT, Jaroszkiewicz EM, Ablett S, Blanshard JMV, Lillford PJ (1992) The glass transition of amylopectin measured by DSC, DMTA and NMR. Carbohydrate Polymers 18(2): 77–88.

Kalichevsky MT, Jaroszkiewicz EM, Blanshard JMV (1993) A study of the glass transition of amylopectin-sugar mixtures. Polymer 34(2): 346–358.

Karathanos VT, Vagenas GK, Saravacos GD (1991) Water diffusivity in starches at high temperatures and pressures. Biotechnology Progress 7(2): 178–184.

Kirby AR, Ollett AL, Parker R, Smith AC (1988) An experimental study of screw configuration effects in the twin-screw extrusion-cooking of maize grits. Journal of Food Engineering 8: 247–272.

Kokini JL, Chang CN, Lai LS (1992) The role of rheological properties on extrudate expansion. In: Kokini JL, Ho CT, Karwe MV (eds) *Food Extrusion Science and Technology*. New York: Marcel Dekker, pp. 631–652.

Lai LS, Kokini JL (1991) Physicochemical changes and rheological properties of starch during extrusion (a review). Biotechnology Progress 7(3): 251–266.

Lelievre J (1974) Starch gelatinization. Journal of Applied Polymer Science 18(1): 293–296.

Levenspiel O (1984) *The Chemical Reactor Omnibook*. Corvallis: Oregon State University Book Stores.

Li PX, Campanella OH, Hardacre AK (2004) Using an in-line slit-die viscometer to study the effects of extrusion parameters on corn melt rheology. Cereal Chemistry 81(1): 70–76.

Liu H, Lelievre J, Ayoung-Chee (1991) A study of starch gelatinization using differential scanning calorimetry, X-ray, birefringence measurements. Carbohydrate Research 210: 79–87.

Maga JA (1989) Flavor formation and retention during extrusion. In: Mercier C, Linko P, Harper JM (eds) *Extrusion Cooking*. St Paul, Minnesota: American Association of Cereal Chemists, pp. 387–398.

Martin O, Averous L, Della Valle G (2003) In-line determination of plasticized wheat starch viscoelastic behaviour: impact of processing. Carbohydrate Polymers 53: 169–182.

Meuser F, Pfaller W, Van Lengerich B (1987) Technological aspects regarding specific changes to the characteristic properties of extrudates by HTST-extrusion cooking. In: O'Connor C (ed) *Extrusion Technology for the Food Industry*. London: Elsevier Applied Science, pp. 35–53.

Mitchell JR, Areas JAG (1992) Structural changes in biopolymers during extrusion. In: Kokini JL, Ho CT, Karwe MV (eds) *Food Extrusion Science and Technology*. New York: Marcel Dekker, pp. 345–360.

Mitchell JR, Fan J, Blanshard JMV (1994) The shrinkage domain. Extrusion Communique March: 10–12.

Mohamed S (1990) Factors affecting extrusion characteristics of expanded starch-based products. Journal Food Processing and Preservation 14: 437–452.

Noguchi A (1989) Extrusion cooking of high-moisture protein foods. In: Mercier C, Linko P, Harper JM (eds) *Extrusion Cooking*. St Paul, Minnesota: American Association of Cereal Chemists, pp. 343–370.

Pai DA, Blake OA, Hamaker BR, Campanella OH (2009) Importance of extensional rheological properties on fiber-enriched corn extrudates. Journal of Cereal Science 50(2): 227–234.

Robin F, Engmann J, Pineau N, Chanvrier H, Bovet N, Della Valle G (2010) Extrusion, structure and mechanical properties of complex starchy foams. Journal of Food Engineering 98: 19–27.

Robin F, Dattinger S, Boire A, et al. (2012a) Elastic properties of extruded starchy melts containing wheat bran using on-line rheology and dynamic mechanical thermal analysis. Journal of Food Engineering 109(3): 414–423.

Robin F, Schuchmann HP, Palzer S (2012b) Dietary fiber in extruded cereals: limitations and opportunities. Trends in Food Science and Technology 28: 23–32.

Senouci A, Smith AC (1988a) An experimental study of food melt rheology. I. Shear viscosity using a slit die and a capillary rheometer. Rheologica Acta 27: 546–554.

Senouci A, Smith AC (1988b) An experimental study of food melt rheology. II. End pressure effects. Rheologica Acta 27: 649–655.

Shogren RL (1992) Effect of moisture content on the melting and subsequent physical aging of cornstarch. Carbohydrate Polymers 19: 83–90.

Tharrault JF (1992) Contribution à l'étude du comportement des farines de blé tendre en cuisson-extrusion bi-vis. Application à l'appréciation de leur valeur industrielle. PhD dissertation, Ecole Nationale Supérieure des Industries Agricoles et Alimentaires, Université Paris XI, France.

Tolstoguzov VB (1988) Creation of fibrous structures by spinneretless spinning. In: Blanshard JMV, Mitchell JR (eds) *Food Structure – Its Creation and Evaluation*. London: Butterworths, pp. 181–196.

Tomka I (1991) Thermoplastic starch. In: Levine H, Slade L (eds) *Water Relationships in Food*. New York: Plenum Press, pp. 627–637.

Vagenas GK, Karathanos VT (1991) Prediction of moisture diffusivity in granular materials, with special applications to foods. Biotechnology Progress 7(5): 419–426.

Vergnes B, Villemaire JP (1987) Rheological behaviour of low moisture maize starch. Rheologica Acta 26: 570–576.

Voisin I (1993) Influence de divers ingrédients alimentaires sur le procédé de cuisson-extrusion et sur les propriétés physiques d'extrudés à base de produits amylacés. PhD dissertation, Université de Technologie de Compiègne, France.

Wang SM, Bouvier JM, Gélus M (1990) Rheological behaviour of wheat flour dough in twin screw extrusion cooking. International Journal of Food Science and Technology 25: 129–139.

Wang SM, Casulli J, Bouvier JM (1993) Effect of dough ingredients on apparent viscosity and properties of extrudates in twin-screw extrusion-cooking. International Journal of Food Science and Technology 28: 465–479.

Wang SS, Chiang WC, Yeh AI, Zhao B, Kim IH (1989) Kinetics of phase transition of waxy corn starch at extrusion temperatures and moisture contents. Journal of Food Science 54(5): 1298–1301.

Wang SS, Chiang WC, Zhao B, Zheng XG, Kim IH (1991) Experimental analysis and computer simulation of starch-water interactions during phase transition. Journal of Food Science 56(1): 121–124, 142.

Wang SS, Chiang WC, Zheng XG, Zhao B, Cho MH, Yeh AI (1992) Application of an energy equivalent concept to the study of the kinetics of starch conversion during extrusion. In: Kokini JL, Ho CT, Karwe MV (eds) *Food Extrusion Science and Technology*. New York: Marcel Dekker, pp. 165–176.

Zeleznak KJ, Hoseney RC (1987) The glass transition in starch. Cereal Chemistry 64(2): 121–124.

Zheng X, Wang SS (1994) Shear induced starch conversion during extrusion. Journal of Food Science 59(5): 1137–1143.

Figure 1.2 Two-stage reactive extrusion equipment designed in 1966 for polymerization. Source: Reproduced with permission of Clextral, France.

Extrusion Processing Technology: Food and Non-Food Biomaterials, First Edition. Jean-Marie Bouvier and Osvaldo H. Campanella.

Figure 1.3 The former extrusion-cooked food products. Source: Reproduced with permission of Clextral, France.

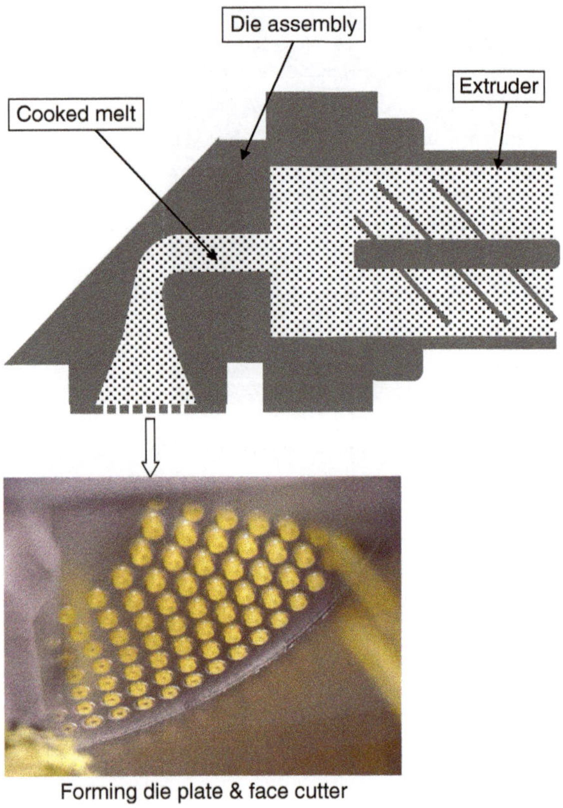

Figure 2.11 Die-face cutting system used in 1D snacks pellet extrusion processing.

Figure 4.26 Biodegradable starch-based loose-fill packaging foams. Source: Reproduced with permission of Clextral, France.

Figure 4.28 Different grades of technical flax fibers. Source: FiMaLin®, 2011. Reproduced with permission of Dehondt Technologies.

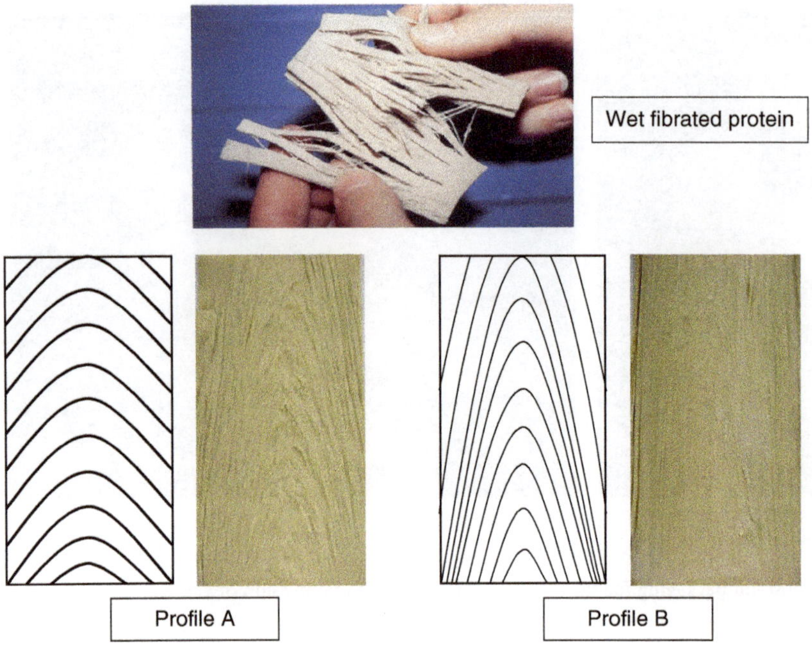

Figure 6.38 Texture profiles of wet fibrated protein.
Source: Reproduced with permission of Clextral, France.

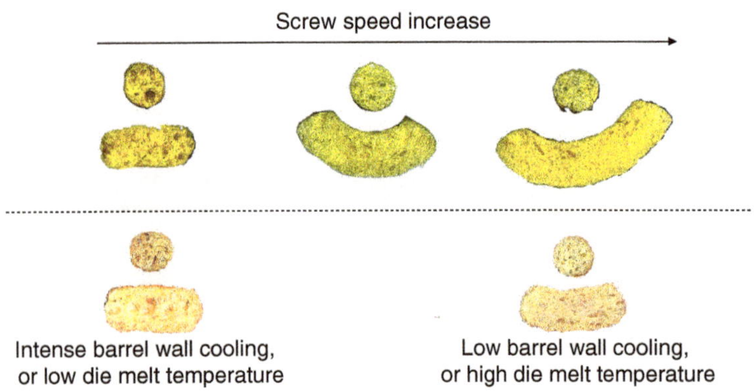

Figure 7.19 Internal structure of extrudates as a function of extrusion processing conditions.

7 Quality Analysis of Extrusion-Textured Food Products

The complexity of food extrusion processing has clearly emerged from the different topics presented in Chapter 6. As a matter of fact, this complexity results mainly from the nature of the raw materials used in extrusion-cooking, the specificities of the conversion process, and the physical characteristics of the final products produced by the process. Food extrusion-cooking deals with natural biopolymeric materials such as starch- and protein-based materials with characteristics that vary extensively depending on many factors such as botanical origin, climate, harvesting, and storage conditions. All these factors significantly affect the physicochemical ability of biopolymeric materials to sustain the extrusion process.

As seen in Chapter 6 (section 6.2.2.3), food extrusion processing is characterized by high temperature and short time cooking conditions that provide severe mechanical and thermal energy inputs, which lead to specific conversion profiles of the biopolymers used in the process. These conversion profiles involving both physical and chemical modifications differ notably from the physical and chemical modifications undergone by the same biomaterials during other conventional processes such as baking and hydrothermal cooking. Regarding product quality, extrusion-cooked products have a unique cellular structure with relatively thin cell walls that confer a crispy texture due to their particular mechanical characteristics and sound effects during consumption.

Since its first introduction to the food industry few decades ago, extrusion technology has led food scientists and engineers to develop or adapt specific instrumental and analytical methods to gain a better understanding of the phenomena occurring during the food extrusion process, and to allow optimum control of the final product texture. The purpose of this chapter is to present specific instrumental and analytical methods used by extrusion practitioners to investigate the extent of thermomechanical cooking undergone by the participating biopolymers (first part of the chapter) and its effects on the final product texture (part two of the chapter). The focus is on methods which are simple and easy to implement in industrial environments as well as for product development activities. The application of the described methods will be illustrated using various practical examples of food produced by extrusion processing. The third part of the chapter deals with assessment of the texture of directly expanded extrusion-textured food products.

7.1 Methods of thermomechanical cooking analysis

In Chapter 6 (section 6.2.2), different methods and equipment were described to investigate the extrusion-cooking process applied to biopolymer-based materials. Some of these methods, such as X-ray diffraction, scanning electron microscopy (SEM), and gel permeation chromatography (GPC), are sophisticated and may require extensive training for their operation and therefore the input of scientific experts for data analysis. Other methods and techniques such as optical microscopy, differential scanning calorimetry (DSC), water solubility index (WSI), water absorption index (WAI), intrinsic viscosity, alkaline viscosity, iodine complexation, and Rapid Visco Analyzer (RVA), to name a few, can be managed successfully in industrial environments; and can be used routinely in industry for quick evaluation of raw and intermediate materials or the extruded final products. These methods are used not only in areas of quality control and quality assessment but also for product development and process/product optimization. Thus, it is worth describing some of these methods to allow extrusion process

Extrusion Processing Technology: Food and Non-Food Biomaterials, First Edition. Jean-Marie Bouvier and Osvaldo H. Campanella.

operators and engineers to implement them in practical applications associated with production and new product development.

7.1.1 Optical microscopy for birefringence analysis

All native starch granules show a hilum which is the nucleation point around which the granule is developed. In polarized light, starch granules are birefringent, showing a black Maltese cross with a crossing point located at the hilum near the center (Figure 7.1). This is indicative of a spherocrystalline arrangement, whereby most of the starch molecules radiate out from the hilum toward the periphery. Full conversion of starch, through either hydrothermal or thermomechanical treatments, results in complete loss of crystallinity as judged by loss of birefringence and the Maltese cross.

The experimental technique for examining native starch granules consists of preparing suspensions of the starchy material by suspending ~10.0 mg starch in 1.0 mL of water in a vortex mixer. The suspension is transferred to a slide, covered with a cover-slip and sealed with nail polish to prevent moisture loss during heating. The sealed specimen is then mounted on a hot stage. For these tests, a Linkam LTS 350 hot stage from Linkam Scientific Instruments, Tadworth, Surrey, UK, can be used to observe specimens under 40× magnification during heating at a heating rate of 5°C/min. Images of native and heated starch granules can be captured using a camera. For example, for the pictures shown in Figure 7.1, a SPOT Insight QE camera from Diagnostic Instruments Inc., Sterling Heights, MI, USA, was used. Photomicrographs of ungelatinized starch granules were observed for Maltese crosses on a petrographic microscope (Olympus BX51) with plane and polarized light at 400× magnification and pictures were taken using a 5 megapixel cooled digital camera attached to the microscope.

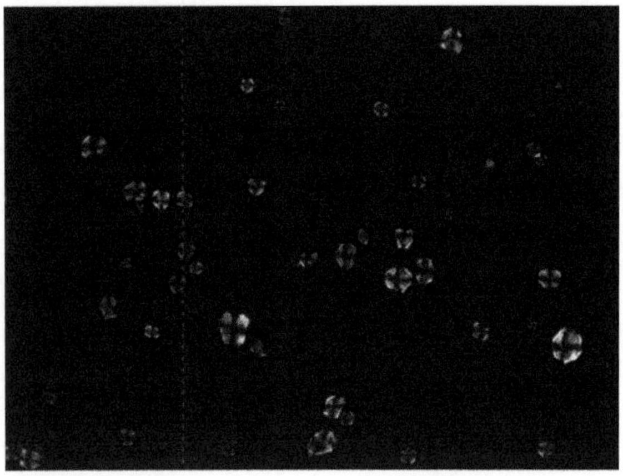

Figure 7.1 Optical microscopy of corn starch granules under polarized light (courtesy of O.H. Campanella, Purdue University, USA).

7.1.2 Water solubility (WSI) and absorption (WAI) indices

Native starch in granular form is not soluble in cold water. On heating and in the presence of excess water, starch granules absorb water and swell. Upon further heating, starch molecules undergo gelatinization. Consequently, structural characteristics of the starch granule architecture are modified and affect its physicochemical behavior when mixed with water to form solutions or pastes. Mixing water with extrusion-cooked starch-based products produces some swelling of the gelatinized starch and, depending on the extrusion conditions, a portion of the starch material is dissolved in water. Hence, the water absorption index (WAI) and water solubility index (WSI) can be used to characterize the extent to which starch structure was modified in the extrusion process. WAI measures the water uptake by the extrudate due to swelling. The hydration of this gelatinized starch produces a "gel" phase in the water–starch system. The WSI determines the amount of soluble materials such as polysaccharides (free or released from the granules), small sugars, undenatured proteins, inorganic ions, etc., which is solubilized in excess of water; this is the "sol" phase of the water–starch system.

The determination of WAI and WSI is based on an experimental procedure developed by Anderson et al. (1969). It consists of grinding a dried sample (native or processed product) and screening the resulting powder in order to keep the fraction that corresponds to a particle size range of 106–422 microns. 0.3 g of that fraction is dispersed with distilled water (10 mL) and vigorously agitated for some seconds, then the suspension is left to stand at ambient temperature overnight. The sample is then centrifuged at about 3000 g for 10 minutes. The supernatant of the centrifuge tube is decanted and dried at 105°C until a constant weight is reached. The WSI is calculated as a percentage of the dry solids in the supernatant per weight of dry sample. Hence, theoretically the WSI should vary from about 0% (unprocessed sample) to 100%, depending upon the extent of structural changes undergone by the starch molecules during the extrusion process. In general, the more severe the process, the higher the measured WSI values. The sediment of the centrifuge tube is weighed and the bound water is removed by drying the sediment at 105°C until a constant weight is achieved. The WAI is calculated as the weight of bound water per 1 g of dry sample; the WAI may also be expressed as a percentage of the bound water per 100 g of dry sample. The application of WAI and WSI measurements is well illustrated in Chapter 6 (section 6.2.2.3, Figure 6.17), where data are collected and correlated to the extrusion-cooking conditions and the extent of starch conversion.

7.1.3 Alkaline viscosity

This method is used to obtain an indication of molecular weight changes of starch due to hydrothermal or extrusion thermomechanical processing. The method gives a relative measurement of the polymer molecular size but not absolute value. When starch is solubilized in an alkaline medium (e.g. potassium hydroxide), comparison of viscosity readings from diluted solutions of known concentrations at a constant shear rate can reveal differences in the molecular weight of the starch molecules.

Alkaline viscosity is measured following the method of Paterson et al. (1996). The dried sample (0.2 g) is mixed with ethyl alcohol (0.5 mL), and 8 mL of 5 M KOH is added to the resulting suspension for solubilization. The sample is stirred overnight and a clear solution is then formed. The solution is diluted with distilled water to 40 mL (1% starch). Aliquots (30 mL) of the resulting solutions are measured using a Couette type rheometer over an increasing shear rate range 1–100 s^{-1} at a constant temperature of 25°C. The sample is equilibrated at 25 +/− 0.1°C for 10 minutes on the rheometer before the first measurement is taken. Data are compared at a constant shear rate (e.g. 10 s^{-1}).

Results of the alkaline viscosity method are illustrated in Figure 7.2, in which the viscosity of starch solutions is plotted as a function of the process SME. The results shown in Figure 7.2 are from solubilization of corn-based and wheat-based extrudates which were subjected to various extrusion-cooking conditions (two different screw profiles, and four combinations of water content and screw speed) to investigate the effect of process SME in the ranges of 370–555 kJ/kg for wheat-based extrudates and 490–720 kJ/kg for corn-based extrudates. For both samples, Figure 7.2 shows clearly that alkaline viscosity decreases almost linearly with increases of process SME values, indicating a significant decrease in the molecular weight of starch molecules with increases of process SME.

7.1.4 Differential scanning calorimetry

As shown in Chapter 6 (section 6.2.2.2), differential scanning calorimetry (DSC) is frequently used to measure phase transitions of biopolymers such as protein denaturation, starch gelatinization and melting (Donovan, 1979), as well as to quantify the extent of structural changes of starch (Wang et al., 1991).

DSC is an instrumental technique in which the difference in the heat flux required to increase the temperature of a sample and an inert reference is measured as a function of temperature at a given heating rate. When the sample goes through a transition, a larger amount of heat is required by the sample than the heat required by the inert reference material. The difference of those heat fluxes is translated as the enthalpy change associated with the transition undergone by the sample. In the DSC instrument, the sample and reference material are supplied with separate heat fluxes in order to maintain the

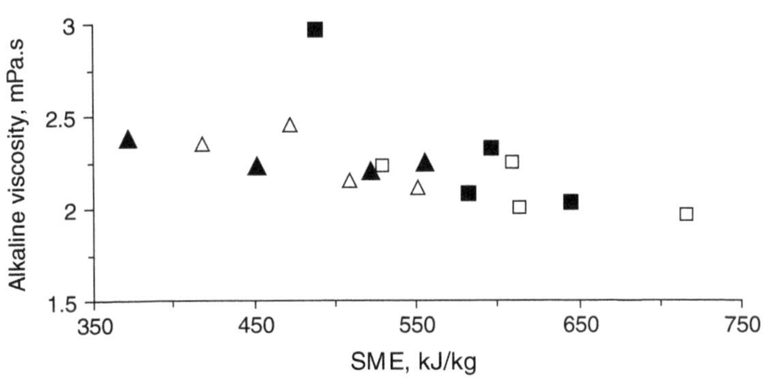

Figure 7.2 Alkaline viscosity as a function of process SME (open symbols: screw profile 1; closed symbols: screw profile 2; □, ■: corn-based extrudates; △, ▲: wheat-based extrudates).

same temperature. If transitions occur in the sample, it will require a different heat input than the inert reference material. As the sample and the reference are maintained at the same temperature, the calibration constant for the instrument is independent of temperature facilitating the use of this technique because the calibration constant is only determined for one standard material. Like any instrumental technique, there are aspects of the DSC operation such as calibration procedures, sample preparation, and measurement protocols that must be carefully taken into account to obtain accurate data.

Differential scanning calorimetry data are obtained in the form of differential heat input (dq/dt) versus temperature (or time since constant heating rates are used). These data can be readily used to obtain temperatures and enthalpy associated with transitions or reactions. The slope of the heating curve is also used to estimate the heat capacity of the sample. When a crystalline sample is heated through its melting point, it will require more energy than the reference material to maintain the same temperature so an endothermic peak is observed in the heat flow/temperature curve, whereas during recrystallization there is a release of energy and an exothermic peak is observed. Glass transition events are not phase or first-order transitions; instead they are second-order transitions involving small enthalpy changes so they are better detected by measuring changes in the material heat capacity. Determination of enthalpy for a transition event requires measurement of the area of the endo/exotherm peak. Integration of the peak area may present some difficulties because often the baseline of the DSC signal is not horizontal and the peak is not well defined. The baseline accuracy and repeatability must be checked regularly.

The typical procedure for determining the baseline for peak integration and enthalpy calculation is shown in Figure 7.3. The baseline on each side of the peak is extrapolated across the peak. A linear portion of each side of the peak is drawn to intercept the baseline on each side of the peak. The intersection of these linear side lines and the baseline gives the peak onset temperature, To, and the completion temperature, Tc, of the transition while the peak temperature, Tp, gives the transition temperature. In the example illustrated in Figure 7.3 (potato starch in excess water; Donovan, 1979), the gelatinization onset, peak and completion temperatures are 64°C, 67.5°C, and 73°C, respectively. The transition enthalpy is calculated from the peak area as calories per gram of dry matter (or J/g). As seen in Chapter 6 (section 6.2.2.4), from estimations of the endothermic peak area it is possible to calculate the extent of starch conversion.

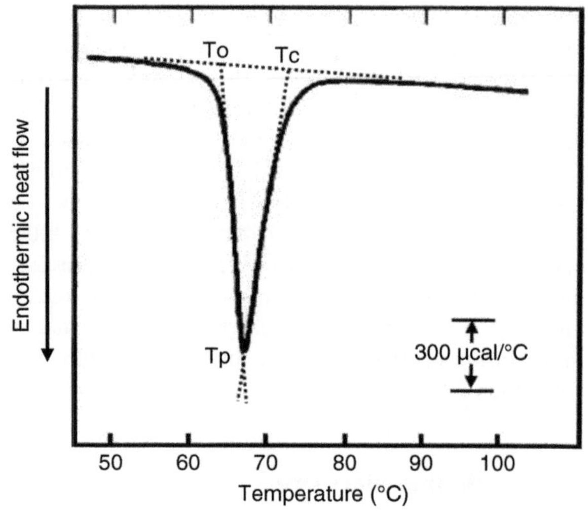

Figure 7.3 DSC thermogram of potato starch obtained at a heating rate of 10°C/min and water volume fraction of 0.81. Source: Donovan 1979. Reproduced with permission of John Wiley & Sons.

Nowadays, DSC is a classic technique extensively used in research, academia, and industry, where technical staff operate DSC equipment routinely. Hence, industrial practitioners who need to use DSC for checking on native or extrusion-cooked biopolymeric materials are encouraged to build a partnership with universities or research centers dealing with food science, and take advantage of their technical skill and knowledge to carry out the desired investigation. New instruments with improved temperature measurement and control have been developed (e.g. Tzero™ technology; Danley et al., 2001) to improve the quality of the baseline. In addition, a new methodology known as modulated differential scanning calorimetry (MDSC) has been widely applied to detect glass transition events in amorphous solids (Schawe, 1995).

7.1.5 Rapid Visco™ Analyzer

The Rapid Visco™ Analyzer (RVA™) is a rotational viscometer/mixer that continuously records the viscosity of a suspension or sample under conditions of controlled temperature (Booth & Bason, 2007). It was first developed in 1984 by the CSIRO Wheat Research Unit and the Bread Research Institute of Australia. In 1986, Newport Scientific took over its development and manufacture, and has continuously promoted it for use in the food industry.

The RVA is typically used to study the pasting characteristics of starches, cereal flours, and derivatives in the grain and grain products industries. It exhibits significant advantages over the traditional viscoamylograph because it requires smaller amounts of sample (3–4), is simpler and faster to operate and gives pasting curves in about 25 minutes. Therefore, the RVA is used for several applications such as the determination of enzyme activity, the suitability of cereals for further processing, the effect of additives or processing conditions on starch-based materials, the characterization of the properties of protein-based materials, the analysis of hydrocolloids, etc. Use of the RVA to assess the pasting characteristics of extruded products has been reported in several publications (Mason & Hoseney, 1986; Paton & Spratt, 1981; Ryu et al., 1993; Whalen, 1998; Whalen et al., 1997). Use of the RVA to monitor the suitability of raw materials for the extrusion-cooking process or to analyze food extruded products and related extrusion processing conditions has been adopted by the cereal industry over the last 10 years. In fact, the viscosity response of suspensions tested by the RVA can help to determine structural changes occurring in the biopolymer structures due to thermomechanical processing. When carrying out food extrusion experiments, there is a lack of methods for analyzing the effects of processing variables on extruded products in real time. Thus, there is a constant search for quick, simple and accurate analytical methods to assist professionals to evaluate the properties of extrudates. In that respect, the RVA method may provide relevant analytical information to guide extrusion researchers and technical personnel in developing new products as well as optimizing food extrusion processes. It must be noted that quick methods help to save time and money in long and complex projects such as those dealing with food extrusion processing.

Readers interested in a detailed description of the RVA and its applications should refer to *The RVA Handbook* (Crosbie & Ross, 2007).

7.1.5.1 RVA composition

The RVA comprises four main components (Figure 7.4).

- The split copper block system which permits temperature control of the sample. Heating is ensured by means of electrical resistance, while cooling is achieved by a chilled coolant (usually water) circulating through channels placed in the block. The instrument controls the temperature through platinum resistance sensors located in the block, and a digital control loop allows the block temperature to be maintained at the desired values.
- A disposable aluminum canister in which the sample is placed and a disposable plastic paddle which stirs/shears the sample. The canister is tightly pressed into the block to ensure perfect contact between the canister and the block to ensure good temperature transfer to the sample. The paddle is attached to a torque-sensing motor.
- The head which holds the torque-sensing electric motor enables the application of torque through the paddle, and the digital tachometer measures the

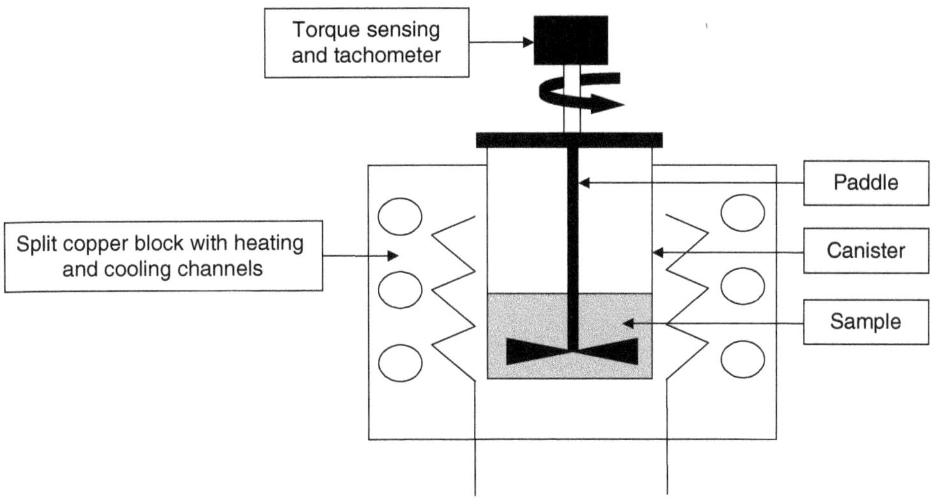

Figure 7.4 RVA instrument with its main components.

rotational speed of the paddle. A closed-loop control system maintains the paddle rotation at the desired speed.
• The Thermocline for Windows™ (TCW) software program controls the instrument and analyzes the temperature and viscosity data generated by the RVA instrument. The paddle ensures two basic functions in the canister: to mix and suspend the sample on one hand, but it also generates a shear rate on the sample. The shear rate applied in the RVA paddle and canister assembly has been analyzed by Lai et al. (2000), who have determined an effective shear rate of 20.1 s^{-1}, corresponding to a rotational speed of 1 revolution per second. Thus, for a typical RVA paddle rotational speed of 160 rpm, the effective shear rate is 53.6 s^{-1}.

7.1.5.2 RVA methodology

The methodology for sample preparation prior to RVA analysis of native and processed starch-based products is schematically illustrated in Figure 7.5. Based on a great deal of research focusing on an experimental procedure for sample preparation in order to obtain highly reproducible RVA profiles, the following methodology has been suggested.
• *Grinding/milling*, when coarse material is used. It is common practice in laboratories to grind various materials prior to analysis. In the case of RVA analysis, this first step should receive close attention as it determines the RVA profile and different RVA pasting curves that can be obtained when different grinding/milling processes are used for the same material. In fact, the grinding/milling mechanism determines various particle properties such as size, size distribution, shape and surface, and possible starch damage. The grinding/milling process depends on the properties of the material to be ground. Two different cases are considered: (i) highly expanded products, such as directly expanded snacks and breakfast cereals, expanded cereal flakes, expanded snacks pellets, etc.; (ii) non-expanded semi-finished extrusion-cooked products, such as coarse raw materials, cereal pellets prior to flaking, snacks pellets prior to expansion, etc. For highly expanded extrudates which are friable, a high-speed coffee-type grinder provides good results with no risks of further starch modification. The amount of material in the grinder as well as the grinding time must be identical for the same type of product. Grinding time usually depends on the hardness of the material; for instance, a grinding time of about 30 seconds is usually applied for directly expanded snacks while about 1 minute is necessary for expanded cereal flakes. For non-expanded extrusion-cooked products which are hard, starch damage may occur during the grinding/milling process; thus, different grinding mechanisms must be applied. In this case, disk mills as well as impeller mills are preferred. With the disk mill, different disk gap settings can be applied while with the impeller mill, screens of different aperture sizes can be used. In both cases, repeated grinding may be necessary. In general, milling products with the impeller mill results in greater starch conversion than when a disk mill is used (Becker et al., 2001a). It must be noted that once a grinding/milling procedure is optimized, the same milling conditions must be used for all samples to be compared.
• *Sieving*. Particle size has great influence in RVA viscosity profiles and thus it must be controlled carefully by sieving the ground material to a defined size fraction.

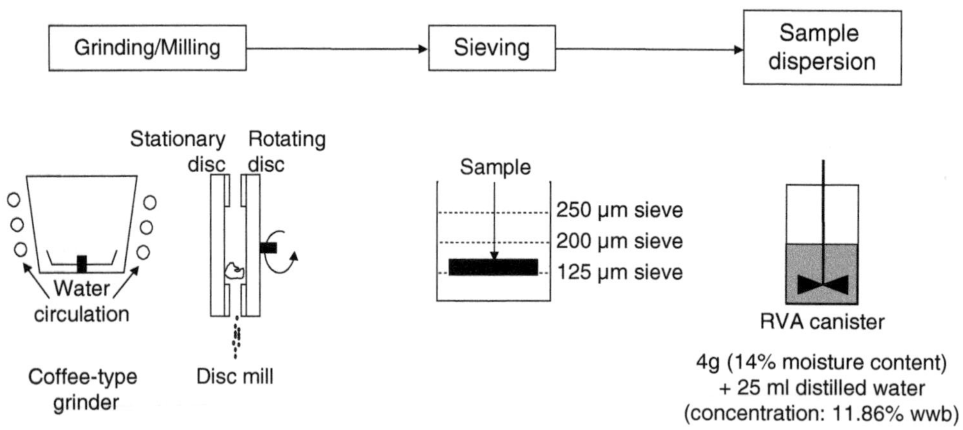

Figure 7.5 Sample preparation for RVA testing.

It is suggested to use a typical sieve shaker equipped with the sieves mentioned in Figure 7.5. The sieve fraction of 125–200 microns is usually maintained in the sample preparation.

- *Sample dispersion.* RVA viscosity is very sensitive to changes in solid concentration, in particular when the close packing fraction is exceeded (Bagley & Christianson, 1982; Doublier, 1981). Thus, a constant solid concentration in the RVA canister must be ensured and its control requires accurate determination of the moisture content of the sample to be analyzed. Most RVA measurements in extrusion processing are performed with 29 g sample dispersion/suspensions with a solid concentration in the canister of 11.86% (wet weight basis) and dry material content of 3.44 g in the dispersion; this corresponds to a 4 g sample at 14% moisture content to which 25 g distilled water is added. For a raw starch-based material with 12% (wwb) moisture content, the sample dispersion in the RVA canister will consist of 3.909 g of ground/sieved material and 25.091 g of distilled water. For an extrudate whose moisture content is 3% (wwb), sample dispersion in the RVA canister will consist of 3.546 g of ground/sieved material and 25.454 g of distilled water.

The RVA testing protocol (Figure 7.6) consists of filling the canister with the appropriate amount of distilled water and then adding the ground sample, putting the paddle into the canister and mixing the solid/liquid suspension manually (to avoid clumping), placing the canister/paddle assembly into the split block system, coupling the paddle to the torque-sensing motor, and finally starting the test. These operations must be done rigorously while respecting the same testing time for all tests. The RVA test comprises a sequence of paddle rotation speeds and temperatures to which the sample is subjected (see Figure 7.6). The following protocols are suggested for extrusion-cooked products.

- *Stirring-time protocol.* The protocol includes first a high-speed mixing step (10 s at 960 rpm) to stir and vigorously disperse the powder particles in the water and eliminate possible clumps. Then, the paddle rotational speed is fixed at 160 rpm for the test.
- *Standard temperature-time protocol.* The protocol starts at 25°C with an initial holding time of 6 minutes to allow cold water swelling and solubilization of the sample before the heating stage starts. Then, heating allows the temperature to be raised from 25°C to 95°C over 4 minutes. A rather long holding time of 6.5 minutes at 95°C is applied to allow full development of a gelatinization peak and trough before cooling starts. Cooling occurs over 4.5 minutes to ensure that the sample is cooling back

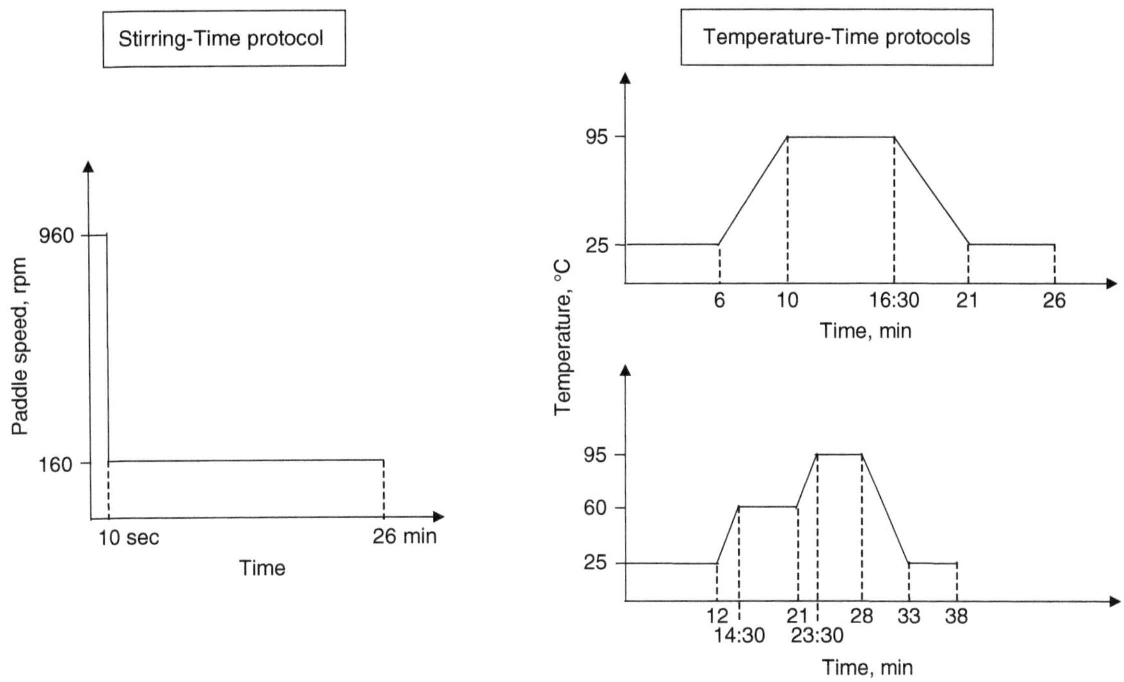

Figure 7.6 RVA testing protocols.

to 25°C. Finally, a holding time of 5 minutes at 25°C is applied to allow determination of the final viscosity of the gelled sample. This protocol can be used for raw starch-based materials and most extrusion-cooked products (directly expanded extrudates, in particular).

- *Two-step heating protocol.* The protocol gradually heats the samples from 25°C to 60°C, and then from 60°C to 95°C, while holding them at 60°C for 6.5 minutes. This allows a more sensitive and accurate study of swelling events below the gelatinization temperature, through the identification of a peak viscosity at 60°C, called the "hot swelling peak". The protocol can be used for raw starch-based materials and extrusion-cooked products produced with low SME.

7.1.5.3 RVA response

The RVA response, which is the recorded viscosity curve as a function of either temperature or time, is directly related to the state of the starch structure and closely related to its functional properties. Viscosity values at characteristic points along the curve are typically used to describe the response for a given sample, and to compare it among different samples. Typical RVA responses of starch-based raw materials and extrusion-processed materials are now presented and discussed.

7.1.5.3.1 RVA response of starch-based raw materials

The RVA response of starch-based raw materials reflects their pasting properties as a result of structural changes occurring during gelatinization and gelation induced by the temperature-time testing protocol. Figure 7.7 shows the RVA responses of unprocessed corn grits and wheat flour obtained when applying the standard temperature-time protocol in the RVA testing (Clextral, 2013a).

When unprocessed starch particles are dispersed in water, a dilute particle suspension is obtained since native starch is not soluble in cold water. The RVA viscosity is very low (close to the viscosity of water) during the first step of RVA testing at 25°C, due to the low volume fraction of particles in the suspension. Native starch does not swell in cold water due to the packing and hydrogen bonding of the starch molecules within the granule. On heating, granules swell but the viscosity remains the same until as a consequence of swelling, the particle phase

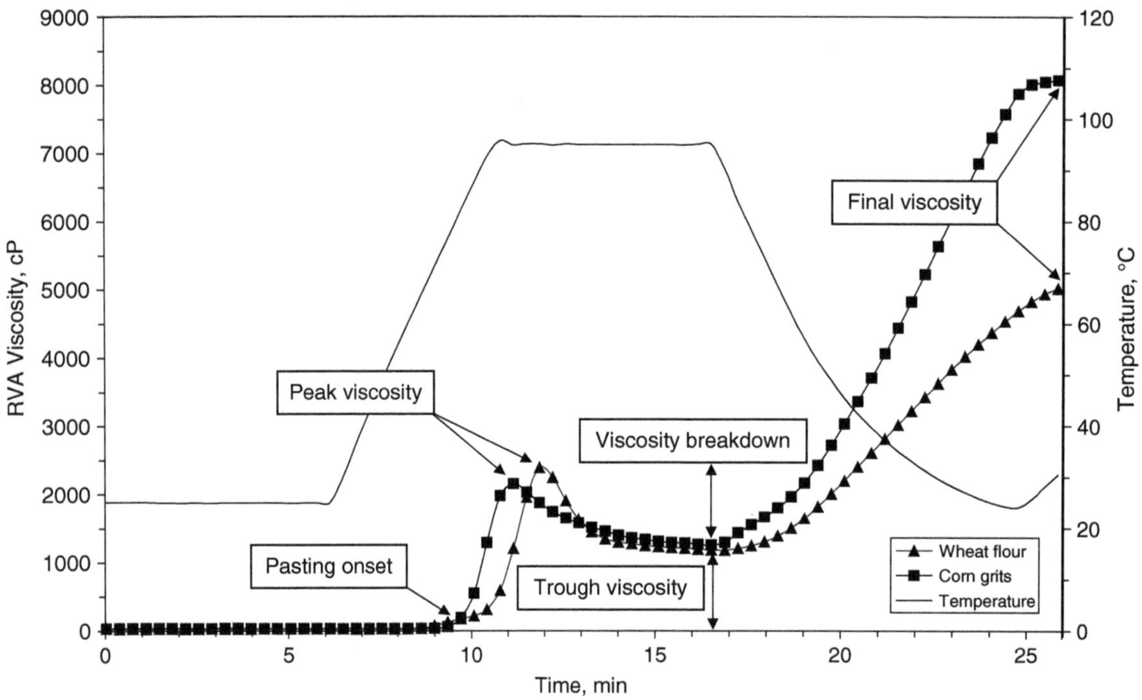

Figure 7.7 Typical RVA responses of cereal starch-based raw materials (Clextral, 2013a). Source: Reproduced with permission of Clextral, France.

volume is sufficiently high to reach the close packing stage. When close packing is reached, at the pasting onset, the viscosity starts to increase and the suspension switches from a dilute to a concentrated state. The time and temperature at which this situation occurs depend on the starch gelatinization temperature, the rate of increase of granule size during swelling, and the packing of the swelling granules as a result of their size distribution and shape. Figure 7.7 shows the pasting onset at 9.35 minutes and 76.6°C for corn grits while it is 9 minutes and 72.5°C for wheat flour. The RVA pasting curve of wheat flour shows a slight increase before the main pasting peak. This occurs often when testing wheat flours, due to the presence of damaged starch.

After the pasting onset, viscosity increases significantly due to continuous particle swelling and the subsequent increase of the volume occupied by the swollen granules. As the gelatinization temperature is exceeded, there is a progressive loss of crystallinity in the granule structure, and the separation of amylose and amylopectin molecules starts, mainly due to leaching of the amylose fraction. Continuous shearing of the swollen granules further enhances amylose leaching. In this concentrated regime, viscosity is mainly determined by the deformability of the swollen granules, and in the continuous phase viscosity depends on the concentration of polymer (mainly amylose) in the solution. The peak viscosity is related to the maximum swelling capacity of the starch granules, and this value is often used to qualify starch pasting characteristics. At peak viscosity, the swollen starch granules are highly shear sensitive and the starch polymers are completely in an amorphous state; the shear sensitivity of the system depends closely on the extent of amylose leaching, and on the swelling capacity of the remaining amylopectin-rich granule ghosts. The balance between maximum swelling of starch granules and breakage induced by shear depends on the starch botanical origin. This balance may also be affected by non-starch components such as lipids or proteins, which reduce amylose leaching. The existence of the peak viscosity at 95°C reveals the presence of non-gelatinized material in the product. Figure 7.7 shows that peak viscosity was reached at 11.2 minutes with a value of 2150 cP for corn grits while it was reached at 11.9 minutes with a value of 2390 cP for wheat flour.

On shearing the sample at 95°C, starch granules, which at that stage are weak and have a fragile structure, are ruptured into fragments and granule remnants, while the amylopectin fraction is released into the solution. The resulting starch paste is then composed of amorphous granule remnants in a polymer solution containing amylose and amylopectin. The disruption of the granule remnants causes an important decrease of the viscosity of the starch paste, due to the decrease of the granule phase volume. This effect is predominant over the simultaneous increase of viscosity in the continuous phase due to the increasing amount of polymers (amylose and amylopectin) in the solution. Additionally, continuous shearing of the starch paste helps to align granules remnants and polymers in the shear field, thus inducing the viscosity breakdown of the paste. Figure 7.7 shows that the trough viscosity was reached at 16.5 minutes with a value of 1255 cP for corn grits while it was reached at 16.9 minutes with a value of 1178 cP for wheat flour, giving a viscosity breakdown of 895 cP and 1212 cP for corn grits and wheat flour, respectively.

On cooling, the viscosity shows a large increase due to the effect of temperature on starch gel viscosity. The final viscosity of the starch paste is determined by the phase volume and deformability of the dispersed remnants, by the extent of amylose gelation (association or retrogradation of amylose and formation of a polymer network) in the continuous phase, and by the interactions of both dispersed and continuous phases. If the starch paste is cooled down to 25°C and held for some time, viscosity increase can be observed due to gelation, and possibly a viscosity decrease as a consequence of the thixotropic effects on the sample induced by continuous stirring. Figure 7.7 shows that the final viscosities of the starch pastes were 8075 cP and 5020 cP for corn grits and wheat flour, respectively. The final viscosity obtained with corn grits is much higher than that obtained with wheat flour.

7.1.5.3.2 RVA response of extrusion-processed materials

The viscosity curve of extrusion-processed starch materials reflects the changes in pasting properties that result from starch structural modifications induced by the process, hence leading to a measure of the extent of starch conversion. Typical RVA responses of extruded corn- and wheat-based products are shown in Figure 7.8. Both samples have been fully gelatinized during extrusion-cooking, as their process SMEs were 614 kJ/kg and 472 kJ/kg for corn- and wheat-based extrudates, respectively (Clextral, 2013a).

When the powder particles are dispersed in excess cold water for testing, starch polymers and possible granular remnants readily hydrate and swell, and a significant fraction of converted polymers dissolves. With extrusion-cooked products, the time required for water migration into the particles, and for granule swelling and

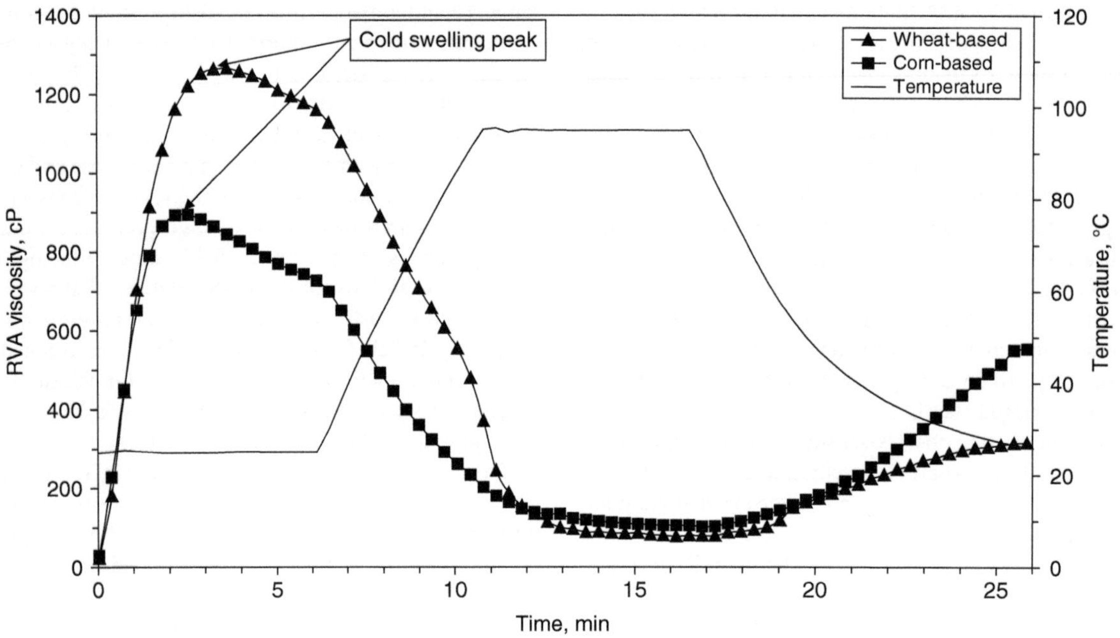

Figure 7.8 Typical RVA responses of extrusion-cooked cereal starch-based materials (Clextral, 2013a). Source: Reproduced with permission of Clextral, France.

dissolution, is quite short. Thus, as soon as the sample is prepared for testing, the dilute sample suspension becomes a concentrated suspension within a short time, resulting in a steep increase of the RVA viscosity. The maximum viscosity, called the "cold swelling peak," is related to the swelling capacity and shear resistance of the particles and granules, and the dissolution rate of the starch polymers. At the cold swelling peak, the viscosity is predominantly influenced by the extent of starch gelatinization and depolymerization during extrusion-cooking. Figure 7.8 shows that the cold swelling peak was reached at 2.5 minutes with a viscosity of 894 cP for corn grits while it was reached at 3.6 minutes with a viscosity of 1265 cP for wheat flour. After the cold swelling peak, viscosity decreases due to the continuous shearing of the starch paste. As the temperature increases from 25°C to 95°C, a significant change in the slope of the viscosity decrease is observed due to the effect of temperature on the paste viscosity. No gelatinization peak viscosity is detected. The trough viscosity of both samples is very low (103 cP and 79 cP for corn- and wheat-based products, respectively). On cooling, increased viscosity is obtained, which is generally much lower than that observed for the unprocessed material. The value of the final viscosity is influenced by the viscosity of the continuous phase and the phase volume of possible remnants. The viscosity of the continuous phase depends closely on the extent of starch depolymerization; it may also be affected by possible amylose retrogradation. Figure 7.8 shows that the final viscosities of the starch pastes were 553 cP and 315 cP for corn- and wheat-based samples, respectively.

7.1.5.4 Effect of thermomechanical cooking on the RVA response

7.1.5.4.1 Effect of process SME

This section presents and discusses RVA responses of extrusion-cooked corn-based material obtained with the same screw configuration, while changing screw speed and water content and hence varying process SME from 115 kJ/kg to 650 kJ/kg. RVA tests used the two-step heating protocol described above. Three categories of RVA responses have been obtained depending upon the range of process SME (Clextral, 2013a).

Low-SME extrudates: SME ≤209 kJ/kg
The extrudates were extrusion-processed at high water content (40% wwb) and different screw speeds (400 rpm, 600 rpm, and 900 rpm), with process SMEs

of 115, 150, and 209 kJ/kg, respectively. The temperature of the resulting melts at the die exit increased from 92°C to 100°C when the process SME increased from 115 to 209 kJ/kg. Figure 7.9a shows no cold swelling peak at 25°C for all samples, meaning that the starch polymers remained insoluble in water. At 60°C, a slight increase of viscosity is observed only for the extrudate processed at 209 kJ/kg SME. Upon heating up to 95°C, RVA responses showed the presence of gelatinization peaks for all samples. These peaks were lower when compared to that obtained during RVA testing of raw corn grits. Unexpectedly, the extent of the gelatinization peak of the extrudate processed at 209 kJ/kg SME was significantly higher than that of the other extrudates processed at lower SMEs (115 and 150 kJ/kg). Finally, on cooling down to 25°C at the end of the test, Figure 7.9a shows high values of final viscosity for all samples. The final viscosity was significantly higher for the corn grits sample while it was lower for the extrudates processed under 209 kJ/kg SME.

Inspection of the starch granules of all extrusion-processed samples under polarized light optical microscopy showed no loss of the granule integrity, but it did show a progressive loss of semi-crystalline structure of starch granules with increasing process SME. The loss of birefringence as estimated by the decrease of the proportion of Maltese crosses was about 20–25% for very low SME extrudates (115 and 150 kJ/kg) while the loss was about 50% for the extrudate processed at 209 kJ/kg. Interestingly, the RVA responses of extrusion-processed samples did not show any sign of cold swelling peak, although part of the starch had lost some of its crystalline structure. The final viscosity progressively decreased with increasing process SME.

According to Becker et al. (2001b), although mechanical and thermal energies are applied in the extrusion process, the lack of water uptake and swelling at 25°C as well as the decrease of the gelatinization peak of samples processed at very low process SME would be related to the formation of an insoluble film of amylose–lipid complexes on the granule surface. In fact, when studying thermal conversion of various starch materials without shear at 50% (db) water content, it was observed that RVA responses of the resulting samples were characterized by decreasing gelatinization peak and final viscosity with longer heating times, and no cold swelling peak was observed for any heat treatment. The authors demonstrated that the presence of amylose–lipid complexes played a key role in the restriction of water penetration in the particles as well as starch granule swelling and solubilization in cold water.

Based on that experimental evidence, it can be considered that during extrusion processing of lipid-rich, starch-based materials (such as cereal starches like maize, wheat, rice, etc.) at low process SME (around 200 kJ/kg and below) and limited amount of water (water content ≤40% wwb), small amounts of amylose are leached from the outer surface of the starch granules but the granular integrity of starch is maintained and a significant amount of starch polymers becomes amorphous. During the extrusion process the temperature is sufficiently high for the amylose molecules to be mobile and able to complex with native lipids and crystallize. Hence, an insoluble film of amylose–lipid complexes is formed at the periphery of the granules which inhibits water uptake and swelling at temperatures below 65–70°C. Due to these structural changes in the system, RVA responses show no cold swelling peak and no hot swelling peak. Gelatinization peak and final viscosity are present but their values are lower than those obtained by testing native starches mainly due to the formation of amylose–lipid complexes in the extrusion-processed materials.

Intermediate range of process SME: 200 kJ/kg < SME <400 kJ/kg

The extrudates were extrusion-processed at different water contents (from 19% to 32% wwb) and screw speeds (from 225 to 600 rpm), leading to process SME ranging from 210 to 400 kJ/kg. At that range, the temperature of the resulting melts at the die exit varied from 110°C to 140°C. Figure 7.9b presents a selection of four RVA responses for extrudates obtained in this intermediate range of process SME. For these samples, the loss of the native crystalline structure increased from about 50% to 100% (as examined by polarized microscopy).

As shown in Figure 7.9b, RVA responses showed cold swelling peaks with values increasing with increases in SME. The extent of mechanical energy input was sufficiently high to allow amylose to leach out of the granules and degrade starch granules. Amylose leaching and granule damage were higher as the SME increased, which allowed a significant part of starch molecules to be water soluble at 25°C. Hence, the cold swelling viscosity of the resulting suspension increased from less than 100 cP to about 1200 cP when the process SME increased from 216 to 379 kJ/kg. Upon heating from 25°C to 60°C, a small viscosity decrease was observed due to the effect of temperature on the viscosity of materials. And when the temperature reached 60°C, a significant increase of viscosity (hot swelling peak) was noted, which revealed some additional water uptake into the particles of the

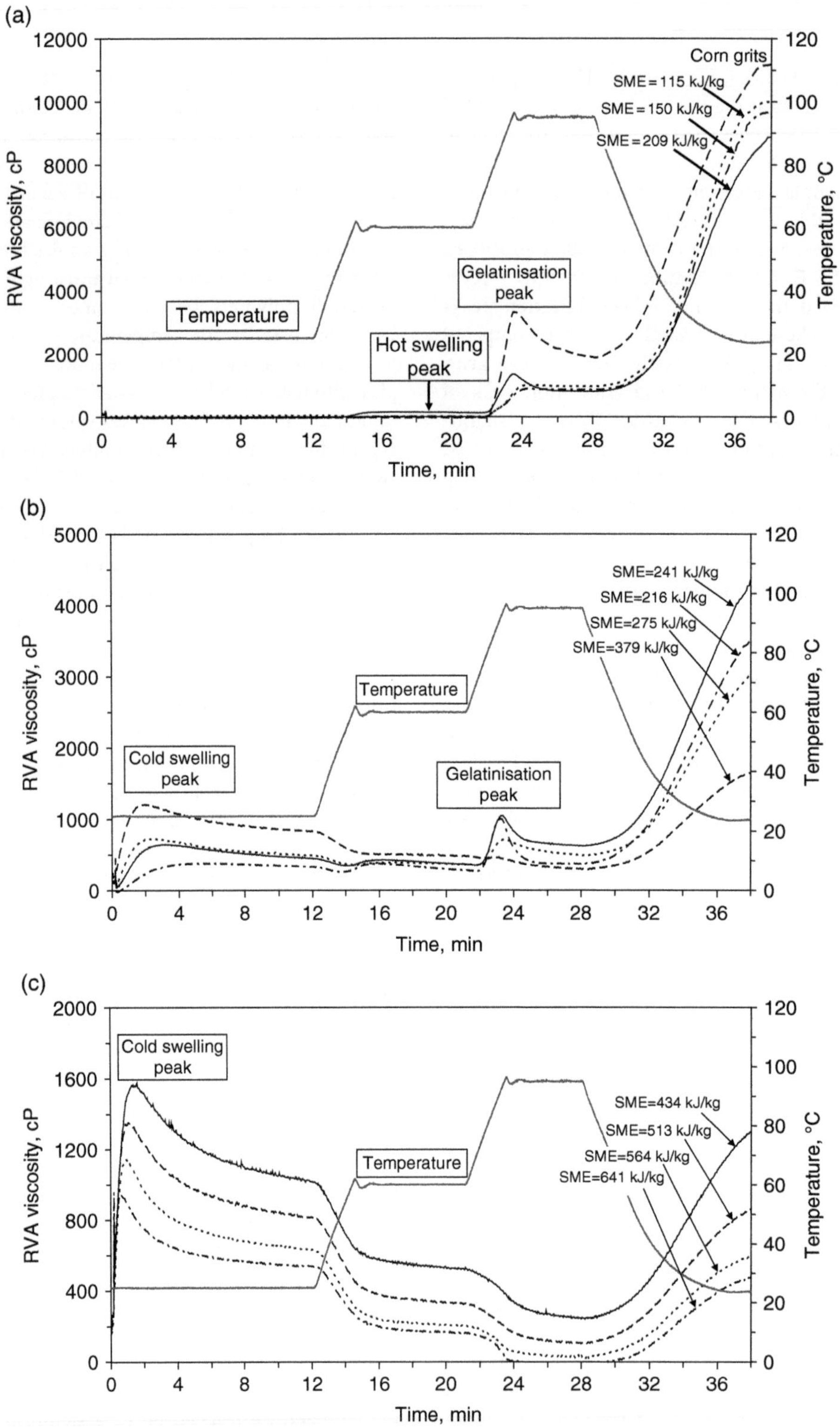

Figure 7.9 (a) RVA responses of low-SME extrusion-processed corn-based material. (b) RVA responses of extrusion-processed corn-based material under intermediate range of process SME (200 kJ/kg < SME <400 kJ/kg). (c) RVA responses of high-SME extrusion-processed corn-based material (Clextral, 2013a). Source: Reproduced with permission of Clextral, France.

suspension, solubilization of extra starch molecules, and swelling of residual undamaged starch granules. Viscosity increases at 60°C as well as the gelatinization peaks were higher as the process SME was low. It turned out that hot swelling peak and gelatinization peak were negligible when SME reached 379 kJ/kg, which indicates that the integrity and native crystalline structure of the starch granule no longer existed at this level of process SME, and that starch polymers were fully converted. As shown in Figure 7.9b, the final viscosity decreased drastically when the process SME increased, leading to a rather low viscosity (about 1600 cP) at 379 kJ/kg, mainly due to a significant shear-induced degradation of starch macromolecules (named dextrinization).

It was observed in the study that the RVA response obtained with an extrudate extrusion-processed with a SME of 399 kJ/kg was identical to that obtained with the material processed with a SME of 379 kJ/kg. Thus, it was concluded that a minimum process SME is necessary for ensuring complete gelatinization of starch material (or complete loss of starch crystallinity) and it is in the range of 380–400 kJ/kg (or 105–110 W.h/kg), which is in good agreement with values observed by other researchers and discussed in Chapter 6 (section 6.2.2.3).

High-SME extrudates: SME >400 kJ/kg

The material was extrusion-processed at different water contents (from 16% to 22% wwb) and screw speeds (from 400 to 1175 rpm), leading to process SMEs ranging from about 400 to 650 kJ/kg. In that range, the temperature of the resulting melts at the die exit varied from 145°C to 155°C. Figure 7.9c presents four RVA responses for extrudates processed with SMEs higher than 400 kJ/kg. Under such conditions, starch has completely lost its native crystalline structure as well as its granular integrity. RVA responses showed an important cold swelling peak with values decreasing with increases in SME. A large part of the starch was water soluble in cold water, as the extent of mechanical energy input was sufficiently high to degrade starch macromolecules. Then, as SME increased in extrusion-cooking, the molecular weight of starch molecules decreased which led to decreasing viscosity of the starch-rich solution.

As the temperature in the RVA increased from 25°C to 60°C, and then from 60°C to 95°C, a significant viscosity decrease was observed due to the effect of temperature in the materials discussed above and described in Chapter 4. Hot swelling peak and gelatinization peak were not present due to the rapid water uptake and swelling of the fully degraded starch molecules which occurs at 25°C, hence allowing starch macromolecules to become fully soluble. As shown in Figure 7.9c, final viscosity peak decreased drastically when the SME increased, due to the increase of shear-induced starch dextrinization. Under high SME extrusion-cooking of starchy materials (over 400 kJ/kg), the cooking process was characterized by a mechanical energy input higher than the thermal energy input inducing starch dextrinization.

It is worth examining the changes in the three main RVA parameters which are the gelatinization peak, the cold swelling peak and the final viscosity as a function of process SME. These parameters can be identified easily from the RVA responses. It should be noted that the gelatinization peak is determined by the difference between the viscosity of the gelatinization peak and the viscosity of the baseline or minimum viscosity before the gelatinization peak starts. Figure 7.10a presents the effect of process SME on the RVA gelatinization peak. The gelatinization peak generally decreases almost linearly with SME in the range 200–400 kJ/kg. Noted in the figure are the two amylose–lipid complexed extrudates produced at very low process SME (SME <200 kJ/kg). The figure shows that complete gelatinization is achieved when process SME reaches values in the range 380–400 kJ/kg, whereas the two extrudates produced at very low process SME (SME <200 kJ/kg) and that promote the complexation between amylose and lipids deviate from the general trend.

As discussed above for these materials, their gelatinization peak is significantly decreased due to the formation of amylose–lipid complex. Figure 7.10b presents the effects of process SME on cold swelling peak and final viscosities. Final viscosity decreases almost linearly from 11200 cP (unprocessed material) to 8900 cP (extrudate obtained at 209 kJ/kg) in the low process SME range of 0–209 kJ/kg. When process SME reaches values of 210–215 kJ/kg, a dramatic decrease of final viscosity is observed (decrease of about 4000 cP). At this level of process SME, there is some discontinuity in the thermomechanical conversion of starch material, which might be interpreted as follows: below 210 kJ/kg SME (water content of 40% wwb), the effect of thermal energy input on the extruded material would be highly predominant with low effect of mechanical energy input whereas above 210 kJ/kg (water content ≤32% wwb), the effect of mechanical energy input on the extruded material would become predominant and produce significant depolymerization of starch molecules. When SME ≥215 kJ/kg, final viscosity decreases continuously but a plateau of viscosity around 1000 cP is reached when SME is about 500 kJ/kg. Cold swelling peak remains

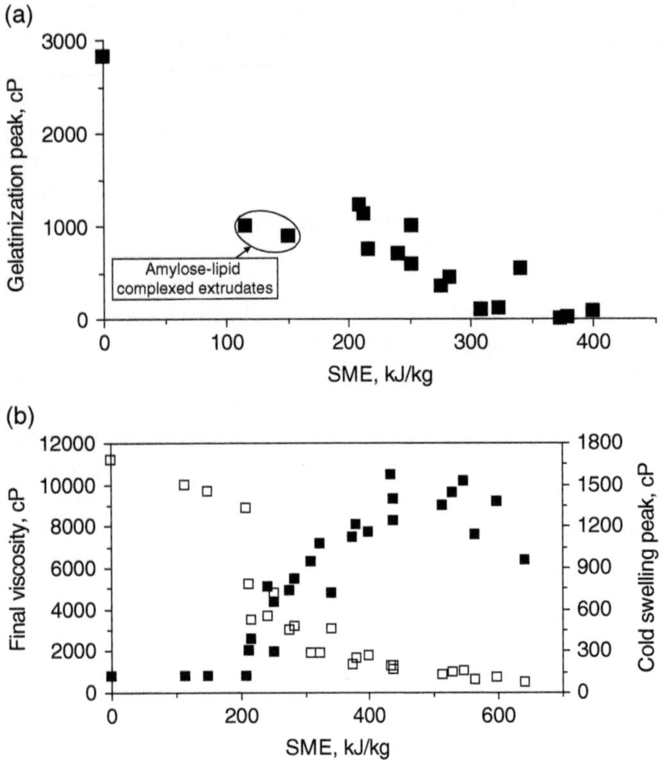

Figure 7.10 (a) RVA gelatinization peak versus process SME (corn-based material). (b) RVA cold swelling viscosity (■) and final viscosity (□) versus process SME (corn-based material).

negligible at low process SMEs (in the range of 0–209 kJ/kg). For SME ≥215 kJ/kg, cold swelling peak then increases sharply with increases in process SME mainly due to the predominant effect of the mechanical energy input at those process SME levels. In the process SME range 400–500 kJ/kg (or 111–139 W.h/kg), maximum viscosity values in the range 1200–1700 cP are observed due to an increasing proportion of gelatinized starch granules reaching complete gelatinization. Above approximately 500 kJ/kg (or 139 W.h/kg), the cold swelling peak decreases slightly due to starch dextrinization.

Data on cold swelling viscosity versus process SME is of paramount importance in practical extrusion-cooking, as it is closely related to the functional properties of many extruded products produced by a wide range of cereal extrusion processes.

Finally, Figure 7.11 illustrates the relationship between cold swelling peak viscosity and final viscosity. Apart from the two extrudates produced at very low process SME (SME <200 kJ/kg) and in which a amylose–lipid complex is formed, Figure 7.11 shows that cold swelling peak viscosity decreases continuously with increases of final viscosity.

Results showing the effect of SME on RVA responses of extrusion-processed corn-based material are in good agreement with published studies focusing on the thermomechanical conversion of starch materials which are presented in Chapter 6 (section 6.2.2.3). This shows that RVA response allows operators to assess and describe accurately structural changes of starch materials during extrusion processing. Consequently, the RVA method can be used to optimize the properties of extruded products as well as the associated extrusion processes. For instance, it can be helpful in building relevant databases incorporating RVA responses and associated functional properties of products such as, for example, the relationship between cold swelling peak viscosity and process SME presented above. That information would help cereal manufacturers to select suitable screw configurations as well as processing variables such as screw speed and water content to produce good-quality, palatable extruded products.

7.1.5.4.2 Effect of screw profile

It is worth illustrating the use of the RVA method to assist process designers in developing suitable screw elements and screw profiles for optimization of food extrusion-cooking processes and products. For that purpose, a typical screw configuration set on a laboratory-scale Clextral twin screw extruder-cooker (25 mm screw diameter) with a L/D of 16 was used. The screw configuration comprised three main processing sections: one compression section (L/D = 14) composed of conjugated right-handed screw elements having decreasing pitches, one cooking section (L/D = 1) and finally one pumping section (L/D = 1) composed of one conjugated right-handed screw element of low pitch. Three different screw designs were tested for the cooking section; Figure 7.12 presents the designs of the screw elements used:

- one conjugated, double-flight, left-handed screw element without slots at the periphery of the flight (pitch of -16.6 mm), hence leading to screw profile 1
- one conjugated, double-flight, left-handed screw element with four half-moon slots at the periphery of the flight (pitch of -16.6 mm), hence leading to screw profile 2
- one bilobe kneading block, neutral 90° staggering, hence leading to screw profile 3.

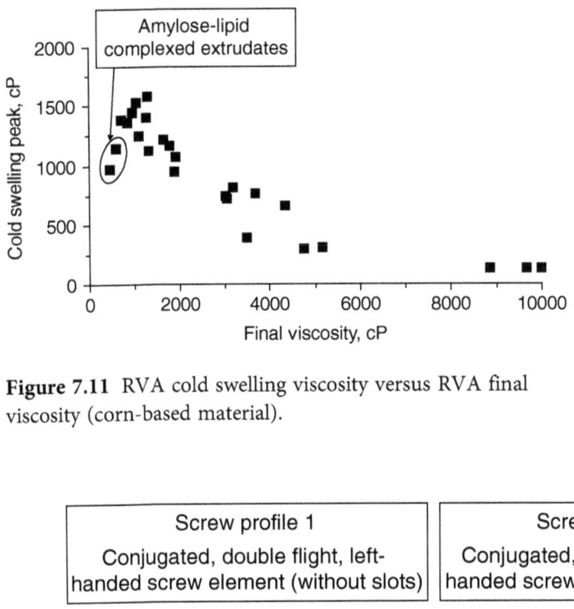

Figure 7.11 RVA cold swelling viscosity versus RVA final viscosity (corn-based material).

The two left-handed screw elements used in profiles 1 and 2 were designed to have the same cross-section offered to the melt flow, meaning that the left-handed screw element of profile 1 had a lower screw diameter than that of the left-handed screw element of profile 2. All screw profiles setting ended with a die assembly able to produce directly expanded extrudates. The experiment consisted of the extrusion-cooking of corn grits utilizing the set screw profiles, under different processing conditions that consisted of varying the water content (14% and 17% wwb) and the screw speed (250 and 400 rpm), hence operating in a range of process SME of 490–790 kJ/kg. In total, 12 different extrudates were analyzed by the RVA method.

Figure 7.13 presents RVA responses of the different extrudates obtained with the three screw profiles at two different screw speeds, but at a constant water content (17% wwb). It is shown that the cold swelling peak viscosity was significantly affected by the screw design in the cooking section: screw profile 2 (left-handed screw element with slots) showed the highest value of the cold swelling peak viscosity whereas screw profile 3 (90° staggering kneading block) showed the lowest value of cold swelling peak viscosity. Values of final viscosity followed the same trend. These results indicate that the intensity of mechanical shearing of the three profiles increases as follows: profile 2 < profile 1 < profile 3. As expected, low screw speed led to higher cold swelling peak and final viscosities. As seen in section 7.1.5.4.1 (see Figure 7.10b), Figure 7.14 also confirms that cold swelling viscosity and final viscosity are almost linearly related, showing decreases in these viscosities with increases in process SMEs, in a range of 490–790 kJ/kg. Thus, the RVA method allows for a quantitative determination of effects that mechanical shearing provided by different screw elements on starch conversion.

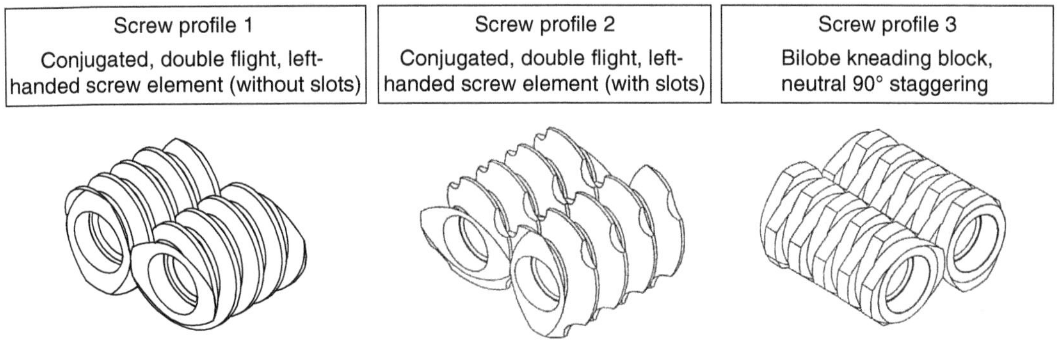

Figure 7.12 Screw designs of the cooking section of the laboratory-scale Clextral twin screw extruder.

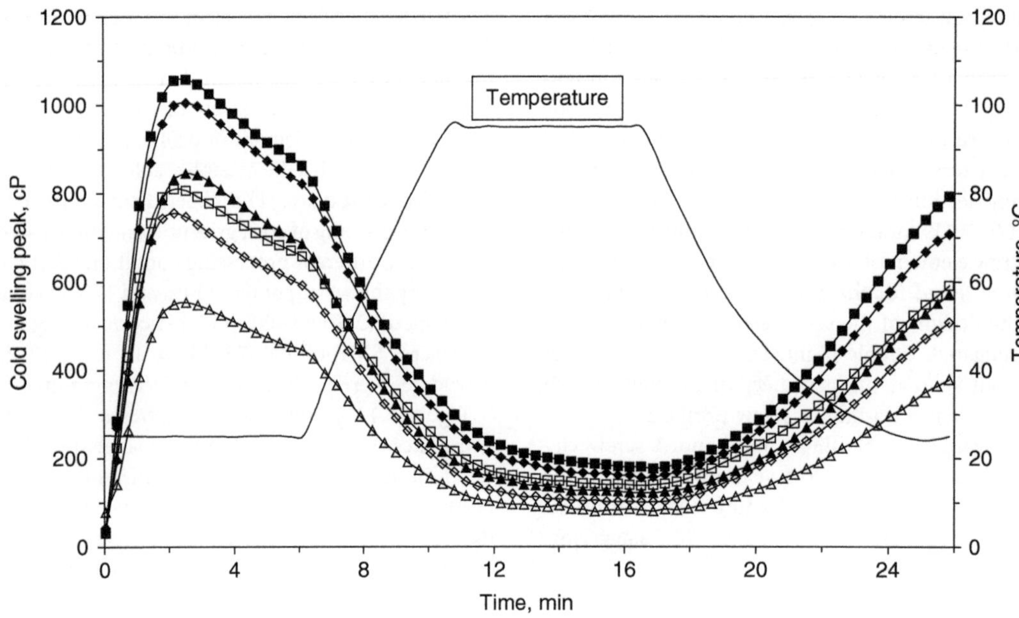

Figure 7.13 Effect of screw design on the RVA response; closed symbols: 250 rpm/17% (◆: profile 1; ■: profile 2; ▲: profile 3); open symbols: 400 rpm/17% (◇: profile 1; □: profile 2; △: profile 3) (Clextral, 2013a).

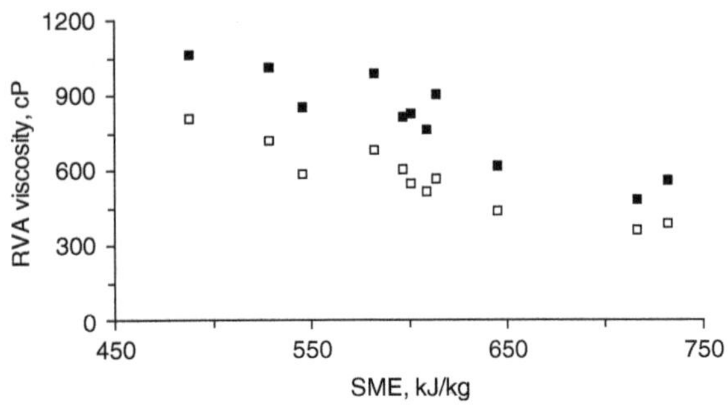

Figure 7.14 RVA criteria versus SME (■: cold swelling peak; □: final viscosity).

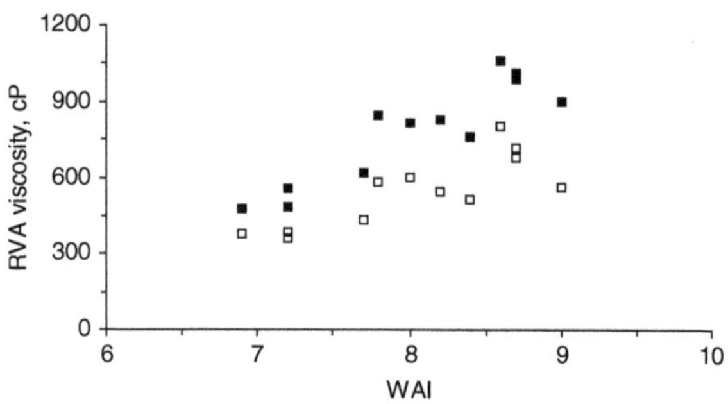

Figure 7.15 RVA criteria versus WAI (■: cold swelling peak; □: final viscosity).

As shown in Figures 7.15 and 7.16, for extrudates produced in a range of process SME of 490–790 kJ/kg, there is a strong correlation between cold swelling and final viscosities and WAI (Figure 7.15) and WSI (Figure 7.16).

Section 7.1.5 has shown that the RVA-based method can very sensitively monitor properties of both raw and processed starch-based materials, that are related to pasting and gel properties which in turn are associated with the thermomechanical treatment to which the sample is subjected. In addition, tests are easy to set, and provide consistent and reproducible results when a rigorous protocol is followed. RVA-based methods can thus be useful for quick analysis and control of starch conversion in food extrusion processing. The use of relevant RVA parameters such as cold swelling viscosity, gelatinization peak viscosity, and final viscosity can help to set some processing criteria to investigate the effect of a wide range of processing conditions on the quality of products, hence giving valuable support to product and process designers.

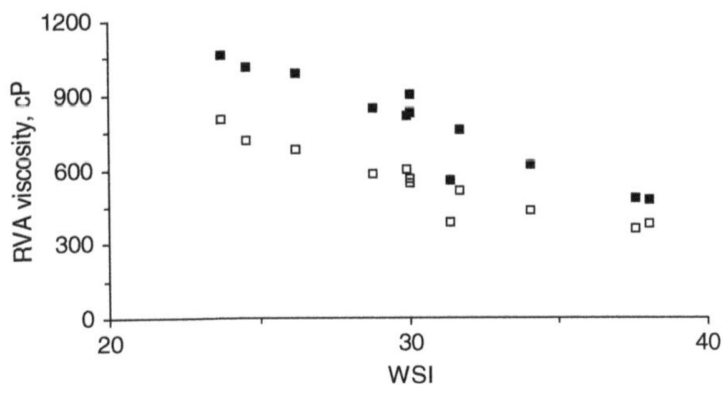

Figure 7.16 RVA criteria versus WSI (■: cold swelling peak; □: final viscosity).

7.2 Methods of characterizing extrudate texture

The introduction of extrusion-cooking in the food industry has led to a new class of food products characterized by a cellular structure and crispy texture. Since the arrival of the first extruded snacks and breakfast cereals several decades ago, the market for extrusion-textured foods is now well established, and numerous new products are launched every year. In fact, this market benefits from increasing consumer demand for extruded foods mainly due to their characteristic crispness, a texture property that is highly appreciated by consumers.

The key sensory value of these products is closely related to their structural characteristics and textural properties. Consequently, when dealing with extrusion-cooking processes, a global methodology must be used for successful development of extrusion-textured food products and consistent control of the manufacturing processes. Figure 7.17 presents a concept for the development of a texturally sound extruded product. As illustrated in the figure, the interconnection between subjective and objective measurements, some of which have not been established yet, can help guide product designers and process engineers to optimize product quality according to consumer acceptance.

In Chapter 6, the mechanism of product texturization was described and discussed. It was shown that the structure of extrusion-textured food products derives from steam-induced texturization of rubbery biopolymer-based melts, through either on-line or downstream processes. An expansion chart was proposed which allows the establishment of relationships between processing conditions and expansion criteria of food products (see section 6.1.2).

The texturization process generates very specific cellular structures with cell sizes and shapes that vary extensively depending upon the rheological characteristics of the extrudate melt and the texturization conditions. After the extrusion process, extrudate products are often dried so that during consumption, they are in a glassy state with a wide range of crispness attributes, which are determined by product characteristics such as physical properties as well as sensory attributes. The challenge then consists of producing products with structures that fit consumer acceptance. But, as consumer acceptance is often associated with highly subjective descriptions of the product texture, the development of correlations between the extrudate structure and consumer acceptance is a critical step in the approach to textural concept illustrated in Figure 7.17. Thus, the objective of the proposed methodology includes the measurement of mechanical and acoustic properties as well as sensory attributes such as crispness profile evaluated by trained sensory panels with the main focus of building correlations between product structure, physical properties, and sensory attributes.

Due to the typical cellular structure and crispy texture of extrusion-textured food products, it is important to describe relevant instrumental methods which would enable food product developers to characterize and control the structure and texture of extrusion-textured food products in relation to the process conditions and sensory attributes. These are not separate areas and they need to be studied using an integrated approach because physical properties, sensory attributes, and consumer descriptors all form the texture profile of extrusion-textured food products.

The purpose of this section is an overall description of relevant methods used to characterize these products while focusing on the measurement of their structural characteristics.

7.2.1 Measurement of product density

In food extrusion processing, developers and product manufacturers measure product density of extrudates as one important physical property to characterize or control the quality of extrudates. Two types of density are determined: bulk density and apparent density. Both are in the typical units for density, i.e. mass per volume. However it is normal practice to express these densities in grams per liter.

Bulk density (ρ_b) is evaluated by weighing (M_b) a known volume V (e.g. a 5 L container) of dried extrudates and calculating the bulk density as $\rho_b = M_b/V$.

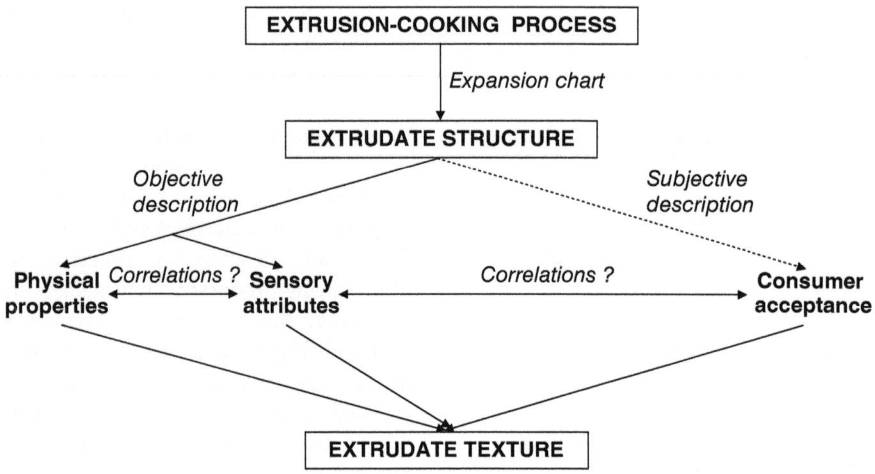

Figure 7.17 Textural concept for extrusion-textured food products.

The volume V includes the pores of the extruded product as well as the voids between extrudate pieces. Consequently, measured bulk density depends on the shape of extrudates and the packing process of extrudates in the container. However, for extrudate pieces of a given shape and following a standardized packing process, measurement of bulk density is very quick, easy, and accurate. Since bulk density may closely reflect the extrusion processing conditions and partly affects the textural properties of the product, its measurement is widely used in product development and quality control. As an illustration, Figure 7.34 (section 7.3.2) presents a typical master curve correlating process SME and bulk density of directly expanded snacks. In this example, several extrudates were produced using a given corn-based formulation, while varying extrusion processing variables such as water content, screw speed, and screw configuration. Regardless of the processing conditions, this curve shows a similar decrease in bulk density as process SME increases from 125 to 215 W.h/kg.

The second parameter, apparent density (ρ_a), is evaluated by weighing (M_o) a known volume V_o of dried extrudates, while adding calibrated microspheres (screened sand particles, glass beads or rapeseeds) to fill in the air voids between the extrudate pieces. Hence apparent density is calculated as $\rho_a = (M_o - M_m)/(V_o - V_m)$, where M_m and V_m are the weight and volume of the microspheres added, respectively. It must be noted that the volume of dried extrudates includes the pores of the extruded product. Consequently, the measured apparent density no longer depends on the shape of extrudates and the packing of extrudates into the container. The measurement of apparent density takes longer but it provides a very accurate determination when applying a rigorous well-established protocol.

7.2.2 Measurement of structural characteristics

Little has been reported on instrumental methods used to analyze the cellular structure of extrudates. It is worth mentioning the use of mercury porosimetry and X-ray tomography techniques to describe the cell structure of extruded products which includes closed and open porosity, cell size distribution, and internal organization of the cell structure. Mercury porosimetry has been widely used to analyze the structure of food products like bread (Datta et al., 2007). Recently, various studies have reported the use of X-ray tomography for describing the three-dimensional cellular structure of extruded starch foams (Agbisit et al., 2007; Babin et al., 2007; Trater et al., 2005). X-ray tomography performs a series of scans that characterize the cellular structure of the sample; the scans are then processed using specialized 3D image analysis software to determine relevant properties of the structure such as cell size distribution, cell wall size distribution, and cell density. This technique has been successfully used to investigate the cellular structure of directly expanded wheat-based extrudates (Robin et al., 2010) prepared under different extrusion processing conditions. In particular, the authors studied the effects of ingredients such as sucrose and sodium bicarbonate and extrusion-cooking conditions (screw speed, moisture content, barrel temperature, die opening) using a laboratory-scale twin screw extruder to explore the effect of process SME in a range of 130–445 kJ/kg (relatively low compared with values currently used in the production of direct expanded products). Although the process SME values lead to relatively low volumetric expansion indices (ranging from 1.6 to 9.1), the authors were able to establish relevant correlations between cellular structure, melt expansion properties (sectional, longitudinal and volumetric expansion indices, in particular), and mechanical properties of extruded products.

Although mercury porosimetry and X-ray tomography provide relevant information on cell structure of extrusion-textured food products, they require highly skilled technical assistance and methods and may take a relatively long time which makes these methods more suitable to a research than to an industrial environment.

This section focuses more on the use of a 2D image analysis method which can be easily integrated and handled in an industrial environment for product development. Image processing techniques have been used since the late 1980s to characterize either the internal structure of extrusion-textured products (Barrett & Peleg, 1992; Moore et al., 1990; Smolarz et al., 1989) or the surface appearance of extrudates (Tan et al., 1994). Smolarz et al. (1989) and Van Hecke (1991) proposed using image analysis to characterize the cell structure of extruded products. The authors developed two different approaches:

- a statistical approach which corresponds to a microscopic vision of the structure, at the pixel scale
- a structural approach which corresponds to a macroscopic vision of the structure, at the cell scale.

This section deals with the structural approach as it provides physical measurement at the cell scale, which is a better fit to the global methodology described in Figure 7.17.

7.2.2.1 Two-dimensional image analysis methodology

An experimental methodology for the direct analysis of clear-cut sections of extrudates is presented. The method uses simple hardware able to provide structural data within a short time (a few minutes, including replicates). As many image analysis instruments are on the market, the development of a methodology for extrudate characterization may require operating adaptations and/or technical arrangements which depend on the type of instrument design.

Figure 7.18 shows a schematic view of an instrument used for image analysis of transversely cut sections of extruded products (Smolarz et al., 1989). The instrument comprises:

- the *measurement unit* which holds an extrudate sample to be analyzed. It is recommended to cut the sample using a high-speed circular saw or a microtome in order to reveal transversely clear-cut sections as close as possible to real internal structures. The sample is placed between two black supports, while its transverse cut is positioned horizontally at the same level as the tops of the supports. A circular neon tube is placed around the sample (about 2 cm above the sample); it delivers an intense light, which allows the effect of room lighting to be eliminated. Thus, such lighting provides a non-diffuse, uniform illumination of the transversely cut section of the extruded sample. The measurement unit is the most important part of the instrument, as its design determines the accuracy of the analysis and the rigorous application of the experimental procedure provides good repeatability and consistency of the structural characterization of samples

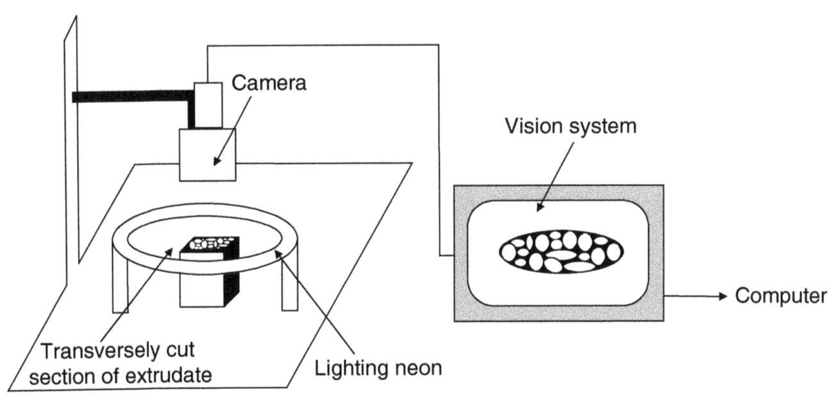

Figure 7.18 Experimental apparatus for image analysis.

- the *classic image processing hardware* which generally consists of a camera (about 40 cm above the sample), a vision system, and a computer. This system allows image digitizing (on 256 gray levels), image conversion (such as binary conversion), and contour extraction (such as contours of the whole transverse cut section and cells) to be carried out. The conversion of digital images into binary images is the simplest procedure and computation is largely reduced. The method consists of identifying the step value of the gray level which better discriminates cells and non-cells in the histogram of gray levels. A binary image is obtained by assigning zero to pixels having a gray level lower than the step value and assigning the value of one the other way. Thus, a digital binary image shows dark cells on a white background.

Various structural parameters can be calculated from digital binary images of cut sections of extruded products (Van Hecke, 1991). According to the global methodology proposed in Figure 7.17, the two following parameters are particularly selected as they reflect simply the internal cell structure of extrusion-textured food products:

- the number of cells per pixel area, or cell density, which is the ratio between the total number of cells and the transversely cut section of extrudate, expressed in mm^{-2}
- the average cell size, which designates the ratio between the sum of cell areas and the total number of cells, expressed in mm^2.

Cell density (named Dc) and average cell size (named Sc) can then be used to investigate the effect of extrusion-cooking conditions on the cell structure of the resulting extrudates.

7.2.2.2 Effect of mix formulation and composition on cell structure

The effect of the formulation on the melt expansion properties (sectional, longitudinal and volumetric expansion indices) has been described extensively in Chapter 6 (section 6.2.3.2.2). Food extrusion practitioners know that the addition of key ingredients used in typical extrusion formulations may significantly affect the cell structure of extrusion-textured food products, and thus their texture. For instance, the addition of sodium bicarbonate, or sodium chloride, results in a significant decrease of cell size and subsequent increase of cell density. However, little has been reported considering a systematic approach to studying the effects of ingredients on the cellular structure of extruded products.

Harper and Tribelhorn (1992) qualitatively compared the internal cell structure of extrudates obtained from regular corn meal (amylopectin content of approximately 70%) and amylose-rich corn (amylopectin content of approximately 30%), and observed that decreasing the amylopectin content in the starch resulted in an important decrease in cell size. Voisin (1993) determined the average cell size and cell density of two types of extrudates (formulated with corn grits and 50/50 wheat flour/potato starch granules mix) using a 2D image analysis methodology. The study included the effects of adding various ingredients such as soy flour, vegetable oil, emulsifiers, sucrose, sodium chloride, sodium bicarbonate, and sodium citrate. Desrumaux et al. (1999) used the same methodology to investigate the effect of free fatty acids on the cell structure of corn-based extrusion-textured products. The results of these two studies are reported in Table 7.1 which presents the relative trends observed on cell density (Dc) and average cell size (Sc) of two types of extrudates (corn grits and 50/50 wheat flour/potato starch granules mix) and after the addition of typical ingredients in starch-based formulations. The following comments can be made in addition to the results shown in Table 7.1.

Whatever ingredient was added to the corn-based formulation, when compared to the extrudate prepared without ingredients it was observed that the cell density of extrudates prepared with ingredients increased while average cell size decreased. The same trend was observed when the sample was formulated with the 50/50 wheat flour/potato starch granules mix, except when adding vegetable oil or sucrose which decreased the cell density and increased the average cell size. It must be noted that the changes in structural parameters when adding ingredients (such as soy flour, vegetable oil, sucrose, sodium bicarbonate, sodium citrate) into wheat/potato-based formulation tend to be more important.

As expected, an inverse relationship between cell density and average cell size is observed. Hence a classic power law relationship can be used to represent the experimental data:

$$Dc = K^* (Sc)^{a*} \tag{7.1}$$

where K^* and a^* are constants known as the cell denseness index and cell size index, respectively. In an ideal case considering highly expanded extrudates and neglecting the volume occupied by the biopolymeric matrix, the Dc–Sc relationship should result in the following parameter values: $K^* = 1$ and $a^* = -1$. Voisin (1993) identified values of both indices, while gathering Dc–Sc data obtained by adding a large range of functional ingredients (such as soy flour, vegetable oil, emulsifiers, sucrose and

Table 7.1 Trends of the evolution of structural properties (Dc and Sc) of starch-based extrudates (Desrumaux et al., 1999; Voisin 1993).

Ingredient added (range of content, in %)	Corn-based formulation		50/50 Wheat/potato formulation		[$] Comments
	Dc, mm^{-2}	Sc, mm^2	Dc, mm^{-2}	Sc, mm^2	
Soy flour (0–5%)	↗	↘	↗	↘	Significant effect with CG Very strong effect with W/P
Vegetable oil (0–5.75%)	↗	↘	↘	↗	Slight effect with CG Strong effect with W/P
Free fatty acids (0.2–0.8%)	↗	↘	-	-	Slight effect with CG Unknown with W/P
GMS (0–1.15%)*	↗	↘	↗	↘	Significant effect with CG Slight effect with W/P
PGSE (0–1.15%)†	↗	↘	→	→	Slight effect with CG No change with W/P
Sucrose (0–11.5%)	↗	↘	↘	↗	Very slight effect with CG Significant effect with W/P
Sodium chloride (0–4.5%)	↗	↘	→	→	Significant effect with CG No change with W/P
Sodium bicarbonate (0–1.15%)	↗	↘	↗	↘	Very important effect with CG and W/P
Sodium citrate (0–1.15%)	↗	↘	↗	↘	Very slight effect with CG Significant effect with W/P

* Emulsifier: glycerol monostearate (GMS).
† Emulsifier: polyglycerol saturated ester (PGSE).
$ CG, corn grits; W/P, 50/50 wheat flour/potato granules mix.

salts –sodium chloride, sodium bicarbonate, sodium citrate) into corn-based and 50/50 wheat/potato-based formulations. The resulting values of K^* and a^* were 0.20 and –0.85, respectively, with a good correlation coefficient of 0.94. As noted, the experimentally determined values of K^* and a^* differed significantly from those given for the ideal case.

It is worth identifying values of the indices K^* and a^* while considering separately different starch-based formulations, and splitting the ingredients in two different groups, the first group with emulsifiers and the second group with vegetable oil, sucrose, and salts. The resulting values of K^* and a^* are reported in Table 7.2. Examining the values of the two constants allows identification of the functional effect of the different ingredients on the extrudate structure. The index K^* depends on matrix-forming biopolymeric materials, whereas the index a^* depends on functional ingredients. The cell denseness index K^* varies in the ranges of 0.170–0.177 and 0.20–0.23 for corn-based and wheat/potato-based matrices, respectively. This means that cell denseness of the wheat/potato-based matrix is significantly higher than that of the corn-based matrix. When adding functional ingredients to the biopolymeric matrix, the cell size index changes significantly depending upon

Table 7.2 Effect of mix formulation on cell characteristics of starch-based extrudates.

Group of ingredients added	Corn-based formulation		50/50 Wheat/potato formulation	
	K^*, $mm^{-2(a^*+1)}$	a^*	K^*, $mm^{-2(a^*+1)}$	a^*
Emulsifiers	0.177	−0.98	0.23	−1.02
Vegetable oil, sucrose and salts	0.170	−0.60	0.20	−0.78

the type of ingredient added: with emulsifiers, index a^* is close to −1 for both matrices, while it ranges from −0.78 to −0.60 for the second group of ingredients. It occurs that when adding emulsifiers, the volumetric expansion of the resulting extrudates varies slightly, as the cell size index is equal to −1, which reflects an ideal cell structuring process caused by addition of the ingredient. However, when adding vegetable oil, sucrose or salts, the volumetric expansion of the resulting extrudates decreases significantly due to either limited bubble growth capacity of the melt or die melt shrinkage, which results in a cell size index notably different from the ideal for cell structuring.

When putting together the evolution of melt expansion properties presented in Chapter 6 (section 6.2.3.2.2) and the cell structure characteristics presented in this section, obvious correlations between sectional expansion index (SEI) and longitudinal expansion index (LEI) (expansion indices) and Dc–Sc (cell parameters) are obtained. For instance, gathering data derived from both biopolymer matrices and all functional ingredients, Voisin (1993) found a linear relationship between cell density and LEI, and a power law relationship between cell density and SEI (with a power law index value close to −1); correlation coefficients were 0.92 and 0.91, respectively. Hence, cell density tended to increase linearly when LEI increased whereas an inverse relationship was obtained between cell density and SEI.

7.2.2.3 Effect of thermomechanical cooking on cell structure

The effect of thermomechanical cooking on melt expansion properties has been discussed extensively in Chapter 6 (section 6.2.3.2.3). Food extrusion practitioners have observed that the cell structure of extrusion-textured food products changes when the extent of thermomechanical cooking varies. For instance, Figure 7.19 shows corn-based extrudates (tranverse cut and longitudinal cut sections) which were subjected to different extrusion processing conditions with varying screw speed and die melt temperature. When screw speed increases, the length of the extruded products increases significantly, while the internal cell structure becomes finer (disappearance of larger cells). When the barrel wall temperature increases, the length of extruded products does not change significantly but the internal cell structure becomes finer, with smaller cell size. These pictures show clearly that cellular structure of extrusion-textured food products is significantly affected by the extrusion-cooking conditions.

Figures 7.20 and 7.21 illustrate the typical trends in the evolution of cell structure properties of starch-based directly expanded extrudates when process SME or die melt temperature is changed. In this example, the raw material consisted of corn grits without ingredients, while the extrusion processing conditions corresponded to typical industrial conditions that have process SME in the range of 430–570 kJ/kg to ensure full starch gelatinization

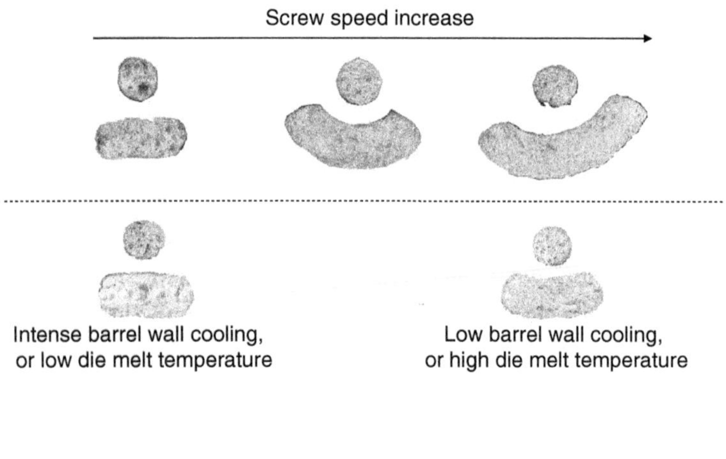

Figure 7.19 Internal structure of extrudates as a function of extrusion processing conditions. For color detail, please see color plate section.

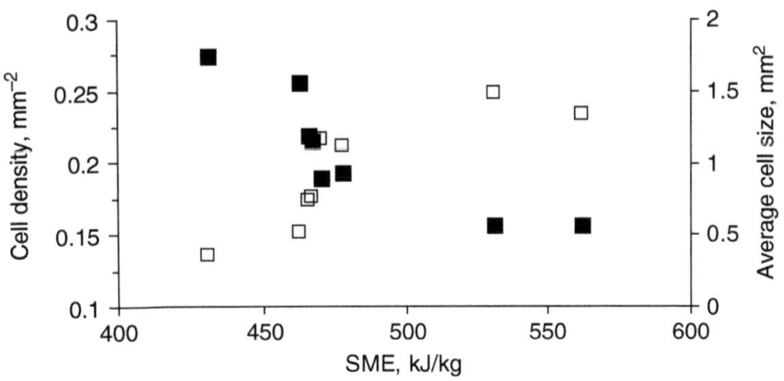

Figure 7.20 Effect of process SME on cell density (□) and average cell size (■) of directly expanded extrudates. Source: Adapted from Desrumaux 1996.

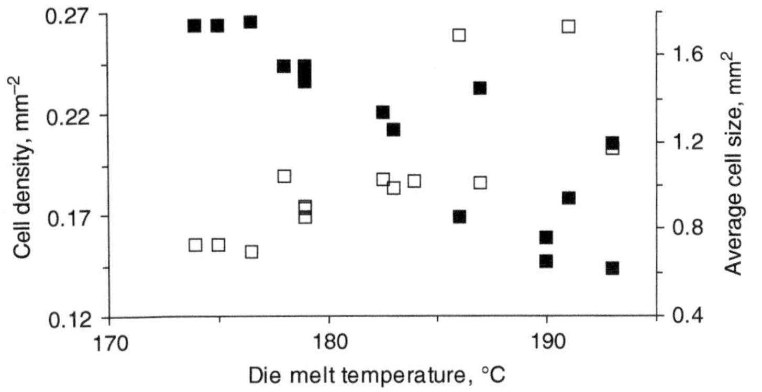

Figure 7.21 Effect of die melt temperature on cell density (□) and average cell size (■) of directly expanded extrudates. Source: Adapted from Desrumaux 1996.

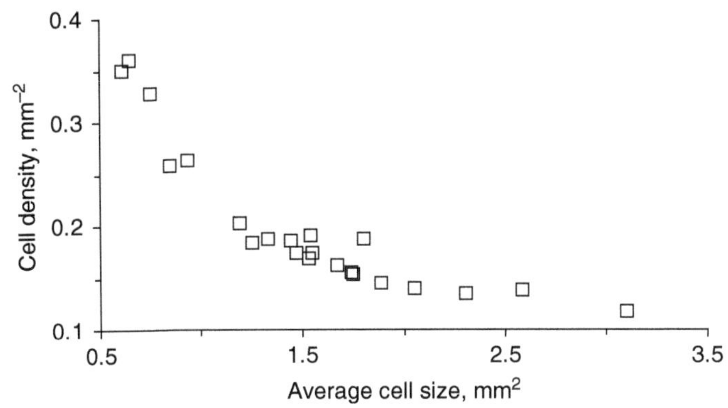

Figure 7.22 Cell density versus average cell size (process SME range of 430–570 kJ/kg). Source: Adapted from Desrumaux 1996.

(Desrumaux, 1996). Figures 7.20 and 7.21 show that cell density increases and average cell size decreases when process SME or die melt temperature increases. Conclusive comments can be made on the effect of extrusion-cooking conditions on extrudate structure.

- For a given starch-based formulation and for practical process SME in the range of 430–570 kJ/kg, the higher the process SME or the higher the die melt temperature, the finer the internal structure of directly expanded extrudates.
- An inverse relationship between cell density and average cell size is observed when thermomechanical conditions of cooking and die texturization are varied. This is confirmed by plotting cell density versus average cell size (Figure 7.22). For these data, Desrumaux (1996) determined the cell denseness index as well as the cell size index by use of the power law equation (Eq. 7.1) as $K = 0.36$ and $a^* = -0.78$ with a reasonable correlation coefficient of 0.89. Compared with the indices

determined by Voisin (1993), the cell size index, a^*, was in good agreement whereas K^* was significantly higher, due probably to the different conditions of thermomechanical cooking and die texturization used in the process. This tends to confirm that the cell denseness index is defined by the characteristics of the biopolymeric matrix – that is, by the nature of starch as well as by the starch conversion.

• When putting together data on melt expansion properties presented in Chapter 6 (section 6.2.3.2.3) and cell structure characteristics presented in this section, reasonable correlations between SEI–LEI (expansion indices) and Dc–Sc (cell parameters) can be observed. For instance, Desrumaux (1996) found that cell density increased almost linearly with increases of LEI whereas cell density decreased almost linearly with increases of SEI. By use of data obtained for different process SME, the author was able to establish a power law relationship between average cell size and LEI on one hand, and between average cell size and SEI on the other hand (Figure 7.23). The power law index values were −1.03 for the LEI–Sc and 1.17 for the SEI–Sc relationships with satisfactory correlation coefficients of 0.92 and 0.85, respectively. This means that the average cell size decreases when LEI increases. The evolution of the average cell size is inversed when SEI increases. It is worth confirming experimentally that melt expansion properties, in this case represented by the LEI and SEI parameters, are closely related to the internal structure of extruded products described by the parameters Dc and Sc.

7.2.3 Measurement of mechanical characteristics

Identification of mechanical properties is important to characterize the texture of extruded food products. The mechanisms that govern mouth-feeling sensation and product acceptance are rather complex. It is known that the mechanical resistance developed by extruded products to the biting process contributes effectively to the assessment of textural acceptance. Consequently, the development of instrumental methods which are able to imitate the biting action would contribute to assessment of mechanical properties that fit the overall concept illustrated in Figure 7.17.

In general, two types of instrumental methods for the measurement of mechanical characteristics of food products are used.

• *Fundamental methods*, which consist of identifying force-deformation responses while deforming food products by bending, compression, or shearing and then determining standard mechanical properties such as the elasticity modulus, breaking stress, shearing modulus, etc. Fundamental methods have been used extensively to investigate the mechanical behavior of the cellular structure of extruded products, as well as to correlate their mechanical characteristics with the extrusion processing conditions utilized for their production. For instance, the three-point bending method was used to test flat bread and cylinder-shaped extruded products and correlate the measured mechanical characteristics resulting from the test with extrusion-cooking parameters or formulation used (Moore et al., 1990; Robin et al., 2010; Van Hecke et al., 1995). Though fundamental methods provide relevant basic mechanical properties, they do not account for the effect of the internal cell structure on the sensory texture of expanded extrudates.

• *Empirical and imitative methods*, which consist of imitating mechanical deformation of food products resembling the biting and masticating process. These methods lead to empirical parameters which are often

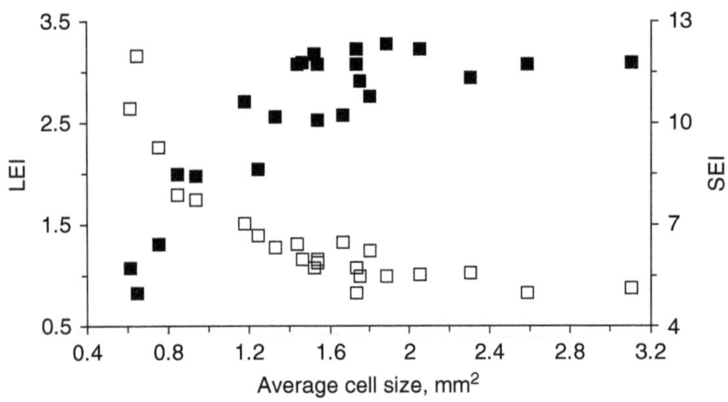

Figure 7.23 LEI (□) and SEI (■) versus average cell size (process SME range of 430–570 kJ/kg). Source: Adapted from Desrumaux 1996.

successfully correlated with sensory attributes of food products but these empirical parameters and processing conditions are generally badly correlated. Suitable empirical or imitative methods that relate to the food extrusion process and conditions used need significant assessment.

The sensory perception of extrudate crispness starts with the first bite in the mouth followed further by mastication which tends to break the product and reduce it into small particles. Thus, the sensory perception of crispness results from a complex process with dynamics controlled by the consumer through regulation of the mechanical stresses applied during mastication. The breaking behavior of crispy extruded products depends on three main factors:

- the mechanical deformation during biting which depends on tooth geometry, masticating speed, and the force developed by the mouth
- the macrostructure of the expanded product which depends on its cell structure (cell density, cell size and dispersion, cell shape, in particular)
- the mechanical properties of the continuous phase of the extrudate matrix (or mechanical resistance of cell walls).

When an expanded extrudate is placed between two teeth and subjected to biting, a combination of compression and shearing deformations is exerted, causing the extrudate to store energy during its deformation until the product is broken. Upon breaking, the stored energy is then released instantaneously in the form of acoustic vibrations mainly due to cell wall relaxation after breakage. Hence, in addition to visual, tactile and mechanical sensations, textural properties of expanded extrudates should also be perceived through a combination of auditory and vibratory sensations in the mouth (Vickers & Bourne, 1976).

Many researchers have conducted acoustic measurements of the noise produced during mastication of crispy, crunchy and crackly food products (Duizer, 2001). In these studies, the authors either measured the perception of air-conducted sounds to establish the contribution of these sounds to the sensation of crispness and crunchiness (Christensen & Vickers, 1981; Vickers, 1981, 1985; Vickers & Christensen, 1980) or they recorded the sounds produced during application of a force to a noisy product to obtain quantitative information related to the crisp, crunchy or crackly sounds (Dacremont et al., 1991; Edmister & Vickers, 1985; Lee et al., 1988; Seymour & Hamann, 1988; Vickers & Bourne, 1976). The authors used transformations such as the Fourier transform to identify frequencies of particular importance associated with a characteristic texture attribute (Dacremont, 1995; Edmister & Vickers, 1985; Seymour & Hamann, 1988; Vickers, 1985; Vickers & Christensen, 1980). Acoustic data have also been analyzed using fractal analysis to evaluate the amplitude-time curves produced during mechanical testing of products. In that respect, Tesch et al. (1995, 1996) showed that fractal dimensions were useful to determine the relative roles of mechanical and acoustic stimuli in the perception of crunchy products. Duizer et al. (1998) used fractal analysis of acoustic data of chewing and observed significant correlations between fractal dimensions of acoustic signatures and sensory characteristics of extruded snacks.

Combining analysis of acoustic recordings with mechanical testing results has been successful for correlating physical properties and sensory characteristics of snacks. However, correlations between textural properties (either physical or sensory properties) and processing conditions are still rare. However, in a study of puncture-generated sounds (Bouvier et al., 1997) generated by crispy foods (including extrusion-textured food products) in the audible frequency range of 0–20 kHz, the authors observed that the power of the frequency 15–16 kHz was correlated with the water content added during extrusion-cooking: the higher the water content, the higher the power.

Mechanical deformation generated by biting can be simulated through a puncturing instrumental method to identify physical criteria that can be correlated with either sensory texture of extruded products or extrusion-cooking processing parameters. This section presents a puncturing instrumental method applied to extrusion-textured food products, and explores its use to link it with the extrusion processing history and sensory attributes of extruded products.

7.2.3.1 Puncture methodology

Based on pioneering investigations by Hutchinson et al. (1987) as well as Hayter and Smith (1988) on testing extrudates by using the puncturing method, several investigators and product designers from research laboratories dealing with extrusion-textured food products have studied this method thoroughly or applied it successfully. The puncture test consists of penetrating a product with a needle/probe at a constant rate of deformation while recording the resulting force-time response. This test can be carried out with standard texture analyzer instruments able to control the conditions of the deformation. Van Hecke (1991) studied puncturing conditions with an eye on the statistical significance of the force-time curve, and has optimized puncture testing conditions for better product discrimination. The conditions include various factors such as the shape and geometry of the needle, the puncture rate, the threshold force which discriminates between signal noise and a real physical event, and the number of replicates needed to obtain suitable data. For instance, for highly expanded extrudates such as snacks or breakfast cereals, Van Hecke (1991) found that optimum puncturing conditions are obtained by using a cylindrical flat-faced needle of about 15 mm^2 cross-section, a puncture rate of 40 mm/min and a threshold force of 0.1 N while 10 replicates are needed to ensure satisfactory significance to discriminate sample properties. Such conditions are generally valid for highly expanded products but might vary notably when testing moderately expanded or harder extrudates. Regardless of the type of sample, it must be noted that optimization of puncturing conditions is required to control the physical significance and repeatability of the puncture test.

Figure 7.24 shows force-time curves recorded during puncture testing of commercial corn-based snack balls stored for different times at ambient conditions.

- The test was conducted right after opening the package, i.e. for a crispy, dry fresh sample. For the fresh sample, it is observed that the force rises to a peak and then falls back. The force-time trace is not smooth but rather fluctuates randomly, sometimes falling to zero and sometimes rising to larger values. The observed peaks in the force-time curve have been attributed to successive cell wall breakage of the extrudate structure while the force drops of each peak remain low. A moderate overall rise of the puncture force is observed, leading to a relatively low value for the integral of the force-time curve, which would be an indication of the energy required to fracture and reduce the sample to small pieces during consumption.
- The test was conducted 2 hours after opening the package. Since the sample is exposed to ambient conditions, a moderate amount of water is absorbed by the sample. The absorption of moisture from the ambient air changes the mechanical properties of the sample and as a consequence the force-time curve shows fewer peaks while the force drops after each peak increase significantly. A significant rise in the overall force is observed leading to a higher value of the integral of the force-time curve, i.e. more energy would be used to consume this sample. It must be noted that at low puncturing time (for approximately the first 2 seconds of puncturing), the force-time curve shows no peak and an important force rise, probably due to the formed moisture gradient between the periphery and the center of the sample.
- The test was conducted 3 days after opening the package. Due to the significant moisture absorption, the sample becomes chewy and fully humidified. In this condition, the water uptake of the sample has reached an equilibrium value. As noted in the figure, the force-time curve shows no peak which indicates that no cell wall rupture occurs. Thus, the sample is no longer crispy but it is chewy. A continuous force over time increase is observed,

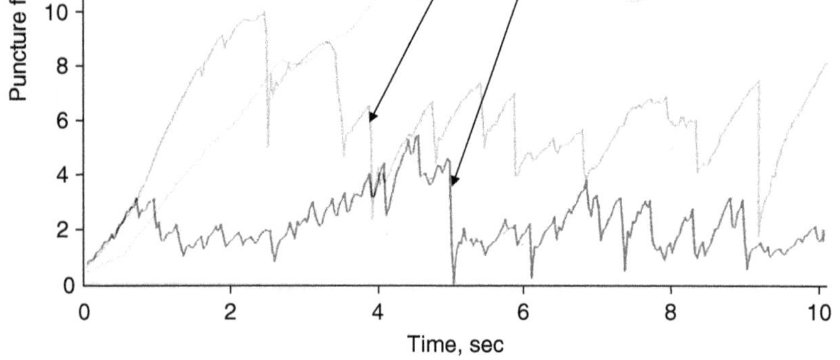

Figure 7.24 Typical force-time puncture curves of commercial corn-based snacks: effect of moisture uptake from ambient air.

leading to the highest value of the integral obtained over the force-time curve, which would indicate the greater energy/work necessary to reduce the sample into small pieces to be able to consume it.

Four relevant puncture parameters can be identified from the puncture force-time response of extrusion-textured food products.

- Average puncture force (F_m), expressed in N. This is the ratio of puncture work (integral of the force-time curve) to puncture time.
- Spatial frequency of the structural ruptures (N_{sr}), expressed in m^{-1}. This is the ratio of the total number of peaks to the distance of puncturing.
- Average force of the structural ruptures (f_{sr}), expressed in N. This is the ratio of the sum of force drops per peak to the number of peaks.
- Crispness work (W_c), expressed in N.m. This is the ratio of the average puncture force (F_m) to the spatial frequency of the structural ruptures (N_{sr}). This parameter may represent a crispness scale, as the higher the crispness, the harder the product with regard to puncturing.

The use of these parameters allows quantitative and instrumental quantification of the physical meaning of crispness. They are discussed in section 7.2.3.2 for evaluating the effect of water activity on the puncture behavior of extrudates.

7.2.3.2 Effect of water activity on puncture response

Water activity of extrusion-textured food products is certainly one of the most important factors affecting its texture. As seen in section 7.2.3.1, premature opening of a package before consumption, or inappropriate choice of packaging materials, may significantly depress the sensory attributes of crispy food products during consumption (the first and second biting, in particular) due to moisture sorption. Katz and Labuza (1981) have evidenced the effect of moisture content on crispness of cereal-based food products. Moisture content lower than approximately 6% (wwb) must be ensured to preserve crispness of such products, and a slight increase of moisture above that level may lead to a drastic decrease of crispness; beyond 10–12% (wwb), crispness may be completely lost (Roudaut et al., 1998; Sauvageot & Blond, 1991). Dry cereal-based food products are highly hygroscopic due to the presence of amorphous starch in their composition, which may absorb moisture from either the ambient air or high moisture fillings.

Van Hecke et al. (1998) investigated crispness loss due to water sorption, while storing commercial corn-based snack balls under different relative humidities ranging from 0% to 100% for several days to ensure moisture uptake and reach equilibrium. Figure 7.25 shows the evolution of the spatial frequency of the structural ruptures (N_{sr}) as a function of water activity. The curve clearly indicates two main domains: a low-slope linear initial decrease of N_{sr} at low water activity ranging from 0 to about 0.25 which corresponds approximately to 5–6% (wwb) water content; then a steep, almost exponential decrease of N_{sr} above 0.25 water activity is observed. Figure 7.26 shows the changes of both average puncture force (F_m) and

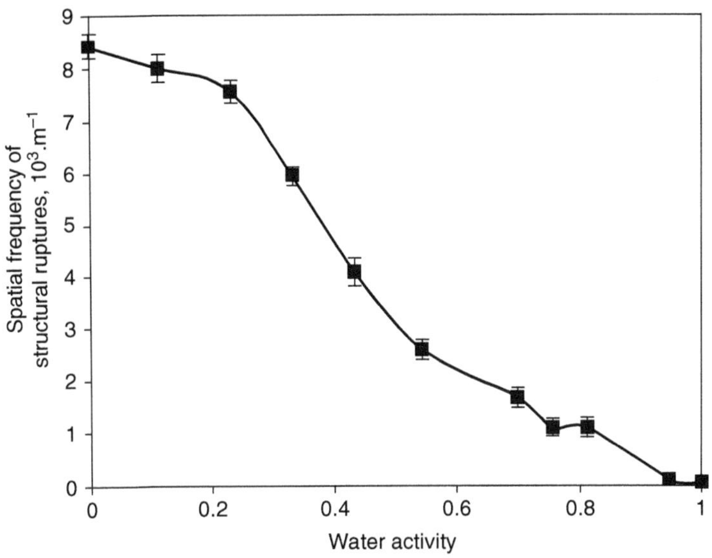

Figure 7.25 Effect of water activity on spatial frequency of structural ruptures. Source: Van Hecke, Allaf & Bouvier 1998. Reproduced with permission of John Wiley & Sons.

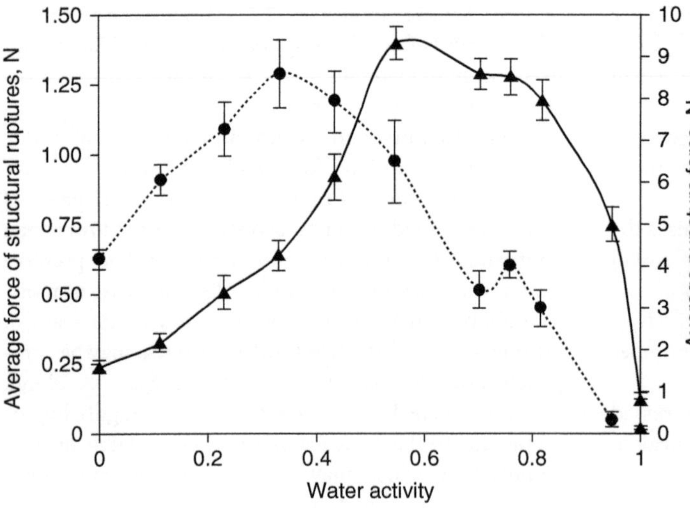

Figure 7.26 Effect of water activity on average puncture force (▲) and average force of structural ruptures (●). Source: Van Hecke, Allaf & Bouvier 1998. Reproduced with permission of John Wiley & Sons.

average force of structural ruptures (f_{sr}) as a function of water activity. Both curves exhibit the same profile but with distinct positions of the maximum values. The figure shows a water activity level at 0.30–0.35 water activity for the maximum value of f_{sr} and at about 0.55 for the maximum of F_m. Results shown in Figures 7.25 and 7.26 can be summarized by the following observations.

- $0 < a_w < 0.35$: in this range of water activity, extrudates show numerous cell wall breakages (N_{sr} >7 breakages per mm of deformation), while the average force associated with cell wall breakages (f_{sr}) tends to increase significantly from about 0.65 N to 1.30 N. The water is strongly bonded to the starch-based matrix, while starch polymers are in a glassy state. As observed by Le Meste et al. (1996), through the change of the apparent Young's modulus of cereal-based extrusion-cooked products (tested by axial compression) versus moisture content, the increase of f_{sr} may be attributed to some apparent moisture stiffening effect of water on the starch network, hence allowing the extrudate to exhibit brittle behavior. The range of 0–0.35 a_w should correspond to the crispness domain.
- $0.35 < a_w < 0.55$: in this range of water activity, N_{sr} and f_{sr} considerably decrease from about 6.5 to two breakages per mm deformation, and from about 1.3 N to 0.75 N, respectively, while F_m increases and reaches its maximum value at approximately 9 N. Extruded products become less and less crispy, but slightly harder as a whole. At 0.50–0.55 water activity, the extrudate has approximately 10–12% (wwb) water content, which might correspond to a condition in which the cell walls undergo a brittle/ductile transition, while starch polymers are still in a glassy state. A critical water activity of 0.50 was noted by Sauvageot and Blond (1991) when evaluating the sensory attributes of breakfast cereals. They assigned this value as the upper limit for accepting such products as crispy. But for commercial snack products, this limit is generally closer to $a_w = 0.35$. Thus, the range of 0.35–0.55 a_w should correspond to a transition domain between crispy and chewy.
- a_w >0.55: in this range of water activity, both parameters N_{sr} and f_{sr} show a continuous decrease down to values close to zero, while F_m also decreases down to about 1 N when water activity increases to values closer to 1. Under these conditions, the starch-based matrix undergoes the glass–rubber transition at 0.7–0.8 water activity, which causes a dramatic decrease in both F_m and f_{sr}. Water activities higher than 0.55 should correspond to a chewiness domain.

The textural domains described (crispness, crispy-chewy transition, and chewiness) of commercial corn-based snack balls conditioned under a wide range of relative humidities (0–100%) can be defined in the puncture chart shown in Figure 7.27. The chart shows the evolution of the puncture parameters F_m and f_{sr} for different values of water activity.

7.2.3.3 Crispness connotation of puncture parameters

Crispness work (W_c) versus average puncture force (F_m)
Physically, F_m corresponds to the overall force required to puncture the whole extrudate, while W_c represents the

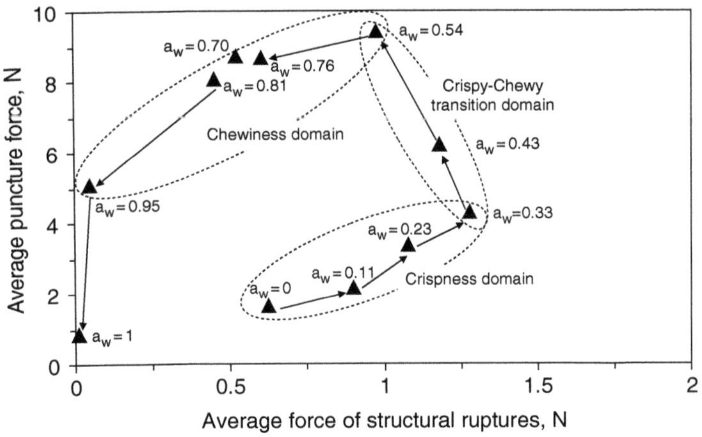

Figure 7.27 Puncture chart of commercial corn-based snacks balls: effect of water activity. Source: Van Hecke, Allaf & Bouvier 1998. Reproduced with permission of John Wiley & Sons.

average work needed for structural rupture. Hence, both parameters should vary proportionally, as the larger the work needed to obtain one rupture, the larger the average puncture force. Using data derived from various experimental sources (including data from Voisin (1993) as well as data obtained when varying thermomechanical processing conditions), and considering crispy, fully converted extrudates (a_w <0.30; process SME >430 kJ/kg), Van Hecke (1991) has shown that the two puncture criteria W_c and F_m are correlated by use of a power law equation (with a correlation coefficient of 0.98), as follows:

$$W_c = 0.08F_m^{1.42} \tag{7.2}$$

According to the experimental sources considered in this study, Eq. 7.2 is valid in a range of $F_m : 2\,\text{N} < F_m < 20\,\text{N}$, and for corn-based and wheat-based and 50/50 wheat/potato-based formulations.

Average puncture force (F_m) versus average force of structural ruptures (f_{sr})

Physically, f_{sr} corresponds to the average force required to obtain one structural rupture; hence both parameters must be correlated. Using the same data given above, Van Hecke (1991) has shown that the correlation between the two puncture parameters F_m and f_{sr} also obeys a power law equation (with a correlation coefficient of 0.93), as follows:

$$F_m = 2.58f_{sr}^{1.72} \tag{7.3}$$

As the power law index is larger than 1, F_m increases more rapidly than f_{sr}, meaning that some force rise occurs during puncturing, depending upon the denseness of the cellular structure of extrudates. Hence, such index might represent a scale of crispness or hardness. The closer to a value of 1, the less force rise is expected during puncturing, and also the crispier the extrudate.

Figure 7.28 reproduces the puncture chart presented above (see Figure 7.27) with cross symbols representing the water activities. The crispness, crispy-chewy transition, and chewiness domains are also mentioned in the chart as well as the curve given by Eq. 7.3 (named crispness curve). It is worth noting that the crispness curve fits well with the F_m and f_{sr} values measured for water activities ranging from 0 to 0.33, which define the crispness domain defined in the puncture chart. In this chart, data obtained by Desrumaux (1996) have been added (closed squares); they were obtained from extrudates prepared by extrusion-cooking of corn grits without adding other ingredients, and under process SME ranging from 430 to 570 kJ/kg. As expected, the crispness parameters position the extrudates well into the crispness domain, on both sides of the crispness curve. As the process SME increases, f_{sr} decreases significantly while F_m increases moderately. This change of puncture parameters and the consequent change of the product location in the crispness domain might be due to the modification of the cellular structure of extrudates caused by an increase of process SME. In fact, as seen in section 7.2.2.3, the higher the process SME, the lower the cell size and the finer the internal structure of directly expanded extrudates, hence leading to a decreasing

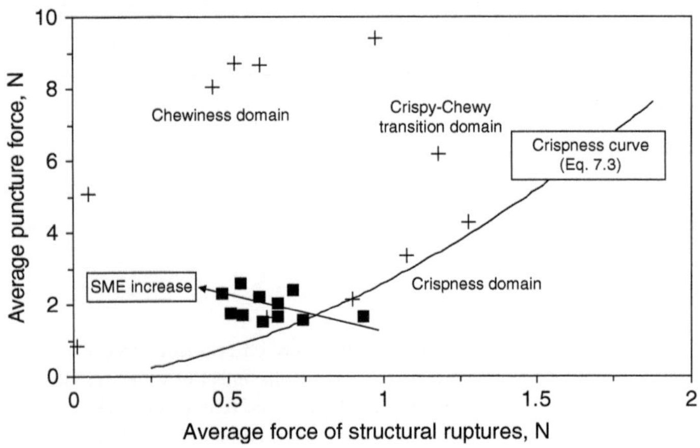

Figure 7.28 Puncture chart of crispy corn-based extrudates: effect of process SME in the range of 430–570 kJ/kg. Source: Adapted from Desrumaux 1996.

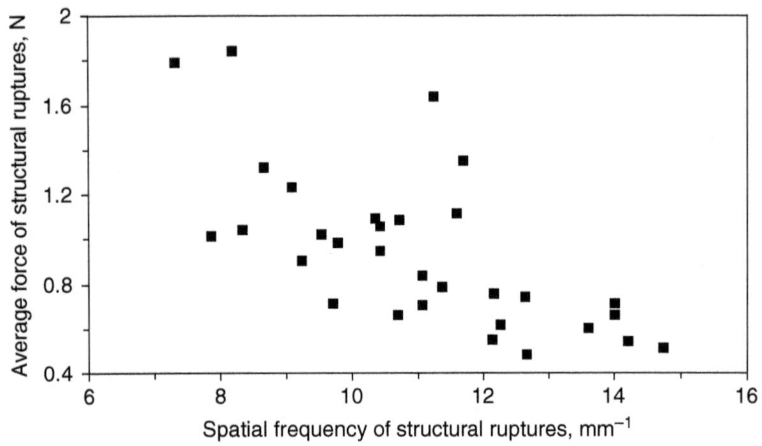

Figure 7.29 f_{sr} versus N_{sr} for crispy corn-based extrudates. Source: Adapted from Desrumaux 1996.

resistance of cell walls (lower f_{sr}), but an increasing force rise during puncturing (larger F_m).

Average force of structural ruptures (f_{sr}) versus spatial frequency of structural ruptures (N_{sr})

Desrumaux (1996) studied the effect of die melt temperature and die pressure on the puncture responses of extrudates prepared by extrusion-cooking of corn grits without using ingredients and under process SME in the range of 430–570 kJ/kg. When die melt temperature increased from about 170°C to 210°C, it was observed that f_{sr} decreased, while N_{sr} increased. But when die pressure increased from about 40 to 140 bar, f_{sr} increased, while N_{sr} decreased. As illustrated in Figure 7.29, although the data appear somewhat scattered, the general trend shows that f_{sr} and N_{sr} are inversely correlated. In fact, an increase of the spatial frequency of structural ruptures means that the number of cells increases; therefore, the thickness of cell walls tends to decrease, leading to a decrease in the average force, which is closely associated with the cell wall rupture. Such relationship results from the mechanism of melt expansion, or product texturization, which controls the cellular structure (cell size and dispersion and thickness of cell wall, in particular), hence influencing the resistance of the extrudate structure to puncturing deformations. Thus, it is worth examining the relationship between puncture parameters and expansion indices.

7.2.3.4 Relationships between puncture parameters and expansion indices

Desrumaux (1996) investigated general trends governing relationships between the puncture parameters f_{sr} and N_{sr} and the expansion indices SEI and LEI over a wide range of values. Samples were prepared by extrusion-cooking of corn grits of different particle size, content of free fatty acids in the formulation, and different thermo-mechanical cooking conditions defined by process SME ranging from 430 to 570 kJ/kg, and die texturization controlled by different die openings. Figure 7.30 presents graphs of f_{sr} and N_{sr} versus SEI. On both graphs, the isotropic expansion transition line is shown (for definition, refer to Chapter 6, Figure 6.21). In the LEI-oriented expansion domain (low values of SEI), when SEI increases from about 3 to 6, both puncture parameters do not change significantly though they might increase very slightly. But in the SEI-oriented expansion domain (SEI >6), for increases in SEI both puncture parameters vary extensively: f_{sr} increases from about 0.4–0.7 N up to 1.7–1.9 N, while N_{sr} decreases from about 14–15 mm^{-1} down to 6–8 mm^{-1}.

Figure 7.31 presents plots of f_{sr} and N_{sr} versus LEI. In the SEI-oriented expansion domain (low values of LEI), when LEI increases both puncture parameters vary extensively: f_{sr} decreases from about 1.8–2 N down to 0.5–07 N, while N_{sr} increases from about 7 to 15 mm^{-1}. But, as extruded products move to the LEI-oriented expansion domain, for increases in LEI both puncture parameters do not change significantly though they might decrease very slightly. Hence, when progressively modifying the direction of die melt expansion from highly SEI-oriented to isotropic expansion at constant volumetric expansion index (VEI) (or constant bulk

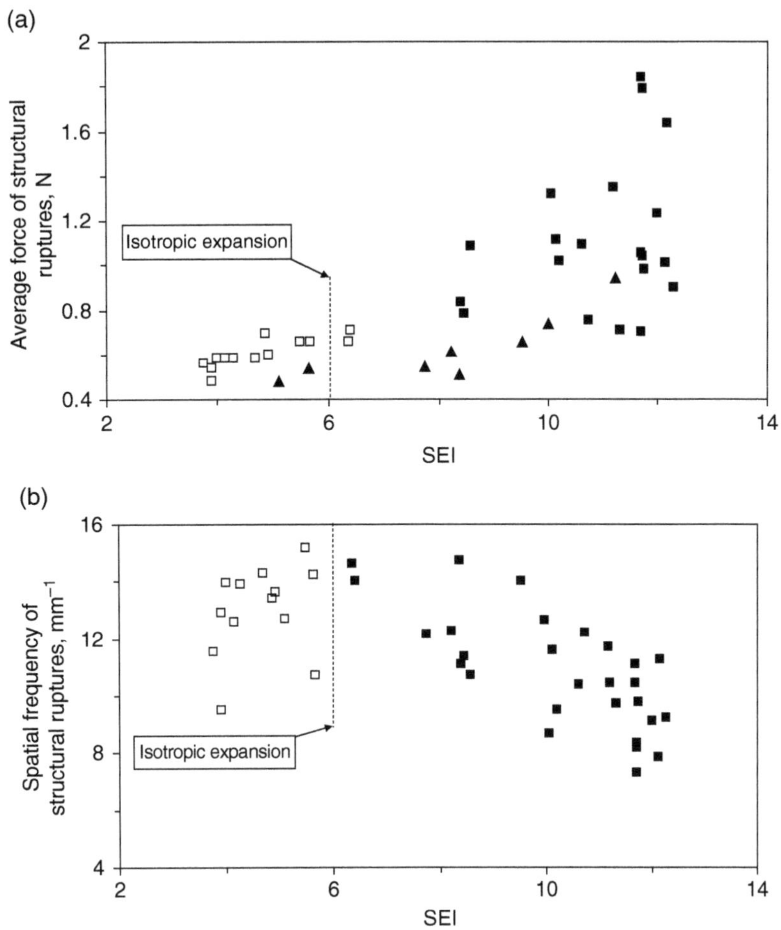

Figure 7.30 Puncture parameters ((a) f_{sr}; (b) N_{sr}) as a function of SEI (□: LEI-oriented expansion; ■: SEI-oriented expansion). Source: Adapted from Desrumaux 1996.

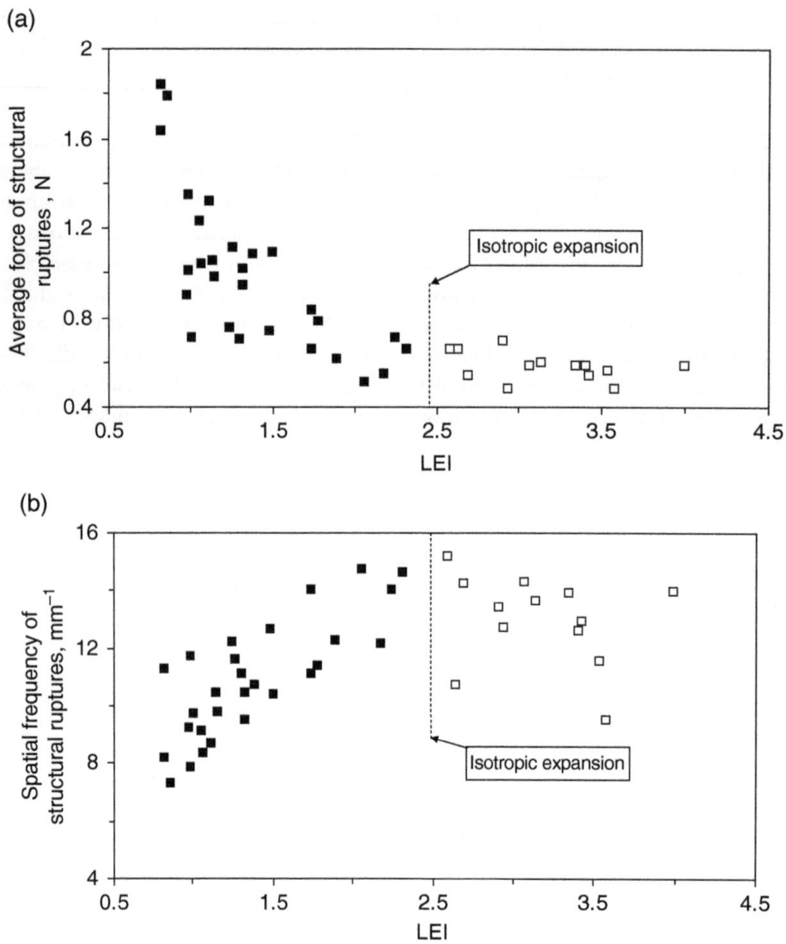

Figure 7.31 Puncture parameters ((a) f_{sr}; (b) N_{sr}) as a function of LEI (□: LEI-oriented expansion; ■: SEI-oriented expansion). Source: Adapted from Desrumaux 1996.

density), by the use of specific ingredients or appropriate thermomechanical cooking conditions the texture profile of the resulting extrudates changes significantly. As far as puncture characteristics of the extrudates are concerned, a wide range of textures can be obtained in the SEI-oriented expansion domain whereas in the LEI-oriented expansion domain, no significant change of texture is observed.

Desrumaux (1996) determined the constants of the power law equations correlating the puncture parameter f_{sr} and the expansions indices SEI and LEI, as follows:

$$f_{sr} = 0.2SEI^{0.66} \quad (7.4)$$

$$f_{sr} = 1.34LEI^{-1.13} \quad (7.5)$$

7.2.4 Physical texture of directly expanded extrudates

It is worth concluding this section by clarifying the main trends relating expansion characteristics, cell structure and puncture properties of directly expanded extrudates. When sectional expansion is predominant (SEI-oriented expansion), the resulting extrudates are characterized by high cell sizes and low cell density, leading to thicker cell walls and giving extrudates which are more resistant to puncture deformation (high average puncture force, high average force of structural ruptures). As the number of cells is low, structural ruptures under puncturing tend to be relatively low (low spatial frequency of structural ruptures). On the other hand, when longitudinal expansion

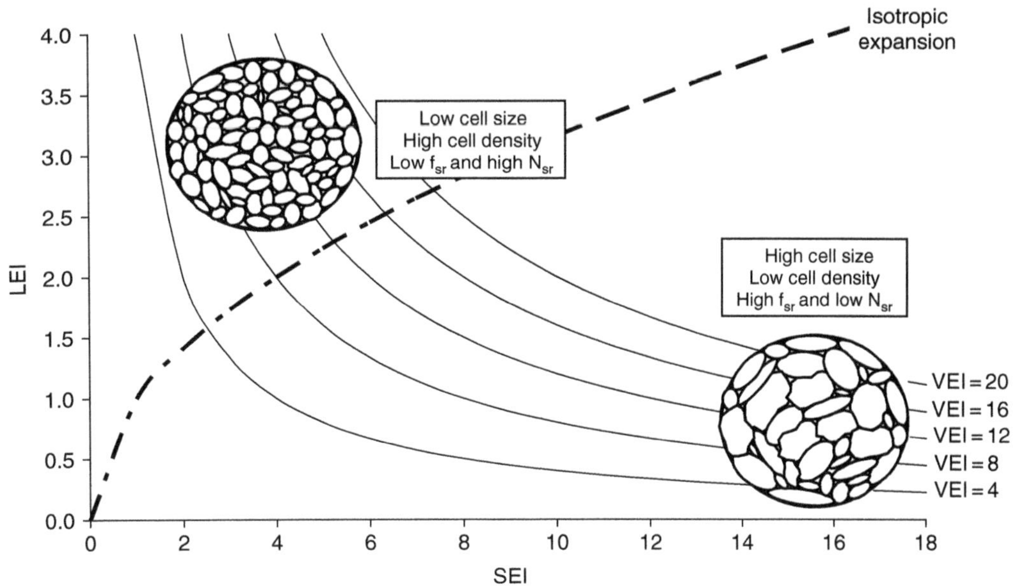

Figure 7.32 Expansion chart of directly expanded, extrusion-textured food products: trends of physical texture.

is favored in contrast to sectional expansion due to the effect of using appropriate process conditions or specific ingredients, the resulting extrudates show a lower cell size and a higher cell density, leading to finer cell walls and giving extrudates which are less resistant to puncture deformation. In the SEI-oriented expansion domain, the change of puncture parameters is particularly important when moving from high SEI-oriented products to isotropic expansion, where f_{sr} decreases while N_{sr} increases when LEI increases. Beyond the isotropic expansion transition line (high LEI-oriented expansion domain), both puncture parameters f_{sr} and N_{sr} do not change significantly. Consequently, the SEI-oriented expansion domain is of great importance from a practical standpoint. Figure 7.32 summarizes the aforementioned trends in the expansion chart of directly expanded, extrusion-textured food products.

7.3 Case study: texture monitoring of directly expanded extrudates

Instrumental methods presented in this chapter can be used for quality analysis of various types of extrusion-textured food products. Though these methods are common practice with directly expanded products (on-line product shaping and texturization, as defined in section 6.1.2.1), they can also provide relevant physical information on indirectly expanded products (downstream product shaping and texturization, as defined in section 6.1.2.2). Such methods have been used extensively for the analysis of starch-based extruded products, for which experimental methodologies are straightforward. However, for protein-based extruded products, significant adaptation of experimental methodologies should be considered for ensuring that consistent and pertinent information is obtained.

It should be noted that long-term use of instrumental methods allows determination of relevant product-related parameters. This enables designers to improve their understanding of the relationships between the extrusion processing history and the product characteristics, and to formalize their long-term experience through the establishment of internal master correlations as presented in Figure 7.32. This section presents master correlations relating textural characteristics of directly expanded extrudates (bulk density and puncture hardness, in particular) and process SME of thermomechanically cooked extrudates, proposing a practical strategy for monitoring their texture.

7.3.1 Main features of process–product relationships

From formulated mixes to the consumer, the process–product context of direct expansion extrusion-cooking

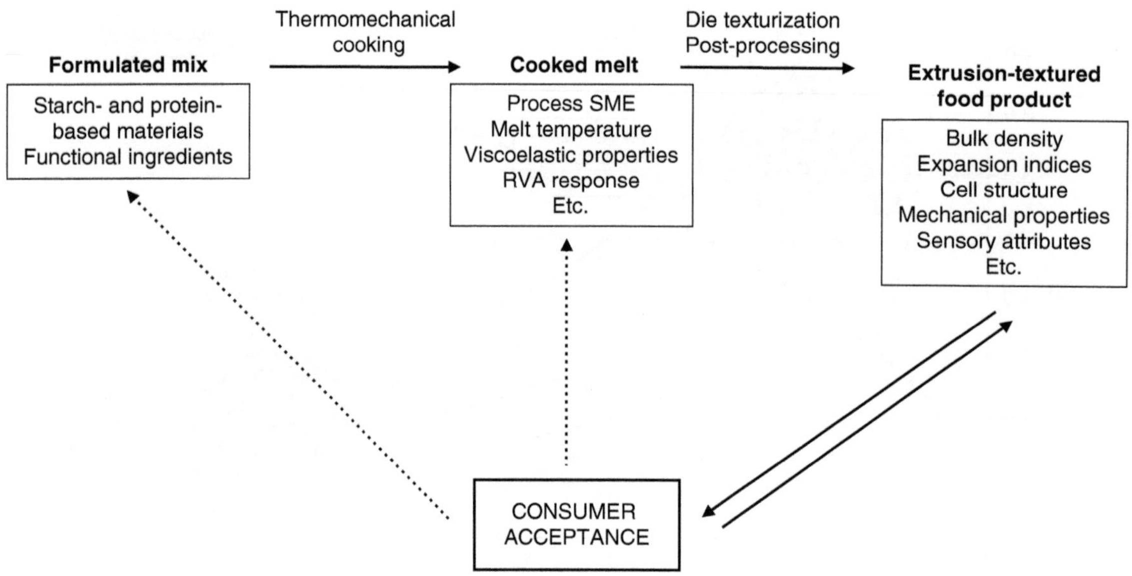

Figure 7.33 From extrusion processing to consumer acceptance: direct expansion extrusion-cooking.

is characterized by three main steps as indicated in Figure 7.33.

- A formulated mix (starch- or protein-based materials with various functional ingredients) is subjected to thermomechanical cooking which is characterized by several independent process variables (such as screw configuration, throughput, moisture content, screw speed, temperature profile, etc.) to produce a cooked biopolymeric melt which is characterized by processing response variables such as process SME, melt temperature, melt viscosity and elasticity, RVA response, etc.
- The cooked melt is then subjected to steam-induced die texturization which is characterized by independent variables such as die design, die temperature, die opening, etc., followed by postprocessing operations such as drying, coating, and packaging to produce a textured crispy product characterized by physical properties such as bulk density, expansion indices, cell structure, mechanical and acoustical properties and sensory attributes such as crispness profile evaluated by trained sensory panels.
- The extrusion-textured crispy product is exposed to consumers whose acceptance depends on many criteria (culture, lifestyle, age, etc.). Though consumer acceptance is highly subjective, product manufacturers aim to analyze texture-related consumer preferences and their evolution, by use of sensory attributes determined by internal sensory panels and physical parameters to investigate and understand what consumer acceptance means from the product design standpoint.

The understanding of consumer acceptance can be aided by use of master correlations between process indicators and textural properties such as process SME and bulk density, process SME and expansion indices, expansion indices and structural characteristics, and finally expansion indices and puncture parameters, as was shown in section 7.2. All this information can be gathered following the systematic approach described in section 7.2 and constituted by the long-term experience accumulated by process and product designers. This allows textural characteristics of end-products to be improved while appropriately adapting the mix formulation and extrusion processing conditions. But in most cases, process and product designers cannot benefit from the use of such master correlations as long-term experience has not been formalized. Thus, a simple methodology for quick monitoring of product texture is needed.

7.3.2 Methodology for texture monitoring

A methodology for texture monitoring of directly expanded products consists of determining only three parameters: the process SME, the bulk density, ρ_b, and the average force of structural ruptures, f_{sr} (or the puncture hardness). The intention is to be able to establish two basic master correlations such as bulk density versus SME, and bulk density versus f_{sr}. As seen in Chapter 6 (section 6.1.1.4), process SME is the key factor which reflects the extent of thermomechanical cooking. Thus, process SME must be considered when building master correlations for process–product fine tuning. Similarly, bulk density is one key product characteristic that results from the cellular structure of expanded extrudates, and closely affects their textural properties. Thus, it is worth introducing bulk density of extrusion-textured food products to set up a process–product relationship. Finally, it is proposed to determine the puncture parameter f_{sr} which reflects the texture of crispy extruded products, thus enabling sensory attributes and puncture hardness to be correlated.

This methodology is simple and easy to implement in practice when defining processing conditions according to textural characteristics of extrusion-textured food products. The proposed methodology is applied below to typical corn-based formulations subjected to different thermomechanical cooking conditions while keeping die texturization and postprocessing conditions constant. Three independent extrusion variables (screw configuration, water content, and screw speed) were selected to vary the thermomechanical cooking conditions while throughput and temperature profile were maintained identical for all extrusion trials. In addition, eight different screw configurations with different levels of shearing intensity were used.

- Profiles A, B, C, D, E, and F which have already been presented in Chapter 6 (section 6.2.3.2.3).
- Profile H which consisted of one single stage of shearing, composed of one conjugated, single-flight, left-handed pitch element (L/D ratio of 1.8) with standard half-moon openings.
- Profile I which consisted of three separate stages of shearing: the first stage was composed of one bilobe kneading block (right-handed 45° staggering; L/D ratio of 0.9); the second stage was composed of one bilobe kneading block (right-handed 45° staggering; L/D ratio of 0.9) followed by one trapezoidal, single-flight, left-handed pitch element (L/D ratio of 0.45) with rectangular shaped openings; and the third stage was composed of one bilobe kneading block (neutral 90° staggering; L/D ratio of 0.9), followed by one trapezoidal, single-flight, left-handed pitch element (L/D ratio of 0.45) with rectangular shaped openings. Obviously, profile I offers the highest intensity of shearing.

This set of eight screw configurations combined with different water contents (ranging from 18% to 20% wwb) and screw speeds (ranging from 400 to 500 rpm) allowed process SME to be varied extensively (in the range of 120–215 W.h/kg); the increase of process SME was in the following order: H < D < B < A < C < E < F < I. The experimental runs provided 40 directly expanded extrudates with different physical characteristics that were determined by the experimental procedures presented above in sections 7.2.1 (bulk density) and 7.2.3 (puncture hardness).

Figure 7.34 shows the master correlation relating bulk density and process SME while Figure 7.35 illustrates the master correlation relating puncture hardness and bulk density. As expected, bulk density decreases when process SME increases, and puncture hardness increases as bulk density increases. It must be pointed out that the observed dispersion of bulk density values at constant process SME is mainly due to significant difference of the processing

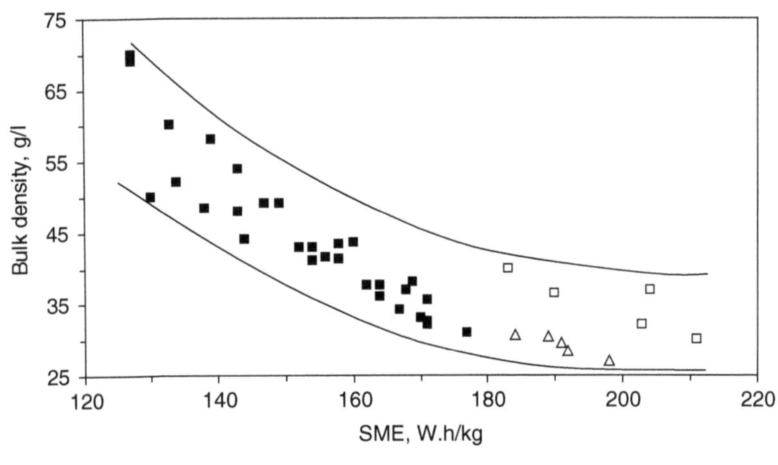

Figure 7.34 Master curve correlating bulk density versus process SME: directly expanded corn-based extrudates (■: screw profiles A, B, C, D, E, H; △: screw profile F; □: screw profile I).

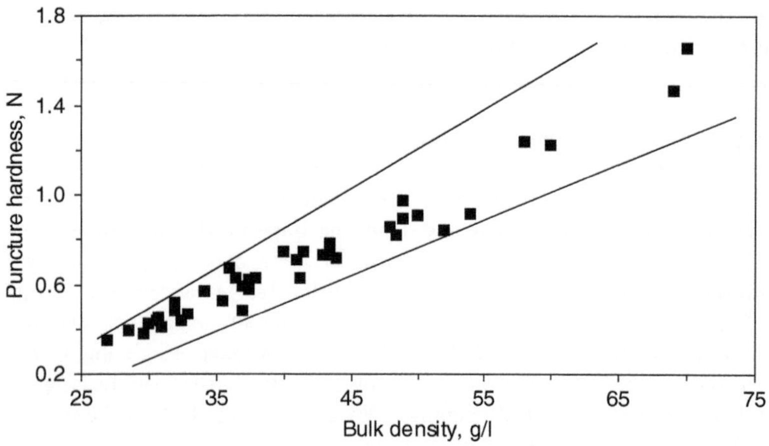

Figure 7.35 Master curve correlating puncture hardness versus bulk density (directly expanded corn-based extrudates).

history used to produce the extrudates. In fact, as seen in Chapter 6 (section 6.1.1.4, Eq. 6.6), at constant apparent melt viscosity, process SME results from the residence time-shear rate history in the screw-barrel assembly. Thus, at constant process SME and melt apparent viscosity, depending upon the combination of screw speed, screw configuration and die opening, the residence time-shear rate history may vary significantly. Therefore, depending upon the relative importance of the average residence time to the average shear rate in the screw-barrel assembly, a significant dispersion of the bulk density of directly expanded extrudates may be observed. Hence, the screw configuration in the shearing sections of the extruder can significantly affect the residence time-shear history, and thus may define the extent of volumetric expansion and bulk density at constant process SME.

For instance, let us compare screw profiles F and I. Both screw profiles hold three separate stages of shearing and they are characterized by similar process SME (in the range of 180–210 W.h/kg). However, screw profile I should show relatively higher average residence time and lower average shear rate in the shearing sections due to the presence of bilobe kneading blocks in its screw configuration, leading to the production of denser extrudates than those produced with screw profile F, though the process SMEs are comparable (see Figure 7.34). These results clearly illustrate the importance of the screw configuration in the shearing sections to vary the residence time-shear history at constant process SME, and hence to fine-tune the expansion characteristics of directly expanded extrudates. Therefore the significant dispersion of values of bulk density and puncture hardness shown in Figure 7.35 results from different residence time-shear history in the shearing sections of the extruder, thus leading to values which are contained between two straight lines of different slopes. The process-induced dispersion of the values of bulk density and puncture hardness is a point of practical importance in food extrusion processing as it gives significant process flexibility, thus providing the possibility to vary the hardness at constant bulk density, and vice versa.

7.3.3 Master correlations between sensory attributes and puncture parameter

Textural properties of food are key quality attributes that contribute to consumer product acceptability. These properties are perceived through a combination of visual, tactile, kinesthetic, and auditory sensations. Low-moisture, directly expanded extrusion-cooked products have a crispy texture and crispness is an important attribute of their quality profile; hence, the loss of crispness is a major cause of consumer rejection.

Sensory evaluation provides a direct measure of food crispness but this method requires important human resources and time when carrying out routine tests. Thus, instrumental methods such as the puncture test may be used conveniently to characterize and control food crispness, in order to provide a quick evaluation of crispness at reasonable cost. Of course, instrumental methods can only be used after demonstrating their robustness from a methodological standpoint and their capacity for product scoring. The purpose of this section is to illustrate the ability of the puncture method to score directly expanded extrusion-cooked food products, and to correlate the score with sensory crispness attributes. Directly expanded extrudates described in section 7.3.2 which were obtained by varying

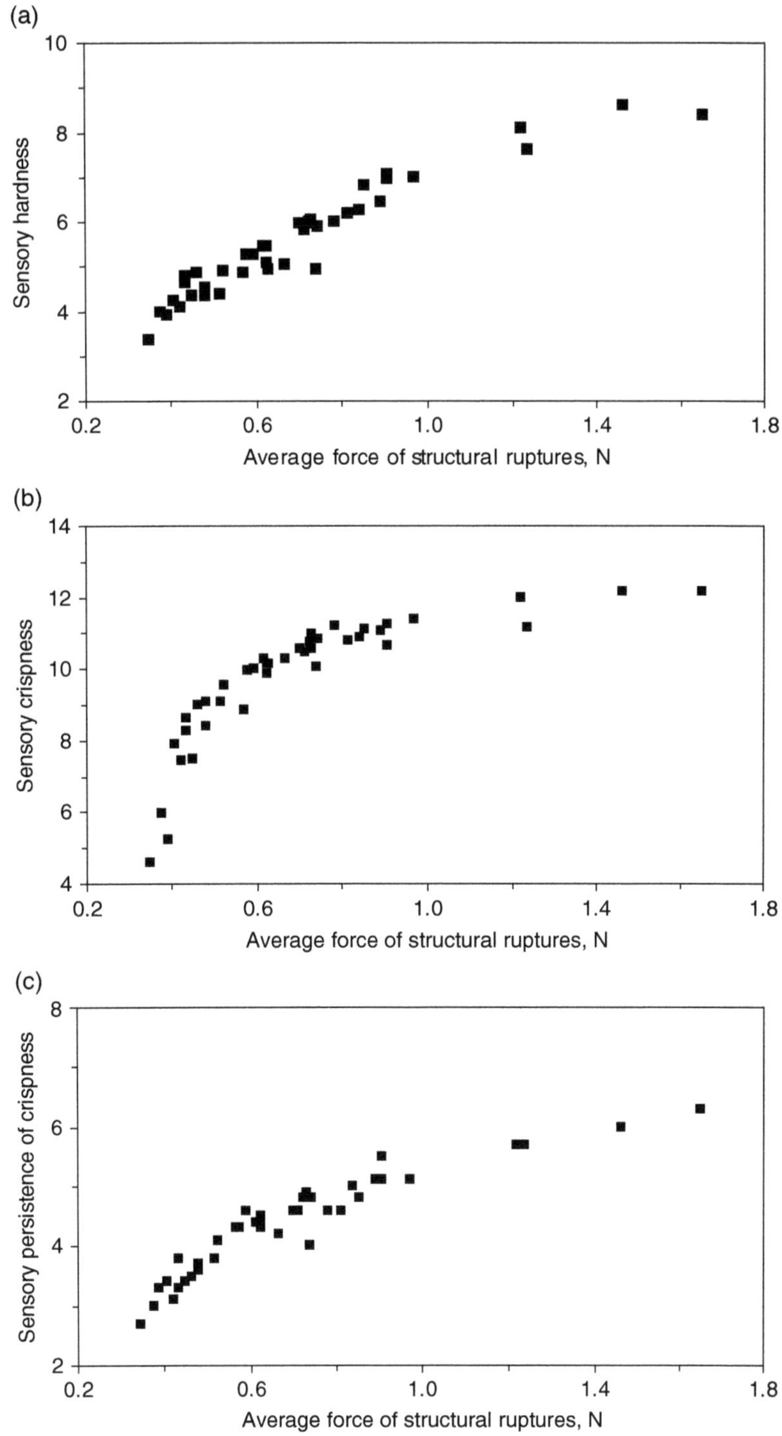

Figure 7.36 Master correlations between sensory attributes and puncture parameter f_{sr} for directly expanded extrusion-cooked extrudates. (a) Sensory hardness. (b) Sensory crispness. (c) Sensory persistence of crispness (Clextral, 2013b). Source: Reproduced with permission of Clextral, France.

extrusion variables were subjected to sensory crispness evaluation, at controlled moisture content (below 3% wwb). Forty different extruded samples were scored by judges trained in texture evaluation of cereal-based crispy food products. Three sensory attributes were measured.

- *Hardness*: sensory scoring resulting from the evaluation of the force required to break the sample between the teeth (first biting).
- *Crispness*: sensory scoring resulting from the evaluation of crispness sensation during the first biting (brittleness, noisiness, in particular).
- *Persistence of crispness*: sensory scoring resulting from the evaluation of crispness sensation during further biting.

Figure 7.36 presents master correlations obtained between the sensory attributes and the puncture parameter f_{sr} (the average force of structural ruptures) (Clextral, 2013b). The figures show that sensory crispness attributes are fairly well correlated with the puncture parameter f_{sr}, illustrating that sensory characteristics can be predicted from the instrumental puncture method. In practice, it appears that texture monitoring of directly expanded food products can be handled by determining only three key process and product indicators: process SME, bulk density, and average force of structural ruptures in puncture.

References

Agbisit R, Alavi S, Cheng E, Herald T, Trater A (2007) Relationships between microstructure and mechanical properties of cellular cornstarch extrudates. Journal of Texture Studies 38(2): 199–219.

Anderson RA, Conway HF, Pfeifer VF, Griffin E.L (1969) Gelatinization of corn grits by roll- and extrusion-cooking. Cereal Science Today 14(1): 4–11.

Babin P, Della Valle G, Rendievel R, Lourdin D, Salvo L (2007) X-ray tomography study of the cellular structure of extruded starches and its relations with expansion phenomenon and foam mechanical properties. Carbohydrate Polymers 68(2): 329–340.

Bagley EB, Christianson DD (1982) Swelling capacity of starch and its relationship to suspension viscosity – effect of cooking time, temperature and concentration. Journal of Texture Studies 35(1): 115–126.

Barrett AM, Peleg M (1992) Extrudate cell structure–texture relationships. Journal of Food Science 57(5): 1253–1257.

Becker A, Hill SE, Mitchell JR (2001a) Milling – a further parameter affecting the Rapid Visco Analyser (RVA) profile. Cereal Chemistry 78(2): 166–172.

Becker A, Hill SE, Mitchell JR (2001b) Relevance of amylose-lipid complexes to the behaviour of thermally processed starches. Starch/Stärke 53: 121–130.

Booth R, Bason M (2007) Principles of operation and experimental techniques. In: Crosbie GB, Ross AS (eds) *The RVA Handbook*. St Paul, Minnesota: American Association of Cereal Chemists, pp. 1–17.

Bouvier JM, Bonneville R, Goullieux A (1997) Instrumental methods for the measurement of extrudate crispness. Agro-Food-Industry Hi-Tech: 16–19.

Christensen CM, Vickers ZM (1981) Relationships of chewing sounds to judgements of food crispness. Journal of Food Science 46: 574–578.

Clextral (2013a) Private communication: the use of RVA instrument for the analysis of raw and extrusion-processed starch-based materials.

Clextral (2013b) Private communication: instrumental and sensory evaluation of extrudate crispness.

Crosbie GB, Ross AS (eds) (2007) *The RVA Handbook*. St Paul, Minnesota: American Association of Cereal Chemists.

Dacremont C (1995) Spectral composition of eating sounds generated by crispy, crunchy and crackly foods. Journal of Texture Studies 26(1): 27–43.

Dacremont C, Colas B, Sauvageot F (1991) Contribution of air- and bone-conduction to the creation of sounds perceived during sensory evaluation of foods. Journal of Texture Studies 22: 443–456.

Danley R, Kelly T, Groh J (2001) Improved DSC performance using tzero technology. International Labmate VI(II): 30–31.

Datta A.K, Sahin S, Sumnu G, Ozge Keskin S (2007) Porous media characterization of breads baked using novel heating modes. Journal of Food Engineering 79(1): 106–116.

Desrumaux A (1996) Comportement technologique des semoules de maïs en cuisson-extrusion. PhD Dissertation, Université de Technologie de Compiègne, France.

Desrumaux A, Bouvier JM, Burri J (1999) Effect of free fatty acids addition on corn grits extrusion cooking. Cereal Chemistry 76(5): 699–704.

Donovan JW (1979) Phase transitions of the starch–water system. Biopolymers 18: 263–275.

Doublier JL (1981) Rheological studies on starch – flow behaviour of wheat starch pastes. Starch/Stärke 33(12): 415–420.

Duizer L (2001) A review of acoustic research for studying the sensory perception of crisp, crunchy and crackly textures. Trends in Food Science and Technology 12: 17–24.

Duizer L, Campanella OH, Barnes GRG (1998) Sensory, instrumental and acoustic characteristics of extruded snack food products. Journal of Texture Studies 29: 397–411.

Edmister JA, Vickers ZM (1985) Instrumental acoustical measures of crispness in foods. Journal of Texture Studies 16: 153–167.

Harper JM, Tribelhorn RE (1992) Expansion of native cereal starch extrudates. In: Kokini JL, Ho CT, Karwe MV (eds) *Food Extrusion Science and Technology*. New York: Marcel Dekker, pp. 653–667.

Hayter AL, Smith AC (1988) The mechanical properties of extruded food foams. Journal of Materials Science 23(2): 736–743.

Hutchinson RJ, Siodlak GDE, Smith AC (1987) Influence of processing variables on the mechanical properties of extruded maize. Journal of Materials Science 22(11): 3956–3962.

Katz EE, Labuza TP (1981) Effect of water activity on the sensory crispness and mechanical deformation of snack food products. Journal of Food Science 46(2): 403–409.

Lai KP, Steffe JF, Ng PKW (2000) Average shear rates in the Rapid Visco Analyser (RVA) mixing system. Cereal Chemistry 77(6): 714–716.

Lee WE, Deibel AE, Glembin CT, Munday Shepherd EG (1988) Analysis of food crushing sounds during mastication: frequency-time studies. Journal of Texture Studies 19: 27–38.

Le Meste M, Roudaut G, Davidou S (1996) Thermomechanical properties of glassy cereal foods. Journal of Thermal Analysis 47(5): 1361–1375.

Mason WR, Hoseney RC (1986) Factors affecting the viscosity of extrusion-cooked wheat starch. Cereal Chemistry 63(5): 436–441.

Moore D, Sanei A, Van Hecke E, Bouvier JM (1990) Effect of ingredients on physical/structural properties of extrudates. Journal of Food Science 55(5): 1383–1387, 1402.

Paterson L, Mitchell JR, Hill SE, Blanshard JMV (1996) Evidence for sulfite induced oxidative reductive depolymerisation of starch polysaccharides. Carbohydrate Research 292: 143–151.

Paton D, Spratt WA (1981) Simulated approach to the estimation of degree of cooking of an extruded cereal product. Cereal Chemistry 58(3): 216–220.

Robin F, Engmann J, Pineau N, Chanvrier H, Bovet N, Della Valle G (2010) Extrusion, structure and mechanical properties of complex starchy foams. Journal of Food Engineering 98(1): 19–27.

Roudaut G, Dacremont C, Le Meste M (1998) Influence of water on the crispness of cereal-based foods: acoustic, mechanical, and sensory studies. Journal of Texture Studies 29(2): 199–213.

Ryu GH, Neumann PE, Walker CE (1993) Pasting of wheat flour extrudates containing conventional baking ingredients. Journal of Food Science 58(3): 567–573, 598.

Sauvageot F, Blond G (1991) Effect of water activity on crispness of breakfast cereals. Journal of Texture Studies 22(4): 423–442.

Schawe JEK (1995) Principles for the interpretation of modulated temperature DSC measurements. Part 1. Glass transition. Thermochimica Acta 261: 183–194.

Seymour SK, Hamann DD (1988) Crispness and crunchiness of selected low moisture foods. Journal of Texture Studies 19: 79–95.

Smolarz A, Van Hecke E, Bouvier JM (1989) Computerized image analysis and texture of extruded products. Journal of Texture Studies 20: 223–234.

Tan J, Gao X, Hsieh F (1994) Extrudate characterization by image processing. Journal of Food Science 59(6): 1247–1250.

Tesch R, Normand MD, Peleg M (1995) Comparison of the acoustic and mechanical signatures of two cellular crunchy cereal foods at various water activity levels. Journal of the Science of Food and Agriculture 70: 347–354.

Tesch R, Normand MD, Peleg M (1996) On the apparent fractal dimension of sound bursts in acoustic signatures of two crunchy foods. Journal of Texture Studies 26: 685–694.

Trater AM, Alavi S, Rizvi SSH (2005) Use of non-invasive X-ray microtomography for characterizing microstructure of extruded biopolymer foams. Food Research International 38(6): 709–719.

Van Hecke E (1991) Contribution à l'étude des propriétés texturales des produits alimentaires alvéolés. Mise au point de nouveaux capteurs. PhD dissertation, Université de Technologie de Compiègne, France.

Van Hecke E, Allaf K, Bouvier JM (1995) Texture and structure of crispy-puffed food products. I: mechanical properties in bending. Journal of Texture Studies 26(1): 11–25.

Van Hecke E, Allaf K, Bouvier JM (1998) Texture and structure of crispy-puffed food products. Part II: mechanical properties in puncture. Journal of Texture Studies 29(6): 617–632.

Vickers ZM (1981) The relationships of pitch, loudness and eating technique to judgements of the crispness and crunchiness of food sounds. Journal of Texture Studies 16: 85–95.

Vickers ZM (1985) Relationships of chewing sounds to judgements of crispness, crunchiness and hardness. Journal of Food Science 47: 121–124.

Vickers ZM, Bourne MC (1976) A psychoacoustical theory of crispness. Journal of Food Science 41: 1158–1164.

Vickers ZM, Christensen CM (1980) Relationships between sensory crispness and other sensory and instrumental parameters. Journal of Texture Studies 11: 291–307.

Voisin I (1993) Influence de divers ingrédients alimentaires sur le procédé de cuisson-extrusion et sur les propriétés physiques d'extrudés à base de produits amylacés. PhD dissertation, Université de Technologie de Compiègne, France.

Wang SS, Chiang WC, Zhao B, Zheng XG, Kim IH (1991) Experimental analysis and computer simulation of starch-water interactions during phase transition. Journal of Food Science 56(1): 121–124 & 142.

Whalen PJ (1998) Detection of differences in corn quality for extrusion processes by Rapid Visco Analyser. Cereal Foods World 43: 69–72.

Whalen PJ, Bason ML, Booth RI, Walker CE, Williams PJ (1997) Measurement of extrusion effects by viscosity profile using the Rapid ViscoAnalyser. Cereal Foods World 42(6): 469–475.

8 The Generic Extrusion Process IV: Thermomechanical Pretreatment and Solid–Liquid Separation

Many processing techniques for solid–liquid separation have been developed independently and applied to a number of industries, thus they have different characteristics although they bear common objectives. Practically, these developments have occurred with little cross-fertilization, and the nature and sizing of equipment depend on the proportions of the phases to be separated, certain physical properties that influence relative movements of liquids and solid particles, as well as economics. Pressure is often the main operating variable used in solid–liquid separations, leading to mechanically or physically driven separation processes.

Mechanical/physical processes used in solid–liquid separation can be broadly classified into three main categories: (i) settling by gravity, centrifugal force, air flotation, dense media flotation and magnetic properties; (ii) filtration on screens by gravity and on filters by application of vacuum, by pressure and by centrifugation; and (iii) expression using batch or continuous presses (screw presses, rolls and disks). Generally, settling, or clarification, is the removal of small amounts of unwanted solids from a valuable liquid. Filtration is also aimed at recovering valuable solids from slurries whereas expression is the removal of relatively small contents of liquids (water or organic solvents) from compressible sludges by mechanical means. Mechanical expression is also extensively used in the processing industry, particularly the agro-food industry, due to significant energy and time savings when eliminating water or solutions from cellular biomaterials such as sugar beet pulp, vegetable pulps, fruits, cellulose pulps, waste water sludges, etc. It is also used in the efficient and economic extraction of oil from oilseeds.

Processes have been developed to push out the liquid occluded in a solid matrix by the application of pressures high enough to rupture the cells containing the liquid and overcome the friction created by the flow of the expressed liquid moving through the porous medium. Owing to the high pressure generated in single and twin screw extruders and the possibility of running continuous processes, screw extrusion-pressing (either single or twin screw extruder-pressing) has been one of the preferred processes for solid–liquid separation. Hence, thermomechanical expression of solid–liquid systems using extrusion-pressing technology is used in the processing industry to continuously separate solutions or liquids from solid materials. It allows concentrated liquids or solutions to be separated in a liquid state, hence avoiding energy-consuming heat-based separation methods (such as distillation or evaporation). It must be noted that extrusion-pressing is considered as a process-intensifying method for solid–liquid separation (see Chapter 10).

The use of screw extruders to thermomechanically express solid–liquid mixtures is presented in this chapter. This process has initiated the development of the Generic Extrusion Process IV (GEP IV), generically named: thermomechanical pretreatment and solid–liquid separation. GEP IV is described qualitatively in the first part of this chapter while the second part gives an engineering analysis on the thermomechanical expression of cellular biological materials. The third part of this chapter presents a modeling approach of the extrusion-pressing process. The fourth part deals with relevant industrial

Extrusion Processing Technology: Food and Non-Food Biomaterials, First Edition. Jean-Marie Bouvier and Osvaldo H. Campanella.

applications of extrusion-pressing technology and typical processing lines.

8.1 The fourth Generic Extrusion Process: continuous mechanical expression

Two main approaches are effectively exploited in extrusion-pressing.

• Direct extrusion-pressing of solid materials (or one-stage extrusion-pressing) to express liquids from solid matrices (e.g. edible oil from oilseeds, liquid water from agricultural by-products, juices from fruits). Regardless of the effectiveness of the operation, the resulting expressed cake always contains a significant amount of residual liquid which needs further separation in a downstream processing unit, through either solvent extraction (for instance, hexane extraction of edible oil from oilseeds) or drying (for instance, water removal from agricultural by-products). This approach is extensively used in industry to eliminate most of the liquid component from the solid in the screw extruder-press stage and hence minimize as much as possible downstream energy-consuming processing such as evaporation for postdrying operations or environmentally and safety questionable solvent extraction for liquid extraction postprocessing operations.

• Solvent-aided extrusion-pressing of solid materials to dissolve solutes and express out the resulting solution from the solid matrices (for instance, water-aided washing of alkaline-digested lignocellulosics and bleached cellulose fibers). Several integrated washing stages can be set up in series on the extruder barrel, depending upon the required final concentration of solutes.

Historically, screw pressing of oilseeds has been a unit operation used with two main purposes. One is a high-pressure operation to expel oil from oilseeds and the other is used as a pre-press operation prior to solvent extraction (Ward, 1976). Figure 8.1 shows the flow diagram of these two operations. As observed in the figure, the typical flow sheet to obtain crude pure oil as the main product of the screw pressing operation is similar for both purposes. The difference is in the treatment of the oil exhausted cake or meal. In one case the meal is stored and used as a feeding material of a second extrusion process, for instance to produce texturized protein (see the case study described in section 8.4). Or the meal is used as the feeding material of a solvent extraction column or as pretreated feed for a second screw pressing operation. The steps indicated in Figure 8.1 include: seed preparation; cooking; screw pressing; and separation of solids from expelled oil and

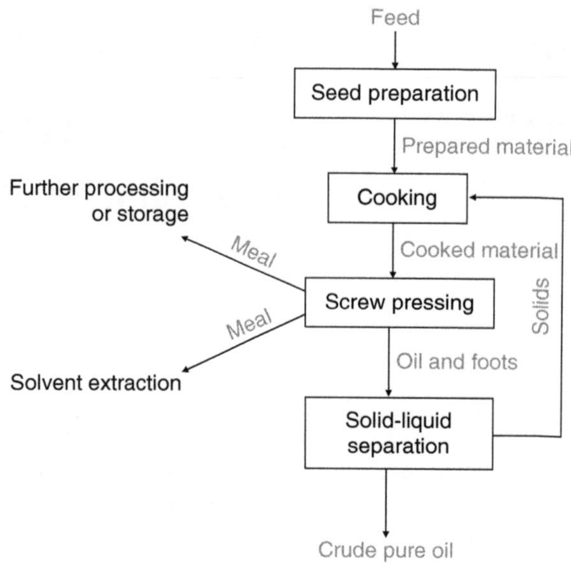

Figure 8.1 Flow chart showing processing steps for the separation of oil from oilseeds including the process of solvent extraction.

return to cooker/screw press (Ward, 1976). Ward highlights the importance of these four operations and states that unless all four steps are properly performed, good results will not be obtained.

In the preparation process, seeds are reduced in size or tempered at a fixed moisture content and temperature. Some processes utilize the whole seed as, for example, during the screw extrusion-pressing operation of soybeans. Tempering and hydration of the seeds serve to achieve raw materials with uniform moisture content going to the cooking process. Working on the expelling of soybean oil, Nelson et al. (1987) reported that pre-extruding (dry extrusion) coarsely ground whole soybeans at 10–14% moisture produces an extrudate emerging from the die in a semi-fluid state, which when immediately pressed in a continuous screw extruder-press yields high-quality oil and press cake. This research clearly showed that proper conditioning of the feed and extrusion prior to liquid expulsion greatly increases the throughput of the expeller over the rated capacity.

The cooking process serves several purposes, one is to modify the rheology of the seeds so an efficient screw pressing operation can be achieved. The other purpose relates to the high temperature-short time characteristics of the extrusion process that reduce the prolonged heating and holding of the material existing in conventional expression operations. Under conditions of high temperature in the screw press operation, besides the production of a meal with good properties in terms of protein content and residual oil (50% protein and 6% residual oil), Nelson et al. (1987) observed a 90% inactivation of trypsin inhibitors. Inactivation of trypsin inhibitors and other antinutritional factors has great relevance when the cake or meal is used as an ingredient for animal feed (Powell et al., 2011; Zhang et al., 1993).

In addition to oil pressing operations, screw extrusion-pressing and other pressing methods have been used in a variety of food and non-food processes such as pressing of whey out of cheese, the production of fruit and vegetable juices, the dewatering of pectin greens, spent brewery mashes, to produce burnable processing wastes from wet material, the recovery of pectin from cooked apple pomace, the expression of sugar from sugar cane, and recovery of protein-containing juices from alfalfa (Schwartzberg et al., 1977). From an economic standpoint, screw pressing or other types of pressing technology are presented as potential alternatives for material dehydration. For instance, for pressed wet burnable wastes, the net fuel value of the waste is increased because only a small amount of energy is required to dehydrate the material when compared with a conventional process that uses heat to evaporate the water out of the material. At the same dehydration level, the energy required for pressure-induced dewatering of materials is significantly lower than the energy required for heat-induced water removal. For example, approximately 7 kJ/kg of energy is necessary to expel water at a pressure of 6.8 MPa (68 bar) against 2300 kJ/kg for heat-assisted evaporation (Schwartzberg, 1997).

One of the most common applications of screw extrusion-pressing is the separation of oil from oilseeds. Table 8.1 shows the increasing trend of oil processing worldwide due to new food and non-food applications. Considering the favorable energy rate of this technology, many applications have been related to the overall theme of energy savings, which is in line with concepts discussed in this book addressing advantages of extrusion and extrusion-related processes from an energy-saving standpoint. Screw pressing operations have also been used to obtain juices and oils from fruits and vegetables such as apple, grapes, and palm fruit (Berger, 1983; Choo et al., 1996; Hunt & Hunt, 1981; Little, 1958; Marin et al., 2002; Ohta et al., 1982; Omobuwajo et al., 1998; Owolarafe et al., 2002).

Screw pressing operations are commonly performed using single screw and twin screw extruder-presses. Single screw extruder-presses use similar mechanical means

Table 8.1 Major oilseeds world production for the years 2009–2013.

Oilseeds	Production (in million metric tons)			
	2009/10	2010/11	2011/12	2012/13
Soybean	260.40	263.92	239.15	267.58
Rapeseed	61.06	60.60	61.17	62.28
Cottonseed	38.91	43.55	46.41	45.33
Peanut	35.92	39.52	37.87	39.93
Sunflower seed	32.14	33.63	40.64	36.31
Palm kernel	12.44	12.88	13.66	14.70
Copra	5.88	6.02	5.66	5.96

Source: Reproduced with permission of the U.S. Department of Agriculture, Food Safety and Inspection Service.

to conventional extruders to pressurize and shear the material, but in single screw extruder-presses the screw turns inside a perforated barrel. The feed is introduced at one end of the barrel and moves along the extruder-press due to the drag flow generated by the velocity gradient created between the turning screw and the stationary barrel. During transit of the material in the press, the pressure increases and the cake compacts because: (i) the screw root diameter progressively increases; (ii) the pitch progressively decreases; (iii) the diameter of the barrel decreases; (iv) the material outflow is throttled by a valve; or (v) a combination of these approaches (Schwartzberg, 1997).

Different types of single screw extruder-press are available and some of the more commonly used are illustrated in Figure 8.2. Figure 8.2a illustrates a typical screw press with perforated barrel used for squeezing soft fruits whereas Figure 8.2b illustrates a heavy-duty screw press for processing oilseeds applying high pressures. Rather than circular orifices in the barrel, heavy-duty screw presses have a barrel-cage made of bars separated by a distance that decreases down the screw channel to prevent plugging the openings with solids (Schwartzberg, 1997). The expelled liquid flows radially outward through the cage but if openings are too small or the compaction of the cake is too large, oil tends to flow back towards the press feed. This condition, known as "oiling up", reduces friction between the cake and the barrel and screw walls so that the cake cannot be conveyed by the drag action and turns *en masse* (Schwartzberg, 1997). Breaker bars that project inward from the barrel wall and fit between gaps in the screw are used to prevent slipping at the screw and barrel surfaces and the cakes from turning *en masse.*

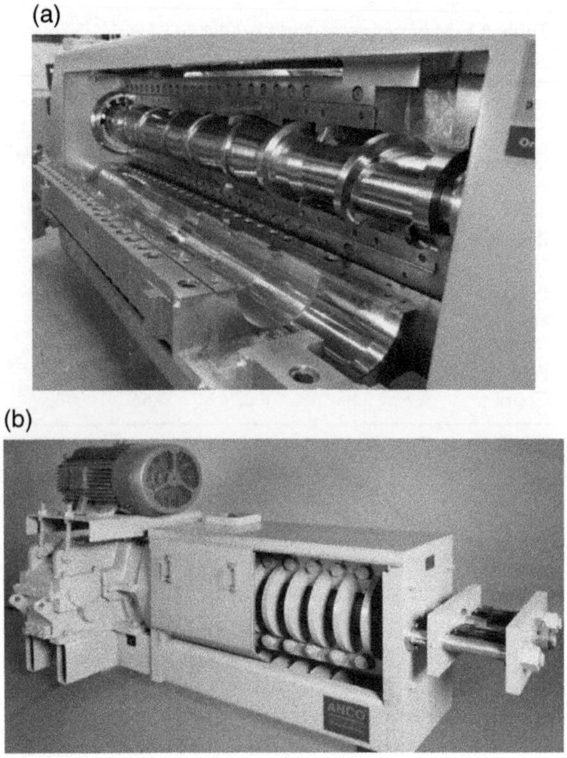

Figure 8.2 Different single screw press technologies used for the screw extrusion-pressing operations. (a) Single screw press (from http://lipidlibrary.aocs.org/processing/expanding/index.htm). (b) Industrial single screw press (oil expeller) using a wedge-bar cage (from http://ancoeaglin.com/images/pdfs/202ScrewPress.pdf). Source: Reproduced with permission of the American Oil Chemists' Society, Lipid Library.

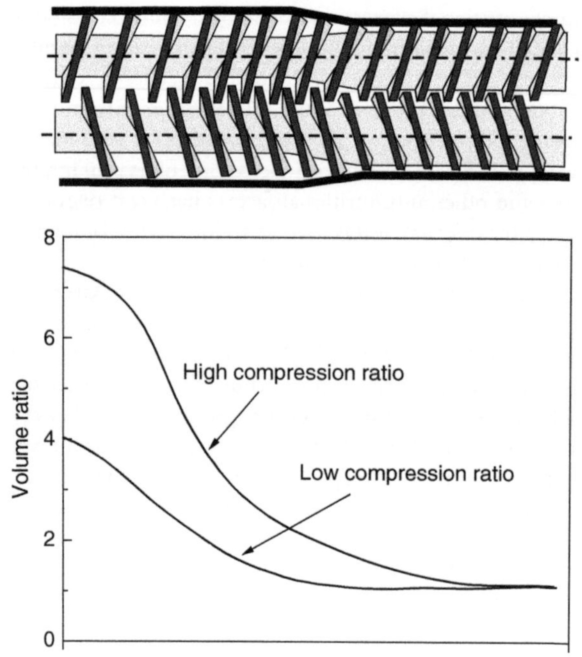

Figure 8.3 Schematic of twin screw extruder-press utilized in screw extrusion-pressing operations along changes in the volume ratio for two different screw profiles.

Cut flight screws such as those shown in Figure 8.2a are also used to improve the conveying of materials with tendency to pack, as well as promoting shear and mixing (Lusas & Watkins, 1988).

The deformation and pressure needed to rupture oil-containing cells is a characteristic associated with the cellular structure of the material being expressed. For instance, the oil point is a bulk property of oilseeds such as rapeseed under compression and is defined as the stage at which compression of the cake has squeezed individual seeds so that the oil is forced out from the interior of the cells (Sukumaran & Singh, 1989). The oil point of seeds such as rapeseeds is affected by the rate of deformation and the strain/deformation of the cake, which clearly shows the viscoelastic nature of cakes formed during processing of oilseeds. Processing parameters such as pressure and moisture also affect the rheological properties of the pressed cake and thus its oil point. Rheological properties of cellular materials are described in section 8.2.4.

Twin screw extruder-presses are also used in screw pressing operations. Most of the technology involves intermeshing counter-rotating twin screws housed in a close-fitting perforated barrel. As illustrated in Figure 8.3, compaction of the cake is generated by a decrease in the barrel casing diameter and an increase in the screw root diameter. The figure also shows the changes in the cake volume along the extruder shaft for two different screw profiles, specifically low and high compression ratios. Twin screw extruder-presses with capacities up to 85,000 kg/h are available, and are used extensively in a variety of applications such as dewatering spent sugar beet pulp, expelling oil and water from cooked oil-rich fish. On a lesser scale, the technology is also used for expelling protein-rich juices from alfalfa, dewatering sugar beet trash, tails and leaves, pressing spent brewers and distillers grains, expelling hydrocolloid juices and

Figure 8.4 View of the interrupted flight twin screw technology used in twin screw extrusion-pressing (courtesy of Vincent Corporation; www.vincentcorp.com/content/twin-screw-press-2). Source: Adapted from Vadke & Sosulski 1988. Reproduced with permission of Vincent Corporation.

pressing fat from cooked dried meat scraps or from heated moist meat scraps in rendering operations (Schwartzberg, 1997).

Depending on the product, high pressures of up to 7 MPa (70 bar) are generated during the expression operation using this technology. In terms of energy rate, Schwartzberg (1997) reports that 1.5–1.9 kWh of energy are required per 1000 kg of feed and up to 3 kWh per 1000 kg of feed may be required for pressing alfalfa. Newer technologies offer high-capacity continuous dewatering of slippery materials such as spent grain, fruits and vegetables, citrus waste, fish, sugar beet pulp, and shrimp waste, and also for preconditioning husk and cob for corn silage (Vincent Corporation, 2013). The introduction of interrupted flight screws (Figure 8.4) improves the slicing action of the press and results in reduced power consumption (Johnston, 2003).

Twin screw extrusion-pressing has also been used for oil expelling of oilseeds using specific screw profiles and long L/D and inserting a filtration section as shown in Figure 8.5. Intermeshing co-rotating twin screw extruders with filtration sections along the barrel wall are also used for extrusion-pressing operations such as pressing oilseeds (Bouvier & Guyomard, 1997; Guyomard, 1994), or water-aided washing of alkaline-digested lignocellulosics and bleached cellulose fibers (see sections 5.3.3 and 10.2.2.3.2). The modularity of twin screw extruders and the use of specific screw geometries described in Chapters 2 and 3 increase the applicability of these

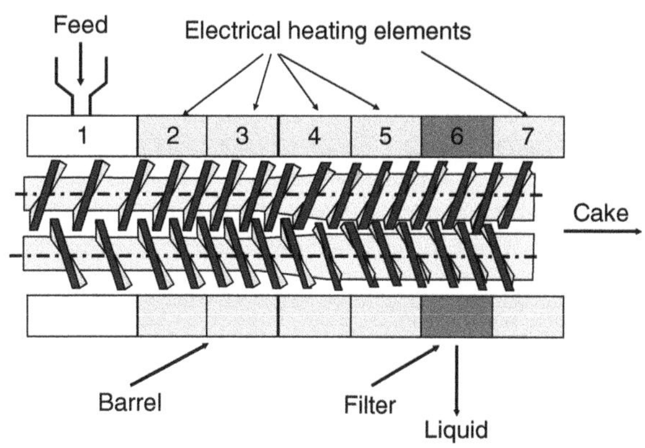

Figure 8.5 Schematic of the twin screw extruder adapted to be operated as a twin screw extruder-press.

extruders for screw extrusion-pressing of a variety of oilseeds for food and non-food applications. For example, oil obtained from jatropha seeds cannot be used in food applications due to the presence of toxic components that have to be removed using expensive processes. Thus, an alternative is to use this oil as an energy source for fuel production, which is a very attractive option given its renewable and environmentally friendly characteristics and the fact that it can be produced locally in many underdeveloped areas (Evon et al., 2013).

A number of studies using twin screw extruders of different sizes have clearly shown that different screw

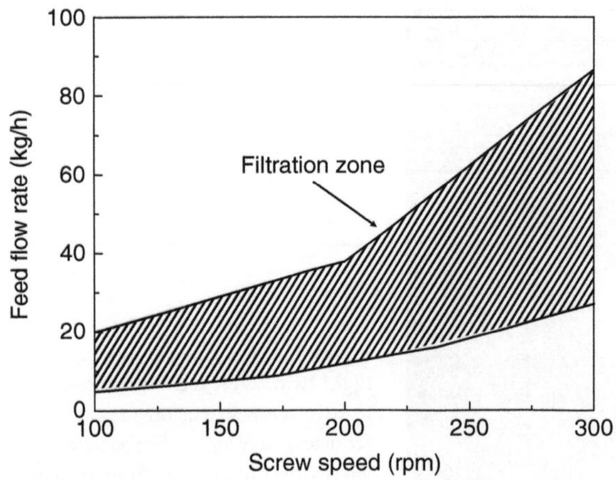

Figure 8.6 Influence of screw speed and mass feed rate on the liquid expression of sunflower seeds in a twin screw extruder modified to work as a twin screw extruder-press (extrusion temperature: 80°C).

geometries have a great effect on oil yield and also its quality. Working with sunflower seeds pressed in twin screw extruders, Dufaure et al. (1999) showed minimal and maximal conditions beyond which the operation was ineffective (Figure 8.6). The effectiveness of the pressing operation of twin screw extrusion-pressing is related to many of the concepts already discussed in Chapter 3. When the feed rate without pretreatment was low, the filling of the extruder was not enough to form a solid plug close to the reverse screw element and the filtration media. Thus, the pressure exerted on the cake was insufficient to express the oil from the solid through the filter media. For these processing conditions, the cake exits the extruder end still containing a large amount of oil (Dufaure et al., 1999). Another problem observed was the presence of solids (defined as foots – see section 8.2.2.1) in the expelled oil, which had to be eliminated using decanters or centrifuges. The selection of an appropriate filter medium could alleviate this problem but the extra pressure created in the system by the finer opening filter could affect the extrusion process. Other studies using modified twin screw extruders as the sole process for oil extraction or as a seed pretreatment before solvent extraction (e.g. using n-hexane) for processing sunflower and other oilseeds can be found in the literature (Bouvier & Guyomard, 1997; Guyomard, 1994; Isobe et al., 1992; Kartika et al., 2006, 2010; Prat et al., 2002; Sriti et al., 2011, 2012).

In all these studies the advantages of using twin screw extrusion technology in terms of its capacity to perform various functions useful for screw pressing operations are discussed. They include better process control and versatility, especially in pumping efficiency, control of residence time distribution, and uniformity of processing; process specialty formulations, which the single screw extruder cannot handle; and machine set-up flexibility, allowing self-cleaning mechanisms and rapid changeover of screw configurations without disassembling the extruder (Evon et al., 2013).

8.2 Engineering analysis of thermomechanical expression

The engineering analysis that describes the behavior of cellular material moving through a screw extruder-press comprises many aspects that are similar to those given in Chapter 3 to describe the thermomechanical treatment of a melt in conventional extruders. In both cases, a semi-solid mass is conveyed along the extruder shaft by the drag action of one or two screws rotating inside an enclosure called a barrel. As a consequence of the semi-solid mass movement, a reduction of the volume in the flow channel and changes in the properties of the moving mass, the pressure of the mass in the channel increases significantly along the shaft. The description of that mass movement into the screw channel and the increase of pressure can be undertaken using mathematical models similar to the ones used in Chapter 3.

However, two main aspects differ in screw extrusion-pressing and conventional extrusion. The first is that the solid mass contains liquids which have to be separated during the pressing to accomplish the solid–liquid separation. In screw pressing, the solid–liquid separation is achieved using perforated barrels that act as a screen or filter that retains the solids in the screw extruder-press while allowing the liquid to be separated. Compared to other pressing approaches, screw extruder-presses have been popular for solid–liquid separation processes due to their ability to generate high pressures able to force the liquid out of the solids using continuous or semi-continuous processes. The other aspect to keep in mind in the engineering analysis of liquid expression is the cellular characteristics of the material being processed. The exudation of the liquid from the semi-solid phase has been analyzed using concepts of fluid mechanics to estimate the pressure in the press volume associated with the filtration process while the liquid is exudated through the filtering media. A model to describe the flow of liquid expressing out of the solid is further complicated by the cellular structure of the solid. There are cases where the liquid in the solid is mostly contained in closed cells which have to be ruptured to facilitate liquid expression. In other cases, the solid has a cellular structure formed by communicating pores and there is no need to induce rupture of cells to express out the liquid from the solid (Schwartzberg, 1983).

Regardless of the structure of the solid cake, the applied pressure on the solid should be high enough to overcome three types of forces: (i) compaction forces or pressure necessary to reduce the volume of the cake; (ii) forces to overcome the pressure drop required to expel the liquid from the pores in the cake; and (iii) forces to overcome the pressure drop caused by the flow of the liquid through the outflow medium (perforated barrel on the screw extruder-press). These forces are strongly influenced by the cellular structure of the cake. Pressures of more than 1000 bars are used in oilseed expellers, more than 400 bars are used to express sugar solutions from sugar canes, and 35–70 bars to dewater spent sugar beets. For fruit juices, pressures are significantly lower, in the order of 15–30 bars, and extremely low pressures are used to express whey from different types of cheese (Schwartzberg, 1983). Due to the variation in operating pressures as a result of the large variation in the nature of the cake and its cellular structures, an engineering analysis must include consideration of the mechanical characteristics of cakes and how they are affected by their cellular structures.

8.2.1 Structure of cellular biological materials

Mechanisms of liquid expression from minerals have been described in the literature and although the process may appear to be similar, it must be kept in mind that liquid expression of cellular biological materials is different due to the inherent compressibility of these materials. Unlike biological material, solids forming mineral cakes are incompressible and during compaction the solid is not deformed. Instead, deformation of the cake is produced by reorientation of the solid particles forming it. In most biological materials, the liquid is occluded in cells (Figure 8.7), where some gas can also be present (Lanoisellé et al., 1996). Under increasing compaction pressure, the gas is released and the large pressures applied may rupture the cells and the contained liquid is expelled along with the gas.

Figure 8.7c illustrates the structure of cellular material and the approximately regular geometry of the cells. The cell wall material surrounding those cells is one of the main components affecting the cellular structure of plants, and is made up of four building blocks: cellulose, hemicellulose, lignin, and pectin (Gibson, 2012). Key literature in the area (e.g. Lanoisellé et al., 1994, 1996; Rebouillat, 2002; Schwartzberg, 1983, 1997) has named those cells "particles." In order to avoid confusion, the word "particle" as used in this chapter concerns cells including the cell wall material (solid) and the liquid and gases contained in the cells. Thus, the volume of a particle will contain solid, liquid, and gas phases. As schematically shown in Figure 8.7c, the cellular biological material is assumed to have two types of porosities. One is associated with the volume of the particle, which could be fully or partially saturated with liquid and include the presence of gases. The other porosity is associated with the interparticle volume. Like the internal porosity in the particle volume, the interparticle volume could be saturated with liquid or unsaturated, containing both liquids and gases. Compaction of the cake generates pressures inside the cells that can rupture them.

The structure of most plants shows specific cellular characteristics that affect their mechanical properties. They may differ in composition, geometry of the cell walls and the presence of liquids (saturation) and air or other gases (unsaturation). The variation in the hierarchical structure of plants in terms of their length and the volume fractions of the four building blocks forming the cell wall, including the presence of intercellular and intracellular liquid and gases, gives rise to a wide range of mechanical properties, which are illustrated in Figure 8.8. The figure shows a plot of the cellular material strength as a function of its Young modulus in double logarithmic co-ordinates. In particular and related to this chapter, it must be noted the mechanical properties of plants such as potato and apples and species having parenchymal tissue, and how they compare with the mechanical properties of other cellular structures such as wooden species. It is clear that the cellular structure and the resulting properties of fruits and vegetables, including oilseeds (not shown in the plot), are weaker in terms of strength and elasticity than their wooden counterparts and thus they are more prone to be destroyed by physical means to separate their liquid fractions. In addition to the structure and composition of the cell wall material, the presence of liquid and the level of saturation, as well as intra- and intercell porosity, and the density of these materials have a profound influence on their mechanical properties. The cellular structure of these materials resembles the structure of dry foams when pores are unsaturated and liquid foams when pores are fully saturated.

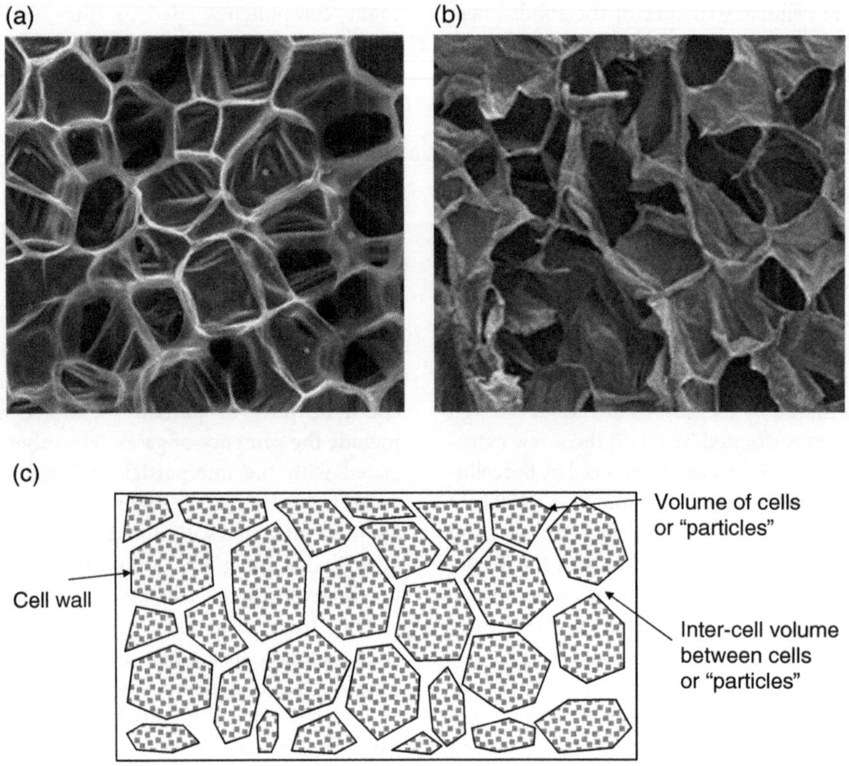

Figure 8.7 Electron scanning microscopy of (a) carrot tissue; (b) apple flesh. (c) Schematic of the structure of cellular materials. Micrograph (a) from Gibson (2012) with kind permission from Cambridge University Press; micrograph (b) from Xiao and Gao (2012) with permission from InTech. Source: Gibson 2012. Reproduced with permission of the Journal of the Royal Society Interface.

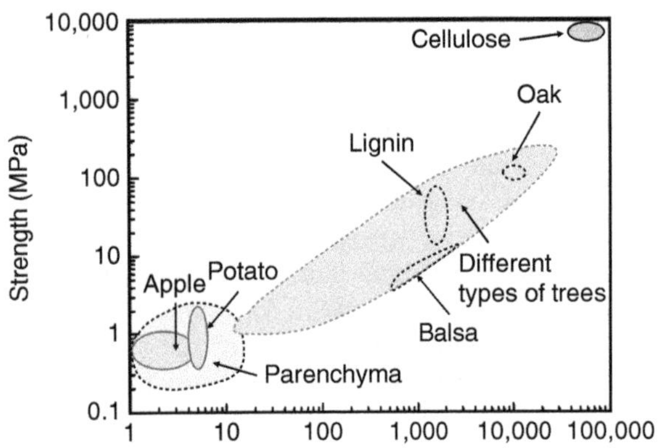

Figure 8.8 Mechanical properties of cellular biological materials. Strength versus Young modulus. Source: Adapted from Gibson 2012.

Gibson and Ashby (1982) proposed an equation relating the mechanical properties of cellular materials with the ratio between the density of the material, ρ, and the density of the cell wall material, ρ_s. By assuming that the solids in the cellular porous structure are those forming the cell wall material and defining the densities of the solid and the cake as $\rho_s = m_s/V_s$ and $\rho = m_s/V$ respectively, the equation proposed by Gibson and Ashby can be written in terms of the volume of the cake V and the volume of the solids in the cake V_s as:

$$\frac{\textit{Mechanical property of foam}}{\textit{Mechanical property of cell wall}} = C\left(\frac{V_s}{V}\right)^n \tag{8.1}$$

where C and n are empirical coefficients. The value of n usually lies between 1.5 and 2 for foams with open cells and between 2 and 3 for foams with closed cells (Schwartzberg, 1997). Thus, this relationship may provide information for estimating how the deformation properties of the cellular material changes for application of forces and pressures. The above equation works well for low relative densities ($\rho/\rho_s \leq 0.1$) a condition that has a large relevance on liquid expression of biological materials. With such materials, the equation holds in a range of relative densities that goes from of 0.01 to 1.

As described in the preceding sections, the nature of the cell wall material and the cellular porous structure dictate the pressure necessary to rupture the cells and express the liquid contained inside them. Plant cells are encased in a permeable lipoprotein membrane known as the plasma and surrounded by a relatively rigid wall made of cellulose fibers embedded in a matrix containing hemicelluloses, pectin and small amounts of glycoproteins (Schwartzberg, 1997). Large amounts of lignin, an amorphous polyphenol, are found in many plant cell walls and cement cells together in wood, which contains 20–35% lignin by weight. Cellular materials with a thicker cell wall (large V_s) exhibit large changes upon compression. Conversely, fruit and vegetable cells are cemented together by protopectin, which is an insoluble form of pectin (the junction region of the cells is called the middle lamella) and the cell walls are thinner and weaker and changes in the mechanical properties are not so significant upon cake compression.

Plant cells also contain an internal, fluid-filled compartment called vacuole, which is surrounded by a second membrane named the tonoplast and occupies most of the interior of cells of juicy plant materials. An excess internal pressure associated with plant cell turgor is present inside biologically active plant cells due to osmotic imbalances between intracellular and extracellular fluid through the membrane. Turgor pressure, which may range from 0.5 bar to 50 bar depending on the plant, provides much of the structural strength of plant cells (Schwartzberg, 1997). Most of the strength of these cell structures under tension is provided by the cell wall but they buckle under compression, especially when the cell loses turgor. If plasma membranes are denatured, e.g. by heating or other means such as chemical treatment, or by using novel technologies such as pulsed electric field (PEF), turgor disappears and cells deform much more easily, thus reducing the pressure necessary to expel the liquid from the solids. In fruit tissues, heating above 70 °C causes pectin to dissolve, affecting their pressability and reducing resistance to the expression of juices and diffusion of water or solutes out of cells. Assisted expression is discussed in section 8.4.

8.2.2 Introduction of the nomenclature

8.2.2.1 Composition in terms of mass and volumes

Scanning electron microscopy (SEM) of carrots and apple tissues are illustrated in Figures 8.7a and 8.7b respectively, whereas Figure 8.7c shows a schematic of the structure of those plant tissues. The volume of the particle is composed of the volume of the cell, the interparticle volume, and the volume of the cell wall material. The collection of those cells or particles forms the structure of the cellular material. The amount of water (or oil in oilseeds) and the solids contained in those materials can be accounted for by a weight fraction defining the amount of water or solids contained in a given mass of the material and defined as:

$$y_w = \frac{\textit{mass of water}}{\textit{total mass of the product}} \tag{8.2a}$$

$$y_s = \frac{\textit{mass of solids}}{\textit{total mass of the product}} \tag{8.2b}$$

Since in expression operations the amount of liquid (water or oil) of the expressed solid (also known as cake) varies, it is more convenient to express these concentrations in a dry solid basis, expressed by a capital letter, which for water is calculated using the following transformation:

$$Y_w = \frac{y_w}{y_s} = \frac{y_w}{1 - y_w} \tag{8.3}$$

It is important to distinguish two types of water in these plant structures. One type is the water that is expressible by the action of mechanical forces or pressure whereas the second type is so tightly bound to the solid that it cannot be separated by the sole action of forces or pressures; this non-expressible water can be separated only by drying (Schwartzberg, 1997). The amount of non-expressible water can vary depending on the material; for example, for fruits and vegetables, cell wall material can adsorb roughly 0.25–0.35 kg water/kg dry solids whereas for wood fibers it can be of the order of 0.40 kg water/kg dry solids. Another parameter of importance in expression operations is the porosity of the material, which is defined as the fraction of volume not occupied by solids. As illustrated in Figure 8.6c, two volumes containing liquid (or a gas) can be considered. One is the volume occupied by the liquid and/or gas inside the cells/particles and the other is the volume of the liquid and/or gas contained in the interparticles space of the cake. Based on those volumes, two porosity parameters can be defined, one based on the cell/particle volume, ε', and the other based on the interparticle volume, ψ, defined by Schwartzberg (1997) as:

$$\varepsilon' = \frac{Liquid/gas\ volume\ in\ cells}{Total\ volume} \quad (8.4a)$$

$$\psi = \frac{Interparticle\ volume}{Total\ volume} \quad (8.4b)$$

The overall porosity or fraction of material volume occupied by pores (involving both the pores inside the cells and the interparticle volume) can be estimated as:

$$\varepsilon = \frac{Volume\ of\ liquid/gases}{Total\ volume} = \varepsilon' + \psi \quad (8.5)$$

whereas the fraction of the solids in the material, known as solidosity, can be determined as:

$$\varepsilon_s = \frac{Solids\ volume}{Total\ volume} = 1 - \varepsilon \quad (8.6)$$

As already stated, the space volume no occupied by the solid can contain liquids and gases. If s is a parameter measuring the saturation of the sample, the fraction of the material occupied by gases can be calculated as (Schwartzberg, 1997):

$$\frac{V_{gas}}{Total\ volume} = (1 - s)\varepsilon \quad (8.7)$$

When the cake is deformed, the liquid moves out of the cells and the interparticle space, and the flow direction depends on the type of cake compaction. In a screw extruder-press, compaction of the solid is produced by the action of a diminishing pitch in the axial direction and also by fitting a restriction to the outflow through a valve, which increases the pressure in the cake, thus expressing out the water (or oil) contained in the inter- and intraparticle pores. The outflow of the liquid is radial and comes out through the opening of the barrel. The semi-solid cake in the screw extruder-press follows spiral trajectories such as those described in Chapters 3 and 5, generating a pressure gradient along these trajectories which produces the radial outflow (Schwartzberg, 1997). As explained in Chapter 3, most of the time the pressure gradient created in the extruder channel is positive, which may induce an axial backflow in the screw extruder-press, especially in zones where pressure gradients are very high, whereas higher radial flow through the barrel perforations occurs in regions where the cake is less compacted. During the pressing operation and as a consequence of the compaction of the cake, the volume of liquid and the volume of the cake itself change. Thus, it is convenient to define variables based on the volume of the non-expressible solids, which are constant in the expression operation. The porosity ratio e defined below expresses the fraction of pores in the cake per unit volume of dry solids:

$$e = \frac{\varepsilon}{1 - \varepsilon} \quad (8.8)$$

During compaction in a screw extruder-press, the volume of cake per kg of dry solids changes from an initial value V_o to a final value V_f, and consequently the corresponding values of the porosity ratios change from e_o to e_f. For materials of biological origin, the liquid being expressed is initiated from both the inside of the cells/particle and the interparticle space so it is difficult to assess what part of the liquid is associated to the internal porosity ψ and what to the interparticle porosity ε'. Hence, to characterize the process, it is preferable to use the volume of the cake per unit of dry mass, which can be easily measured. Regardless of the parameters used to characterize the materials obtained in the expression process, they are related by equations that can be derived considering simple mass balances. Schwartzberg (1997) gives relationships to interconvert the different parameters used in expression operations, which are reported in Table 8.2. For example, if it is necessary to estimate

Table 8.2 Inter-conversion between parameters used to characterize changes in the properties of the cake during expression of liquid laden solids. Adapted from Schwartzberg (1997).

Parameter	ε	e	V
ε	-	$\frac{\varepsilon}{1-\varepsilon}$	$\frac{\varepsilon}{(1-\varepsilon)\rho_s}$
e	$\frac{e}{1+e}$	-	$\frac{1+e}{\rho_s}$
V	$\frac{V\rho_s - 1}{V\rho_s}$	$V\rho_s - 1$	-

Source: Adapted from Schwartzberg 1997.

the overall porosity ε from the volume of the cake per kg of dry solids V, the derivation of the equation starts with the basic definition of the parameter as:

$$V = \frac{V_{cake}}{m_s} = \frac{V_{pores} + V_s}{m_s} \tag{8.9}$$

But the porosity is defined as:

$$\varepsilon = \frac{V_{pores}}{V_{cake}} = \frac{V_{pores}}{V_{pores} + V_s} \tag{8.10}$$

By combining Eqs 8.9 and 8.10, the following relationship is obtained:

$$\varepsilon = \frac{Vm_s - Vs}{m_s V} = \frac{V\rho_s - 1}{\rho_s V} \tag{8.11}$$

which corresponds to the expression given in Table 8.2 to obtain the cake porosity ε (ε - column from Table 8.2) from the value of the cake volume per kg of dry mass V (V - row of Table 8.2) and ρ_s is the density of the solid (m_s/V_s).

Typical values of volume for different types of cakes at the start and end of the expression process as well as the corresponding porosity ratios are given in Table 8.3. Other volumes can be estimated based on these initial parameters. For instance, the volume of the solids per kg of dry matter in the particles can be calculated as:

$$V_p = V_o(1 - \varepsilon') \tag{8.12}$$

where V_p is the volume of the solids forming the particle wall material per kg of dry matter. As mentioned, these particles also contain liquids and gases. During compaction of the cake, the gas is first expelled until the cake becomes fully saturated. The volume of the cake per kg of dry mass at that saturation condition is designated as V_c, and depends on the saturation level s. In particular, $V_c = V_o$ when the fractional saturation is $s = 1$, i.e. when the particle is initially fully saturated. Liquids tend to stick in drained cakes and in general $V_c > V_p$ and also $V_c < V_o$. As mentioned, some materials may contain quantities of non-expressible water, or oil in oilseeds. The volume of the solids plus the non-expressible water (or oil) per kg of mass of dry solids, $V_{non\text{-}exp}$, can be calculated by the following equation:

$$V_{non\text{-}exp} = V_{avg} + \frac{Y_{non\text{-}exp}}{\rho_w} \tag{8.13}$$

where V_{avg} is an average volume of the solids in the cake per unit mass of the dry solids, $Y_{non\text{-}exp}$ the mass fraction of non-expressible water based on the dry solids and ρ_w the density of pure water at the processing temperature. For oilseeds, the water mass fraction is replaced by the oil mass fraction and the density of water by the oil density. Values of non-expressible water, which can be considered as the residual water remaining in the solid after the expression operation, are also included in Table 8.3. To facilitate further discussions, some definitions and abbreviations have been defined by Schwartzberg (1997) and will be utilized in this chapter. Insoluble solids that cannot pass through the filter media are called fixed solids. It is important to differentiate these solids from the ones present in the expressed liquid, which can contain dissolved solids and also finer solids not retained in the filter media. Those solids that

Table 8.3 Typical values of the cake volume per unit of dry mass and the corresponding porosity ratio for a variety of expressible solids (adapted from Schwartzberg, 1997).

Material	V_o	V_f	e_o	e_f	$V_{non\text{-}exp}$
	m^3/kg dry solid			-	m^3/kg dry solid
Fruits	$19–60 \times 10^{-3}$	$1.7–11 \times 10^{-3}$	28–90	1.5–15	$1.1–1.25 \times 10^{-3}$
Oilseeds	$3.3–6 \times 10^{-3}$	$0.74–0.83 \times 10^{-3}$	4–8	0.06–0.16	0.74×10^{-3}
Sugar cane	$10–11.3 \times 10^{-3}$	3.8×10^{-3}	15–17	4.7	–
Waste treatment sludge	$9–16 \times 10^{-3}$	$3.5–6 \times 10^{-3}$	13–23	3–5.3	–
Kraft pulp	$4–4.3 \times 10^{-3}$	2.2×10^{-3}	5–5.5	2.3	$1–1.85 \times 10^{-3}$
Moist soils	$0.7–0.9 \times 10^{-3}$	$0.5–0.6 \times 10^{-3}$	0.7–1.3	0.25–0.5	0.5 × 10–3
Swellable clays	$1.1–2.4 \times 10^{-3}$	$0.6–1.3 \times 10^{-3}$	1.7–5	0.45–2.3	–

Source: Adapted from Schwartzberg 1997.

are not retained by the media, but that can be retained if a filtering medium with finer openings is used, are called foots.

8.2.2.2 Pressures and stresses in the cake and the expressible liquid

During compaction of the expressible biological material, the cells containing liquids and gases deform and eventually rupture, releasing the liquid that flows through the cake and eventually through the filtering media. The deformation of these cells, their rupture and the flow of liquids through the cake are governed by two different pressure drops that are associated with the solid and the liquid phases and designated as Δp_s and Δp_f respectively. The total pressure drop in the system can be obtained from the addition of these specific pressure drops as:

$$\Delta p_T = \Delta p_f + \Delta p_s \tag{8.14}$$

Based on a conceptual visualization developed by Terzaghi on the consolidation of soil and the flow of liquid during uniaxial compression, Schwartzberg (1997) explained the relationship between these pressure drops generated by the applied force of a Universal Testing Machine (UTM) on a bed of a saturated solid. The liquid-saturated solid is represented by a series of perforated plates, which are separated by liquid layers (representing the liquid phase) that contain parallel springs representing the solid phase of the material (Figure 8.9). Plates are perforated by small and constant-diameter circular channels communicating with the liquid layers. When a force is applied, initially it is supported by the liquid, and a pressure drop Δp_f is generated to force the liquid to flow through the channels in the plates and out of the bed. While the bed is saturated, the counterpressure provided by the springs does not change, i.e. $\Delta p_s = 0$ and $\Delta p_T = \Delta p_f$. When part of the liquid leaves the space between plates, further increases of the compression force start to deform the springs, generating a pressure drop Δp_s that balances the total pressure, i.e. at any instant after the liquid starts to flow, continuous increases in Δp_s and decreases in Δp_f are observed, which are predicted by Eq. 8.14. When all the liquid contained in the solid is expressed out, $\Delta p_f = 0$ and the total pressure is now given by the deformation of the springs, thus $\Delta p_T = \Delta p_s$. In unsaturated solids, initially $\Delta p_f = 0$ but not Δp_s, and when the gas contained in the solid is expressed out the value of the Δp_f starts to increase. Other more complex rheological models based on springs and dashpots representing the pure solid and liquid phases respectively have been used to describe the viscoelastic behavior of cellular biomaterials (e.g. Lanoisellé et al., 1996; Rebouillat, 2002; Schwartzberg, 1997).

Equation 8.14 applies in all parts of saturated cakes, but local pressures in both the solid (p_s) and liquid (p_f) phases change during liquid flow and solid compaction as a consequence of the external load; and at any instant $p_T = p_s + p_f$. It is important to note that pressures in liquids are isotropic, i.e., local p_f in the interparticle liquid acts uniformly on the particle surfaces. Conversely, stresses in the solid are not isotropic and depend on the direction. Specifically, p_s is the axial component of the stress acting on the solid particles present in the cake with a transverse component $p_h = Bp_s$, where B is defined as an analog to the Poisson ratio (Schwartzberg,

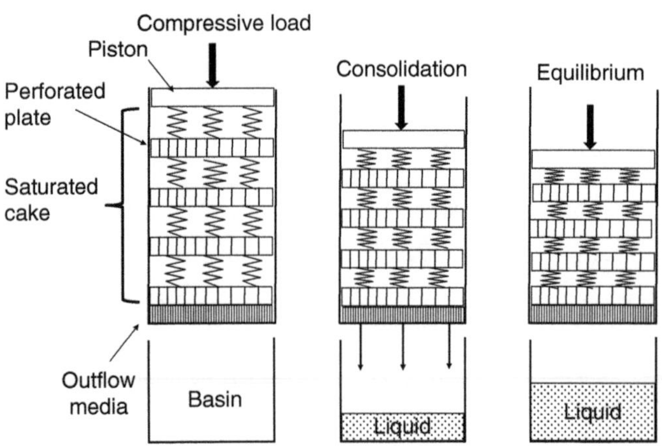

Figure 8.9 Schematic illustration of the Terzaghi–Peck model to describe the consolidation of cellular materials. Source: Adapted from Schwartzberg 1997.

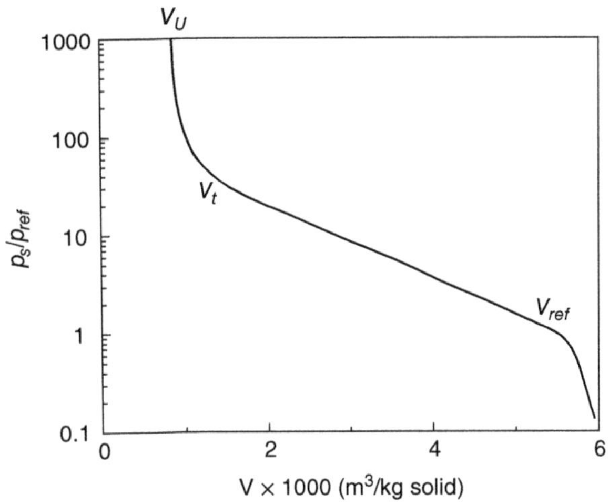

Figure 8.10 Relationship between the solid pressure in the cake and its volume. V_U is the upper limit volume at which the solid pressure becomes very large. V_t is the volume of the cake at which the relationship becomes linear. V_{ref} and p_{ref} are reference values for volume and pressure respectively.

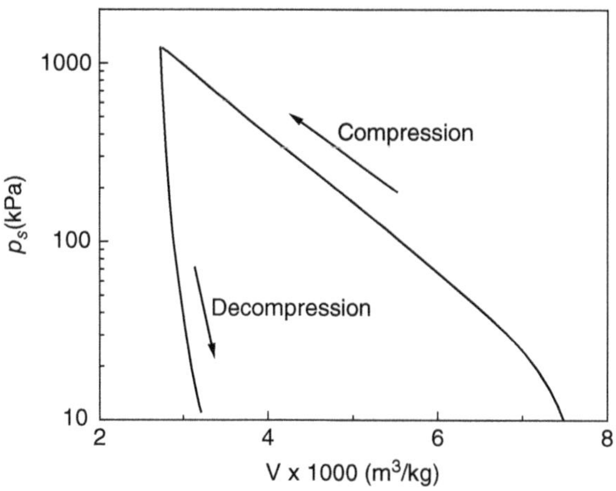

Figure 8.11 Hysteresis phenomenon observed in compaction and expansion of a cellular material cake. Source: Adapted from Schwartzberg 1997.

1997). Since the solid particles in the cake occupy a volume $1-\varepsilon'$ and:

$$1-\varepsilon' = \frac{V_s}{V_{total}} \approx \frac{A_s}{A_{total}} \tag{8.15}$$

the axial pressure corrected by the area of the solid particles, or true axial stress, can be then calculated as:

$$\frac{p_s A_{total}}{A_s} = \frac{p_s}{1-\varepsilon'} \tag{8.16}$$

Local true axial stresses act locally on the particles and when the particle surfaces press on one another, the true axial stress is larger than $p_s/(1-\varepsilon')$. The axial stress component p_s changes when the cake is compressed and its volume decreases following the non-linear trend illustrated in Figure 8.10. The trend can be represented by the following equation (Schwartzberg, 1997):

$$\ln \frac{p_s}{p_{ref}} = K\left(V_{ref} - V\right) \tag{8.17}$$

Although Eq. 8.17 has been used to describe the deformation of many cellular biological materials such as coffee grounds, protein curds and sugar beets by the action of pressures, it does not apply to all the pressing range. At extremely low cake volume, identified as V_U in the figure, the axial stress p_s increases more rapidly than what is predicted by Eq. 8.17. Similarly, for initially no compacted cakes i.e. $p_s = 0$ when $V = V_o$, Eq. 8.17 would not provide a reasonable prediction. Thus, Eq. 8.17 can predict the variation of the axial stress p_s in a range of cake volume between V_t and V_{ref} (volumes indicated in the figure), whereas real experimental values extending out of that range, which are not predicted by equation 8.17, are also included in the figure.

The other singular characteristic of cellular biomaterials compared to mineral materials is that the compression of cellular biomaterials is not a reversible process. Figure 8.11 schematically illustrates changes in the axial stress component when the volume of the cake is subjected to compaction and expansion showing significant hysteresis. Hysteresis which is attributed to friction between particles during cake compaction (Schwartzberg, 1997).

8.2.3 General description of the filtration and consolidation processes

In many ways the pressing operation of biological materials is similar to the filtration of highly compressible solids. In both cases, the liquid flows through a solid cake, following patterns that can be described by the same fluid mechanics governing equations. In the filtration operation, the flow-induced pressure drop causes compaction of the cake in the flow direction and an increase in cake filtration resistance. As a consequence, large pressure drops are necessary to overcome the high resistance of the cake. A similar mechanism occurs during liquid expression where a larger pressure is necessary to express the liquid from a more compacted cake.

During liquid filtration, the liquid flows through the solid cake and there is significant friction, which is assumed to be proportional to the velocity of the liquid relative to the solid. Tiller and Cooper (1960) developed a model to estimate a variable liquid flow through the cake during filtration. However, it was assumed that the solids in the cake do not move, which often is not the case, especially in the filtration operation. Shirato et al. (1969) extended this model to include cases in which the solid can also move during formation of the cake and the consolidation process. Tiller and Shirato (1964) related these internal flows with the pressure drop in the filtration operation. These filtration models were used to describe the expression of liquids from slurries and compactable beds of particles which are individually incompressible but that can accommodate during compaction (Shirato et al., 1970, 1971). A mechanism describing the formation of the cake was thought to be initiated by a filtration stage created by the displacement of a concentrated slurry by a moving piston. At the point where the slurry was no longer existing, i.e. when the piston had achieved the surface of the deposited cake, further displacement of the piston generated a consolidation stage during which the liquid was squeezed out and solid particles were reaccommodated in the bed. The liquid expression process is different; rather than dealing with a slurry, the cake is a cellular material that has little interstitial liquid as all the liquid is contained in the cells. The cake must go through compaction to be able to rupture the cell and fill the interstitial space with the liquid before the liquid can be expelled from the cake.

The other important difference between filtration and expression is that the systems modeled by the filtration theories developed by Tiller and Shirato apply to incompressible solids, whereas biological materials are compressible and deformable. Deformability of the material is essential to promote the flow of liquid from the interior of the cells to the interstitial space and eventually through the outflow media. Thus, the overall axial stress is highly influenced by the force required to break the cells and produce the liquid exudation, which is a dynamic process that sometimes cannot be detected by equilibrium consolidation pressure measurements such as those performed by Shirato and collaborators (Schwartzberg et al., 1977).

Although filtration and liquid expression exhibit some similarities and differences, often they are complementary unit operations because after a cake is produced by filtration, the liquid is then further expressed by mechanical pressure. Thus, the final cake structure of the filtration process is the initial condition for the expression process (Tiller & Yeh, 1987). Mathematically, the difference between filtration and liquid expression is related to the process conditions. In filtration, it can be assumed that the local flow rate is a sole function of time and independent of distance, i.e. the rate of change of porosity with time is small, which results in ordinary equations governing the flow of liquid through the cake whose solutions enable the derivation of relationships between time, filtrate volume, cake thickness, porosity, and pressure (Tiller & Yeh, 1987). In contrast, in the liquid expression process the squeezing of the compressible cake is essential for the exudation of the liquid, but cake porosities vary with position in the cake. As a result, the equations governing become complex.

The compaction and flow processes during liquid expression have been examined by studying the compaction and liquid flow of a cake contained in a cylinder with impermeable side walls and pressed between an impermeable flat piston and a flat filter medium in a uniaxial compression test. Figure 8.12 illustrates different compression stages from an initial height Z_o to a final height Z_f. Both the volume and the height of the cake vary during the process; assuming that foots are not lost with the expressed liquid, the mass of fixed solids per unit of cross-sectional area, designated by W, remains constant. Thus, an equation relating the height of the cake can be considered if the specific volume of the cake or the volume of the cake per kg of fixed solid V is used. The relationship between the height and the volume of

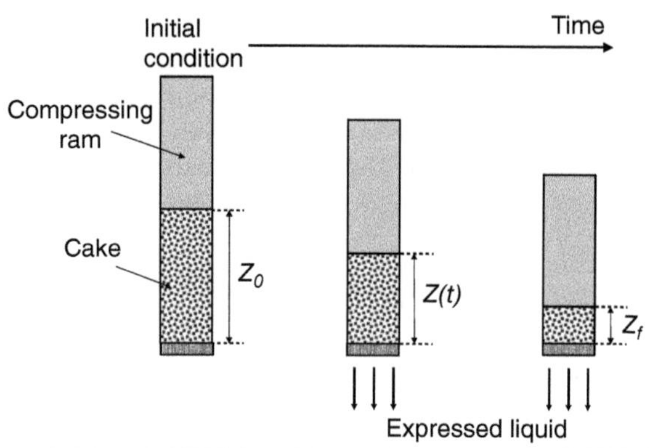

Figure 8.12 Schematic of the liquid expression from a cellular material undergoing uniaxial compression.

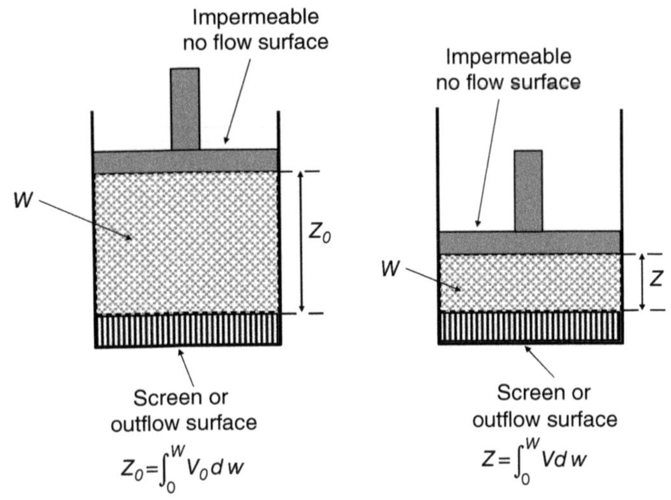

Figure 8.13 Changes in the height Z during the compression of a cellular material. Relationship between the Z and the W co-ordinates.

the cake is expressed by the following equations (Figure 8.13):

$$Z_0 = \int_0^W V_o dw \quad \left(\frac{cake\ volume\ m^3}{kg\ dry\ solid} \cdot \frac{kg\ dry\ solid}{cake\ area\ m^2} = cake\ height\ m \right) \tag{8.18a}$$

$$Z = \int_0^W V\, dw \quad \left(\frac{cake\ volume\ m^3}{kg\ dry\ solid} \cdot \frac{kg\ dry\ solid}{cake\ area\ m^2} = cake\ height\ m \right) \tag{8.18b}$$

An average specific volume for the cake, $\bar{V}$, can be defined as:

$$\bar{V} = \frac{Z}{W} \tag{8.19}$$

By substituting Eq. 8.19 into Eq. 8.18b, the following equation is obtained:

$$\bar{V} = \frac{1}{W} \int_0^W V dw \tag{8.20}$$

During a deformation test where either a constant compaction rate is set by a UTM crosshead speed or by setting a constant pressure (or load) test, the compaction rate dZ/dt can be evaluated by the following equation:

$$\frac{dZ}{dt} = \int_0^W \frac{\partial V}{\partial t} dw \tag{8.21}$$

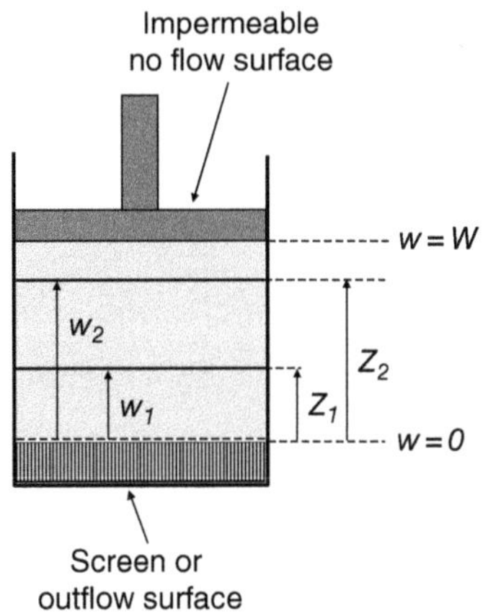

Figure 8.14 Schematic illustrating the definition of the variable w, which is measured from the outflow surface to a fixed position in the cake.

The choice of the variable w can be made using a Lagrangian approach to study the consolidation stage of the expression process, in which the deformation of a "marked" portion of the solids is followed during the process. Figure 8.14 schematically illustrates the definition of the variable w, which is measured from the outflow surface to a fixed position in the cake. Thus, $w = 0$ at the outflow surface and $w = W$ at the impermeable no-flow surface (see Figure 8.14). These Lagrangian co-ordinates can be used to consider the balance of mass in a differential element of the cake dz (or dw) where only the liquid flow into a defined fixed solid

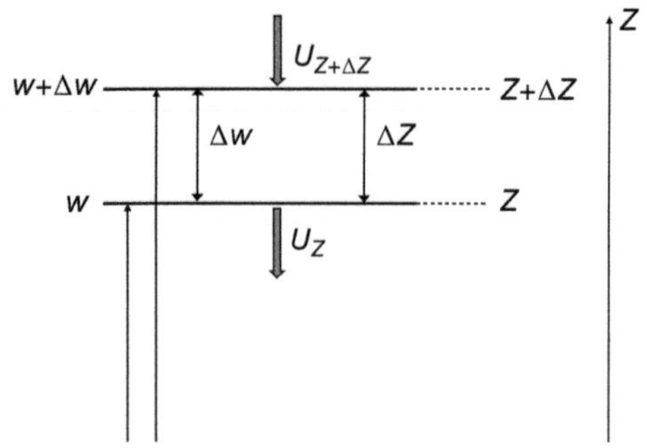

Figure 8.15 Differential region of the cake where balance of mass for the liquid is considered.

region is considered. Figure 8.15 schematically illustrates a differential region of the cake where a balance of mass for the liquid is considered.

$$(AU_{Z+\Delta Z} - AU_Z)\Delta t = \Delta V_{liquid} = A(Z + \Delta Z - Z) \quad (8.22)$$

Considering $\Delta t \rightarrow 0$ and $\Delta Z \rightarrow dZ$ the above equation becomes:

$$dU = \frac{dZ}{dt} \quad (8.23)$$

Substitution of Eq. 8.18b into Eq. 8.23 yields:

$$\frac{\partial U}{\partial w} = \frac{\partial V}{\partial t} \quad (8.24)$$

U is the local velocity of the fluid relative to the fixed solids, which can be obtained by integration of the above equation at a particular location defined by the variable w:

$$U = \int_0^w \frac{\partial V}{\partial t} dw \quad (8.25)$$

The solid pressure p_s and the fluid pressure p_f can be estimated as a function of the co-ordinate w. It has been assumed that these pressures are constant along a line of constant w (see Figure 8.14), i.e. they do not vary in the lateral direction. The assumption is fairly realistic provided there is negligible friction between the cake and the wall of the container around it (Tiller & Lu, 1972). The pressure change in the fluid in terms of the variable w can be written as (Schwartzberg, 1997):

$$\frac{dp_f}{dw} = \alpha\mu U \quad (8.26)$$

where α is the specific local resistance of the cake having units of m/kg and μ is the viscosity of the expressed liquid. The above expression is equivalent to the Darcy's law applicable to the flow of a liquid through a porous medium. The equation predicts a positive pressure drop in the liquid phase at increasing distances from the outflow surface (i.e. surface $w = 0$ in Figure 8.14). The balance of forces $dp_f = -dp_s$ enables to obtain a similar equation for the changes in pressure in the solid phase as:

$$\frac{dp_s}{dw} = -\alpha\mu U \quad (8.27)$$

By partially differentiating the above equation with respect to the variable w, the following equation is obtained:

$$\frac{\partial U}{\partial w} = -\frac{1}{\mu}\frac{\partial}{\partial w}\left[\frac{1}{\alpha}\frac{\partial p_s}{\partial w}\right] \quad (8.28)$$

which after substitution of Eq. 8.24 yields:

$$\frac{\partial V}{\partial t} = -\frac{\partial}{\partial w}\left[\frac{1}{\alpha\mu}\frac{\partial p_s}{\partial w}\right] \quad (8.29)$$

As discussed before and illustrated in Figure 8.10, p_s is assumed to be a sole function of the specific cake volume V, hence Eq. 8.29 can be changed to:

$$\frac{\partial V}{\partial t} = -\frac{\partial}{\partial w}\left[\frac{1}{\alpha\mu}\frac{dp_s}{dV}\frac{\partial V}{\partial w}\right] = \frac{\partial}{\partial w}\left[C\frac{\partial V}{\partial w}\right] \quad (8.30)$$

where the coefficient C is defined as:

$$C = -\frac{1}{\alpha\mu}\frac{dp_s}{dV} \quad (8.31)$$

Similar equations were derived by Körmendy (1974), substituting the cake volume V by the cake porosity ε and the porosity ratio e, both defined above, and showing that all the equations have the same form. Thus, solutions of these equations (e.g. Eq. 8.30 for volume or their equivalent equations if V is replaced by ε or e) would provide information on local changes in the cake characteristics, in terms of their volume, porosity and porosity ratio, as a function of the time under which a load is applied. For the special case where the parameter C is constant, Eq. 8.30 simplifies to:

$$\frac{\partial V}{\partial t} = C\frac{\partial^2 V}{\partial w^2} \tag{8.32}$$

Equation 8.32 is identical to the equation describing conductive heat and diffusive mass transfer in solids so solutions to these problems can be adapted to solve the above equation. Schwartzberg (1997) gives the solution of the above equation when the volume of the cake is initially uniform and equal to V_o, the coefficients involved in the boundary conditions at $w = 0$ and $w = W$ are constants, and the resistance of the media to liquid flow is negligible:

$$\frac{V_0 - V}{V_0 - V_\infty} = 1 - \sum_{i=1}^{\infty} \frac{8}{\pi^2 (2i-1)^2} \exp\left(\frac{(2i-1)^2 \pi^2 Ct}{4W^2}\right) \tag{8.33}$$

where V_∞ is the equilibrium volume of the cake.

A similar equation can be obtained to describe the local solid pressure as a function of time. By assuming that p_s is only a function of the cake volume, i.e. $p_s(V)$, Eq. 8.29 becomes:

$$\frac{\partial p_s}{\partial t} = -\frac{dp_s}{dV}\frac{\partial}{\partial w}\left[\frac{1}{\alpha\mu}\frac{\partial p_s}{\partial w}\right] = C\left[\frac{\partial^2 p_s}{\partial w^2} - \frac{1}{\alpha}\left(\frac{d\alpha}{dp_s}\right)\left(\frac{\partial p_s}{\partial w}\right)^2\right] \tag{8.34}$$

For the case where the specific resistance of the cake α does not depend on the pressure and C is constant, Eq. 8.34 simplifies to:

$$\frac{\partial p_s}{\partial t} = C\frac{\partial^2 p_s}{\partial w^2} \tag{8.35}$$

The assumption of constant values of C is a good approximation for non-biological solids, but it may fail when applied to biological materials where 20–50-fold changes are observed in values of C for pressing casein curds, cranberries, and spent sugar beet pulps (Rebouillat et al., 1993; Schwartzberg, 1997). For applications in which the parameter C is not constant, the solution of Eq. 8.30 has to be performed numerically provided the local variation of the coefficient C with position is known. Schwartzberg (1997) summarized numerical methods used to solve the relevant equations as well as relationships used to characterize the specific cake resistance with solid pressure and cake volume for a number of biological materials.

8.2.4 Rheological properties of cellular biological materials and their characterization

The structure of cellular materials, including selected biological systems such as fruits and vegetables, was reviewed by Gibson (2012) and has been associated with their mechanical behavior. The structure of those materials shows the presence of regular polyhedral cells with thin walls. These cells are filled with water or solutions and are densely packed together, forming a structure resembling liquid-filled closed-cell foams. Stress–strain curves obtained from compression tests have shown typical non-linear behavior and compression-relaxation tests displayed the viscoelastic nature of these materials. Results shown in Figure 8.9 indicating the deformation (volume) of the cake as a function of solid pressure resulting from an externally applied load show a behavior similar than that observed in porous solids when stress is plotted as a function of dimensionless cake volume defined as $(V-V_o)/V_o$.

A typical plot, known as Ashby's map, of these cellular solids is shown in Figure 8.16. The figure clearly shows a flat response in terms of stress at the intermediate deformation region that depends on the ratio between the volume of the solids and the initial volume of the cake (V_s/V_0), which can be described by Eq. 8.1. Cells with thicker cell walls are characterized for larger V_s/V_0, which are the ones showing the initial steepest change in axial stress upon compression. Materials having high V_s/V_0 ratios also exhibit an almost elastic behavior whereas those with lower V_s/V_0 ratios exhibit either a plastic or viscoelastic behavior. Materials with intermediate V_s/V_0 values exhibit cell bulking and densification during compression (Figure 8.16).

The viscoelastic behavior of cellular materials has been studied using compression-relaxation tests performed in universal testing machines. The test is based on compression of the sample at a given rate to achieve a deformation or strain. After the set strain is achieved, it is maintained for a given time to study the stress relaxation of the sample. Purely solid materials do not show relaxation of the stress, i.e. the stress is maintained as long as the applied strain is applied. Conversely, purely liquid materials have an instantaneous stress relaxation, clearly indicating that liquids cannot sustain the applied strain. In practical terms, it can be stated that the components forming the material relax to an equilibrium state immediately. Viscoelastic materials exhibit a behavior situated between these two ideal behaviors. More liquid-like viscoelastic materials relax faster to an equilibrium or a zero stress but still within a finite time. Compression-relaxation tests of cellular materials and

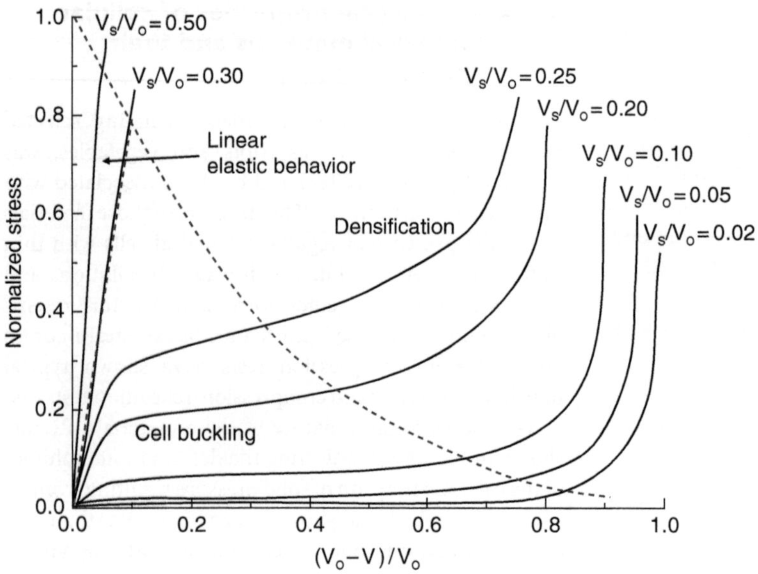

Figure 8.16 Modified Ashby compressive stress–strain map for cellular materials. Source: Adapted from Schwartzberg 1997.

particularly biomaterials demonstrate this characteristic behavior.

The relaxation phenomenon observed with cellular materials is not well understood but has been attributed to two different mechanisms. One is the decay of pressure inside the cells as the liquid is coming out and the other is an equalization of internal stresses and strains in the press-cake due to cake movement within the testing cylinder (Schwartzberg et al., 1977). Although no quantitative explanation has been obtained to explain the stress relaxation phenomenon, different materials exhibit specific stress relaxation behaviors as illustrated in Figure 8.17 during testing of a sponge material, coffee grounds, and apples. Different relaxation behaviors have also been observed in materials such as peat, in which the dewatering process was able to modify not only the relaxation behavior of the material but also its compressibility curve, observed in terms of the density of the sample as a function of the applied normal stress (Rebouillat et al., 1987).

During constant stress/load tests, deformation of the expressible and compressible cake exhibits a typical viscoelastic behavior known as creep which has been described in the literature using various mechanical analogies such as springs (representing solid behavior) and dashpots (representing liquid behavior). The same approach has been utilized in classic rheology to study the mechanical properties of viscoelastic materials. However, when characterizing materials that will be deformed in liquid pressing operations, the loss of the expressed

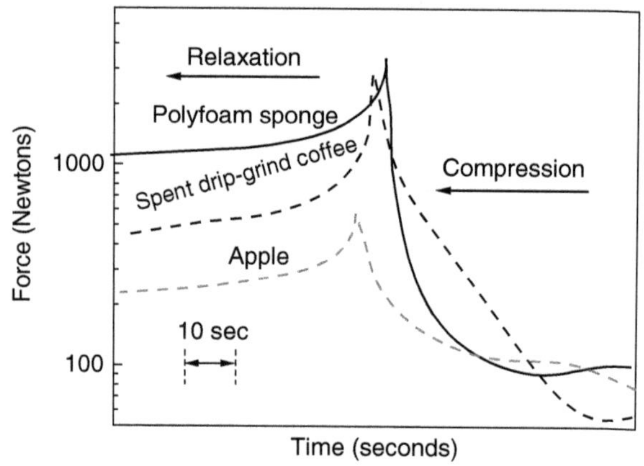

Figure 8.17 Rheological properties of cellular biological materials. Compression-relaxation cycles performed in a Universal Testing Machine. Source: Adapted from Schwartzberg et al. 1977. Reproduced with permission of John Wiley & Sons.

liquid from the material must be taken into account. Unlike non-biological materials, compaction of cellular biological materials causes considerable liquid outflow from the particles. This liquid outflow causes a larger reduction, or deformation, of the cake volume, which must be considered. Thus, for modeling the viscoelastic behavior of cellular biological materials, springs and dashpots models have been combined with mechanical elements that consider the liquid outflow, e.g. using perforated dashpots (Schwartzberg, 1997).

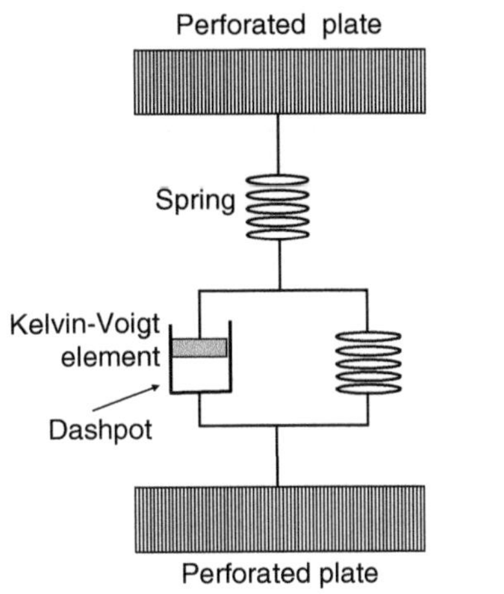

Figure 8.18 Rheological model to describe a cellular material. Kelvin–Voigt modification of Terzaghi–Peck model. Perforated plates are used to simulate the outflow of the expressed liquid.

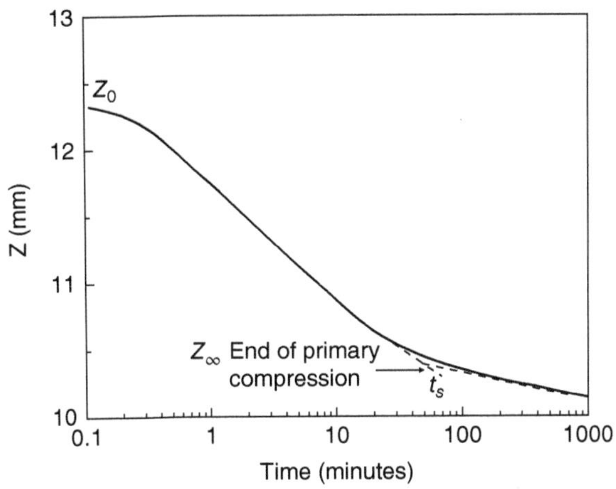

Figure 8.19 Typical plot of creep behavior of a cellular material in terms of changes in the cake height as a function of time.

Figure 8.18 illustrates a typical mechanical model to describe the viscoelastic behavior of cellular biological materials where the liquid outflow through the plates deforming the material is incorporated. The model consists of a spring, having a constant k_1, in series with a Kelvin–Voigt element consisting of a spring with a constant k_2 and a dashpot with viscosity η in paralell. The mathematical expression describing the model in terms of changes in the cake volume with the applied load is given by the following equation:

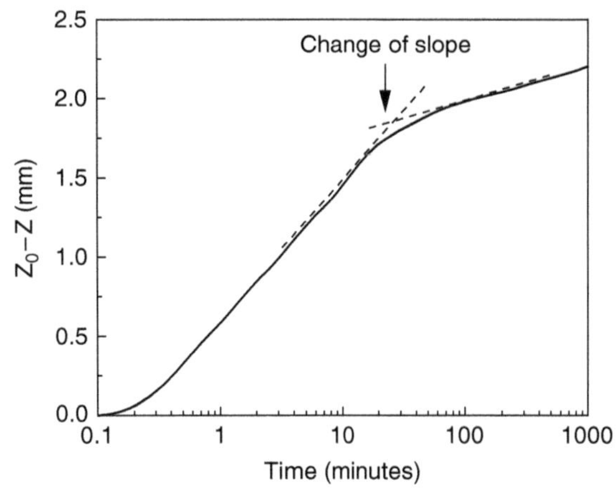

Figure 8.20 Transformed creep data in terms of $Z_o - Z$ versus time.

$$V_{ref} - V = \ln\left(\frac{p_s}{p_{ref}}\right) \cdot \left[\frac{1}{k_1} + \frac{1 - \exp\left(-\frac{k_2 t}{\eta}\right)}{k_2}\right] \quad (8.36)$$

Creep analysis has been used in association with some of the models explained above to describe the cake deformation and liquid outflow due to the applied load in the cake. Figure 8.19 shows the typical plot creep behavior of a cellular material in terms of the changes in cake height as a function of time upon imposition of a constant load or pressure. Data can also be presented as the change in cake height, i.e. Z_o–Z, versus time (Figure 8.20). In both cases, the change in the slope indicates the end of the primary compression of the material and the start of a secondary consolidation stage produced by the creep of the material.

The creep behavior of cellular biological materials can be studied using some of the solutions already presented. For example, Eq. 8.33 also applies to the height of the cake. Thus, the equation could be written as:

$$\frac{Z_0 - Z}{Z_0 - Z_\infty} = 1 - \sum_{i=1}^{\infty} \frac{8}{\pi^2 (2i-1)^2} \exp\left(\frac{(2i-1)^2 \pi^2 C t}{4W^2}\right) \quad (8.37)$$

The equation is written in terms of a series which often does not provide results which are practical for analysis and interpretation of rheological data from creep tests applied to these materials. An equivalent, simpler equation has been derived, which provides an estimation within 1% of the above equation when $\frac{Z_0-Z}{Z_0-Z_\infty}<0.62$:

$$\frac{Z_0-Z}{Z_0-Z_\infty}=\frac{V_0-V}{V_0-V_\infty}=2\left[\frac{Ct}{\pi W^2}\right]^{1/2} \tag{8.38}$$

Thus, for these situations the analysis of creep data often involves plotting $\frac{Z_o-Z}{Z_0-Z_\infty}$ as a function of $t^{1/2}$ to visualize departure of the linearity which is generally associated with changes in the mechanisms of consolidation and deformation or other events such as expulsion of entrapped air prior to liquid outflow.

This viscoelastic creep as well as the stress relaxation behaviors discussed above have relevance for liquid pressing operations, especially when screw extruder-presses are used. Viscoelastic effects have a time component that can affect the behavior of the materials when different rates of stresses and strains are applied. In processes like screw extrusion-pressing, stresses and strains are applied suddenly and also change significantly during the course of the process, so they will be largely affected by the viscoelastic properties of the cake material.

8.3 Process modeling

8.3.1 The fluid mechanics of the process and determination of relevant parameters

8.3.1.1 The process model

Vadke et al. (1988) developed a simplified model to describe the expression of oil from oilseeds. The model is based on the schematic diagram shown in Figure 8.21 and a number of simplifying assumptions that include the following:

- the oilseed mass is macerated in the feed section where expulsion of any air contained in these seeds is also achieved
- the development of pressure in the extruded seeds starts at the beginning of the compression section
- the temperature of the extruded seeds remains constant because the heat generated by the shearing action is removed efficiently through the perforated barrel.

The transport of the material in single screw extruders has been discussed in Chapter 3, and for Newtonian melts

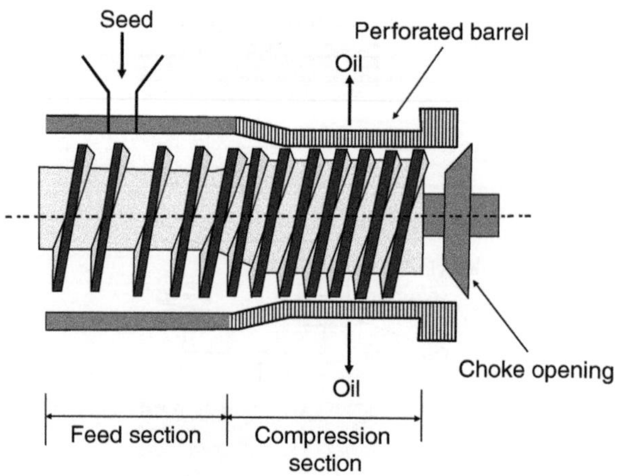

Figure 8.21 Schematic of a single screw extruder-press system.

Eqs 3.81–3.84 were used to describe it. These equations, suitable to describe one-dimensional flow in single screw extruders, can also be applied to describe the flow mechanics in the screw extruder-press operation when they are combined with equations describing the filtration of the expelled liquid through the perforated barrel. The equation describing the transport of the extruded material derived in Chapter 3 is given in a simplified form as:

$$Q_z=\alpha N-\frac{\beta}{\mu_{app}}\frac{dp}{dz} \tag{8.39}$$

where expressions to estimate the parameters α and β are given for Newtonian fluids by Eqs 3.82–3.84 and power law fluids by Vadke et al. (1988); μ_{app} is the apparent viscosity of the cake. Equation 8.39 can be combined with the filtration of the expressed liquid through the cake and the perforated barrel.

Figure 8.22 shows a schematic of the flows involved in the expression operation. A mass balance in a differential section of the screw dz is written as:

$$Q_z\rho_c=(Q_z-dQ_z)\rho_c+Q_r\rho_l \tag{8.40}$$

and:

$$dQ_z\rho_c=Q_r\rho_l \tag{8.41}$$

The above equation assumes that the density of the cake, ρ_c, remains constant in the differential region considered; Q_z and Q_r are the volumetric flows of the cake and the expressed liquid respectively in that region, both

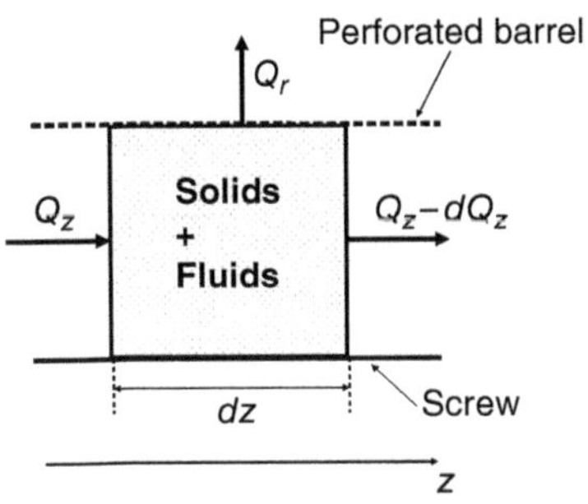

Figure 8.22 Control volume in the channel of the screw extruder-press to analyze the mass balance in that differential volume region.

measured in m^3/s, and ρ_l is the density of the expressed liquid in units of kg/m^3. The radial flow of the expressed liquid can also be estimated from the linear velocity of the liquid u_r through the perforated barrel as:

$$Q_r(z) = u_r A_b = u_r \pi D dz \tag{8.42}$$

That linear velocity can be estimated using the theory of filtration (Geankoplis, 2003) that relates the liquid flow with the applied pressure, the viscosity of the expressed liquid, and the specific filtration of the cake α_c (in units of m/kg):

$$u_r = \frac{\Delta p}{\alpha_c \mu_l m_s} \tag{8.43}$$

where m_s is the mass of solids per unit of barrel area and Δp is the pressure gradient through the barrel. Substituting Eqs 8.41 and 8.43 into Eq. 8.42, the following equation is obtained:

$$\frac{dQ_z}{dz} = \frac{\rho_l}{\rho_c}\pi D\left(\frac{\Delta p}{\alpha_s \mu_l m_s}\right) \tag{8.44}$$

The pressure generated in the screw extruder-press channel, p, increases from the atmospheric pressure at the feed to the final pressure at the die, p_{die}, which can be controlled by a valve or choking the flow of the cake leaving the screw extruder-press (Figure 8.2). At the die, the pressure drop between the die pressure and the atmospheric pressure outside was derived in Chapter 3 (Eq. 3.87) and can be written as:

$$\Delta p_{die} = \frac{1}{K'} m Q_{die}^n \tag{8.45}$$

where m and n are the consistency and the flow indices of the cake assumed to behave rheologically as a power law fluid, and K' is a geometric constant that depends on the geometry of the die-choking system which for different geometries is given in Table 3.1 (refer to Chapter 3). Under steady-state conditions, $Q_{z\_at\ die} = Q_{die}$ and as done for the water component of the cake in Eqs 8.2a and 8.2b, it is possible to define the fraction of oil and the solids in the cake as:

$$y_{oil} = \frac{mass\ of\ oil}{total\ mass\ of\ the\ product} \tag{8.46a}$$

$$1 - y_{oil} = \frac{mass\ of\ solid}{total\ mass\ of\ the\ product} \tag{8.46b}$$

By assuming no loss of foots in the expressing process, the mass of solids in the cake remains constant during the expression process, which is expressed by the following equation:

$$Q_{z,0}\rho_{c0}\left(1 - y_{oil,0}\right) = Q_z \rho_c \left(1 - y_{oil}\right) \tag{8.47}$$

and the oil fraction in the cake at a given axial location of the screw extruder-press defined by the distance z can be obtained from:

$$y_{oil} = 1 - \frac{Q_{z,0}\rho_{c0}}{Q_z \rho_c}\left(1 - y_{oil,0}\right) \tag{8.48}$$

where Q_{z0} is the mass flow rate of the cake at the screw extruder-press entrance having a density ρ_{c0} and an initial mass fraction of oil content $y_{oil,0}$. Q_z and ρ_c are the mass flow rate of the cake and its corresponding density at a given location z.

The solution of the mathematical model is essentially a boundary value problem which Vadke et al. (1988) solved using a trial and error approach. The approach consisted in solving Eqs 8.39, 8.44 and 8.48 progressively in incremental regions of length Δz to calculate values of Q_z, p_z and y_{oil}. A Runge–Kutta fourth-order method was used to simultaneously solve Eqs 8.39 and 8.44 and obtain the values of p and Q_z; the oil fraction y_{oil} was calculated by Eq. 8.48. As viscosity, filtration resistance, and density of the cake are functions of the oil fraction and cake pressure, they were calculated using values of oil fraction and pressure estimated by the numerical integration. The procedure was repeated until the end of the screw extruder-press was reached and the end pressure was used to calculate Q_{die}, using Eq. 8.45. This volumetric flow rate was compared with the volumetric flow rate at the end of the screw extruder-press estimated with the numerical approach. The interaction was finalized when the values of the two volumetric flow rates converged within 2%.

The apparent viscosity of the cake was assumed to be dependent on the shear rate (power law model), the oil content measured by y_{oil}, and the absolute temperature of the process, T, as expressed in the following relationship already discussed in Chapter 4:

$$\mu_c = m\dot{\gamma}^{n-1} \cdot e^{-a y_{oil}} \cdot e^{\frac{b}{T}} \tag{8.49}$$

where a and b are empirical constants. The density of the cake was assumed to be a function of the oil content and estimated by the following relationship:

$$\rho_c = 1451 - 700 y_{oil} \tag{8.50}$$

In the work of Vadke et al. (1988), the mathematical model was validated with experimental results and one of the processing variables used was the die pressure which in practice is controlled with a valve or by the choke opening at the screw extruder-press die. Thus, it is necessary to establish a relationship between the choke opening and the pressure at the die. That relationship between the die pressure and the flow annular area can be estimated by the following equation (Vadke & Sosulski, 1988):

$$\Delta p = \frac{2mL}{R} \cdot \left(\frac{2n+1}{n\pi R^3}\right)^n \cdot \left(\frac{R}{d}\right)^{n'} \cdot Q_z^n \cdot e^{-a y_{oil}} \cdot e^{\frac{b}{T}} \tag{8.51}$$

where L is the length of the choke, R the outer ratio of the choke (annular orifice), d (defined as the choke opening) is the width of the annulus, and n' is an empirical constant. Experiments using canola seeds and a small oilseed press running at different screw speeds, different choke openings (die pressure), and temperatures enabled determination of the empirical parameters involved in Eqs 8.49–8.51 to yield the following operating equations:

$$\mu_c = 2.5 \cdot 10^5 \cdot \dot{\gamma}^{-0.87} \cdot e^{-18.4 y_{oil}} \cdot e^{\frac{1476}{T}} \tag{8.52}$$

$$\Delta p = 2.4 \cdot 10^6 \cdot \left(\frac{R}{d}\right)^{0.27} \cdot Q_z^{0.13} \cdot e^{-18.4 y_{oil}} \cdot e^{\frac{1476}{T}} \tag{8.53}$$

Expressions to estimate shear rates in single screw extruders were derived in Chapter 3 (see Eqs 3.212 and 3.213) and found to depend on the position of the channel. Also for a 1D flow in the direction z, the relevant shear rate can be estimated by Eq. 3.213, which is clearly dependent on the position of the flowing material in the channel and the back pressure (given by the ratio Q_p/Q_d). Vadke et al. (1988) considered a constant shear rate which can be calculated as:

$$\dot{\gamma} = \frac{\pi N D \cos\theta}{H} \tag{8.54}$$

For nomenclature and geometric considerations, refer to Figure 3.17 in Chapter 3. Other physical parameters involved in this mathematical model are those associated with Eq. 8.44 that describes the filtration of the expressed liquid through the cake and the barrel. They include the specific filtration resistance, α_s, the mass of solids in the cake per unit of barrel area, m_s, in addition to the density and viscosity of the expressed oil, designed as ρ_{oil} and μ_{oil} respectively. For canola oil, the density was assumed constant and equal to 910 m^3/kg, whereas its viscosity was assumed to be Newtonian and dependent on temperature as indicated in the equation below:

$$\mu_{oil} = 4.6 \cdot 10^{-3} \cdot e^{\frac{200.9}{T}} \tag{8.55}$$

8.3.1.2 Estimation of specific filtration resistance α_s

Schwartzberg (1997) describes an approach to measure the specific filtration resistance of pressed solids. The apparatus used, called a compression permeability cell, consisted of a hollow cylinder of internal diameter D and cross-sectional area A, where the cake to be tested is placed. The cake is supported by a perforated bottom plate, covered with a filtering medium, which is connected to a container to collect the drained liquid (Figure 8.23). The cylinder can be placed on a press stand where a known load is applied through an impermeable plate. The set-up can be installed on the base of a UTM to apply either a constant load or a constant deformation rate. A sample of a measured weight of solids, M, is placed inside the cylinder and saturated with a liquid. Alternatively, the cylinder can be filled with a slurry containing a weight M of solids. A close-fitting piston fitted with a porous tip and flow channel is inserted into the cylinder and pushed downward until it reaches the top of the cake. Liquid supplied from a constant source is fed into the piston, flows out through its porous tip and the cake, and then out of the cell through the perforated plate and outflow channel to be collected in a graduated cylinder. The test consists of measuring the volume of permeate,

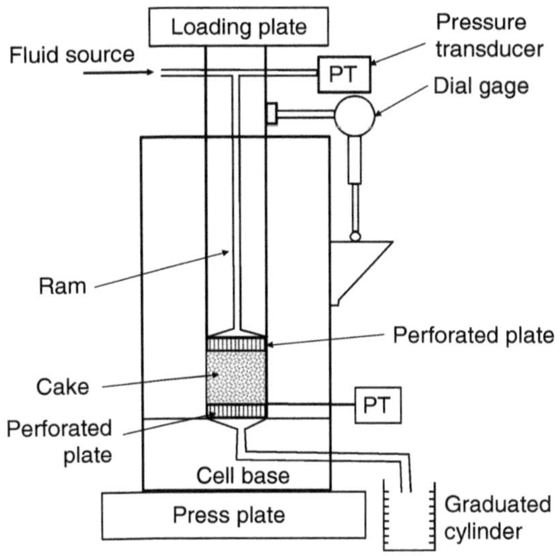

Figure 8.23 Schematic of the compression permeability cell. Source: Adapted from Schwartzberg 1997.

V, over time t at different pressures until the rate $\Delta V/\Delta t$ becomes constant. The basic equation for the filtrate rate in a batch process can be written as (Geankoplis, 2003):

$$\frac{dV}{A\,dt} = \frac{1}{A}Q = \frac{\Delta p}{\mu_{oil}\left(\alpha_s m_s + R_m\right)} \tag{8.56}$$

where $\Delta p = \Delta p_s + \Delta p_f$, R_m is the resistance of the media, $m_s = M/A$ and $Q = \Delta V/\Delta t$. Hence, after rearranging Eq. 8.56 an expression to calculate an average specific resistance for the cake as a function of the measured variables can be obtained:

$$\alpha_s = \frac{\Delta p A}{\mu_{oil} m_s Q} - \frac{R_m}{m_s} \cong \frac{\Delta p A^2 \Delta t}{\mu_{oil} M \Delta V} - \frac{R_m A}{M} \tag{8.57}$$

If the load on the cake is given by the height h of a liquid of density ρ_l, the average specific filtration resistance is calculated as:

$$\alpha_s \cong \frac{\rho_l g h A^2 \Delta t}{\mu_{oil} M \Delta V} - \frac{R_m A}{M} \tag{8.58}$$

In general, relatively small liquid heights are usually used for estimation of the cake specific filtration resistance, e.g. 0.5 m of water, and gas pressurization of the liquid feed is used to provide higher Δp when the value of α_s is very large.

In such cases, Eq. 8.57 is used (Schwartzberg, 1997). The resistance of the media R_m is determined by running the experiment without the cake, i.e. $\alpha_s = 0$ in Eq. 8.56, and thus:

$$R_m = \frac{\Delta p}{\mu_{oil} Q_m} = \frac{\rho_l g h \Delta t A}{\mu_{oil} \Delta V_m} \tag{8.59}$$

where Q_m is the volumetric flow rate without cake present and V_m the corresponding permeate volume. Depending on the applied pressure, cakes can be compressible. For example, for expression of canola oil, Vadke et al. (1988) found that due to cake compressibility, the specific filtration resistance is a function of the applied pressure described by the following relationship:

$$\alpha_s = 4.3 \times 10^9 + 0.8 \times 10^3 p \tag{8.60}$$

During screw extrusion-pressing, the oil must overcome two sources of resistance. One is provided by the individual cell walls through which the oil has to diffuse out, or break the cell to be exuded out. The other resistance is that offered by the matrix of solid particles to the flow of oil through the interparticle porous cake (Vadke et al., 1988). The specific filtration resistance calculated by the method described above only measures the second resistance. Since the first resistance is very difficult to estimate, Vadke et al. (1988) proposed using an effective specific filtration resistance affected by a parameter known as the permeability factor f_{per} defined as:

$$f_{per} = \frac{\alpha_{eff}}{\alpha_s} \tag{8.61}$$

8.3.1.3 Estimation of the mass of solids in the cake per unit filtration area m_s

From the definition of the porosity of the cake, it is possible to estimate the volume of solids in the cake as:

$$Volume\,solids = V(1-\varepsilon) \tag{8.62}$$

Thus, the mass of solids per unit of filtration area is calculated as:

$$m_s = \frac{M}{A} = \frac{\rho_s H_c A(1-\varepsilon)}{A} = \rho_s H_c(1-\varepsilon) \tag{8.63}$$

In terms of the porosity ratio e, defined by Eq. 8.8, the above equation is transformed to:

$$m_s = \frac{\rho_s H_c}{1+e} \tag{8.64}$$

and the porosity ratio can be calculated as:

$$e = \frac{y_{oil}}{1-y_{oil}} \frac{\rho_s}{\rho_{oil}} \tag{8.65}$$

Solution of the model using the material properties defined in this section was obtained by Willems et al. (2009), who were able to show the pressure profile in the screw extruder-press as well as the residual oil content (Figure 8.24). As illustrated in the figure, close to the end of the screw extruder-press the residual oil content becomes negative. Willems et al. (2009) attributed this problem to the fact that Vadke's model assumed that the residual oil content decreased continually as long as pressure is applied to the cake and does not consider an equilibrium condition under which the oil cannot be removed despite the pressure applied. Willems et al. (2009) extended Vadke's model by incorporating an equilibrium residual oil content once the cake reaches the consolidation regime. The consolidation regime was defined based on the operation of a hydraulic press and modified for a screw extruder-press as:

$$U_c(t) = \frac{Z_o - Z(t)}{Z_o - Z_\infty} = (1-B)\cdot\left[1-e^{-\frac{\pi^2 C_e}{4\omega_o^2}t}\right] + B\cdot\left[1-e^{-\frac{E}{G}t}\right] \tag{8.66}$$

where $Z(t)$, Z_o and Z_∞ are the local, initial and equilibrium heights of the cake, B is an empirical parameter, ω_o is the volume of the solids per unit area, and E/G is a rheological property of the cake known as the creep constant, which can be determined from a creep test such as that described in section 8.2.4. The parameter C_e and the time t can be calculated from the following equations:

$$C_e = \frac{1}{\mu_{oil}\rho_s\alpha_s\frac{\partial e}{\partial p}} \tag{8.67a}$$

$$t = \int_0^z \frac{HW}{Q_z}dz \tag{8.67b}$$

where H and W are the screw extruder-press channel height and width respectively. In screw extruder-presses, however, the height of the cake is not determined by the nature of the material or the pressure; instead it is determined by the geometry of the screw. Thus, in the screw extruder-press the consolidation ratio can be determined from the porosity of the cake as:

$$U_c(z) = \frac{(\varepsilon_{eq}-1)\cdot(\varepsilon(z)-\varepsilon_o)}{(\varepsilon_{eq}-\varepsilon_o)\cdot(\varepsilon(z)-1)} \tag{8.68}$$

Solution of the model was done using data from Vadke and Sosulski (1988) on canola seeds having an initial oil content of 43% and a moisture content of 4.3%. The geometry of the screw extruder-press incorporated to the model was that of a small oilseed press (Simon-Rosedowns mini 40 screw press) with a nominal capacity

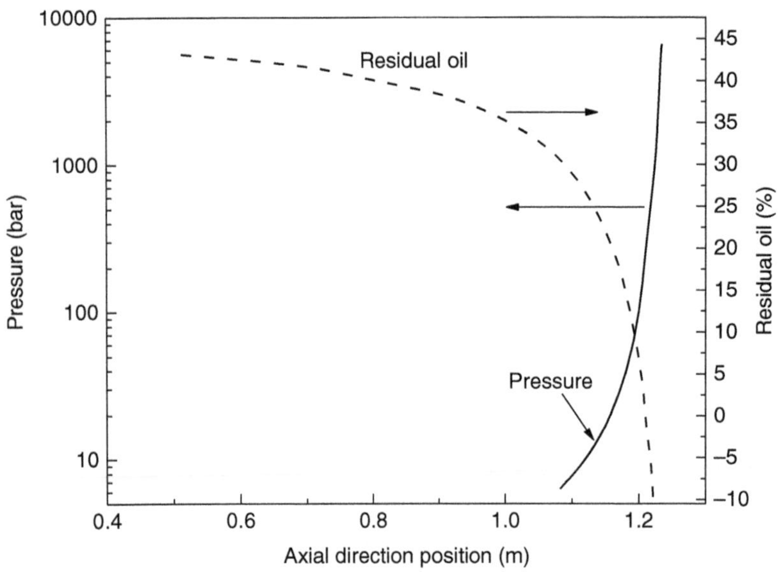

Figure 8.24 Simulated pressure and residual oil profiles in a screw extrusion-pressing operation using Vadke's model. Properties used to solve the model were taken from a screw pressing operation of canola oil in a small oilseed press (Simon-Rosedowns mini 40 screw press) with a nominal capacity of 40 kg seeds/h that was used by Vadke and Sosulski (1988). Source: Adapted from Willems et al. 2009. Reproduced with permission of Elsevier.

of 40 kg seeds/h that was used by Vadke and Sosulski (1988) in the experimental trials, and also used to predict results illustrated in Figure 8.24. Results of the modified Vadke's model are illustrated in Figure 8.25, which clearly shows that at the position near 0.8 m the consolidation of the cake increases and no more oil flow is observed, likely due to the fact that the porosity of the cake reaches a constant value. However, values of throughput used in the simulations (0.01 kg/h) were significantly lower than the nominal capacity of the screw press. The use of more realistic values would help to better assess the capabilities of these mathematical models.

8.3.2 Effects of material properties on the process yield

The properties of the material being pressed have relevance to its processing. Of special interest in oilseed processing is the comparison of using whole or flaked seeds as feed due to the impact that the addition of a flaking process may have on the overall cost of the process. Flaking of seeds may require some preheating and/or tempering/drying steps prior to compressing them between rolls to minimize shattering of seeds during the operation. Thus, it is necessary to assess whether the extra energy used in the flaking process will be compensated by greater efficiency in the pressing process when flakes are used.

As illustrated in Figure 8.26, pressure, throughput, oil content, and residual oil were measured in a pilot screw press operated at different temperatures and using flaked canola and whole seeds as the feed (Vadke & Sosulski, 1988). For all processing temperatures, pressures and throughout were higher when flakes were processed instead of whole seeds (Figure 8.26a). The same behavior was observed, at all temperatures, in the oil mass flow extracted from the screw extruder-press, where the oil extraction was always higher for flaked canola seeds than for whole canola seeds. Both increased with temperature, likely due to the decrease in oil viscosity with temperature increases. In addition, at higher temperatures the cake was likely to be closer to its glass transition temperature, being more deformable and thus increasing the extraction yield with temperature. Residual oil in the cake was, however, higher in processed flakes than in processed whole seeds despite the higher pressures obtained, at all temperatures, during processing.

This somewhat surprising behavior was attributed to a shorter residence time of the flakes in the extruder-press due to higher throughputs (Figure 8.26a). The difference in cake residual oil is more apparent at lower temperatures and tends to reduce when the processing temperature increases. In commercial pressing plants, canola seed is flaked and then heated to 85–90°C prior to screw extrusion-pressing. As shown in Figure 8.26, these treatments would allow for higher throughputs as well as higher oil extraction (Vadke & Sosulski, 1988). However, a critical cost analysis should include

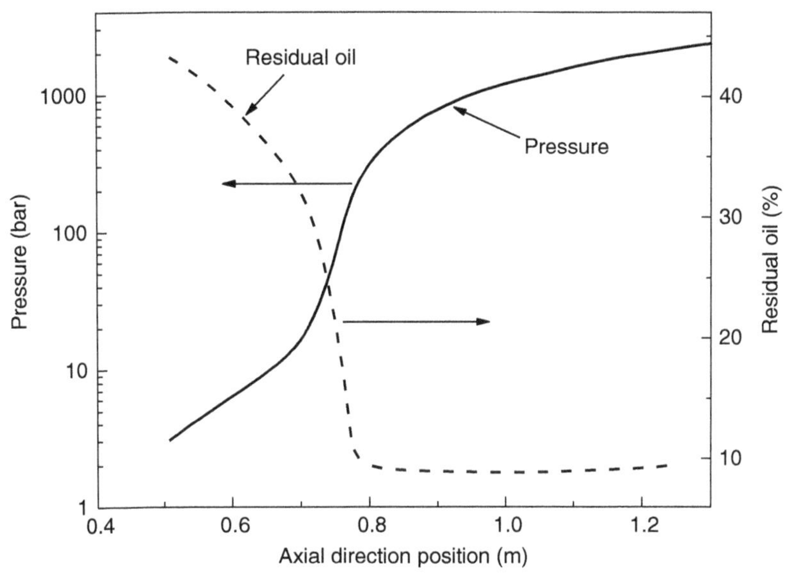

Figure 8.25 Simulated pressure and residual oil profiles in a screw extrusion-pressing operation. Properties used to solve the model were taken from a screw pressing operation of canola oil in a small oilseed press (Simon-Rosedowns mini 40 screw press) with a nominal capacity of 40 kg seeds/h used by Vadke and Sosulski (1988) in experimental trials. Source: Adapted from Willems et al. 2009. Reproduced with permission of Elsevier.

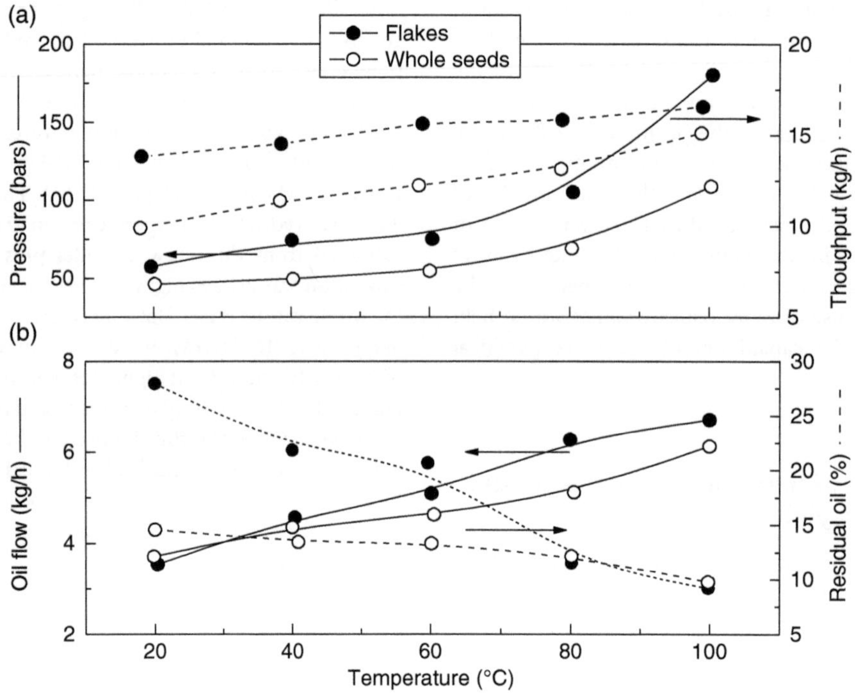

Figure 8.26 Experimental values showing (a) die pressure and throughput as a function of temperature for flaked and whole canola seeds; (b) oil flow and residual oil in the cake as a function of the temperature for flaked and whole canola seeds. Source: Vadke & Sosulski 1988. Reproduced with permission of Springer Science+Business Media.

expenses involved in the flaking process to establish a true comparison. In addition, the use of high-temperature pressing has been associated with quality issues, such as oil discoloration, and often cold pressing is preferred despite a lower oil flow is obtained. For many cases, increases of temperature due to friction could be around 20°C or more, so proper cooling of these pressing systems is essential. Hence, for modeling more real-life systems it would be necessary to include the heat transfer equation associated with the motion equation (see Chapter 3 for similar approaches to solving the equation of motion and the energy equation simultaneously comprising what is known as a true multiphysics model).

The model developed by Vadke et al. (1988) and described in this section provides results of effects of temperature on the screw extrusion-pressing of crushed canola seeds. Although the model predicts values within the experimentally observed range, results appear to show a different trend than the ones experimentally observed and reported in Figure 8.26 where the oil mass flow and residual oil increase and decrease with temperature, respectively. It must be noted that the choke opening is different (opening 0.42 mm for the data shown in Figure 8.26) than the value of 0.62 mm, which was used-for the simulations shown in Figure 8.27. Higher temperatures were also used in the simulations, thus complicating a true comparison with experimental results. Increases in oil extraction have been observed with barrel temperature during oil extraction of sunflower seeds in an intermeshing co-rotating twin screw extruder-press (Kartika et al., 2010). Other applications include the extraction of cocoa butter, which relies principally on expression processes solely using filter press and screw extruder-presses.

Concerning other mathematical models, several studies have been conducted on the modeling of oilseed expression incorporating the cell structure of the material (Gros et al., 2003; Lanoisellé et al., 1996; Mrema & McNulty, 1985; Toscano & Pedretti, 2007; Venter et al., 2007). Models based on the cellular structure of the material can provide good insight and better prediction of the expression process but they require knowledge of the cell structure and its rheological characteristics.

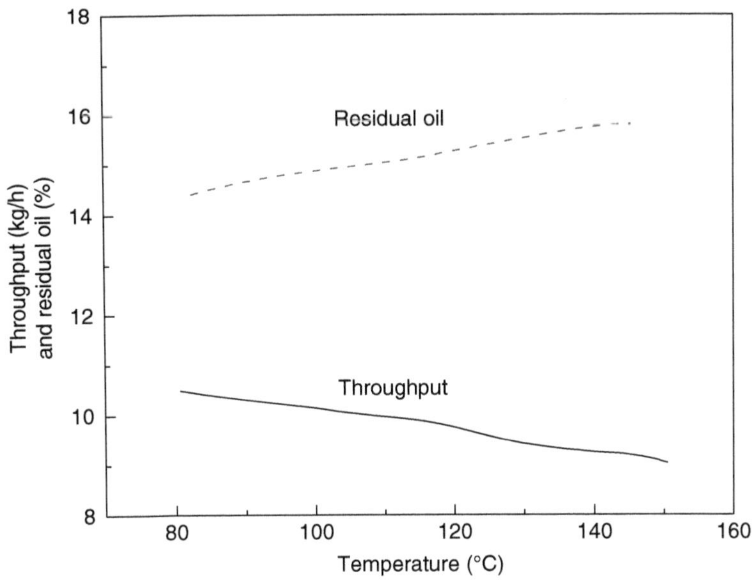

Figure 8.27 Simulated values of throughput and residual oil as a function of temperature. Material data were those of canola seed and canola oil; choke opening was 0.62 mm and temperature 120°C. Source: Adapted from Vadke et al. 1988. Reproduced with permission of Springer Science+Business Media.

However, models including the complex rheological behavior of the cellular material are scarce. As described in this section, some studies have focused on processing variables that may influence the rheological properties of the expressed cake. The example discussed above which illustrates the effects of either feeding flakes or whole oilseeds in the process shows the impact of material properties on the expression operation.

One processing variable that has a profound effect on material properties is temperature, especially when processing temperatures closer to the glass transition range are used. In many instances, results of the model do not seem to agree experimental results (for example, results shown in Figures 8.26 and 8.27). Although no reasons for such discrepancies are given, it could be hypothesized that the failure of the model is due to the lack of inclusion of the glass transition phenomenon, especially when processing temperatures are closer to that region. Close to glass transition temperatures, the compressed cake may experience large deformations that cannot be predicted by simple viscoelastic models like the ones described in section 8.2.4. At temperatures near the glass transition region, the deformability and mobility of the cellular structure increase significantly so an increase in oil flow extraction and lower residual oil extraction would be practically expected with increases of temperatures, which is shown in the experimental results (Figure 8.26) but not predicted by Vadke's model (Figure 8.27). The decrease in throughput with increases in temperature appears to be logically predicted by Eq. 8.39. That effect was also discussed in Chapter 3, as increases of temperature lower the apparent viscosity of the cake and increase the pressure flow component, thus decreasing the net flow or throughput. However, the equation assumes that the extruded material behaves rheologically as a true melt, which is not a true representation of the rheological behavior of pressed cakes.

Other approaches based on the Shirato's model (Shirato et al., 1974, 1986) have been intended to describe the expression operation, although they were not focused exclusively on screw extrusion-pressing. Venter et al. (2007) proposed a model based on Shirato's work to predict the expression of cocoa butter from cocoa nibs. Development of the model, which is specific to screw pressing operations, results in an equation similar to Eq. 8.34 able to predict the pressure on the solid cake as a function of the location in the press and time. The model takes into account physical characteristics of the expressed cake but only in terms of its porosity, permeability and compressibility, whose dependence with pressure are expressed by empirical relationships such as:

$$\alpha_s = \alpha_0 \left(1 + \frac{p_s}{P_a}\right)^{n_\alpha} \tag{8.69}$$

$$(1-\varepsilon) = (1-\varepsilon_0)\left(1 - \frac{p_s}{P_a}\right)^{\beta} \tag{8.70}$$

where α_0 , P_a, n_α, ε_0 and β are empirical parameters of the model. It is important to recognize that these equations, including Eq. 8.34, assume that the material is elastic, i.e. that the material behavior is time-independent and that instantaneous changes to p_s will cause instantaneous changes in the porosity and filtration resistance (Venter et al., 2007). This assumption is in clear conflict with the experimentally observed rheological behavior of cellular biomaterials described in section 8.2.4.

Extension of mathematical models incorporating viscoelasticity has been done by assuming that the pressed cake behaves as a Kelvin–Voigt solid (Venter et al., 2007). However, these rheological models are only applicable for conditions of linear viscoelasticity, i.e. for small deformations. Besides the rheological properties, the consolidation and permeability properties of oilseed material vary with temperature (Bargale et al., 2000; Tiller & Li, 2003). It can also be seen that for oilseeds, the contribution of secondary consolidation to the expression process is much smaller at higher temperatures. This is because at high temperatures, the pressed cakes are weaker and more deformable than at lower temperatures, thus favoring their fracture. Thus, temperature influences the nature of the solid structure and its behavior during expression, especially when the processing temperature is near or at the glass transition region. It must also be recognized that processing temperature affects the viscosity of the expressed liquid but that is accounted for in the existing models (e.g. using Eq. 8.55). Enhancement of oil extraction with a heating pretreatment to modify the mechanical properties of the feed has also been observed with other oil-containing biological materials such as corn germs, in which microwave and oven heating promoted the extraction of residual oil (Moreau et al., 2005).

Moisture content of oilseeds is a key factor controlling solid–liquid expression because it has a great impact on the expressed cellular material rheological properties. Although based on practical experience, press manufacturers usually provide processors with information on acceptable seed moisture ranges to be used with their equipment (Savoire et al., 2013). It is generally observed that increases in moisture content lead to decrease of oil yield. That inverse correlation between moisture content and oil yield has been observed for a great number of oilseeds, including crambe (Singh et al., 2002), canola seed (Vadke & Sosulski, 1988), flaked cooked cuphea (Evangelista & Cermak, 2007), flaxseed (Zheng et al., 2003, 2005), sesame and and linseed (Willems et al., 2008). For some seeds, an optimum moisture content can be found. For example, for walnuts, increasing moisture content from 2.4% to 7% results in an increase in the oil yield from 61% to 83.5%, after which the yield decreases. However, the process was assisted with supercritical CO_2 (Martínez et al., 2008), a process that is described in section 8.4.4. In addition to the oil yield, moisture content affects other pressing operating variables such as the capacity of the press and parameters such as barrel pressure and oil quality. Savoire et al. (2013) have reviewed most of the work performed in this area.

8.3.3 Effects of processing conditions and screw geometry on pressure build-up and liquid expression

8.3.3.1 Barrel pressure and screw speed

Figure 8.28 illustrates experimental results showing the effect of choke opening and screw speed on the maximum pressure in an extruder-press as well as the throughput (Figure 8.28a) and the flow of extracted oil and residual oil in the cake (Figure 8.28b).

Figure 8.28a clearly shows that, at all screw speeds, as the choke opening is reduced, the maximum pressure in the barrel increases and the throughput decreases. This experimentally observed behavior is anticipated by Eq. 8.39, which predicts a decrease in throughput when the pressure-induced flow increases. It must be noted that increases in screw speed also contribute to the drag flow and throughput should increase with increases in screw speed. The effect of screw speed on pressure shows that larger screw speeds, at a given temperature and choke opening, reduces the barrel pressure. As discussed in Chapter 3, this is due to the fact that at lower screw speeds, the residence time of the product in the screw extruder-press is longer and the degree of filling in the extruder channel is higher resulting in higher barrel maximum pressures (Figure 8.28a). The increase of maximum pressures in the barrel is slow for a choke opening range 0.4–0.8 mm but increases more sharply in the range 0.2–0.4 mm. This behavior is almost paralleled by the decrease in throughput with decreases in the choke openings.

Oil extraction flow output remains unchanged until an intermediate choke opening around 0.40 mm is reached, but decreases significantly for smaller choke openings. In parallel, the residual oil in the press-cake decreases with decreases in the choke opening (Figure 8.28b). It should be noted that there is a strong relationship between these process parameters and the residual oil, which is expected to be inversely related to oil output and directly related to the throughput.

8.3.3.2 The length of the compression zone

The length of the compression zone of the screw extruder-press can be varied and its effects on the screw extrusion-pressing process, in terms of throughput and residual oil, can be studied using the mathematical model developed in this section. Figure 8.29 illustrates these results for a temperature of 100°C and a choke opening of 0.8 mm, while screw geometry was maintained constant. Simulations show that throughput increases and residual oil decreases for screw extruder-presses having a longer compression zone. Thus, predictions of the model show that longer screw extruder-presses have a better extraction performance, which is in agreement with reports stating that longer screw extruder-presses give higher throughputs (Ward, 1976). The model described in this section also has the capacity to analyze other situations which may include geometrical parameters such as barrel diameter, choke length, height and width of flights. It would also show the effects of individual parameters on extruder-press performance, thus providing not only a useful insight into screw extrusion-pressing flow characteristics but also guidance on extruder-press design (Vadke et al., 1988).

Figure 8.30 shows a comparison between experimental results and those predicted by the model at a screw speed of 120 rpm. Both the predicted and theoretical values show a decrease in the residual oil at smaller choke openings. The close agreement between the experimental results and the results predicted by the model is clearly illustrated (Figure 8.30a). The model also correctly predicts that the residual oil is lowered at slower shaft rotation speeds (Figure 8.30b) as a consequence of the longer residence time caused by the slower conveying action. However, the predictions of the model and the experimental results at different screw speeds do not appear to fully match (Figure 8.30b). Another limitation of the model can be seen in Figure 8.31, where it is clear that the model fails to predict changes in throughput with changes in choke opening at low rotation speeds. As before, this limitation is explained on the basis of changes in material characteristics, during the cake passage through the extruder-press, which are not accounted for in the model (Vadke et al., 1988). In all these simulations in which temperature was maintained constant, the actual values of extrusion-pressing temperature as

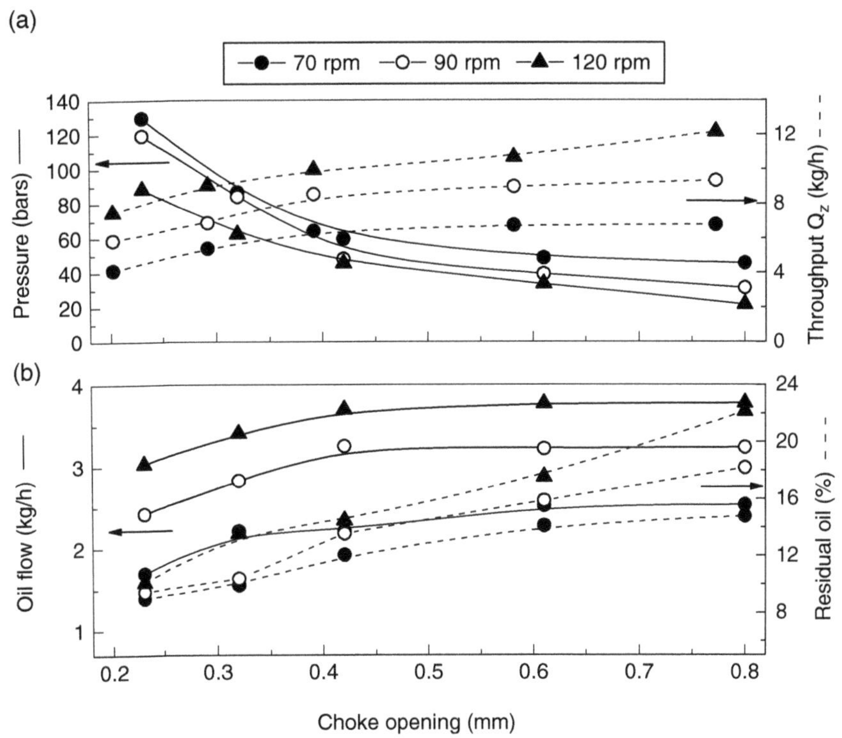

Figure 8.28 Experimental values showing the effects temperature and choke opening on (a) barrel maximum pressure and throughput, and (b) oil extraction flow and residual oil during cold screw pressing of canola seeds. Source: Adapted from Vadke & Sosulski 1988. Reproduced with permission of Springer Science+Business Media.

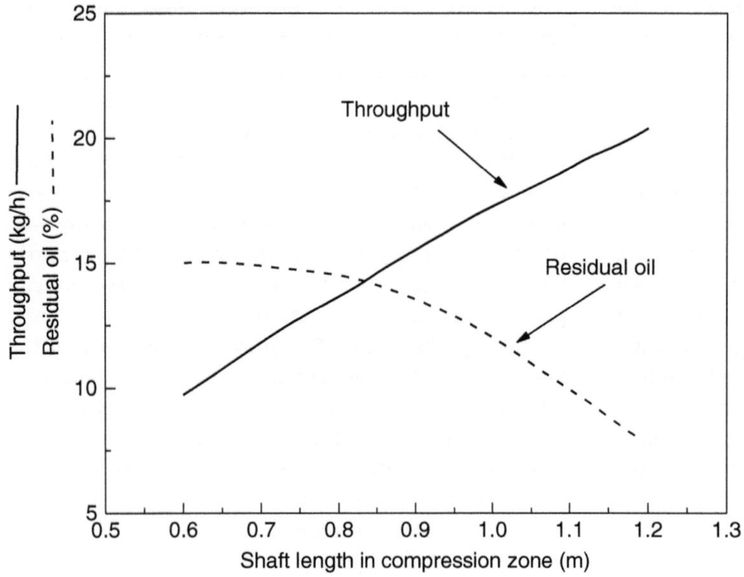

Figure 8.29 Throughput and residual oil in the press cake as a function of the length of the shaft in the compression zone of the screw extruder-press. Other conditions are screw speed 120 rpm, temperature 100°C, and choke opening 0.80 mm.
Source: Adapted from Vadke et al. 1988. Reproduced with permission of Springer Science+Business Media.

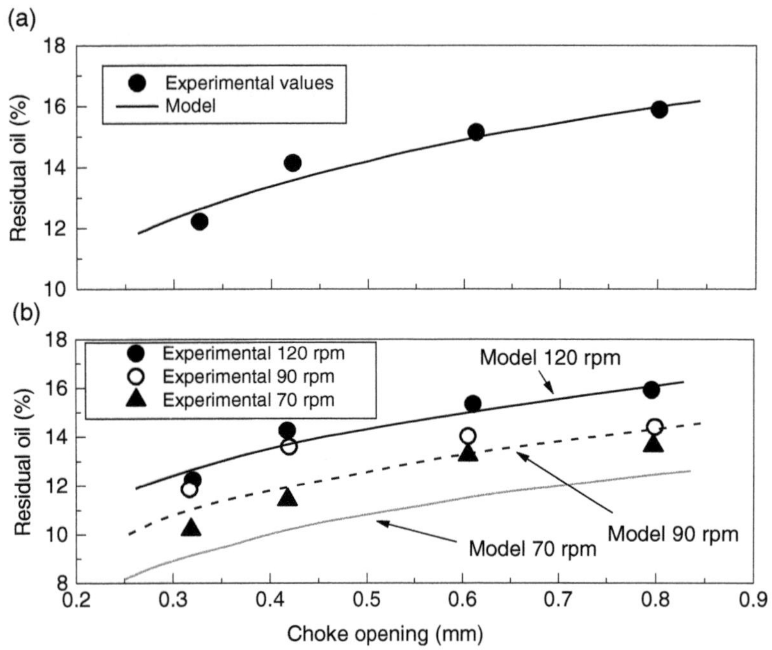

Figure 8.30 Comparison between experimental and predicted values of the model described in section 8.3 in terms of residual oil in the pressed cake as a function of the choke opening at different temperatures and screw speeds. (a) Comparison at a screw speed of 120 rpm; (b) comparison at different screw speeds.

measured in the pressing experiments were used. These temperatures varied with the screw speed and choke opening and values for the screw extrusion-pressing operation are given in Table 8.4 (Vadke & Sosulski, 1988).

Although the model described in this section is simple and uses a number of approximations, it was able to predict trends on the effects of processing parameters of relevance to the screw extrusion-pressing operation. The approach taken to develop the model, which includes superimposition

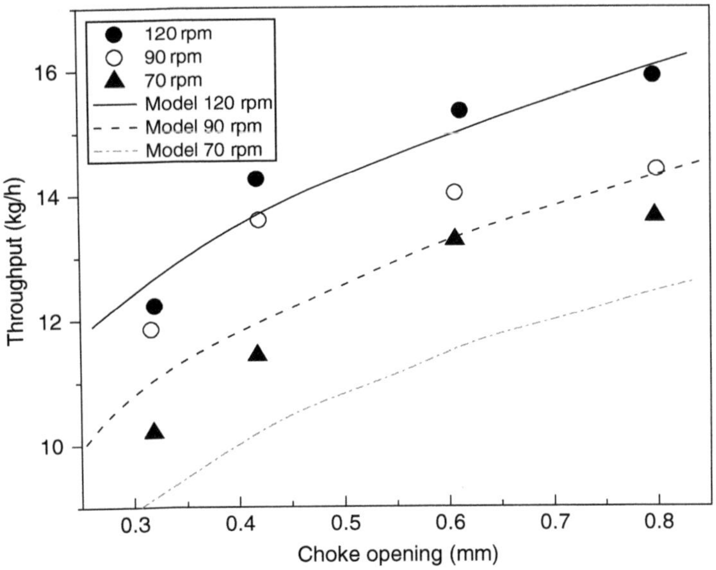

Figure 8.31 Experimental results and model prediction showing the effect of the choke opening and screw speed on the pressed cake throughput. Source: Adapted from Vadke et al. 1988. Reproduced with permission of Springer Science+Business Media.

Table 8.4 The effect of shaft speed and choke opening on measured barrel temperature during the screw extrusion-pressing of canola seeds. The temperatures were used to analyze model predictions which are illustrated in Figures 8.30 and 8.31.

	Shaft speed		
	120 rpm	90 rpm	70 rpm
Choke opening (mm)	Temperature (°C)		
0.23	133	128	124
0.32	116	116	112
0.42	112	108	105
0.61	108	105	104
0.80	104	102	101

Source: Adapted from Vadke & Sosulski 1988. Reproduced with permission of Springer Science+Business Media.

of filtration to the screw extrusion theory developed in Chapter 3, appears to be valid. With the advent of more computer power which can handle more sophisticated numerical approaches, extension of the 1D flow model to 2D or even 3D flow models is a direct result of these new modeling efforts. As noted, expanding the model by incorporating more accurate and complete rheological models describing the flow of the semi-solid pressed cake as well as the effects of temperature on the structure and rheology of the cake would improve its predictability, especially when processing temperatures are near the glass transition region.

8.4 Case studies: examples of industrial applications

As a preamble, it is worth mentioning the use of the intermeshing co-rotating twin screw extruder-press in water-aided washing of alkaline-digested lignocellulosics and bleached cellulose fibers in the extrusion pulping of non-wood fibers. This process may lead to important water savings compared to conventional washing processes. This relevant application is described in Chapter 5 (section 5.3.3) whereas the process-intensifying capability of twin screw extrusion-pressing process is developed in Chapter 10 (section 10.2.2.3).

This subsection focuses mainly on applications relative to oil separation from oilseeds which use solid–liquid extrusion-pressing processes. One of the pioneer publications reviewing the processing of a range of oilseeds using continuous screw extruder-presses is that of Ward (1976). The study provides a description of the process, as well as definitions of the terminology used in the area which has been covered in section 8.1 of this chapter. Continuous screw extrusion-pressing in the oil industry has been used for two main purposes. One is a high-pressure operation to lower the residual oil content of oilseeds. The other application is as a pre-press operation prior to solvent extraction. Figure 8.1 illustrates a schematic flow sheet showing the two possible avenues that a continuous screw extrusion-pressing process can take. Ward (1976) gives details on the processing of hard, fibrous seeds which include palm kernels and copra and also groundnuts and sunflower meals which in comparison with the former can be considered soft materials. Thus, two different cases in which the main difference is the texture and hardness of the oilseed are considered.

8.4.1 Continuous screw extrusion-pressing of copra, a hard cellular material

Copra is the name given to the dried meal of coconuts produced by coconut palms. It is abundant in countries such as the Philippines, Papua New Guinea, Vanuatu, Mozambique, Malaysia, and the Pacific Islands. After coconuts are cracked, their internal meal contains about 50% water and 30–40% oil; about 30 nuts provide 4.5 kg of copra. More uniform quality copra is produced by hot-air drying with kernels up to 4–5% moisture and 63–70 % oil. The oil from copra is commonly separated by continuous screw extrusion-pressing that includes the following processing steps.

• *Cleaning*: used to remove sand which could affect the screw extrusion-pressing operation.

• *Coarse grinding*: this reduces meal coming from the coconut halves into pieces about 20 mm, normally done in a hammer mill without a screen or grid.

• *Fine grinding or breaking*: reduces the broken copra to pass an 8 mesh screen. The moisture content of the copra is extremely important for the grinding operation. Dry cobra is brittle and can be handled in the grinders. Conversely, high moisture content copra has a rubbery texture that is difficult to grind. For high moisture content seeds, grinding temperatures may soften the structure of the cell material so oil can exude and blind the screen.

• *Cooking*: a necessary step to reduce the moisture content as low as possible. A feed with a moisture content of 1% is preferred for better continuous extrusion-pressing operation conducted at a temperature range of 115–120°C (Ward, 1976). With low moisture content, the maximum pressure at the screw extruder-press increases significantly. It has been observed in products such as canola seeds that the throughput and oil extraction increase and residual oil decreases with decreases in the feed moisture content, to reach an optimum moisture of 5%, which is the moisture used by the industry for a good screw extrusion-pressing operation (Vadke & Sosulski, 1988).

• *Continuous screw extrusion-pressing*: given the hardness of the material, the screw extruder-press must be manufactured with alloys (e.g. cobalt-chromium) to resist wear under harsh pressure and temperature conditions. Barrel drainage slots must be wide at the feed end of the extruder-press to allow the large quantity of oil to be drained as rapidly as possible. Since copra forms a fibrous drainage matrix within the extruder-press barrel, the quantity of foots produced is not high if the seed is properly cooked. Shaft cooling is necessary to prevent burning of the cake on the shaft side (Ward, 1976). Cooling of the screw, shaft and barrel is useful to maintain the flavor of the oil.

A slightly different process can be applied to copra when it is used as a pre-pressing step for solvent extraction. Variations include: (i) the fine grinding is taken further; (ii) the cooking is eased to feed a product to the press at 105–110°C and higher moisture content (3–4% moisture). These conditions make the cake rubbery and deformable and its structure is less fractured during the continuous screw extrusion-pressing operation. Finer grinding tends to free oil in the cooker, so the grinding and cooking stages must be balanced to eliminate this problem. Due to the conditions of the cake in terms of high moisture and high temperature, the pre-pressing process is not as severe as the standard screw extrusion process. Hence, the pre-pressing process provides conditioning of the cake for good solvent extraction.

8.4.2 Continuous screw extrusion-pressing of groundnuts/peanuts, a soft cellular material

Groundnut oil is another name for peanut oil, which is commonly used in foods due to its good fat composition. It contains a high percentage of monounsaturated fats, but more saturated fats than canola oil and more polyunsaturated fats than olive oil. Processing steps are similar to those used for the processing of copra, with some variations in the screw extrusion-pressing conditions.

• *Cleaning*: groundnuts need to be cleaned due to the presence of sand, which can seriously wear the screws. Groundnuts also contain a fiber-rich shell that is highly abrasive and needs to be separated before the screw extrusion-pressing operation.

• *Preparation*: each half of a groundnut is virtually one oil cell. Breaking of the seed will free oil in the cooker, yet such a stage is necessary. Ward's opinion, however, is that groundnuts should not be broken prior to high-pressure pressing (Ward, 1976).

- *Cooking*: used to coagulate some of the protein present in the nut. Overstirring groundnuts in the cooker is avoided. The temperature of the product going to the screw extruder-press should be around 115°C with moisture content of 3.5–4.5%. Injection of steam in the cooker may improve the pressing operation (Ward, 1976). To keep the same moisture content in the cooked product, superheated steam is injected into the cooker. Ward (1976) did not find a logical explanation that justifies the use of steam injection in the cooking process but it is likely that an increase of heating rate may favor the distribution of temperature and moisture in the cooking mass.
- *Screw extrusion-pressing*: groundnuts have a low friction coefficient so slippage is a potential problem in the screw extrusion-pressing operation. In addition, the mechanical strength of the groundnut seeds is low so pressure build-up should be applied gradually. Hence, low shaft speeds and long screw extruder-presses are preferred. Since the soft nature of groundnuts may generate a large amount of foots (closely related to the feeding rate), it is recommended to use relatively low feed rates and thus minimize the presence of foots in the expressed oil. Foots in the oil are commonly returned to the process, which represents a severe unbalancing effect on the screw extrusion-pressing operation as feed rates may become erratic and the loading of the extruder-press is inconsistent (Ward, 1976).

The cases presented in this section illustrate the pressing operation of two oilseeds with different textures and it is clear that the processing steps do not change substantially except the conditions used for the screw extrusion-pressing and whether the operation is used to reduce the oil content of the seed or as a preparation step for solvent extraction.

Groundnuts can also be pre-pressed as a conditioning step prior to oil solvent extraction using similar conditions to those used for the standard screw extrusion-pressing operation. The main concern in the pressing of groundnuts is the rupture of the seeds mainly in the cooking and screw extrusion-pressing steps. Breaking of the seeds is essential to produce an easily extractable cake, without large pieces of whole nut embedded in it. The recommendation is that temperatures in the cooking step be lowered so the cooked mass is fed to the screw extruder-press at temperatures near 105°C.

8.4.3 Soybean processing

Before the development of solvent extraction, the oil extraction industry was based on mechanical methods such as hydraulic and continuous screw extrusion-pressing. Out of 1.9 million tonnes of soybeans processed for oil in the US in 1940–41, 77% was extracted by mechanical methods (Nelson et al., 1987). Over the years, solvent extraction technology has replaced mechanical methods of oil extraction, especially in developed countries. However, with current trends of using oils as a source of renewable energy, oil pressing has become a viable process that is receiving considerable attention in both research and the industry, especially for potential applications in many developing countries characterized by small-scale processors (Evangelista & Cermak, 2007; Evon et al., 2013; Kartika et al., 2006, 2010; Okoye et al., 2008; Omobuwajo et al., 1998; Owolarafe et al., 2002, 2007; Singh & Bargale, 2000).

Currently, there are two competitive processes used to extract oil from soybeans: continuous screw extrusion-pressing and extraction by solvent. Screw extruder-press and other pressing technologies have advantages because they can be used with almost any kind of oilseeds and nuts. They can be adapted to small production and to multi-purpose plants built to process different types of raw materials, which is beneficial for application in developing countries. In addition, the process is relatively simple and not capital intensive. Conversely, the smallest solvent extraction plant would have a processing capacity of 100–200 tonnes per day, whereas screw extruder-presses and other pressing equipment are available for much smaller capacities, from a few tonnes per day and up.

The main disadvantage of screw extrusion-pressing is its relatively low yield of oil recovery compared to solvent extraction. In general, oil pressing technology cannot reduce the level of residual oil in the press-cake below 3–5%. However, most of the oil left in the cake can be recovered by a stage of solvent extraction (Lusas & Watkins, 1988). Such two-stage processes (pre-press/solvent extraction) are now widely applied. In the case of soybeans, however, a 5% residual oil level in the cake represents an oil loss of about 25%, but solvent extraction of the cake would not be economical, because the bulk of material which must be processed is very large. Screw extrusion-pressing followed by solvent extraction processes are therefore not applied to soybeans. In addition, there are safety concerns about solvent traces in the cake when it is intended for animal or human consumption.

World soybean oil production is around 42.4 million tonnes, with China the main producer followed by the USA and Argentina (Table 8.5). Thus, there is continuous

Table 8.5 Soybean oil: production, trade and consumption (million tonnes) in 2011–2012 (information extracted from lipidlibrary.aocs.org/market/soybean.htm).

	Production	Exports	Imports	Consumption
World	42.40			41.76
China	10.91	-	1.50	11.94
USA	8.95	0.66	-	8.31
Brazil	7.09	1.88	-	5.21
Argentina	6.84	3.79	-	3.07
EU-27	2.22	0.75	0.38	1.98
India	1.71	-	1.17	2.75

Source: Reproduced with permission of the American Oil Chemists' Society, Lipid Library.

interest in technologies used to produce soybean oil more economically. Environmental concerns in terms of energy usage and waste have also increased the pressure to find more sustainable processes for oil extraction.

Solvent extraction to produce soybean oil includes drying, tempering, cleaning, and classification, followed by the processing steps given below (Berk, 1992).

- *Cracking*: the purpose of this operation is to break the seeds into smaller particles in preparation for flaking. If the beans have been dried to 10% moisture and tempered, cracking also loosens the hulls and permits their separation by aspiration. Ideally, the seeds should be broken to 4–6 pieces of fairly uniform size.
- *Heating/conditioning*: heated to between 60°C and 88°C to increase the plasticity of the cracked beans in preparation for flaking.
- *Flaking*: cooked cracked beans are flaked by passing them through the nip of two rotating rolls. Flaking machines consist of a pair of horizontal counter-rotating smooth steel rolls. Typical roll sizes are in the range of 60–80 cm diameter. The rolls are pressed against one another by means of heavy springs or by controlled hydraulic systems. Cooked cracked beans are fed between the rolls and are flattened as the rolls rotate. The roll-to-roll pressure can be regulated and determines the average thickness of the flakes. The main purpose of flaking is to increase the contact surface between the oilseed tissues and the solvent, and to reduce the distance that the solvent and the extract will have to travel in the process of extraction.
- *Solvent extraction*: flaked beans are solvent-extracted with hexane.
- *Refining*: the oil is then refined, blended for different applications, and sometimes hydrogenated. Most of the remaining residue (soybean meal) is used as animal feed.

Regardless of the method of oil extraction, the oil expelling industry is expanding and geared to producing oil for edible and industrial purposes. While in the past the spent cake (meal) was used predominantly for livestock feed, in recent times, there has been interest in its use as a source of edible protein for human food (Nelson et al., 1987). However, this has raised food safety concerns with the possibility that cakes may contain traces of the solvent. Also, as discussed above, the scale economy of the solvent extraction process is very large and it would not be suitable for small processors in developing countries.

So there is interest in developing processes that do not use solvent extraction. From that perspective, screw extrusion-pressing offers a unique processing tool for these developments. It is important to note that when the press-cake from the oil extraction process is intended for human food and livestock feed, it must undergo a heat treatment to reduce the levels of antinutritional factors (e.g. trypsin inhibitor) minimizing protein damage. It is not uncommon to find the use of screw extruder-presses with several passes, which increase oil recovery but at the same time result in excessive heating of the cake, causing deleterious changes in its color and flavor and significantly affecting the quality of the oil.

The soybean oil is deposited within the bean tissue cells in bodies called spherosomes. Although screw extrusion-pressing operations are not common for processing soybeans, the reasons mentioned above have prompted some new research and development in this area. For standard screw extrusion-pressing of soybeans, they are cracked and subjected to dry heating to reach temperatures in the range 116–132°C. As a consequence of that heating, the moisture content of the cracked beans reaches a range of 2–5%, which is preferred for commercial screw extrusion-pressing operations. The heated beans are maintained at high temperatures (in a relatively long tempering process) to allow for moisture equilibration before they are fed to the screw extruder-press. The screw extruder-press breaks the oil cells so oil is released hot inside the matrix. As a consequence of the increase in pressure, when the cake moves down the screw extruder-press the oil is forced out through the perforated barrel.

Nelson et al. (1987) proposed a method that involves use of an extruder to break the oil cell and prepare the press-cake for further oil expression in the screw extruder-press. This approach has some advantages as the heating step is done in an extruder with a residence time of 30 seconds and the temperature is raised to 135°C from ambient

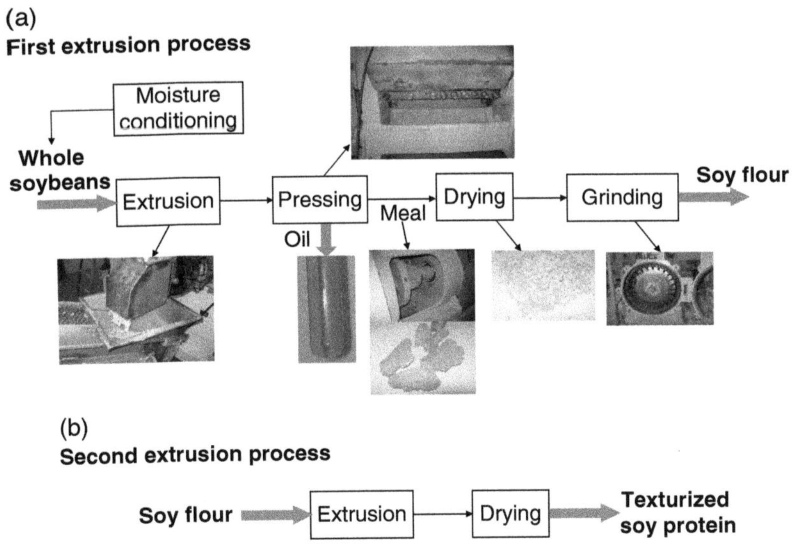

Figure 8.32 (a) Coupled extrusion and the screw extrusion-pressing processes for the production of soybean oil from whole soybeans. (b) Second extrusion process to produce texturized protein from soy flour. The extruder used is the same with some slight changes in the screw geometry.

Table 8.6 Effect of extrusion temperature on the oil yield obtained after pressing hot and cold extrudates in a Carver press.

	Oil yield (%)		
Extrusion temperature (°C)	Hot extrudate	Cooled extrudate	Residual oil in hot extrudate
107	57.8	45.6	9.5
127	56.3	51.8	9.7
132	60.7	50.3	8.6
135–141	65.5	46.1	6.7

Source: Adapted from Nelson et al. 1987.

temperature. The high temperature-short time treatment (HTST) can destroy antinutritional factors in soybeans while retaining the nutritional value of the proteins. This process makes use of relatively low-cost extruders, which generates heat by friction, thus eliminating the need for external heating. The other advantage is that the extrusion pre-process can handle the feed of whole soybeans, resulting in considerable savings of energy that was previously used in the cracking process.

The proposed processing approach is illustrated in Figure 8.32.

- *Whole soybean conditioning*: moisture of the whole beans is adjusted to the desired level and allowed to equilibrate; about 11% moisture content was used in the work of Nelson et al. (1987), but whole soybeans can be extruded using this technology with a moisture range 11–14%.
- *Extrusion*: used to break the oil cells. High shear forces and temperatures disrupt cells containing the oil and release it into the matrix. The high temperature also reduces oil viscosity which is partially separated from the extrudate exiting the die. Once the extrudate emerges from the die, flash evaporation of moisture occurs, reducing the temperature of the extrudate due to the evaporative cooling effect. At lower temperatures, the extrudate changes from a semi-liquid material to a granular mass that can reabsorb some of the separated oil. In addition to the extruder screw configuration, the extrusion temperature has a significant effect on the oil yield and the residual oil in the cake.
- *Oil pressing*: to optimize processing temperatures, Nelson et al. (1987) performed trials in which the extrudate was hot- and cold-pressed using a carver press after the first extrusion process. Results of these tests are given in Table 8.6. They show that when extrusion temperature increases, the yield of oil in the press also increases, and residual oil in the cake reduces in correspondence. Using a number of screw presses, Nelson et al. (1987) compared the conventional process using only a screw extruder-press to process conditioned whole soybeans and the dual process illustrated in Figure 8.32 that uses a pre-extrusion process followed by the screw extrusion-pressing operation. Results showed a significant improvement in the dual process in terms of oil yield and residual oil in the cake. In the dual process after the extrusion process,

Table 8.7 Composition of the raw material (whole soybean), extrudates, and screw-pressed cake. Destruction of antinutritional factors is also included.

Chemical component	Whole soybean	Extrudate	Screw-pressed cake
Protein, %	41.9	42.6	50.1
Oil, %	20.3	21.0	6.5
Carbohydrate, %	33.0	31.2	37.4
Total dietary fiber, %	17.3	16.9	19.4
Destruction of trypsin inhibitors, %	0	91.1	93.8
Protein dispersibility index, %	88.4	15.0	12.7

Source: Adapted from Nelson et al. 1987.

the extrudate is fed into a screw extruder-press to separate the oil and the meal (Figure 8.32a). Nelson et al. (1987) compared a number of commercially available screw extruder-presses. Table 8.7 illustrates the composition of the whole soybean, the extrudate and the screw-pressed cake, indicating that a low fat content cake suitable for food could be obtained. Figure 8.32a shows that the cake coming from the press can be dried and milled to reduce its particle size and produce defatted soy flour. The defatted soy flour can be further extruded in the same extruder to produce texturized soy protein (Figure 8.32b). Other recent studies using dry extrusion and expelling of the oil of other oilseeds include those of Evangelista (2009) and Riaz and Cheewapramong (2009).

Tindale and Hill-Haas (1976) surveyed most of the screw extruder-presses available at that time, illustrating technical issues. Some of those machines may not be manufactured any longer, while others are still manufactured or have introduced new technologies (e.g. http://frenchoil.com; http://www.vincentcorp.com). Other reviews concerning oilseed processing using screw extruder-presses and other pressing technologies are given by Khan and Hanna (1983) and Schwartzberg (1997). The latter, in addition, contains information concerning the pressing of other plant materials, including biomass and wastes. It also provides methodologies to characterize the materials and the pressing operations from an engineering standpoint, many of which were considered in this chapter. Another complete source of available equipment for oilseed processing is provided by Williams (2010) in which screw extruder-press technologies are described along with other technologies and processes. A recent review (Savoire et al., 2013) on mechanical continuous oil expression from oilseeds describes newer developments and the working principles of the available technologies. These characteristics are summarized in Table 8.8. In addition, it provides a summary of typical and new pretreatments to improve the screw extrusion-pressing performance of oilseeds.

8.4.4 Feed pretreatments

In section 8.4.3 a process involving extrusion pretreatment of whole soybeans to improve the efficiency of screw extrusion-pressing was discussed. Other studies concerning soybean processing include those investigating the effects of screw speed and feed pretreatments on the quantity and quality of the soybean oil and cake produced in the process. Pretreatments included soybean crushing, soy grits crushing, and crushing of soy grits pre-extruded at 135°C (Patil & Ali, 2006).

Although mechanical liquid pressing offers advantages from an economy of scale (suitable for small operations) and environmental standpoints, its efficiency in terms of oil recovery is inferior to solvent extraction. Hence, other pretreatments or technological developments that could enhance oil yield are being investigated. Figure 8.33 illustrates some of the processes currently used for oil refining and as shown, most of them consist of a combination of mechanical expression and solvent extraction. Indicated in the figure are also conventional pretreatments, which include thermal, size reduction and mechanical sieving. Some of these have been described in the case studies. However, there are other treatments that could improve the oil yield, which are now briefly discussed.

8.4.4.1 Gas-assisted mechanical expression (GAME process)

Gas-assisted mechanical expression is a relatively new approach that could recover high-quality oil from oilseeds that have low yields. The aim sought in the GAME process is to replace high oil yield solvent extraction that uses toxic solvents and requires expensive purification processes because the extraction is not selective and along the oil undesirable components such as volatile substances responsible for questionable taste and flavor are also removed (Venter et al., 2006). The GAME process utilizes the solubility of the oil in supercritical CO_2 to enhance the extraction yields of mechanical expression (Goodrum & Kilgo, 1987).

Table 8.8 Types and characteristics of screw presses used in oilseed processing.

Type of screw press			
	Expeller		
Expander	Farm use	Industrial use	Twin screw systems
Forced entry	Feeding by gravity on demand from the process	Forced entry	Forced entry
Variable flight-to-flight distance and screw root diameter	Constant flight-to-flight distance and screw root diameter	Variable flight-to-flight distance and screw root diameter	Variable flight-to-flight distance and screw root diameter
Discharge through a perforated plate	Discharge through a die or choke opening	Discharge through an annular space	Discharge through an annular space
Unperforated barrel; perforated close to the die	Barrel perforated	Barrel formed with bars	Unperforated barrel; perforated close to the die
Possible intermediate injection of steam or liquids		Possible refrigeration of barrel and screw	Possible intermediate injection of steam or liquids

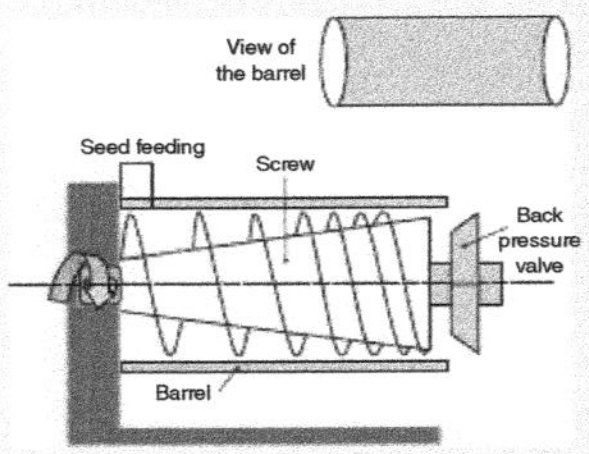

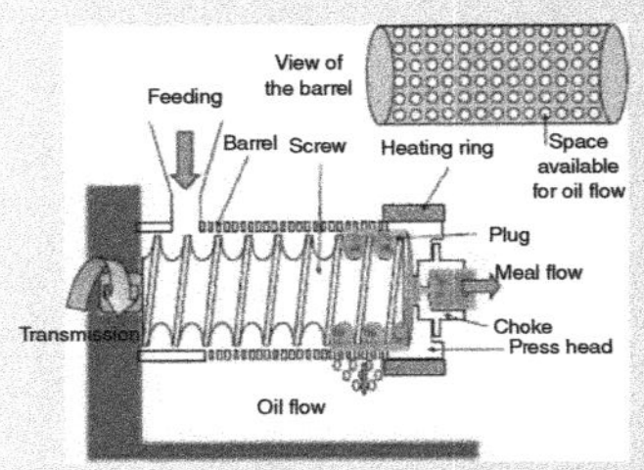

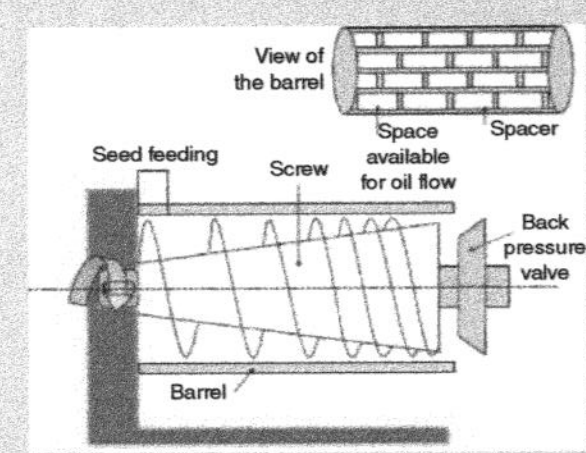

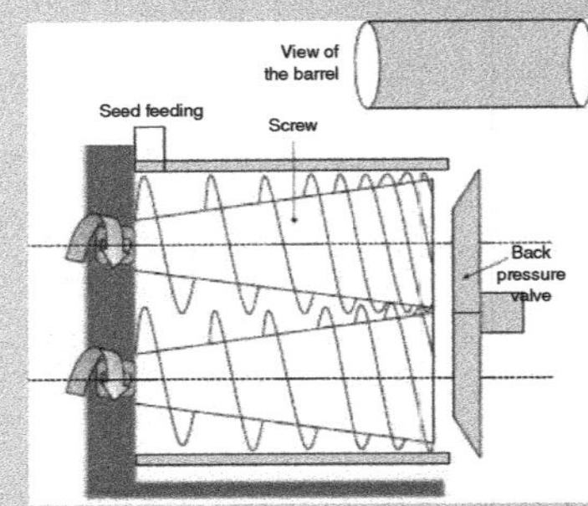

Source: Adapted from Savoire et al. 2013. Reproduced with permission of Springer Science+Business Media.

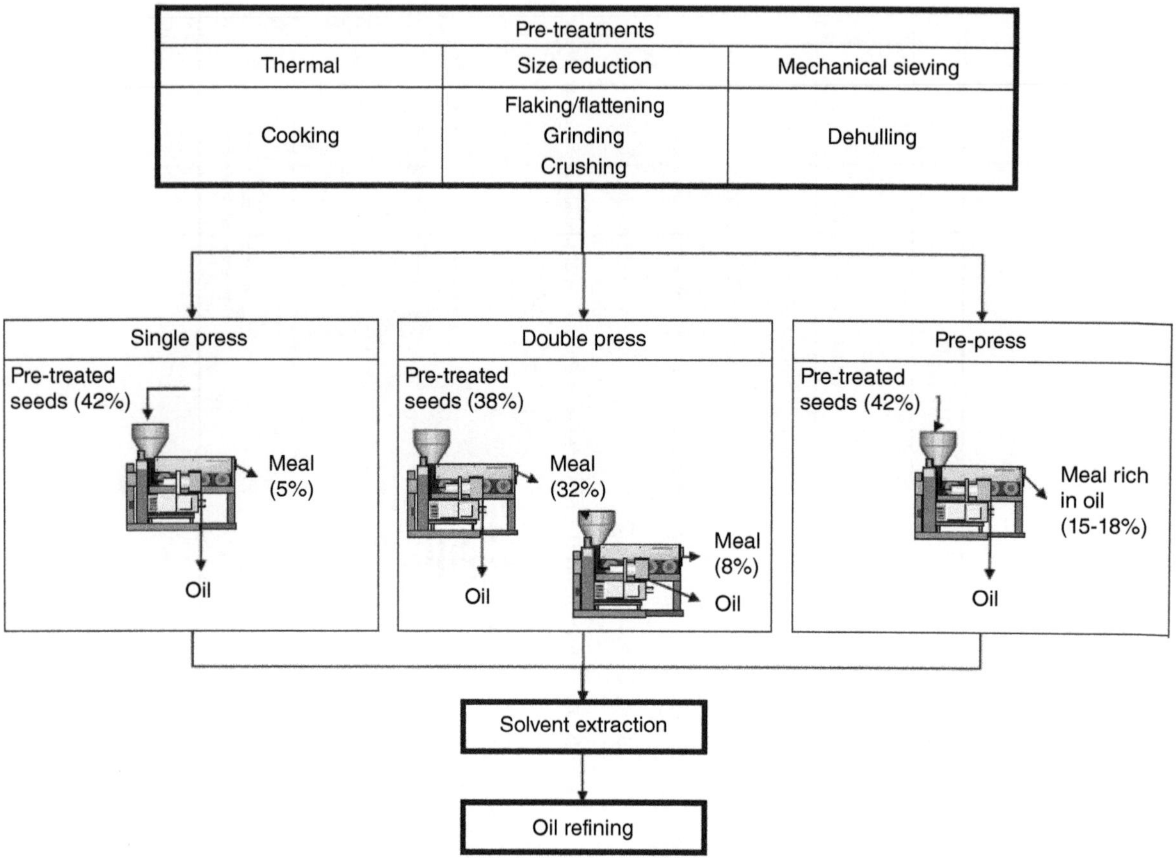

Figure 8.33 Different processes to obtain pure oil from oilseeds. Source: Savoire et al. 2013. Reproduced with permission of Springer Science+Business Media.

As discussed in order to improve the oil yield, solvent extraction is used in conjunction with extrusion pretreatment. In developing countries farms where the availability and cost of solvents such as hexane are prohibitive, water is used as the solvent but the yield is extremely low. The use of enzymes for weakening the cell wall and thus enhancing the removal of oil is a practical alternative but often the cost of the enzymes is prohibitive. The use of supercritical CO_2 as a solvent for oil extraction is another alternative. The solubility of the oil is dependent on the type of oil as well as the processing temperature and pressure, but generally it is in the order of only 1–5 wt% under normal extraction conditions at a range of pressure of 250–300 bar and temperatures of 40–60°C (Venter et al., 2006). Given the moderate solubility of most oils in supercritical CO_2, the use of large volumes of carbon dioxide is required and consequently the process is only economically feasible for high-value products like essential oils and flavors.

Hence, there is need for new processes that combine the advantages of current industrial technologies while overcoming their disadvantages. In that respect, the positioning of the GAME process in terms of product quality, oil yield and suitability for commercial use in the large-scale production of cocoa is shown in Table 8.9. As illustrated, the GAME process better equates to an ideal oil extraction technology. Other studies using a hydraulic press investigated the effect of process parameters such as moisture content, pressing duration, pressure profile, and press load, and it was found that only the moisture content, affecting the solid structure of the cellular cake, had an influence on the pressing operation (Venter et al., 2007).

Willems et al. (2009) developed a model incorporating the GAME process to compare the effect of extruder designs on the residual oil screw extrusion-pressed cakes. The study specifically investigated the influence of CO_2

Table 8.9 Comparison of oil recovery processes in terms of product quality, oil yield and commercial use, with regard to liquid expression of cocoa.

Process	Oil quality	Cake quality	Oil yield	Commercial use
Mechanical expression	++	++	+	++
Aqueous extraction	+	+	-	-
Organic solvent extraction	-	-	++	+
Supercritical CO_2 extraction	++	++	++	-
Ideal process	++	++	++	+ to ++

+ Good, ++ excellent, – poor.
Source: Adapted from Venter et al. 2006. Reproduced with permission of Elsevier.

in the process. For modeling purposes, it was assumed that the CO_2 gas was contained in the extruder-press regardless of the pressure at the extruder barrel. Four cases were studied.

- Conventional screw extruder-press.
- GAME screw extrusion-pressing in which the feed cake is fully saturated with CO_2 before entering the extruder.
- A two-step conventional screw extrusion-pressing process in which the cake from the first pass is fed to the second step of screw extrusion-pressing.
- A pretreatment of the feed with the extruder-press, the cake saturated with CO_2 and fed again to the screw extruder-press.

The model used is similar to the one described in section 8.3. In all cases, the extruder-press geometry and process conditions were kept constant, in addition, the screw speed was assumed to be 90 rpm, the feed temperature 40°C, and in both GAME cases the CO_2 pressure was assumed to be 100 bar. The mass fraction of oil y_{oil} defined in Eq. 8.46a, which is used to estimate the properties of the expressed cake (e.g. the viscosity by Eq. 8.52), represents the total liquid content, i.e. the amount of oil and CO_2 in the feed instead of only the oil content for conventional processes that do not use supercritical CO_2. Results obtained from the model are given in Table 8.10. In the conventional cases, the liquid content is equal to the residual oil content, whereas in the GAME cases the liquid content is the CO_2-saturated oil content; for comparison the final liquid and oil contents are shown in the table. Although the effect of the CO_2 pressure shows only small differences in the residual oil, results showed that the conventional process followed by GAME was the most efficient in terms of reducing the residual oil.

Table 8.10 Final liquid and oil contents for the four systems analyzed.

Process	Liquid content (%)	Oil content (%)
Conventional	19	19
GAME	47	40
Two-step conventional	17.5	17.5
Conventional followed by GAME		
- 80 bars	19.0	15.2
- 100 bars	19.0	14.9
- 150 bars	18.9	14.3

Source: Adapted from Willems et al. 2009. Reproduced with permission of Elsevier.

8.4.4.2 Pulsed electric field (PEF) treatment

Although PEF technology has been seen primarily as an alternative non-thermal treatment for microbial inactivation, recent studies have looked at the permeabilization of cell membranes with the aim of enhancing mass transfer from the inner part of the cells (Donsì et al., 2010). PEF treatment is based on the electroporation phenomenon, which is not fully understood and several theories have been proposed to explain it. A common feature of all these theories is the importance of the cell membrane in amplifying an applied electric field as the electrical conductivity of an intact membrane is several orders of magnitude lower than the extracellular media. By application of an external high electrical field, a transmembrane potential is created and increases as a result of the charging process at the membrane interface. When a critical value of the field strength is applied, a critical transmembrane potential can be induced, leading to the formation of reversible or irreversible pores in the membrane.

Pretreatments using PEF may cause plasmolysis, cellular damage, or permeabilization of plant cells and enhance the yield of solid–liquid separation using extrusion-pressing processes. Donsì et al. (2010) listed liquid pressing processes that use PEF as a pretreatment step to enhance the expressed liquid yield. Although the surveyed studies do not include a continuous screw extrusion-pressing operation, positive results observed with other types of pressing systems would indicate the potential of a combined process involving PEF as a pretreatment of the feed to be continually screw extrusion-pressed. Fruits and vegetables with textures particularly affected by thermal treatments are potential candidate products to be processed using these new technologies. Extraction of valuable components from fruits and vegetables is another application considered in Donsì's review. High-voltage electrical discharges have been successfully used for linseed oil extraction with subsequent pressing in a hydraulic press (Gros et al., 2003). Other approaches using PEF treatments and other physical methods such as ohmic heating, high-intensity ultrasound, and microwaves to enhance the extraction of valuable solutes and assist liquid pressing operations are reviewed by Vorobiev and Chemat (2010).

References

Bargale PC, Wulfsohn D, Irudayaraj J, Ford RJ, Sosulski FW (2000) PH Postharvest technology: prediction of oil expression by uniaxial compression using time-varying oilseed properties. Journal of Agricultural Engineering Research 77(2): 171–181.

Berk Z (1992) Technology of production of edible flours and protein products from soybeans. FAO corporate document repository. Information available through: www.fao.org/docrep/t0532e/t0532e00.htm#con.

Berger KG (1983) Production of palm oil from fruit. Journal of the American Oil Chemists' Society 60(2): 206–210.

Bouvier JM, Guyomard P (1997) Procédé et installation d'extraction en continu d'un liquide contenu dans une matière première. EP 0914243 B1.

Choo YM, Yap SC, Ooi CK, Ma AN, Goh SH, Ong ASH (1996) Recovered oil from palm-pressed fiber: a good source of natural carotenoids, vitamin E, sterols. Journal of the American Oil Chemists' Society 73(5): 599–602.

Donsì F, Ferrari G, Pataro G (2010) Applications of pulsed electric field treatments for the enhancement of mass transfer from vegetable tissue. Food Engineering Reviews 2(2): 109–130.

Dufaure C, Leyris J, Rigal L, Mouloungui Z (1999) A twin-screw extruder for oil extraction: I. Direct expression of oleic sunflower seeds. Journal of the American Oil Chemists' Society 76(9): 1073–1079.

Evangelista RL (2009) Oil extraction from lesquerella seeds by dry extrusion and expelling. Industrial Crops and Products 29(1): 189–196.

Evangelista RL, Cermak SC (2007) Full-press oil extraction of Cuphea (PSR23) seeds. Journal of the America Oil Chemists' Society 84(12): 1169–1175.

Evon Ph, Kartika IA, Cerny M, Rigal L (2013) Extraction of oil from jatropha seeds using a twin-screw extruder: feasibility study. Industrial Crops and Products 47: 33–42.

Geankoplis CJ (2003) *Transport Processes and Separation Process Principles*, 4th edn. New Jersey: Prentice-Hall.

Gibson LJ, Ashby MF (1982) The mechanics of three-dimensional cellular materials. Proceedings of the Royal Society of London. Series A, Mathematical and Physical Sciences 382: 43–59.

Gibson LJ (2012) The hierarchical structure and mechanics of plant materials. Journal of the Royal Society Interface 9: 2749–2766.

Goodrum JW, Kilgo MB (1987 Peanut oil extraction using compressed CO_2. Energy in Agriculture 6(3): 265–271.

Gros C, Lanoisellé JL, Vorobiev E (2003) Towards an alternative extraction process for linseed oil. Chemical Engineering Research and Design 81(9): 1059–1065.

Guyomard P (1994) Etude de faisabilité d'un extrudeur bivis en pressage-extrusion de graines oléoprotéagineuses. PhD thesis, Université de Technologie de Compiègne, France.

Hunt AJ, Hunt AJ Jr (1981) Screw press with continuous slope feed screw. US Patent 4,266,473.

Isobe S, Zuber F, Uemura K, Noguchi A (1992) A new twin-screw press design for oil extraction of dehulled sunflower seeds. Journal of the American Oil Chemists' Society 69(9): 884–889.

Johnston RB (2003) Twin screw press with interrupted flights. US Patent 6,550,376 B2.

Kartika IA, Pontalier PY, Rigal L (2006) Extraction of sunflower oil by twin screw extruder: screw configuration and operating condition effects. Bioresource Technology 97(18): 2302–2310.

Kartika I.A, Pontalier PY, Rigal L (2010) Twin-screw extruder for oil processing of sunflower seeds: thermomechanical pressing and solvent extraction in a single step. Industrial Crops and Products 32(3): 297–304.

Khan LM, Hanna MA (1983) Expression of oil from oilseeds – a review. Journal of Agricultural Engineering Research 28(6): 495–503.

Körmendy I (1974) Contribution to the three-dimensional pressing theory and its one-dimensional application. Acta Alimentaria 3: 93–110.

Lanoisellé JL, Vorobyov EI, Bouvier JM (1994) Modélisation du pressage à pression constante. Cas des produits à structure cellulaire. Entropie 186: 39–50.

Lanoisellé JL, Vorobyov EI, Bouvier JM, Piar G (1996) Modeling of solid/liquid expression for cellular materials. AIChE Journal 42(7): 2057–2068.

Little TH (1958) Method of producing juices from vegetables including fruits. US Patent 2,823,126.

Lusas EW, Watkins LR (1988) Oilseeds: extrusion for solvent extraction. Journal of the American Oil Chemists' Society 65(7): 1109–1114.

Marin FR, Martinez M, Uribesalgo T, Castillo S, Frutos MJ (2002) Changes in nutraceutical composition of lemon juices according to different industrial extraction systems. Food Chemistry 78(3): 319–324.

Martínez ML, Mattea MA, Maestri DM (2008) Pressing and supercritical carbon dioxide extraction of walnut oil. Journal of Food Engineering 88(3): 399–404.

Moreau RA, Johnston DB, Hicks KB (2005) The influence of moisture content and cooking on the screw pressing and pre-pressing of corn oil from corn germ. Journal of the American Oil Chemists' Society 82(11): 851–854.

Mrema GC, McNulty PB (1985) Mathematical model of mechanical oil expression from oilseeds. Journal of Agricultural Engineering Research 31(4): 361–370.

Nelson AI, Wijeratne WB, Yeh SW, Wei TM, Wei LS (1987) Dry extrusion as an aid to mechanical expelling of oil from soybeans. Journal of the American Oil Chemists' Society 64(9): 1341–1347.

Ohta H, Tonohara K, Watanabe A, Iino K, Kimura S (1982) Flavor specificities of satsuma mandarin juice extracted by a new-type screw press extraction system. Agricultural and Biological Chemistry 46: 1385–1386.

Omobuwajo TO, Ige MT, Ajayi AO (1998) Theoretical prediction of extrusion pressure and oil flow rate during screw expeller processing of palm kernel seeds. Journal of Food Engineering 38(4): 469–485.

Owolarafe OK, Faborode MO, Ajibola OO (2002) Comparative evaluation of the digester-screw press and a hand-operated hydraulic press for palm fruit processing. Journal of Food Engineering 52(3): 249–255.

Owolarafe OK, Osunleke AS, Oyebamiji BE (2007) Effect of hydraulic press parameters on crude palm oil yield. International Agrophysics 21(3): 285–291.

Okoye CN, Jiang J, Hui LY (2008) Design and development of secondary controlled industrial palm kernel nut vegetable oil expeller plant for energy saving and recuperation. Journal of Food Engineering 87(4): 578–590.

Patil RT, Ali N (2006) Effect of pretreatments on mechanical oil expression of soybean using a commercial oil expeller. International Journal of Food Properties 9(2): 227–236.

Powell S, Naranjo VD, Lauzon D, Bidner TD, Southern LL, Parsons CM (2011) Evaluation of an expeller-extruded soybean meal for broilers. Journal of Applied Poultry Research 20(3): 353–360.

Prat L, Guiraud P, Rigal L, Gourdon C (2002) A one dimensional model for the prediction of extraction yields in a two phases modified twin-screw extruder. Chemical Engineering and Processing: Process Intensification 41(9): 743–751.

Rebouillat S (2002) Contribution à l'étude fondamentale de la déformation de milieux poreux saturés. Thèse présentée pour obtenir le grade de Docteur ès Science. Institut National Polytechnique de Lorraine, Nancy, France.

Rebouillat S, Campanella OH, Peleg M (1987) Mechanical characteristics of raw and processed peat. Powder Technology 51: 273–275.

Rebouillat S, Schwartzberg HG, Leclerc D (1993) The expression of biological and food materials under constant rate of strain. Journal of Chemical Engineering of Japan 29: 29–36.

Riaz MN, Cheewapramong P (2009) Characterization of partially defatted peanut flour using dry extruder and screw pressing. International Journal of Food Properties 12(2): 427–437.

Savoire R, Lanoisellé JL, Vorobiev E (2013) Mechanical continuous expression from oilseeds: a review. Food Bioprocess Technology 6: 1–16.

Schwartzberg HG, Rosenau JR, Richardson G (1977) The removal of water by expression. AIChE Symposium Series 73: 177–190.

Schwartzberg HG (1983) Expression-related properties. In: Peleg M, Bagley EB (eds) *Physical Properties of Foods*. Connecticut: AVI Publishers, pp. 423–472.

Schwartzberg HG (1997) Expression of fluid from biological solids. Separation and Purification Methods 26: 1–213.

Shirato M, Sambuichi M, Kato H, Aragaki T (1969. Internal flow mechanism in filter cakes. AIChE Journal, 15(3): 405–409.

Shirato M, Murase T, Negawa M, Senda T (1970) Fundamental studies of expression under variable pressure. Journal of Chemical Engineering of Japan 3: 105–112.

Shirato M, Murase T, Negawa M, Moridera H (1971) Analysis of expression operations. Journal of Chemical Engineering of Japan 4(3): 263–268.

Shirato M, Murase T, Tokunaga A, Yamada O (1974) Calculations of consolidation period in expression operations. Journal of Chemical Engineering of Japan 7(3): 229–231.

Shirato M, Murase T, Iwata M, Nakatsuka S (1986) The Terzaghi–Voigt combined model for constant-pressure consolidation of filter cakes and homogeneous semi-solid materials. Chemical Engineering Science 41(12): 3213–3218.

Singh J, Bargale PC (2000) Development of a small capacity double stage compression screw press for oil expression. Journal of Food Engineering 43(2): 75–82.

Singh KK, Wiesenborn DP, Tostenson K, Kangas N (2002) Influence of moisture content and cooking on screw pressing of crambe seed. Journal of the American Oil Chemists' Society 79(2): 165–170.

Sriti J, Talou T, Faye M, Vilarem G, Marzouk B (2011) Oil extraction from coriander fruits by extrusion and comparison with solvent extraction processes. Industrial Crops and Products 33(3): 659–664.

Sriti J, Msaada K, Talou T, Faye M, Kartika IA, Marzouk B (2012) Extraction of coriander oil by twin-screw extruder: screw configuration and operating conditions effect. Industrial Crops and Products 40: 355–360.

Sukumaran CR, Singh BPN (1989) Compression of a bed of rapeseeds: the oil-point. Journal of Agricultural Engineering Research 42(2): 77–84.

Tiller FM, Cooper HR (1960) The role of porosity in filtration: IV. Constant pressure filtration. AIChE Journal 6(4): 595–601.

Tiller F.M, Li W (2003) Radial flow filtration for super-compactible cakes. Separation Science and Technology 38 (4): 733–744.

Tiller FM, Lu WM (1972) The role of porosity in filtration. VIII: cake nonuniformity in compression-permeability cells. AIChE Journal 18(3): 569–572.

Tiller FM, Shirato M (1964) The role of porosity in filtration: VI. New definition of filtration resistance. AIChE Journal 10(1): 61–67.

Tiller FM, Yeh CS (1987) The role of porosity in filtration. Part XI: Filtration followed by expression. AIChE Journal, 33(8): 1241–1256.

Tindale LH, Hill–Haas SR (1976) Current equipment for mechanical oil extraction. Journal of the American Oil Chemists' Society 53(6): 265–270.

Toscano G, Pedretti EF (2007) Evaluation of a mathematical model for oil extraction from oleaginous seeds. Journal of Agricultural Engineering 38: 11–20.

Vadke VS, Sosulski FW (1988) Mechanics of oil expression from canola. Journal of the American Oil Chemists' Society 65(7): 1169–1176.

Vadke VS, Sosulski FW, Shook CA (1988) Mathematical simulation of an oilseed press. Journal of the American Oil Chemists' Society 65(10): 1610–1616.

Venter MJ, Willems P, Kuipers NJM, De Haan AB (2006) Gas assisted mechanical expression of cocoa butter from cocoa nibs and edible oils from oilseeds. Journal of Supercritical Fluids 37(3): 350–358.

Venter MJ, Hink R, Kuipers NJM, De Haan AB (2007) The influence of process parameters on gas assisted mechanical expression (GAME) of cocoa nibs. Innovative Food Science and Emerging Technologies 8(2): 172–179.

Vincent Corporation (2013) http://www.vincentcorp.com/brochures.

Vorobiev E, Chemat F (2010) Principles of physically assisted extractions and applications in the food, beverage and nutraceutical industries. In: Rizvi SSH (ed) *Separation, Extraction and Concentration Processes in the Food, Beverage and Nutraceutical Industries*. Cambridge: Woodhead Publishing, pp. 71–108.

Ward JA (1976) Processing high oil content seeds in continuous screw presses. Journal of the American Oil Chemists' Society 53(6): 261–264.

Willems P, Kuipers NJM, De Haan AB (2008) Hydraulic pressing of oilseeds: experimental determination and modeling of yield and pressing rates. Journal of Food Engineering 89(1): 8–16.

Willems P, Kuipers NJM, De Haan AB (2009) A consolidation based extruder model to explore GAME process configurations. Journal of Food Engineering 90(2): 238–245.

Williams MA (2010) Separation technologies in oilseed processing. In: Rizvi SSH (ed) *Separation, Extraction and Concentration Processes in the Food, Beverage and Nutraceutical Industries*. Cambridge: Woodhead Publishing, pp. 396–429.

Xiao HW, Gao ZJ (2012) The application of scanning electron microscope (SEM) to study the microstructure changes in the field of agricultural products drying, scanning electron microscopy. InTech. Available from: www.intechopen.com/books/scanning–electron–microscopy/the–application–of–scanning–electron–microscope–sem–to–study–the–micro-structure–changes–in–the–field.

Zhang YE, Parsons CM, Weingartner KE, Wijeratne WB (1993) Effects of extrusion and expelling on the nutritional quality of conventional and kunitz trypsin inhibitor-free soybeans. Poultry Science 72(12): 2299–2308.

ZhengYL, Wiesenborn DP, Tostenson K, Kangas N (2003) Screw pressing of whole and dehulled flaxseed for organic oil. Journal of the American Oil Chemists' Society 80(10): 1039–1045.

Zheng Y, Wiesenborn DP, Tostenson K, Kangas N (2005) Energy analysis in the screw pressing of whole and dehulled flaxseed. Journal of Food Engineering 66(2): 193–202.

9 The Generic Extrusion Process V: Thermophysical Micromixing and Material Porosification

The field related to the manufacture of powders, granules and finely divided solids is vast and of considerable importance in industries such as food, chemicals, pharmaceuticals and metallurgy, among others. Many processing industries involved in forming (compression in metallurgy, compaction in galenical applications, etc.), preservation (drying techniques in the food industry), or product functionalization (encapsulation and instantizing in the food industry and galenical applications) require materials to be in a finely divided solid state.

The international scientific community has been very interested in this field and, as early as 1988, the major scientific societies created multidisciplinary research groups on the subject of powder science and technology. The subject is at the crossroads of several different fields of knowledge and expertise in material science, process, formulation design and metrology, which is shared with the industrial world, giving an interdisciplinary approach that has benefited the research and development of novel materials.

While the starting material (liquid, liquid-solid, solid) of a powder can be characterized relatively easily and accurately by its basic physicochemical or biochemical properties, the resulting powder in the finely divided state is already a sophisticated, complex or even a composite product.

In general, starting from either a liquid or a solid as well as a liquid-solid mix, the process that leads to a powder state gives rise to a product with specific properties that depend on the operating conditions and physical mechanisms employed in its production. Thus, the determination, interpretation (product–process relationship) and assessment of consequences (end-use properties) of these properties are complex and often difficult to control. Table 9.1 illustrates the specific features and complexity of powders and finely divided solids, from conventional powder-forming processes to end-use properties.

It is not necessary to dwell upon the difficulties of controlling the relationship between the operating conditions for the powder-forming process and the physical properties of the finely divided solid state and also controlling the relationship between the process and the powder's end-use properties. This is an area where technical expertise makes a difference, and such expertise is more effective when it is based on a methodological approach.

Porous powders constitute an especially complex knowledge-dependent category of powder that is essential to many processing industries, in particular the food industry. For these powders, physical texture is paramount, as it serves as an interface between the powder-forming process and the end-use properties.

Porous powders alone and in bulk do not consist only of a solid phase, they also have a mobile gaseous phase located in the space between the particles and also in the internal pores of the powder particles. The main factors used for describing a powder's physical texture are the density and specific surface area, and for porous powders the pore shape, size and size distribution. These properties are crucial in the case of porous powders, where the end-use value is highly linked to their porosity. Some examples of end-products are catalysts, ceramics and metal powders where the specific surface area can reach several hun-

Extrusion Processing Technology: Food and Non-Food Biomaterials, First Edition. Jean-Marie Bouvier and Osvaldo H. Campanella.

Table 9.1 Process/product properties in powder technology.

Starting material	Powder-forming process	Powder properties	End-use properties
Solid	Grinding	Physical properties:	Flow properties
	Separation-classification	Shape, size, size distribution	Fluidity
	Freeze drying	Textural properties:	Caking
Liquid	Concentration	Density, specific surface area, porosity, etc.	Mixability/
Liquid-solid	Crystallization	Mechanical and rheological properties:	demixability
	Precipitation-decantation	Angle of repose, angle of friction, angle of slide, etc.	Compactibility
	Spray drying	Electrical properties:	Compressibility
	Drum drying	Conductivity, resistivity, permittivity,	Cohesiveness
	Freeze drying	chargeability, etc.	Fluidizing ability
		Surface properties:	Dispersibility
		Roughness, surface geometry, surface interaction and reactivity, etc.	Wettability
			Solubility
			Flammability
			Explodability etc.

dreds of $m^2.g^{-1}$. In the food industry, specific surface areas are smaller ($<10\ m^2.g^{-1}$), but they are still significant in terms of quality, since they determine the technical performance of porous powders in relation to unit operations such as drying, dissolution and extraction, which depends on the particle's surface/mass ratio.

Worldwide, the food industry produces millions of tonnes of powders each year, either to save cost by eliminating water for storage and transport or for industrial processing purposes. These powders are rehydrated and then used or consumed at different levels of the industrial production chain and the human and animal food chains. Rehydration is dependent upon a category of functional properties that includes wettability, sinkability, dispersibility, and solubility. The physical texture and therefore the porosity of these instant powders are mainly what determine the powders' end-use values. Some of the most widespread of these instant products are dairy powders (whole milk, skim milk, milk substitutes, functional dairy ingredients, infant formula), coffee, cocoa, flavorings, extracts, instant drinks, yeast, probiotics, etc, with uses that require quick rehydration.

Instant powder production technologies must be able to handle dehydration of the starting material and texturization of the resulting powder. This is currently done using well-known technologies such as drum drying, spray drying, and freeze drying. However, there is a need for alternative technologies which allow dehydration to be intensified and the texture of the resulting dried powder to be improved. The ability of twin screw extrusion technology to process and texturize viscous media led researchers to design a new Generic Extrusion Process V, called "Extrusion-Porosification," which aims at intensifying drying and generating high-porosity powders. The extrusion-porosification process uses the two main process functions of GEP V: thermophysical micromixing of gas-liquid systems and porosification of the resulting foams which lead to porous powders after water (or solvent) removal. GEP V is considered a breakthrough in the field of instant powders production. The technology is still in the development phase with regard to process optimization and end-use applications. Although it is not industrialized yet, it is worth presenting this new generic extrusion process as well as its potential applications.

After presenting and discussing drying technologies which are classically used for the production of instant powders and the main drivers of instant powders processing, the innovative extrusion-porosification process is discussed. The second part then gives an engineering analysis of the main process functions of GEP V: the vacuum evaporation of a viscous medium, the thermophysical micromixing of concentrated gas-liquid systems and the intensified spray drying. The purpose of the third part of this chapter is to present the industrial perspectives of the extrusion-porosification process, i.e. eligible industrial applications with a focus on dairy products such as milk (whole and skim milk).

9.1 The new generic extrusion-porosification process

9.1.1 Typical drying processes for instant powders

9.1.1.1 Drum drying

Drum drying is a relatively old process that has been used to dry highly viscous liquids or pastes (Bonazzi et al., 1996). This technology is not commonly used by the dairy industry for drying milk except in the production of milk products with high content of fat, dairy products having enhanced flavor arising from the Maillard reaction for potential uses in the chocolate industry and special caseinate powders. Table 9.2 illustrates the properties of skim milk powder dried by drum drying, and for comparison also includes the properties of powders produced by spray drying. The table clearly indicates that the amount of heat treatment that the skim milk receives during the drum-drying process is significantly higher than that received during the spray-drying process. This is evidenced by the better functional properties of the spray-dried milk in terms of solubility and dispersibility. The Whey Protein Nitrogen Index (WPNI), developed by the American Dairy Products Institute, quantifies how much whey proteins are denatured by heat. It assesses the amount of undenatured whey protein present in the powder. Thus, high values are associated with less denatured proteins. Table 9.2 indicates that, due to higher processing times and higher temperature gradients, the drum-drying process produces skim milk with a large amount of denatured proteins that may affect their functional properties. Despite these differences, both methods produce skim milk powders considered to be low heat dairy ingredients.

Table 9.2 Comparison of quality of skim milk powder produced by drum and spray drying.

Property	Skim milk powder dried by drum drying	Skim milk powder dried by spray drying
Solid content (%)	94.6	> 99.5
Solubility index (%)	94.6	98.9
Dispersibility index (%)	88.0	98.9
WPNI (mg N/g of powder)	7.0	9.2

WPNI, Whey Protein Nitrogen Index.
Source: Adapted from Schuck 2011.

Drum dryers are commonly used in the corn wet milling industry for the drying of corn gluten separated from the corn kernel after stepping and grinding processes to reach final product moisture contents of about 10%. However, the process is not very efficient and requires about 0.07 kg of natural gas/kg of water evaporated (Ramirez et al., 2008). Despite the relatively low energy efficiency, drum drying is a practical alternative for the drying of products with high content of solids and high viscosity because its energy efficiency, although low, is comparable to other drying techniques but has the advantage of being able to handle larger solid concentrations in the feed. With regard to dairy applications, the consumption of steam is 1.3–1.5 kg of steam per kg of evaporated water, which translates to a specific consumption of thermal energy of about 990 kWh per tonne of water evaporated (Schuck, 2011), whereas the specific evaporation rate of the process varies from 10–30 $kg/m^2.h$ of water evaporated for products difficult to dry to 40–50 $kg/m^2.h$ of water evaporated for materials easy to dry (Daud, 2006).

Figure 9.1 illustrates a schematic of the drum-drying technology that uses a single drum. The feed can be injected into the rotating drum and spread by the use of one or two applicator rolls (Figure 9.1a). The product thickness in the rotating drum is affected by the applicator roll and drum rotation as well as the properties of the feed. An alternative feeding system uses the principles of dip coating where the rotating drum picks the feed (Figure 9.1b). It has been demonstrated that the thickness of the product can be predicted as a function of the drum rotational speed and the properties of the feed. The analysis was done for Newtonian as well as non-Newtonian fluids (Campanella & Cerro, 1984; Campanella et al., 1986).

Another alternative in drum drying is using twin counter-rotating dryers fed through a pool formed at the nip of the two drums (Figure 9.2). The thickness can be estimated using models developed to describe the calendering of plastic processes (Daud, 1986; Levine et al., 2002; Mitsoulis, 2009). Other feeding and process arrangements are detailed by Daud (2006).

As schematically illustrated in Figures 9.1 and 9.2, essentially the process consists of the evaporation of the water contained in the product by passing it through one or more rotating drums heated internally by steam. The internal temperature of the drum is controlled by the pressure of the steam utilized in the process (Trystram & Vasseur, 1992). The feed of the material to

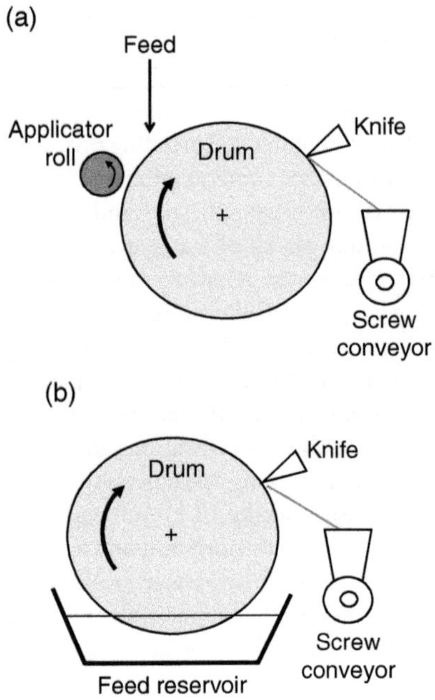

Figure 9.1 Schematic of the drum-drying system using only one drum. (a) System utilizing an applicator roll. (b) System using a feed reservoir where the feed is picked by the rotation of a roll in a process known as dip coating.

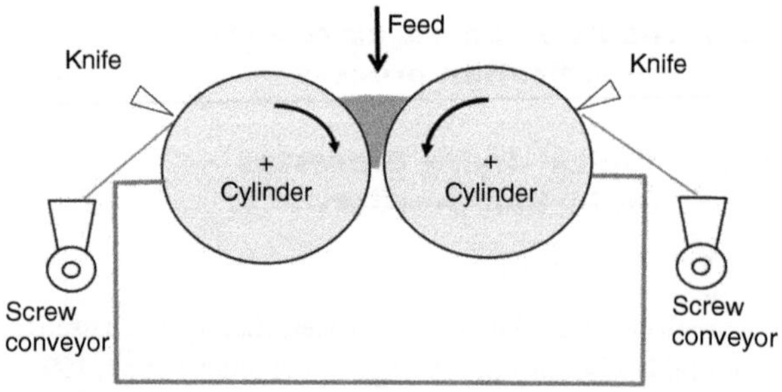

Figure 9.2 Schematic of a drum-drying system using two drums and feeding through the twin roller nip.

the dryer is a viscous slurry or paste that is spread on the heated cylinder(s), forming a thin sheet from which the water is evaporated (Daud, 2006). The first stage of the drum-drying process, when the content of water in the product is high, is characterized by boiling the water in the product at temperatures higher than the water normal boiling point (due to the boiling point elevation) if the drying is operating as an open system, i.e. at atmospheric conditions. This drying stage, known as evaporative drying, is characterized by very high and fairly constant drying rates. When the moisture content of the product decreases and the surface of the product becomes more dried, the drying rate decreases significantly because the movement of water is controlled by its diffusion through a porous internal medium and an outer dried layer, which is a crust formed on the surface of the product. At this stage, boiling decreases significantly but the porous structure is already set due to the low moisture of the product. An important option for heat-sensitive products is that evaporation can be achieved at temperatures lower than the normal boiling point of the solvent. For those applications the drum-drying system is enclosed and vacuum is applied.

A schematic of the temperature profiles in the product, drum and steam as well as the overall product moisture content in a cycle of the drum rotation is illustrated in Figure 9.3. The sudden exposure of the feed to the high temperature of the drum and the high heat flux present in the system quickly raises the temperature of the water present in the feed to a temperature higher than the normal boiling of water in open drying systems and remains constant whereas the product has free moisture. During the boiling stage, most of the free moisture is evaporated with a constant drying rate (Daud, 2006), which is noted by the almost linear changes of moisture in the product with the position in the drum. Since the boiling of the water requires a significant amount of heat, the temperature of the drum surface decreases due to the removal of that heat. When the free moisture in the product is evaporated, the temperature of the product increases in correspondence with the temperature of the drum surface (Figure 9.3). During that stage, the drying rate is significantly reduced. The different temperatures and moisture content profiles enable the classification of different zones in the drying process identified as the feed application zone, the boiling zone and the slow drying zone, which are indicated in Figure 9.3. Given the intensive heat flux, the temperature gradient and the boiling of the water of the product in the feed application and boiling zones, significant porosity is generated in the product, whereas in the slow drying zone the structure of the product is set before it is scraped off by the knife. It is assumed that the drying process occurs in about 75% of a drum revolution, with a residence time in a range of 30–35 seconds.

The heat flux (q) required for the evaporation of water in the product under steady-state conditions can be estimated by the following equation:

$$q = UA(T_s - T_{air}) \tag{9.1}$$

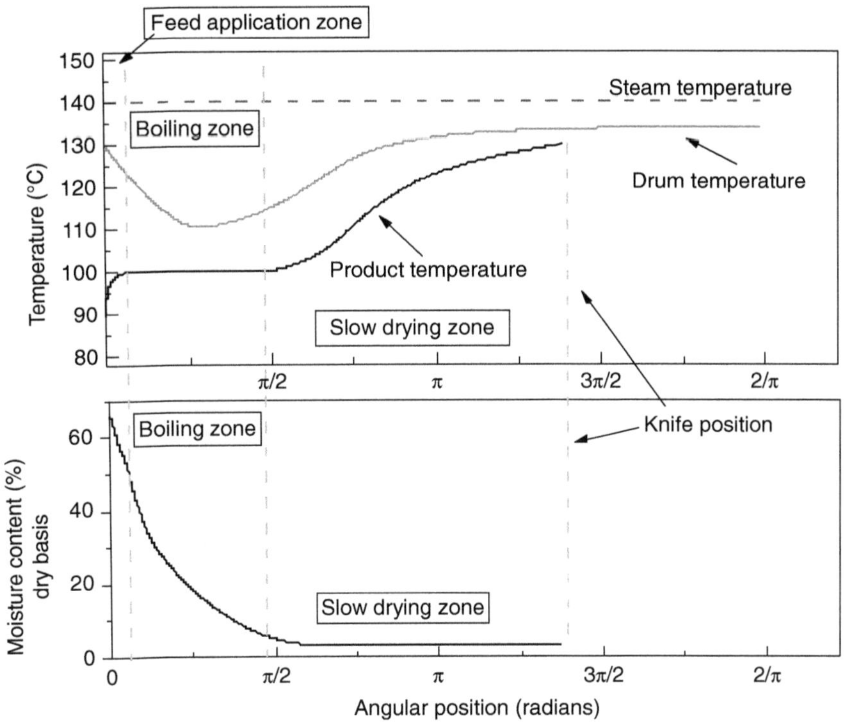

Figure 9.3 Temperature and moisture profiles at different positions in the drum. Temperatures of the steam, the drum and the product are also shown.

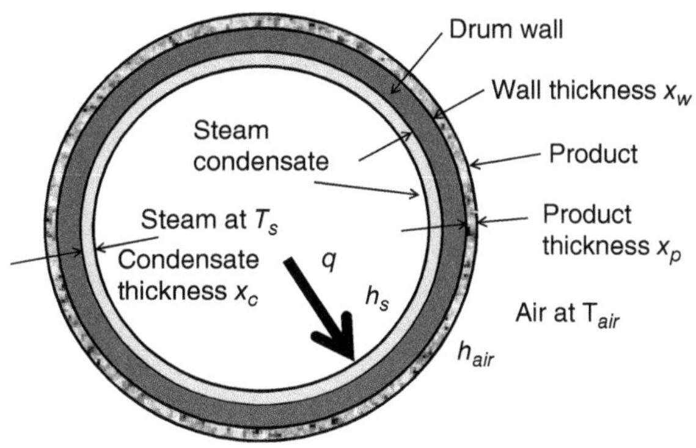

Figure 9.4 Heat transfer mechanisms and heat transfer resistances in a drum internally heated with steam.

where U is the overall heat transfer coefficient, which can be calculated by considering the heat flux resistances to convection in the steam side, the layer of steam condensate formed on the inside of the drum, the drum wall, the product layer, and the convection outside the product as schematically illustrated in Figure 9.4 (Okos et al., 2007). A is an average cylindrical area and T_s and T_{air} are the temperatures of the steam and the air outside, respectively.

By neglecting the curvature of the cylindrical surface, a condition that applies for large drum diameters, the overall resistance to heat transfer, $1/U$, can be approximated as (Fritze, 1972; Incropera & DeWitt, 2002):

$$\frac{1}{U} = \frac{1}{h_s} + \frac{x_c}{\lambda_c} + \frac{x_{wall}}{\lambda_{wall}} + \frac{x_p}{\lambda_p} + \frac{1}{h_{out}} \tag{9.2}$$

where h_s and h_{out} are the convection heat transfer coefficients inside and outside the drum, respectively; x_c, x_{wall} and x_p are the thickness of the condensate, the drum wall and the product, respectively; and λ_c, λ_{wall}, and λ_p are the thermal conductivities of the steam condensate, the drum wall and the product, respectively.

Although all these heat flow resistances are involved in the calculation of the overall heat transfer coefficient, many of them, such as the resistance of the layer of steam condensate and the drum wall, are often considered negligible because the heat transfer of condensing steam is high, also either the thickness of the drum wall is small or its thermal conductivity is large, or both. Thus, control of the heat transfer process is mostly governed by the resistance exhibited by the layer of the product. The amount of heat transferred through the drum is used to evaporate the water and dry the product. Models developed to describe this heat transfer include other effects involved in the process such as the presence of non-condensable gases that can accumulate within the drum affecting the steam temperature, and the presence of a pool of condensate at the bottom of the drum that is agitated by its rotation and affects the heat transfer. All these effects are incorporated into a local heat transfer coefficient in the steam side (Trystram & Vasseur, 1992; Vasseur et al., 1991).

Although the analysis based on Eqs 9.1 and 9.2 assumes conditions of steady state, the heat transfer through the product is essentially an unsteady state process in which the thickness of the product layer changes with time/position in the drum as well as its temperature, which is schematically illustrated in Figure 9.3. As a consequence of the different thermal conditions to which the product is subjected at different locations of the drum, microstructural characteristics such as porosity and product moisture content vary along the drum surface. This unsteady state heat transfer process has been described utilizing the equation for heat conduction as:

$$\frac{\partial T_p}{\partial t} = \frac{\lambda_p}{\rho_p c_p} \nabla^2 T_p \tag{9.3}$$

where c_p and ρ_p are the specific heat and density of the product and ∇^2 is the Laplace operator applied to the temperature of the product T_p. These properties as well as the thermal conductivity of the product, λ_p, are functions of the porosity of the material (Saravacos & Maroulis, 2001), which as discussed changes during the drying process. Due to the increase of porosity during drying, Trystram and Vasseur (1992) reported that the value of the product thermal conductivity decreases by a factor of 2–3 while drying progresses. A simplified model was developed by Vasseur et al. (1991) using the above equation to describe the unsteady state heat transfer in the product, heat flows in the steam side and outside of the drum as boundary conditions, and assuming constant values for the thermophysical properties. Results of the model showed trends similar to those depicted in Figure 9.3.

A model that describes the drying kinetics was developed by Trystram and Vasseur (1992), considering that the drum temperature is different from the temperature of the steam and varies with time during the drum revolution. The model also considers heat convection due to the drag flow of product created by the rotation of the drum. Major results of the model are presented as parametric curves of vapor temperatures versus drum rotation speed needed to achieve specific final moistures in the product. Variables considered in the model are the dry matter loading in the drum dryer, the feed moisture content and the use of one or two auxiliary rollers to distribute the feed on the drying drum. Other models used to describe drum drying have been published (Daud, 1991; Kostoglou & Karapantsios, 2003).

Convection induced by the product flow caused by the drum rotation is an aspect of singular importance not only for predicting accurate temperatures in the drum and drying rates but also to answer practical questions about the process in terms of product quality. Daud and Armstrong (1988) reported, simply by visual observation, that in the drum-drying process the product film has zones of mixing and stagnation upstream of the feed application point, specifically when the application is done by smaller rollers or when the feed is introduced in the nip formed by the position of the two rollers in twin drum systems. Knowledge of the flow pattern of the product in that zone, a problem that has been discussed in the calendering of non-Newtonian fluids (Daud, 1986; Levine et al., 2002; Mitsoulis, 2009), can be used to predict the mean residence time of the product, which is of practical importance to avoid exposing heat-sensitive products to prolonged heating. The approach of Cholette and Cloutier (1959), originally applied to stirred tank reactors, was used to estimate the percentage of stagnation and mixing zones in a drum-drying process (Daud, 1986). Results of the model that were supported by experimental validation showed a significant high distribution of the product residence time in the stagnation zone that can cause product overexposure to heat and represents a serious drawback for drying heat-sensitive products.

As mentioned, products that are suitable for drying with this technology are high-viscosity suspensions and pastes, including chemicals and foods (Daud, 2006). In addition to drying the product, drum dryers have been used to gelatinize cereal starch slurries for producing instant pregelatinized and weaning foods (Anastasiades et al., 2002; Kozempel et al., 1986; Onayemi, 1978; Supprung & Noomhorm, 2003; Vallous et al., 2002).

For non-starch materials, earlier applications of drum drying for food materials include tomato puree, milk, skim milk, and whey (Daud, 2006). Studies have focused on the effect of operational variables and also the type of dryer on the physicochemical and nutritional characteristics of the dried product, including starches (Abdel-Aal et al., 1996; Anastasiades et al., 2002; Vallous et al., 2002) and proteins (Arrage et al., 1992; Occeña et al., 1997). For starch-based products, properties such as moisture content, morphological properties of the product, texture and color have been studied. Moisture absorption and desorption isotherms were employed to indirectly study the porous structure of the dried material and how it was affected by processing conditions.

9.1.1.2 Spray drying

9.1.1.2.1 Principle

Spray drying is used in the dairy industry for the production of dairy powders including skim milk, different types of whey proteins, caseins, etc.. Additionally it is used to dry other food and non-food products such as instant coffee, coffee lightener, instant tea, eggs and baby foods, biochemicals, protein concentrates, and pharmaceuticals. Over 25,000 spray drier processes are estimated to be in commercial use for drying agrochemicals and biotechnology products, chemicals, dairy products, and pharmaceuticals (Filková et al., 2006). The convenience of spray drying is evidenced not only by the significantly higher throughput but also the notable differences observed in quality. Those benefits have influenced the dairy industry to adopt the spray-drying process for the drying of products such as skim milk and whole milk (Schuck, 2011).

During the spray-drying operation, a pumpable liquid suspension feed is atomized and sprayed into the stream of a hot drying medium (usually air but in some cases it can be a hot inert gas) moving inside a stainless steel chamber. For some low-capacity laboratory spray dryers, drying chambers made of glass are utilized. Heat is transferred from the hot air to the droplets, causing evaporation of the water contained in them. Although water is the solvent present in most food applications if flammable solvents such as ethanol are used, the air is replaced by an inert gas such as nitrogen. As a consequence of the evaporation of water from the product, the humidity of the hot air increases and its temperature decreases. Since the droplets formed have diameters of the order of microns and the rate of evaporation is extremely fast, the temperature of the droplet is low even though the air may have a high temperature. That relatively low temperature in the product combined with the short residence times of the droplets in the drying chamber makes spray drying an excellent alternative for heat-sensitive products. However, the process is less energetically efficient than drum drying.

A list of advantages and disadvantages of spray drying includes the following (Filková et al., 2006).

Advantages

- Product properties and quality can be effectively controlled.
- It allows for the drying of heat-sensitive products.
- It allows for large-capacity processes, up to several tonnes per hour.
- Equipment corrosion is largely reduced because of the lower moisture, or solvent, in the product when it enters in contact with the equipment surfaces.
- It produces uniform and spherical particles with the same solid composition as the solid composition in the feed.

Disadvantages

- It is limited to pumpable, not extremely viscous suspensions.
- It does not produce products of relatively high bulk density.
- It is not flexible. Design of the spray-drying system is targeted to a product with specific properties that are difficult to modify only by changing processing conditions.
- Large, more expensive, water evaporation by drying is required because the feed has to be pumpable so preconcentration by evaporation has practical limits in the maximum permissible solid concentration of the feed.
- Product recovery and fine and dust collection increase the cost of the process.

Spray drying is considered a one-step unit operation with three different stages: (1) atomization; (2) spraying a liquid-air (or inert gas), mix; and (3) separation of dry product from the air (or inert gas).

Depending on the movement of the air flow with respect to the movement of the droplets, spray-drying processes can be classified as either co-current (Figure 9.5a) or counter-current (Figure 9.5b). The co-current arrangement is usually used when foods and biochemicals are spray-dried because it minimizes the exposure of the product at high temperatures and reduces its impact on heat-sensitive materials. This is because when the liquid is atomized and contacts the hot dry air, the temperature of the droplet quickly approaches the wet bulb temperature of the air. At the end of the drying,

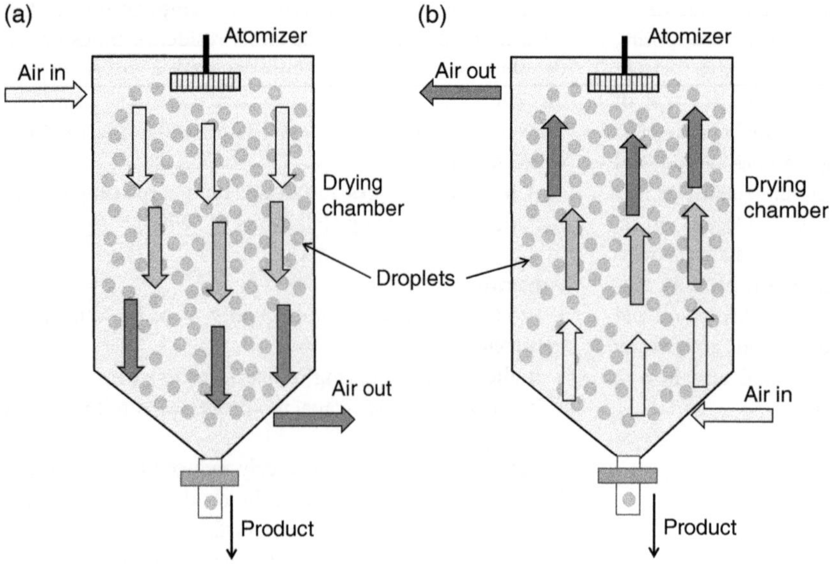

Figure 9.5 Schematic showing main components of a spray-drying process. (a) Co-current operation and (b) counter-current operation.

the temperature of the droplet approaches the dry temperature of the air but at that time, the temperature of the air has decreased significantly due to loss of its thermal energy that is used for the evaporation of the water or solvent in the product. Simple calculations using the psychometric chart can show that for an air temperature of about 290°C and an inlet absolute humidity of 0.01 kg water/kg dry air, the initial drying temperature of the droplet would be about 60°C whereas for those air inlet conditions, the outlet air temperature is about 120°C. Usually the product temperature is about 15–20°C below that temperature, i.e. in the 100–105°C range, but only for extremely short times. For heat-sensitive products the air temperature can be reduced to a 50–70°C range. In addition to the system schematically illustrated in Figure 9.5, spray-drying systems include air flow distributors, liquid feed tanks, pumps to transfer the liquid from the tanks to the atomizer, a heater to heat the incoming air (or inert gas), fans and blowers to supply the air to the drying chamber, and filters to remove dust particles in the incoming and exiting air. Often, the air flow arrangement used permits most of the dried powder to separate from the air due solely to the action of gravity or centrifugal forces. However, more frequently, cyclones or bag filters are used to recover the dry products. Fines or dust separated with the air in cyclones can be recovered.

For the counter-current spray-drying process, the droplets reach high temperatures so it is not convenient for heat-sensitive materials although the process achieves a greater thermal efficiency than co-current spray drying.

9.1.1.2.2 Atomization process

This is the more important operation in the process because it determines the size and size distribution of the droplets as well as their trajectories and speed which largely influence the size and physicochemical characteristics of the final powder. Three types of atomizers are available commercially. The two most commonly used in terms of practicability and simplicity of design are the rotary wheel, in its diverse designs, and the pressure nozzle single-fluid atomizer, whereas pneumatic two-fluid nozzles are used to atomize high-viscosity complex non-Newtonian fluids (Filková et al., 2006; Walzel & Furuta, 2011).

In rotating wheel atomizers, the liquid is fed by a metering pump in the center of a wheel rotating at a speed in a range of 10,000–30,000 rpm. These atomizers are used for high-viscosity fluids and slurry feeds and the rotating speed is a key variable to control the size distribution of the formed droplets (Walzel & Furuta, 2011). Guide vanes in the wheel cause the liquid to accelerate until moving at the same speed as the vanes. The liquid also moves radially outward due to centrifugal forces. The most common design of the wheel atomizer has radial vanes but in general the number and shape of the vanes differ according to the product characteristics and capacity requirements (Filková et al., 2006). Recent studies have shown the influence of the spraying system on the droplet formation and spray pattern (Walzel, 2011). For high-capacity applications, the number and height of the vanes are increased to maintain the same liquid height thickness in each vane;

rotary wheel atomizers with capacities up to 200 tonne/h are being commercially used (Filková et al., 2006). Rotary wheels with curved vanes instead of straight ones are used by the dairy industry because they produce powders with about 15% more bulk density.

During the atomization process, the liquid forms first a film and string-like streams, but quickly breaks up into droplets shortly after leaving the wheel. The formation of droplets in Newtonian fluids has been theoretically and experimentally studied and in general, droplets of uniform size and a narrow distribution are formed. However, when non-Newtonian or viscoelastic liquids and also liquids containing surfactants are sprayed, the formation of smaller satellite droplets or fluid strings negatively affects the particle size distribution (Bhat et al., 2010; Déchelette et al., 2011a; Notz & Basaran, 2004; Xue et al., 2008). In addition to these well-developed theories of liquid flow in atomizers, various empirical correlations have been proposed for calculating the mean drop size (the Sauter diameter, $D_{3,2}$) produced by rotary wheel atomization (Masters, 1979):

$$D_{3,2} = 1.31 d x 10^5 \left(\frac{\dot{M}_P}{Nd^2}\right)^{0.6} \left(\frac{\mu}{\dot{M}_P}\right)^{0.2} \left(\frac{\sigma(n_v b)\rho}{\dot{M}_P^2}\right)^{0.1} \quad (9.4)$$

$$D_{3,2} = 14x10^4 \left(\frac{1}{N}\right)^{0.6} \left(\frac{1}{\rho}\right)^{0.5} \left(\frac{\mu \dot{M}_L}{d}\right)^{0.2} \left(\frac{\sigma}{n_V b}\right)^{0.1} \quad (9.5)$$

$$D_{3,2} = 1.4x10^4 \cdot \frac{\dot{M}_L^{0.24}}{N^{0.83} \cdot d^{0.83} \cdot (n_V b)^{0.12}} \quad (9.6)$$

where $D_{3,2}$ (μm) is the Sauter diameter of the particle defined as $D_{3,2} = \frac{\Sigma_1^i f_i D_i^3}{\Sigma_1^i f_i D_i^2}$, whereas $\dot{M}_L$ is the mass liquid feed rate (in kg/h), $\dot{M}_P$ is the vane liquid loading ($\dot{M}_P = \frac{\dot{M}_L}{n_V b}$, kg/h.m), N is the wheel rotation speed (in rpm), d is the wheel diameter (m), n_v is the number of vanes in the wheel and b its height (m), ρ is the density of the fluid (kg/m^3); μ and σ are the viscosity (in cP) and the surface tension (in dynes/cm) of the liquid, respectively. For non-Newtonian fluids, which is the case commonly encountered during the drying of biomaterials, the viscosity of the fluid should be replaced by the apparent viscosity at the corresponding shear rate applied in the atomizer. For very large shear rates, the viscosity approaches the Newtonian behavior (refer to Chapter 4, section 4.4) so if that information is accessible, that value should be used in the above equations. For intermediate shear rates, an approximated calculation for the shear rate along with the power law rheological model described in Chapter 4 will allow for the calculation of the apparent viscosity as (Filková et al., 2006):

$$\mu_a = k\left[\left(\frac{r\omega^2}{m}\right)^2 \frac{V_V}{b} \frac{2n+1}{n}\right]^{\frac{n-1}{2n+1}} \quad (9.7)$$

where μ_a is the apparent viscosity of the liquid (Pa.s), ω the angular velocity (in rad/s), V_V the volumetric feed rate per vane (m^3/s), m (Pa.s^n) and n (dimensionless) are the power law rheological parameters of the liquid, named the consistency index and flow index, respectively. There is no preferred equation to predict the Sauter diameter but estimations given by Eqs 9.5–9.7 are within 30% of the experimental results (Filková et al., 2006).

In pressure-driven centrifugal nozzles, specifically single-fluid pressure atomizers, the liquid to be atomized is pumped through the vanes, causing the liquid to spin and entering as swirls inside the nozzle. The spray is created by conversion of pressure to kinetic energy as the liquid passes through the nozzle under pressures in the range of 20–70 bars. In forming the spray, the liquid leaves the orifice of the nozzle, with a diameter that ranges from 0.4 to 4 mm, as a sheet in the form of a hollow cone with an angle that varies from 40° to 140° (Filková et al., 2006). Single-fluid pressure atomizers have limited capacity and in general, one nozzle cannot exceed a capacity of 100 L/h. For large processing facilities, a system with several nozzles forming a manifold/multi-nozzle is used. In these cases, the energy consumption required to produce the droplets is smaller than that used in rotating wheel atomizers.

To estimate the droplet size from a single-fluid pressure nozzle, the pressure drop across the nozzle is required, and a rough estimation of the diameter of the droplets is given in the following equation (Okos et al., 2007):

$$D_a = \frac{9575}{\Delta P^{1/3}} \quad (9.8)$$

where D_a is the average droplet diameter (μm) and ΔP (Pa) is the pressure across the nozzle. Pressures ranging from 2 to 200 bars are used to operate these nozzle atomizers.

Pneumatic nozzles are also known as two-fluid atomizers in which either compressed air or steam is used to atomize the liquid. The feed is mixed with the air outside the nozzle and approximately 0.5 m^3 of air at pressures ranging from 1.5 to 8 bars are used per kg of liquid, which represents a significant operating cost. Maximum capacities of 1000 kg/h can be achieved using this system and the Sauter mean diameter can be predicted by the following equation (Filková et al., 2006):

$$D_{3,2} = \frac{535x10^3 \sqrt{\rho/\sigma}}{v_{rel}} + 597\left(\frac{\mu}{\sqrt{\rho\sigma}}\right)^{0.45} \cdot \left(\frac{1000\,\dot{V}_{FL}}{\dot{V}_{AIR}}\right) \quad (9.9)$$

The liquid density, viscosity and surface tension in this equation are expressed in SI units, that is, kg/m^3, Pa.s and N/m, respectively. $\dot{V}_{FL}$ (m^3/s) and $\dot{V}_{AIR}$ (m^3/s) are the volumetric flow rates of the liquid and air, respectively. More elaborated theories and mechanisms to predict the mean diameter and size distribution of atomized droplets as a function of the type of atomizer, and properties of the fluids, mostly for applications with Newtonian fluids, as well as different types of spray systems are given by Déchelette et al. (2011b) and Omer and Ashgriz (2011). The diameter of the dried particle, d_p, can be calculated from the diameter of the formed droplets, d_d, the solid mass fraction of the feed, x_s, the densities of the solid, ρ_s and the liquid feed, ρ_l, as well as the porosity of the particle by the following equation (Walzel & Furuta, 2011):

$$d_p = \frac{d_d}{\left[(1-\varepsilon)\left(\frac{1-x_s}{x_s}\cdot\frac{\rho_s}{\rho_l}+1\right)\right]^{1/3}} \quad (9.10)$$

When designing a spray dryer, one of the primary objectives is to select a shell diameter large enough to prevent high moisture wet droplets striking the hot dryer wall. At high moisture contents, the glass transition of the material forming the droplet (glass transition concepts are discussed in Chapter 4, and in more detail by Champion et al., 2000) is significantly lower than the temperature of the dryer wall so the droplets touching the wall stick to the wall, not only affecting the quality of the product due to its thermal degradation but also creating a fire and explosion hazard. Figure 9.6 indicates the evolution of a typical dairy-based droplet temperature in the last stages of the spray-drying process superimposed with the corresponding glass transition temperature of the droplet as a function of its water content. The figure clearly shows the regions where sticking of the dried particles to the walls of the drier would occur. It has been stated that if the temperature of an amorphous

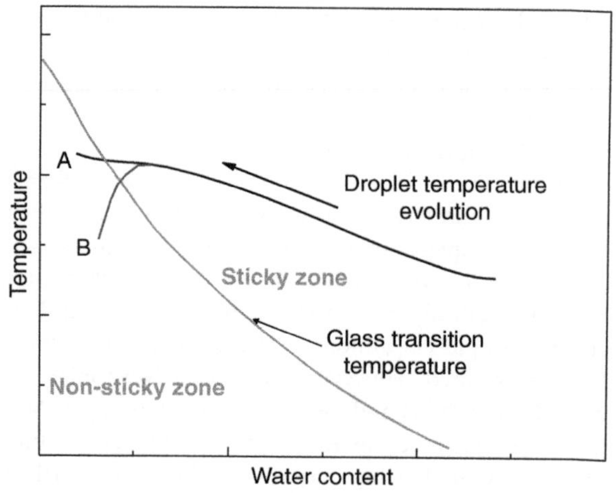

Figure 9.6 Evolution of droplet temperature in the drying process along the glass transition temperature demarking sticky and non-sticky zones.

material is above 10–20K, its glass transition, the material will be sticky (Adhikari et al., 2005). As illustrated in the figure, in the latest stages of the drying process, i.e. when a high-viscosity droplet becomes a semi-solid particle, the particle temperature is below its glass transition temperature and it will not stick to the dryer wall (point A). By changing the drying conditions, the particle temperature can be lowered even more, for example to the state indicated by the letter B, to minimize undesirable results that may affect the final product quality (Walzel & Furuta, 2011).

Most spray dryer systems have a cylindrical shell mounted on top of a conical surface where the dried droplets are conveyed to the exit valve. Rotary wheel atomizers produce droplets with high horizontal velocity so cylinder shells with a height to diameter ratio of 0.6/1 to 1/1 are used. Conversely, nozzle atomizers provide droplets with large downward vertical velocities and small horizontal velocities. For these cases, cylindrical shells with a height to diameter ratio of 3/1 to 4/1 are used, specifically for co-current spray dryers. Table 9.3 lists guidelines used in the design of industrial spray dryers (Masters, 1979).

9.1.1.2.3 Characteristics of the dried product

The physical properties of spray-dried powders, such as particle and bulk densities, as well as functional properties such as powder dispersibility and rehydration are related to their structural and physicochemical characteristics. It has been demonstrated using scanning electron

Table 9.3 Guidelines used to select size of spray-drying chambers.

Droplet diameter (μm)	Maximum evaporation capacity (kg/h)	Spray drying chamber (m)
40	20	1.0
60	150	2.0
80–100	1000	4.0
100–120	1500	5.0
120–150	1500	6.0

Source: Masters 1979. Reproduced with permission of John Wiley & Sons.

microscopy (Buma & Henstra, 1971a, 1971b) that most of the products produced by spray drying are hollow spheres with a density that is greatly reduced from that of a solid particle (about 1600 kg/m^3) to that of a hollow particle (about 300 kg/m^3). Typical spray-dried powders have densities within that range, which vary with the drying conditions.

In a series of publications, Verhey (1972a, 1972b, 1973) describes the mechanisms that explain the generation of the hollow and porous structure of spray-dried powders. It is hypothesized that their vacuoles and porous characteristics are induced by nucleation of the air entrained in the liquid feed during atomization in a kind of foaming process. The foaming is more pronounced when rotary atomizers are used and the feed has a low viscosity. During the drying process and due to evaporation of water from their surfaces, the droplets develop a high-viscosity layer that eventually forms a rigid crust (Verhey, 1973). While water is evaporated from the droplet, its temperature is maintained at low levels due to the evaporative cooling effect and the volume of the droplet is reduced due to the evaporation of water. But once the surface crust is formed and the amount of water in the droplet is low, further shrinkage of the droplet is drastically limited. Under these conditions, the solid network forming the droplet, especially the rigid crust, is at its glassy state and the little water evaporated produces expansion of the air bubbles inside the droplet, leaving behind vacuoles and increasing the porosity of the dried particles. Based on a no-air atomization process, Verhey (1972b) showed that when the droplets do not contain entrained air, shrinkage only occurs on the particle surface and the presence of vacuoles is not observed. Because of the occurrence of cracks due to shrinkage and the formation of the superficial rigid crust, the surface of the dried particles shows some wrinkling and denting and not a homogeneous structure (Verhey, 1973).

This seminal work has led to the conclusion that the porosity generation of spray-dried particles comprises two stages, the first in which air is introduced in the feed during the atomization process and a second and more complex mechanism in which formation of the rigid crust at its glassy state produces a drastic reduction of water diffusion to the droplet surface and a significant increase in droplet temperature, all affect in the morphology of the final powder. In addition to the atomization process, the drying conditions also contribute to providing a porous structure to dairy powders. For instance, increasing the drying air temperature has a significant and positive effect on the increase of the vacuole volume present in the hollow particles of skim milk powders whereas the viscosity and total solids concentration of the feed decrease the volume of these vacuoles (Verhey, 1972a). Further research using different gases (e.g. CO_2 and steam) and processing conditions that promoted air-free atomization clearly showed that vacuoles and the porosity of spray-dried powders are originated by the entrance of air during the atomization process. By using different atomization processes, Verhey (1972b) showed that the densities of dried powders were always significantly smaller than the densities of particles produced by no-air atomization conditions. Calculations used to estimate temperatures of the atomized droplets and temperatures of the drying air based on simple heat and mass transfer equations as well as an analysis using the theory of spontaneous or homogeneous nucleation confirmed the hypothesis that the porosity of the dried particles is originated by the entrance of air during the optimization (Verhey, 1973).

Morphological features of spray-dried milk particles have been studied by scanning electron microscopy (Aguilera & Stanley, 1999). Figure 9.7 shows a spray-dried skim milk particle exhibiting deep surface folds (Figure 9.7a) and a section of the particle showing the presence of vacuoles (Figure 9.7b).

Sano and Keey (1982) developed a mathematical model that describes the formation of hollow particles during the spray-drying process. It was assumed that when the particle reaches the drying falling rate period, its temperature rises above the wet bulb value. As a consequence, the vapor pressure of the moisture in the particle becomes greater than the pressure in the dryer and the droplet inflates at a rate that depends on the rheological properties of the material forming the particle. The inflation of the particle that generates the hollow sphere is initiated in air nuclei introduced during the atomization process, but

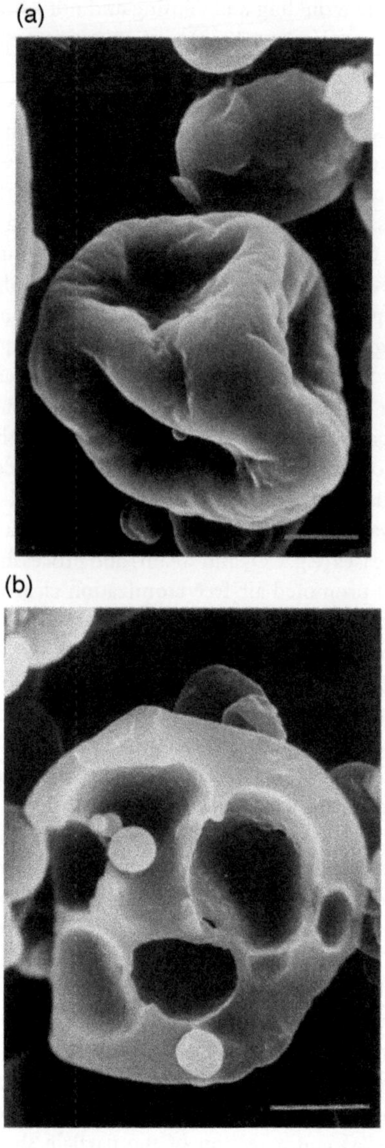

Figure 9.7 Scanning electron microscopy (SEM) of skim milk particles. Source: Aguilera & Stanley 1999. Reproduced with permission of Springer Science+Business Media.

these nuclei can also be provided by solutes or solids present in the droplets. At a certain stage of the inflation process, the particle bursts, releasing some moisture vapor through cracks formed in the external shell which leads to the shrinking of the droplets.

Sano and Keey (1982) described schematically the mechanism for deformation and formation of hollow particle spheres during spray drying, illustrated in Figure 9.8.

Several studies have been conducted focusing on the drying of individual droplets under different drying conditions and their effects on the morphology of the final dried particle (Walton, 2000; Walton & Mumford, 1999a, 1999b). Based on these results, Walzel and Furuta (2011) provide a complete description of the different particle morphologies obtained under different spray-drying conditions as well as possible mechanisms to form them. In agreement with previous work, the generation of dried particles with porous structures is due to the formation and expansion of bubbles inside the drying droplets as well as the formation of the crust.

Among the mechanisms assumed to affect the morphology of the particles are considered the possible formation of solid non-porous particles, especially for conditions in which the drying rate is slow. Although different morphologies can be obtained for successive expansion inflation-collapse-inflation cycles that depend on the drying conditions, the largest effect on particle morphology is attributed to the nature of the solute/solids contained in the liquid feed, which can be classified as: (1) solutions of low molecular weight substances; (2) solutions of polymers; (3) suspensions containing small solid particles; and (4) suspensions containing large particles. For feeds containing small molecular weight solutes, for example during the spray drying of fruit and vegetable juices or liquids containing sugar-rich components, the morphology of the dried powders, especially their surfaces, is affected by crystallization of solutes that reach the solubility limit as a consequence of the evaporation of water from the droplet surface. The supersaturation conditions of these solutes and their crystallization contribute to formation of the crust. Depending on the permeability of the crust, part of the solution and the surrounding air can penetrate inside the droplets to increase the inside pressure that expands the droplets into hollow spheres. For cases in which the solute does not crystallize, a very concentrated high-viscosity solution, probably at the rubbery state, forms the external shell of the particles. Since it comprises small molecular weight solutes, the glass transition of the shell material is well below the drying temperature and the powders become very sticky (see Figure 9.6), generating serious practical problems. The drying temperature, which is affected by the solvent evaporation from the droplet surface, has a significant effect on the morphology of the dried powder surface. Scanning electron microscopy of mannitol dried at low (80°C) and high (130°C) exhaust temperatures showed significant differences in surface morphology; the product dried at low temperatures had smooth

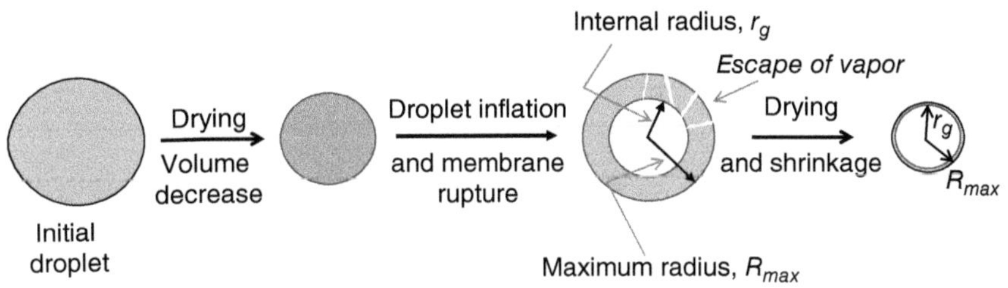

Figure 9.8 Schematic of the evolution of a liquid droplet during the spray-drying process.

surfaces whereas the product dried at high temperatures exhibited rough surfaces (Walzel & Furuta, 2011).

Many commercial applications of spray drying, however, are intended to dry products that contain large molecules in solutions or colloidal suspensions such as for example in the production of dairy powders. As discussed in Chapter 4 (see section 4.4), the rheological properties of many macromolecular solutions or colloidal suspensions are non-Newtonian and many of them exhibit viscoelasticity. Viscoelastic effects not only complicate the formation of droplets in the atomization process but also have a strong effect on the morphology of the dried powders. There exist practical rules that are used to atomize these fluids; for example, the spraying process has been limited to feeds with viscosities smaller than 200 mPa.s because higher viscosities cause the formation of fiber-like droplets combined with spherical droplets, and for viscosities larger than 10,000 mPa.s only fibers or filaments are formed, seriously limiting the spray-drying process (Walzel & Furuta, 2011). Due to the evaporation of water/solvent from the surface of the droplets, the solutes in the surface become more concentrated but because these solutes are large molecules, their diffusion to the center of the droplet is strongly impaired. Thus, during drying an external shell is formed and the interior of the droplet is liquid. When the surface of the droplet has little water/solvent content, the evaporative cooling effect ceases and the temperature of the droplet increases beyond the boiling temperature of the solvent, generating evaporation of water inside the droplet with the consequent formation of bubbles, which may penetrate the droplet shell and form blowholes in the droplet skin. The resulting dried powders are hollow and porous in the interior and on the particle shell; some typical particles of this kind are described in the article by Walzel and Furuta (2011).

Regarding spray-dried dairy powders, the drying process is similar to the one described above. Their diameters change during the drying process and initially the droplet shrinks rapidly because of the large evaporation of water in the early stages of the process. For liquids that have large amounts of readily crystallizable solutes such as lactose, shrinkage of the droplet may continue during the entire drying process. During the drying of whole and skim milk, the air desorbed in the droplets or entered by the atomization process starts to form nuclei in the center of the droplets. These nuclei become bubbles that grow as the air and water vapor pressure inside the bubble increase due to the increase of droplet temperature as the drying proceeds; while above 100°C bubble expansion/inflation starts. When the temperature of the droplet is higher than the T_g of the solid material forming the droplet, the expansion of these bubbles is favored and the droplets can reach up to 90% expansion. However, expansion is halted when the moisture of the droplets is significantly decreased and the T_g of the solid material forming the droplets is increased beyond the particle temperature, which makes the solid material become rigid. Under high temperatures and high initial moisture contents, the droplets can collapse, which makes their surface more irregular. The consequence of this process is the production of spray-dried milk powders that are hollow spheres with densities significantly smaller than the density of the solid itself, for example 350 kg/m^3 versus 1600 kg/m^3. These powders of low density are fluffy and do not wet or sink readily when brought in contact with water or any other polar liquid.

Lactose and other sugars present in milk powders do not crystallize during drying but instead are transformed to an amorphous material (Alexander & King, 1985; Bhandari & Howes, 2005; Buma & Henstra, 1971a, 1971b; Jayasundera et al., 2009). Amorphous lactose is highly hygroscopic and the presence of water promotes recrystallization during storage and cements the powder particles into large lumps, producing caking. The change in lactose from amorphous to crystalline releases water and causes microstructural collapse, favoring non-enzymatic browning of whey powders (Saltmarch & Labuza, 1980) and also skim milk powders (Morgan et al., 2005). Using scanning electron microscopy, Buma and Henstra

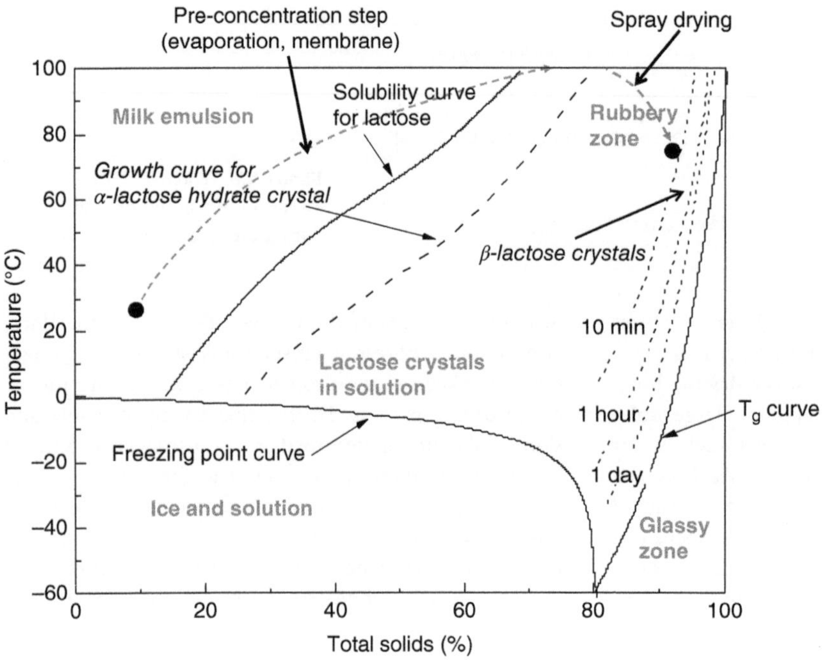

Figure 9.9 Temperature and solid content evolution of a dried whole-milk particle along with its state diagram.

(1971a, b) demonstrated the hollow sphere structure of spray-dried dairy powders as well as the distribution of free fat on the powder surface. The shells of these hollow particles have a glassy structure, primarily formed by amorphous lactose. Further work using scanning electron microscopy showed the fate of lactose when milk is dried and its relationship to quality losses during the storage of milk powders (Aguilera & Stanley, 1999). Crystallization of lactose can also occur if the feed (whole and skim milk and whey protein) has very high solid contents and processing temperatures are low. That crystallization can cause problems during the spray-drying operation.

Figure 9.9 provides the state diagram for whole milk describing its states during the preconcentration and spray-drying processes (Vuataz, 2002). It also includes changes during storage of the powder indicating the zones where the crystallization of lactose occurs and the glass transition curve to indicate the conditions where lactose eventually reaches the glassy state. During the preconcentration by evaporation, it is important to avoid crystallization of lactose and the evaporation process should be conducted at temperatures higher than the solubility of lactose. Other components of dairy powders and their chemical composition are of importance and affect their functionality. Specifically, the free fat located on the powder surface has a large influence on quality and industrial applications.

Fäldt et al. (1993) discuss the advantages and disadvantages of having fat on the surface of dairy powders. Fat molecules are prone to oxidation and their presence on the powder surface is a serious drawback when the powder is in contact with air. In addition, the natural hydrophobicity of fat molecules negatively affects the dispersibility of dairy powders in water. From a different point of view, for example in the manufacture of chocolate products, whole-milk powders with a high surface fat content are preferable because they have important functional properties in these applications.

Quantification of free fat in powders, defined as the amount of fat which can be extracted by an adequate organic solvent, presents some problems because during the extraction process not only fat in the surface of the powders is extracted. Fat in the interior of the particles, which is accessed by the organic solvent (transferred to the interior of the particle through capillaries, cracks and pores) can also be extracted. Fäldt et al. (1993) used the electron spectroscopy for chemical analysis (ESCA) technique to study fat coverage on the surface of spray-dried emulsions. With this method, the sample is exposed to an X-ray beam of known wavelength which is able to eject electrons from the sample. The spectrum of emitted electrons is analyzed and compared with a spectrum generated by the individual chemical components forming the sample. The wavelength of the X-ray beam is selected

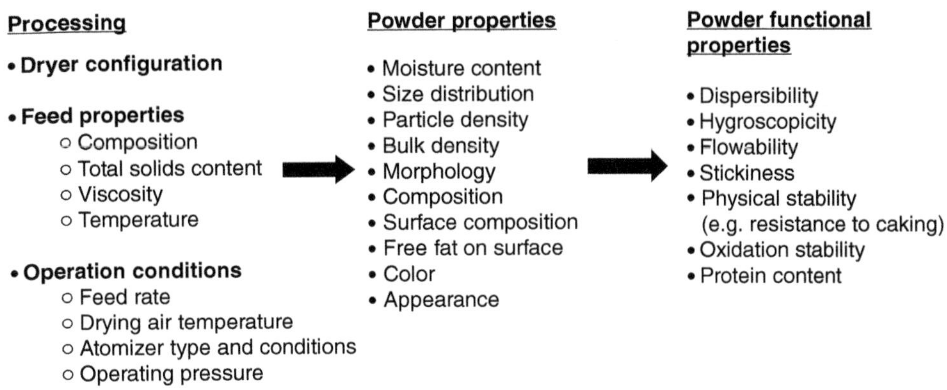

Figure 9.10 Interrelationship between processing conditions, powder properties and functional properties important for uses of dairy powders.

to assure that the emission of the electron spectrum is only from, or near, the surface of the sample (about 10 nm of the surface).

In a study described in a series of articles, Kim et al. (2002, 2003, 2005a, 2005b, 2009a, 2009b, 2009c) used the ESCA technique to characterize the chemical composition of the surface of different spray-dried dairy powders, the distribution of fat between the surface and the interior of dairy powder particles as well as the influence of the chemical composition of these powders on their functionality such as flowability, wetting, and stability during storage. The work also describes mechanisms of surface formation along with the effects of drying conditions and the chemical composition of the dried powders. Many interrelationships between processing parameters, the physicochemical properties of the dried powders and their functional properties are given in Figure 9.10. Kim et al. (2009b) provide a schematic of droplet evolution with time in terms of moisture content and temperature and morphology for dairy powders.

Figure 9.11 illustrates the changes in water content and temperature of a drying whey protein droplet for three different temperatures of the drying air (low, medium and high temperatures) as well as a schematic of the particle morphologies obtained during the drying process. Especially at higher air drying temperatures, two clear drying regimes are noted. In the first period, the drying rate is constant and the water evaporated from the surface is replenished rapidly by diffusion of water from the interior of the droplet to keep the surface saturated. Thus, the temperature of the droplet surface remains approximately constant and equal to the wet bulb temperature of the drying air. The evaporative cooling phenomenon prevents the temperature of the droplet rising which is clearly observed in the first drying period for all three drying temperatures. As a consequence of the water evaporation, the volume of the droplet decreases. As the drying progresses, the total solids content of the droplet increases and the diffusion of water from the droplet interior decreases significantly and is not sufficient to keep the surface of the droplet wet, so a rigid crust is formed on the droplet surface. The temperature of the surface increases to values that depend on the air drying temperature whereas the drying rate is significantly decreased, indicating the start of the second drying period (Bernard et al., 2008; Kim et al., 2009b). In the second drying period, the crust layer increases in size and the water from the interior of the droplet evaporates, leaving the hollow structure described before for general non-dairy particles (see Figure 9.11). Due to possible crystallization of some molecules, collapse and rupture of the crust and escape of vapor through the cracks the final structure of the droplet varies. In general these changes are affected by the drying conditions and the nature of the product to be dried. Regardless of the fine details of the final morphology, the general characteristic of the powder is its appreciable porosity and a chemical composition on the surface that depends on the processing conditions.

9.1.1.2.4 Mass and heat transfer mechanisms in spray drying

A mass and energy balance for the spray-drying operation is given by Langrish (2009). Heat energy enters the dryer primarily through the hot air used for drying the product. Some of the energy is also entered with the feed. Figure 9.12 illustrates the spray-drying process and the balance of energy is given by the following equation:

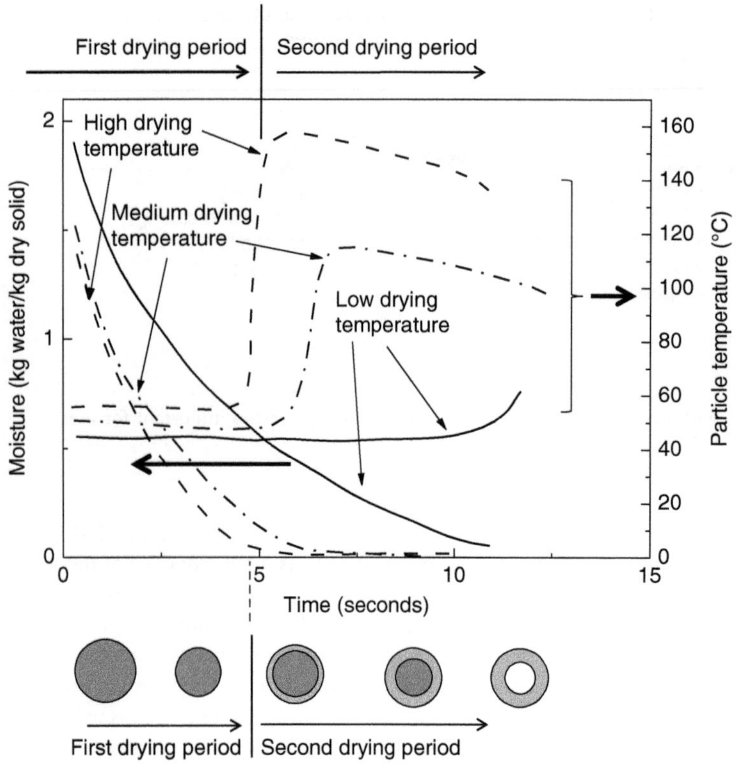

Figure 9.11 Moisture, temperature and morphology evolution of a dairy drying particle at three different drying temperatures (low, intermediate and high).

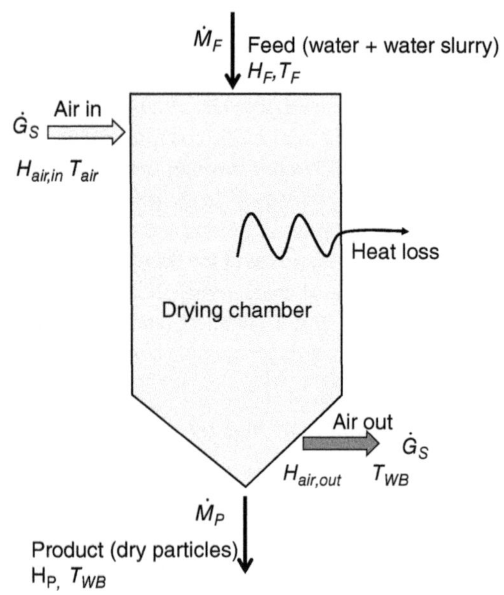

Figure 9.12 Schematic of a spray-drying system indicating mass and energy exchange of the system with the ambient.

$$\dot{G}_S\left(H_{air,in} - H_{air,out}\right) = \dot{M}_S\left(H_P - H_F\right) + Heat\,Loss \tag{9.11}$$

where $\dot{G}_S$ is the mass flow of air in kg dry air/s, $H_{air,in}$ and $H_{air,out}$ are the enthalpies of the air entering and leaving the dryer respectively, in kJ/kg dry air, $\dot{M}_S$ is the mass flow of dry solids entering the dryer in kg dry product/s, and H_F and H_P are the enthalpies of the feed and product respectively, in kJ/kg dry product. Enthalpy of air can be expressed in terms of the heat capacities of air and water vapor, the latent heat of vaporization of water at a reference temperature and the air temperatures as:

$$H_{air} = \left(\bar{C}_{air} + Y\bar{C}_{w,v}\right)\left(T - T_{ref}\right) + Y\Gamma_{ref} \tag{9.12}$$

where $\bar{C}_{air}$ and $\bar{C}_{w,v}$ are the average heat capacities of air and water vapor in a range of temperatures between the reference temperature T_{ref} and the air temperature T, Y is the absolute humidity of the air (kg water/kg dry air), and Γ_{ref} the water latent heat of vaporization at the

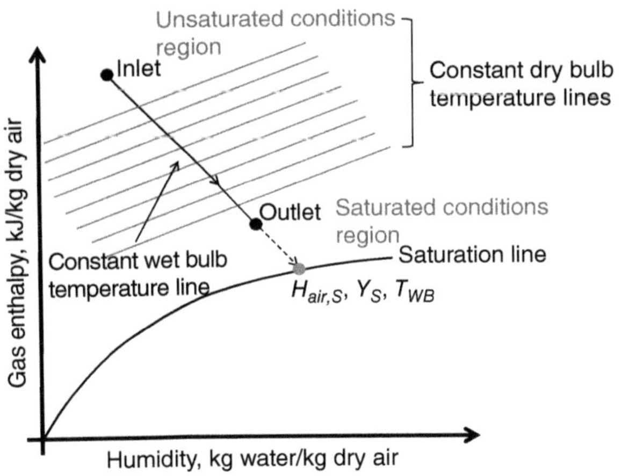

Figure 9.13 Evolution of the drying gas conditions from inlet to outlet in terms of the gas enthalpy and its absolute humidity.

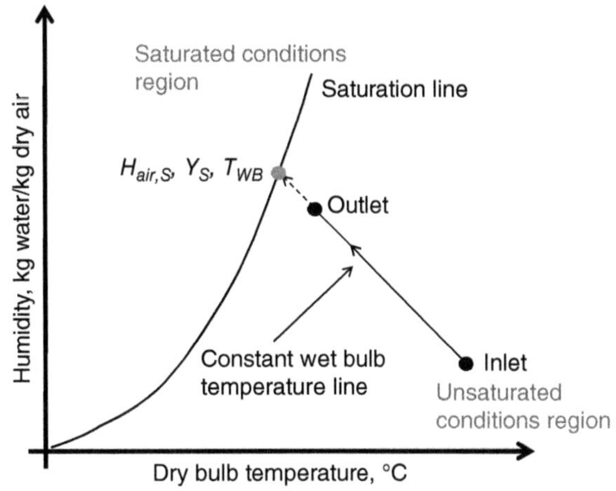

Figure 9.14 Evolution of the drying gas conditions from inlet to outlet in terms of the gas humidity and its temperature (dry bulb temperature).

reference temperature T_{ref}, in kJ/kg water. The enthalpy of the feed and the product can be expressed as:

$$H = (\bar{C}_S + X\bar{C}_w)(T - T_{ref}) \tag{9.13}$$

where $\bar{C}_S$ and $\bar{C}_w$ are the average heat capacities of the solids to be dried and liquid water in the range of temperature T_o - T, in $\frac{kJ}{kg\,dry\,solid \cdot K}$ and $\frac{kJ}{kg\,water \cdot K}$ respectively. X is the amount of water in either the feed or the product in a dry basis, i.e. $\frac{kg\,water}{kg\,dry\,solid}$. The mass flow of feed and the product on a wet basis can be calculated as:

$$\dot{M}_F = \dot{M}_S(1 + X_F) \tag{9.14a}$$

$$\dot{M}_P = \dot{M}_S(1 + X_P) \tag{9.14b}$$

Figure 9.13 is an enthalpy-humidity plot, which shows the evolution of the drying air in the dryer. It moves from unsaturated condition to saturated condition regions. For an adiabatic drying process where heat losses are considered negligible, the evolution of the drying air from the unsaturated to the saturated condition region follows an isenthalpic process, which for the air/water system approximately coincides with the line of constant wet-bulb temperature. Figure 9.14 shows the evolution of the air in a humidity-temperature plot.

The efficiency of the spray-drying process is defined as the ratio of the heat used in the evaporation of the water in the product to the total heat input in the process (Filková et al., 2006), calculated as:

$$\eta = \frac{\dot{M}_w \Gamma_{water}}{\dot{G}_S C_{p,air}(T_{air,in} - T_{air,out}) + \dot{M}_S C_{p,F}(T_{F,in} - T_{F,out})} \cdot 100 \tag{9.15}$$

The equation indicates that the total heat input is used to heat the air and the feed from inlet to outlet conditions. With the assumption that the air is out of the dryer at saturation conditions of humidity and temperature and that the product temperature is equal to the temperature of the air, which is equal to the temperature of the wet bulb T_{WB}, Eq. 9.15 simplifies to the following equation:

$$\eta = \frac{\dot{M}_w \Gamma_{water}}{\dot{G}_S C_{p,air}(T_{air,in} - T_{WB}) + \dot{M}_S C_{p,F}(T_{F,in} - T_{WB})} \cdot 100 \tag{9.16}$$

where Γ_{water} and $\bar{C}_{p,air}$ are the latent heat of vaporization of water and the specific heat for air at the average drying temperature, respectively, and $\dot{M}_w$ is the water evaporation rate.

For an adiabatic drying process the above equation further simplifies to:

$$\eta = \left(\frac{T_{air,in} - T_{air,out}}{T_{air,in} - T_{amb}}\right) \cdot 100 \tag{9.17}$$

where T_{amb} is the ambient temperature.

To establish comparisons between efficiencies of spray-drying systems utilizing the above equation, it should be kept in mind that the residual moisture of the product should be the same because the liquid moisture carried out with the product may have a significant influence on the balance of energy (Bellinghausen et al., 1994). The above equation also indicates, theoretically, that in order to achieve higher efficiencies, the spray-drying process should operate with a maximum difference between the inlet and outlet temperatures in the air, i.e. with high temperature in the incoming air and low temperature in the leaving air. Those options are often not practical due to the effects that the high drying temperature may have on the quality and morphology of the dried powder, as discussed in the previous section. Furthermore, drying processes with low outlet air temperatures produce powders of high moisture content, which bring serious quality concerns.

Increase in the amount of solids in the feed has proved to be a convenient option to improve the efficiency of the drying process because it is more economical to evaporate water from the product in single or multiple stage evaporation operations than in a drying process. It has been reported that an increase in the feed solid content from 10% to 25% results in a 67% reduction in the supplied heat. Filková et al. (2006) provide a comparison between the energy necessary (in kJ) to evaporate 1 kg of water in the product in several processes (Table 9.4). As indicated in the table, spray drying is an energy-demanding process compared to other processes so the approach followed by the dairy industry has been to preconcentrate, by using multiple effect evaporators, the feed going to the dryer as much as possible and within practical limits under which the concentrated feed does not affect the quality of the milk powder or the process itself. Studies on the effects of the rheological properties of milk (skim and whole-fat milk) on the atomization process and quality of the milk powder recommend a solid content in the range of 44–48% for an efficient spray-drying operation (Baldwin et al., 1980; Bloore & Boag, 1981; Trinh et al., 2007). However, it is generally accepted that the amount of solids in the feed will affect the heat consumption for the production of powder. Table 9.5 illustrates the dependence between energy consumption and concentration of the feed (Filková et al., 2006).

Table 9.4 Comparison of energy efficiency for different preconcentration processes used in the dairy industry and spray drying.

Process	Heat consumption for evaporation of 1 kg of water (kJ)
Membrane (UF or reverse osmosis)	140
Evaporator 1 stage	2600
Evaporator 2 stages	1300
Evaporator 6 stages	430
Evaporator 6 stages with thermocompression	370
Evaporator 6 stages with mechanical compression	400
Spray drying	Up to 6000

Table 9.5 Effect of solid content in the feed on the energy consumption for the evaporation of water in the spray-drying process.

Solids in the feed (%)	Energy consumption (kJ/kg powder)
10	24,000
20	10,500
30	6200
40	4000
50	2700

9.1.1.3 Freeze drying

9.1.1.3.1 Principle

In the freeze-drying process, a material is dried from a frozen state and the solvent sublimates directly from the surface of the solid network. Figure 9.15 illustrates the pressure-temperature diagram for water. In the figure, the changes undergone by the water during a freeze-drying process are included. The process starts with the water contained in the product at room temperature and atmospheric pressure (state A), the product is frozen at constant pressure and most of the freezable water becomes ice at temperatures below the freezing point (state B). Although the initial and final states of the freezing process are considered equilibrium states, the freezing rate has a significant effect on the quality of the final product. For instance, it has been found that quick-frozen foods (such as meat and coffee) reached a whiter color than the same foods frozen slowly (Flink, 1975; Karel, 1975; Ratti, 2008). This is attributed to the smaller ice particles formed during the quick-freezing process which impart a finer pore distribution in the product after it is freeze dried. This results in better reflection of light and a product with a whiter appearance in contrast with the color of products that are slow frozen and have a coarse pore distribution.

After the product is frozen, the pressure is decreased at constant temperature to the state indicated by C, below the water triple point. Afterwards, the water in the product is sublimated at a constant pressure and the residual non-freezable water in the product reaches the temperature of the dried product, commonly at room temperature (state D). For comparison, phase changes and states

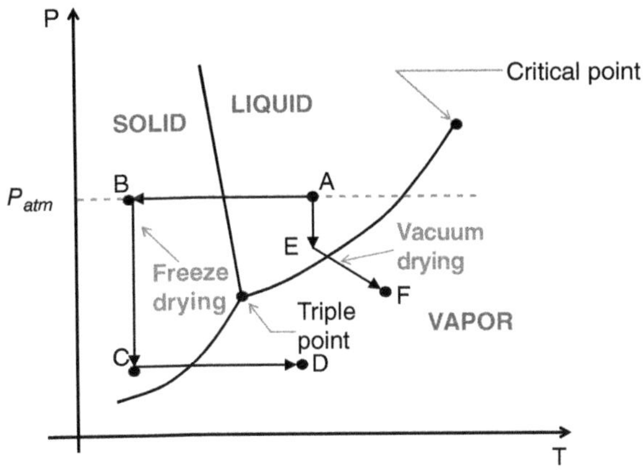

Figure 9.15 P-T diagram for water, including the different conditions of the water substance in a freeze-drying process and also for vacuum drying.

undergone by water when the product is dried using a vacuum-drying process are included in the diagram (states E and F). As observed in the figure, temperatures in the process are higher than those used in the freeze-drying process, and often could be higher than the glass transition temperature of the dried solids, which as will be discussed could cause collapse on the product structure, thus impairing its quality.

Products that are freeze dried from solutions of solutes such as salts, sugars, and macromolecules have a freezing point lower than the freezing point of the pure solvent. However, proper sublimation generally occurs below the eutectic point, which is the lowest freezing point reached by the solution (Kessler, 2002). For most biological materials the solvent is water, which is desorbed from zones of high concentration of solutes where there is also a certain amount of unfreezable water. A schematic of the freeze-drying cycle profile is illustrated in Figure 9.16. The three stages indicated in the figure, freezing, primary drying, and secondary drying, constitute the entire process. In the first stage, the product is frozen and after the phenomenon known as thermal arrest (indicated in the figure), nucleation and formation of ice start and the temperature of the process is maintained constant until all freezable water is frozen. The characteristics of the freezing process are important because they have an important effect on the drying rate occurring during the primary and secondary drying stages. If freezing is performed at higher rates, the sizes of ice crystals are small and although as discussed above this may have a beneficial effect on the

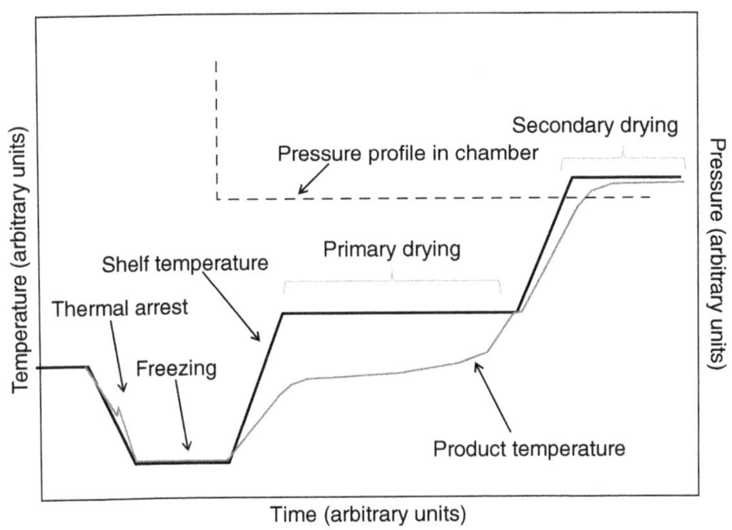

Figure 9.16 Temperature profile in a freeze-drying process including the temperature in the product. Source: Liu 2010. Reproduced with permission of John Wiley & Sons.

properties and quality of the dried matrix, the remaining porous structure after the ice crystals are sublimated has a small size distribution, which increases resistance to the passage of the water vapor through the dried layer during primary and secondary drying. Conversely, if the freezing is conducted at a slow rate, the ice particles formed are large and although this may negatively affects the properties of the dried solid matrix, the porous structure left after ice sublimates has a large size distribution, which results in larger primary and secondary freeze-drying rates.

During the initial phase of the drying process all free or freezable water, in the form of ice crystals, is progressively removed by sublimation. And at the end of this stage, known as the primary drying phase, some bound or unfreezable water can still be removed from the porous structure left after the ice is sublimated by using higher temperatures. Thus, higher shelf temperatures are used in the primary drying stage. At the last stage of the process, known as secondary drying, only desorption takes place and the product is no longer in the frozen state. Freeze-drying operations are generally conducted under vacuum conditions ranging from 100 to 1500 milliTorr. Primary and secondary drying stages are indicated in Figure 9.16 in which typical shelf and product temperatures as well as the pressure profile used in the drying chamber are also schematically shown.

A schematic of the basic set-up of a simple vacuum freeze-drying unit is illustrated in Figure 9.17. The product is placed in a perforated tray positioned between two heating plates at temperature T_P, which transfer heat mainly by radiation to the outer surface of the product that is at temperature T_o. Other arrangements incorporate a heated bottom plate on which the sample rests and heat is transferred by conduction (Kessler, 2002). In order to sublimate the formed ice crystals in the freezing process, the chamber pressure is reduced to at least below the vapor pressure of ice at the product temperature, while the temperature of the radiant or conductive plates is increased to supply the energy for the sublimation process (Liu, 2010). In industrial dryers, the heating plates are internally heated by a heating medium whereas in small laboratory dryers the heating is performed by electrical resistances. Most of the water sublimated during drying condenses on the condensers whose low temperatures are maintained by a refrigeration unit. Along with the water vapor, the drying chamber may contain small amounts of non-condensable gases produced by leaks in the system or gases derived from the product that have to be eliminated by the vacuum pump. The vacuum pump also maintains the vacuum in the chamber. It has to be designed to evacuate the chamber at the start of the process, but the pumping rate used to remove non-condensable gases during operation is significantly smaller. Thus, often two pumps are used, one for carrying up the initial evacuation and a second one, a holding pump, to remove non-condensable gases. Practically,

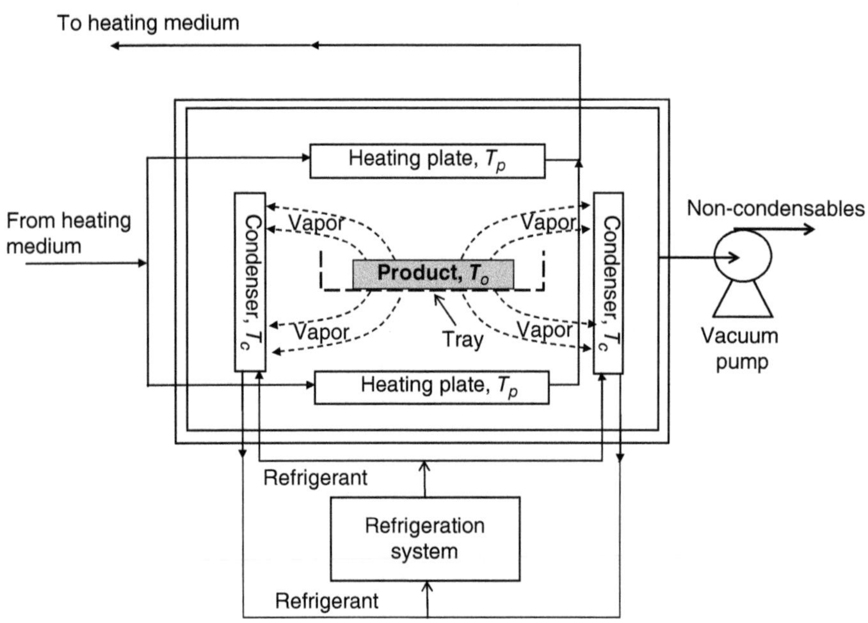

Figure 9.17 Typical freeze-drying set-up.

the pressure to be reached in the vacuum chamber, p_o, is slightly higher than the vapor pressure of ice at the condenser temperature, p_c^*. Hence, the condenser temperature, T_c, which is set and controlled by the refrigeration system, is a key process parameter to control the dryer pressure.

Food products that are freeze dried commercially include beverage extracts such as soluble coffee, vegetables, fruits to be added to ready-to-eat breakfast cereals, meals including vegetables and meats for camping and military rations due to their light weight and capacity for easy hydration and re-expansion to their normal size and shape. Biological, pharmaceutical and biochemical products are also freeze dried to keep their viability during storage. These materials are usually confined in vials that are partially closed with elastomeric closures (stoppers). After secondary drying, an inert gas, e.g. nitrogen, is introduced inside the chamber and when the pressure is increased to full atmospheric pressure, the vials are fully stoppered within the product chamber by hydraulic compression of the shelf stack and stored (Liu, 2010).

Advantages of freeze drying include: (1) low drying temperatures, so thermal damage is minimized; (2) good retention of volatiles and aromas; (3) the formation of very porous materials having fast rehydration; (4) no shrinkage or collapse of product because it is dried at temperatures below its glass transition temperature; (5) the biological activity of the product is maintained during storage. Disadvantages are associated with: (1) the high cost of the process; (2) susceptibility of the products to freeze-induced damage; (3) the tendency of freeze-dried products to rapidly pick up moisture due to their high porosity unless they are packed and maintained in low relative humidity; (4) an enhanced photosensitivity of some freeze-dried products, specifically when the process is applied to dry biotechnological products (Greiff & Rightsel, 1965).

9.1.1.3.2 Mass transfer

Heat and mass transfer mechanisms existing in the freeze-drying process are schematically visualized in Figure 9.18. The simplest mathematical analysis of the process is done through a slab of thickness *2a* placed in a perforated tray located between the heating plates. During the course of drying, the sublimation interface recedes within the product, which in many instances can be fairly uniform, showing a relatively sharp demarcation between the frozen central core and an outer porous "dried" layer. The sublimating water from the ice in the frozen layer tends to leave channels that promote the porosity of the dried layer. As a consequence of the porous structure and the low content of water in the dried layer, its color is pale and contrasts with the darker color of the frozen layer, which makes the sublimation interface visible. Although the use of a slab is an oversimplification to visualize this complex process, in the freeze drying of materials with irregular forms, a sublimation interface that is curved but still a relatively distinct line between the dried and frozen zones is visible.

Based on Figure 9.18, simplified models that describe the heat and mass processes can be developed. The flow of vapor through the porous dried layer per unit of area cross-sectional to the flow, G_s, can be estimated by defining a permeability coefficient, K_p, which is multiplied by the pressure gradient in the porous dried layer to give the flow of vapor through the porous network:

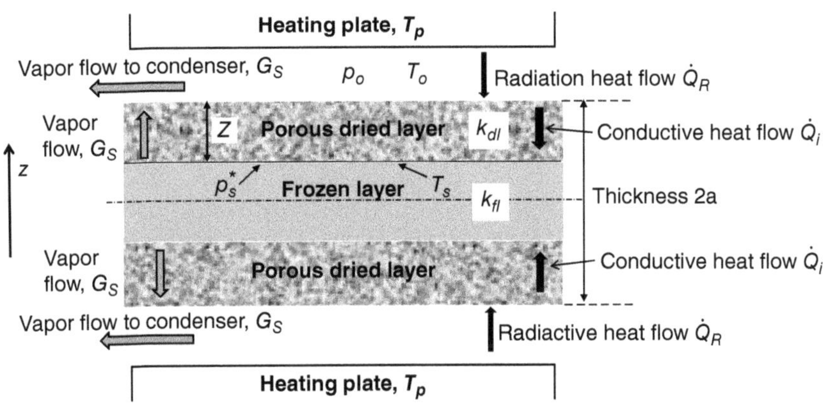

Figure 9.18 Product being freeze dried, indicating its structure changes due to the process as well as mass and heat transfer phenomena involved in the process.

$$G_S = K_p\left(\frac{p_s^* - p_o}{Z}\right) \quad (9.18)$$

As indicated in Figure 9.18, Z is the thickness (in meters) of the dried layer, which changes during the process, and G_s is the mass flow rate of sublimed water per square meter (kg/m^2.s). The pressure inside the drying chamber is p_o whereas p_s^* is the vapor pressure of water at the sublimation interface (in Pascal). Since the units of mass flow vapor G_s, in the SI system, are in $\frac{kg\,of\,vapor}{m^2.s}$, the units of permeability should be in seconds (s). However, as the pressures used in freeze-drying operations are very small and conveniently measured in μHg (microTorr), values of K_p are expressed in units of kg/m.s. μHg. If the initial and final moisture contents of the product are expressed in a dry basis (kg of water/kg of dry product) and designated as X_o and X_f, respectively, and ρ is the density of the frozen undried product, the mass flow of water vapor can also be calculated as:

$$G_S = \rho\left[\frac{X_o - X_f}{1 + X_o}\right]\frac{dZ}{dt} \quad (9.19)$$

By combining Eqs 9.18 and 9.19 the following equation is obtained:

$$\rho\left[\frac{X_o - X_f}{1 + X_o}\right]\frac{dZ}{dt} = K_p\left(\frac{p_s^* - p_o}{Z}\right) \quad (9.20)$$

If p_s^*, p_o and K_p are assumed to be approximately constant, a simple integration of the above equation allows for estimation of the drying time for the slab, as the time under which $Z = a$ by:

$$t_D = \frac{\rho}{2K_P}\cdot\left[\frac{X_o - X_f}{1 + X_o}\right]\cdot\frac{a^2}{(p_s^* - p_o)} \quad (9.21)$$

Typical values of the permeability for freeze-dried food materials such as apple, carrot, potato, fish, beef, and turkey vary in the range 0.5×10^{-9}–10×10^{-9} kg/m.s. μHg (King, 1970; Quast & Karel, 1968). For freeze-dried pharmaceutical products, values of water vapor permeability of the dried layer are in the order of 0.5–1×10^{-3}m^2/s (Nakagawa et al., 2011), whereas for other foods (e.g. beef, coffee) values are an order of magnitude larger and highly dependent on the porosity of the material (Sagara, 2001). It is important to note that it is not possible to establish a direct comparison among permeabilities of foods and pharmaceuticals because two different phenomenological models are used to define these parameters. The reported permeabilities of water for foods are based on Eq. 9.18 and have units of vapor flux over pressure gradient. Those units simplify to units of time (e.g. in seconds) as noted in the transformation of units given in the equation:

$$\frac{kg}{m\cdot s\cdot \mu Hg} = \frac{kg}{m\cdot s\cdot \mu Hg}\cdot\frac{1000\,\mu Hg}{1\,mmHg}\cdot\frac{760\,mmHg}{101{,}300\,Pa} = 7.5s \quad (9.22)$$

Instead, permeabilities of pharmaceuticals are calculated based on molecular diffusion in the Knudsen regime and they are in very good agreement with experimental results (Nakagawa et al., 2011). To gain an idea of the magnitude of water vapor permeability values in the porous dried layer and its transport from the sublimation line, the values are of the same order of magnitude as the permeability of water vapor through plastic food packaging material such as HDPE and LDPE, PP, PS, PVC, and PET, which are reported by Saravacos and Maroulis (2001). Thus, due to the low values of water vapor permeability for most freeze-dried foods and pharmaceuticals, experimental drying times are excessively long. Measured values of drying times are in general agreement with estimations done by Eq. 9.21. From this equation it can be seen that one way to reduce drying times would be by increasing the sublimation pressure p_s^*. Values of sublimation pressure close to 1000 μHg could be used for permeability values of the order 1×10^{-9} kg/m.s. μHg. The importance of using sublimation pressures close to the maximum tolerable values can be noted by applying Eq. 9.21 to calculate the time to dry a sample with an initial moisture content of 75% (w/w) to a final moisture of 5% (w/w); the initial density (ρ) of a slab-shaped product of 20 mm thickness, a, can be assumed to be 1070 kg/m^3.

If K_p and $(p_s^* - p_o)$ are expressed in $\frac{kg}{m\cdot s\cdot \mu Hg}$ and μHg respectively, the drying time is expressed in seconds. For calculations, the initial and final moisture contents of the sample are more conveniently expressed on a dry basis as:

$$X_0 = \frac{x_0}{1 - x_0} = \frac{0.75}{1 - 0.75} = 3 \quad (9.23)$$

$$X_f = \frac{x_f}{1 - x_f} = \frac{0.05}{1 - 0.05} = 0.053 \quad (9.24)$$

where x_o and x_f are the initial and final moisture contents of the sample, expressed on a wet basis. Substituting these values in Eq. 9.21 yields:

$$t_D = \frac{\rho}{2K_P}\left[\frac{X_o - X_f}{1+X_o}\right]\cdot\frac{a^2}{(p_s^* - p_o)} = \frac{1070}{2\cdot K_p}\left[\frac{3-0.053}{1+3}\right]$$

$$\times\frac{(10x10^{-3})^2}{K_p\cdot(p_s^* - p_o)} = \frac{0.039\,kg/m}{K_p\cdot(p_s^* - p_o)} \qquad (9.25)$$

By assuming a value for the permeability of the sample of 1.0×10^{-9} kg/m.s. μHg and a pressure in the drying chamber (p_o) of 100 μHg, Eq. 9.25 predicts that the drying time t_D changes from approximately 72 hours to 12 hours when the sublimation pressure changes from 250 to 1000 μHg. Regardless of the sublimation pressure used, within tolerable ranges, calculated and experimental drying times are excessively long and there are no practical approaches to implement a freeze-drying process as an economic alternative to handle large production volumes.

9.1.1.3.3 Heat transfer

Heat is required in the process to sustain sublimation. If the heating plates are not in contact with the material and the heat required for sublimation is coming solely from radiant energy transferred from the heating plate, a balance of energy yields:

$$\dot{Q}_R = \dot{Q}_i = G_S\Gamma_{sub} \qquad (9.26)$$

where $\dot{Q}_R$ is the radiant energy from the heating plates, $\dot{Q}_i$ is the internal heat flow conducted through the porous dried layer of thermal conductivity k_{dl} and thickness Z, and Γ_{sub} is the average latent enthalpy of water sublimation (see Figure 9.18). The heat transferred by conduction in the dried layer, assuming unidirectional flow in the direction z, can be calculated as:

$$\dot{Q}_i = -k_{dl}\frac{dT}{dz} = k_{dl}\frac{T_o - T_s}{Z} \qquad (9.27)$$

Substituting Eq. 9.19 into Eqs 9.26 and 9.27, the following equation is obtained:

$$k_{dl}\frac{T_o - T_s}{Z} = \frac{\rho(X_o - X_f)}{(1+X_o)}\Gamma_{sub}\frac{dZ}{dt} \qquad (9.28)$$

If T_o, T_s and Γ_{sub} are assumed constant during the process, integration of Eq. 9.28 gives:

$$t_D = \frac{\rho}{2k_{dl}}\cdot\frac{(X_o - X_f)}{(1+X_o)}\cdot\frac{\Gamma_{sub}a^2}{(T_o - T_s)} \qquad (9.29)$$

The above equation shows that drying time is proportional to the density of the frozen product, the difference in moisture, calculated on a dry basis, of the fresh and dried product, the square of the product size represented by the half thickness, a, in the simplified slab geometry used, and inversely proportional to the driving force for the heat flow, which is the difference between the temperature of the chamber T_o and the temperature of the sublimating surface T_s, whereas that for calculating the vapor transfer, the driven force is given by the difference of pressures in the same location. The latent enthalpy of sublimation in Eq. 9.29 varies from approximately 2835 kJ/kg at temperatures close to 0°C to 2839 kJ/kg at −40°C, and since it should include the enthalpy for vapor desorption, the value of Γ_{sub} to be used to estimate drying time by Eq. 9.29 should be within the range of 2840–2850 kJ/kg (Kessler, 2002). Eqs 9.21 and 9.29 provide drying times estimated using mass and heat transfer mechanisms, respectively. By equating them the following is obtained:

$$(p_s^* - p_o) = -\frac{k_{dl}}{K_p\Gamma_{sub}}(T_s - T_o) \qquad (9.30)$$

Equation 9.30 is the equation of a straight line if p_s^* and T_s are considered the variables of the equation, and p_o and T_o are constants and specified parameters of the drying process. The slope of the straight line is related to the thermal conductivity, k_{dl}, the permeability, K_p, of the dried layer and the enthalpy of sublimation of the solvent Γ_{sub}. Kessler (2002) gives values of thermal conductivity and vapor permeability in the porous dried layer of a range of dried foods, which are reproduced in Table 9.6. As shown in Eq. 9.30, which applies to the arrangement in which heat is supplied by radiation from the heating plates located above and below the sample (see Figure 9.18), the thermal conductivity and vapor permeability of the dried layer are important material properties that control the drying rate. These transport properties are not only dependent on the pressure and temperature of the drying process but also on the structure of the drying layer, notably the porosity.

Figure 9.19 illustrates the thermal conductivity of freeze-dried milk as a function of the pressure in the drying unit for materials having different porosity. It can be noted that the thermal conductivity of the freeze-dried milk increases with the pressure and decreases with the porosity. The distribution of the pores in the dried layer as well as its structure plays an important role in the measured value of thermal conductivity. Predictive models based on the structure of the materials have been

Table 9.6 Values of thermal conductivities and mass permeabilities in the porous dried layer of selected food products.

Product	Thermal conductivity k_{dl} [W/mK]	Mass permeability K_P (seconds) $\times 10^{-9}$
Whole milk	0.02–0.08	20–40
Apple	0.035–0.11	25
Beef	0.04–0.05	5–11
Peach	0.015–0.04	33
Tomato juice Brix 6	0.035–0.05	78
Tomato concentrate Brix 15	0.043–0.065	46
Tomato concentrate Brix 22	0.08–0.11	16
Coffee powder	0.033	–
Avocado powder	0.055	–
Carrot	–	42
Spinach	–	15–30
Potato	–	10
Mushroom	–	10
Banana	–	7
Fish	–	65

Source: Adapted from Kessler 2002.

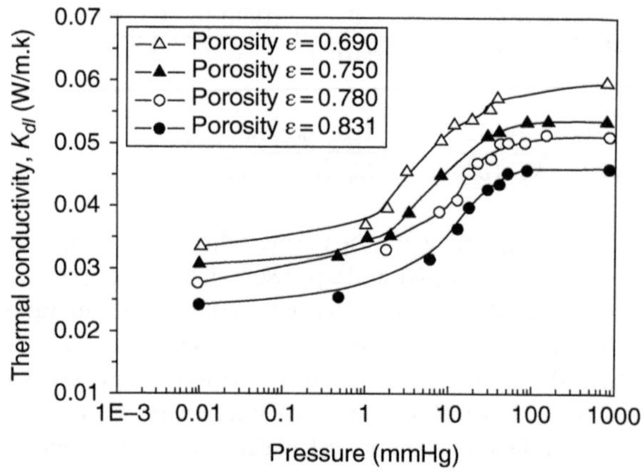

Figure 9.19 Thermal conductivity of freeze-dried milk as a function of the pressure in the chamber and the porosity. Source: Ratti 2008. Reproduced with permission of John Wiley & Sons.

successfully used to predict transport properties of biomaterials and foods (Chen, 2008; Saravacos & Maroulis, 2001). Figure 9.20 shows values of the thermal conductivity of dried beef and turkey meat and vapor permeability for turkey meat as a function of pressure in the drier.

These results clearly show that porosity, ε, is one of the main characteristics of freeze-dried products. However, there are conflicting reports indicating either negligible or important effects of the freeze-drying operational variables, e.g. drying temperature, on the transport properties of dried food products (Ratti, 2008). Some researchers have reported that processing variables significantly affect the structure of the product with a general increase of porosity during freeze drying (Rahman et al., 2002), whereas other researchers have reported no effects (Tsami et al., 1999) or a decrease of the sample porosity with increases in the heating plate temperature (Krokida et al., 1998).

The sublimation temperature is a very important variable to control during the process because if it exceeds the glass transition temperature zone of the product (discussed in Chapter 4), the solid matrix collapses. It has been demonstrated that to control collapse of the material structure and thus the quality of freeze-dried products, two temperatures must be considered. One is the temperature of the frozen core that has to be below the ice melting temperature to provide good-quality products and the other has to be lower than the glass transition temperature of the dry solids (Ratti, 2008). Collapse occurs when the temperature of the dried material is higher than the glass transition temperature range of the dried solid and it is a phenomenon that can be visually observed as a decrease in the porosity and color of the sample after drying. After collapse of the dried material, the vapor permeability through the drying layer is significantly reduced because of the sealing of capillaries. Conversely, the thermal conductivity of the dried material may increase after the material has collapsed due to reduction of its porosity. Figure 9.20 clearly shows those trends of increase in thermal conductivity and decrease in vapor permeability with increasing pressures in the chamber, which are associated with higher product temperatures and significantly higher collapse of the solid matrix structures. Thus, collapse and the limiting temperatures to avoid it are governed by the drying process operating conditions, in particular the chamber pressure and the shelf temperatures (Flink, 1975; Karel, 1975; Karel et al., 1994; King et al., 1968; Levi & Karel, 1995; Pikal & Shah, 1990; Ratti, 2001; To & Flink, 1978a, b, c).

A schematic of the evolution of the product temperature during drying for freeze, vacuum and classic air drying are depicted in Figure 9.21 in which the glass transition of the dry material is also included. It is clear that the temperature of the product for air and vacuum

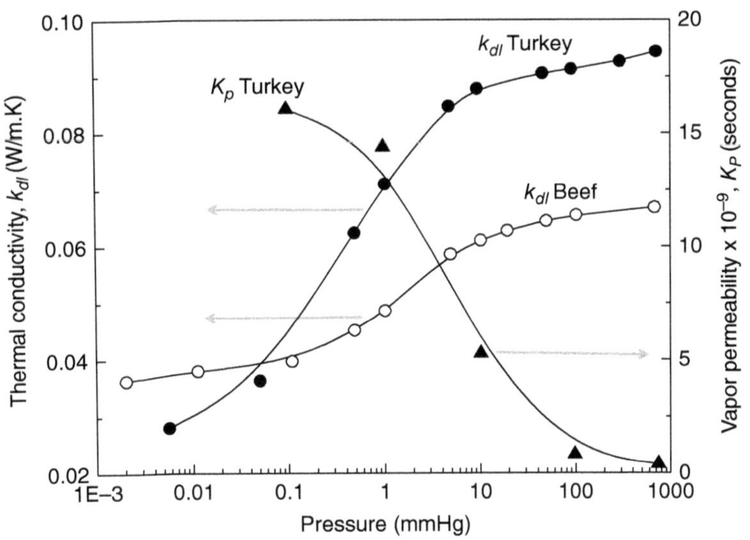

Figure 9.20 Thermal conductivities of beef and turkey meat and vapor permeability of turkey meat as a function of the drying chamber pressure. Source: Ratti 2008. Reproduced with permission of John Wiley & Sons.

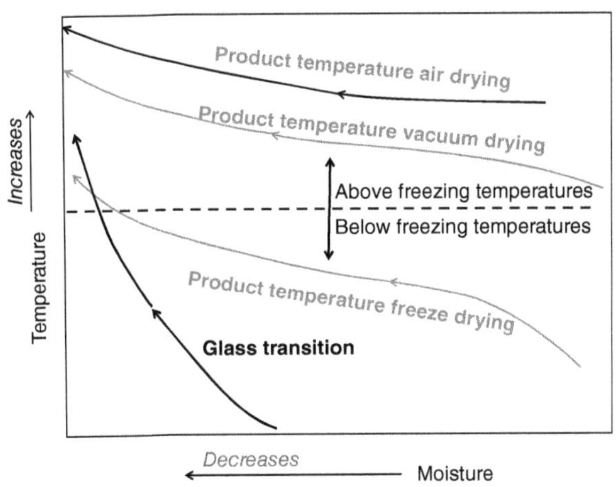

Figure 9.21 Temperature and moisture evolution of a product in conventional air drying, vacuum drying and freeze drying and their relationship with the glass transition of the product.

drying is significantly higher than the glass transition temperature of the drying product, which promotes collapse of dried products with higher densities and, depending on the type of material, poorer quality (Ratti, 2001). Conversely, the product temperature is lower than the glass transition temperature of the drying product in the freeze-drying process. Contrary to what would be expected, the product temperature in the freeze-drying process is below T_g only at the end of the process. However, it should be noted that during the process, the drying product is divided into porous dried and frozen layers by the receding surface illustrated in Figure 9.18 (Ratti, 2001). Thus, the part of the product in contact with warmer air has a glass transition temperature that is higher than the temperature in that part of the product and it remains rigid without collapsing.

Illustrated in Figure 9.22 is a different set-up in where the drying material is resting on a bottom heating plate which transfers heat by conduction through the frozen layer in order to sublimate water from the receding surface separating the frozen from the porous dried layer. The flow of vapor is then transferred through the porous dried layer and an equation relating pressures and temperatures similar to Eq. 9.30 can be obtained as (Ratti, 2008):

$$\left(p_s^* - p_o\right) = -\frac{k_{fl}}{K_p \Gamma_{sub}}\left(T_s - T_{cond}\right) \qquad (9.31)$$

In this case the transport property involved in the process is the thermal conductivity of the frozen layer whose value is less dependent on the porosity and structure of the material.

9.1.2 Main drivers of instant powder drying

During recent decades, much attention has been paid to the development of novel, highly functional ingredients

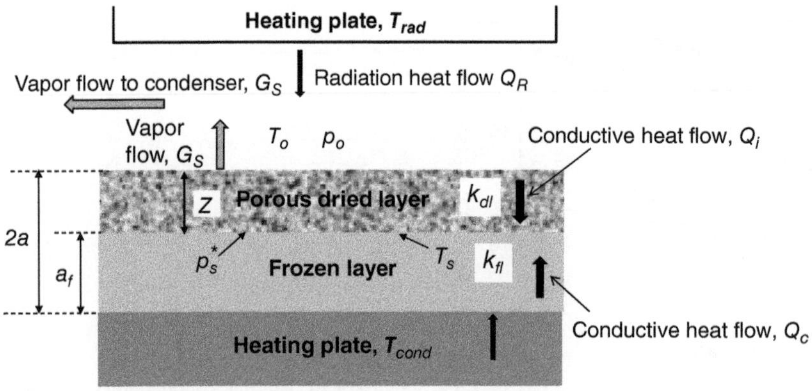

Figure 9.22 Schematic of heat and mass transfer mechanisms during a freeze drying process in which the heat is supplied from only a heating plate.

for food and non-food applications. Most of these ingredients are dried to stabilize, preserve and enhance the functions in the products in which they are used (e.g. minimization of microbial deterioration, reduction of lipid oxidation, preservation of original structures, etc.), to reduce storage and transport costs, and to make handling of the material easier. It must also be noted that high added-value heat-sensitive ingredients are developing quickly so methods to minimize the exposure of products to extensive heat and high temperatures are being investigated.

As an illustration, it is worth mentioning the rapid growth of dairy ingredients, i.e. the production of functional ingredients with good functional properties to promote sensory-related functions such as texturization, gelation, foaming, and emulsification as well as to retain the bioavailability and nutritional properties found in many dairy ingredients such as β-casein, α-lactalbumin, β-lactoglobulin, and bioactive peptides. To retain their bioactive properties, such ingredients generally require processing conditions that involve relatively low time-temperature treatments during their preparation, fractionation and drying processes.

A large number of functional ingredients are prepared from aqueous media (homogeneous solutions, heterogeneous liquid mixtures such as o/w emulsions, or liquid-solid suspensions), so the final drying process to form powder particles with specific end-use properties (such as flowability, stickiness, wettability, rehydration, etc.) consists of removing water under diverse conditions. An important class of functional powders consists of instant powdery ingredients or products produced by the spray-drying process, which was discussed previously in terms of production volume and quality. Spray drying is a process commonly used to dry and functionalize powdered materials. It must be noted that the production of spray-dried instant powders needs significant improvement in relation to:

- upgrading of product functionality, particularly its rehydration behavior, in order to meet the technological requirements of many end-applications
- increase of process flexibility, to be able to process various products of varying physicochemical properties and heat sensitivity
- decrease of processing costs, the capital investment and energy consumption in particular.

These three product- and process-related constraints constitute key drivers for the manufacture of instant powders.

9.1.2.1 Product functionality

Quick and complete rehydration of dried powders is of paramount importance in industrial practices and is an essential quality criterion of instant powders. The dispersion of powders in liquids involves four important attributes that define the quality of these powders.

9.1.2.1.1 Wettability

The wetting of a solid depends on the surface state and is a key step involved in dispersion, dissolution, and floating of the powder. For a given liquid and solid in equilibrium, wettability can be defined as the ability of the liquid to spread spontaneously on the surface of the solid, or as the ability of a bulk powder to absorb the liquid under the influence of capillary forces. This ability is governed by the surface free energy of the solid. Wettability is favored by large particles, high porosity and low contact angle (Freudig et al., 1999). For a given powder, instant properties are improved when the powder consists of porous agglomerates, to which a surfactant (e.g. lecithin for food products) is added to reduce the contact angle of the liquid on the surface. However, wettability is also affected by the surface composition of the powder which depends mainly on drying conditions as well as the type of drying process. For instance, for dairy products, high lactose and/or low lipid content at the surface of the particles improves their wettability.

As discussed above, the distribution of lipids on the surface of the particle is strongly associated with the operating conditions in a spray-drying process (Kim et al., 2005a, b, 2009a, b, c). Gaiani et al. (2009, 2010) found that different proteins exhibit different wettability. For example, powders of whey protein isolate (92.7% protein content) showed poor wettability when compared with the wettability of powders of native phosphocaseinate (87.6% protein content), probably due to the much higher heat sensitivity of whey proteins which may have been affected during drying. Droplet size during atomization and the time-temperature history in the drier affect the powder size distribution and powder surface composition. During spray drying, the larger the droplet size during atomization, the higher the expected size of the finished product, which has a positive effect on the particle wettability. However, high droplet temperatures induce modifications on the particle surface composition, such as high lipid content, and the presence of denatured proteins which impair product wettability. This is a clear example of two antagonistic effects in the production of powders using a spray-drying process, i.e. the production of large particles favors wettability but higher temperatures that negatively affect surface composition are needed to achieve a specific drying process.

Thus, in the drying process, there is a need to avoid the development of concentration and temperature gradients in the atomized droplets and later in the dried particle during the last stages of drying, whatever the size is. One way to achieve this would consist of generating dried particles with a high internal porosity to favor moisture transfer during the falling rate period of the drying process (last stage of the drying process), which would allow the temperature and moisture content in the particle to be controlled and approximately uniform. In addition, porous dried particles provide quick absorption of liquid in the rehydration process.

9.1.2.1.2 Sinkability

This is the ability of wetted particles to sink. Sinkability depends on particle size and density. Difficulties often arise if very fine particles are present due to abrasion in postprocessing operations such as particle separation in cyclones, or when the particles are hollow due to air trapped in the center of the particles during drying. As explained, such difficulties may occur with spray-dried powders, especially for high drying temperatures that invariably lead to surface case hardening, which inhibits air or moisture transfer and allows the air inside the particle to expand. For materials whose bulk density equals approximately 1500 kg/cm^3, the average particle size must exceed 100 μm to allow sinking of the dried powder.

9.1.2.1.3 Dispersibility

This is the ability of powders to disintegrate quickly into primary particles throughout the liquid, followed by relatively low settling in order to allow complete dissolution. Sediment formation during rehydration should be avoided with high-porosity particles.

9.1.2.1.4 Solubility

This is the ultimate quality attribute of instant powders. It corresponds to the complete separation of the molecules that compose the particles. Full dissolution can be achieved with particles dried with relatively low time-temperature processes that minimize physicochemical changes and exhibit slow settling during rehydration. For dried milk powders, the solubility index, measured by the percentage of the dissolved material, must exceed 99%.

Most instant powders are spray dried owing to the ability to produce powders with satisfactory rehydration behavior (solubility > 99%), while physicochemical damage of powder components remains acceptable. Given its disadvantages from an energy standpoint, drum drying is rarely used, except when formulations show no heat sensitivity. For instance, in the dairy industry, it can be used for drying high free fat mixes for chocolate manufacturing, which requires Maillard reaction products to be developed, whereas freeze drying is required for heat-sensitive materials, for example for the production of probiotics. However, as discussed, freeze drying is an expensive process and may not be suitable for high-capacity processing. In addition, it must be noted that functionality of instant powders still needs to be enhanced owing to the technical requirements of end-applications, namely quicker and easier rehydration. The rehydration behavior of instant powders could be significantly enhanced if high-porosity products could be generated in the drying process, in order to reduce wetting time and control sinkability and dispersibility, hence leading to quick and full solubilization of the powder material.

9.1.2.2 Process flexibility

Process flexibility is a common quality sought in the processing industry, to be able to meet variable demands and evolving conditions with regard to changes of raw materials and formulations, as well as the physicochemical characteristics of processed products. Hence, any drying process producing instant powders must offer high process and product flexibility, and be able to:

- process a large variety of product compositions such as protein-rich mixes (protein concentrates and isolates resulting from various types of proteins), high fat content formulations, carbohydrate-rich recipes (polysaccharides, oligosaccharides, in particular), heat-sensitive materials (such as biofunctional ingredients, probiotics, etc.), and complex formulations (such as ice cream mixes, coffee whiteners, etc.). Such materials may consist of either homogeneous single-phase solutions or heterogeneous two-phase emulsions and suspensions with an extensive range of viscosities and rheological behaviors
- vary extensively the functional properties of dried powders depending upon the required end-use quality profile.

Rehydration behavior of instant powders depends on three main physical characteristics: the size of particles (average size and size distribution), the agglomeration structure of primary particles (or interparticle texture), and the internal structure of primary particles (or intraparticle texture). For a given formulation, physical characteristics of instant powders are adjusted appropriately by acting on independent processing variables with operating ranges depending upon the level of process flexibility: the higher the operating range, the higher the process flexibility.

Among existing drying technologies used for the production of instant powders, spray drying presently offers the highest flexibility with regard to its ability to process products with various compositions and to handle product functionality. But, although spray drying is well and widely accepted, it has been recognized that its flexibility for standard and novel powders still needs significant enhancement to process a larger range of formulations, including those which cannot be dried by the use of existing drying technologies, such as high fat content formulations and complex multi-component mixes, and to produce powders with improved rehydration behavior.

The atomizing systems of spray dryers are technically capable of delivering variable-size droplets in a wide range of size distributions. The material is apparently flexible enough, although in practice, the process flexibility is limited in particular by the drying mechanism of the liquid droplets into solid particles. As discussed in section 9.1.1.2, when a particle dries in a spray-drying process, the drying time in the falling drying rate regime (diffusion-governed regime) is proportional to the square of particle radius: the larger the particle size, the longer the drying time. Hence, spray-drying processes are limited to rather low particle sizes (with an average diameter generally smaller than 100 μm). An ideal drying process should enable end-users to vary independently the three main surface or textural properties of instant powders in a wide range of values, as indicated in the schematic diagram below.

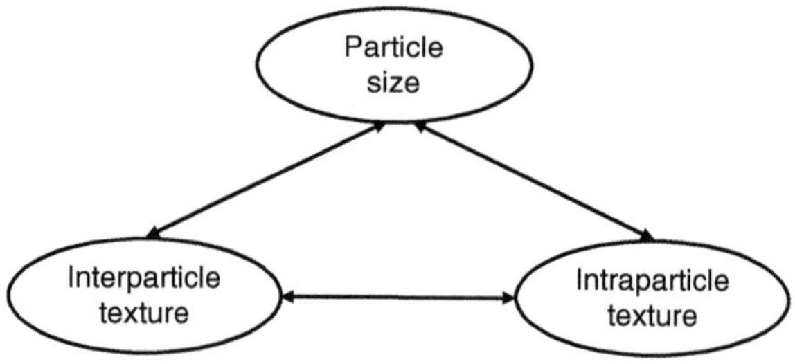

In addition, process flexibility also relates to the ability of a drying technology to handle various formulations on the same unit, with easy and quick adaptation of processing conditions to reduce product loss and production downtime. Existing drying technologies are generally designed and optimized for a specific manufacturing process, and are not very adaptable to product variations.

9.1.2.3 Cost reduction

The cost structure of drying units for the production of instant powders is mostly affected by the capital investment (which defines the amortization cost) and the consumption of energy, owing to the high value of the latent heat of vaporization of water or the solvent that must be evaporated from the material. Hence, cost reduction primarily consists of reducing capital investments and energy consumption.

For a given production capacity, capital investment of typical drying technologies for instant powders increases in the following order: drum drying < spray drying < freeze drying. Drum drying requires much lower capital investment, while freeze drying (a batch process) shows a higher capital investment. Spray-drying units operate very large volumes due to the fact that a somewhat longer contact between the product to be dried and the drying air is required and due to the implementation of a sophisticated liquid atomization system. Adding these elements to existing facilities (buildings, height in particular, and utilities) requires a relatively high investment cost. For instance, spray drying alone accounts for 40–50% (according to the capacity of the unit) of the investment cost of a complete milk powder production unit.

However, the main drawback of drying operations is their consumption of energy. Removing water or a solvent from a material by state change (from liquid to gas state through evaporation, or from solid to gas state through sublimation) consumes a lot of energy. In the case of water removal, thermal drying requires approximately 2500 kJ to evaporate 1 kg of water. In practice, the energy consumption may be twice or three times that value depending upon the drying process and the type of drier. Drum drying consumes approximately 1.3 kg of steam per kg of evaporated water, which corresponds to a consumption of specific thermal energy of 900 W.h per kg of evaporated water, or 3240 kJ/kg of evaporated water. Energy consumption of spray drying is roughly twice that value. In industrial practice, an energy consumption of 2.5–3.8 kg of steam per kg of evaporated water is observed, depending upon the nature of the products involved and particularly the heat sensitivity of its constituents. With low heat-sensitive products such as caseinates, energy consumption ranges from 1.8 to 2.5 kg of steam per kg of evaporated water whereas for highly heat-sensitive products such as whey concentrate, which is rich in soluble proteins, energy consumption ranges between 3.2 and 3.8 kg of steam per kg of evaporated water.

The energy consumption of the spray-drying unit of a milk powder manufacturing unit may represent 75% of the total energy consumption. So, energy consumption for spray drying is generally a significant line item in the operating costs of instant powder manufacturing units. For freeze-drying technology, the energy consumption is extremely high, 2.5–5 times higher than that of spray drying. For a given product or formulation, energy consumption of typical drying technologies for instant powders increases in the following order: drum drying < spray drying < freeze drying. Besides, there is increasing pressure to reduce the energy consumption of drying processes, owing to new environmental challenges.

This brief discussion of the two main factors that impact the cost of typical drying technologies for the production of instant powders indicates that product designers and process developers involved in the development of instant powders need to look for alternative drying technologies in which capital and energy costs are reduced significantly compared to those of existing drying technologies, and of spray drying in particular.

9.1.3 The extrusion-porosification process

When examining key drivers for instant powder processing, it is worth discussing the limitations of existing powder drying processes in their capability to meet emerging process requirements with regard to product functionality, process flexibility, and cost reduction.

9.1.3.1 Limitations of existing drying processes to produce instant powders

Basic principles as well as the advantages and disadvantages of drying processes to produce powders were discussed in section 9.1.1. The first process described was drum drying and although this presents a good choice for the drying of products with high solids content and high viscosity, the long residence time and the relative high temperatures used generally limit its application only to products which are not heat sensitive.

Although the origins of drying dairy products are related to the introduction of a double roll drier in a patent by Just (1902), its use to dry dairy products has been limited because the high temperature and long time processing conditions produce products with scorched flavors and low solubility due to protein denaturation (Hall & Hedrick, 1971). Solubility and dispersibility of powders produced by drum drying are inferior to those of powders produced by spray drying (see Table 9.2), whereas final moisture and protein denaturation are larger. Applications of drum drying in the dairy industry are limited to products in which high temperatures and long drying times are sought to produce flavor- and color-enhancing characteristics derived from the Maillard reactions. Other uses of drum drying have been limited to the production of confectioneries, bakery products, and feed blends. Problems associated with the high drying temperature have been eliminated by using a vacuum in which drying takes place at temperatures around 60°C in applications to dry whey (Henning et al., 2006). Most of the applications of drum driers have been oriented to the drying of cereals, legume seeds, and flours as the processing conditions are very favorable for promoting starch gelatinization (Vallous et al., 2002).

Studies have compared the effects of drying processes on volatile compounds such as flavors. Reineccius and Coulter (1969) compared retention of the flavor component diacetyl added to skim milk and dried by drum drying, spray drying, and freeze drying. Whereas similar proportions of added diacetyl (about 60%) were retained during spray and freeze drying, the drum-drying process hardly retained this flavor compound. The lack of flavor retention was attributed to the characteristics of the drum-drying process where the vigorous boiling of the moisture impairs the formation of the characteristic shell/crust observed in the dried droplets during spray drying, thus retaining volatiles. In the freeze-drying process water is sublimated below the triple point of water so the low vapor and low temperature are not large enough to carry volatiles.

Encapsulation to entrap sensitive ingredients, e.g. β-carotene, is generally based on the emulsification of the sensitive ingredient (for lypophilic ingredients dissolved usually in oil) in a solution containing the material that will form upon drying a coating of the oil phase containing the active ingredient to protect it from oxygen and water. Spray drying is the most common method of drying encapsulated material as the cost this process is significantly less than freeze drying, being the latter most commonly used in the flavor industry (Desobry et al., 1997). However, spray drying promotes a significant increase in the material surface area which may enhance oxidation if the coating material does not provide a good oxygen barrier. Increase in the oxygen barrier can be achieved by increasing the thickness of the coating layer but that reduces product delivery efficiency (Moreau & Rosenberg, 1996). Instead of oils, maltodextrins, which have bland flavors and exhibit low viscosities at high solids content, are a good potential alternative for use as encapsulation agents. Because of their relatively low molecular weight, maltodextrins have the tendency of lowering the glass transition of formulations, which as discussed in Chapter 4 and also in section 9.1.1 may complicate spray- and freeze-drying processes, producing very hygroscopic products and products with collapsed and dense matrices that can cake and become sticky during storage (Bhandari & Howes, 1999, 2005; Chuy & Labuza, 1994). By comparing spray drying, freeze drying and drum drying, it can be noted that time-temperature regimes are different and can lead to differences in structure, texture, physicochemical characteristics, and stability during storage.

The hydration, dispersibility, and solubility properties of dairy ingredients are of primary interest for the formulation of a variety of foods and are not only dependent on the physicochemical make-up, but are also influenced by the drying process (Modler, 1985). As discussed, spray drying, drum drying, and freeze drying offer different options with specific benefits and drawbacks but any of these processes by themselves can provide the necessary versatility to handle the large number of applications and products relevant to the dairy area. Freeze drying is an excellent choice if the primary objective is to preserve the quality of the final product. But a downside of freeze drying is its high operational cost. Although in terms of nutritional quality, freeze-dried products exhibit very little degradation produced by thermal effects, often their physical characteristics are affected by freezing damage which leads to a loss of color in the product. The high porosity of freeze-dried products gives them a fragile structure that may impair the packaging of the product due to the production of crumbs. Owing to the high porosity, freeze-dried materials tend to adsorb great quantities of water so they must be stored or packed in low-humidity environments. Spray drying is the most commonly used process for dairy and non-dairy foods but it is limited only to feeds that are pumpable and can be atomized into small droplets without excessive pressure loss. In addition to the feed high-viscosity limitation, spray-drying processes have serious problems in the atomization of solutions or suspensions that exhibit strong viscoelastic behavior because they produce powders with fiber-like shapes, which have poor dispersibility when compared to powders with more spherical shapes. The drum-drying process can handle feeds with very high solids content and there is no limitation with regard to their rheological behavior. However, this process lacks versatility and produces powders of inferior quality.

9.1.3.2 Process presentation

The technologies more widely used to produce powders show real limitations for meeting emerging technical and economical requirements of instant powder drying. Combining these limitations with the increasing demand for functional ingredients, nowadays product designers and process developers are challenged to propose alternative technologies for instant powder production. Thus, the main question which interests extrusion experts and process engineers is the following: is the extrusion processing technology able to offer an alternative processing method for instant powder production? Because of its ability to continuously micromix and texturize single and multiphase concentrated viscous media with a high level of flexibility, twin screw extrusion technology has the potential to functionalize instant powder products, more cost-effectively than traditional processes and technologies. In this challenging context, in 2004, Maxwell Scott (Max Scott Consulting Pty, Australia, expert in dairy processing technology) and Jean-Marie Bouvier (Clextral, France, expert in extrusion processing technology) joined their expertise and creativity to design an innovative generic extrusion process, called the "extrusion-porosification" process, aimed at cost-effectively drying and functionalizing instant powders.

The innovative extrusion-porosification process is based on two ideas driven by principles of the process intensification concept defined in Chapter 10 (section 10.1.4).

- Operate continuously high solids feeds, in order to intensify the instant powder drying process, hence leading to reduced processing volumes and energy consumption, and consequently decreased operating cost (capital investment and energy cost) and increased process flexibility.
- Generate high porosity textures in instant powders, in order to decrease the characteristic time of a very important transport phenomenon, which is moisture transfer in the drying and rehydration processes, in particular, in order to decrease the time-temperature loads in the drying process and consequently to enhance instant powder functionality.

High solids processing is fundamental in extrusion-porosification. As discussed in section 9.1.3.1, in conventional spray drying the apparent viscosity of the concentrate determines the instant behavior of the product owing to its effect on the size of the sprayed droplets (Déchelette et al., 2011a, b; Schuck et al., 2005): larger droplet size may lead to degradation of rehydration behavior. For optimal concentrate spraying, the viscosity of the concentrate should not exceed 200 mPa.s; this corresponds to approximately 48–52% total solids for skim milk and 20–25% total solids for proteins isolates, for instance. In extrusion-porosification, the viscosity of feeds is much higher, typically ranging from 2000 to 20,000 mPa.s, which enables a significant increase of the concentrate total solids (from 40% to 75% increase, depending upon the nature and composition of the media to be dried). High solids processing thus permits the use of energy-efficient vacuum evaporation to be extended to solids concentration much higher than those used in conventional spray-drying processes, thus extracting more benefit from the energy efficiency of vacuum evaporation and concomitantly decreasing the energy consumption of the whole drying process.

This new route for instant powder processing requires a real breakthrough from both process and technical standpoints. In fact, high solids processing requires the use of processing equipment that is able to handle highly viscous concentrated media while controlling the possible physicochemical modifications which may result from thermomechanically oriented processing. Besides, high solids spraying and porous powders formation necessitate appropriate texturization of the concentrated feeds prior to final spray drying; hence choosing foam processing as the central phase of the extrusion-porosification process allows the spray drying to be intensified, the porosity of powders to be controlled and the whole process to be flexible (that is, adapted to different types of formulations and able to vary functional properties of instant powders). The inventors of the extrusion-porosification process have chosen extrusion processing technology (twin screw extrusion technology, in particular) to meet all these processes and technical requirements appropriately. As a matter of fact, because of its ability to handle micromixing and texturization of single- and two-phase viscous media, twin screw extrusion technology is the relevant equipment to process and create porosity (porosify) within matrix-forming concentrates for instant powder products.

Figure 9.23 shows a simplified flowsheet of the extrusion-porosification process which consists of three main processing units: high solids vacuum evaporation, twin screw extrusion-aeration, and intensified spray drying.

9.1.3.2.1 High solids vacuum evaporation

The starting materials of the process typically consist of relatively low solids media (solutions, emulsions, liquid-solid suspensions) resulting from classic preprocessing units such as evaporation (atmospheric evaporation, vacuum evaporation), membrane separation (ultrafiltration,

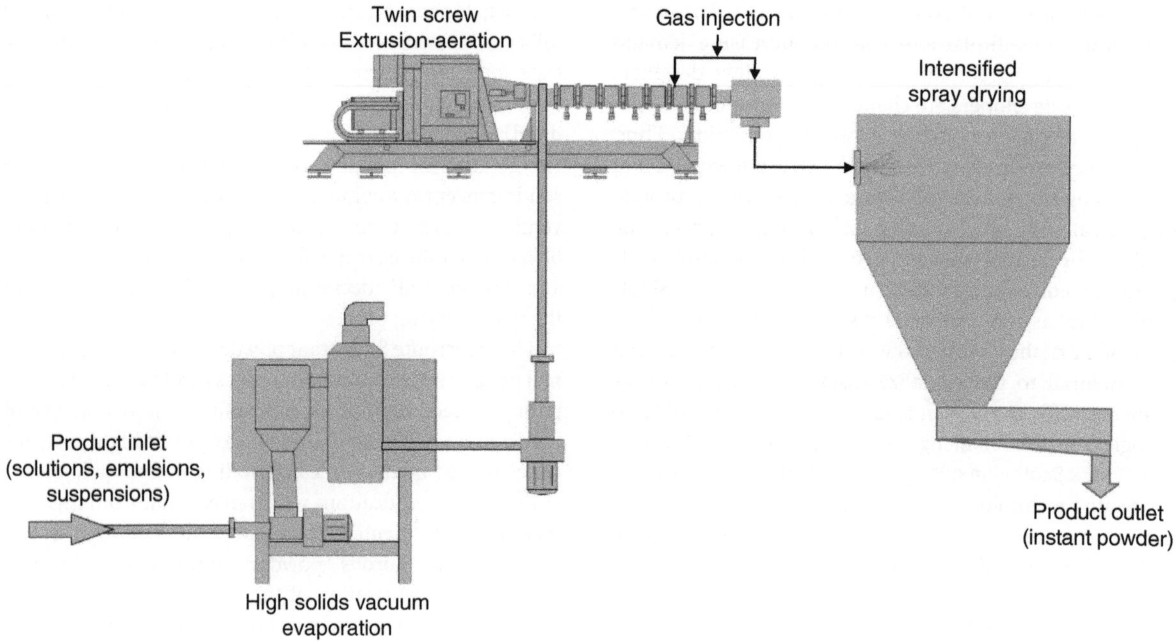

Figure 9.23 Simplified flowsheet of the generic extrusion-porosification process. Source: Reproduced with permission of Clextral, France.

Figure 9.24 A 38% dry solids milk protein concentrate exiting a high solids vacuum evaporation unit.
Source: Reproduced with permission of Clextral, France.

microfiltration, nanofiltration), premixing (diluted premixes), etc. The product to be dried is fed into a high solids vacuum evaporation unit to concentrate it, while taking as much benefit as possible from the energy efficiency of the evaporation technique (approximately 1.05 kg steam per kg of evaporated water), and therefore being able to minimize the energy consumption of the whole process: the higher the amount of water removed through vacuum evaporation, the higher the energy efficiency of the whole process. According to the nature and heat sensitivity of the product to be concentrated, it is of paramount importance to select the most appropriate design of the evaporator to optimize the performance of the evaporation unit from technical and economical standpoints (refer to the review presented in section 9.2.1). The viscosity of the concentrate typically ranges between 2000 and 20,000 mPa.s. Figure 9.24 shows a 38% dry solids milk protein concentrate (85% protein content, dry basis) exiting the high solids vacuum evaporation pilot unit. The evaporation unit may not be necessary when the concentrate results directly from premixing the components of the mix to be dried (ice cream mixes or coffee whiteners, for instance).

9.1.3.2.2 Twin screw extrusion-aeration

The concentrate is fed into an intermeshing co-rotating twin screw extruder and mixed with a gas (carbon dioxide, for instance) to produce a textured foam, while controlling the time-temperature load in the screw-barrel assembly depending upon the process and product requirements. The extrusion-aeration process is of paramount importance, as it determines both the spray-ability of the foam in the spray drier and the porous structure of the final powder. The foaming process of a given mix depends on the composition of the concentrate (presence of surfactants, emulsifying components, texturing agents, in particular), on the rheology of the concentrate (Newtonian or non-Newtonian behavior), and on the mixing performance of the twin screw extruder-aerator (that is, how the two phases, gas and viscous liquid, are brought into contact). The foaming ability of a raw material can be described by the dimensionless Weber number, which corresponds to the ratio of viscous to surface tension forces. The Weber number, further described below, then allows one to evaluate the influence of the rheological behavior and process parameters on the characteristics of the foam (e.g. bubble diameter and bubble bursting). Foam texturization is ensured by the twin screw extruder-aerator, the design of which consists of an innovative configuration for intense and intimate mixing of gas-liquid systems. This configuration is based on an appropriate screw configuration and an innovative aeration section ahead of the screw-barrel assembly. Figure 9.25 shows the 38% dry solids textured foam obtained from milk protein concentrate (85% protein content, dry basis). The use of an intermeshing twin screw extruder brings high levels of process flexibility owing to the ability to add secondary feed streams separately along the screw-barrel assembly, which may be solids or liquids, depending upon the product requirement.

Figure 9.25 Extrusion-textured foam obtained from a 38% dry solids milk protein concentrate. Source: Reproduced with permission of Clextral, France.

9.1.3.2.3 *Intensified spray drying*

The twin screw extruder-aerator delivers the high solids textured foam which is sprayed easily through air pressure spray nozzles into a spray-drying chamber. The air pressure (pressure range of 6–12 bars) allows variable-sized droplets to be atomized. In the drying chamber, atomized particles of high solids textured foam are dried by direct contact with hot air with inlet and outlet temperatures typically in the ranges of 140–220°C and 65–80°C, respectively, depending upon the type of product. As shown in Figure 9.26, the resulting dried powder is highly porous owing to the starting textured foam. The extrusion-porosified powder particle shows numerous pores with communicating channels which are created by the release of gas during drying. The porous structure has an important impact on the drying process as well as on the functional properties of the products (the rehydration characteristics, in particular). The porous structure is controlled by the initial structure of the foam, and by the drying conditions.

9.1.3.3 Potential benefits of the extrusion-porosification process

Compared with conventional instant powder processing, the innovative extrusion-porosification process brings relevant benefits for the production of instant powders from various food and non-food concentrates.

- Processing high solids concentrates or mixes.
- Macromixing and micromixing viscous, multi-component formulations with in-line addition of gases, liquids or solids.
- Reducing energy consumption.
- Controlling the time-temperature-shear history in processing units, which allows the integrity of heat- or shear-sensitive materials to be preserved.
- Reducing processing volumes, and hence unit footprint.
- Enhancing process flexibility.
- Improving product quality.
- Reducing the capital and operating expenditures of processing units.

9.2 Engineering discussion of process functions

As discussed in section 9.1, the extrusion-porosification process allows the handling of feeds with larger solids content, which permits eliminating more water through

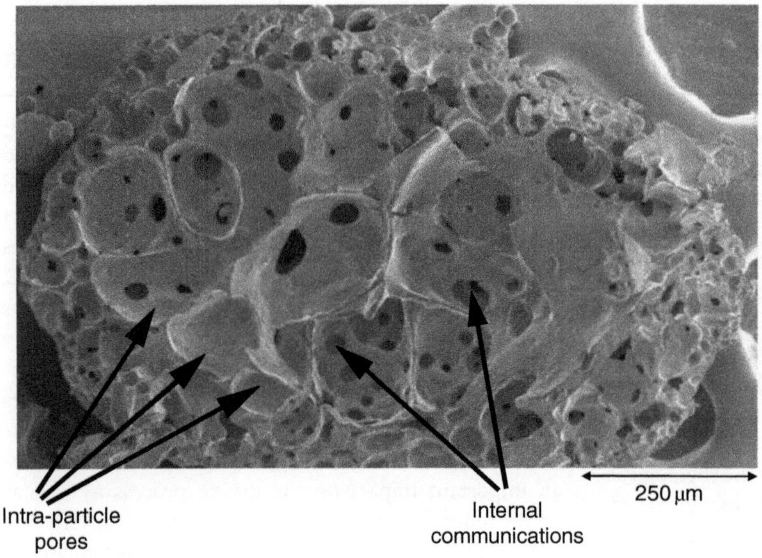

Figure 9.26 Typical SEM picture of an extrusion-porosified milk powder particle. Source: Reproduced with permission of Clextral, France.

vacuum evaporation, hence providing a significantly more economical alternative from an energy consumption standpoint. Given the limitations on the viscosity and solid content in the feed that most of the current drying processes have, the overall extrusion-porosification process makes use of the less expensive evaporation technologies to evaporate most of the water in the feed. Evaporation technologies and particularly vacuum evaporation are discussed in section 9.2.1. The concentrate which results from the evaporation unit is then fed into the extrusion-aeration unit, where the product characteristics and process phenomena of interest to this unit operation are related to the formation of foams, their physicochemical properties and the transport through the twin screw geometry, aspects that are described in section 9.2.2. The last unit operation of relevance to the overall extrusion-porosification process is described in section 9.2.3 (Intensified spray drying).

9.2.1 Vacuum evaporation

Evaporation is a unit operation widely used in many technological areas but of singular importance for the processing of foods and biomaterials. The dairy industry in particular employs evaporation for concentration of whole milk, skim milk, whey ultrafiltration permeates, and buttermilk. Other food applications include concentration of extracts, from sugar beet and sugarcane for the production of sugar, coffee, tea, fruits, and gelatine (Kessler, 2002). Typical preconcentration operations used in the dairy industry change the water content of whole milk from 87% to approximately 50–60% wet basis for products that are subsequently dried.

There are several ways of expressing the water or solvent content of the product in an evaporation process. One is to express it on a wet basis. By convention, in this chapter lower-case letters are used for wet basis concentrations, for example moisture content of a product:

$$x_w = \frac{m_w}{m_w + m_s} \tag{9.32}$$

A similar expression could be used to express the content of solids in the product:

$$x_s = \frac{m_s}{m_w + m_s} = 1 - x_w \tag{9.33}$$

where m_w and m_s are the mass of water (or solvent) and solids (fat plus solids non-fat) in the product. The same amount of water (or solvent) expressed on a dry basis, by convention using capital letters, is defined as:

$$X_w = \frac{m_w}{m_s} \tag{9.34}$$

Concentrations of the components in the product (solids and solvent) expressed on a wet basis give a more physical sense of the effect of these components on the physical characteristics of the solution, e.g. density and viscosity.

Furthermore, expressing percentages of water or solid content on a wet basis provides an indication of the relative amounts of each component in the liquid. However, since during the evaporation process the composition of water and solid varies, it is more convenient to express the water and solid content on a dry basis. Regardless of the basis used to estimate the amount of water in the solution, it is simple to transform concentrations from a wet basis to a dry basis by the following equation:

$$X_w = \frac{x_w}{1 - x_w} \tag{9.35}$$

For assessment of energy requirements in an evaporation process, it is necessary to estimate the percentage of water that has to be evaporated in the process, which can be calculated as follows:

$$percentage\ water\ evaporated = \frac{m_{wo} - m_{w1}}{m_{wo} + m_s} \cdot 100$$

$$= \frac{X_{wo} - X_{w1}}{1 + X_{wo}} \cdot 100 \tag{9.36}$$

where X_{wo} and X_{w1} are the initial moisture contents of the feed and the concentrate expressed on a dry basis. The percentage of residual water to be removed by drying after preconcentration by evaporation is then calculated from the moisture content of the feed and the concentration expressed on a dry basis as:

$$Residual\ water\ to\ be\ removed\ by\ drying$$
$$= \frac{m_{w1}}{m_{wo} + m_s} = \frac{X_{w1}}{X_{wo} + 1} \cdot 100 \tag{9.37}$$

Equations 9.35– 9.37 can be used to calculate the percentage of water to be removed by a preconcentration step and the remaining moisture content in the product, some of which is removed by the drying process. Concerning dairy products, for preconcentrating skim milk from about 8% total solids (92% moisture content) to a final total solids content of 45% (55% moisture content) on a wet basis, the percentage of water to be removed by evaporation is 80%, leaving remaining moisture to be evaporated by spray drying of about 11%. Similar estimations could be done for other commercially important dairy products such as whole milk with a 13% total solids (87% moisture content) to be concentrated by evaporation up to 45% solids and whey proteins with a 6% total solids initial concentration (94% moisture) to be concentrated to 50% total solids content (50% moisture content), all expressed on a wet basis. Figure 9.27 illustrates the percentages of water to be removed by evaporation during the preconcentration process and the percentage of water remaining in the product going to the drying process.

Since the energy required for removal of 1 kg of water from the product by drying is about 10–20 times greater

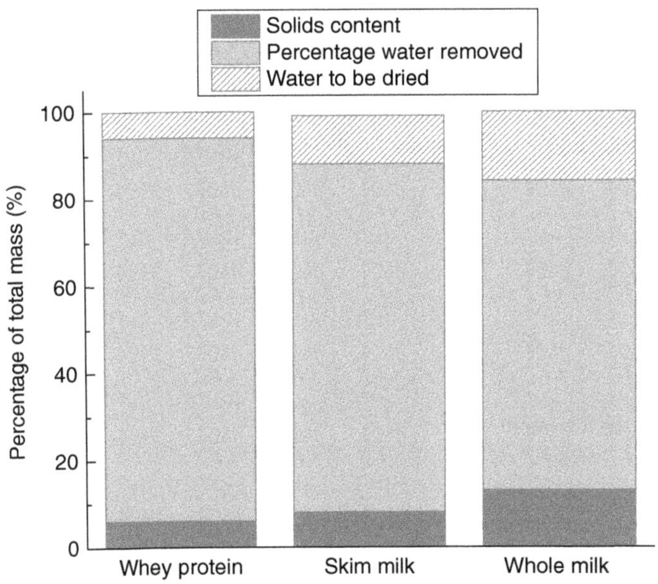

Figure 9.27 Comparison of the amount of water to be evaporated in the production of dairy powders. A differentiation between the water to be evaporated by evaporation and drying processes is made. Source: Adapted from Kessler 2002.

than the energy required for removing the same amount of water by evaporation, it is important to perform this type of analysis to optimize the process and minimize processing costs (Kessler, 2002). It is clear from Figure 9.27 that from an economical standpoint, it is always more convenient to attain the lowest permissible water content in the feed entering the dryer. However, there are limitations inherent to the process concerning the minimum water content in the dryer feed that will allow an efficient operation. That is particularly important in spray-drying processes due to the relevance of the feed atomization on the efficiency of the process (discussed in the previous section) and the strong effect that the water content in the feed has on the efficacy of the atomization process, which imposes limitations on the minimum content of water permissible in the feed.

Kessler (2002) describes a number of evaporators used in the dairy industry and Geankoplis (2003) provides details of evaporation technologies used in a number of applications including juice and sugar as a preconcentration of crystallization operations. In the dairy and food industry multiple falling film evaporators have been used due to their relative simplicity, energy savings, and short hold-up times that minimize product degradation, aspects that are discussed in section 9.2.1.1.

9.2.1.1 Falling film evaporation

9.2.1.1.1 Single effect evaporators

Balance of mass

Typical components of a falling film evaporator are illustrated in Figure 9.28. The feed is introduced via an inlet located in the upper part of the evaporator body, from which it is uniformly distributed to the tubes which, depending on the design, have a range from 4 m to 12 m (Jebson & Iyer, 1991; Kessler, 2002). Inside the tube, the liquid feed flows down its interior surface, forming a falling film which is heated from the outside with steam. During the flow in the interior of the tube, the solvent (water in most bioapplications) is evaporated and the solid content increases down the tube, as well as its viscosity. In the food industry, the evaporated liquid is water but some other volatile components present in the liquid can be partially or totally evaporated. Exiting the tube is a mixture of vapor and concentrate; the vapor is stripped out in a separator.

Basically an evaporator consists of: (1) a heat exchanger, also known as a steam chest, to raise the temperature of the product to its boiling point at the processing pressure and vaporize part of the solvent of the solution; (2) a separator where the vapor is separated from the concentrate; (3) a condenser to condense the vapor and remove the condensate from the evaporator; and

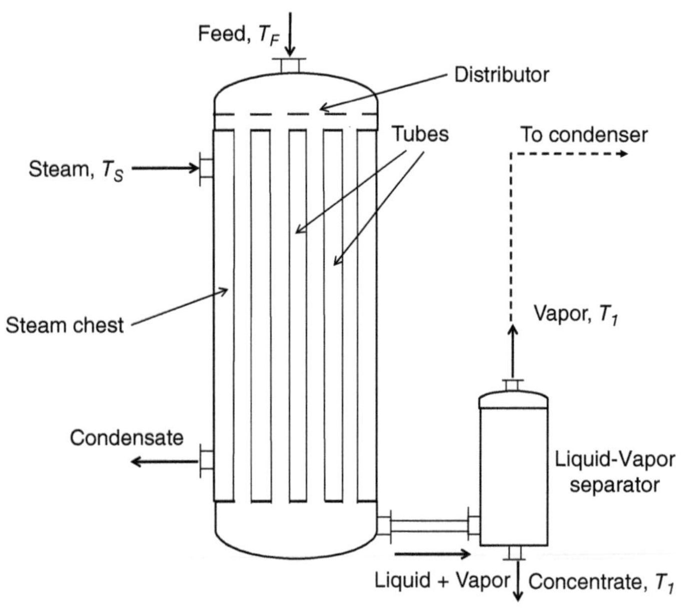

Figure 9.28 Schematic of a falling film evaporation system.

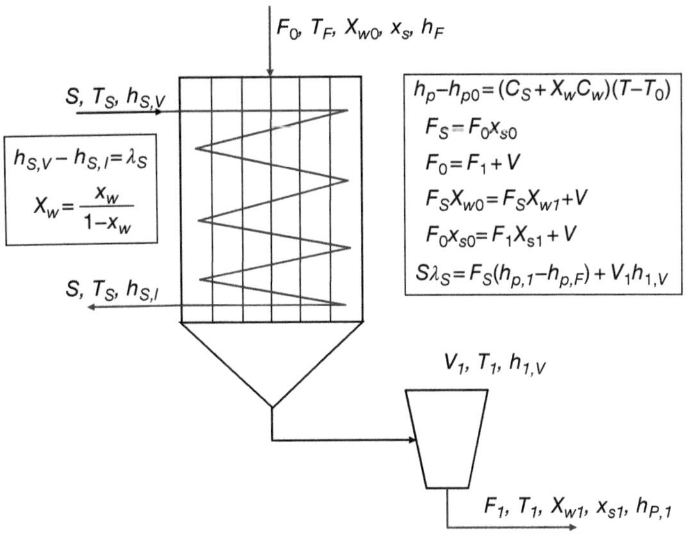

Figure 9.29 Schematic of a single-step falling film evaporation system where the mass and energy balances are performed.

(4) a vacuum system to withdraw non-condensable gases and maintain a constant evaporation temperature. The steam used to heat and vaporize the solvent, or water, in the solution condenses and exits at the bottom of the evaporator.

Before writing the mass and energy balances relevant to the evaporation process, it is convenient to define flow rates and concentrations on a basis that facilitates the calculations, for instance dry solids, which remain constant during the evaporation process. Mass balances can be applied to the evaporator schematically illustrated in Figure 9.29 where all the relevant variables are included as well as the main balance equations.

$$\text{Total balance:} \quad F_0 = F_1 + V_1 \tag{9.38}$$

$$\text{Solids balance:} \quad F_0 x_{s0} = F_1 x_{s1} \tag{9.39}$$

$$\text{Water/solvent balance:} \quad F_S X_{w0} = F_S X_{w1} + V_1 \tag{9.40}$$

where F_0, F_1 and F_S are the feed, the concentrate and the dry solids flow rates, respectively. V_1 is the flow rate of vapor coming out of the separator and x_{s0} and x_{s1} are the solid contents in the feed and concentrate expressed on a wet basis, whereas X_{w0} and X_{w1} are the water content in the feed and the concentrate expressed on a dry basis, respectively.

The dry solids flow rate in the evaporator can be calculated from the flow rate of liquid entering the evaporator as:

$$F_S = F_0(1 - x_{w0}) = F_0 x_{s0} = F_1(1 - x_{w1}) = F_1 x_{s1} \tag{9.41}$$

Balance of energy

The balance of energy in the evaporator involves the enthalpies of the liquid calculated at the feed and concentrate conditions as well as the latent heat of condensation of the heating steam. With the assumption of steady state and adiabatic process (no heat losses), the balance of energy in the evaporation process yields:

$$S(h_{S,V} - h_{S,l}) = S\Gamma_S = F_S(h_{P,1} - h_{P,F}) + V_1 h_{1,V} \tag{9.42}$$

where $h_{P,F}$ and $h_{P,1}$ are the enthalpies of the feed and the concentrate calculated at their respective temperatures T_F and T_1, S is the mass flow rate of steam, with enthalpies for the vapor and liquid phases $h_{S,V}$ and $h_{S,l}$ respectively Γ_S the latent heat of vaporization of water at temperature T_S and $h_{1,V}$ the enthalpy of the vapor at T_1. The latent heat for steam can be obtained from steam tables with the saturation temperature T_s, whereas the enthalpies of the feed and concentrate should be obtained from enthalpy versus concentration data for the liquid product, which often are not known. For these cases, an expression to estimate the enthalpy of the solution as a function of the concentration can be used. As noted in Eq. 9.43, for convenience, these enthalpies are expressed on a dry basis and it is assumed that the heat capacities of the solid and water are independent of temperature:

$$h = (C_S + X_w C_w).T \tag{9.43}$$

where h is the enthalpy of the product in kJ/kg dry solids, C_s and C_w are the heat capacity of the solids and water expressed in kJ/K.kg solids and kJ/K.kg water, respectively, and T is the temperature of the liquid. The equation assumes that enthalpy is zero at a reference temperature of 0°C. In the equation it is also assumed that the steam is saturated and condenses completely after giving up its latent heat. By using the expression of the enthalpy given in Eq. 9.43 along with Eq. 9.40, the balance of energy simplifies to:

$$S\Gamma_S = F_S(C_S + X_{w0}C_w)(T_1 - T_F) + V_1\Gamma_1 \qquad (9.44)$$

where Γ_1 is the latent heat of vaporization of water at the temperature of the concentrate. It should be noted that concentrated solutions exhibit a boiling point elevation and the temperature of the concentrate T_1 is higher than the boiling point of pure water (or the solvent in non-aqueous solutions) at the evaporation pressure. The boiling point elevation of simple solutions of non-ionic solutes can be calculated by an expression that depends on the concentration of the solute and the type of solvent in the solution:

$$\Delta T_b = (T_b - T_{b0}) = K_b \cdot m \qquad (9.45)$$

where T_b and T_{b0} are the boiling points of the solution and pure water (or solvent) at the evaporation pressure, K_b is a constant that depends on the solvent (e.g. 0.51 K/mol/kg solvent for water, 1.22 K/mol/kg solvent for ethanol, and 2.93 K/mol/kg solvent for acetic acid), whereas m is the concentration of the solute expressed as molality, i.e. mol of solute/kg of solvent. For ionic solutes the above equation is modified by using the Van't Hoff factor, i, as (Chang, 2000):

$$\Delta T_b = (T_b - T_{b0}) = K_b \cdot m \cdot i \qquad (9.46)$$

Heat transfer in falling film evaporators

The flow of liquid films on surfaces, frequently observed in nature, is the key phenomenon that governs heat transfer in falling film evaporators. This phenomenon has been studied in great detail and specifically has been used to estimate the overall heat transfer coefficient that controls the transfer of heat from the heating medium (steam) to the liquid product. Another phenomenon, involving the action of centrifugal forces instead of gravity to form the falling film, has been utilized in the development of evaporators suitable for high-viscosity materials. Evaporators that use rotating surfaces were first described by Blass (1979) and are presented in section 9.2.1.2 (High solids evaporation).

The balance of energy mathematically represented by Eq. 9.44 clearly illustrates that the energy supplied by the heating steam provides the energy to vaporize part of the solvent of the solution and increase the enthalpy of the solution. In that sense, the equation provides information on how much energy is involved in the process but it does not give information on how fast this energy is provided. In order to estimate the energy transport in terms of rates that may affect the process and temperature in the product, Eq. 9.44 needs to be expanded and includes a term that takes into account that heat transfer rate. An equation taking into account the heat transfer rate is useful from a design standpoint because it provides an estimation of the area of heat exchange. The amount of energy involved in the evaporation process can be combined with the energy transfer rate by the following equation (Geankoplis, 2003):

$$S\Gamma_S = Q = UA(T_s - T_p) \qquad (9.47)$$

Several considerations need to be taken into account when using Eq. 9.47 to estimate the heat transfer Q. The first consideration is related to the geometry and the area A, which is used in calculations of the overall heat transfer coefficient U. As discussed, the film of the flowing liquid is formed in the interior and flows down the tube, part of which is evaporated. Calculations based on the properties of water at normal evaporation temperatures have shown that the velocity of the vapor inside the tube can be higher than 50 m/s, which promotes a turbulent regime for the evaporating vapor and a propelling action on the falling liquid film. As a consequence of the water (or solvent) evaporation, the thickness of the liquid film decreases down the tube. The average thickness of the product can be calculated assuming that the mass flow is uniformly distributed in the wall of the interior tube with a wetted surface πD by the following equation:

$$\delta = \left(\frac{3\dot{m}\nu_{film}}{\rho_{film}\pi D g}\right)^{1/3} \qquad (9.48)$$

where D is the internal diameter of the tube, $\dot{m}$ (kg/s) is the mass flow rate of the product in a tube, ν_{film} (m^2/s) and ρ_{film} (kg/m^3) are the kinematic viscosity and the density of the product, respectively, and g is the gravity constant. Equation 9.48 is applicable for Reynolds numbers (Re) of up to 400, i.e. within the laminar regime, which can be calculated by the following equation (Kessler, 2002):

$$Re = \frac{v_{film} \cdot \rho_{film} \cdot \delta}{\mu_{film}} = \frac{v_{film} \cdot \delta}{\frac{\mu_{film}}{\rho_{film}}} = \frac{v_{film} \cdot \delta}{\nu_{film}} \qquad (9.49)$$

The velocity of the film can be calculated from the mass flow rate in the tube, the density of the liquid film and the area of flow that is calculated with the average thickness and the wetted area:

$$v_{film} = \frac{\dot{m}/\rho_{film}}{A_{film}} = \frac{\dot{m}}{\rho_{film}\pi D\delta} \tag{9.50}$$

Using the above equations, it can be demonstrated that the thickness of the film is very small compared to the tube diameter. For example, assuming that milk with 3.5% fat is evaporated at 60°C, from tables (Kessler, 2002) it can be obtained that ρ_{film}= 1011.4 kg/m^3 and μ_{film}= 0.86 × 10^{-3} Pa.s, which gives a kinematic viscosity of $\nu_{film} = \mu_{film}/\rho_{film}$= 0.85 × 10^{-6} m^2/s. In addition, considering a tube of 25.4 mm (1 inch) internal diameter and a mass flow rate per tube, $\dot{m}$, of 40 kg/h (1.1 × 10^{-2} kg/s) and by substituting these values into Eq. 9.48, the following is obtained:

$$\delta = \left(\frac{3 \times 1.1 \times 10^{-2} \times 0.85 \times 10^{-6}}{1011.4 \times \pi \times 25.4 \times 10^{-3} \times 9.81}\right)^{1/3} = 3.3 \times 10^{-4}\, m = 0.33\, mm \tag{9.51}$$

Thus, the thickness of the film is small compared with the diameter of the tube and the curvature of the cylindrical geometry as illustrated in Figure 9.30 can be ignored and represented a the rectangular (slab) geometry. Hence, the overall heat transfer coefficient is calculated as:

$$\frac{1}{U} = \frac{1}{h_{steam}} + \frac{x_{wall}}{k_{wall}} + \frac{x_{fouling}}{k_{fouling}} + \frac{1}{h_{film}} \tag{9.52}$$

The above equation describes the overall resistance (*1/U*) to the heat transferred from the heating steam to the falling film inside the tube and includes the resistance to heat transfer due to convection in the film of the condensing steam, the resistance to conduction through the wall and the fouling layer if present, and the resistance to heat through the product film, which for simplicity is approximated as $h_{film} = k_{product}/\delta$. This coefficient incorporates the heat transfer by conduction in the liquid film and the convection through the vapor phase in the product side, which is high in comparison with the heat transfer mechanism in the product film (see Figure 9.30).

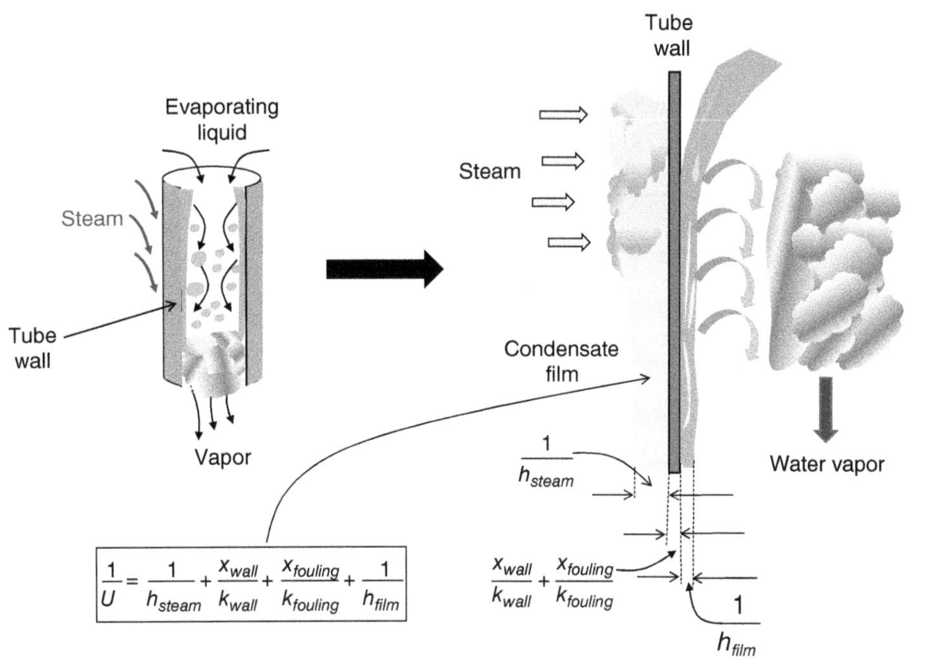

Figure 9.30 Actual (cylindrical) and approximated (slab) geometries used to describe and calculate the overall heat transfer coefficient in a falling film evaporation system.

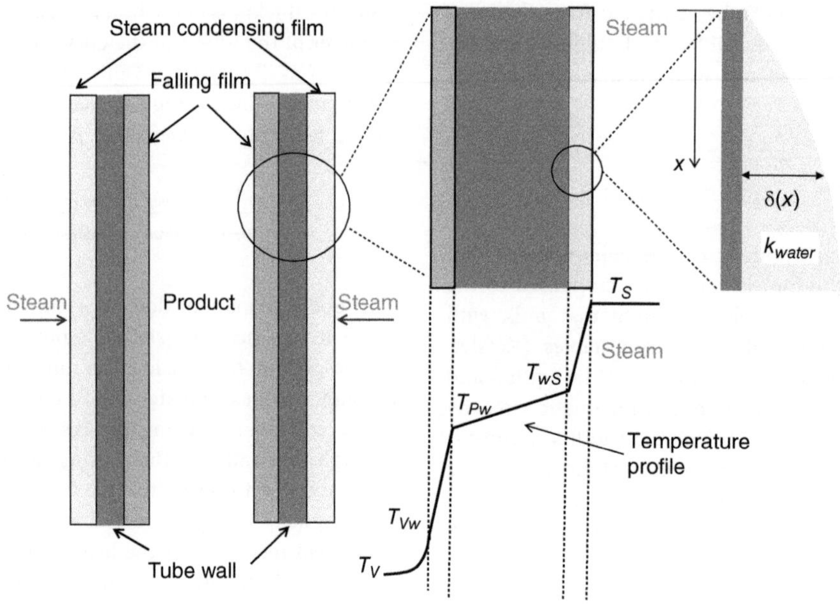

Figure 9.31 Temperature profile in one tube of a falling film evaporator.

Steam condensation heat transfer coefficient h_{steam}
The heat transfer coefficient in the steam side is generally high, with values up to 6 kW/m^2.K. Geankoplis (2003) describes an approach used to predict this coefficient, which is based on the Nusselt theory that assumes the formation of a variable thickness film through which the heat released by the condensing steam is transferred (see Figure 9.30). Since the thickness of the condensing steam film increases with the distance down the tube, the heat transfer coefficient varies with position and is calculated as a local coefficient (Geankoplis, 2003):

$$h_{steam}(x) = \left(\frac{k_l^3 \cdot \rho_l \cdot (\rho_l - \rho_v) \cdot g \cdot \Gamma_S}{4 \cdot \mu_l \cdot (T_S - T_{wS}) \cdot x}\right)^{1/4} \quad (9.53)$$

Thermal conductivity (k_l), density (ρ_l), and viscosity (μ_l) correspond to liquid water whereas ρ_v is the density of water vapor, which is significantly lower than the density of the liquid water and can be dropped from the equation. The thermal conductivity and the density of water are weak functions of temperature and both can be considered constant, whereas the viscosity of liquid water is affected by temperature and has to be calculated at the steam temperature. The distance measured from the tube entrance is designated as x, and T_S and T_{wS} are the temperatures of the steam and the interface between the film and the tube exterior wall (Figure 9.31). The assumed temperature profile is schematically depicted in Figure 9.31 without the fouling layer. When the product fouls the tube wall, a resistance to the heat transfer due to that fouling layer is incorporated in the equation to estimate the overall heat transfer coefficient. To simplify calculations, the local heat transfer coefficient can be integrated along the tube length (L) to obtain an average value:

$$h_{steam} = 0.943\left(\frac{k_l^3 \cdot \rho_l \cdot (\rho_l - \rho_v) \cdot g \cdot \Gamma_S}{\mu_l \cdot (T_S - T_{wS}) \cdot L}\right)^{1/4} \quad (9.54)$$

The above equations apply to a laminar regime but depending on the evaporation conditions and the pressure and temperature of the steam, other regimes can exist and thus prediction of the heat transfer coefficient in the steam side can underpredict experimental results.

If a Reynolds number is defined as $Re = \frac{4\dot{m}}{\pi D \mu_l}$

For turbulent flow Re > 1800

$$Nu = \frac{h_{steam} L}{k_l} = 0.0077\left(\frac{g \rho_l L^3}{\mu_l^2}\right)^{1/3} Re^{0.4} \quad (9.55)$$

To get the value of heat transfer coefficient in the film, h_{film}, the solution of the above equation is by trial and error since a value of Re must first be assumed (Geankoplis, 2003).

Heat transfer coefficient in the liquid product

Heat transfer coefficients in the falling film are more difficult to predict because two different evaporation regimes are found. One is close to the tube entrance, i.e. when the solid content in the liquid is low and there is vigorous boiling of water. The other regime is close to the bottom of the tube where boiling of water is prevented due to the high solids content of the liquid (Geankoplis, 2003).

Three heat transfer mechanisms are identified in the boiling regime: (1) natural-convection boiling; (2) nucleate boiling; and (3) film boiling. In the natural convection boiling mechanism, heat transfer through the liquid film occurs by conduction and natural convection. Evaporation of the water occurs only at the vapor–liquid interface (identified by temperature T_{Vw} in Figure 9.31). In the nucleate boiling mechanism, the presence of bubbles may affect the heat transfer process. At low nucleation, the presence of a few bubbles does not greatly affect the heat transfer whereas for high rate nucleation boiling the presence of many bubbles strongly affects both the flow of the film and the heat transfer coefficient. The vigorous mixing action due to the presence and motion of many bubbles increases the heat transfer coefficient. For high temperatures in the liquid, the evaporation rate can increase greatly and a blanket of vapor is formed at the interface of the liquid product which significantly decreases the heat transfer coefficient due to the low heat conductivity of the vapor. Predictive equations for this boiling regime are less accurate because of the large variation in the heat transfer coefficient which is highly dependent on the difference between the film–vapor interface temperature and the temperature of the bulk vapor. Depending on that temperature difference, values of this coefficient vary from 1 to 30 kW/m^{2}K as illustrated in Figure 9.32. As indicated in the figure, the maximum heat transfer coefficient can be reached for temperature differences of about 45–50°C. For higher temperature differences, the blanketing effect reduces the heat transfer significantly.

The heat transfer coefficient in the non-boiling zone can be calculated by an equation proposed by Geankoplis (2003):

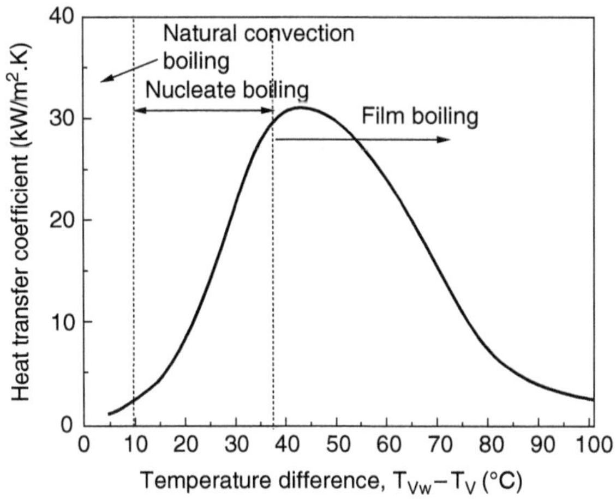

Figure 9.32 The effect of the difference between the heating steam temperature and the temperature of the product vapor on the falling film heat transfer coefficient.

$$Nu = \frac{h_{film}D}{k_l} = 0.028 Re^{0.8} Pr^{1/3} \left(\frac{\mu_b}{\mu_w}\right)^{0.14} \qquad (9.56)$$

where μ_b and μ_w are the viscosities of the product at the evaporation and wall temperatures, respectively. If these temperatures vary along the tube length the viscosities should be evaluated at the log mean temperatures calculated from the top and bottom temperatures. *Re* and *Pr* are the Reynolds and the Prandtl numbers defined as:

$$Re = \frac{Dv\rho}{\mu} \qquad (9.57)$$

$$Pr = \frac{C_p\mu}{k} \qquad (9.58)$$

To calculate these dimensionless numbers, the properties of the product must be evaluated at the evaporation temperature. If the product temperature varies along the tube length, a mean value should be used to estimate the properties of the product.

9.2.1.1.2 *Multiple effect evaporators*

The increased cost of energy in the evaporation process prompted the development of more energy-efficient evaporation systems, in which, among many modifications done on the basic initial evaporation unit, the incorporation of a larger number of effects has resulted as one of the most effective. The analysis to demonstrate the energy savings with the introduction of more effects is simple. By inspection of Eq. 9.44 and considering that the sensible heat utilized to raise the temperature of the product from the feed temperature up to the evaporation temperature is about 10–15% of the latent heat of vaporization, which is slightly dependent on the temperature, it can be easily demonstrated that to evaporate a certain amount of water from the product, approximately the same amount of steam must be used. Thus, if the *specific steam consumption* is defined as the amount of steam used in the process divided by the amount of water evaporated, its value for single effect evaporators is close to 1 (or 100%). However, if the vapor removed in the first effect is used as the heating medium for a second effect, the specific steam consumption is reduced to approximately 50%. If three effects are used, the specific steam consumption would be further reduced to 33% and so forth.

Thus, it appears that, at least theoretically, by increasing the number of effects larger energy savings could be obtained. However, in practice, there are limits to the maximum number of effects that can be economically used. One of the more serious problems is that in each effect, the pressure and temperature of the vapor are reduced and therefore a larger and more expensive vacuum is required to achieve product evaporation temperatures lower than the vapor temperature of the previous effect used to heat the product. In addition, to set an optimum number of effects, other modifications have been introduced in the evaporation process with the main objective of saving energy. These modifications include preheating of the feed, reduction of heat losses using better insulation, flashing of condensate passed from one effect to the next, the use of inter-effect steam to preheat the feed, and better regulation of non-condensable gases purging from the evaporator. Other significant changes have been the use of thermal and mechanical recompression across two or three effects rather than a single effect.

Figure 9.33 illustrates a multiple effect system with N effects. As indicated in the figure, which describes a feed-forward operation, the vapor removed from the concentrate in the liquid-vapor separator of each effect is used as the heating medium for the next effect. It should be noted that during vapor transfer from one effect to the next, the pressure and temperature of the vapor decrease. The pressure and temperature of the vapor in the last effect, which control the pressures and temperatures of all the effects, are the lowest obtainable in the multiple effect system. The pressure, and consequently the temperature, of the vapor entering in the last effect are controlled by conditions in the condenser, mainly the temperature of the cooling water. In the feed-forward operation the effects are numbered from *1* to N starting with the effect where steam is introduced whereas in the backward feed operation, the fresh feed enters the last and coldest effect.

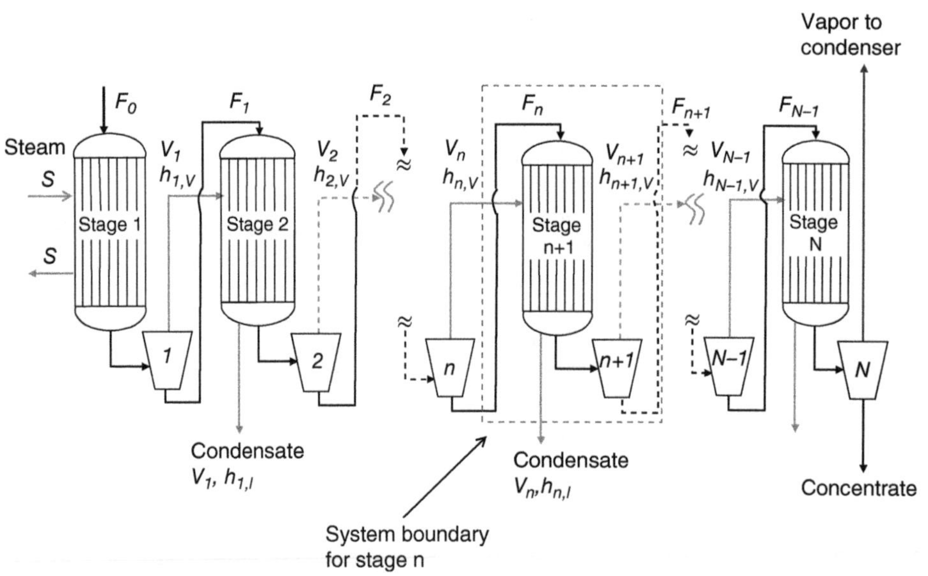

Figure 9.33 Schematic of a multi-effect falling film system working in a co-current mode.

The feed-forward operation is used when the feed is hot or to minimize heat damage on the final concentrate. Conversely, the backward feed operation is used when the feed is cold but pumps are necessary to transport the product against an increasing pressure profile (Geankoplis, 2003). The balance of mass and energy is similar in these two operation modes so the balance of mass and energy will be done only for the feed-forward operation.

The calculation methods for multiple effect evaporators are fairly involved because in order to analyze the performance of a N-effects evaporation system, it is necessary to deal with N mass and energy balances, N heat transfer equations, N equations relating local heat transfer coefficients to local flow rates in each effect, and N equations relating local boiling point elevations to local concentrations in each effect whereas many other variables such as feed rate, its temperature and concentration, the desired final concentration, the last effect pressure, the maximum tolerable product temperature and the conditions of the available steam are specified. Schwartzberg (1983) estimated that for a multiple effect evaporation system with N-stages, there are *6N+6* variables to be determined or specified and *5N+7* specified quantities or equations available for determining these *6N+6* variables. Thus, *N–1* additional or design criteria must be used to determine all the variables that define the problem. One approach that has been used for design purposes is to specify the heat transfer area in each effect. This approach is described by Geankoplis (2003) and is based on the assumption that all the heat transfer areas are equal. Regardless of whether the assumption is suitable for commercial practice and the simplification it brings to the calculations, the approach still requires a cumbersome trial-and-error procedure. Schwartzberg (1983) proposed shortcut calculations that not only can be used to design these systems for specific applications but also enable energy efficiency estimations.

Considering the multiple effects with N stages evaporation system depicted in Figure 9.33, the mass and energy balances for the stage *n+1* are:

$$\text{Total balance: } F_n = F_{n+1} + V_{n+1} \quad (9.59)$$

$$\text{Solids balance: } F_n x_{s,n} = F_{n+1} x_{s,n+1} \quad (9.60)$$

$$\text{Water/solvent balance: } F_S X_{w,n} = F_S X_{w,n+1} + V_{n+1} \quad (9.61)$$

$$\text{Energy balance: } V_n(h_{n,V} - h_{n,l}) = F_S(h_{p,n+1} - h_{p,n}) + V_{n+1} h_{n+1,V} \quad (9.62)$$

In terms of the enthalpy for the liquid product defined by Eq. 9.43 and substitution of the expression $X_{w,n+1} = X_{w,n} - V_{n+1}/F_S$, from Eq. 9.61, the energy equation results in the following:

$$V_n(\Gamma_n + C_V\phi_n) = V_{n+1}\left[\Gamma_{n+1} - (C_w - C_V)\phi_{n+1}\right] - F_S(C_S + X_n C_w)\cdot\left[(T_n + \phi_n) - (T_{n+1} + \phi_{n+1})\right] \quad (9.63)$$

Equation 9.63 introduces concepts of importance in evaporation technology that relate to the boiling point rise of solutions, which in the above equation is designed as ϕ. Although the elevation of the boiling point is small for diluted solutions, it becomes appreciable when the solution becomes concentrated in the process. It must also be noted that due to the boiling point rise of the solution at the exit of each effect, the vapor separated and used as a heating medium for the next effect is superheated and its latent heat of vaporization is estimated as:

$$h_{n,V} - h_{n,l} = \Gamma_n + C_V\phi_n \quad (9.64)$$

where C_V and ϕ_n are the heat capacity at constant pressure of the vapor and the boiling point rise of the solution leaving the stage *n*, respectively. Γ_n and Γ_{n+1} are the latent heat of vaporization of water at T_n and T_{n+1}, which are the normal boiling points of water corresponding to the vapor pressure existing in stage *n* and *n+1*, respectively. ϕ_{n+1} is the boiling point rise of the solution leaving the stage *n+1*. The derivation of Eq. 9.63 also assumes the realistic condition that all the heat capacities at constant pressure included in the equation are independent of temperature. Equation 9.63 provides a good description of the balance of energy including the boiling point rise phenomenon mentioned above. The term on the left is the heat given up by the condensation of the vapor V_n in the effect *n +1* whereas the first term on the right side of the equation is the energy required to produce the vapor V_{n+1}, which is superheated by the boiling point rise ϕ_{n+1}. The second term on the right side corresponds to the heat released when the liquid from the *n* stage cools from $T_n + \phi_n$ to $T_{n+1} + \phi_{n+1}$ in the next *n+1* stage.

9.2.1.1.3 Energy savings approaches

In order to improve the efficiency of the evaporation process, part of the condensate extracted in each stage (Figure 9.34) could be used in the following effect operating at lower pressure. As a consequence of the condensate moving from a high pressure to a low pressure condition, when the multiple effect system is operating in the feed-forward mode, part of the condensate will flash off. The amount of condensate that is converted to vapor, x, as a consequence of the pressure decrease can be estimated from a balance of energy across a valve where the pressure of the liquid condensate is decreased from p_{n-1} to p_n (inset in Figure 9.34). Assuming 1 kg of condensate coming from stage n-1, the amount of vapor flashing off per kg of condensate and the energy associated are calculated as:

$$x(h_{n,V}-h_{n,l})=(h_{n-1,l}-h_{n,l})=C_w(T_{n-1}-T_n) \quad (9.65)$$

The amount of condensate flashing off in the effect n+1 is $\Sigma_1^{n-1}\dot{V}_C$, and to calculate the energy savings, the above equation should be multiplied by the flow of condensate flashing off. Including that energy saving in Eq. 9.63 yields:

$$V_n(\Gamma_n+C_V\phi_n)+\left(\sum_1^{n-1}\dot{V}_C\right)\cdot C_w(T_{n-1}-T_n)=V_{n+1}\big[\Gamma_{n+1}-(C_w-C_V)\phi_{n+1}\big]-F_S(C_S+X_nC_w)\cdot\big[(T_n+\phi_n)-(T_{n+1}+\phi_{n+1})\big] \quad (9.66)$$

The second term on the left side is the contribution of the vapor flashing off to the energy necessary for water evaporation of the product in the following stage. Another energy saving considered in these systems is the preheating of the feed using part of the vapor coming from other effects (Schwartzberg, 1983). A schematic of the setting for the stage n+1 is illustrated in Figure 9.35, showing that the feed is preheated using parts of the vapors from effects N through n through heat exchangers positioned in the steam chests of the effects. The temperature of the feed entering the stage n+1 is $T_{n+1}-\delta_{n+1}$, whereas the temperature of the concentrate leaving the same stage is $T_n-\delta_n$, where δ_n is the temperature difference between the feed temperature leaving the stage n+1 and the temperature of the heating vapor in the chest of the stage. Thus, the preheating of the feed in the stage n+1 can be calculated as:

$$preheating=F_S\cdot(C_S+C_wX_0)\cdot\big[(T_n-\delta_n)-(T_{n+1}-\delta_{n+1})\big] \quad (9.67)$$

To take into account the preheating of the feed in the balance of energy, it is necessary to subtract Eq. 9.67 from the left side of Eq. 9.66 to yield:

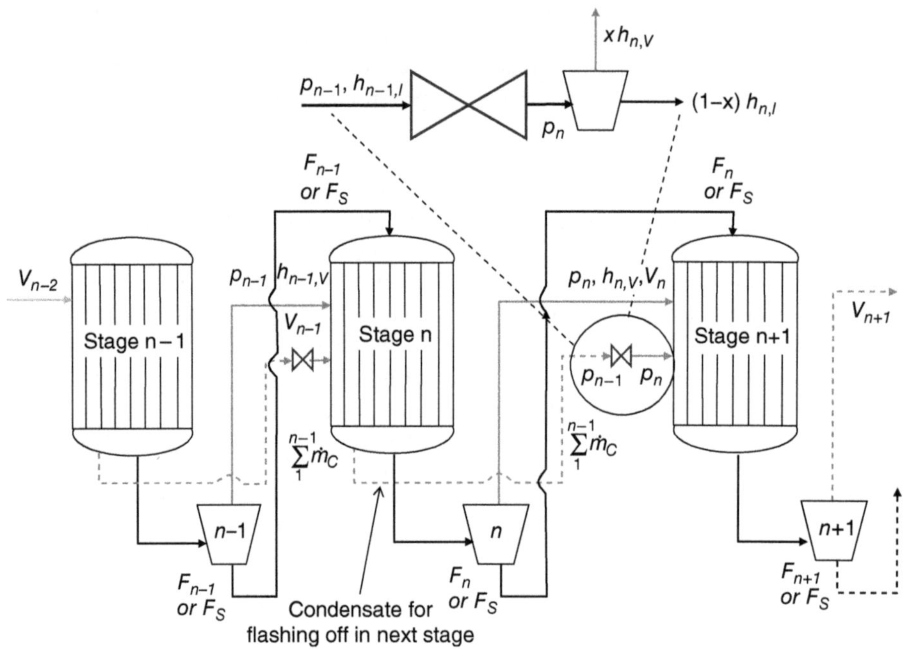

Figure 9.34 Schematic of a multi-effect falling film system working in co-current mode and with the incorporation of the condensate flashing-off going to the next effect.

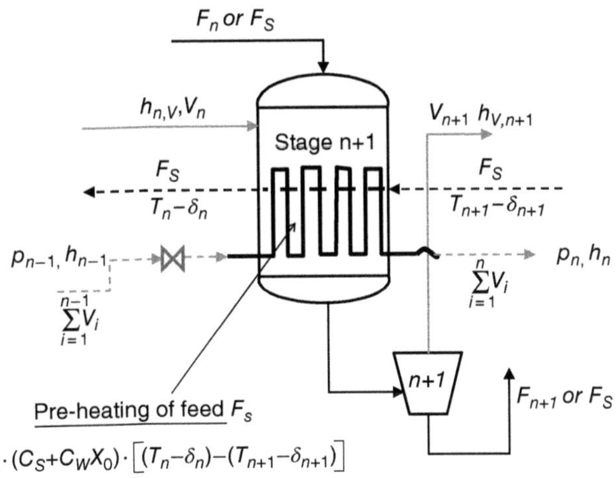

Figure 9.35 Multi-effect falling film evaporation system including preheating of the feed.

$$V_n(\Gamma_n + C_V\phi_n) + \left(\sum_{1}^{n-1}\dot{V}_C\right)\cdot C_w(T_{n-1} - T_n) = V_{n+1}[\Gamma_{n+1} - (C_w - C_V)\phi_{n+1}] + F_S(C_S + X_0C_w)\cdot[(T_n - \delta_n) - (T_{n+1} - \delta_{n+1})] - F_S(C_S + X_nC_w)\cdot[(T_n + \phi_n) - (T_{n+1} + \phi_{n+1})] \tag{9.68}$$

Equation 9.68 gives a complete balance of energy for multiple effect evaporators and provides a practical tool to analyze "what if" situations related to the saving of energy during the process. In order to simplify the equation to use it for shortcut calculations, a few assumptions must be made. One assumption is concerning the boiling point elevation, which is important to determine the heat transfer but its contribution to the balance of energy is relatively small and can be ignored. With the assumption that all ϕ_i for i 1→n are zero, several cases can be considered.

1. No flashing off condensate and no feed preheating:

$$\frac{V_{n+1}}{F_S} = \frac{V_n}{F_S}\cdot\left(\frac{\Gamma_n}{\Gamma_{n+1}}\right) + \frac{(C_S + X_nC_w)(T_n - T_{n+1})}{\Gamma_{n+1}} \tag{9.69}$$

To express the moisture content of the concentrate leaving the effect n, X_n, a mass balance between the initial effect and the effect n can be used:

$$X_n = X_0 - \frac{\sum_{1}^{n} V_i}{F_S} \tag{9.70}$$

Substituting Eq. 9.70 into Eq. 9.69 yields the following:

$$\frac{V_{n+1}}{F_S} = \frac{V_n}{F_S}\cdot\left(\frac{\Gamma_n - \left(\sum_{i=1}^{n} V_i\right)C_w(T_n - T_{n+1})}{\Gamma_{n+1}}\right) + \frac{(C_S + X_0C_w)(T_n - T_{n+1})}{\Gamma_{n+1}} \tag{9.71}$$

2. Flashing off condensate and no feed preheating:

$$\frac{V_{n+1}}{F_S} = \frac{V_n}{F_S}\cdot\left(\frac{\Gamma_n - C_w(T_n - T_{n+1})}{\Gamma_{n+1}}\right) + \frac{(C_S + X_0C_w)(T_n - T_{n+1})}{\Gamma_{n+1}} - \frac{\left(\sum_{i=1}^{n-1} V_i\right)\cdot C_w[(T_n - T_{n+1}) - (T_{n-1} - T_n)]}{F_S\Gamma_{n+1}} \tag{9.72}$$

3. Flashing off condensate and feed preheating with $\delta_n = \delta_{n+1}$:

$$\frac{V_{n+1}}{F_S} = \frac{V_n}{F_S}\cdot\left(\frac{\Gamma_n - C_w(T_n - T_{n+1})}{\Gamma_{n+1}}\right) - \frac{\left(\sum_{i=1}^{n-1} V_i\right)\cdot C_w[(T_n - T_{n+1}) - (T_{n-1} - T_n)]}{F_S\Gamma_{n+1}} \tag{9.73}$$

9.2.1.1.4 Shortcut equations

Mass balances

Further assumptions can be used to simplify the general or specific equations given above. These assumptions include (Schwartzberg, 1983):

- $T_n - T_{n+1} = T_{n-1} - T_n = ... = \frac{T_S - T_N}{N}$

  or $T_n - T_{n+1} = T_{n-1} - T_n = ... = \frac{T_1 - T_N}{N-1}$

where T_1 is a specified upper limit for the product.

- $\Gamma_{n+1} = \Gamma_n + (C_w - C_V)(T_n - T_{n+1})$
- $\Gamma_n = \bar{\Gamma}$, which is assumed to be the average latent heat of vaporization over the range of temperatures used in the evaporator.
- A water balance in the effect n given by Eq. 9.61 and with the assumption of a uniform water distribution in each effect: $\frac{V_n}{F_S} = (X_{w,n-1} - X_{w,n}) = \frac{X_0 - X_N}{N}$

By using the above assumptions in Equations 9.71 to 9.73 to each effect and summing all the V_n/F_S terms with the additional condition $\sum_{i=1}^{N} V_i/F_S = X_{w,0} - X_{w,N}$, Schwartzberg (1983) obtained the following shortcut equations:

1. No flashing off condensate and no feed preheating (shortcut equation):

$$\frac{V_1}{F_S} = \left(\frac{X_{w0} - X_{wN}}{N}\right) \cdot \left[1 - \frac{T_1 - T_N}{(N-1)\bar{\Gamma}} \cdot [(C_V - C_w) - NC_w]\right] - \frac{(T_1 - T_N)(C_S + X_{w0} C_w)}{(N-1)\bar{\Gamma}} \tag{9.74}$$

2. Flashing off condensate and no feed preheating (shortcut equation):

$$\frac{V_1}{F_S} = \left(\frac{X_{w0} - X_{wN}}{N}\right) \cdot \left[1 + C_V \cdot \frac{T_1 - T_N}{(N-1)\bar{\Gamma}}\right] - \frac{(T_1 - T_N)(C_S + X_{w0} C_w)}{(N-1)\bar{\Gamma}} \tag{9.75}$$

3. Flashing off condensate and feed preheating with $\delta_n = \delta_{n+1}$ (shortcut equation):

$$\frac{V_1}{F_S} = \left(\frac{C_V(T_1 - T_N)}{(N-1)\bar{\Gamma}}\right)\left(\frac{X_{w0} - X_{wN}}{1 - r^N}\right) \tag{9.76}$$

where the factor r is defined as:

$$r = \left(1 - \frac{C_V(T_1 - T_N)}{(N-1)\bar{\Gamma}}\right) \tag{9.77}$$

Energy balance

From an economic standpoint, a key parameter in the evaporation process is the thermal energy required in the first effect per unit mass of dry solids, i.e. Q_1/F_S. Two different cases can be considered.

1. Without feed preheating, which can be estimated as:

$$\frac{Q_1}{F_S} = \frac{V_1}{F_S}\Gamma_1 + (C_S + X_{w0} C_w)(T_1 - T_0) \tag{9.78}$$

2. With feed preheating:

$$\frac{Q_1}{F_S} = \frac{V_1}{F_S}\Gamma_1 + (C_S + X_{w0} C_w) \cdot \delta_1 \tag{9.79}$$

where δ_1 is the approach temperature between the feed and the preheated vapor drawn from the first effect. The steam requirements corresponding to a value of Q_1 can be calculated as:

$$\frac{S}{F_S} = \frac{Q_1/F_S}{\Gamma_S} \tag{9.80}$$

Either Eq. 9.79 or Eq. 9.80 could be used with the different values of V_1/F_S for the cases considered and estimated by Eqs 9.74–9.76. For a four effect evaporator with T_1 = 100°C and T_4 = 40°C, initial moisture content on a wet basis of 90% (X_{w0} = 9) and a final moisture content of 50% wet basis (X_{wN} = 1), two cases are considered: (1) the same ΔT in each effect; and (2) $\Delta T_4 = 2.33\ \Delta T_2$, $\Delta T_3 = 1.53\ \Delta T_2$ and $\Delta T_2 = \Delta T_1$. Other values assumed are C_w = 4.18 kJ/kg.K, C_V = 1.68 kJ/kg.K, Cs = 1.4 kJ/kg.K, $\bar{\Gamma}$= 2332 kJ/kg, and δ_1= 6°C. Results obtained from the calculations are given in Table 9.7.

It can be noted that by using multiple effect evaporators and some processing modifications, the energy can be decreased up to 30%. The greater the reduction in the moisture content of the sample, the more efficient will be the overall drying process. However, as discussed before, there are limitations on the maximum amount of solids in the feed that atomizers of spray-drying systems can handle. With regard to the use of multiple effect evaporation systems, theoretically more effects improve the efficiency of the evaporation system, although 4–6 effects provide high efficiency and performance. Limitations for using a large number of effects are related to

Table 9.7 Results of shortcut calculations for some of the energy-saving options discussed in section 9.2.1.1.4.

ΔT	Q_I/F_S (kJ/kg)		
	No flashing No preheating	Flashing No preheating	Flashing Preheating
Same ΔT in each effect	7050	6892	5155
$\Delta T_4 = 2.33\Delta T_2$ $\Delta T_3 = 1.53\Delta T_2$	7090	6980	5172

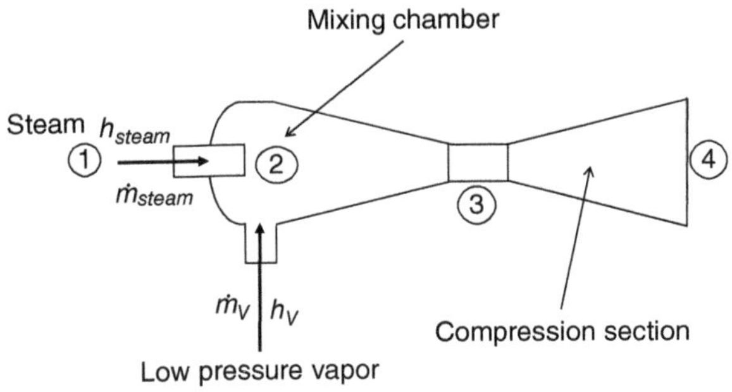

Figure 9.36 Thermocompressor system used for thermocompression of low-pressure vapors in thermal vapor recompression (TVR).

the low pressures of the vapor generated in the latest effects. Further improvement of the efficiency of the evaporation process can be achieved by using thermal vapor recompression (TVR) and mechanical vapor recompression (MVR). In both cases, part or all of a vapor from an effect is compressed to elevate its temperature and further used as a heating medium for the same or a different effect (Kessler, 2002). In TVR, recompression of the vapor is carried out using steam-jet compression in a nozzle system as illustrated in Figure 9.36. Direct steam of pressure in the range 8–12 bars is used and mixed with low-pressure vapor coming from any of the effects. A balance of energy and momentum in the steam-jet compressor, assuming isentropic conditions, yields (Kessler, 2002):

$$\frac{\dot{m}_V}{\dot{m}_{steam}} = \sqrt{\frac{h_{steam} - h_2}{h_4 - h_3}} - 1 \qquad (9.81)$$

For real-life processes, losses due to friction and turbulence will decrease the enthalpy difference h_{steam}-h_2 and will increase the enthalpy difference h_4-h_3, decreasing the theoretical ratio to about 80–85% of the theoretical

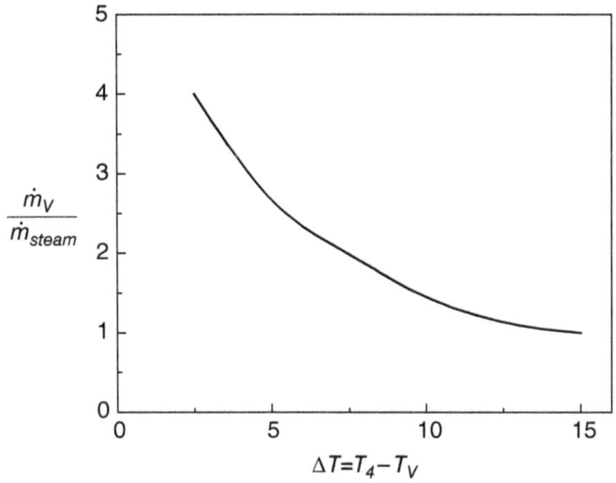

Figure 9.37 Steam used as a function of the temperature rise from the vapor temperature T_V to the exit temperature of the thermal vapor recompression system T_4.

values. Equation 9.81 allows estimation of the mass flow rate of steam for a given vapor mass flow rate and specific temperatures at the steam-jet compression exit. Results of the flow rates ratio as a function of the temperature difference $T_4 - T_V$ are illustrated in Figure 9.37. Schwartzberg (1977) discusses saving of energy using these systems and shows that the use of thermocompression yields the thermal efficiency of a *N+1* effects in a *N*-effect evaporator.

Mechanical recompression is used to increase the pressure and temperature of vapor obtained in some effects. It can theoretically provide greater energy savings than thermocompression. Mechanical recompression is achieved by using axial turbocompressors and the overall mechanical recompression lowers the primary energy consumption to about 200 kJ per kg of removed water (Kessler, 2002). The change of enthalpy in the low-pressure vapor can be calculated by the following equation:

$$\Delta h = \frac{n}{n-1} \cdot p_1 v_1 \cdot \left[\left(\frac{p_2}{p_1} \right)^{\frac{n-1}{n}} - 1 \right] \qquad (9.82)$$

n = C_p/C_v with a value of approximately n = 1.3 corresponding to superheated steam

9.2.1.2 High solids evaporation

From an energy standpoint, the previous section has shown that the solution being dried should be preconcentrated as much as possible without exceeding the solubility of the solute. Furthermore, the concentration of solids should be such that the concentrate remains pumpable in the dryer atomizer. Falling film evaporators are also limited when the viscosity of the concentrate becomes too high to flow in the evaporator tubes. Under these circumstances, the flow is extremely slow and a thick film is formed in the tubes, which significantly decreases the film heat transfer coefficient and the evaporation rate. Thus, the use of high solids evaporators ahead of vacuum falling film evaporators must be considered in order to increase the solids content of the concentrates. Various options can then be proposed such as plate evaporators, mechanically scraped surface evaporators, or rotating surface evaporators, which typically operate under vacuum.

Plate evaporators utilize the climbing-falling film principle within a plate heat exchanger. In this device for the concentration of products, concentrates and steam flow through alternate passages defined by the plate assembly. High liquid velocities lead to good heat transfer with short residence times, making these evaporators suitable for the concentration of heat-sensitive materials. However, they show real limitations when the viscosity of the concentrates is high.

Scraped surface evaporators, widely used in the food processing industry, consist of a jacketed cylinder with an internal cylinder concentric to the first and fitted with a rotor which is equipped with scraper blades. The concentrate flows in the central cylinder with the outer heat transfer surface constantly scraped while the steam circulates through the jacketed cylinder. These devices are particularly well adapted to high-viscosity concentrates, with relatively small evaporation loads. Their ability to process heat-sensitive products is moderate.

As for rotating surface evaporators, it has been recognized that if a centrifugal force can be generated in a rotating surface, the film thickness is decreased and the film heat transfer coefficient is greatly enhanced. Evaporation of heat-sensitive materials at high concentrations requires short residence times which cannot be achieved in falling film evaporators. Initial work, including some industrial practices, started with thin film evaporators with rotating wipers but residence times were of the order of minutes and not a few seconds which are more suitable for liquids containing heat-sensitive materials. Evaporators that supply residence times of a few seconds for liquids with viscosities of 20 mPa.s or more are not available among standard technologies (Billet, 1989). Rotating the surface on which evaporation is taking place appeared to be the logical way to handle these requirements. Some of the first rotating surface evaporators date from 1935 when Hickman introduced the first rotating cone evaporator. Rotating surface evaporators have many important features such as: (1) low residence times which are important for evaporating heat-sensitive materials such as antibiotics, enzymes, protein solutions, etc.; (2) high solids concentrations; (3) high overall heat transfer coefficient; (4) ability to handle concentrates from intermediate to high apparent viscosity. Rotating surface evaporators have been used to produce a wide range of products that includes fruit juices, coffee and tea extracts, egg, milk products, etc. (Bouman & Waalewijn, 1994a, 1994b; Chen et al., 1997; Jebson et al., 2003).

Although technical modifications, or adaptations, are needed to better control the processing conditions (such as the residence time distribution), the above-mentioned high solids evaporators can be used in the innovative extrusion-porosification process. Compared with falling film evaporators the energy consumption of which is approximately 0.1 kg steam per kg evaporated water, these evaporators are characterized by moderate energy efficiency as the steam consumption reaches approximately 1.05 kg steam per kg evaporated water. This energy efficiency is still higher than that of spray dryers, and hence high solids evaporators may contribute significantly to the energy savings of the whole extrusion-porosification process.

9.2.2 Twin screw extrusion-aeration

9.2.2.1 Physical chemistry of foaming

Foams are two-phase systems, in which gas (usually air) bubbles are dispersed in a continuous liquid phase. The bubbles are separated by thin liquid films, so the liquid volume fraction of foams is very small (Kraynik, 1988). Foams are widely used in chemical and food applications, including detergents, shaving creams, ice cream, whipped cream, and beverages, to name a few. In general, air or other gases forming the non-continuous phase are dispersed in an aqueous surfactant solution (the continuous phase). However, there are other speciality foams in which the continuous phase can be a hydrocarbon-based fluid or a polymeric liquid (Heller & Kuntamukkula, 1987). The production of foamed plastics for a variety of non-food applications such as cushioning and insulation materials, and impact absorbers has seen continuous growth (Lee, 2000). Thus, there is continuous interest in industry as well as academia researching not only new technologies to produce stable and good-quality foams but also in a better understanding of the fundamentals and mechanisms of foam formation to assist new development of foam-based materials.

Foams with high liquid content are known as liquid foams and they have a gas to liquid volume ratio less than 90–95%. Liquid foams are very unstable and the action of gravity on the continuous phase could be severe. The Nestlé company has announced that it is using zero gravity research to understand the effect of gravity in the destabilization of high liquid content foams with the aim of developing perfect foam-like products (Kuntz, 2012). Foams with a gas to liquid volume ratio larger than 90%–95% are called dry foams. Extrudates or porous powders, which are foams of low liquid content and known as solid foams, are more stable and have a longer lifetime. Regardless of the lifetime of foams, it should be kept in mind that they are thermodynamically unstable systems and given time, they will inevitably break down (Campbell & Mougeot, 1999).

The instability of foams lies in the characteristics of their microstructure of a dispersed phase embedded in a continuous medium. As a consequence of the large surface area due to the dispersed phase, foams have high free energies which tend to move to equilibrium states with lower free energies and reduced surface area, resulting in a collapsed foam system. Thus, physicochemical forces and thermodynamic phenomena mean that the dispersed phase cannot remain as a fine dispersion and eventually will separate as a separate phase or collapse as happens in foam systems (Wilde, 2000).

One way to control the unavoidable path to breakdown of foams is by controlling their rheological properties. Figure 9.38 provides a qualitative description of the relationship between the mechanical properties of foams and their shelf-lives. These mechanical properties are closely associated with the content of liquid in foam systems; for instance, beverages have a high liquid content and low viscosities. It can be observed that bubbles present in beverages have a very short life, as for example in wine, beer, champagne, and cappuccino coffee. In many other cases, the lifetime of foams is stabilized by processes such as drying (e.g. in the production of meringue) or freezing (e.g. in ice cream); that is, stabilization of the foam is achieved by modifying its rheology through processing means such as drying or freezing.

The lifetime of foams can be also stabilized by the addition of ingredients such as surfactants that affect the interfacial properties of the bubbles, or gums that increase the viscosity of the continuous medium, thus retarding the coalescence of the bubbles. It must be noted that although the unavoidable fate of foams is phase separation and breakdown, the kinetics of breakdown can be altered within the limits of the applicability of foam-based products.

9.2.2.1.1 Foam structure and stability

Based on their microstructure, various types of foams exist and two in particular are described in the literature in great detail. The first type is the so-called quasi-spherical foam which comprises a system of nearly spherical bubbles dispersed in a continuous liquid phase, as in emulsions. The second type is the polyhedral foam, which

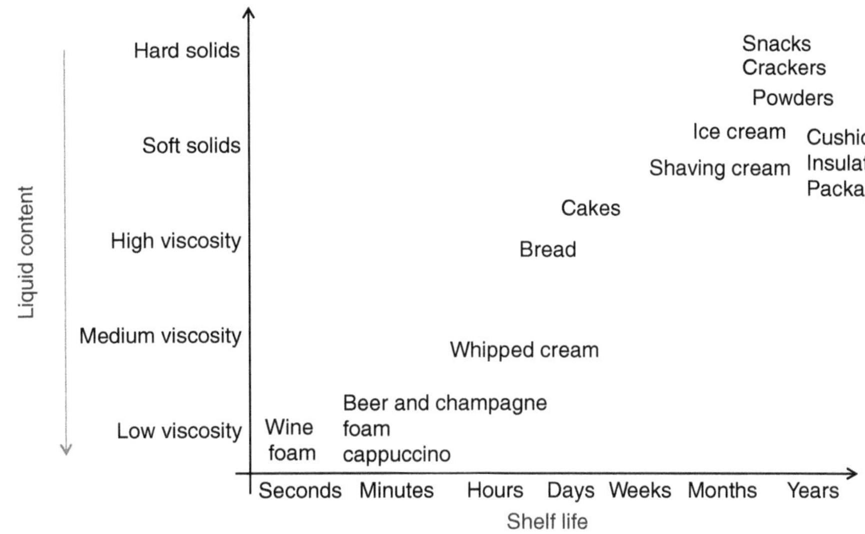

Figure 9.38 Relationship between mechanical properties of foams and their shelf-lives.

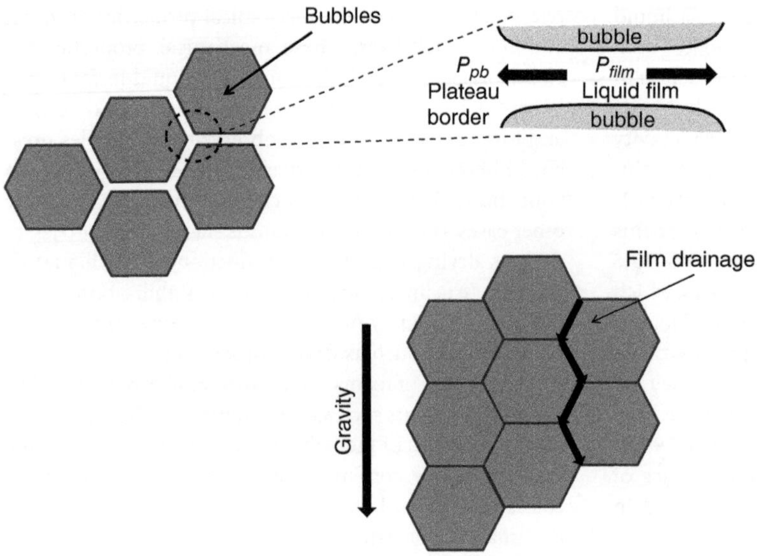

Figure 9.39 Structure of foams formed by polyhedral bubbles forming the Plateau border zones; equilibrium of forces acting on films forming the Plateau borders; action of gravity on liquid foams.

is a system of coarsened bubbles separated by thin liquid films, called lamellae (Kornev et al., 1999). As illustrated in Figure 9.39, due to the regular shape of the polyhedra cells, three films intersect in a zone known as the Plateau border, named in recognition of the French scientist Joseph Plateau who described these structures (Weaire & Hutzler, 1999). Since for this type of cellular structure the surface tension forces acting along each film must balance at the Plateau border, each film forms an angle 120° from the other. The pressure in the liquid film contributes to the stability of the foam, which is greatly affected by the liquid film drainage. Due to the curvature of the Plateau border, there is a pressure difference between the liquid close to the Plateau border and the liquid in the film, which can be calculated by the following equation:

$$P_{film} - P_{Pb} = \frac{\gamma_f}{R} \tag{9.83}$$

where P_{film} and P_{Pb} are the pressures of the liquid in the film and the Plateau border respectively, γ_f is the liquid surface tension and R is the radius of curvature of the interface between the bubble and the liquid (see Figure 9.39). Due to this difference in liquid pressure, there is a natural drainage of liquid from the film to the Plateau border. Decrease in surface tension by the addition of surfactants can contribute to the decrease of gravity-induced drainage that promotes the instability of the foam. It is thought that by populating the film/bubble interfaces with surfactant molecules, the flow within the interface is retarded (Heller & Kuntamukkula, 1987).

9.2.2.1.2 *Formation of bubbles in foams*

During mixing, whipping, beating or shaking of the continuous media, the gas (e.g. air) enters and is incorporated into the liquid continuous phase. But due to this mechanical action bubbles can also be broken into bubbles of smaller sizes. It is assumed that shear forces and turbulence contribute to the dispersion of the gas in the continuous phase. In the case of dispersion of a gas, shear forces are those that deform bubbles and eventually reduce their size. The Weber number, We, provides a measure of the shear forces applied to the bubble and also the property of the bubble to resist deformation induced by the shear forces, which is controlled by the surface tension of the liquid. The Weber number is thus defined as follows:

$$We = \frac{\tau d}{\sigma} \tag{9.84}$$

where τ is the shear stress of the continuous phase, a parameter directly related to its rheology, d is the diameter of the bubble and σ the surface tension of the liquid. If the Weber number exceeds a critical value, We_c, the bubble will reduce in size (Campbell & Mougeot, 1999). For non-Newtonian liquids an apparent viscosity of the foam is defined, by using the generalized Newtonian model defined in Chapter 4, and the Weber number could be expressed in terms of the rheology of the foam, characterized by the apparent viscosity and the average or equivalent shear rate applied in the process, $\dot{\gamma}_e$ (Djelveh et al., 1999).

The critical Weber number allows one to estimate the maximum diameter of the bubble in the foam as:

$$d_{\max} = \frac{\sigma We_c}{\tau} \tag{9.85}$$

9.2.2.1.3 *The effect of the foam characteristics on its rheological properties*

Foams are considered structured fluids with a structure generated by a large volume of gas (non-continuous phase) in a small volume of liquid or solid medium (continuous phase). The principal distinguishing characteristic of foams is the large volume fraction of the non-continuous phase, which can be about 74% for foams with uniform cell sizes and even more for polydisperse foams (Heller & Kuntamukkula, 1987). It is the presence of these structure-forming bubbles that increases the viscosity of the liquid by several orders of magnitude in comparison with the viscosity of the liquid. The bulk rheology of foams is extremely important because it has a profound effect on the foam application. Thus, the rheology of foams has attracted the attention of researchers, specifically in trying to understand how the bubble characteristics (size, morphology, structure) can be modified by processing and formulation, and how they affect the functional properties and stability of foams.

The rheological behavior of foams under shear deformations has been schematically described by Weaire (2008) through a shear stress versus shear strain diagram as that illustrated in Figure 9.40. Although, as discussed by Weaire, understanding the rheology of foams is still a challenging goal, it is clear from this schematic diagram that for small strains and for stresses below the yield stress, the behavior of foams is that of elastic or viscoelastic solids whereas for large strains foams flow. The elastic/viscoelastic behavior of foams at small strains can be characterized by bulk and shear moduli. An important property of foams when compared to solid materials is that the range of shear strain under which the material behaves elastically is larger. The elastic/viscoelastic properties of foams are mainly significantly affected by cell size distribution and structure of the gas cells, which remains approximately stable for stresses below the yield stress except for some local rearrangements that can be described by the Plateau rules applied to three films

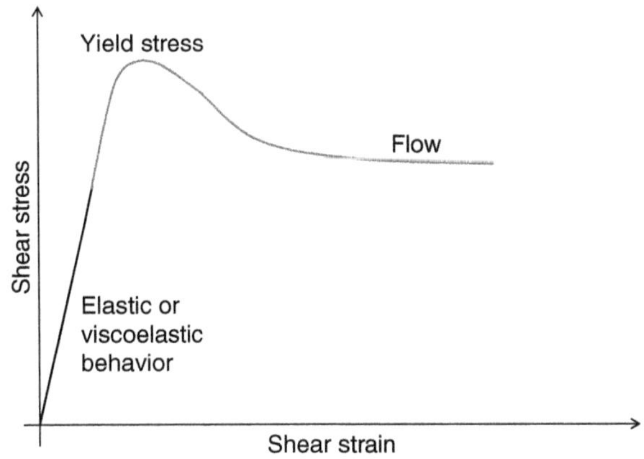

Figure 9.40 Shear stress versus shear strain behavior of foams.

converging in a 2D representation of the foam, or six films converging in a 3D representation of the foam (Weaire, 2008). An example of these rearrangements is the hop phenomenon, which is described in Figure 9.41.

The relationship between foam structure and foam rheological properties has been described by Princen in a series of publications (Princen, 1983, 1985; Princen & Kiss, 1989) and a critical review of the foam rheology literature performed by Heller and Kuntamukkula (1987). The volume fraction of the gas dispersed phase, the bubble radius, the interfacial tension, and the thickness of the films separating bubbles are all considered influential variables in determining the rheological properties of foams, including their yield stress, elastic modulus, and shear viscosity. The effect of surfactants is also important because the yield stresses of foams as well as their shear moduli are proportional to the interfacial tension, which is affected by the presence of surfactants. Most experimental data on foam rheology have shown that yield stress, viscosity, and shear modulus of foams are affected mainly by the air volume fraction and the bubble size distribution, which increase with increases in the gas volume fraction and decrease with the polydispersity of the bubble size distribution.

Kraynik (1988) describes the flow of foams and critically reviews the models used to describe the rheology of foams during flow. Most of those models are based on emulsions in which the properties of the non-continuous and continuous phases are well differentiated. The other aspect of importance discussed in Kraynik's review is the relevance of the bubble/cell dimensions (typically in the 10 μm–1 cm range) and their influence on the complex bulk rheological behavior exhibited by foams. Based on many models used to describe the rheology of emulsions, the bulk viscosity of dilute gas-bubbles foams could be estimated by the following equation (Kraynik, 1988):

$$\mu_d = \mu_o \left[\frac{4}{3\phi_d} \left(1 + \frac{\kappa}{\mu_o a} \right) - 1.733 \right] \quad (9.86)$$

where μ_d is the bulk viscosity of the foam, μ_o the viscosity of the liquid, κ the intrinsic surface dilational viscosity, a the bubble diameter and ϕ_d the dispersed gas volume ratio. Unlike shear-based rheological properties, it can be noted that the bulk viscosity of foams decreases with the gas volume ratio, and the rheological properties of the bubble–air interface positively affect the bulk rheological properties of foams. By considering monodisperse 2D hexagonal cells, no liquid drainage due to gravity, negligible liquid hold-up in the liquid films (gas volume fraction equals unity), an affine motion of the cells with the flow, and a Lagrangian approach to describe the deformation of the foam in terms of a position-independent deformation gradient relating non-deformed and deformed states, Khan and Armstrong (1986) obtained expressions to describe the deformation of the foam cells in terms of their undeformed dimensions and the specific applied strain. These authors also develop an expression to estimate the stress tensor in terms of the deformation experienced by the foam microstructure. A complete stress tensor was obtained by averaging the stresses existing in the foam microstructure over a volume V that is representative of the foam (Khan & Armstrong, 1986). The expression for the stress obtained in dimensionless form was a function of the capillary number, Ca, a dimensionless number defined by the following expression:

$$Ca = \frac{\mu \dot{\gamma}_c a}{\sigma} \quad (9.87)$$

where μ is the viscosity of the liquid phase, $\dot{\gamma}_c$ a characteristic shear rate, a the bubble size and σ the interfacial tension of the liquid forming the film. Capillary numbers tested ranged from values of 10^{-7} to 0.1. The equations obtained were tested to describe simple shear tests and were suitable to describe low deformations of foams such as that schematically described in Figure 9.41 for a 2D foam cell structure. The sides of the cells of size a are labeled with numbers from 1 to 3 as indicated in the unstrained conformation illustrated in Figure 9.41a. When shear is applied, the gas cells deform but the three liquid films still meet, forming angles of 120° as illustrated in the strained conformation indicated by Figure 9.41b.

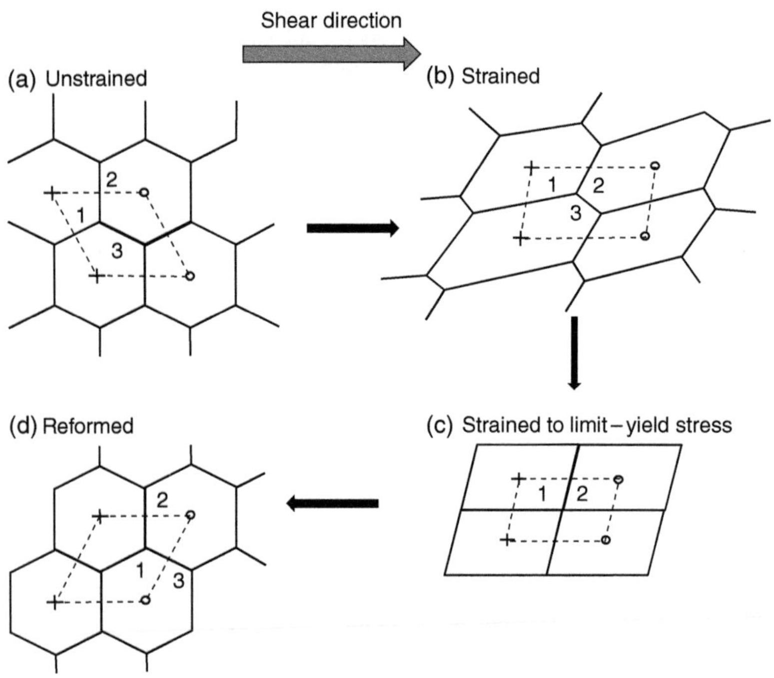

Figure 9.41 Sequence showing changes in the structure of a monodisperse hexagonal foam due to the action of a shear deformation. (a) Unstrained foam; (b) strained; (c) strained to limit; (d) reformed. Source: Adapted from Khan et al. 1988.

Due to the affine motion of the cells with the shear flow, side "3" shrinks while sides "1" and "2" elongate (Figure 9.41b). With further application of the shear deformation, the length of side "3" decreases and eventually reaches the value of 0 (Figure 9.41c). Due to a condition of non-equilibrium induced by the stress localized in that area of the foam, which corresponds to the yield stress τ_0 of the foam, the cellular structure reforms with a different configuration (Figure 9.41d). The phenomenon illustrated schematically in Figure 9.41, known as a "hop" of the foam structure, has enabled establishment of relationships to estimate the yield stress of the foam as a function of the air phase fraction (ϕ), the size of the gas cells expressed as the Sauter mean size R_{32}, the interfacial tension (σ), and an empirical parameter, $Y(\phi)$, which depends on the specific foam and is commonly obtained by fitting. A typical equation is given below (Foegeding et al., 2006):

$$\tau_0 = \frac{\sigma}{R_{32}} \phi^{1/3} Y(\phi) \tag{9.88}$$

Other equations to estimate the yield stress of foams are similar but the interfacial tension is replaced by the interfacial dilational elastic modulus E' (Yang et al., 2009).

In addition to the yield stress, the rheological properties of foams under shear flows are important to estimate their behavior during flow and transport. Experimental characterization of foams to determine their shear-based rheological properties is complicated given their inherent instability and the relative fragility of the foam structure. In addition, the presence of a liquid film can induce slippage on the wall of the testing geometry (Khan et al., 1988). Shear-based rheological properties of liquid foams have been tested using a variety of instruments and geometries that include Couette and parallel plate geometries as well as capillary viscometry, in which the presence of slippage may invalidate the testing if slippage is not properly recognized and accounted for. In addition to slippage, a major experimental problem observed with a rotational viscometer is destruction of the sample structure before a steady state can be achieved to obtain the measurement, thus the rheological properties are not those of the original sample.

As a consequence of these problems, much effort has been focused on approaches to minimize and/or correct artifacts produced by slippage during the testing of foam systems (Kraynik, 1988; Princen, 1985; Yoshimura & Prud'homme, 1987). In order to minimize slippage, vane geometry has been used to characterize the rheology of food (Pernell et al., 2000, 2002) and non-food (Zhang et al., 1998) foams. The compressibility of some foam systems also affects the data analysis obtained from capillary viscometry because the pressure gradient along the channel cannot be assumed to be constant and as a result, the standard equations used to calculate shear stresses and wall shear rate are no longer applicable and require corrections (Heller & Kuntamukkula, 1987). The other aspect of importance lies in the size of the foam cells with respect to the size of the testing geometry because if they are of the same order of magnitude, the measured rheological properties will depend on the type of geometry used for the testing, information that is often not reported in publications (Heller & Kuntamukkula, 1987). Foam systems with bubble sizes comparable to the dimensions of the testing geometries would violate the continuum approach used in rheological characterization of materials. In addition, the assumption of fluid isotropy would be compromised when foams are rheologically tested using geometries that have dimensions comparable to their bubble size.

Thus, two critical requirements have been proposed as standard for the rheological characterization of foams. One refers to the reproducibility of the data and the other is related to the uniqueness of the measurement regardless of the instrument and geometry used (Heller & Kuntamukkula, 1987). The stability of the foams subjected to the strains applied during loading of the sample and testing may seriously affect the reproducibility of the data. For example, liquid drainage, coalescence and rupture of the cells may affect foam cell size distribution and the resulting measured rheology. Heller and Kuntamukkula (1987) give a list of factors that might influence the flow behavior of foams and help to interpret and analyze the measured experimental data. Even considering the great variability in rheological behavior and values reported in the literature, the rheology of liquid foams under shear deformations has been described for non-Newtonian models such as the power law model, or the plastic Bingham and Herschel–Bulkley models due to the presence of a yield stress. The plastic behavior and/or the presence of slippage are likely to be the cause of the plug flow observed during the flow of foams in tubes and channels when the applied stress is smaller than the yield stress of the foam. For aqueous foams, the plug flow rate of the foam is calculated as a function of a shear water layer δ_w formed near the tube wall and the viscosity of water η_w as:

$$Q_{plug} = \frac{\pi R^3 \delta_w \Delta P}{2L\eta_w} \tag{9.89}$$

If the apparent shear rate at the wall is estimated as $4Q_{plug}/\pi R^3$, the calculation of the apparent shear viscosity of foams takes into account the shear layer and can be estimated as (Heller & Kuntamukkula, 1987):

$$\eta_{foam,app} = \left(\frac{\tau}{\dot{\gamma}_{app}}\right)_{wall} = \frac{\frac{R\Delta P}{2L}}{\left(\frac{4}{\pi R^3}\right)\cdot\left(\frac{\pi R^3 \delta_w \Delta P}{2L\eta_w}\right)} = \frac{\eta_w R}{4\delta_w} \tag{9.90}$$

Thus, under conditions of small shear stresses the pressure drop required to produce a foam flow through a pipe, annulus or channel is determined by the shear layer thickness and the viscosity of the liquid forming the foam rather than the bulk properties of the foam. The cell size distribution of the foam significantly influences the magnitude of the shear layer thickness which can be a few microns thick, leading to foams with high apparent viscosities flowing in pipes. Once the stress applied exceeds the yield stress of the foam, the apparent viscosity of the foam decreases substantially.

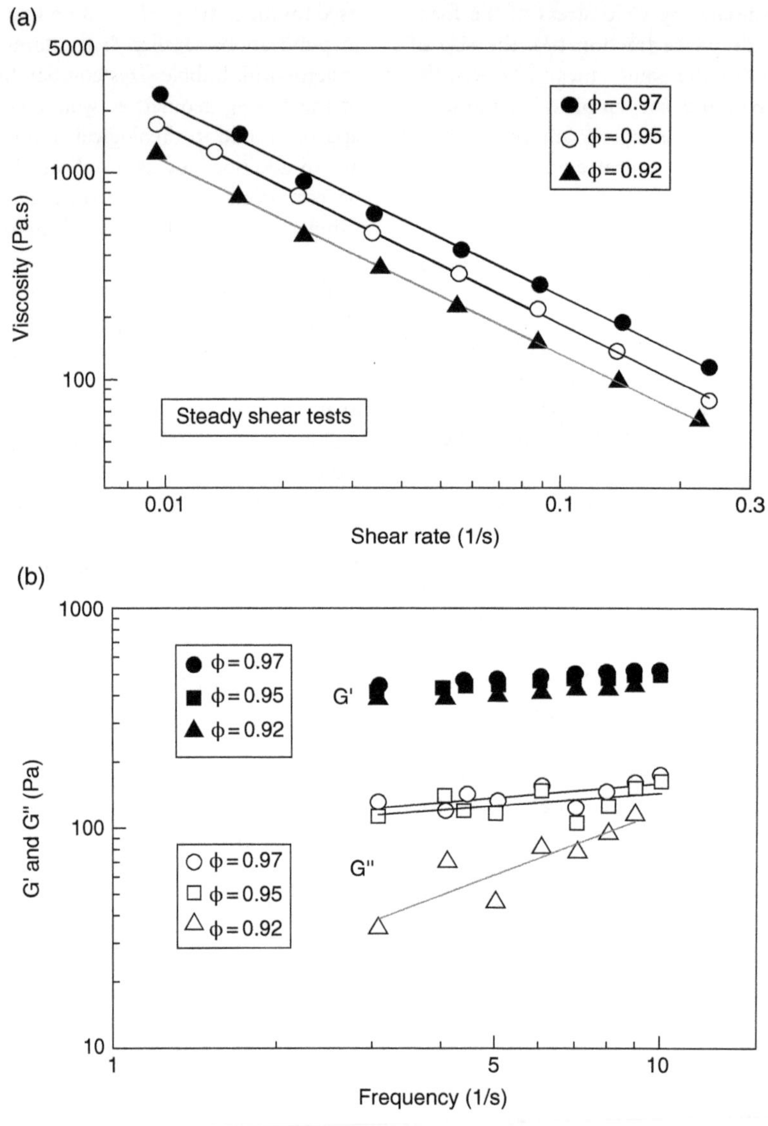

Figure 9.42 Steady shear viscosities and viscoelastic properties of foams having different air volume ratio. (a) Steady shear viscosity versus shear rate; (b) viscoelastic properties storage and loss modulus versus frequency. Storage Modulus G' (filled symbols) and Loss Modulus G" (open symbols).

The transition of the rheological behavior of foams from an approximately Newtonian behavior for shear stresses below the yield stress to a strong shear-thinning behavior for shear stress exceeding this yield stress was observed in experiments conducted in transparent tubes utilized to estimate the yield stress of foams (Heller & Kuntamukkula, 1987).

Khan et al. (1988) obtained shear-based rheological properties of foams tested with parallel plate geometry, and artifacts such as slippage were minimized using sandpaper attached to the plates. Steady shear tests as well as transient tests such as oscillatory dynamic rheology were performed to determine the viscous and elastic properties of stable foams specially prepared for the measurements. Foam drainage was decreased by using fresh samples for each test. Results of the steady shear and oscillatory dynamic tests performed on foams prepared with polyethylene oxide and two types of alcohol mixed with a commercial surfactant serving as the continuous phase, at three different gas volume fractions, are shown in Figure 9.42. In Figure 9.42a, it can be clearly observed that the foams exhibit a shear-thinning behavior that could be described by the power law model in a small range of shear rates. Higher shear rates were not used because they affected the structure of the foam and probably the validity of results. It is also clear in the figure that foams prepared with a larger air volume ratio had a larger viscosity. Viscoelastic results obtained from the oscillatory dynamic tests for the same foams are illustrated in Figure 9.42b. For all the foams studied, the viscoelastic behavior observed was one of an elastic material in which values of the storage modulus were significantly greater than the corresponding loss modulus. There was a small effect of the air volume ratio on the elastic properties of the foams, which was more noticeable in the loss modulus, but in general the trend was that more elastic foams are obtained with higher air volume ratios.

As discussed, yield stress is an important functional property of foams that is closely related to their structure and used often as a parameter of quality in terms of stability. In addition to the use of vane geometry and the observation of the flow in transparent tubes to determine yield stress of foams, an approach based on the stress/torque relaxation after the imposition of a low constant shear or rotation rate was presented by Khan et al. (1988). A schematic of the approach and results obtained with the method are illustrated in Figure 9.43. The measurement principle is based on the properties of liquid and solid materials. After a known shear rate is imposed, for liquid materials the stress or torque would fall to zero when the rotation is halted (dashed lines in Figure 9.43 left) whereas it would relax to a finite value (solid line Figure 9.43 left), which is assumed to be the yield stress of the foam. Integrity of the foam structure was assessed by running tests with different constant shear rates and observing how long the sample could sustain a constant torque reading. As shown in figure 9.43 (right), the effect of the shear rate on the determination of the yield stress using this method is almost negligible. It is also shown in the figure that to avoid the permanent destruction of the foam structure, only small shear rates should be used.

9.2.2.2 Governing factors in foam processing

As discussed above, foams are thermodynamically unstable systems that require that the dispersed gaseous phase remains as a fine dispersion for a period of time equal to or larger than their expected shelf-lives. Thus, foams have to be stabilized by surface-active agents such as proteins and other surfactants.

Proteins are amphiphilic molecules with polar and non-polar groups which are good candidates to stabilize food foams. Given their chemical characteristics, proteins can adsorb at the gas–liquid interface of foams, reducing the interfacial tension and easing the process to prepare them. The kinetics of adsorption of proteins at air–water interfaces is an important factor to be considered in the preparation of foams and it has been studied extensively because the effectiveness of proteins as surface-active agents is related to their rate of adsorption at the interface. The kinetics of adsorption of proteins (and other surfactants) at air–liquid interfaces is often determined using a Langmuir through/Wilhelmy plate apparatus to measure changes of interface pressure and surface concentration of protein as a function of time. Graham and Phillips (1979) discussed the effect of protein size on the adsorption rate; large and complex proteins adsorb at a slower rate compared with small and simple proteins, so the latter have better foaming capacity. Other studies have shown that the adsorption of proteins into air–liquid interfaces is influenced by other protein molecular properties beyond size which include shape, surface hydrophobicity, conformation, and charge as well as conditions in the bulk solution such as concentration, pH, and ionic strength (Cho et al., 1996, 1997). The presence of surfactants not only produces long-lasting foams but also promotes their formation due to reduction of the liquid surface tension and consequently the energy needed to produce them.

To produce foams, it is necessary to mix a gas, a liquid, and a suitable surfactant and supply energy to increase the free energy of the system by the increase in the

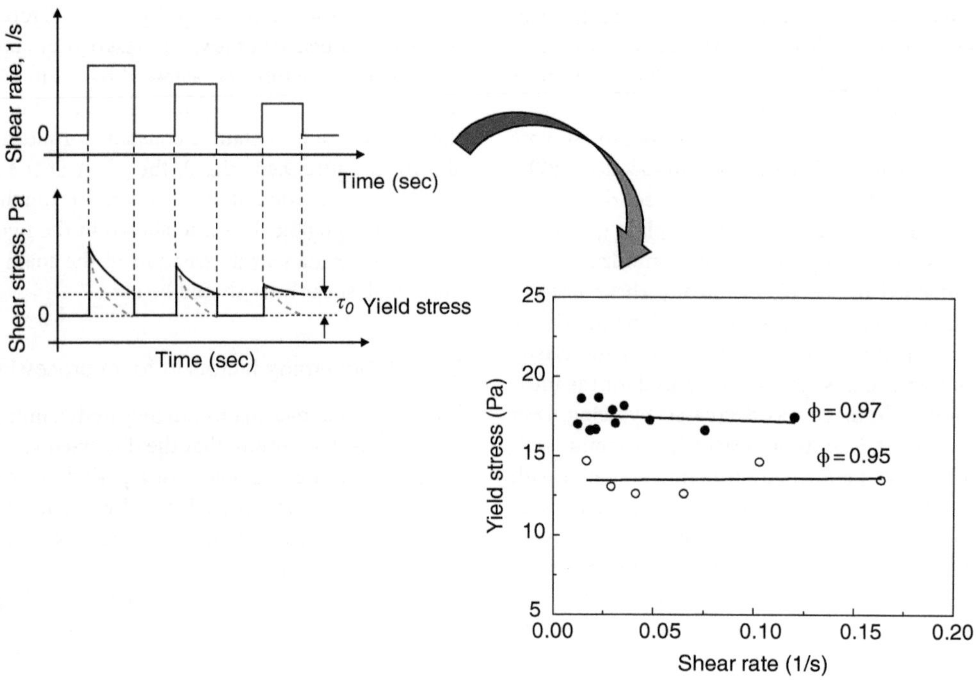

Figure 9.43 Transient rheological method to determine the yield stress of the foam. The method uses the relaxation of the stress upon application of a constant shear rate.

interfacial area between the liquid and the gas. Several methods have been developed to produce foams in food and non-food materials, and they are described by many publications and books (e.g. Campbell & Mougeot, 1999; Lee, 2000; Walstra, 2003). They mainly differ in the manner in which the gas is dispersed in the continuous phase. One traditional method for preparing foams is by dispersing the gas (e.g. air) through mechanical action (shaking, impelling, and whipping) in a liquid medium containing a surfactant. The method is not particularly suitable for industrial production because the amount of gas incorporated in the solution is difficult to control (Pu, 1996). A special type of agitation is cavitation induced by ultrasonic waves and ultrasonic generators (Walstra, 2003). In the cavitation phenomenon, vapor cavities are formed in the liquid due to negative pressures generated by the ultrasonic waves and the collapse of these cavities. For small-scale foam production, introducing a flow of gas through a small opening, for instance a perforated sheet, a process known as sparging of a gas, is a suitable and controllable manner of producing foams. However, a well-established procedure for preparing foams is by nucleation of gas bubbles, based on the supersaturation of the liquid with the gas, which is released from the solution by lowering the pressure (Pu, 1996; Walstra, 2003). Supersaturation of the liquid can be achieved by either injecting the gas into the liquid or producing the gas inside the liquid by a chemical or fermentation reaction. In soft drinks, the fizzing sensation is produced by bubbles formed due to the loss of solubility of CO_2 when the can is open and subjected to a pressure (atmospheric) lower than the pressure in the closed can under which the CO_2 was completely soluble. In other beverages, for example, beer bubbles are produced *in situ*. The structure of dough and bread, which can be considered as solid foam, is also created by generating the gas *in situ* by fermentation.

Nucleation and bubble formation may occur through a homogeneous or heterogeneous process. For the heterogeneous nucleation process, nuclei and bubbles originate from "seeds" that can be gas pockets or solids present in the solution. This is particularly applicable to the formation of the cellular structure of extrudates during the expansion process created by the evaporation of water from the extrudate coming out from the die. Injection of gases such as CO_2 in fully cooked doughs flowing in extruders has been performed to improve expansion (Rizvi & Mulvaney, 1995). Under the high-pressure conditions prevalent in an extruder, CO_2 dissolved in the water present in the extrudate melt is near or at critical conditions and can enhance the solubilization of flavors, colors and other soluble materials. When the material exits the extruder, in an extrusion operation carried out at temperatures lower than 100°C to minimize puffing by water evaporation, the die pressure is reduced and the dissolved CO_2 vaporizes, expanding the product, but flavor and color components stay in the water/solid phase forming the extrudate. Thus, the solubility of the gas in the continuous medium has a strong effect on the structure and formation of the foam. For cases where the continuous phase is a liquid, the solubility of gases can be calculated by Henry's law:

$$p = HC \tag{9.91}$$

where p is the partial pressure of the gas, C is the gas concentration in the liquid phase and H is the Henry's constant. Bubbles can shrink or grow if the gas inside the bubble expands or contracts and the volume change due to this contraction/expansion can be approximately calculated by the gas ideal equation:

$$V = \frac{nRT}{p} \tag{9.92}$$

where n is the number of moles of gas, T the absolute temperature and R the universal gas constant, which is equal to 8.31 kJ/mol.K when pressure p is expressed in kPa. The bubble can expand or contract due to gas diffusing in or out of the bubble (Campbell & Mougeot, 1999). The transfer of gas in and out of the bubbles can be estimated by a typical mass transfer equation:

$$N_{gas} = kA(C_{\infty} - C_{sol}) \tag{9.93}$$

where N_{gas} is the flux of gas, A is the area normal to the flux, C_{∞} the concentration of gas in the liquid phase and C_{sol} the solubility of the gas; k is a mass transfer coefficient related to the diffusivity of the gas in the liquid, D, and the diameter of the bubble, d, through the Sherwood number, Sh, as:

$$Sh = \frac{kd}{D} \tag{9.94}$$

Hailemariam et al. (2007) developed a model that describes the dynamics of a bubble growing in a viscoelastic medium, where mass transfer from the medium to the bubble was considered. Parameters used in the model that predicts bubble growth were taken from the study of Rizvi and Mulvaney (1995) on extrusion of dough saturated with CO_2. Results of the model in terms of the final equilibrium radius were compared to experimental results reported by Alavi et al. (2003) on the growth of a CO_2 bubble in bread dough during isothermal supercritical fluid extrusion-expansion at ambient temperature.

The size distribution of the bubbles is also an important characteristic of foams because different bubble sizes can promote a phenomenon called disproportionation, by which the gas is transferred from smaller to larger bubbles, leading to the cannibalization of the small bubbles. The reason is because the chemical potential of the gas inside bubbles is proportional to the surface tension of the liquid and inversely proportional to the diameter of the bubble (Larson, 1999). Thus, the chemical potential of the gas inside a small bubble is larger than that of the gas contained in a larger bubble, leading to the flow of gas from small bubbles to large ones. If the area of the bubble, which is proportional to its square diameter, is used to calculate the gas flow per unit of the area, it can be demonstrated that the rate of change of volume of a bubble is proportional to its diameter. This argument leads to the theoretical conclusion that the average diameter or volume of a bubble grows with time, following a relationship $t^{1/2}$, which agrees well with the found experimental value of $t^{0.45}$ (Larson, 1999).

During the production of foams, several physicochemical properties need to be measured and controlled and they are significantly affected not only by the type of process used to prepare the foams but also the processing conditions used. Physicochemical properties of importance in the production of foams are given in the following subsections.

9.2.2.2.1 Gas volume fraction or ratio ϕ

This is defined as the volume of gas over the total volume of the foam. This is an important factor to control because it has an important effect on the rheology of foams. In many processes, e.g. ice cream manufacture, the volume of gas in the foam is measured by the overrun, which is defined as the percentage increase in volume due to the incorporation of the gas. The relation between overrun and gas volume fraction is given by the following equation (Walstra, 2003):

$$\%overrun = 100 \cdot \frac{\phi}{1-\phi} \quad (9.95)$$

For example, foams with a gas volume fraction of 0.5 would give an overrun of 100%, meaning that the volume of the liquid is increased by 100% as a consequence of the incorporation and dispersion of the gas in the liquid.

9.2.2.2.2 Particle size distribution

Bubble size and bubble size distribution have an important effect on the stability of foams. From an average size standpoint, it is known that the smaller the bubbles, the more stable is the foam. However, consideration must be given to the process used to produce foams with tight particle distributions to minimize the phenomenon of cannibalization of small bubbles by the larger ones, described above. The particle size distribution and mean particle diameter d (estimated as d_{32} from the distribution) are associated with the interfacial area of the foams A calculated by the following relationship, which assumes monodisperse spherical bubbles of average diameter d:

$$A = \frac{6\phi}{d} \quad (9.96)$$

Equation 9.96 allows for simple calculations to estimate some parameters relevant to foam systems. Walstra (2003) used the relationship to estimate the interfacial area of a foam system produced from 200 mL of liquid whipped for 2 minutes with a small kitchen beater using about 50 watts of power to reach an overrun of 100%. From Eq. 9.96, for a foam having bubbles of average diameter 200 μm, the surface area is 15,000 m^2 per m^3 of foam. Thus, the interfacial area of the 400 mL foam produced from the initial volume of liquid with a 100% overrun would be 6 m^2. That information can be used not only to estimate the amount of surfactant used but also to obtain a rough estimation of the energy utilized to prepare the foam. Assuming that the surface tension of the liquid in the presence of the surfactant is 20 dynes/cm (0.02 N/m), the energy used to prepare the foam would be equal to 0.02 N/m × 6 m^2 = 0.12 joules, which is significantly lower (0.002%) than the energy available of 50 W × 120 s = 6000 joules. This simple example clearly shows that using beating of liquids produces foams of variable quality and the efficiency of the process in terms of energy use is poor. Thus, new methods that improve product quality and process efficiency are needed.

9.2.2.2.3 Film and Plateau border characteristics

Thickness, composition, and the presence of surfactants all have a great influence on the stability and properties of foams. As discussed above, there is natural drainage of liquid from the foam due to the action of gravity (see Figure 9.39), resulting in a liquid gradient with a dry foam at the top and a wet foam at the bottom. The size of the surfactant plays a key role in the characteristics of the Plateau border and liquid drainage. It has been demonstrated that the use of proteins reduces drainage of the liquid in foams when compared with the drainage observed in foams prepared with small molecule surfactants.

The extrusion-porosification process takes advantage of the very efficient mixing capability of twin screw extrusion technology (see Chapter 5) to aerate viscous concentrates resulting from the high solids vacuum evaporation unit, and convert the concentrates into stable viscous foams. The aeration of concentrates is realized by injection of gases such as CO_2. Part of the CO_2 is dissolved into the continuous phase (water), depending upon the pressure conditions in the extruder-aerator. It is worth noting that the twin screw extrusion-aeration unit requires major innovations from equipment and process standpoints to be able to produce stable, homogeneous foams from a wide diversity of concentrates.

9.2.3 Intensified spray drying

High solids foams produced by the twin screw extrusion-aeration unit are dried in a spray-drying chamber, the design of which exhibits special technical features owing to the physicochemical and rheological characteristics of the foams. Extrusion-porosification is an emerging generic process which does not yet have long-term experience. Thus, the fundamentals of the foam spray-drying process are still unknown, and complete engineering analysis of foam spray drying cannot be described extensively. However, it must be noted that the main specificity of the spray-drying chamber relates to its ability to intensify the drying process, hence leading to significant operational benefits.

It is worth reviewing the key factors which allow foam spray drying to be considered as a process-intensifying method according to the process intensification concept presented in Chapter 10 (section 10.1.4). In fact, when compared with conventional spray drying of concentrates, the spray-drying operation of the generic extrusion-porosification process exhibits two main process-intensifying characteristics.

- Continuous and flexible processing of high solids concentrates, the viscosity of which ranges from 2000 to 20,000 mPa.s. It must be noted that conventional spray drying operates with relatively low solids concentrates, the viscosity of which is less than 200 mPa.s. High solids foam drying considerably reduces the amount of water to be removed, reducing the energy required as well as the processing volume of the chamber. Besides, foam processing allows a larger range of concentrates to be processed and powder properties to be improved, leading to high levels of process-product flexibility.
- Acceleration of transport phenomena, which reduces the characteristic times of heat and mass transfers in the drying operation. The drying of foam droplets is mainly governed by the transfer of moisture from the droplet to the drying air, following the two classic moisture transfer regimes: the constant drying rate regime (or convective regime) and the falling drying rate regime (or diffusion regime) which determines the residence time in the drying chamber. In the falling drying rate regime, according to the diffusion mechanism, the drying time is inversely proportional to the diffusion coefficient of moisture within the particle. With wet porous products, moisture diffusivity generally increases when product porosity increases. For instance, Waananen and Okos (1996) have shown that effective moisture diffusivity of pasta products depends closely on its pore structure: the higher the porosity, the larger the effective moisture diffusion. With highly porous particles, the diffusion coefficient for moisture transfer tends to be significantly larger compared to that of bulk particles or low-porosity particles, which often result from conventional spray drying. Hence, moisture transfer in highly porous particles is faster, and consequently the drying time is shorter: in foam spray drying, it varies in the range of 8–15 seconds depending upon the size of particles, which is approximately 10 seconds shorter than the residence time of conventional spray drying of liquid concentrates. Besides, faster moisture transfer during drying allows the outlet temperature of the particles to be reduced significantly: in foam spray drying, it varies in the range of 45–55°C, which is approximately 20°C lower than the outlet temperature observed in conventional spray drying. The decrease of both residence time and powder temperature allows the time-temperature loads in the drying process to be reduced drastically, hence enhancing the quality of instant powders.

According to the principles of process intensification, it is worth pointing out that continuous spray drying of high solids textured foams belongs effectively to the class of intensified spray drying, due to the following drying conditions: important reduction of the amount of solvent (water) to be removed; significant increase of moisture transfer in drying particles. For a given capacity, such conditions lead to significant operational benefits such as reduction of the processing volume of the drying chamber, energy saving, and decrease of the time-temperature loads. Hence, the capital and operating expenditures of the drying unit are reduced, and the quality of instant powders is improved.

9.3 Perspectives on industrial applications

Extrusion-porosification technology has not been industrialized yet. It is still undergoing very active industrial development under the control of Clextral, France (supplier of twin screw extrusion processing technology).

As a new generic process, extrusion-porosification, is suitable to a wide range of applications which rely on the drying and texturization of food and non-food formulations to produce functional powders; hence, the food industry and chemical industry are primary candidates to use this technology as an alternative to conventional drying technologies (drum drying, spray drying, freeze drying) which were described in this chapter, or as a potential technology for product innovation. For the time being, the owners of the technology have given priority to the development of food instant powders (such as milk powders for high value-added functional ingredients, among others). This field of application is discussed in the following sections.

9.3.1 Range of applications

The use of twin screw extrusion technology for the production of functional powders is still fairly rare. One real example of its industrial use is the crystallization of food ingredients such as organic acid salts or hydrogenated sugars. Another example in the dairy industry is in casein-to-caseinate processing through a reactive extrusion process as formerly presented by Wagner and Sillard (1989); the conversion of high solids casein to caseinates by extrusion processing is described in Chapter 5 (section 5.5.2). At the time of its development, the process was a significant technological breakthrough, considering the unique conditions for the chemical reaction of casein with alkali: chemical reaction in the molten state, at high temperature and short residence time. In that example of high solids caseinate manufacturing, twin screw extrusion was used as an alternative to the spray-drying technology used in the traditional process. In this particular case, it led to substantial savings in investment cost and energy consumption without sacrificing the end-use properties of the end-products, owing notably to the advances made in twin screw extrusion technology (better temperature and shear control, in particular), and in extrusion processing expertise. This successful industrial experience applied to a relatively heat-sensitive dairy product has encouraged the owners of the extrusion-porosification technology to explore the potential of this innovative process in the field of instant porous powders.

Hence, extrusion-porosification can be applied to various raw materials, formulated mixes, or functionalized ingredients, as an alternative to conventional drying technologies to produce instant powders from:

- multicomponent raw materials such as whole and skim milk concentrates, to reduce the capital investment and energy consumption, in particular. This analysis is developed in section 9.3.2
- technofunctional ingredients such as dairy protein-based fractions, to enhance product quality (rehydration behavior, protein functionality). Milk protein concentrates (up to 85% protein content on a dry basis) are successfully extrusion-porosified at 38–42% dry solids concentration. Figure 9.44a presents scanning electron microscopy (SEM) pictures of milk protein concentrate powder which has been subjected to extrusion-porosification at 38% dry

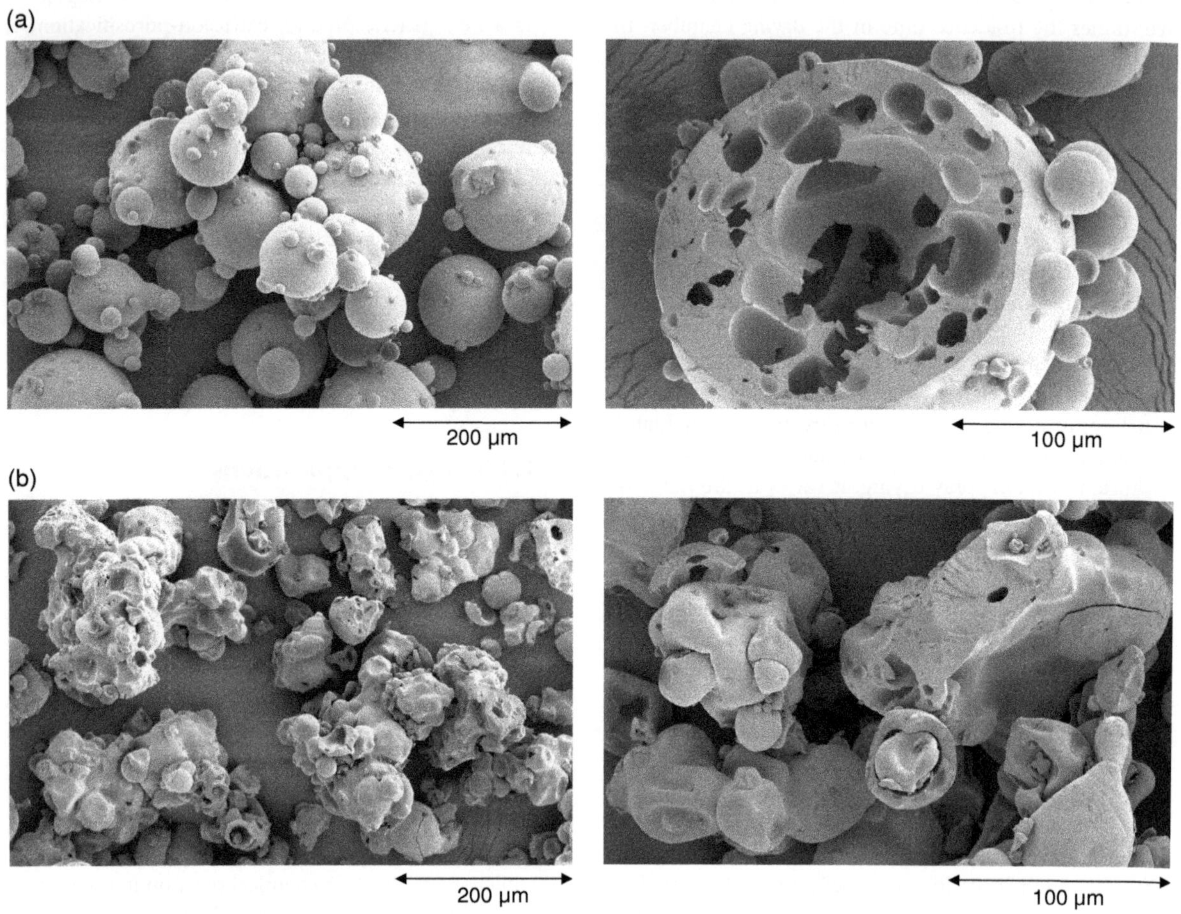

Figure 9.44 (a) SEM pictures of a 38% dry solids extrusion-porosified milk protein concentrate. (b) SEM pictures of a conventionally spray-dried milk protein concentrate. Source: Reproduced with permission of Clextral, France.

solids. The resulting size of the powder particles ranges from approximately 15 to 120 μm; they are perfectly spherical, highly porous with fine particles agglomerated to the largest ones. In contrast, conventional spray-dried milk protein concentrates (Figure 9.44b) which have been spray dried at approximately 28% dry solids show irregular-shaped, non-porous particles with poor agglomeration. Functional analysis of extrusion-porosified powders (no lecithin added) showed much quicker and better rehydration (improved wettability and cold water solubility) than conventional spray-dried powders

- formulated mixes such as ice cream mixes, coffee whiteners, instant drinks, infant formula, etc., to enhance mix homogeneity, add value to end-products, increase process flexibility, and reduce the process carbon footprint
- high fat content materials (such as buttermilk), up to 80% fat content, which cannot be dried with conventional drying technologies, for product innovation and product functionality enhancement
- highly heat-sensitive materials (such as probiotics, biofunctional ingredients, etc.), to enhance product functionality, increase process flexibility and reduce operating costs
- extracts from plants, roots or seeds (such as coffee concentrates), to enhance product functionality and reduce operating costs
- chemically or enzymatically modified materials (such as protein hydrolyzates, enzyme-modified cheese, etc.), to enhance product functionality, increase process flexibility and reduce operating costs.

The extrusion-porosification process can also be combined with other generic extrusion processes such as polymer compounding, extrusion-cooking and reactive extrusion, thus taking advantage of twin screw extrusion technology to combine various process functions in a single equipment. For instance, in an intermeshing twin screw extruder, flavor encapsulation (first processing section) can be associated with extrusion-porosification (second processing section) to produce instant encapsulated flavors (or fragrances) with improved functionality and reduced operating cost. Extrusion-cooking (first processing section) can be combined with extrusion-porosification (second processing section) to produce instant cereal-based powders (such as instant drinks, for instance). Reactive extrusion (first processing section) and extrusion-porosification (second processing section) can be carried out in the same extruder, or in two extruders in series, to produce instant chemically modified powders (aromas or flavors, for instance resulting from the Maillard reaction).

All potential applications concern food as well as non-food products.

9.3.2 Case study: extrusion-porosification of dairy products

Besides its many constituents having energy and nutritive value, milk also contains large amounts of water, that is 87–88% (wet basis). Eliminating this water opens up possibilities for preserving and storing the milk, which is crucial from an economic point of view. The water can be removed by evaporation, drying, or by using membrane separation techniques (in particular for the functional protein bases). The initial step in the process, called milk cracking, has become more and more complex over time, giving rise to a variety of products of different biochemical composition (different fat contents and different protein types and contents, in particular) and thus to a whole range of milk powders and by-products.

The following section is dedicated to milk powder processing (whole and skim milk powders), illustrating steps used in the current process as well as the steps used in the extrusion-porosification process.

9.3.2.1 Milk powder processing: the current situation

Figure 9.45a shows a simplified milk powder manufacturing process. The raw material is liquid milk with about 12% dry solids content. After reception, the milk is cleaned, clarified, refrigerated and then stored at about 4°C. Depending on whether whole or skim milk powder is being manufactured, part or all of the milk's fat content is centrifuged out in the form of cream. For whole milk powder, the fat content is standardized at 28–29% (total solids basis); for skim milk powder, it is lower than 1.9% (total solids basis). Skimming is currently done at 40–60°C. The milk then undergoes a heat treatment process (pasteurization and possibly sterilization). This has several objectives:

- inactivating all pathogenic bacteria to reduce the overall bacterial count
- inactivating undesirable enzymes, in particular lipases
- activating the thiol groups in the β-lactoglobulin to make the powder more oxidation stable during storage.

Relatively high temperature treatment and low residence times are preferred. Usually, pasteurization is carried out at 72°C for 15 seconds whereas sterilization conditions require either 120°C for 3 minutes or 140°C for 3 seconds.

Although Figure 9.45a does not include membrane separation processing units (reverse osmosis, nanofiltration, ultrafiltration, microfiltration), it is worth mentioning

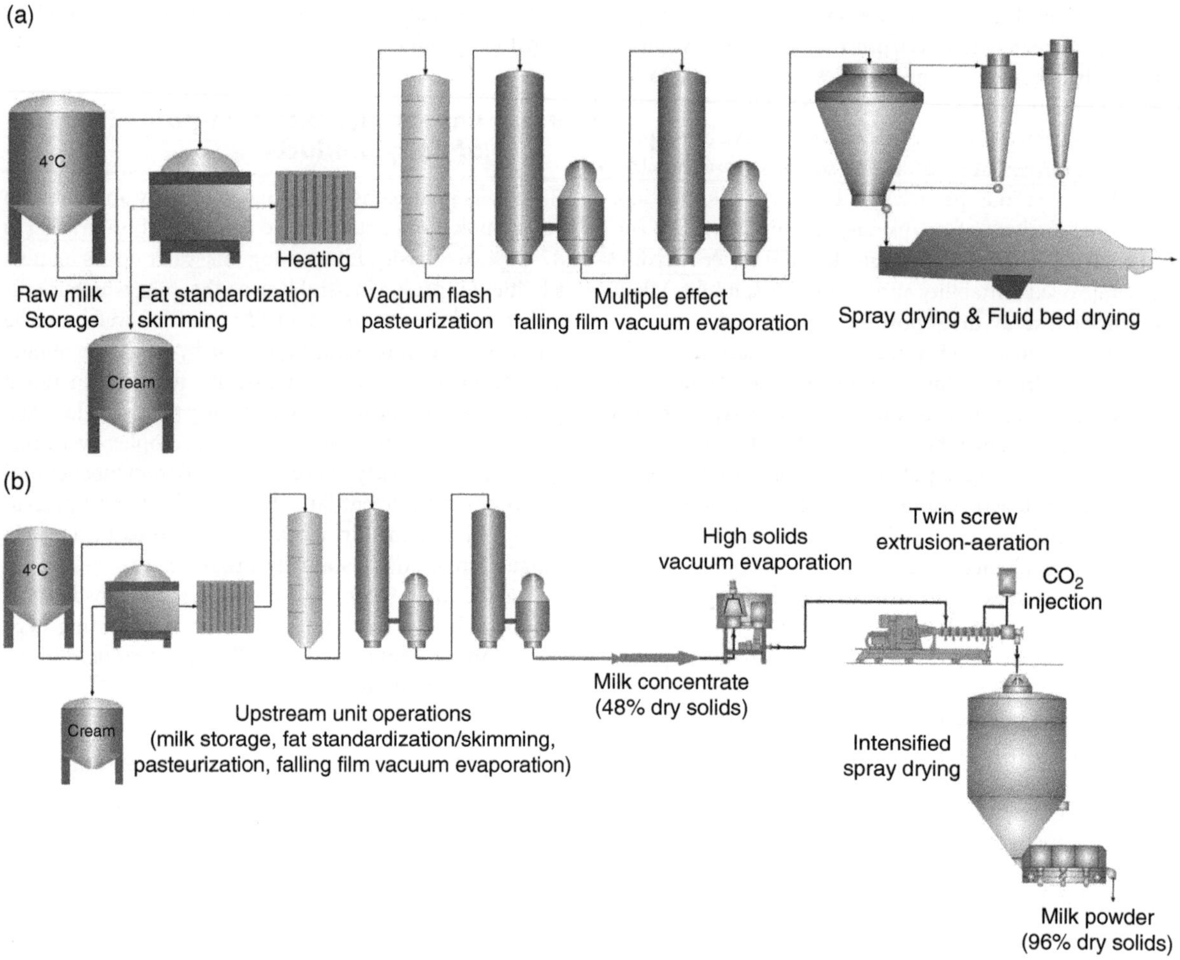

Figure 9.45 (a) Simplified flowsheet for the current milk powder manufacturing process. (b) Simplified flowsheet for the extrusion-porosified milk powder manufacturing process.

membrane separation technologies, which were introduced in the 1980s for the production of functional protein matrices from milk fractionation operations. As a matter of fact, membrane separation technologies minimize physicochemical modifications of the milk components owing to low processing temperatures, in the range of 10–50°C. In addition, in the process the water is removed with low energy consumption (in the range of 3–10 kW.h per tonne of permeate depending upon the processing conditions).

After standardization and heat treatment, the milk is concentrated by vacuum evaporation at low temperature (under 70°C); vacuum evaporation typically delivers a milk concentrate at 48% dry solids content. A great portion of the water in the milk is eliminated during this step; to reach a 48% solid content, 85% of the water originally contained in the milk must be eliminated. This requires the use of an energy-efficient technology. The most widely used technology in the dairy industry is a multiple effect falling film evaporator system, which was described in section 9.2.1.1. Approaches to reduce energy are included in that section, including the use of thermal or mechanical vapor recompression. As discussed, these approaches reduce the heat-energy (steam) consumption considerably; steam consumption with this type of evaporator is approximately 0.10 kg of steam per kilogram of evaporated water.

The concentrated milk is then dehydrated in the drying unit to dry solids content from 48% to 96%. A spray-drying process is used to dry milk concentrate from 48% to approximately 90% dry solids content, followed by fluid bed drying to finalize powder drying from 90% to 96% dry solids content. The concentrate is sprayed into the top of the drying tower in the form of small round droplets (50–100 microns) into a current of hot air (150–250°C) to produce a powder almost instantly. The water evaporates quickly which cools the air abruptly and limits the temperature increase for the particles to approximately 65–75°C. This temperature is critical for the physicochemical properties of soluble milk proteins, which would be denatured if subjected to higher temperatures, thereby affecting the quality of the milk powder.

As seen earlier (sections 9.1.1.2 and 9.1.2.3), the heat-energy consumption of spray drying varies whithin a wide range of steam consumption: it is roughly in the range of 2.5–3.8 kg of steam per kg of evaporated water, depending upon the heat sensitivity of the product. For whole and skim milk powders, the energy consumption of spray drying is approximately 2.8 kg of steam per kg of evaporated water. Energy consumption for fluid bed drying varies between 1.2 and 1.5 kg of steam per kg of evaporated water. For whole and skim milk powders, the energy consumption of fluid bed drying is approximately 1.4 kg of steam per kg of evaporated water.

As the powder leaves the fluid bed, it is separated out from the air using a cyclone separator, sieved, and packaged.

The spray drying is evidently the most critical operation in the current process. This is not surprising, as it is the point in the process where the milk makes the transition from a liquid state to a dry solid state, which was schematically illustrated in Figure 9.9. Control of this transition remains a challenge because the milk is heat sensitive and the quality requirements are strict. In this context, although spray drying is suitable and has been widely adopted in the dairy industry for the production of instant powders, as seen earlier (section 9.1.3.1), it shows process-product limitations (such as high energy consumption, constraints in particle size management, low process and product flexibility, high capital investment) that need to be addressed. Hence, extrusion-porosification technology can bring significant and relevant process and product enhancements in milk powder manufacturing.

9.3.2.2 Milk powder processing: the extrusion-porosification process

The proposed process (Durand et al., 2007) is presented in Figure 9.45b. It keeps the upstream unit operations from the current process but replaces the conventional spray drying process with the twin screw extrusion-porosification process. The objective is to guarantee the transition from liquid to solid powder state during the concentration process, while preserving product quality (nutritional and functional properties in the case of milk powders) and adding significant cost and process flexibility advantages.

Based on experimental developments carried out by the owners of the extrusion-porosification technology, the alternative process includes the following unit operations.

- Raw milk pretreatment operations such as fat standardization/skimming, heating, and pasteurization (identical to the current process).
- Milk concentration through vacuum evaporation; the unit is designed to concentrate the milk from a dry solids content of 12% to approximately 65%. As for the current process, multiple effect falling film vacuum evaporation is used to concentrate the milk from dry solids of 12% to approximately 48% while as discussed in section 9.2.1, high solids vacuum evaporation enables milk concentration from dry solids of 48% to approximately 65%. According to the heat sensitivity of milk components (particularly protein fractions, and possibly phosphocalcium salts), process developers have shown that the upper limit of dry solids is around 65%, so as not to damage milk components and be able to preserve functionality (that is, rehydration behavior) of the final powder. As seen in sections 9.2.1.1 and 9.2.1.2, it must be noted that the energy consumptions of falling film vacuum evaporation and high solids vacuum evaporation are 0.10 and 1.05 kg of steam per kg of evaporated water, respectively.
- The milk concentrate is fed into a twin screw extruder-aerator; the objective is to convert the milk concentrate into a textured foam, with good control of temperature and shear. Carbon dioxide is currently used to aerate the concentrate in two different stages (one gas injection port in the screw-barrel assembly to initiate concentrate aeration, followed by a second gas injection in the aeration unit ahead of the screw-barrel assembly). With milk concentrates, twin screw extrusion-aeration is carried out at low processing temperature and short residence time; the product temperature remains below 25°C and the residence time is in the range of 40–45 seconds, in order to maintain the time-temperature load as low as possible. It must be noted that the specific mechanical energy of the extrusion-aeration unit is quite low, in the range of 12–16 W.h/kg of milk concentrate (65% dry solids).

• The textured foam is dried (from 65% to 90% dry solids content) using an intensified spray drier (low processing volume, simplified air pressure spray nozzles, low product temperature). Based on the experimental data obtained in a pilot plant unit, the design of the intensified spray drier leads to an energy consumption of approximately 2.6 kg of steam per kg of evaporated water.
• The porous powder is finally dried in a fluid bed drier, to obtain a dry solids content of 96%, with an energy consumption of approximately 1.4 kg of steam per kg of evaporated water.

Milk powder characteristics are highly sensitive to time-temperature loads, as well as the shear history in the whole process, from high solids concentration to final fluid bed drying. Hence, the control and optimization of residence times, product temperatures and shear stresses in the three unit operations require special care to preserve powder properties.

Figure 9.46 shows typical porous structures of both whole and skim milk powders obtained from the extrusion-porosification process. Table 9.8 presents typical characteristics of extrusion-porosified powders obtained from whole milk. It shows satisfactory results (functional properties as well as sensory attributes) compared with standard values for whole milk powder. It must be noted that free fat content may show higher values than the standard, due to the inadequacy of the standard extraction method used for the analysis of porous products. However, it is worth mentioning that extrusion-porosified milk powders (whole and skim milk powders) show good rehydration properties in hot water as well as in cold water. Sodium dodecyl sulfate polyacrylamide gel electrophoresis (SDS-PAGE) analysis has confirmed no protein denaturation (Clextral, 2013).

9.3.2.3 Technological relevance of the extrusion-porosification process

In the innovative extrusion-porosification process, both the process approach and the configuration of the twin screw extruder-aerator are the result of a reassessment of the extrusion concepts used in standard applications

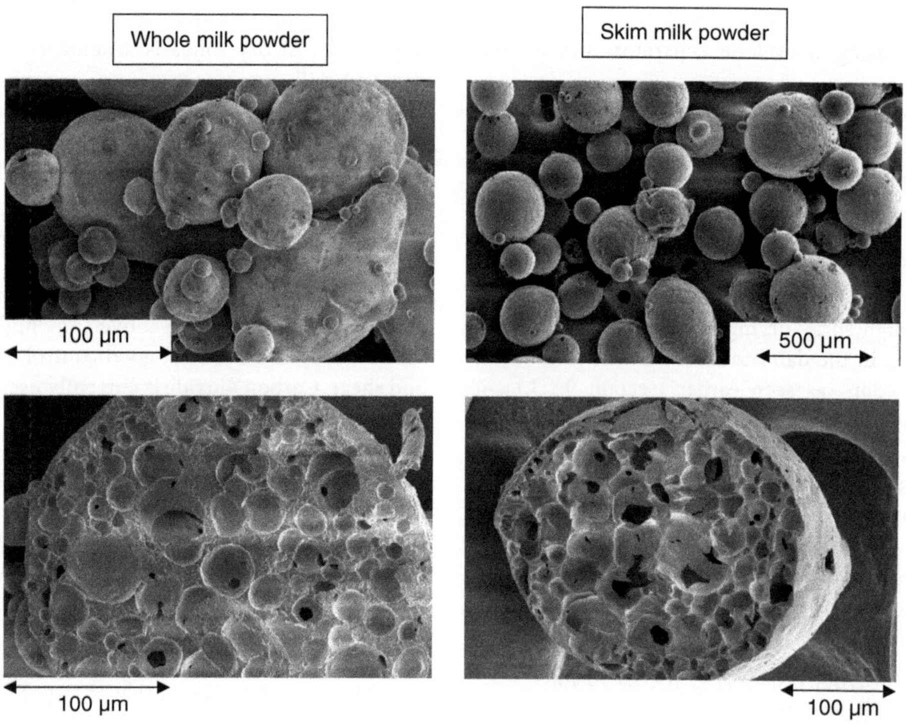

Figure 9.46 Typical SEM pictures of extrusion-porosified instant milk powders. Source: Reproduced with permission of Clextral, France.

Table 9.8 Characteristics of extrusion-porosified whole-milk powder.

Characteristics	Extrusion-porosified powder	Standard values
Moisture (%)	2.5–3.9	<4
Fat (%)	27–28	27–29
Insolubility index (mL)	0.1–0.2	<0.5
Titratable acidity (%)	0.1	<0.15
Flavor and odor	Good	Good
Appearance and color	Normal	Normal
pH (10% solution)	6.73–6.77	>6.6
Free fat (%)	3.2–6.7	<5
Peroxide value (meq. O_2/kg fat)	0.3	<1
Water activity	0.22–0.27	<0.5
Density (kg/L)	0.41–0.44	>0.4

Source: Reproduced with permission of Clextral, France.

such as polymer processing or food extrusion-cooking. The purpose of the process is to convert the liquid concentrate of a heat-sensitive raw material into high solids foams and from that, texturize and produce a functional powder whose properties must be consistent with a group of more or less restrictive end-use properties as a function of the final use of the powder. Compared to conventional drying technologies, the extrusion-porosification process aims to offer relevant process and product benefits, which are reviewed and discussed below.

9.3.2.3.1 Product functionality

In general, product functionality is largely dependent on the processing history, owing to heat- and shear-sensitive components. Heat sensitivity relates to the intensity of heat treatments and the associated residence times. Here, the consequences for the product's physicochemistry are fairly well known and it is possible to get a proper handle on the process conditions. The more critical changes mostly involve the denaturation of soluble proteins and the insolubilization of phosphocalcium salts. In the extrusion process, mechanical stresses must also be considered owing to the shear imposed on the material inside the screw-barrel assembly; this means adding the concept of product shear sensitivity (sensitivity of the material to local shear rates and to the deformation history as it goes through the extruder-aerator) to the concept of heat sensitivity.

The effect of shear on milk constituents is due to two factors: the direct mechanical effect on the material components' structure and conformation, and a secondary mechanical effect, which occurs as a local temperature increase due to viscous dissipation. Viscous dissipation depends on the product's viscosity and on the shear rate distribution in the screw-barrel assembly. While the induced mechanical effect can be easily understood for all practical purposes because its mechanism has been properly described, not much is known about the direct effect of mechanical forces on the material constituents, and in this case specifically milk. Does it have an effect on the conformation of soluble proteins, on the distribution of phosphorus and calcium between the continuous and colloidal phases, on the colloidal phase structure or on the chemical reactivity of the functional groups? So far, there is no scientific knowledge giving an insight into these fundamental questions to guide process developers in configuring the extruder-aerator and setting up the processing parameters.

The process and equipment of extrusion-porosification have been designed based on know-how and experiments, as well as end-use properties of resulting powders. It must be noted that the conversion of high solids milk concentrates into textured foams in the extruder-aerator system allows for the shear to be controlled fairly well. In addition, high solids textured foams permit the production of porous structures after spray drying, which considerably enhances the functionality of milk powders. Furthermore, the high heat transfer capacity of twin screw extrusion (refer to Chapter 10, section 10.2.2.1.2, and Table 10.2) associated with short residence times maintains time-temperature loads within acceptable limits, so as not to damage the constituents of milk products. Thus, as seen in Table 9.8, the concept of extrusion-porosification applied to milk products gives milk powders the functionality and properties that satisfies all the standard values.

Recently, Bouvier et al. (2013) applied the extrusion-porosification process to dry and texturize milk protein concentrates (so-called MPC85 that contains 86.7% w/w milk proteins) and compared the physical and rehydration properties of resulting powders with those of conventional spray-dried MPC powder particles. Extrusion-porosified MPC particles showed spherical shapes and highly porous matrices with numerous pores and communicating channels whereas conventional spray-dried MPC particles had irregular shapes, rather dense shells, and large internal pores. It is well known that conventional spray-dried milk protein concentrates exhibit low solubility and dispersibility and require very long rehydration times (several hours). The study clearly proved that the use of extrusion-porosification to

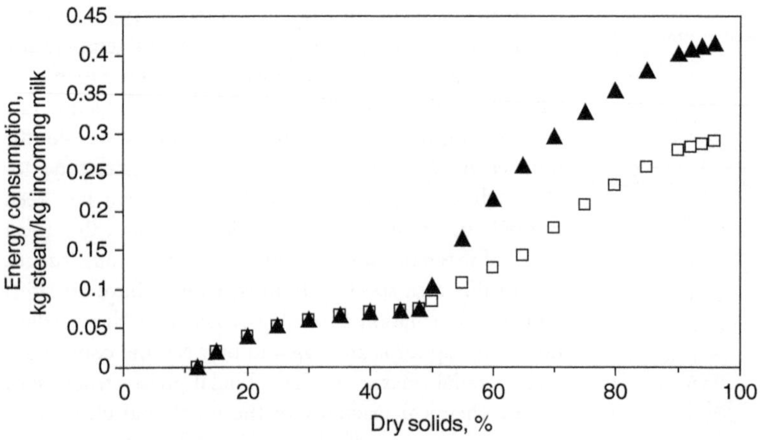

Figure 9.47 Energy consumption versus dry solids content for milk powder manufacturing (▲: current process; □: extrusion-porosification process).

texturize MPC85 material significantly upgraded the hydration behavior of MPC powder particles. Extrusion-porosified MPC particles had a very high dispersibility index of 96%, while spray-dried MPC particles showed typically a much lower dispersibility index of 38%. The authors have investigated the rate of powder rehydration by measuring particle size reduction in the dispersed phase versus reconstitution time. Complete dispersion/dissolution of extrusion-porosified MPC powder particles was obtained within 2 h of rehydration, while conventional spray-dried MPC powder particles required over 8 h for rehydration. The study demonstrated that extrusion-porosification processing had a very positive effect on the microstructure of MPC powder particles that considerably enhanced the porosity of powder particles and hence remarkably improved powder reconstitutability, with complete dissociation of submicron casein micelles within a relatively short time.

9.3.2.3.2 Energy consumption

The annual production of dairy powders is roughly 10 million tonnes worldwide. The energy consumption of drying operations in the dairy industry amounts approximately to 30 TeraW.h worldwide, of which spray drying alone may account for about 65%. Hence, it is worth estimating the potential benefit of the extrusion-porosification technology to decrease this considerable amount of energy consumption.

Based on process conditions (dry solids contents) and energy consumptions of the different unit operations of the two milk powder processes, the current process (section 9.3.2.1) and the extrusion-porosification process (section 9.3.2.2), the energy consumptions for a manufacturing unit can be estimated and compared while dry solids content of milk increases from 12% (raw material) to 96% (finished powder). Figure 9.47 presents the energy consumptions for both processes; the energy consumption is based on 1 kg of incoming milk (12% dry solids) and is then expressed in kg steam per kg of incoming milk.

It is observed that the total energy consumption required for milk concentration and drying from 12% to 96% of dry solids content using the extrusion-porosification process is about 0.29 kg of steam per kg of incoming milk, while the current process consumes 0.414 kg of steam per kg of incoming milk. This is due to the extrusion-aeration unit which permits the use of cost-effective vacuum evaporation at higher dry solids contents. Hence, the extrusion-porosification process allows 30% of the energy to be saved, which is considerable. This estimation is quite low as it does not account for the water removed by the flash-carrying effect of carbon dioxide during the earlier stages of drying in the spray-drying chamber. It must be noted that the specific mechanical energy of the extrusion-aeration unit, which is not accounted for in this estimation, is low compared to the consumption of thermal energy; as mentioned in section 9.3.2.2, it is in the range of 12–16 W.h/kg of milk concentrate (65% dry solids content).

9.3.2.3.3 Process flexibility

Given its highly efficient mixing capacity, the extruder would allow operators to process a wide range of raw materials and new product-oriented formulations, and take advantage of all the benefits of extrusion being used to serve product innovation, as is the case with today's applications (polymer processing, food extrusion, etc.).

The ability to vary the structure of dried powders through foam texturization gives high levels of flexibility regarding powder functionality. Because of the twin screw extruder's reduced or confined operating volume, the periods of downtime due to unproductive regimes (ramp-up, shut-down, cleaning operations, product change) are short, so the efficiency, or productivity, is relatively high. This advantage also allows the manufacture of a range of products using the same unit without sacrificing line productivity.

These quality advantages (access to a variety of raw materials and formulations, ability to vary powder structure, high process productivity, rapid shift from one product to another), which can be classified under the term "process-product flexibility," are appreciable and are part of the long-term requirements for instant powders manufacturing units. Many standard industrial applications, notably extrusion-cooking in the food industry and reactive extrusion or complex mixing in the chemical and paper industries, are reaping the benefit of the flexibility which twin screw extrusion technology offers, due to its reduced operating volume.

9.3.2.3.4 Capital investment

Projections on investment costs for new units from 5.5 to 9 t/h show substantial savings in favor of the very compact extrusion-porosification process, with a 20–30% reduction in the investment cost depending on the capacity and the processing line. The reduction is due mainly to two factors: the equipment, which is less expensive; and the building (less space occupied) and utilities. The impact of these economic advantages on the cost of the end-product is clearly significant; the savings on costs of milk powders produced using the extrusion-porosification process is assessed roughly at 25%.

The results obtained to date and the economical and technical potential of this new extrusion-porosification technological breakthrough lead its owners to intensify its development and promotion, for the manufacture of high value-added instant powders (functional food and feed ingredients in particular). This consists of pursuing process scale-up and optimization, finalizing the complete engineering line (including sanitation and hygienic aspects of the processing equipment), and installing mid-size industrial units in the near future (Clextral, 2013).

At this stage of its development, extrusion-porosification technology opens doors on new horizons. Indeed, beyond the dairy industry, porous powders are present in food ingredients, baby foods, instant drinks, cosmetology, pharmaceuticals, fine chemistry, etc.

References

Abdel-Aal ESM, Sosulski FM, Shehata AY, Youssef MM (1996) Nutritional, functional and sensory properties of wheat, rice and fababean blends texturized by drum drying. International Journal of Food Science and Technology 31(3): 257–266.

Adhikari B, Howes T, Lecomte D, Bhandari BR (2005) A glass transition temperature approach for the prediction of the surface stickiness of a drying droplet during spray drying. Powder Technology 149: 168–179.

Aguilera JM, Stanley DW (1999) Microstructural aspects of a fluid food: milk. In: *Microstructural Principles of Food Processing and Engineering*, 2nd edn. Maryland: Aspen Publishers, pp. 293–324.

Alavi SH, Rizvi SSH, Harriot P (2003) Process dynamics of starch based microcellular foams produced by supercritical fluid extrusion. I: Model development and II: Numerical solution and experimental evaluation. Food Research International 36: 309–330.

Alexander K, King CJ (1985) Factors governing surface morphology of spray-dried amorphous substances. Drying Technology 3: 321–348.

Anastasiades A, Thanou S, Loulis D, Stapatoris A, Karapantsios TD (2002) Rheological and physical characterization of pregelatinized maize starches. Journal of Food Engineering 52: 57–66.

Arrage JM, Barbeau WE, Johnson JM (1992) Protein quality of whole wheat as affected by drum-drying and single-screw extrusion. Journal of Agricultural and Food Chemistry 40: 1943–1947.

Baldwin AJ, Baucke AG, Sanderson WB (1980) The effect of concentrate viscosity on the properties of spray-dried skim milk powder. New Zealand Journal of Dairy Science and Technology 15(3): 289–297.

Bellinghausen R, Esper GJ, Gehrmann D (1994) Design of spray dryers on the basis of straightforward laboratory experiments and computer simulation. Chemie Ingenieur Technik 66(6): 825–828.

Bernard C, Broyart B, Vasseur J, Granda P, Relkin P (2008) Enhancement of protein structure-forming properties in liquid foams by spray drying. Dairy Science and Technology 88: 65–80.

Bhandari BR, Howes T (1999) Implication of glass transition for the drying and stability of dried foods. Journal of Food Engineering 40: 71–79.

Bhandari BR, Howes T (2005) Relating the stickiness property of foods undergoing drying and dried products to their surface energetics. Drying Technology 23(4): 781–797.

Bhat PP, Appathurai S, Harris MT, Pasquali M, McKinley GH, Basaran OA (2010) Formation of beads-on-a-string structures during break-up of viscoelastic filaments. Nature Physics 6(8): 625–631.

Billet R (1989) *Evaporation Technology: Principles, Applications, Economics.* Chichester: VCH Ellis Horwood Ltd.

Blass E (1979) Gas/film flow in tubes. International Chemical Engineering 12: 183–195.

Bloore CG, Boag IF (1981) Some factors affecting the viscosity of concentrated skim milk. New Zealand Journal of Dairy Science and Technology 16(2): 143–154.

Bonazzi C, Dumoulin E, Raoult Wack AL, et al. (1996) Food drying and dewatering. Drying Technology 14(9): 2135–2170.

Bouman S, Waalewijn R (1994a) Concentration of dairy products with rotating evaporators. 1. Theory and experimental methods. Milchwissenschaft 49: 190–193.

Bouman S, Waalewijn R (1994b) Concentration of dairy products with rotating evaporators. 2. Results and conclusions. Milchwissenschaft 49: 253–256.

Bouvier JM, Collado M, Gardiner D, Scott M, Schuck P (2013) Physical and rehydration properties of milk protein concentrates: comparison of spray-dried and extrusion-porosified powders. Dairy Science and Technology 93(4): 387–399.

Buma TJ, Henstra S (1971a) Particle structure of spray-dried milk products as observed by a scanning electron microscope. Netherlands Milk Dairy Journal 25: 75–78.

Buma TJ, Henstra S (1971b) Particle structure of spray-dried caseinate and spray-dried lactose as observed by scanning electron microscope. Netherlands Milk Dairy Journal 25: 278–281.

Campanella OH, Cerro RL (1984) Viscous flow on the outside of a horizontal rotating cylinder: the roll coating regime with a single fluid. Chemical Engineering Science 39(10): 1443–1449.

Campanella OH, Galazzo JL, Cerro RL (1986) Viscous flow on the outside of a horizontal rotating cylinder. 2. Dip coating with a non-Newtonian fluid. Chemical Engineering Science 41(11): 2707–2713.

Campbell GM, Mougeot E (1999) Creation and characterisation of aerated food products. Trends in Food Science & Technology 10: 283–296.

Champion D, Le Meste M, Simatos D (2000) Review: Towards an improved understanding of glass transition and relaxations in foods: molecular mobility in the glass transition range. Trends in Food Science & Technology 11: 41–55.

Chang R (2000) *Physical Chemistry for the Chemical and Biological Sciences.* California: University Science Books.

Chen H, Jebson RS, Campanella OH (1997) Determination of heat transfer coefficients in rotating cone evaporators. Part I. Transactions IChemE 75, part C: 17–22.

Chen XD (2008) Food drying fundamentals. In: Chen XD, Mujumdar AS (eds) *Drying Technologies in Food Processing.* Oxford: Blackwell Publishing, pp. 1–54.

Cho D, Narsimhan G, Franses EI (1996) Adsorption dynamics of native and alkylated derivatives of bovine serum albumin at air-water interfaces. Journal of Colloid and Interface Science 178: 348–357.

Cho D, Narsimhan G, Franses EI (1997) Adsorption dynamics of native and pentylated bovine serum albumin at air-water interfaces: surface concentration/surface pressure measurements. Journal of Colloid and Interface Science 191: 312–325.

Cholette A, Cloutier L (1959) Mixing efficiency determinations for continuous flow systems. Canadian Journal of Chemical Engineering 37(3): 105–112.

Chuy LE, Labuza TP (1994) Caking and stickiness of dairy-based food powders as related to glass transition. Journal of Food Science 59: 43–46.

Clextral (2013) Private communication: update on twin screw extrusion-porosification for the production of instant powders.

Daud WRW (1986) Calendering of non-Newtonian fluids. Journal of Applied Polymer Science 31: 2457–2465.

Daud WRW (1991) Thermal dynamics of a drum dryer. Drying Technology 9: 463–478.

Daud WRW (2006) Drum dryers. In: Mujumdar AS (ed) *Handbook of Industrial Drying*, 3rd edn. Boca Raton, Florida: CRC Press, pp. 203–21.

Daud WRW, Armstrong WD (1988) Conductive drying characteristics of gelatinized rice starch. Drying Technology 6: 655–674.

Déchelette A, Campanella OH, Corvalan C, Sojka PE (2011a) An experimental investigation on the breakup of surfactant-laden non-Newtonian jets. Chemical Engineering Science 66(24): 6367–6374.

Déchelette A, Babinsky E, Sojka PE (2011b) Drop size distributions. In: Ashgriz N (ed) *Handbook of Atomization and Sprays – Theory and Applications.* New York: Springer, pp. 479–495.

Desobry SA, Netto FM, Labuza TP (1997) Comparison of spray-drying, drum-drying and freeze-drying for β-carotene encapsulation and preservation. Journal of Food Science 62: 1158–1162.

Djelveh G, Cornet JF, Gros JB (1999) Combined effects of substrate and process parameters in food foaming processes. In: Campbell G, Webb C, Pandiella S, Niranjan K (eds) *Bubbles in Food.* St Paul: Eagan Press, pp. 75–82.

Durand D, Bouvier JM, Maller G, Scott M, Stevenson SR, Roberts SJ (2007) Procédé de préparation d'un produit en poudre. French Patent, FR 2890572.

Fäldt P, Bergenståhl B, Carlsson G (1993) The surface coverage of fat on food powders analyzed by ESCA (Electron Spectroscopy for Chemical Analysis). Food Structure 12: 225–234.

Filková I, Huang LX, Mujumdar AS (2006) Industrial spray drying systems. In: Mujumdar AS (ed) *Handbook of Industrial Drying*, 3rd edn. Boca Raton, Florida: CRC Press.

Flink JM (1975) The influence of the freezing conditions on the properties of freeze-dried coffee. In: Goldblith S, Rey L, Rothmayr W (eds) *Freeze-drying and Advances in Food Technologies*. London: Academic Press.

Foegeding EA, Luck PJ, Davis JP (2006) Factors determining the physical properties of protein foams. Food Hydrocolloids 20: 284–292.

Freudig B, Hogekamp S, Schubert H (1999) Dispersion of powders in liquids in a stirred vessel. Chemical Engineering and Processing 38: 525–532.

Fritze H (1972) The use of drum dryers in the human food industry. International Symposium of Heat and Mass Transfer Problems in Food Engineering, Waggeningen, The Netherlands: pp. 1–23.

Gaiani C, Scher J, Schuck P, Desobry S, Banon S (2009) Use of turbidity sensor to determine dairy powder rehydration properties. Powder Technology 190: 2–5.

Gaiani C, Morand M, Sanchez C, et al. (2010) How surface composition of high milk proteins powders is influenced by spray-drying temperature. Colloids and Surfaces B: Biointerfaces 75: 377–384.

Geankoplis CJ (2003) *Transport Processes and Separations Process Principles*, 4th edn. New Jersey: Prentice-Hall.

Graham DE, Phillips MC (1979) Proteins at liquid interfaces: I. Kinetics of adsorption and surface denaturation. Journal of Colloid and Interface Science 70: 403–414.

Greiff D, Rightsel WA (1965) An accelerated storage test for predicting the stability of suspensions of measles virus dried by sublimation in vacuo. Journal of Immunology 94: 395–400.

Hailemariam L, Okos M, Campanella O (2007) A mathematical model for the isothermal growth of bubbles in wheat dough. Journal of Food Engineering 82: 466–477.

Hall CW, Hedrick TI (1971) *Drying of Milk and Milk Products*, 2nd edn. Connecticut: AVI Publishing.

Heller JP, Kuntamukkula MS (1987) Critical review of the foam rheology literature. Industrial & Engineering Chemistry Research 26(2): 318–325.

Henning DR, Baer RJ, Hassan AN, Dave R (2006) Major advances in concentrated and dry milk products, cheese, milk fat-based spreads. Journal of Dairy Science 89: 1179–1188.

Incropera FP, DeWitt DP (2002) *Fundamentals of Heat and Mass Transfer*, 5th edn. New York: John Wiley.

Jayasundera M, Adhikari B, Aldred P, Ghandi A (2009) Surface modification of spray dried food and emulsion powders with surface-active proteins: a review. Journal of Food Engineering 93: 266–277.

Jebson RS, Iyer M (1991) Performance of falling film evaporator. Journal of Dairy Research 58: 29–38.

Jebson RS, Chen H, Campanella OH (2003) Heat transfer coefficients for evaporation from the inner surface of a rotating cone, II. Transitions IChemE, Part C, 81: 293–302.

Just JA (1902) Preserving milk in dry form. US Patent 712,545.

Karel M (1975) Heat and mass transfer in freeze drying. In: Goldblith S, Rey L, Rothmayr W (eds) *Freeze-drying and Advances in Food Technologies*. London: Academic Press, pp. 177–201.

Karel M, Anglea S, Buera P, Karmas R, Levi G, Roos Y (1994) Stability-related transitions of amorphous foods. Thermochimica Acta 246: 249–269.

Kessler HG (2002) *Food and Bio Process Engineering - Dairy Technology*, 5th edn. Munich: Verlag A. Kessler.

Khan SA, Armstrong RC (1986) Rheology of foams: I. Theory for dry foams. Journal of Non-Newtonian Fluid Mechanics 22: 1–22.

Khan SA, Schnepper CA, Armstrong RC (1988) Foam rheology: III. measurement of shear flow properties. Journal of Rheology 32: 69–92.

Kim EHJ, Chen XD, Pearce D (2002) Surface characterization of four industrial spray-dried dairy powders in relation to chemical composition, structure and wetting property. Colloids and Surfaces B: Biointerfaces 26(3): 197–212.

Kim EHJ, Chen XD, Pearce D (2003) On the mechanisms of surface formation and the surface compositions of industrial milk powders. Drying Technology 21(2): 265–278.

Kim EHJ, Chen XD, Pearce D (2005a) Melting characteristics of fat present on the surface of industrial spray-dried dairy powders. Colloids and Surfaces B: Biointerfaces 42: 1–8.

Kim EHJ, Chen XD, Pearce D (2005b) Effect of surface on the flowability of industrial spray-dried dairy powders. Colloids and Surfaces B: Biointerfaces 46: 182–187.

Kim EHJ, Chen XD, Pearce D (2009a) Surface composition of industrial spray-dried dairy powders. 1. Development of surface composition during manufacture. Journal of Food Engineering 94(2): 163–168.

Kim EHJ, Chen XD, Pearce D (2009b) Surface composition of industrial spray-dried milk powders. 2. Effects of spray drying conditions on the surface composition. Journal of Food Engineering 94: 169–181.

Kim EHJ, Chen XD, Pearce D (2009c) Surface composition of industrial spray-dried milk powders. 3. Changes in the surface composition during long-term storage. Journal of Food Engineering 94: 182–191.

King C.J (1970) Freeze-drying of foodstuffs. CRC Critical Reviews in Food Technology 1: 379–451.

King CJ, Lam WK, Sandal OC (1968) Physical properties important for freeze drying poultry meat. Food Technology 22(10): 1302–1308.

Kornev KG, Neimarkb AV, Rozhkov AN (1999) Foam in porous media: thermodynamic and hydro dynamic peculiarities. Advances in Colloid and Interface Science 82: 127–187.

Kostoglou M, Karapantsios TD (2003) On the thermal inertia of the wall of a drum dryer under a cyclic steady state operation. Journal of Food Engineering 60: 453–462.

Kozempel MF, Sullivan JF, Craig JC, Heiland WK (1986) Drum drying potato flakes – a predictive model. Lebensmittel Wissenschaft und Technologie, 19: 193–197.

Kraynik AM (1988) Foam flows. Annual Review of Fluid Mechanics 20: 325–327.

Krokida MK, Karathanos VT, Maroulis ZB (1998) Effect of freeze-drying conditions on shrinkage of dehydrated agricultural product. Journal of Food Engineering 35: 369–380.

Kuntz LA (2012) Tiny bubbles in my food. Editor's page. Food Product Design August: 10.

Langrish T (2009) Applying mass and energy balances to spray drying. Chemical Engineering Progress 105: 30–34.

Larson RG (1999) *The Structure and Rheology of Complex Fluids*. New York: Oxford University Press.

Lee ST (2000) Introduction. In: Lee ST (ed) *Foam Extrusion – Principles and Practices*. Lancaster, PA: Technomic Publishing.

Levi G, Karel M (1995) Volumetric shrinkage (collapse) in freeze-dried carbohydrates above their glass transition temperature. Food Research International 28: 145–151.

Levine L, Corvalan CM, Campanella OH, Okos MR (2002) A model describing the two-dimensional calendering of finite width sheets. Chemical Engineering Science 57(4): 643–650.

Liu J (2010) Freeze-drying, pharmaceuticals. In: Flickinger MC (ed) *Encyclopedia of Industrial Biotechnology – Bioprocess, Bioseparation, Cell Technology*. New York: John Wiley, pp. 2549–2567.

Masters K (1979) *Spray Drying*, 3rd edn. London: John Wiley.

Mitsoulis E (2009) Calendering and coating of polymers. In: Thomas S, Weimin Y (eds) *Advances in Polymer Processing – From Macro- to Nano-scales*. Cambridge: Woodhead Publishing, pp. 312–351.

Modler HW (1985) Functional properties of non-fat dairy ingredients – a review. Modification of products containing casein. Journal of Dairy Science 68: 2195–2205.

Moreau DL, Rosenberg M (1996) Oxidative stability of anhydrous milkfat microencapsulated in whey proteins. Journal of Food Science 61: 39–43.

Morgan F, Appolonia-Nouzille C, Baechler R, Vuataz G, Raemy A (2005) Lactose crystallization and early Maillard reaction in skim milk powder and whey protein concentrates. Lait 85: 315–323.

Nakagawa K, Hottot A, Vessot S, Andrieu J (2011) Short Communication. Modeling of freezing step during vial freeze-drying of pharmaceuticals – influence of nucleation temperature on primary drying rate. Asia-Pacific Journal of Chemical Engineering 6: 288–293.

Notz PK, Basaran OA (2004) Dynamics and breakup of a contracting liquid filament. Journal of Fluid Mechanics 512: 223–256.

Occeña LG, Bennink MR, Uebersax MA, Chung YS (1997) Evaluation of drum-dried meals prepared from split beans (phaseolus vulgaris l.): protein quality and selected antinutritional factors. Journal of Food Processing and Preservation 21: 335 –344.

Okos MR, Campanella OH, Narsimhan G, Singh RK, Weitnauer AC (2007) Food dehydration. In: Heldman D, Lund D (eds) *Handbook of Food Engineering*, 2nd edn. New York: CRC Press, pp. 601–719.

Omer K, Ashgriz N (2011) Spray nozzles. In: Ashgriz N (ed) *Handbook of Atomization and Sprays – Theory and Applications*. New York: Springer, pp. 497–579.

Onayemi O (1978) Studies on the chemical properties of cowpea powders supplemented and drum-dried with methionine. Food Chemistry 3: 265–270.

Pernell CW, Foegeding EA, Daubert CR (2000) Measurement of the yield stress of protein foams by a vane rheometry. Journal of Food Science 65: 110–114.

Pernell CW, Foegeding EA, Luck PJ, Davis JP (2002) Properties of whey and egg white protein foams. Colloid and Surfaces A: Physicochemical and Engineering Aspects 204: 9–21.

Pikal MJ, Shah S (1990) The collapse temperature in freeze drying: dependence on measurement methodology and rate of water removal from the glassy phase. International Journal of Pharmaceutics 62: 165–186.

Princen HM (1983) Rheology of foams and highly concentrated emulsions: I. Elastic properties and yield stress of a cylindrical model system. Journal of Colloid and Interface Science 91(1): 160–175.

Princen HM (1985) Rheology of foams and highly concentrated emulsions. II. Experimental study of the yield stress and wall effects for concentrated oil-in-water emulsions. Journal of Colloid and Interface Science 105(1): 150–171.

Princen HM, Kiss AD (1989) Rheology of foams and highly concentrated emulsions: IV. An experimental study of the shear viscosity and yield stress of concentrated emulsions. Journal of Colloid and Interface Science 128(1): 176–187.

Pu RJ (1996) Foaming, foam films, antifoaming and defoaming. Advances in Colloid and Interface Science 64: 67–142.

Quast DG, Karel M (1968) Dry layer permeability and freeze-drying rates in concentrated fluid systems. Journal of Food Science 33: 170–175.

Rahman S, Al-Amri OS, Al-Bulushi IM (2002) Pores and physico-chemical characteristics of dried tuna produced by different methods of drying. Journal of Food Engineering 53: 301–302.

Ramirez EC, Johnston DB, McAloon AJ, Yee W, Singh V (2008) Engineering process and cost model for a conventional corn wet milling facility. Industrial Crops and Products 27: 91–97.

Ratti C (2001) Hot air and freeze drying of high-value foods: a review. Journal of Food Engineering 49: 311–319.

Ratti C (2008) Freeze and vacuum drying of foods. In: Chen XD, Mujumdar AS (eds) *Drying Technologies in Food Processing*. Oxford: Blackwell Publishing, pp. 225–251.

Reineccius GA, Coulter ST (1969) Flavor retention during drying. Journal of Dairy Science 52: 1219–1223.

Rizvi SSH, Mulvaney SM (1995) Supercritical fluid extrusion process and apparatus. US patent 5,417,992.

Sagara Y (2001) Structural models related to transport properties for the dried layer of food materials undergoing freeze drying. Drying Technology 19: 281–296.

Saltmarch M, Labuza TP (1980) Influence of relative humidity on the physicochemical state of lactose in spray-dried sweet whey powders. Journal of Food Science 45: 1231–1236.

Sano Y, Keey RB (1982) The drying of a spherical particle containing colloidal material into a hollow sphere. Chemical Engineering Science 37: 881–889.

Saravacos GD, Maroulis ZB (2001) *Transport Properties of Foods*. New York: Marcel Dekker.

Schuck P (2011) Modifications des propriétés fonctionnelles des poudres de protéines laitières: impact de la concentration et du séchage. Innovations Agronomiques 13: 71–99.

Schuck P, Méjean S, Dolivet A, Beaucher E, Famelart MH (2005) Pump amperage: a new method for monitoring viscosity of dairy concentrates before spray drying. Lait 85: 361–367.

Schwartzberg HG (1977) Energy requirements for liquid food concentration. Food Technology March: 67–76.

Schwartzberg HG (1983) Optimization short cuts for multiple-effect evaporators for liquid foods. Personal communication.

Supprung P, Noomhorm A (2003) Optimization of drum drying parameters for low amylose rice (kdml105) starch and flour. Drying Technology 21: 1781–1795.

To EC, Flink JM (1978a) Collapse, a structural transition in freeze-dried carbohydrates. I. Evaluation of analytical methods. Journal of Food Technology 13: 551–565.

To EC, Flink JM (1978b) Collapse, a structural transition in freeze dried carbohydrates. II. Effect of solute composition. Journal of Food Technology 13: 567–581.

To EC, Flink JM (1978c) Collapse, a structural transition in freeze-dried carbohydrates. III. Prerequisite of recrystallization. Journal of Food Technology 13: 583–594.

Trinh B, Trinh KT, Haisman D (2007) Effect of total solids content and temperature on the rheological behaviour of reconstituted whole milk concentrates. Journal of Dairy Research 74: 116–123.

Trystram G, Vasseur J (1992) The modelling and simulation of a drum-drying process. International Chemical Engineering 32(4): 689–705.

Tsami E, Krokida MK, Drouzas AE (1999) Effect of drying method on the sorption characteristics of model fruit powders. Journal of Food Engineering 38(4): 381–392.

Vallous NA, Gavrielidou MA, Karapantsios TD, Kostoglou M (2002) Performance of a double drum dryer for producing pregelatinized maize starches. Journal of Food Engineering 51: 171–183.

Vasseur J, Abchir F, Trystram G (1991) Modelling of drum drying. In: Mujumdar AS, Filková I (eds) *Drying '91*. Amsterdam: Elsevier, pp. 121–129.

Verhey JPG (1972a) Vacuole formation in spray-dried powder particles. Part 1: Air incorporation and bubble expansion. Netherlands Milk Dairy Journal 26: 186–202.

Verhey JPG (1972b) Vacuole formation in spray-dried powder particles. Part 2: Location and prevention of air incorporation. Netherlands Milk Dairy Journal 26: 203–224.

Verhey JPG (1973) Vacuole formation in spray-dried powder particles. Part 3: Atomization and droplet drying. Netherlands Milk Dairy Journal 27: 3–18.

Vuataz G (2002) The phase diagram of milk: a new tool for optimising the drying process. Lait 82: 485–500.

Waananen KM, Okos MR (1996) Effect of porosity on moisture diffusion during drying of pasta. Journal of Food Engineering 28(2): 121–137.

Wagner J, Sillard M (1989) Making caseinate by extruder. Food Engineering International, March.

Walstra P (2003) *Physical Chemistry of Foods*. New York: Marcel Dekker.

Walton DE (2000) The morphology of spray-dried particles a qualitative view. Drying Technology 18(9): 1943–1986.

Walton DE, Mumford CJ (1999a) Spray dried products – characterization of particle morphology. Chemical Engineering Research and Design 77(1): 21–38.

Walton DE, Mumford CJ (1999b) The morphology of spray-dried particles: the effect of process variables upon the morphology of spray-dried particles. Chemical Engineering Research and Design 77(5): 442–460.

Walzel P (2011) Influence of the spray method on product quality and morphology in spray drying. Chemical Engineering & Technology 34(7): 1039–1048.

Walzel P, Furuta T (2011) Morphology and properties of spray dried particles. In: Tsotsas E, Mujumdar AS (eds) *Modern Drying Technology Volume 3: Product Quality and Formulation*. London: John Wiley, pp. 231– 294.

Weaire D (2008) The rheology of foam. Current Opinion in Colloid & Interface Science 13(3), 171–176.

Weaire D, Hutzler S (1999) *The Physics of Foams*. New York: Oxford University Press.

Wilde PJ (2000) Interfaces: their role in foam and emulsion behaviour. Current Opinion in Colloid & Interface Science 5: 176–181.

Xue ZJ, Corvalan CA, Dravid V, Sojka PE (2008) Breakup of shear-thinning liquid jets with surfactants. Chemical Engineering Science 63(7): 1842–1849.

Yang X, Berry TK, Foegeding EA (2009) Foams prepared from whey protein isolate and egg white protein. 1. Physical, microstructural, interfacial properties. Journal of Food Science 74(5): E259–E268.

Yoshimura AS, Prud'homme AK (1987) Response of an elastic Bingham fluid to oscillatory shear. Rheologica Acta 26: 428–436.

Zhang XD, Giles DW, Barocas VH, Yasunaga K, Macosko CW (1998) Measurement of foam modulus via a vane rheometer. Journal of Rheology 42: 871–889.

10 Extrusion Technology and Process Intensification

When the time comes to analyze human history in the period ranging from the end of the second millennium to the beginning of the third millennium, historians will probably distinguish the following three key events.

- The Industrial Revolution in the 19th century, which fundamentally transformed national economies, politics, and society.
- The globalization of environmental changes in the 20th century, due to the exponential increase of emissions to air, water and soil.
- Sustainability in the 21st century through which everyone from government to citizen has successfully managed overriding priorities in order to meet the needs of humans while respecting the limits of the planet.

Although the challenge on sustainability is yet to be fully taken up, in general it is now recognized that our cultural development must account for its environmental impact. We must thus consider the concept of sustainable development as the fundamental direction to work on the two largest world problems which are "unlimited exploitation of nature and the growing inequity within and among nations" (Harmsen et al., 2004). This concerns all parts of society, and challenges choices and solutions that relate to environmental and socioeconomic problems. In particular, it concerns challenges and technological solutions that currently do not take into account social and economic aspects of sustainable development. Growing consciousness about sustainable development will undoubtedly drastically influence the use of technology; that is to say, it will accelerate the disappearance of invasive (high environmental impact), unsustainable technologies, and favor the emergence and implementation of new, sustainable technologies.

In that context, processing technologies are of prime importance as they play a determinant role in human activities and contribute extensively to meet the needs of the present generation. As this book deals with extrusion processing technology that serves a variety of processing industries, the main question which may interest process developers and designers is the extent to which extrusion processing technology contributes to sustainable development. In other words, is extrusion processing a sustainable technology? Is it destined to capture new industrial applications or new market domains, based on its suitability for sustainable technology?

This chapter aims to provide information to process developers and designers with regard to the contribution of extrusion processing technology to sustainable development. In the first section, the emergence of sustainable development is explained and the concept of process intensification is discussed. In the second section, the capacity of extrusion processing technology for process intensification is assessed regarding its contribution to sustainability, while focusing on three important process functions: reactive extrusion, extrusion-cooking, and solid-liquid extraction. In that section, the sustainability of extrusion processing technology is illustrated while reviewing the characteristics and performances of existing extrusion processes and applications. Finally, the third section is dedicated to case studies aimed at facilitating understanding of extrusion process-intensifying methods.

10.1 From sustainable development to process intensification

Before analyzing and commenting on the contribution of extrusion processing technology to sustainability through process intensification, it is worth describing mainstream thought about sustainable technology, and deriving a general view on sustainable technology as well as on process intensification.

Extrusion Processing Technology: Food and Non-Food Biomaterials, First Edition. Jean-Marie Bouvier and Osvaldo H. Campanella.

10.1.1 The IPAT equation

In the early 1970s, Ehrlich and Holdren (1971) proposed the following equation to estimate the effects of human activities (economy, culture, and technology) on the total environmental impact:

$$I = P.A.T \quad (10.1)$$

where:

- I is the environmental Impact, which may be expressed in terms of resource depletion or waste accumulation
- P is the Population, or the number of people. The United Nations predict that human population will increase tremendously from 6.8 billion today to about 9.2 billion by 2050, and might stabilize around 9 billion by 2100. Thus, since the world population is set to increase for the next decades, this factor will likely keep increasing environmental impact (increase of land use, resource use, and pollution) in the near future
- A is the Affluence, or the average amount of technology units per person. Affluence refers to the consumption of the population. Production and consumption have been rising steadily over the last few centuries, and will continue to increase in the near future. Thus, the Affluence factor will increase environmental impact (increase of resource use and pollution)
- T is the Technology, or the amount of environmental impact per unit of technology. Technology refers to the processes used to exploit planetary resources, as well as to create, transport, and disperse the goods, services, and amenities. Thus, improvements in process efficiency will likely reduce resource intensiveness, hence reducing the technology factor to the benefit of environmental impact.

The IPAT equation can be considered as a heuristic approach for analyzing the interactions between society and environment. It clearly evidences that the scale of environmental impacts generally grows with the scale of human activities (the multiplicative contribution of factors P and A), but that societies can compensate for their impacts by modifying the way they use resources (the technology factor T).

To illustrate this, let us take the example of worldwide carbon emissions over the past 100 years, ranging from 1890 to 1990 (Hoffert et al., 1998). As shown in Table 10.1, the total carbon emissions in 1990 were 7.4 times as much as they were in 1890. The changes in environmental impact over time can be examined as a collation between factors P and A (which have increased) and factor T (which has decreased, thanks to changes of energy sources). However, it is clear that the increase in technological efficiency does not keep up with the multiplicative contribution of P and A. Therefore, radical change in one or more of the parameters of the IPAT equation is necessary to stabilize carbon emissions.

When writing it as a single product, the IPAT equation is somewhat imperfect because people are not equally affluent or do not use similar technologies, due to the huge cultural and social differences between nations in the world. Thus, it was proposed to split the IPAT equation into the sum of three terms which aims to distinguish three different groups of nations in the world, depending upon their level of industrialization, such as:

$$I = \sum_{i=1}^{i=3} (P.A.T)_i \quad (10.2)$$

where $i = 1$ refers to the highly industrialized nations (such as USA, Western Europe, Japan, Oceania, etc.), $i = 2$ to nations in rapid development (such as BRIC: Brazil, Russia, India and China; Mexico, former communist European countries, etc.), and $i = 3$ to poor nations. The population of group 1 is no longer a problem as it is stable, or slightly declining, but it is characterized by high levels of consumption (high A_1), and to a lesser extent by its technology which is not always clean (relatively high T_1). Group 2 generally uses dirty technology (high T_2). Although its population will increase significantly for the time being, it will become stable soon with the fast increase in education level and numbers of middle-class people (relatively high P_2). As for its affluence, it is presently moderate but it will increase quite quickly (relatively high A_2). Group 3 has very little affluence or technology; thus its contribution to environmental impact is presently negligible compared to the other two groups.

Table 10.1 Carbon emissions: data relative to the period range of 1890–1990.

Factor of IPAT equation	1890	1990
P, in billion	1.7	5.3
A, unit of energy per person per year	640	2050
T, kg carbon emission per unit of energy	0.75	0.56
I, billion tons of carbon emission per year	0.82	6.08

Source: Hoffert et al. 1998. Reproduced with permission of Nature Publishing Group.

Graedel and Allenby (1995) have proposed a *master equation* which is a variation of the IPAT equation, where affluence, *A*, and technology, *T*, are expressed as the Gross Domestic Product per person (GDP_i), and the environmental impact per unit GDP, respectively. By use of the master equation, it is straightforward to calculate the increase in environmental impact for the population, or for the increase in GDP_i with the same technology. In contrast, a decrease in environmental impact should be expected when using alternative, sustainable technology. It must be noted that the global use of the master equation must account for interactions between the three factors, such as for instance:

- interaction between *P* and GDP_i: population growth is a function of GDP_i. For instance, in Western countries where GDP_i is relatively high, the population growth is zero or even less; hence, the population growth in developing countries will certainly decrease when GDP_i increases
- interaction between GDP_i and *T*: beyond a certain level of GDP_i, the environmental impact emission per person tends to decrease. Jackson (1993) showed that on average, above 5000 GDP_i, the hazardous waste per GDP_i is reduced; for instance, this is the case for sulfur emission (kg SO_2 per capita) which decreases from about 250 down to 135 when GDP_i increases from 5000 to 10,000 units. Besides, *T* may vary significantly from nation to nation, depending upon human behavior and societal culture. For instance, this factor is definitely lower in Scandinavia than in North America.

The IPAT equation is useful for policy development by evidencing the variable contribution of different factors, and indicating where resources might be best directed to reduce environmental impact. Although it does not allow sustainable limits (or the level of impact that exceeds a sustainable scale) to be identified, it provides relevant understanding of the general factors that increase or decrease environmental impact. It also shows that different nations might focus on different factors to reduce their overall impact.

In general, there are two interrelated critical factors, affluence and technology, in highly industrialized and rapidly developing nations ($A_1.T_1$ and $A_2.T_2$). The long-term challenge for highly industrialized nations consists of reducing the effect of A_1 (that is, focus on essential needs, reduce consumption of unsustainable goods and services, promote temperateness in consumption, etc. while preserving and improving quality of life) and decreasing T_1 (that is, striving to make all technologies greener, and developing new sustainable technologies). As for the rapidly developing nations, the long-term challenge consists of urgently managing the ineluctable exponential increase of A_2 (that is, curbing increasing affluence, through appropriate actions depending upon the social, cultural, and economic context of the nation) and decreasing T_2 (that is, promoting green technologies). Hence, drastic cultural and technological changes must urgently be implemented in these two groups of nations to move from unlimited exploitation of nature to sustainability. With the third group of nations, sustainable cooperation between poor nations and the other two groups needs to be increased in order to accommodate poor countries in their legitimate yearning for well-being from the social, economic, and ecological standpoints.

In conclusion, the IPAT equation clearly indicates that current technological solutions and subsequent environmental problems suffer from a lack of vision and long-term anticipation. Another way of thinking, and another direction, must therefore lead our development and guide our decisions in the future. Hence, the essential way of thinking is based on the concept of sustainable development.

10.1.2 Sustainable development

Invention of the term "sustainable development" is normally attributed to the World Commission on Environment and Development, or WCED. The most common definition is given in the WCED report (Brundtland, 1987): "*Sustainable development is development that meets the needs of the present without compromising the ability of future generations to meet their own needs.*" This definition contains key concerns, such as:

- the notion of needs and the idea of intergenerational justice
- the acknowledgment that there are ecological and societal limits to growth
- the importance of intragenerational justice, i.e. justice within the present generation.

The concept of sustainable development originated from the attempt to reconcile the ecological consequences of human activities and the sociopolitical concerns of human development issues. In the second part of the 1900s, several publications and debates led to the establishment of the WCED in 1983, and then to the concept of sustainable development in 1987. In particular, it is worth mentioning the publication of *Limits to Growth* (Meadows et al., 1972) ordered by the Club of Rome, in which gloomy predictions of humanity's future being hindered by limits such as resource scarcity, pollution

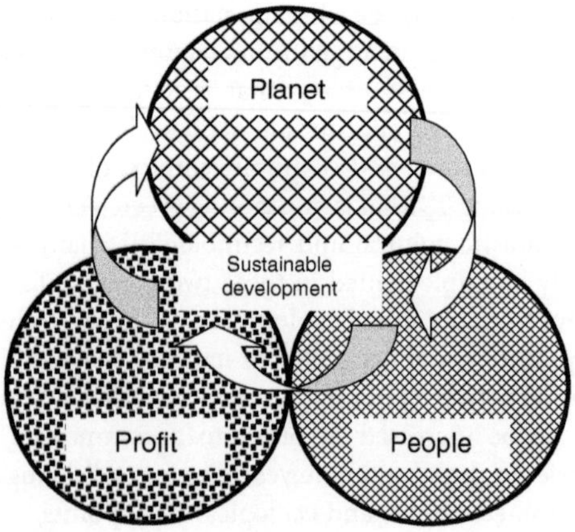

Figure 10.1 The three dimensions of sustainable development.

and/or lack of food production were discussed. The report provoked a wave of protests, especially from developing countries that perceived it as a neocolonial attempt by the industrialized world to hinder their economic growth.

The principles of sustainable development are based on the will and challenge to link what has been separated for a long time: society, environment, and economy. This can be simply translated into the well-balanced interaction of three circles that represent its three dimensions: social equity, ecological integrity, and economic efficiency, or people, planet, and profit (Harmsen et al., 2004). The three dimensions of sustainable development are now examined hereafter (Figure 10.1).

10.1.2.1 People (or social dimension)

This relates to the satisfaction of societal needs while respecting social equity, solidarity, social and cultural diversity. This dimension drives the essential idea that people are not limited to production and consumption.

10.1.2.2 Planet (or ecological dimension)

Throughout history, humans have shown incredible ingenuity in finding strategies to survive. These strategies have been formed by the co-evolution of social institutions, production strategies, and the natural environment. Thus impacts on and alterations of the environment are nothing new or unique to our society. What is different is the global scale and magnitude of the perturbations. The major global environmental challenges that we face today are (not in order of relevance):

- availability of safe drinking water
- land degradation
- spreading of chemicals, which accumulate over many years in the environment as well as in human or animal tissue (i.e. persistent organic pollutants)
- loss of biodiversity which is a big concern due to its link with human well-being. One of the most important concerns is the rapid decline of fish populations, since fish is a major source of protein for many people
- climate change: atmospheric CO_2 concentrations reached 379 ppm in March 2004; this should be compared to an average of 280 ppm in the preindustrial era. The increase of the global mean temperature has reached approximately 0.6°C in the past century which most probably has caused the rise of sea level of 10–20 cm over the same time period
- extensive use of non-renewable resources such as fossil fuels, rich metal ores, etc.

10.1.2.3 Profit (or economical dimension)

This relates to the economic feasibility of sustainable solutions (processes, goods, products, services, amenities, etc.) under the notions of efficiency and responsibility. This means that sustainable solutions: (a) must be profitable to be accepted; (b) must include external costs to account for the value of world ecosystem services, which can be very high; for instance, the price of fossil fuels should depend not only on exploration and transformation costs, but also on the costs involved in the production of CO_2 and resultant adverse climatic effects. This dimension supports the view that finance serves the economy which serves people, and not the opposite.

As Diamond (2005) points out in his book *Collapse*, humans can choose to react and handle the challenges of sustainable development. But is there enough will? When people are more and more in agreement with the fundamentals of sustainable development, the questions that interest the majority are as follows.

- What governance is there to manage interfaces and synergies coherently between the three dimensions of sustainable development, and move from concept to reality in human activities?
- Will governments work together and co-operate with enough belief in the long-term future, and finally take responsibility at the right level to the benefit of sustainable development?
- Will the bank system adapt its strategy and policy to support the emerging sustainable economy that is to

provide the required funds at the right time to those new or existing enterprises which invest in sustainable technology?

10.1.3 Sustainable technology

As seen in section 10.1.1, the IPAT equation clearly underlined the need for new, sustainable technologies coherently with the concept of sustainable development. In the following section, the notion of sustainable technology is examined.

Sustainable technology aims to account for the three dimensions (people/social, planet/ecological, profit/economic) of sustainable development and their interactions. This section reviews the key items which allow the contribution of technology to sustainable development to be assessed, by use of the "sustainable technology scorecard" derived by Harmsen et al. (2004; see pp. 506–512). For each item reported in the scorecard, the relevance to extrusion processing technology is noted hereafter, based on learning curves from existing generic extrusion processes and relative industrial applications, as presented in previous chapters. Scorecard items are the following.

10.1.3.1 People/social dimension

- *Provide for the needs of the poor?* Positive potential of extrusion processing technology: to be defined later (section 10.2.3).
- *Fair distribution of wealth, power, and knowledge?* Extrusion processing technology is accessible to all levels of technical education.
- *Social acceptance?* Positive impact of extrusion processing technology: to be defined later (section 10.2.3).
- *Safety; smell; occupational health?* Positive impact of extrusion processing technology: to be defined later (section 10.2.3).
- *Noise?* Extrusion processing technology allows legal limits to be satisfied.
- *Plot area impact; skyline impact?* Positive impact of extrusion processing technology: to be defined later (section 10.2.3).

10.1.3.2 Planet/ecological dimension

- *Sensitivity to world-scale nature and ecology?* No impact identified.
- *Depletion of abiotic resources?* Extrusion processing technology may have a positive impact: to be defined later (section 10.2.3).
- *Depletion of biotic resources; biodiversity?* No impact identified.
- *Desiccation or dehydration?* No impact identified.
- *Global warming; photochemical ozone pollutants?* Positive impact of extrusion processing technology: to be defined later (section 10.2.3).
- *Human toxicity?* Extrusion processing technology allows legal limits to be satisfied.
- *Acidification and thermal pollution?* Positive impact of extrusion processing technology: to be defined later (section 10.2.3).
- *Ecotoxicity; eutrophication; radiation; waste?* No impact identified.

10.1.3.3 Profit/economic dimension

- *Scarce non-renewable resource depletion? Fossil fuel depletion?* Positive impact of extrusion processing technology: to be defined later (section 10.2.3).
- *Drinking water resource depletion?* Positive impact of extrusion processing technology: to be defined later (section 10.2.3).
- *External costs (or future costs, that lead to exploitation costs of ecosystem via taxation or otherwise)?* No concern identified.
- *Capital and operational expenditures?* Positive impact of extrusion processing technology: to be defined later (section 10.2.3).
- *Decommissioning cost?* No concern identified.
- *Profitable over total lifecycle?* Positive impact of extrusion processing technology: to be defined later (section 10.2.3).

The sustainable technology scorecard is a useful tool to evaluate and enhance the sustainability of extrusion processing technology, through qualitative assessment of scorecard items for all three dimensions (people/social, planet/ecological, profit/economic). On one hand, it can be used to identify those items which need real improvements to increase the sustainability of existing extrusion processing applications. On the other hand, the scorecard may allow the substitution potential of extrusion processing technology to conventional processing systems to be evaluated, with regard to sustainability. However, it will be worth re-examining extrusion-concerned scorecard items in a later conclusive section (section 10.2.3), once the process-intensifying capability of extrusion processing technology is thoroughly presented and discussed in sections 10.2.2.1 through 10.2.2.3, to realistically validate its sustainability.

The sustainable technology scorecard can also be utilized to manage the development of processing technologies, while accounting for the principles of sustainable technologies in the decision-making procedure. For that purpose, Figure 10.2 proposes a decision-making graph which consists of plotting technology sustainability versus market attractiveness for a given generic processing application, or a given processing line. To grade the processing technology, a four-class segmentation is suggested depending upon the respective levels of technology sustainability and market attractiveness.

- Mature sustainable processing technology (high sustainability-high attractiveness).
- Emerging sustainable processing technology (high sustainability-low attractiveness).
- Threatened processing technology (low sustainability-high attractiveness).
- Doomed processing technology (low sustainability-low attractiveness).

Technology sustainability can be evaluated and scored by use of the sustainable technology scorecard whereas market attractiveness is defined by the following market-driven items:

- market volume and growth
- market sensitivity to sustainability
- and possibly, fit with the enterprise's business model.

Estimation of technology sustainability and market attractiveness results from a four-level semi-quantitative scoring of the different items, leading to values which allow the processing technology to be evaluated and graded in the graph. Scoring values can be used to give either a global evaluation of the processing technology to compare it with competitive technology, or a partial evaluation to analyze strengths and weaknesses.

This decision-making procedure can be simply applied to extrusion processing technology, in order to evaluate either generic extrusion applications (such aa food extrusion-cooking, reactive extrusion, etc.) or specific extrusion processes (such as polymer compounding, breakfast cereal extrusion processing, paper extrusion milling, etc.). For a given enterprise dealing with processing technology (equipment engineering and manufacturing, process designing and engineering, etc.), this graph can be used to appropriately manage the technology survey, communication, R&D, commercial development, etc.

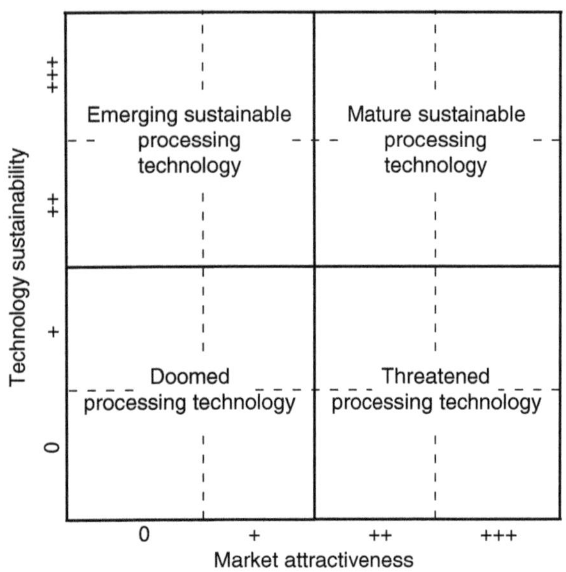

Figure 10.2 Decision-making graph for sustainability-driven technology development.

10.1.4 Concept of process intensification

Process intensification is currently one of the most significant trends in processing technology. It is a totally new approach to processing and plant design, development, and implementation. This concept was originally pioneered in the early 1980s by Colin Ramshaw and his co-workers at ICI New Science Group, when describing the application of centrifugal fields in distillation processes (Ramshaw, 1983). Ramshaw defined process intensification as a strategy to reduce drastically the size of process plant and processing equipment; volume reduction by two or three orders of magnitude was mentioned which can be considered as a real technology breakthrough. Then, from Ramshaw's definition, the concept of process intensification has been adopted by several scientists involved in process engineering, as well as by chemical industries (in Western Europe, USA, Japan, and China, in particular). Readers who are interested in a detailed description of process intensification should refer to the two relevant books of Stankiewicz and Moulijn (2004), and Reay et al. (2008).

In the early 2000s, interest in process intensification grew in both academia and industry, particularly due to more incentives for the development of sustainable technology. The definition of process intensification is no longer restricted to plant or equipment size reduction but covers a wider vision of process engineering, as proposed by Stankiewicz and Moulijn (2000): "*Novel equipment, processing techniques, and process development methods that, compared to conventional ones, offer substantial improvements in (bio)chemical manufacturing and processing.*" This definition accounts for several possible beneficial effects in accordance with the three dimensions of sustainable development. These beneficial effects can be grouped into three main principles or drivers, as follows.

10.1.4.1 First principle: promote continuous processing instead of batch or semi-batch processing

Higher throughputs with equivalent or even smaller volume devices should result from continuous processing, leading to higher process productivity, smaller plants, safer processing, and reduced capital and operating costs. It must be noted that a large majority of industrial processing plants (in chemical and pharmaceutical industries, in particular) are predominantly based on batch or semi-batch processing units.

10.1.4.2 Second principle: develop and promote multifunctional processing equipment

This consists of integrating several process functions in one unit whenever possible, hence reducing process steps, increasing process productivity, and leading to safer and smaller processing plants.

10.1.4.3 Third principle: accelerate fundamental process phenomena such as physical transfer phenomena and chemical reactions

This allows specific energy consumption, environmental impact, and utilities costs to be reduced significantly. Furthermore, smaller quantities of solvents, or even their absence, should be expected, hence leading to safer processing. Finally, higher selectivity for similar or even higher conversion rates in reactive systems should be obtained, leading to reduced consumption of raw materials and reagents, and less waste or by-products.

In sum, process intensification is intended to develop smaller, cheaper, safer, cleaner processing technologies, hence contributing directly to the desired outcomes of sustainable processing technology, and thus to the benefit of sustainable development, as summarized in Figure 10.3. In fact, in processing activities, process intensification provides tools and methodologies to reach a well-balanced intersection of the three dimensions of the sustainable development: cleaner processing to the benefit of planet, safer processing to the benefit of people, cheaper processing to the benefit of profit. Intersections of two circles are favored by the following trends: compact processing units to serve both planet and profit, low wastes and emissions to serve both planet and people, flexible and convivial processing (that is, easy to adopt in all economies) to serve both people and profit. The right long-term strategy then aims to use process intensification to increase as much as possible the area of intersection of the three circles.

The whole field of process intensification is generally divided into two complementary domains, as proposed by Stankiewicz and Drinkenburg (2004).

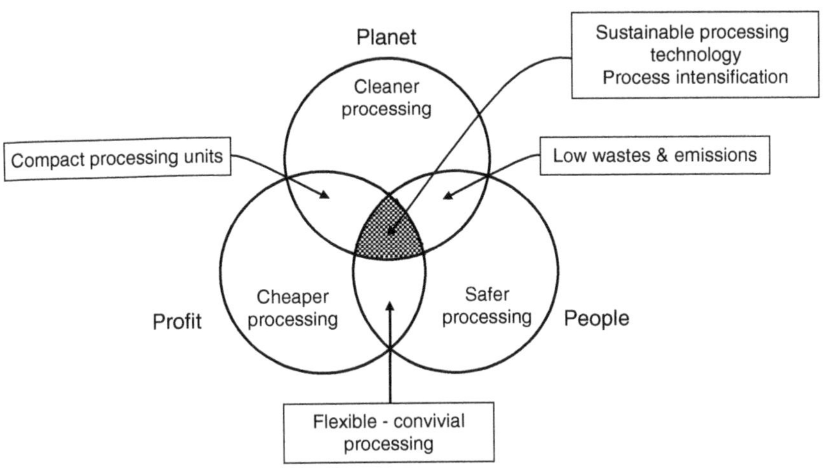

Figure 10.3 Process intensification to the benefit of sustainable development.

10.1.4.4 Process-intensifying equipment

- Novel reactors (spinning disk reactor, microreactor, static mixer reactor, etc.).
- Devices for intensive non-reactive operations (static mixer, compact plate and spiral heat exchangers, centrifugal adsorber, etc.).

10.1.4.5 Process-intensifying methods

- New processing methods which integrate the function of chemical reaction and one or more process functions (such as separation, heat exchange, phase transition, in particular), in so-called multifunctional reactors. For instance, multifunctional reactors include: reactive distillation (where a catalyst-packed column operates as a reactor and distillation column), reactive extraction (combination of reaction and liquid-liquid phase separation), reactive crystallization, reactive absorption, etc.

Stankiewicz and Drinkenburg (2004) have mentioned reactive extrusion as a process-intensifying method which aims to combine chemical reaction (single- or two-phase reactions) and phase transition, in reactive extruders (twin screw extruders, in particular), which are particularly well adapted to the chemical conversion of highly viscous materials without requiring large amounts of solvent, as is the case in conventional stirred-tank reactors. The authors also noted that "*twin-screw extruders offer effective mixing, the possibility of operation at high pressure and temperatures, plug-flow characteristics, and capability of multistaging.*" These process characteristics have been developed and illustrated in this book (Chapter 5).

- Hybrid separations which integrate membranes and another separation technique (such as membrane distillation, membrane adsorption and stripping).
- Alternative sources of energy (such as centrifugal fields, ultrasound, solar energy, microwave heating, electric fields, etc.) which allow intensification of physical and chemical phenomena (chemical reactions, mixing, heat and mass transfer, etc.) to proceed effectively.
- Miscellaneous methods such as the use of supercritical fluids to enhance processing operations, the use of dynamic operating or periodic operating (through flow or concentration pulsing, for instance).

It must be noted that process-intensifying equipment and methods are highly interactive and overlapping. In fact, new methods may require the development of novel designs of equipment, and vice versa. This is particularly true in the case of extrusion processing technology.

Of course, this review of process-intensifying equipment and methods cannot be exhaustive, as it is largely inspired by the evolution of chemical engineering and chemical industry. Other industries (such as food industry, pulp and paper industry) could provide additional relevant process-intensifying equipment and methods, such as extrusion-cooking or solid-liquid extrusion-pressing in the specific field of extrusion processing technology. These methods would definitely enrich the aforementioned list of process-intensifying methods. Besides, new developments are regularly proposed by scientists and technologists worldwide. However, it is comforting to underline the extraordinary dynamics of process intensification which can already provide sustainable equipment and methods to processing industries, hence contributing effectively to sustainable development.

10.2 Process intensification in extrusion processing technology

Processing systems involve processes or operations designed to transform raw materials into desired products through appropriate physical or chemical changes which occur in the processing units (such as reactors, distillation columns, heat exchangers, etc.). Process engineering design then involves coupled phenomena, including chemical reactions, hydrodynamics, and heat and mass transfer, which define the performance and efficiency of the process. These phenomena interact at different space and time scales. The slowest phenomena exhibit the longest characteristic times and determine process performances such as productivity, energy consumption, process selectivity, product quality, etc. Such limitations can be partially overcome through intensification of transfer phenomena.

Basically, process-intensifying equipment and methods involve one or more of the process characteristics which have been described above in the three main principles (section 10.1.4):

- continuous processing
- multifunctional processing units
- intensification of transfer driving forces

to the benefit of volume reduction, productivity increase, energy saving, reduction of solvent-aided carriers for heat and mass transfer, cleaner and safer processing, etc., leading to important reductions of capital expenditure (CAPEX) and operational expenditure (OPEX) of processing plants.

The global performance of processes often results from a competition between physical transfer phenomena (mixing, heat or mass transfer, etc.) and chemical reaction kinetics (homogeneous and heterogeneous reactions). Under steady-state conditions, these physicochemical events proceed at different rates. Process intensification then aims at identifying the characteristic times of the phenomena involved and acting on the limiting phenomenon to decrease its characteristic time in order to accelerate the rate of the global process.

10.2.1 Characteristic times of process phenomena

Processes are governed by fundamental phenomena (such as physical transfer phenomena and chemical reactions) whose characteristic times determine the rate and performance of the global process, depending upon the space time, or the residence time, of the medium in the processing device.

Transfer phenomena, such as mass transfer, heat transfer and momentum transfer, describe the transfer of a particular extensity that can diffuse in a medium. Hence, a general expression of their characteristic times can be written as follows (Matlosz et al., 2009):

$$t_{transfer} = \frac{R^2}{D_{extensity}.N_{\dim}} \quad (10.3)$$

where R is the characteristic dimension of the processing device (such as radius of a stirred tank, radius of a tubular channel, hydraulic radius of a particular geometry), $D_{extensity}$ is the diffusivity of the transferred extensity through the medium (such as mass diffusivity, D_{mass}; heat diffusivity, $\alpha_{heat} = \lambda/\rho.\ C_p$, where λ, ρ and C_p are thermal conductivity, density and specific heat capacity, respectively; and momentum diffusivity, $D_{momentum}$, or kinematic viscosity $\upsilon = \mu/\rho$, where μ and ρ are dynamic viscosity and density, respectively), and $N_{\dim}$ is a dimensionless number when convection effects are involved in the transfer process (such as the Nusselt number for convective heat transfer and Sherwood number for convective mass transfer). It turns out that in processes where the transfer is purely diffusive or conductive, or when dimensionless numbers are considered constant (fully developed velocity and extensity profiles, in steady-state conditions), the characteristic times of transfer phenomena are proportional to R^2, the square of the characteristic dimension of the processing device. Hence, reducing the characteristic dimension of the processing device will strongly accelerate the transfer phenomena. Besides, as the characteristic dimension decreases, the surface-to-volume ratio increases concomitantly.

The case of chemical reactions is far more complex, as most reactions of interest may exhibit complex kinetics, such as heterogeneous reactions. In fact, knowledge of the intrinsic reaction rate may be confused with the apparent, or observed, reaction rate due to interaction of physical phenomena such as diffusion. Nevertheless, the expression of characteristic times of chemical reactions, $t_{reaction}$, results from the relationship between the concentration of a reactant at initial conditions, C_0, and the rate of its consumption at the same initial conditions, r_0, as follows:

$$t_{reaction} = \frac{C_0}{r_0} \quad (10.4)$$

For nth-order homogeneous reactions (single phase reacting systems):

$$t_{\hom\ react} = 1/k.C_0^{n-1} \quad (10.5)$$

where k is the reaction rate constant. Hence, homogeneous reactions show no dependence on the characteristic dimension of the processing device, regardless of reaction order.

For an apparent first-order heterogeneous reaction (multiple phase reacting system):

$$t_{het\ react} = R/2k_a \quad (10.6)$$

where k_a is the apparent rate constant of the first-order heterogeneous reaction. Hence, heterogeneous reactions exhibit a linear dependence of their characteristic time on the characteristic dimension of the processing device, when they are limited by reaction rate.

Dependence on characteristic dimensions of the processing device varies strongly with the phenomena involved in the process. The characteristic times methodology then allows the process phenomena to be classified, and process intensification to be performed through appropriate design of the characteristic dimension of the processing device.

The characteristic times of the fundamental phenomena of the global process can be compared to the space time of the processing device, t, or the average residence time:

$$t = V/Q \quad (10.7)$$

where V and Q are the volume of the processing device and the volumetric flow rate, respectively.

The characteristic times of process performance refer to individual phenomena (such as heat transfer, mass transfer, chemical reaction, etc.). However, the global performance of a real process often results from the coupling of the fundamental phenomena involved, leading to a global characteristic time, t_{global}, which is a combination of the individual characteristic times. Such combination may be rather complex as it depends upon the nature of the coupled fundamental phenomena and the interactions between them (Commenge et al., 2005; Matlosz et al., 2009). For instance, simple coupling such as two phenomena acting in parallel (say, two competitive homogeneous reactions) or two in-series phenomena (say, first-order catalyzed heterogeneous reaction where mass transfer and chemical reaction act in series) is governed by the following global characteristic times.

For in-parallel phenomena:

$$\frac{1}{t_{global}} = \frac{1}{t_1} + \frac{1}{t_2} \qquad (10.8)$$

For in-series phenomena:

$$t_{global} = t_1 + t_2 \qquad (10.9)$$

where t_1 and t_2 are the characteristic times of the two coupled phenomena. When the process is governed by complex couplings of three or more phenomena, the global characteristic time is then derived from the appropriate combination of simple couplings. For instance, the global characteristic time of an in-series/in-parallel combination (say, two competitive first-order catalyzed heterogeneous reactions) will be given by:

$$t_{global} = t_1 + \frac{1}{1/t_2 + 1/t_3} \qquad (10.10)$$

where t_1 is the characteristic time of mass transfer while t_2 and t_3 are the characteristic times of the two competitive reactions.

The number of operation units (NOU) of the process (or NUT, the number of unit transfer for transfer phenomena) is then defined as the ratio of the space time of the fluid in the processing device to the global characteristic time, as follows:

$$NOU = t/t_{global} \qquad (10.11)$$

and the efficiency of the global process, η, is defined by:

$$\eta = 1 - exp(-NOU) \qquad (10.12)$$

The process efficiency is a non-dimensional parameter; it generally varies from 0 to 1. η may correspond to the conversion ratio of a chemical reaction or to the flux ratio of an extensity transferred into its environment.

The objective of process intensification is then decreasing the global characteristic time of the process by acting appropriately on the limiting fundamental phenomenon, in order to reduce the space time or the volume of the processing device at fixed flow rate, while keeping constant the NOU or efficiency of the process. The use of the characteristic time analysis of a process (as proposed by Matlosz et al., 2009) may help when introducing the process-intensifying capability of extrusion processing technology.

10.2.2 Process-intensifying methods in extrusion

A question arises on whether extrusion processing technology involves one or more process-intensifying characteristics presented in section 10.1.4. Said in another way, are single and twin screw extruders process-intensifying equipment? Are generic extrusion technologies process-intensifying methods? There is currently no published work analyzing and discussing the process intensification capability of extrusion processing technology. So far, only a few papers mention reactive extrusion as a process-intensifying method (Stankiewicz & Drinkenburg, 2004). Based on their process characteristics, generic extrusion processes presented in previous chapters show real process-intensifying potential. Actually, the Generic Extrusion Process (GEP) concept can be considered as a basis for defining extrusion process-intensifying methods, leading to:

- four existing extrusion process-intensifying methods which are already used in industry: polymer plasticating (GEP I; Chapter 4), reactive extrusion (GEP II; Chapter 5), extrusion-cooking (GEP III; Chapter 6), and solid-liquid extrusion-pressing (GEP IV; Chapter 8)
- and one emerging extrusion process-intensifying method: extrusion-porosification (GEP V; Chapter 9).

Polymer plastication is very well known and used in the plastics industry as so far there is no alternative process to continuously melt, mix and pump high-viscosity polymer mixes. Actually, polymer plastication fully satisfies the principles of process intensification, and thereby can be classified as a process-intensifying method.

The purpose of the following sections of this chapter is to discuss process-intensifying characteristics within extrusion processing technology through the examination of three extrusion processes: reactive extrusion, extrusion-cooking, and solid-liquid extrusion-pressing.

10.2.2.1 Reactive extrusion

The characteristic time analysis presented above clearly shows that the volume of processing units can be reduced significantly while maintaining constant process efficiency and productivity, as it showed that a reduction of the characteristic dimension of the unit may promote fundamental phenomena associated with the process to be accelerated (transfer phenomena, rate of heterogeneous reactions). Consequently, process intensification mainly has consisted of miniaturizing or microstructuring processing units. Even for homogeneous reactions for which the characteristic time is independent of the characteristic dimension of the processing unit, miniaturization may bring significant benefits through local intensification, resulting from the effects of the characteristic dimensions on local transfer characteristics or local operating conditions. And a number of innovative applications have been developed and implemented in the area of micro-reaction technology, in the chemical and pharmaceutical industries. However, microstructured systems cannot be generalized to all industrial processes, since processing units based on microchannel designs may show real operational or technical limitations when complex fluids (such as highly viscous fluids, solid-liquid media, etc.) are processed. In practice, complex fluids represent a very important domain in processing industries.

As seen in section 10.1.4, Stankiewicz and Drinkenburg (2004) proposed a relevant overview of different process-intensifying equipment and methods which allow processes to be intensified depending upon the process phenomena involved and the couplings between process phenomena. In that overview, the authors mentioned reactive extrusion as a very promising process-intensifying method. In reactive extrusion, examples of homogeneous reactions may include bulk polymerization, chemical modifications of polymers (such as grafting, cross-linking, functionalization, depolymerization, etc.) when reacting species are fully miscible. Examples of heterogeneous reactions may also be mentioned; they include liquid-liquid reacting systems (immiscible reacting species with a macromolecular chemistry) and solid-liquid reacting systems (e.g. alkaline digestion of lignocellulosic materials, cellulose bleaching, chemical modification of cellulose fibers, biomass hydrolysis, etc.).

10.2.2.1.1 Process specificities

Reactive extrusion has been described in Chapter 5, as a fundamental generic extrusion process. In this section, it is worth underlining the key process characteristics which allow reactive extrusion to be classified as a process-intensifying method as defined by Stankiewicz and Drinkenburg (2004).

- Continuous processing, particularly well adapted to chemical conversions in high-viscosity reaction media (homogeneous and heterogeneous reactions) and solid-liquid media.
- Intense laminar mixing, which allows reactive extrusion to intensify transfer phenomena (momentum transfer, mass transfer).
- Bulk reaction, without solvents or diluents, with the overall objective of avoiding further separation of solvents or diluents to minimize contamination, saving energy and consumables, and reducing emission of volatiles.
- Multifunctional processing devices (co-rotating twin screw extruders, in particular) which may handle several process functions in one unit (such as solids conveying, melting, mixing, reacting, venting, and pressurizing). In addition, reactants (solids, liquids, and possibly gases) can be added on line (side-stream addition), particularly when using modular screw-barrel designs.
- Low processing volume, hence reducing footprint and cost of reactive extrusion units. Effective processing volumes of industrial reactive extrusion units amount to a few liters, which makes screw extruders highly productive reactors; consequently the space time of fluids in extruder-reactors is relatively short (order of minutes), which may limit the application range to reactions with low characteristic times.

These process characteristics allow reactive extrusion to be very competitive from technical and economic standpoints. In fact, compared with conventional reactors (such as batch or continuous stirred tank reactors) which require the use of solvents or diluents and large processing volumes, extruder-reactors enable significant improvements in technical performance (product quality, process productivity and safety, etc.) as well as important cost reductions (lower CAPEX and OPEX) to be obtained when performing chemical reactions. Hence, reactive extrusion offers an important potential to replace conventional reactive methodologies, when appropriate processing conditions are applied.

10.2.2.1.2 Process characteristics of extruder-reactors

The key aims of reactors are to: (i) achieve good mixing of the reacting medium to ensure micromixing, meaning that concentrations of all species are the same throughout; (ii) provide the required residence time according to the reaction kinetics; (iii) achieve the heat exchange that provides the temperature required by an optimum reaction. For the study of extruder-reactors, it is worth recalling the process characteristics of the two ideal flow behaviors of continuous-flow chemical reactors in which the flow dynamics show extreme time and mixing histories.

- *Continuous stirred tank reactor* (or CSTR). Characterized by complete longitudinal mixing (complete backmixing), spatial uniformity of composition and highly dispersed residence times (exit probabilities of all species are identical and independent of the time spent in the device). Stirred tank reactors show relatively large diameters (from about 1 m to a few meters) with low length-to-diameter ratio (L/D) and specific area (or surface-to-volume ratio, A/V). L/D is close to unity, while A/V ranges from 1.5 m^{-1} in large reactors to 5 m^{-1} for pilot-scale reactors (Table 10.2). The A/V ratio of cylindrical tank reactors of diameter, *D*, and height, *L*, is generally given by 5/D. The overall heat transfer coefficient, *U*, may vary between 50 and 400 $W/m^2.K$ (see Table 10.2), depending upon the Reynolds number both in the reacting media and in the double jacket. It must be noted that stirred tank reactors are currently used for batch reaction processing. The key issue of CSTR is the scalability of mixing, which soon will become prohibitive from technical (mixing capability decreases as tank diameter increases) and economical (required power for mixing increases drastically as it is proportional to tank diameter to the power of 5) standpoints.
- *Continuous plug flow reactor* (or CPFR). Characterized by no backmixing, only radial uniformity of composition and identical residence time for all material. Continuous plug flow reactors show relatively small diameters (from a few millimeters to a few centimeters) with high L/D and A/V ratios: L/D >> 1, while A/V ratio may range from 50 to 500 m^{-1} depending upon the characteristic dimension of the reactor (diameter or hydraulic diameter); A/V ratio of such reactors of diameter, *D*, is given by 4/D. Hence, for a typical 5 cm diameter tubular reactor, the A/V ratio is 80 m^{-1} while *U* may vary between 250 and 500 $W/m^2.K$. However, for a microtubular reactor of 400 μm diameter, A/V ratio will be 10^4 m^{-1}, while the overall heat transfer coefficient might reach values up to 2.10^4 $W/m^2.K$. Based on previous designs and heat transfer characteristics, it appears that the heat exchange density, HED (or the heat exchange per unit volume, as expressed in $kW/m^3.K$), of continuous plug flow reactors is approximately 50–100 times higher than that of a stirred tank reactor, making linear plug-like flow reactors potentially much more attractive than stirred tank reactors for handling reactive systems.

Ideal CST and CPF reactors, the latter discussed in Chapter 5, are abstractions, as none of the actual industrial reactors shows flow characteristics which are perfectly represented by the two ideal systems. Actual industrial continuous flow reactors tend to belong to either the CSTR or CPFR model in terms of their geometric designs as well as mixing and time histories and heat exchange characteristics. For instance, as discussed in Chapter 5, continuous extruder-reactors resemble more continuous plug flow reactors, as their L/D and A/V ratios are generally in the ranges of 25–50 and 50–500 m^{-1}, respectively. The appropriate question at this point is: what mixing capability, time history and heat exchange characteristics, and consequently what process-intensifying drivers, do characterize extruder-reactors?

Cylindrical stirred tank vessels can provide adequate mixing at viscosities as high as 10^2–10^3 Pa.s, if well-designed impellers are used. They include paddle designs (anchor and helical ribbon designs, in particular) which promote good peripheral mixing near the tank walls or its bottom. But as the viscosity of reacting fluids increases, viscous stagnant regions may develop in the tank, leading to increased degrees of segregation through the tank. This phenomenon is aggravated when processing non-Newtonian fluids. Besides, plug-like flow devices,

Table 10.2 Heat exchange characteristics of typical reactors and extruder-reactors.

Characteristics	Stirred tank	Plug-like flow	SSE	TSE
A/V, m^{-1}	1.5–5	50–500	50–500	40–400
U, $W/m^2.°K$	50–400	250–500	50–160	300–500*
HED, $kW/m^3.°K^{\dagger}$	0.6–2	25–250	8–80	20–200

* U = 300 $W/m^2.°K$ corresponds to positive conveying sections with low filling degree (30%) while U = 500 $W/m^2.°K$ corresponds to shearing or mixing sections with filling degree of 100% (right-handed screw element followed by left-handed screw element or kneading block) (Mottaz & Bruyas, 2001).
† HED: the ranges of values in the table were obtained by multiplying the highest values of U with the lowest and highest values of A/V.
SSE, single screw extruder; TSE: twin screw extruder.

or tubular devices, are generally driven by pressure applied at the inlet, limiting their practical use due to large pressure drops and possibly plugging phenomena when processing highly viscous reactive media. Thus, beyond a viscosity of about 10^3 Pa.s, conventional units such as stirred tank or pressure-driven plug-like flow devices are not practically convenient for chemical reactions in viscous media like pastes, doughs and melts, often found in the plastics, rubber, and food industries. They are even more limited when dealing with concentrated solid-liquid media, such as those found in green chemistry, the pulp and paper industry, etc. Thus, more suitable technologies such as extruder-reactors are needed for handling these systems giving rise to reactive extrusion.

As seen in Chapter 3, screw extruders operate on the principle of drag flow with one moving surface (either one or two rotating screws) which allows pressure to be generated longitudinally when processing highly viscous media. The net volumetric flow rate of a screw extruder is the algebraic sum of the two following flow components: drag flow and pressure flow. It has been demonstrated that the drag flow component increases with increases in the cross-sectional area of the screw channel and the screw speed, while the pressure flow component, which is opposed to the drag flow, increases with decreases in the fluid viscosity. In other words, for a given pressure rise generated down the channel, the net throughput of extruder-reactors increases when the viscosity of the reaction media increases. Hence, the good pumping capabilities of highly viscous materials make extruder-reactors definitely more suitable for handling high-viscosity reaction media.

The time history of the elements in an extruder-reactor as described by the residence time distribution shows that it is rather close to plug flow. Figure 10.4 presents residence time distribution functions $F(t)$ as a function of reduced time, $t/\bar{t}$, for different types of flow: the ideal CSTR and CPFR flows, the plane Poiseuille flow (or Poiseuille flow between infinite parallel plates) and the flow in a single screw extruder calculated from the velocity distribution (Tadmor & Klein, 1970). Actually, the residence time distribution in a single screw extruder lies between that obtained in plane Poiseuille flow and ideal plug flow. The residence time distribution is expected to be even closer to plug flow as the fluid exhibits a non-Newtonian shear-thinning behavior. The flow behavior in a single screw extruder is closer to that observed in an ideal continuous plug flow reactor mainly because of the transverse flow, owing to the helix angle of the screw. For co-rotating twin screw extruders, the residence time distribution is believed to be even closer to ideal plug flow reactor behavior with self-wiping flights and tightly intermeshing screws. Thus, in reactive extrusion, extruder-reactors would show a continuous change of the concentration of reacting species in the axial flow direction, depending upon the flow dynamics in the processing device and the reaction kinetics.

With regard to their heat exchange capabilities, considering that they deal with high-viscosity reacting media, extruder-reactors show satisfactory heat transfer characteristics (see Table 10.2). For a typical screw design of the metering section of single screw extruders, the A/V ratio can be estimated over one complete turn of the helix of the screw channel. Hence, considering the

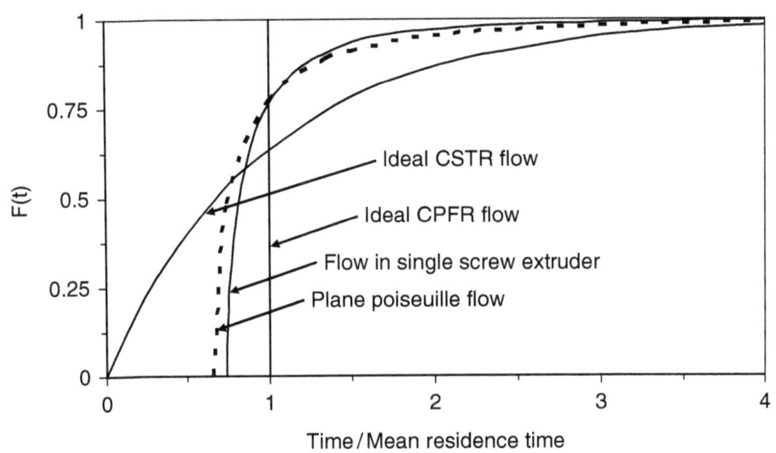

Figure 10.4 Residence time distribution function $F(t)$ versus reduced time $t/\bar{t}$.

nomenclature presented in Chapter 2 (see Figure 2.4), the A/V ratio of the metering section can be defined as follows:

$$\frac{A}{V} = \frac{\pi.D.B}{(H+\delta).W.Z} \qquad (10.13)$$

Equation 10.13 shows that A/V depends on geometric parameters of the screw channel. As a first approximation, let us assume that the metering section of a single screw extruder is equipped with a single-flight square-pitched screw, which means that the helix angle $\theta = 17.66°$ and the screw pitch $B = D_s \cong D$ (as seen in Chapter 2, section 2.1.2.1). In addition, assuming that $(H+\delta) \cong 0.1D$ and neglecting e as compared to $B\cos\theta$, Eq. 10.13 simplifies to:

$$\frac{A}{V} \cong \frac{10}{D} \qquad (10.14)$$

As expected, Eq. 10.14 shows that A/V is inversely proportional to the barrel diameter D. Hence, for a given axial distance of the metering section (constant L/D ratio), heat exchange capability decreases when barrel diameter increases. However, Eq. 10.14 gives a rough estimation of A/V. According to Eq. 10.13, it must be noted that A/V depends closely on the key geometric parameters at constant D, such as the height of screw flight, H (the larger the height, the smaller the A/V), the helix angle of the screw channel, θ (the larger the helix angle, the larger the A/V), and possible additional mixing elements in the metering section (such as mixing pins).

When a reactive extrusion process proceeds in the metering section of single screw extruders, for a given internal barrel diameter D, A/V would be roughly twice that of stirred tank and plug-like flow reactors. A/V typically ranges from 50 m^{-1} (in large single screw extruders: 200 mm screw diameter) to 500 m^{-1} (in laboratory-scale single screw extruders: 20 mm screw diameter). The overall heat transfer coefficient, U, typically varies between 50 and 160 $W/m^2.K$, depending on the processing conditions (Davis, 1992), leading to a HED of 8–80 $kW/m^3.K$ (see Table 10.2).

According to Martelli (1983), considering the nomenclature presented in Chapter 2 (see Figure 2.9), the A/V ratio for intermeshing co-rotating twin screw extruders (assuming that the screw profile of the metering section is based on conjugated, single-flight, right-handed pitch screw elements, and that the metering section is fully filled) can be defined as follows:

$$\frac{A}{V} = \frac{\pi.D_s(\pi.D_s - \sqrt{2D_sH}).sin\phi}{\frac{1}{2}\pi^2.D_s^2.H.tan\phi} \qquad (10.15)$$

Equation 10.15 shows that A/V depends on geometric parameters of the screws, such as D_s, H, and ϕ in particular. As a first approximation, let us assume that $D \cong D_s$, $H \cong 0.2D$ and $\phi < 10°$; hence, Eq. 10.15 simplifies to:

$$\frac{A}{V} \cong \frac{8}{D} \qquad (10.16)$$

As for single screw extruders, Equation 10.16 shows that A/V is inversely proportional to the barrel diameter D. Hence, heat exchange capability decreases when barrel diameter increases. Equation 10.16 also gives a rough estimation of A/V. According to Eq. 10.15, A/V decreases when the height of the screw flight, H, and the helix angle of the screw channel, ϕ, increase. In addition, A/V depends on the screw design (such as the number of flights: a double-flight screw offers larger reactive volume than a single-flight screw, tending to decrease A/V) and the screw configuration (for instance, adding kneading blocks which offer smaller reactive volume tends to increase A/V). In general, A/V of intermeshing co-rotating twin screw extruders tends to be slightly lower than that of single screw extruders, due to the significantly higher height of the screw flight in particular.

As seen in Table 10.2, A/V typically ranges from 40 m^{-1} (large twin screw extruders: 200 mm screw diameter) to 400 m^{-1} (laboratory-scale twin screw extruders: 20 mm screw diameter) while U generally varies from 300 to 500 $W/m^2.K$, depending upon the operating conditions and the design of the screw-barrel assembly (Mottaz & Bruyas, 2001), leading to HED values of 20–200 $kW/m^3.K$ (see Table 10.2). It appears that HED of intermeshing co-rotating twin screw extruders is equivalent to that of plug-like flow reactors, although the viscosity of the reacting media in twin screw extruder-reactors is two or three orders of magnitude higher. Moreover, HED of twin screw extruders is approximately twice that of single screw extruders, because twin screw extruders are characterized by a higher overall heat transfer coefficient owing to the more intense laminar mixing which leads to a higher convective heat transfer coefficient between the viscous media and the barrel wall.

This simple analysis of the heat transfer characteristics in extrusion shows that single and twin screw extruder-reactors have satisfactory heat exchange capabilities, particularly twin screw extruder-reactors which are better suited to reactive extrusion with high-viscosity reactive media. As barrel diameter increases, surface-to-volume ratio decreases concomitantly; hence, in practice, the use of an intermediate barrel diameter will be favored, particularly with exothermic reactive systems. The channel depth, H, is an important factor in reactive extrusion; small values of H facilitate the removal of reaction heat but viscous dissipation will rise concomitantly. Thus, an optimal value of channel depth must be defined carefully, as it affects the performances of extruder-reactors.

10.2.2.1.3 Process-intensifying conditions

Process limitations may be detrimental to the productivity of processing systems, but can be partially eliminated through intensification of heat and mass transfer phenomena. Such intensification is generally obtained by dissipating an excess of mechanical energy in the system, through various methods such as forced convection, mechanical stirring, fluidization of particles, etc. In reactive extrusion processing, intensification results from dissipating mechanical energy in highly viscous reaction media through drag flow. In addition, the helix angle of the screw(s) induces an effective transverse flow that enhances mixing. It is worth analyzing the mixing value-added or mixing capability of continuous drag flow extruder-reactors, and hence discussing the process-intensifying conditions of reactive extrusion.

As seen in previous chapters (Chapters 3 and 4 in particular), the screw-barrel assembly of screw extruder-reactors can be approximated by a flat plate (the unwound barrel) which is moving at a certain velocity over a shallow rectangular channel due to the helix angle (say θ). A drag flow is developed and driven by the moving barrel wall in the shallow rectangular channel. When the helix angle $\theta=0$, all wall motion is dedicated to down-channel flow whereas when $\theta > 0$, a component of wall motion proportional to $\sin\theta$ is directed toward cross-channel flow, which induces transverse convective mixing. As discussed in Chapter 5 (section 5.2.1.2.1), the use of characteristic times associated with the different process phenomena occurring in the process and of appropriate dimensionless numbers such as the first and second Damköhler numbers, Da_I and Da_{II}, which relate the chemical reaction time scale to other phenomena occurring in the reactive system, is of paramount importance in the analysis and comprehension of the limiting factors of the global process. For laminar flows in continuous plug flow reactors, when $Da_{II} > 1$ (relatively fast reactions), a microsegregation regime prevails which means essentially no transverse convective mixing, whereas when $Da_{II} < 1$ (relatively slow reactions), perfect mixing (or a Taylor diffusion regime) predominates which means essentially complete transverse mixing.

Reactive extrusion, like most chemical reaction systems, involves three main process steps which occur simultaneously: transverse convective mixing, diffusion, and reaction. When feeding the feed stream(s) (solid, liquid, melt, or possibly gas) in an extruder-reactor, the most important process step is the mixing. In practice, mixing in extruder-reactors includes blending two or more fluids to produce either a homogeneous phase (miscible fluids) or a heterogeneous phase (immiscible fluids with passive or active interfaces) such as in polymer reaction processing (mixing of immiscible polymeric melts, mixing of miscible fluids having widely differing viscosities and volume ratios such as high-viscosity polymeric melt(s) and low-viscosity reactant(s)). In reactive extrusion, the mixing step may also include blending of solids with liquid reactant(s) (such as lignocellulosic materials and liquid alkali, or raw cellulose fibers and liquid oxidizing agent) in case of alkaline lignin digestion of lignocellulosics and cellulose bleaching (for more information refer to Chapter 5, section 5.3.3).

The process of mixing a small reacting volume of an element of fluid (that can be a liquid droplet, a blob of melt, a dispersed solid particle, or a gas bubble, soluble or insoluble in the continuous phase) with its surroundings includes both reducing the size of the fluid element to as small as possible (dispersive mixing) and the spatial redistribution of resulting fragments as uniformly as possible (distributive mixing) at the molecular level, in order to produce a mix of desired uniformity and mixture quality, with a uniform concentration of reactant(s). Basically, extruder-reactors are characterized by laminar mixing which is composed of the following basic mechanisms.

- *Laminar shear*, where relative motion between streamlines in the flow enables deformation of the fluid elements, hence increasing interfacial areas between the liquids being mixed and reducing striation thicknesses. Laminar shear is extensively developed in the screw channel of extruder-reactors, owing to the drag flow induced by screw rotation.
- *Elongation strains*, or extensional flow, which relate to the stretching effect on fluid elements due to changes in fluid flow, or unsteady-state flow dynamics, which allow striation thicknesses to be reduced and hence mixture quality to be improved. Extensional flows are deliberately generated in extruder-reactors when changing the screw design or installing restrictive screw elements and kneading disks in the screw configuration. In addition,

extensional flows are particularly important in the intermeshing area of co-rotating twin screw extruders.

- *Distributive mixing*, or the spatial redistribution of fluid elements among contiguous regions as uniformly as possible. Distributive mixing aims at producing stream-splitting and recombination effects, hence reducing striation thickness and increasing the interfacial area between components. In extruder-reactors, distributive mixing is favored by the use of appropriate mixing elements such as pin mixers in single screw extruders or kneading disks in intermeshing, co-rotating twin screw extruders.
- *Molecular diffusion*. In viscous liquids, the molecular diffusion coefficients are small (in the range of 10^{-10}–10^{-14} $m^2.s^{-1}$). Hence, according to the expression of the characteristic time for mass diffusion (Eq. 10.3), mixing at the molecular level (or micromixing) is slow and only becomes important once the mechanisms of shear, elongation and distributive mixing have significantly increased the interfacial area or reduced striation thickness down to a sufficiently low level. Thus molecular diffusion acts over a much lower characteristic dimension, and allows the intensity of segregation or the differences in concentration between components of the mixture to be reduced.

In reactive extrusion, proper mixing of screw extruder-reactors means mixing at the molecular level. As discussed in Chapter 5 (section 5.2.2.1), a characteristic of the mixing of viscous polymers in laminar flow is the production of layered or lamellar mixtures that can be conveniently represented by the striation thickness, l, that is related to the interfacial area, S, by the following expression (the same equation is also given as Eq. 5.44 in Chapter 5):

$$l = \frac{V}{S/2} \tag{10.17}$$

where V is the total volume of the system. When two viscous liquids are mixed, the interfacial area between the two components increases whereas the striation thickness decreases.

When a minor component (say, a reactive polymer melt A) is initially introduced as randomly oriented blobs of melt into a major component (say, a reactive polymer melt B), and assuming that the minor and major components have identical rheological properties, it can be demonstrated that the striation thickness is related to the shear strain by the following equation (Tadmor & Gogos, 2006):

$$\frac{l}{l_o} = \frac{2}{\gamma} = \frac{2}{\dot{\gamma}.t} \tag{10.18}$$

where l_o is the initial striation thickness; γ is the total shear strain; $\dot{\gamma}$ and t are the shear rate and the residence time of the processing unit, respectively. Tadmor and Gogos (2006) have shown that the striation thickness is directly proportional to the initial size of melt blobs of the minor component and inversely proportional to the total shear strain and volume fraction of the minor component. The larger the size of melt blobs, the smaller the volume fraction of the minor component, hence a larger total strain is needed to achieve the required striation thickness. It must be noted that it is more difficult to mix by laminar shearing small volumes of high-viscosity minor component into a major component than it is to mix equal volumes of equal-viscosity components. Eq. 10.18 can be used to estimate the strain needed to reduce the striation thickness to a level where ultimate homogenization will be ensured by molecular diffusion for a given strain rate within the residence time of the mixing unit.

Tadmor and Gogos (2006) indicate that a good rule of thumb for adequate mixing in the mixing unit is a total strain of the order 10,000–25,000 shear units, which implies a four order of magnitude reduction in striation thickness. Hence, screw extruder-reactors should be able to target that mixing level through an efficient transverse convective mixing due to the helix angle of the screw. Then, the key factor for process intensification in reactive extrusion results from the capacity of screw extruder-reactors to effectively decrease the striation thicknesses, or to increase the interfacial area between reacting species, or to reduce the scale of segregation, and make possible the mass transfer and chemical reaction within the time available in the processing unit.

Biesenberger (1992) defined a dimensionless number, the segregation number for mixing S_m, in terms of a ratio of the corresponding characteristic time for diffusion and the residence time of the processing unit. For a shallow rectangular channel in which laminar flow is developed and no transverse convection occurs, the characteristic time for transverse diffusion is L_c^2/D_{mass} where L_c and D_{mass} are the characteristic length of the processing unit (that is, one-half the channel depth) and the mass diffusion coefficient, respectively. In this particular type of processing unit, transverse mixing must occur entirely by diffusion and the segregation number S_m is given by:

$$S_m = \frac{L_c^2}{D_{mass}.\bar{t}} \tag{10.19}$$

where $\bar{t}$ is the average residence time of the processing unit. For the existence of a perfect mixing (or Taylor diffusion regime), S_m must be less than some critical value which can be determined either by experimentation or by computation.

For a laminar flow in a screw extruder where a transverse convective mixing occurs in the screw channel, the characteristic length now refers to the reacting volume of an element of fluid which is therefore a variable function of time. From the laminar mixing theory (refer to Chapter 5, section 5.2.2.1), this characteristic length can be defined as the striation thickness which is reduced from l_o within the residence time of the processing unit. Hence, the characteristic length is equated to the striation thickness, l, after time $\bar{t}$ resulting from transverse convective mixing in the screw channel, and the segregation number S_m becomes (Biesenberger, 1992):

$$S_m = \frac{l^2}{D_{mass}.\bar{t}} \tag{10.20}$$

Expressions for estimating the decrease of l as a function of residence time depend on flow fields in the screw channel of screw extruder-reactors. Based on laminar mixing theory, Biesenberger (1992) proposed linear functions for plain screw extruder channels and the segregation number becomes:

$$S_m = \frac{l_o^2}{D_{mass}.\bar{t}} \times \frac{1}{(m_i.Rc)^2} \tag{10.21}$$

where Rc is a dimensionless recirculation number which is a measure of the relative rates of cross-channel versus down-channel convection as follows: $Rc = (2L_{ch}/H)\tan\theta$; L_{ch} and H are the length of unwound rectangular channel and the depth of screw channel, respectively; m_i relates to the effect of initial orientation of interface between two adjacent fluid layers ($m_i = 1.15$ for an initially vertical interface, almost perpendicular to the flow streamlines; $m_i = 0.1$ for an initially horizontal interface, almost parallel to the flow streamlines). For detailed information on the foundations of striation thinning in a screw extruder channel, refer to the article by Chella and Ottino (1985).

In reactive extrusion, most mixing processes are associated with chemical reactions which involve the three simultaneous process steps mentioned above: convective mixing, diffusion, and reaction. By analogy to Eq. 10.20, Biesenberger (1992) introduced another dimensionless number, the segregation number for reaction, S_r, defined as follows:

$$S_r = \frac{l^2}{D_{mass}.t_{reaction}} \tag{10.22}$$

where $t_{reaction}$ is a characteristic time for the reaction; hence S_r is considered to be a generalization of the second Damkhöler number, Da_{II}. For a required molecular mixing, it is expected that S_r must be less than a critical value which can be determined either by experimentation or by computation. For a plain screw channel of an extruder-reactor, by analogy to Eq. 10.21, Biesenberger (1992) proposed the following expression for the segregation number for reaction:

$$S_r = Da_{II}\frac{1}{(m_i.Rc)^2} \tag{10.23}$$

where the second Damköhler number relates to the segregation number for reaction in a processing unit with no transverse convective mixing, and is conventionally expressed as follows: $Da_{II} = L_c^2/D_{mass}.t_{reaction}$, in which L_c is the characteristic dimension of the processing unit.

Among the three main principles which characterize process intensification, it is worth focusing on the third principle which relates to the acceleration of fundamental process phenomena such as physical transfer phenomena and chemical reactions. In reactive extrusion, the purpose of screw extruder-reactors is to accelerate the mass transfer of reactants within the viscous reaction media through the presence of transverse convection in the screw channels. Hence, the key factor for classifying reactive extrusion as a process-intensifying method relates to the mixing value-added or mixing capability of screw extruder-reactors, which results from their capacity to effectively decrease striation thickness, or to increase the interfacial area between reacting species, or to reduce the scale of segregation at the required extent of completion. The quantity $(m_i.\,Rc)^2$ of Eqs 10.21 and 10.23 expresses the mixing value-added of screw extruder-reactors when processing complex homogeneous and heterogeneous, viscous reaction media. It must be noted that the mixing value-added depends on the orientation of initial layer interfaces through the parameter m_i; and on L_{ch}/H ratio and helix angle θ through the recirculation number Rc.

Based on typical values of geometry parameters generally encountered with screw channels, the effect of transverse convection in screw channels is huge as the extent of the quantity $(m_i.\,Rc)^2$ would lie between 10^3 and 10^4 depending upon the design of the reaction section. For instance, for an initially vertical interface entering into a screw channel of a single screw extruder-reactor with $L_{ch}/H = 100$ and a helix angle of 17.66°, the quantity $(m_i. Rc)^2$ is equal to approximately 5300. Hence, transverse convective mixing in screw extruder-reactors provides a determinant processing advantage in reactive extrusion as it allows mass transfer of reactive species to be effectively accelerated and chemical reactions to occur within short residence times in screw extruder-reactors. Although the quantity $(m_i. Rc)^2$ gives a rough estimation of the mixing value-added of screw extruder-reactors, it allows us to objectivize the transverse convective mixing in the processing unit, and hence the process-intensifying capability of reactive extrusion, and the quantity $(m_i. Rc)^2$ can be considered as a process-intensifying factor in reactive extrusion.

As shown by Eq. 10.18, the extent of striation thickness decrease is inversely proportional to the total shear strain regime imposed on the initial striation thickness. Then, the significance of the total shear strain, $\gamma = \dot{\gamma}.t$, in the laminar shearing process has been used as a measure of mixing. In particular, the calculation of the weighted average total strain (or WATS), which accounts for the distribution of strain histories among the fluid elements, can provide a quantitative measure of mixedness in the extruder-reactor (Tadmor & Klein, 1970). By combining the total strain distribution with the residence time distribution, WATS is defined as follows:

$$WATS = \bar{\gamma} = \int_{t_o}^{\infty} \gamma f(t) dt \qquad (10.24)$$

where t_o is the minimum residence time and $f(t)$ is the residence time distribution function; $\bar{\gamma}$ can be calculated by numerical integration. WATS as a measure of mixing is useful in systems for which the flow field is well defined. For a single screw extruder, as shown by Tadmor and Klein (1970) as well as in Chapter 5 (refer to section 5.2.2.2, Figures 5.20 and 5.25), WATS is a function of the pressure flow to drag flow ratio and of the helix angle of the screw channel. Higher average strain and hence mixture quality are achieved when the pressure flow becomes significant compared with the drag flow; hence, when pressure flow attains values above 50% of the drag flow, a sharp increase of $\bar{\gamma}$ with this ratio is observed. Thus, the increase of mixing implies an increase of pressure at the die, which consequently depresses the net throughput of the extruder-reactor. Besides, the mixing is affected by the helix angle of the screw channel, θ. In the limiting cases of 0° and 90°, $\bar{\gamma}$ tends to infinity but when the helix angle increases from 10° to 30° (usual range of helix angles), $\bar{\gamma}$ decreases although the cross-channel mixing which is proportional to $\sin\theta$ increases. In fact, when the helix angle increases, the output per screw turn increases, hence reducing residence time. The reduction in residence time takes precedence over the effect of improved cross-channel mixing when helix angle increases from 10° to 30°.

Based on predictions by Tadmor and Klein (1970), for practical processing conditions in single screw extruders, typical values of strain $\bar{\gamma}$ would lie between *20 L/H* and *30 L/H* where L and H are the axial screw length and channel depth, respectively. Noting that typical values of geometry parameters generally encountered with single screw extruders would be $L/D = 25$ and $D/H = 10$, the resulting value of strain $\bar{\gamma}$ would range between 5000 and 7500. It can be demonstrated that the strain achieved for a Poiseuille flow in a long circular pipe is $\bar{\gamma} = 16L/3D$ where L and D are the length and diameter of the pipe, respectively (refer to section 10.3.3, Exercise 3), hence leading to much lower values of strain.

It should be noted that high values of WATS will not always be associated with a good-quality mixture. Although large strains can be imposed in single screw extrusion through the flow kinematics described in Chapter 3, during melting the fluid elements become aligned with the planes of shear and hence the shearing effect in the metering section does not improve the mixing significantly. As a result of this orientation effect, very large values of WATS can be provided without improving the mixing, unless additional mixing elements are installed in the metering section to disturb fully developed streamlines. Therefore, WATS as a measure of mixing is pertinent and useful as long as the flow field in the processing device is known, and the streamlines of fluid elements are redistributed along the screw channel. Although WATS gives an imperfect analysis of the mixing capability of screw extruder-reactors, it allows the process-intensifying capability of reactive extrusion to be evaluated and can also be considered as a process-intensifying factor in reactive extrusion.

The laminar shear mixing in screw extruders can only be analyzed if the exact flow kinematics is known. As shown in Chapter 3, this can be done fairly well for simplified geometry of the screw channels, hence considering simple rectangular flow channels and simple assumed rheology for the melt. In this case, the analysis of the mixing capability of extrusion-based processing units and identification of limiting process phenomena in reactive extrusion are acceptable. However, such simplification may neglect the effects of various phenomena on flow patterns, such as leakage flow over small flight clearance δ, multiflight screw profiles, or the possible difference in flow properties of fluid components, if any, thus limiting the validity and applicability of results of this mixing analysis.

It should be noted that in single screw extrusion, the lack of effective orientation of the fluid elements at the screw flights causes the bulk of the interfacial area to be oriented parallel to the bottom in a relatively short down-channel distance, regardless of initial orientation of the components to be mixed. Consequently, distributive mixing in single screw extruders is generally poor. Moreover, there is no dispersive mixing capability that can rupture agglomerates except for leakage flow. Hence, specially designed mixing elements (such as mixing pins or fluted mixing sections) are incorporated to redistribute the streamlines and reorient the interfacial area. As for intermeshing, co-rotating twin screw extrusion with regular right-handed pitch conveying screw elements, there is an important difference if compared with single screw extrusion owing to the nip regions, i.e. the zone where the two screws intermesh. In fact, at the intermeshing area, one can observe:

- change of direction of the shearing planes, hence reorienting the interfacial area for better distributive mixing
- relatively high shear rates, due to the high relative velocity at the interface between the screw root of one screw and the tip of flights of the other screw.

In intermeshing co-rotating twin screw extruders, the fluid moves through regions of high to low shear rates which confer a redistributive mixing component. Besides, as shown in Chapter 2 (section 2.2.2), the extensive use of kneading disks in the screw profile allows excellent dispersive as well as distributive mixing to be generated with a high level of flexibility, depending upon the process requirements. It is generally considered that these extruders are capable of providing better mixing through dispersive and redistributive actions than single screw extruders, suggesting that intermeshing, co-rotating twin screw extruder-reactors are more suitable for reactive extrusion processes. However, quantitative analysis of mixing sections in such extruder-reactors is complex. Therefore, the development of mixing elements and the design of screw configurations for reactive extrusion remain an empirical exercise. In practice, mixing requirements encountered in reactive extrusion are not as simple as the mixing of a minor component into a major component both having identical rheological properties. Feed streams may be added separately or premixed on a macroscopic scale or on a molecular scale; they may be solids (pellets, powders, fibers), liquids (low-viscosity reagents), melts or possibly gases. Hence, design of the screw configuration for adequate mixing (dispersive and distributive mixing) is largely based on empirical knowledge, depending upon the physical state of the feeds and their introduction into the barrel of the extruder-reactor.

10.2.2.1.4 Process-intensifying capability

The technology of mixing has made significant progress during the last 25 years, at least for liquid-liquid and gas-liquid systems. This was achieved by developing static mixers in particular. As for single- or two-phase viscous media, laminar mixing capability of screw extruders has been greatly improved by the development of screw designs dedicated to the control of dispersive and/or distributive mixing. Furthermore, heat transfer capability of mixing systems has also progressed significantly, enabling processes in which simultaneous mixing and intensive heat removal are necessary.

Intermeshing co-rotating twin screw extruder-reactors are good examples of process-intensifying equipment, able to successfully perform reactive extrusion. They are multifunctional reactors which may integrate several process functions such as phase transition of polymeric materials, chemical reaction and heat transfer, in particular. They enable reactive processing of highly viscous materials without requiring the large amounts of solvents that stirred tank reactors need. Furthermore, extruder-reactors can achieve effective mixing and operate at high pressures and temperatures. They exhibit plug-type flow characteristics, and are capable of multistaging operations.

Reactive extrusion involves laminar mixing which generally governs the performance of the whole process, in reducing striation thicknesses and/or increasing interfacial areas to allow diffusion and hence chemical reaction to occur within the residence time of the extruder-reactor. It is thus important to define the process-intensifying capability (PIC) of the extruder-reactor, in order to rank the PIC of equipment based on the governing phenomenon of the whole process. In reactive extrusion, the governing phenomenon is definitely linked to the mixing capability of the equipment in relation to the design of screw configuration and operating parameters, in particular. Thus, the process-intensifying factors mentioned above, the quantity $(m_i.\,Rc)^2$ or WATS (which expresses the extent of shear strain applied in the laminar mixing process), can be used to define the PIC in screw extruder-reactors. As an illustration, the use of WATS

(Eq. 10.24) is proposed to estimate the PIC of an extruder-reactor, as follows:

$$PIC = \frac{(WATS_{process} - WATS_{ref})}{(WATS_{target} - WATS_{ref})} \quad (10.25)$$

where $WATS_{process}$ is WATS of the actual mixing process; $WATS_{ref}$ is WATS of a reference process – arbitrarily, the Poiseuille flow in a long circular pipe (essentially no transverse convective mixing) can be chosen ($\bar{\gamma} = 16L/3D$); and $WATS_{target}$ is the target WATS needed to achieve adequate mixing required for the reaction. An adequate mixing process implies a reduction in striation thickness by several orders of magnitude (e.g. three or four orders of magnitude), which corresponds to a total shear strain of several thousand units. $WATS_{target}$ may then be expressed as a number of times (e.g. p times) the value of $WATS_{ref}$; hence, Eq. 10.25 can now be written as:

$$PIC = \frac{(WATS_{process}/WATS_{ref} - 1)}{p-1} \quad (10.26)$$

Thus, the PIC represents a dimensionless criterion which varies from 0 (no transverse convective mixing) to 1 (adequate mixing); it can also be expressed in %. Knowing $WATS_{target}$ according to the requirements of the reacting system, p can then be selected and hence the PIC of any extruder configuration can be determined and used for optimizing and scaling up screw configuration.

10.2.2.2 Extrusion-cooking

The extrusion-cooking process deals with the physicochemical conversion of food biopolymers (starch and proteins, in particular) through the application of thermal and mechanical energy inputs. As seen in Chapter 6, appropriate thermomechanical conversion of food biopolymers requires a certain level of energy input (or SME input; see section 6.2.2.3) and proceeds according to complex reaction kinetics (see section 6.2.2.4). The characteristic time analysis presented in section 10.2.1 obviously applies to thermomechanical extrusion-cooking of food and feed mixes. The present section emphasizes the process-intensifying characteristics of extrusion-cooking through the intensification of kinetic characteristics of biopolymer cooking as well as the intensification of transfer phenomena in processing sections of extruder-cookers (in the cooking section in particular).

10.2.2.2.1 Process-intensifying characteristics

Extrusion-cooking is a fundamental generic extrusion process whose processing characteristics are in good agreement with the fundamentals of process intensification. As an approximation, although the study of biopolymer-based food mixtures subjected to extrusion-cooking is complex, the transformation of these biomaterials can be approached by multiple reactions, which can be simplified leading to two consecutive homogeneous reactions (zero- or first-order single-phase reactions), as follows:

$$A \xrightarrow{k_1} B \xrightarrow{k_2} C \quad (10.27)$$

where A is the uncooked, native food biopolymer (mainly starch or protein macromolecules); B is the converted food biopolymer resulting from order-to-disorder transitions (gelatinized/melted starch and denatured proteins, in particular); C is the cooked food biopolymer (gelatinized and partially depolymerized starch, aggregated and cross-linked proteins, in particular); while k_1 (zero-order kinetics) and k_2 (first-order kinetics) are the rate constants for the first and second reactions. As seen in Chapter 6 (section 6.2.2.4), Wang et al. (1992) have suggested that thermal and shear energies involved in the process are thermodynamically additive and interactive, hence proposing that rate constants result from a combination of two components: a thermal-induced rate constant and a shear-induced rate constant. Both components follow an Arrhenius type of relationship to account for the effects of temperature and shear stress; hence the higher the temperature and shear stress, the faster the conversion rates. In such a simplified reaction scheme which considers homogeneous reactions for cooking (Eq. 10.27), the characteristic times of the two reactions show no dependence on the characteristic dimension of the extruder-cooker (Eq. 10.5).

However, although no relevant kinetic model and quantitative data are available, it turns out that k_1 and k_2 rate constants are highly dependent on temperature and shear stress, which provide a way for cooking intensification of biopolymer-based food mixes through particular processing conditions such as high temperature and shear stress processing. In fact, the process operates at high temperature (over 110°C) and with shear energy inputs which allow the rate constants to be increased and characteristic time of cooking conversion to be substantially decreased. When compared with semi-batch, hydrothermal cooking processes which operate at lower temperatures (below 110°C) without shear energy input, the residence time of extrusion-cooking process is much shorter, leading to intensified cooking rate through high temperature, high shear and short time processing.

In addition, as shown in Eq. 10.3, characteristic times of transfer phenomena such as mass, heat and momentum transfers are proportional to the square of the characteristic dimension of extruder-cookers, providing an opportunity for process intensification through the design of screw-barrel assembly (relatively small characteristic dimension, that is the screw channel depth) and appropriate thermomechanical processing conditions. Extrusion-cooking operates at relatively low water content which generally leads to high-viscosity biopolymer-based melts. As in polymer plasticating, extrusion-cooking takes advantage of this operating condition to reinforce cooking intensification through the acceleration of temperature build-up of the biopolymeric melt in the screw channel of the metering section. After complete melting in the melting section, the biopolymer-based melts then enter into the metering section where the bulk melt temperature increases from the melting temperature to the required cooking temperature, to ensure proper physicochemical conversion of food mixes. The temperature increase may range approximately from 10°C to 40°C depending upon the extrusion-cooking process and related operating conditions, as summarized in Chapter 6 (see Table 6.3).

Hence, it is worth estimating the characteristic times of transfer phenomena in the metering section of extruder-cookers (viscous heat generation and convective heat transfer) which control the temperature build-up of the melt in the screw channel. According to Eq. 10.3, characteristic times for viscous heat generation (t_{visc}) and convective heat transfer (t_{conv}) are given by Equations 10.28 and 10.29, respectively:

$$t_{visc} = \frac{H^2}{\upsilon.Ec} \qquad (10.28)$$

$$t_{conv} = \frac{H^2}{\alpha_{heat}.Nu} \qquad (10.29)$$

where H is the characteristic dimension of the screw channel that is the height of screw flight; Ec is the non-dimensional Eckert number which expresses the relationship between kinetic energy and enthalpy of melt flow, and is used to characterize viscous dissipation (Dorfman, 2010). Ec is defined by Eq. 10.30 as follows:

$$Ec = \frac{V_f^2}{C_p.\Delta T} \qquad (10.30)$$

where V_f is a characteristic velocity of the melt: V_f can be simply defined as the linear velocity of the top of the screw flight, that is equal to $\pi.D_s.N$; C_p is the constant-pressure specific heat of the melt; and ΔT is a characteristic temperature difference of the flow that is the difference between the barrel wall temperature and the bulk melt temperature. Nu is the non-dimensional Nusselt number, which expresses the ratio of convective to conductive heat transfer in the screw channel and is defined by Eq. 10.31 as follows (Brodkey & Hershey, 1988):

$$Nu = h.H/\lambda \qquad (10.31)$$

where h and λ are convective heat transfer coefficient in the screw channel and thermal conductivity of the melt, respectively. According to melt flow in the screw channel, the theoretical value of 3.66 is assumed for estimation of the Nusselt number for fully developed laminar velocity and temperature profiles (Matlosz et al., 2009).

Accounting for typical extrusion-cooking conditions such as those classified in Chapter 6 (see Table 6.3), V_f varies approximately in the range of 0.5 $m.s^{-1}$ (small screw diameter extruder and relatively low screw speed) to 4.5 $m.s^{-1}$ (large screw diameter extruder and relatively high screw speed). Values of C_p are in the range of 2400 $J.kg^{-1}.K^{-1}$ (relatively low water content mixes) to 2600 $J.kg^{-1}.K^{-1}$ (relatively high water content mixes) (Della Valle & Vergnes, 1994). The kinematic viscosity, υ, ranges from 10^{-1} $m^2.s^{-1}$ (low-viscosity biopolymer-based melts, such as melts resulting from dry pet food and aquafeed mixes) to 1 $m^2.s^{-1}$ (high-viscosity biopolymer-based melts, such as melts resulting from directly expanded snacks and directly expanded breakfast cereals mixes) (Della Valle & Vergnes, 1994); while the heat diffusivity, α_{heat}, is approximately equal to 0.6 10^{-7} $m^2.s^{-1}$ for most extrusion-cooked biopolymer-based melts (Della Valle & Vergnes, 1994). ΔT typically varies from 10–20°C (relatively high water content and low cooking temperatures, as for dry pet food processing, aquafeed processing, etc.) to 30–40°C (relatively low water content and high cooking temperatures, as for directly expanded snacks processing, directly expanded breakfast cereals processing, etc). Thus, for most industrial extrusion-cooking conditions the Eckert number varies between approximately 5.10^{-5} (directly expanded snacks processing and directly expanded breakfast cereals processing, for instance) to 5.10^{-4} (dry pet food processing and aquafeed processing, for instance). Based on the aforementioned physical properties and processing conditions, according to Eq. 10.28, the characteristic time for viscous heat generation, t_{visc}, typically ranges from 1–5 s (relatively high-viscosity melts, high temperature, high SME processing and intermediate screw channel depth) to 15–20 s (relatively low-viscosity melts, low temperature, low SME processing and large screw channel depth). And according to Eq. 10.29, in industrial conditions the characteristic time for convective heat transfer, t_{conv}, varies approximately from 10^3 s to 6.10^3 s depending upon the screw diameter. The larger the screw diameter, the larger the screw channel depth and hence the higher the characteristic time for convective heat transfer.

The characteristic time analysis of transfer phenomena in the screw channel of the metering section clearly shows that the characteristic time of viscous heat generation is two or three orders of magnitude smaller than that of convective heat transfer. Hence, thermomechanical extrusion-cooking of biopolymer-based food mixtures is strongly intensified due to the beneficial input of viscous dissipation in the screw channel, which accelerates the temperature build-up in the cooking section and makes possible very short time extrusion-cooking. It must be noted that the ratio of the characteristic time of convective heat transfer to that of viscous heat generation leads to the modified Brinkman number, *Br*, as defined by Eq. 10.32:

$$Br = t_{conv}/t_{visq} = Pr.Ec/Nu \qquad (10.32)$$

where *Pr* is the non-dimensional Prandtl number which expresses the ratio of momentum diffusivity, or kinematic viscosity, to thermal diffusivity, that is $Pr = \upsilon/\alpha$ (Brodkey & Hershey, 1988). In the cooking section of industrial extruder-cookers, the modified *Br* would vary in the approximate range of 10^2–10^3.

It is worth recalling that thermomechanical extrusion-cooking operates according to the processing characteristics of screw extruders already discussed in section 10.2.2.1.2, such as:

- narrow residence time distribution to the benefit of homogeneous physicochemical conversion of food mixtures
- relatively high heat exchange capability to the benefit of accurate temperature control in the different processing sections of the screw-barrel assembly (the cooling sections in particular).

Considering these basic processing characteristics, it is worth reviewing the key factors which allow thermomechanical extrusion-cooking to be considered as a process-intensifying method with regard to the three main principles of process intensification (see section 10.1.4).

- Continuous processing, particularly well adapted to the cooking of high-viscosity biopolymer-based mixtures (starch and/or protein raw materials), with a high degree of flexibility.
- Multifunctional processing equipment, which results from the integration of several process functions in one extruder-cooker (such as melting, laminar mixing, cooking, venting, cooling, and product texturization). In addition, food ingredients (such as solids, liquids, and steam) can be added on line when using modular screw-barrel designs.
- Acceleration of fundamental process phenomena, through the reduction of characteristic times of the cooking rate and heat transfer, leading to an intensified method of food cooking characterized by high temperature, low water content, and short time processing. Basically, cooking intensification is made possible by use of mechanical energy input in addition to thermal energy input, leading to thermomechanical cooking as an alternative to semi-batch, non-intensified hydrothermal cooking. Furthermore, the input of shear energy in the process implies an intense laminar mixing, which allows mass transfer to be intensified when miscible or immiscible minor ingredients are added into biopolymer-based food mixtures (such as sugars, emulsifiers, lipids, dietary fibers, steam, etc.). It is worth noting that the capability of laminar mixing in screw extruders presented in section 10.2.2.1.3 and in Chapter 5 applies effectively to thermomechanical extrusion-cooking processes. That is why co-rotating twin screw extruders, which have better mixing ability, are often preferred when complex, multi-ingredients formulations are extrusion-cooked. Besides, as already mentioned in Chapter 6, a high level of mixing positively influences the quality of product texturization, and thus the quality of end-products.

Based on these factors, thermomechanical extrusion-cooking fully satisfies the principles of process intensification. Consequently, compared with semi-batch, hydrothermal cooking, processing volumes of extrusion-cooking are significantly reduced, which leads to compact and flexible processing lines, and more productive processes (higher yields of material and less energy use, and lower downtime, in particular). It must be noted that low water content cooking, or bulk cooking, enables important energy saving due to the lower amounts of water to be removed in the downstream drying process, which is always required in food biopolymer-based processing; hence processing volumes occupied by downstream unit operations of extrusion-cooking lines are also reduced significantly. Energy saving is discussed in more detail in the following section. In summary, compared with semi-batch hydrothermal cooking, intensified thermomechanical extrusion-cooking of food biopolymer-based mixtures is very competitive from an economic standpoint, with important CAPEX and OPEX reductions.

10.2.2.2.2 *Energy saving in thermomechanical extrusion-cooking*

Thermomechanical extrusion-cooking operates at relatively low water contents which require shear energy input in the cooking step; therefore, the amount of water to be removed downstream is substantially reduced, leading to significant energy saving that is one of the main beneficial aspects of extrusion-cooking. In this section, we quantify this energy saving when comparing thermomechanical extrusion-cooking (thermal and shear energy inputs at low water contents) and hydrothermal cooking (only thermal energy input at relatively high water contents) of biopolymer-based food mixes.

By physicochemically converting and texturizing food biopolymer-based raw materials (such as starch- or protein-based food mixes), the core process generally includes two key unit operations that consist of a cooking operation (such as semi-batch hydrothermal cooking or continuous extrusion-cooking) followed by a drying operation (belt drying, belt drying and toasting, fluidized bed drying, drum drying, etc.). The amount of energy used in the whole process mainly relies on the choice of processing conditions which drive cooking and drying operations. In particular, the water content and temperature of the cooking step importantly affect the energy balance of the whole process. The section gives a rough estimate of the energy consumption of an oversimplified, hypothetical cooking-drying process, while varying water content and temperature of the cooking step according to requirements to produce basic physicochemical changes in cereal-based materials, namely gelatinization and melting. Thus, let us consider a mixture of starch and water which is subjected to:

- a cooking operation with water content and temperature given by the evolution of DSC-peak temperatures when water content decreases from 60% (gelatinization mechanism) to lower values where a dual mechanism of gelatinization and melting occurs (as defined by Wang et al., 1991; Chapter 6, Figure 6.14)
- a drying operation of the cooked starch material from a water content used during the cooking operation down to a water content compatible with long-term, stable storage conditions of resulting cooked cereal-based food products (say, an arbitrary value of 5% wwb).

For such a hypothetical cooking-drying process, the specific energies (or the energy consumed for the conversion of 1 kg of dry starch) involved in the process can be estimated as follows.

For the cooking step, the specific cooking energy (SCE) is defined by:

$$SCE = \left[\frac{Q_{heat}}{(1-X_{water})} + Q_{cooking}\right] / \rho_c \qquad (10.33)$$

where Q_{heat} is the sensible heat needed to raise the temperature of 1 kg of starch-water mix from ambient temperature (say, 20°C) to the cooking temperature, T_c, hence: $Q_{heat} = \left[\bar{C}p_{mix}(T_c - 20)\right]$, where $\bar{C}p_{mix}$ is the average heat capacity of the starch-water mix in the ranges of water content and temperature considered in the process ($\bar{C}p_{mix} \cong 1500\, J.kg^{-1}.°C^{-1}$) (Della Valle & Vergnes, 1994); X_{water} is the weight fraction of water of the starch-water mix (wwb); $Q_{cooking}$ is the gelatinization/melting heat ($Q_{cooking} \cong 10\, kJ/kg$ of dry starch) (Della Valle & Vergnes, 1994); ρ_c is the energy yield of the cooking operation (non-dimensional).

For the drying step, the specific drying energy (SDE) is defined by:

$$SDE = \left[\frac{Q_{drying}}{(1-X_{water})}\right] / \rho_d \qquad (10.34)$$

where Q_{drying} is the heat needed to evaporate the water of the cooked starch-water mix from that of the cooking step down to 5% (wwb), hence: $Q_{drying} = \Gamma_{water} \cdot (X_{water} - 0.05)$, where Γ_{water} is the latent heat of vaporization for water ($\Gamma_{water} = 2257 kJ.\, kg^{-1}$); ρ_d is the energy yield of the drying operation (non-dimensional).

The global specific energy (GSE; the energy consumed for cooking and drying 1 kg of dry starch) of the hypothetical cooking-drying process is then given by:

$$GSE = SCE + SDE \qquad (10.35)$$

Energy yield of the cooking operation, ρ_c, obviously depends on the principle and design of the processing device. Hydrothermal cooking devices (such as batch stirred tank steam cookers) are generally characterized by relatively low energy yields (in the range of 40–70%) due to the significant loss of energy in transient states of cooking cycles whereas energy yields of continuous extruder-cookers may vary between 60% and 90% which accounts mainly for the power consumption for pressure build-up in the restrictive sections of the screw-barrel assembly (approximately 10% of the total power input; Rauwendaal, 2001) as well as thermal energy losses to the ambient (in the range of 10–20%). As for the energy yield of the drying operation, ρ_d, dryers generally exhibit an energy efficiency of 60–90%, depending upon their design and processing conditions.

To simplify, energy yields of 70% will be arbitrarily considered for both ρ_c and ρ_d in the calculation of SCE and SDE, as defined by Eqs 10.33 and 10.34, respectively. It turns out that SDE is much larger than SCE, particularly at relatively high water contents: as water content decreases from 60% to 15% (wwb), SDE decreases from approximately 4450 kJ/kg to 380 kJ/kg, while SCE increases from approximately 290 kJ/kg to 440 kJ/kg. At water contents of 30–15% (wwb), SCE is in the range of 400–440 kJ/kg (or 110–120 W.h/kg), which is in good agreement with the SME of most cereal-based extrusion-cooking processes, except when significant amounts of fat are included in the mixtures (such as pet food and fish feed formulations) which leads to significantly lower values of SME.

Figure 10.5 presents the evolution of the SDE, the SCE and GSE (Eq. 10.35) of the hypothetical cooking-drying process, while water content in the cooking step varies between 15% and 60% (wwb). GSE increases quasi-exponentially when the water content increases, due to the effect of SDE in particular. Hence, it can be observed that the GSE of thermomechanical extrusion-cooking (or process-intensifying biopolymer cooking) is 2–4 times lower than that of hydrothermal cooking; hence, compared with semi-batch hydrothermal cooking, thermomechanical extrusion-cooking allows significant energy savings to be obtained, and thus the operating costs of the manufacturing process are reduced significantly. Actually, energy savings combined with other important technical and economic benefits such as high process productivity, flexibility and compactness, product innovation, reduced CAPEX and OPEX have contributed to the development of thermomechanical extrusion-cooking in the cereal-processing industry during the last decades. And the increasing interest in sustainable technologies in the food-processing industry should drive its development even more strongly in the coming decades. That is to say that thermomechanical extrusion-cooking has an important potential to the detriment of conventional semi-batch hydrothermal cooking.

In practice, however, replacing hydrothermal cooking by extrusion-cooking is not straightforward, because extrusion-cooked food products often show different functional characteristics and end-use properties, owing to the shear energy input which affects the conversion of cereal biopolymers, hence effectively restraining the technology substitution. However, product issues can be overcome from a technical standpoint, through process innovation and optimization, and successful process substitution of conventional hydrothermal cooking by thermomechanical extrusion-cooking can then be obtained. It is worth mentioning the following examples of successful substitution: crispy flat bread (as an alternative to Scandinavian flat bread, or toasted "biscottes" and breads), extrusion-cooked cereal flakes (as an alternative to hydrothermally cooked flakes), extrusion-cooked cereal flours for instant drinks (as an alternative to hydrothermally cooked and drum-dried cereal flours), in particular.

10.2.2.3 Solid-liquid extrusion-pressing

The bioprocessing industry has numerous applications involving solid-liquid mixtures with components that

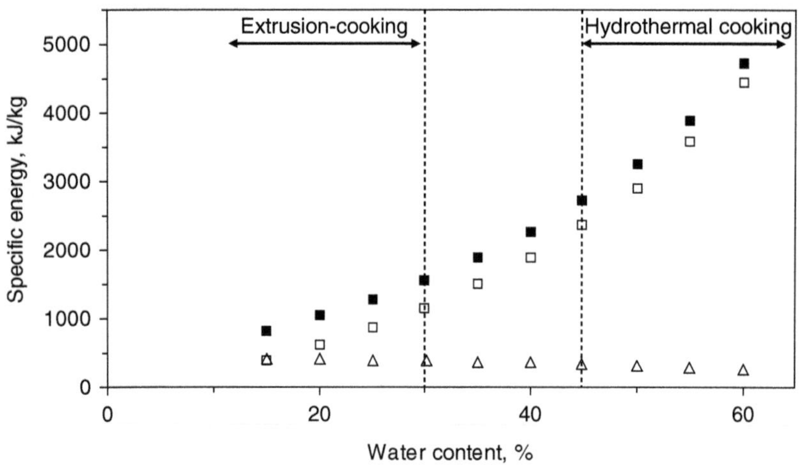

Figure 10.5 Specific energies versus water content for an arbitrary cooking-drying process for the conversion of starch-based food mixes (□: specific drying energy; △: specific cooking energy; ■: global specific energy).

need to be separated. One method of separation consists of the introduction and thorough mixing of a new phase such as a liquid or a solvent, allowing the components of the original raw material to distribute between the original phase and the newly created phase. The components to be separated are distributed between the phases in accordance with the equilibrium coefficients which give the relative concentrations in each phase once equilibrium has been reached. The two phases are then separated by simple physical methods such as gravity settling (or draining) or centrifugation. This process of contacting the two phases can be carried out in a batch processing mode through a single contacting stage or continuously through a series of contacting stages to enhance the separation. Examples of contact-equilibrium separation, or contact-equilibrium extraction, are found in the edible oil industry in which oil is extracted from oilseeds by use of a solvent in which oil, but not the other components of the seeds, is soluble. This is the case for soybean oil, where flaked soybeans are extracted with a solvent such as hexane. Examples of contact-equilibrium separation are also found in the sugar industry where sucrose is removed by water extraction from sugar cane or sugar beet, in the pharmaceutical industry where various active constituents are solvent-extracted from natural plants, and in the paper milling industry where wood or non-wood cellulosic fibers are water-washed to remove residual chemicals. Factors controlling the separation are the area of contact between the phases, the time of contact, the properties of the materials (which determine the equilibrium distribution of the transferred component), and the number of contacting stages. In solid extraction or washing, the transfer of the desired component may be governed by diffusion, hence controlling the rate of extraction or washing.

Contact-equilibrium separation processes are solvent and energy consuming due to the large volume of solvent required for an effective yield of extraction or washing, and the energy needed to recycle or eliminate the solvent; in addition, long processing times and large volumes are required. Hence, solvent-aided solid-liquid separation units are characterized by relatively low productivity and high CAPEX and OPEX. When the solid phase is the value-added component, another method of separation aims at simply evaporating the liquid. Examples are found in the agro-food industry and in water waste processing plants where dewatering of diluted by-products is required (sugar beet pulp, vegetable pulps, fruits, waste water sludges, etc.). However, water evaporation from solid materials is also highly energy and time consuming. Consequently, it is worth intensifying solid-liquid separation processes in order to reduce significantly the use of extracting solvents, reduce energy consumption, and enhance process productivity.

Thermomechanical expression of solid-liquid systems using extrusion-pressing technology enables effective intensification of solid-liquid separation to be carried out; hence it can be considered as a process-intensifying method for solid-liquid separation. Thermomechanical expression of solid-liquid mixtures is described in Chapter 8, as a fundamental generic extrusion process able to separate continuously solutions of solutes or liquids from solid materials. It allows concentrated liquids or solutions to be separated in a liquid state, hence avoiding energy-consuming separation methods (such as distillation or evaporation). So far, two main process options are effectively exploited at the industrial level.

- Direct extrusion-pressing of solid materials (or one-stage extrusion-pressing) to express liquids from solid matrices (edible oil from oilseeds, liquid water from agricultural by-products, juices from fruits, for instance) without solvent. In general, the resulting expressed cake still contains significant amounts of residual liquid which needs further separation in downstream processing units, through either solvent extraction (for instance, hexane extraction of edible oil from oilseeds) or drying (for instance, water removal from agricultural by-products). This process option is extensively used in industry to eliminate most of the liquid component and hence to minimize as much as possible the extent of downstream energy-consuming evaporation.
- Solvent-aided extrusion-pressing of solid materials to dissolve solutes and express the resulting solution from solid matrices (for instance, water-aided washing of alkaline-digested lignocellulosics and bleached cellulose fibers). Several integrated washing sections can be set up in series on the extruder barrel, depending upon the required final concentration of solutes. This process option is much less used in industry though it offers an important capability for process intensification.

The purpose of this section is to present and discuss the process-intensifying characteristics and capabilities of the extrusion-pressing method.

10.2.2.3.1 Process-intensifying characteristics

As presented and discussed in Chapter 8, the mechanism of solid-liquid expression shows variable complexity depending upon the nature of the materials to be processed. For instance, with wet mineral cakes which consist of incompressible solid particles saturated with liquid, the mechanism of solid-liquid expression is only controlled by the reorientation of individual solid particles under the effects of pressure; hence, the filtration/consolidation theory for soils applies satisfactorily, while accounting for two periods of consolidation that are the primary and the creep (or secondary) consolidation (Shirato et al., 1986). But with highly structured materials such as fibrous or cellular materials, which consist of compressible solid particles, the mechanism of solid-liquid expression is much more complex. With such structured materials, liquid expression is controlled by the reorientation of the liquid-containing particles and the evolution of the particle microstructure under the effect of pressure; hence, a four-step consolidation model describes expression performances to account for the effect of inter-particle, extracellular and intracellular volumes on liquid flow (Lanoisellé et al., 1996).

Consolidation periods of solid-liquid expression are governed by characteristic times. For instance, for constant-pressure solid-liquid expression, the characteristic time of the primary consolidation is defined by h_o^2/b (Vorobiev & Lebovka, 2006) where h_o is the characteristic dimension of the cake (or the initial thickness) and b is the consolidation coefficient whose dimension is equivalent to the dimension of $m^2.s^{-1}$. The parameter b depends on the structure of the material to be processed; with fibrous or cellular raw materials, physical as well as chemical pretreatments (such as mechanical crushing or grinding, thermal heating or freezing-thawing, enzymatic or chemical hydrolysis, pulsed electric field treatment, etc.) allow cellular structures to be denatured and the characteristic time of the primary consolidation to be reduced, facilitating further liquid expression (Bouzrara & Vorobiev, 2003; Grimi et al., 2010). At relatively low loading pressure (below 10 bar), typical values of the consolidation coefficient of pretreated fibrous or cellular materials should be in the range of 10^{-6}–10^{-9} $m^2.s^{-1}$, depending upon the raw material and the type of pretreatment. In continuous solid-liquid expression, accounting for the values of b, a relatively low filter cake thickness is required so that reasonably short characteristic times can be obtained (say, a few seconds). Besides, the consolidation coefficient is pressure dependent, as an increase of the loading pressure leads to higher values of b; b may increase from one to three orders of magnitude when loading pressure increases from 1 to 80 bar, depending upon the pretreated materials.

Screw extrusion-pressing is an efficient generic method for industrial application of solid-liquid expression. Presently, single screw extruder-presses are more extensively used than twin screw extruder-presses and most industrial applications use direct extrusion-pressing of solid materials (or one-stage extrusion-pressing) with no use of solvent. In a single screw extruder-press, the solid-liquid mixture is transported and continuously crushed (or macerated) by the rotating screw in the barrel. The screw design determines the filter cake thickness in the active section of the screw (that is, the compression section with restrictive screw elements up to the die), and hence the extent of pressure build-up which causes the liquid to be expressed from the solid material along the filtering barrel. Consequently, the screw design combined with the porosity-pressure relationship, which reflects the consolidation behavior of the solid material, are of paramount importance as they determine the expression performances of the extrusion-pressing process. Operating parameters such as screw speed and temperature also affect expression performance, as they define the pressing time and the liquid viscosity, respectively.

Based on the processing characteristics of solid-liquid expression in screw extruder-presses, it is worth reviewing the key factors which allow screw extrusion-pressing to be considered as a process-indentifying method with regard to the three main principles of process intensification (section 10.1.4).

- Continuous processing, particularly well adapted to mechanical expression of structured solid-liquid mixtures such as oilseeds, fruits, pulps, waste water sludges, etc.
- Multifunctional processing equipment, which results from the integration of several process functions in one screw extruder-press (such as solid crushing or maceration, continuous homogenizing of the solid-liquid mixture and texturization of the press cake).
- Acceleration of fundamental process phenomena, through the important reduction of the characteristic time of primary consolidation by affecting favorably both consolidation coefficient, b, and cake thickness, h_o. On one hand, effective crushing and maceration of solid material during mechanical expression frees the liquid from the solid owing to the shearing effect in the screw channel; in addition, continuous mixing of solid-liquid mixture in the rotating screw channel reduces concentration gradients and breaks up the filter cake. Screw crushing and mixing help to increase the consolidation coefficient, b, and hence to decrease the characteristic time of primary consolidation. On the other hand, due to the design of the screw-barrel assembly, the characteristic dimension of the pressing section is small (of the

order of millimeters) which helps to keep the filter cake thickness small. Hence, both process effects allow high expression yields with short residence times.

Based on these factors, thermomechanical extrusion-pressing fully satisfies the principles of process intensification. Compared with conventional processes (such as continuous solvent-aided solid-liquid extraction as for oil extraction from oilseeds by use of hexane, or continuous thermal drying as for sugar beet pulp), processing volumes and times of extrusion-pressing are greatly reduced, which leads to more compact and productive processes. Furthermore, specific energy consumption and solvents are significantly reduced, leading to safer and cleaner processes. Consequently, intensified extrusion-pressing allows important reductions in CAPEX and OPEX of solid-liquid separation processing units.

10.2.2.3.2 Process-intensifying capability of solid-liquid separation and washing

As mentioned in the introduction to section 10.2.2.3, conventional contact-equilibrium solid-liquid separation processes, which generally consist of separating a desired component from a solid matrix by use of an appropriate solvent, are characterized by high consumption of solvent and energy. Similarly, conventional washing processes (batch or continuous), which generally consist of removing an unwanted component from a solid-liquid mixture by use of washing solvent or water, also consume large quantities of solvent and energy. For both solid-liquid extraction and washing processes, real progress towards process intensification has already been made, particularly when implementing continuous counter-current contact processing which provides significant solvent and energy savings. However, performance of continuous processing can be enhanced through process-intensifying extrusion-pressing methods. Hence, it is of paramount importance to analyze the PIC of thermomechanical extrusion-pressing for solid-liquid separation and washing processes.

For such an analysis, a hypothetical multistage contacting process is considered, as illustrated by the generalized system shown in Figure 10.6, in which there are *n* contacting stages in series and two contacting phases: the solid phase and the liquid phase. Each contacting stage (or processing stage) of the process generally aims at mixing the solid and liquid phases followed by phase separation (through settling, draining, centrifugation or mechanical expression) prior to passing to the next stage. The analysis of the process depends upon establishing the equilibrium condition, that is the distribution of the required solute between the solid and liquid phases, and the operating conditions. Considering the extraction of a solute from a solid matrix, it is assumed that the whole of the solute is dissolved in the liquid phase in one stage, which accomplishes the desired separation. However, it is not possible then to separate all of the liquid from the solid phase because some solution is retained with the solid matrix and this solution contains solute. As the solid retains solution with it, the content of solute in this retained solution must then be progressively reduced by a series of contacting stages. In general, the equilibrium conditions are simple when it is assumed that the concentration of solute is the same in the solution that can be separated as it is in the solution that remains with the solid matrix. As for the operating conditions, as long as the same quantity of solid material passes from stage to stage, and the same quantity of liquid is retained by the solid matrix after each solid-liquid separation, the analysis is straightforward as the flow rates of liquid and solid phases are the same in all contacting stages.

In the first contacting stage (see Figure 10.6), the solid phase entering into the process consists of the solid matrix with the maximum concentration of solute to be extracted or washed while the liquid phase entering into the process

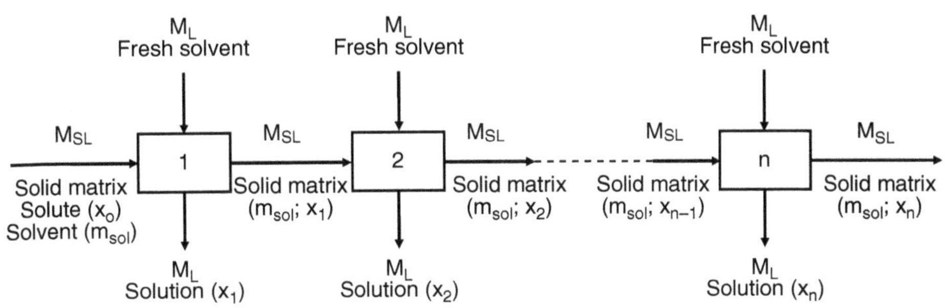

Figure 10.6 Hypothetical multistage contacting process.

is fresh solvent. In the other stages of the process, the solid phase (M_{SL}) which enters into the contacting stage is composed of the solid matrix with the retained solution (m_{sol}: solvent with solute) while the liquid phase (M_L) which enters into the contacting stage is fresh solvent; an equal amount of fresh solvent is used for each stage. The process function of the contacting stages is to continuously mix the solid and liquid phases to allow the solute to be transferred from the solid phase to the liquid phase, and drain the resulting solution concomitantly. A constant proportion of solution (m_{sol}) is retained by the solid phase in each stage and consequently, a constant amount of solution (equal to M_L) is drained from each stage. In the first contacting stage, it is assumed that an amount of fresh solvent equivalent to m_{sol} is virtually included into the solid phase M_{SL}, so that material balances of all contacting stages can be treated identically. According to the equilibrium conditions and operating conditions discussed previously, the mass flow rates of the solid phase (M_{SL}) and the liquid phase (M_L) are the same in all contacting stages. In this hypothetical process, it is assumed that the whole of the solute is dissolved in the first stage. Because the solid matrix retains solution with it, as the extraction or washing operation proceeds from stage 1 to stage n, the mass solute concentration (x_o, x_1, x_2,..... x_n, weight fraction) in the solid phase decreases progressively down to the process requirement.

In the first contacting stage, based on the aforementioned assumptions, the solute which was formerly contained in the solid phase (solid matrix + m_{sol} solvent) is transferred into a larger volume of solvent equal to $m_{sol} + M_L = m_{sol}(k+1)$, where k is the ratio of fresh solvent added ($k = M_L/m_{sol}$). Then, the mass concentration of solute falls by the ratio of solvent volumes, that is to $x_o[m_{sol}/m_{sol}(1+k)]$, which is the mass concentration of solute (x_1) in the two streams exiting the first contacting stage (that are the drained solution and the solution retained with the solid phase):

$$x_1 = x_o\left[\frac{1}{1+k}\right] \qquad (10.36)$$

In the two streams exiting the second contacting stage, the mass concentration of solute becomes:

$$x_2 = x_o\left[\frac{1}{1+k}\right]^2 \qquad (10.37)$$

And finally, when exiting stage n of the multistage extraction or washing process, the mass concentration of solute is given by:

$$x_n = x_o\left[\frac{1}{1+k}\right]^n \qquad (10.38)$$

The yield of extraction or washing of a contacting stage process, ρ_s, is expressed as the ratio of the amount of extracted solute, or washed solute, to the initial amount of solute present in the raw solid material:

$$\rho_s = \frac{x_o - x_i}{x_o} \qquad (10.39)$$

where x_i is the mass concentration of solute in contacting stage i (with $i = 1, 2,.... n$); hence, ρ_s should vary from 0 (raw solid) to 1 (solute-free solid, through complete extraction or washing). The performance of an extraction or washing process having n contacting stages will be:

$$\rho_s = 1 - \left[\frac{1}{1+k}\right]^n \qquad (10.40)$$

Hence, the performance of the hypothetical multistage contact-equilibrium extraction or washing process increases when the number of contacting stages, n, and the ratio of fresh solvent, k, increase.

Figure 10.7 shows the relationship between the process yield (ρ_s) and the fresh solvent ratio (k) for a number of contacting stages varying from 1 to 4. It clearly shows that to achieve the same process yield, continuous multistage contacting process consumes less solvent than single batch contact processing. For instance, an extraction or washing yield of 90% ($\rho_s = 0.90$) is obtained with $k = 9$ and $k \cong 1.15$ for one and three contacting stages, respectively. Thus, significant process intensification is obtained when increasing the number of contacting stages, knowing that the total amount of solvent needed for extracting or washing 90% of the solute contained in raw solid is proportional to k through the quantity $n.\,k.\,m_{sol}$. Hence, it is worth attempting to express the PIC of an actual contact-equilibrium process with regard to the impact on solvent consumption, as follows:

$$PIC = \frac{Ms_{ref} - Ms_{process}}{Ms_{ref}} \qquad (10.41)$$

where, for a given process yield ρ_s, Ms_{ref} and $Ms_{process}$ are the quantities of solvent consumed with a reference process and the actual extraction or washing process, respectively. Suppose the reference process refers to a single batch contact extraction or washing process with gravity settling (or draining) associated with a standard solid

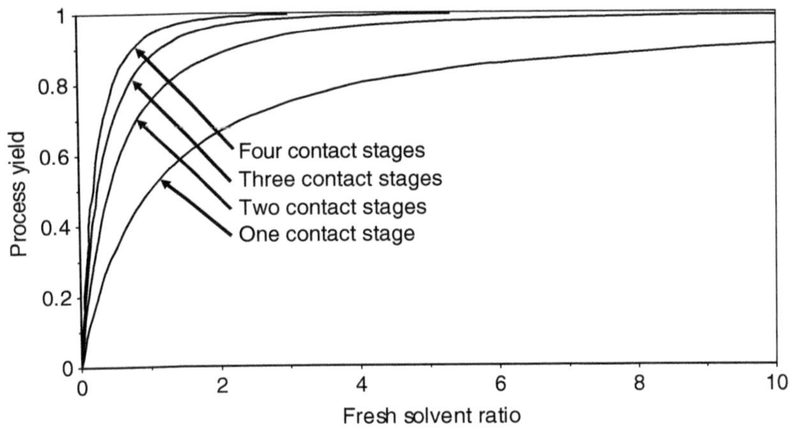

Figure 10.7 Process yield (ρ_s) versus fresh solvent ratio (k) for conventional contact-equilibrium extraction or washing.

material (for instance, a native solid material which has been subjected to standard milling or grinding procedure) from which a solute is extracted, Eq. 10.41 then becomes:

$$PIC = \frac{m_{sol}.k_{n=1} - m_{sol}.(n.k)_{n\geq 1}}{m_{sol}.k_{n=1}} = \frac{k_{n=1} - (n.k)_{n\geq 1}}{k_{n=1}} \quad (10.42)$$

where $k_{n=1}$ and $(n.k)_{n\geq 1}$ are given for an equal process yield, ρ_s. An actual process will consist of intensifying solute extraction or washing through multistaging contact-equilibrium with gravity settling (or draining). The PIC is a non-dimensional criterion which generally varies from 0 (non-intensified, single batch contacting stage process) to 1 (ideal, solvent-free processing); it can also be expressed in %. Figure 10.8a presents the evolution of the PIC as a function of process yield (ρ_s), when the number of contacting stages of the actual extraction or washing process varies from 2 to 4. For instance, for a process yield of 90% ($\rho_s = 0.90$), the PIC of a continuous three stages process will be 61.5%. Figure 10.8a shows that the PIC increases significantly when moving from two to three stages in series but the fourth contacting stage does not bring much PIC, compared to the three first stages.

Process intensification of multistage contact-equilibrium extraction or washing can be further enhanced through screw extrusion-pressing. This is possible by use of a screw extruder-press whose design consists of one or more contacting stages. A contacting stage is a section of the screw-barrel assembly with a configuration that is specially designed (appropriate combination of right-handed pitch screw elements and restrictive screw elements, filtering barrel) for ensuring continuous solid-liquid mixing, liquid expression, and draining. Contact-equilibrium extraction or washing through screw extrusion-pressing takes advantage of the shearing and pressure build-up effects of the screw profile to enhance processing yields in short processing times. Shear- and pressure-aided contact-equilibrium extraction or washing then allows the proportion of the solution retained by the solid phase (m_{sol}) to be reduced significantly, and the extraction or washing performance to be increased for a given amount of solvent.

In screw extrusion-pressing, the proportion of solution retained by the solid phase is determined by the porosity change of the filter cake as a function of pressure. For instance, Vadke and Sosulski (1988) established an experimental porosity-pressure relationship for oil expression of canola seeds processes in a single screw extruder-press without solvent. The authors observed that the porosity of the cake decreased exponentially when pressure increased, and they found residual oil contents of approximately 10% (weight basis) in the press cake. Thus, in screw extrusion-pressing of oilseeds, the liquid retained by the solid phase may be as low as 10% (weight basis), or even lower when processing with a solvent whose viscosity is significantly lower than that of vegetable oil. It must be noted that in conventional solvent-aided contact-equilibrium extraction of oil from oilseeds, the solution retained by the solid phase is typically in the range of 30–45% (weight basis), depending upon the protein matrix. Hence, compared with conventional contacting stage extraction, screw pressing extraction will allow an important reduction of the liquid retained by the solid phase. Westenbroek (2000) found similar results when studying the compressibility behavior of non-wood cellulosic fibers (such as hemp fibers) by use of a uniaxial

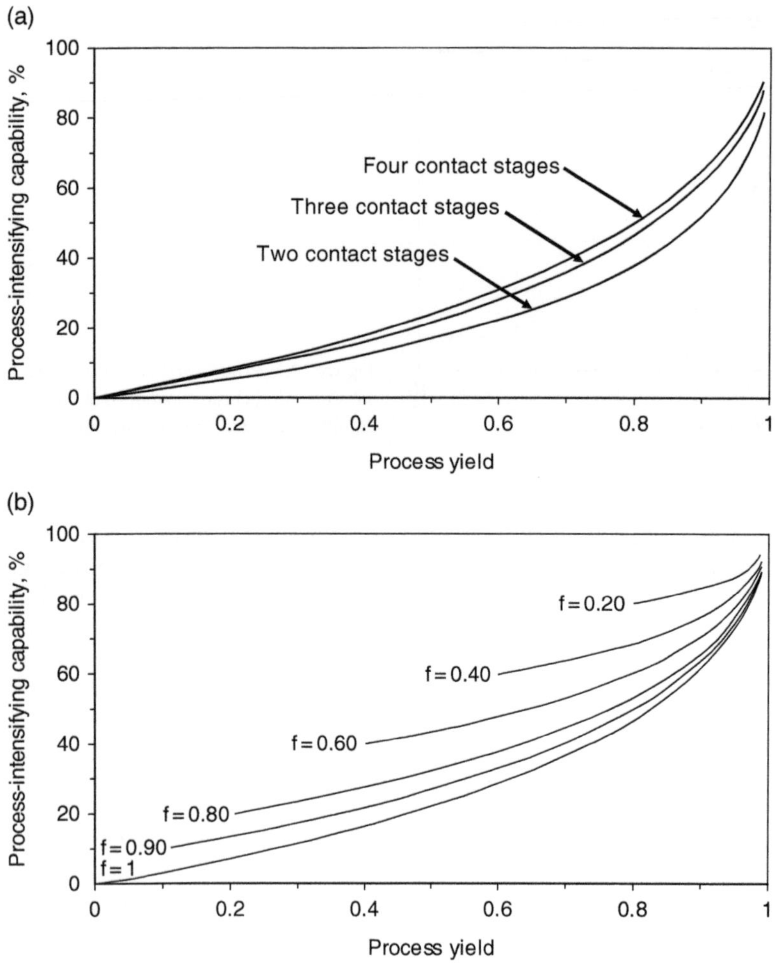

Figure 10.8 (a) Process-intensifying capability (*PIC*) as a function of process yield (ρ_s) for conventional contact-equilibrium extraction or washing. (b) Process-intensifying capability (*PIC*) as a function of process yield (ρ_s) for a three-contact stages extraction or washing process ($0.2 \leq f \leq 1$).

compression cell. With hemp bast fibers, after several hours soaking in water, the author observed that the residual liquid content decreased from 70–75% to 30–35% (weight basis) when the applied pressure increased from atmospheric pressure to approximately 6 MPa. Though there are very few relevant data on the porosity-pressure relationship of structured solid materials in extrusion-pressing, the pressure-induced reduction of the cake porosity and hence of the liquid retained by the solid phase is very important. Based on the aforementioned examples, this reduction may range from 10% to 80% depending upon the physical characteristics of the solid matrix and the pressure applied in the expression process.

For contact-equilibrium extraction or washing through screw extrusion-pressing, a non-dimensional factor is introduced to account for the reduction of the solution retained by the solid phase in a given contacting stage *i*, and consequently for the reduction of the amount of fresh solvent added into the contacting stage. Let f (with $f \leq 1$) be this non-dimensional factor: the lower is f, the higher the reduction of cake porosity and the lower the amount of solution retained by the solid phase. Assuming that pressing conditions are identical for all contacting stages (then f is constant), the stream flows entering into a given contacting stage i will consist of a solid phase (that is, the solid matrix plus $f.m_{sol}$ retained solution with a solute concentration of x_{i-1}) and a liquid phase (that is, M_L fresh solvent, with $M_L = f.m_{sol}(1+k)$). And the stream flows exiting stage i will consist of a solid phase (that is, the solid matrix plus $f.m_{sol}$ retained solution with a solute concentration of x_i)

and a liquid phase (that is, M_L drained solution with a solute concentration of x_i). Based on these assumptions, the mass solute concentration in the drained solution that exits contacting stage i is then given by: $x_i = x_{i-1}[1/(1+k)]$. Eqs 10.36–10.38 are still valid for a contact-equilibrium solid-liquid extraction or washing process carried out in a screw extruder-press; hence the mass solute concentrations in the drained solution will be pressure independent. However, the process yield and PIC are f-dependent as they are both significantly enhanced when processing in a screw extruder-press, due to the reduction of fresh solvent required for each contacting stage as opposed to that needed for the reference process (single batch contact process). Eq. 10.40 and Eq. 10.42 then become:

$$\rho_s = 1 - f\left[\frac{1}{1+k}\right]^n \tag{10.43}$$

$$PIC = \frac{m_{sol}.k_{n=1} - f.m_{sol}.(n.k)_{n\geq 1}}{m_{sol}.k_{n=1}} = \frac{k_{n=1} - f.(n.k)_{n\geq 1}}{k_{n=1}} \tag{10.44}$$

Equations 10.43 and 10.44 are general relations which allow the process yield and PIC of contact-equilibrium solid-liquid extraction or washing process to be estimated. The non-dimensional factor f accounts for a process parameter which may affect the amount of liquid retained by the solid phase during phase separation in the extraction or washing process. As discussed earlier, f is strongly affected by pressure but it will also be affected by structural modifications of the solid material which may result from physicochemical changes through preprocessing (such as thermal heating, microwaving, freezing-thawing, chemical or enzymatic modifications, etc.), and consequently modify phase separation compared to that obtained with a standard milling or grinding process. Hence, f reflects the draining behavior of the solid-liquid mixture and defines the proportion of liquid retained by the solid phase: $f = 1$ corresponds to the conventional single batch contact-equilibrium solid-liquid extraction or washing with a standard solid material while $f < 1$ corresponds to process parameters (such as pressure and/or solid preprocessing) which enhance liquid separation from the solid matrix. According to Eq. 10.44, the PIC varies from $(1 - f)$ to 1. It will be equal to zero when operating in a single batch contacting stage process with $f = 1$ (that is, the reference process). But the PIC will take values higher than zero when operating in a single continuous contacting stage process with $f < 1$ (for instance, by use of screw extrusion-pressing method) with no addition of fresh solvent in the contacting stage ($M_L = 0$). In that case, the PIC benefits from process enhancement aimed at decreasing the amount of liquid retained in the solid matrix, hence leading to collection of an amount of solution equal to $(1 - f).\ m_{sol}$.

Figure 10.8b presents the evolution of the PIC as a function of process yield (ρ_s), for a three contacting stages extraction or washing process, with f ranging from 0.2 to 1. As expected, the PIC increases when process yield increases and when f decreases, which means increases of liquid expression in each contacting stage, leading to important process intensification and higher process performance. It is worth pointing out the effectiveness of the solid-liquid extrusion-pressing option ($f < 1$), that is comparing it with the conventional solid-liquid separation ($f = 1$) by use of the PIC difference between the two process conditions (say, ΔPIC). Figure 10.8b shows that ΔPIC decreases progressively when process yield increases; the relative decrease of ΔPIC becomes important at high process yields ($\rho_s > 0.9$). Hence, the PIC of the screw extrusion-pressing option is less effective at high process yields (say, above 0.9) and the benefit of the extrusion-pressing option at process yields higher than 0.95 is small. This result should be accounted for when designing a solid-liquid extraction or washing process.

10.2.2.4 Benefits of process-intensifying methods: industrial examples

As seen in sections 10.2.2.1 through 10.2.2.3, extrusion process-intensifying methods allow process performance to be greatly enhanced compared with conventional, non-intensified processes (batch or semi-continuous, in particular). The purpose of this section is to describe operational benefits from an industrial standpoint. Hence, operational characteristics of intensified extrusion processes will be compared with those of conventional processes, while considering industrial applications which have been described extensively in previous chapters.

Tables 10.3a and 10.3b include the benefits of the reactive extrusion method while considering two different examples: the well-known casein-to-caseinate processing (Table 10.3a; see also Chapter 5, section 5.3.2) and an innovative alginate alkaline extraction (Table 10.3b; Vauchel et al., 2008, 2009). In casein-to-caseinate processing, Table 10.3a shows clearly that continuous reactive extrusion improves the efficiency of the process in terms of handling high solid concentrations and short

Table 10.3a Casein-to-caseinate processing: reactive extrusion versus conventional batch processing.

Process performances	Conventional processing (batch processing)	Extrusion processing (reactive extrusion)
Processing conditions:		
• Solids concentration, % (wwb)	18–25	75–82
• Temperature, °C	70	120–130
• Processing time, min	30–120	<5
Operational benefits:		
• Relative energy consumption	100	≅ 20
• Process flexibility	+	+++
• Relative footprint	100	≅ 50
• Relative capital cost	100	≅ 50

Source: Data from Chapter 5, section 5.3.2.

Table 10.3b Alginate alkaline extraction: reactive extrusion versus conventional batch processing.

Process performances	Conventional processing (batch processing)	Extrusion processing (reactive extrusion)
Extraction yield, %	33 ± 2	39 ± 2
Reagent consumption, kg/kg dry algae	0.5	0.2
Water consumption, l/kg dry algae	25	10
Processing time, min	60	5
Purity, %	97 ± 1	96 ± 1
Intrinsic viscosity, l/g	0.3 ± 0.05	0.8 ± 0.1
Average molecular weight, Da	36 000	105 000

Source: Adapted from Vauchel et al. 2008 and 2009. Reproduced with permission of Elsevier.

processing time, that result in significantly CAPEX and OPEX savings. As for alginate alkaline extraction, the conventional process consists of five steps: acidification, batch alkaline extraction, solid-liquid separation, precipitation, and drying. Batch alkaline extraction is the main step, as it generally requires high quantities of chemical reagent (Na_2CO_3) and water, as well as long processing times (several hours) to achieve optimum extraction yield. It must be noted that long processing time may negatively affect the rheological properties of the extracted alginates; thus, reducing processing time would lead to alginates with enhanced quality.

Hence, Vauchel et al. (2008, 2009) replaced batch extraction by a continuous reactive extrusion process, in order to enhance process performance and product quality; alkaline extraction of alginate through reactive extrusion is presently under development, and yet not industrialized. Based on data obtained on a laboratory scale twin screw extruder, Table 10.3b indicates the benefits of reactive extrusion with regard to process productivity (improved extraction yield, short processing time), consumables savings (chemical reagent and water), and product quality (purity of alginic acid and average molecular weight). When compared with conventional batch extraction, it is observed that reactive extrusion allows higher productivity to be obtained, and much less reagent and water to be used. Purity of alginic acid is equivalent for both processes. Alginates extracted by reactive extrusion show much higher intrinsic viscosity and average molecular weight: both properties are approximately three times larger than those resulting from conventional batch processing. This is particularly due to the lower processing time in the reactive extrusion process which significantly reduces depolymerization, leading to higher alginate quality. According to these preliminary data from a laboratory-scale unit, cost savings should be expected when scaling up the new process, hence favoring its industrial implementation.

Table 10.4 points out the benefits of a fully continuous extrusion-cooking process in terms of processing time, solid concentration when compared to the semi-continuous conventional process, processing conditions and process performance for the production of cereal flakes. Conventional cereal flakes processing generally consists of six or seven key processing steps: batch steam cooking of coarse grain grits (hydrothermal cooking for approximately 2 hours, with water content ranging from 35 to 40% wwb), drying, tempering (moisture equilibration within grain particles for approximately 8 hours), flaking, toasting/drying, possibly sugar coating/drying, and cooling. The pellet-to-flake extrusion-cooking process was presented in Chapter 6 (see section 6.3.1.2). As extrusion processing operates at higher solids content and temperature, a great reduction of the whole processing time is observed. Consequently, extrusion processing lines for cereal flakes show much better

Table 10.4 Cereal flakes processing: extrusion processing versus conventional processing.

Process performances	Conventional processing	Extrusion processing
Cooking conditions:		
• Solids concentration, % (wwb)	60–65	74–78
• Temperature, °C	100–115	130–145
Whole processing time, h*	≅ 12	≅ 0.25
Operational benefits:		
• Relative energy consumption	100	55–70
• Process/product flexibility	+	+++
• Relative footprint	100	≅ 50
• Relative capital cost	100	≅ 50

* Processing time from raw material feeding in cooking unit operation to final product cooling (before packaging).

Table 10.5 Pulping of non-wood fibers: extrusion processing versus conventional processing.

Process performances	Conventional processing	Extrusion processing
Cooking conditions:		
• Solids concentration, % (wwb)	10–20	≅ 35
• Temperature, °C	140–160	≅ 90
Whole processing time, h	20–24	≅ 2.5
Operational benefits:		
• Fiber pulp yield, %	≅ 50	76–82
• Relative energy consumption	100	60–80
• Relative	100	70–80
consumption		
of chemicals	100	≅ 50
• Relative water		
consumption	+	+++
• Process flexibility	100	≅ 50
• Relative footprint		

performance with regard to process productivity and energy consumption, as well as unit compactness. In addition, an extrusion cereal flakes processing line shows much higher process and product flexibility, as it is able to process various types of formulations (large variety of starch-based raw materials, hence allowing cost of formulations to be minimized) and to produce an extensive range of cereal products (such as traditional corn flakes, oat flakes, multigrain flakes, flakes with inclusions, etc., as well as directly expanded cereals and co-filled cereals). Process and product flexibility is highly beneficial for cereals manufacturers for business development.

Thus, extrusion processing is definitely more advantageous from an economic standpoint (lower CAPEX and OPEX), compared to conventional cereal processing. But, as seen in Chapter 6 (section 6.2.2), starch cooking in conventional cereal processing and extrusion cereal processing is significantly different. In fact, conventional cereal processing is characterized by hydrothermal cooking (thermal energy input at relatively high water content), where starch is fully gelatinized with negligible dextrinization whereas extrusion cereal processing is characterized by thermomechanical cooking (thermal and mechanical energy inputs at relatively low water content), where starch dextrinization may be significant depending upon the processing conditions. As starch dextrinization affects the physical and textural characteristics of the resulting cereal products, the sensory profile of extrusion-processed cereal flakes may show perceivable differences when compared with hydrothermally cooked cereal flakes.

Table 10.5 shows the benefits of a fully continuous extrusion pulping process compared with the conventional semi-continuous pulping approach, based on process conditions and performances presented in Chapter 5 (section 5.3.3). It is worth mentioning that a twin screw extrusion pulping process uses the two following extrusion process-intensifying methods: reactive extrusion (impregnation of chemicals) and solid-liquid extrusion-pressing (pulp washing). Thus, compared with conventional processing, process conditions and performance of extrusion processing benefit from both process-intensifying methods: higher process productivity (important increase of fiber pulp yield, much shorter processing time), consumables savings (significant decrease of the consumption of chemical reagents, energy and water), process flexibility (ability to process a large range of raw materials), and unit compactness, leading to environmentally friendly pulping of lignocellulosics.

10.2.3 Sustainability of extrusion processing technology

From the discussion on extrusion process-intensifying methods and the examples given, it is worth concentrating on the potential of extrusion processing technology for sustainable development. In section 10.1.3, some examples of sustainable technology were identified with regard to their relevance to extrusion processing. In the following section, the potential of extrusion processing for sustainable development is commented upon.

10.2.3.1 People/social dimension

10.2.3.1.1 Provide for the needs of the poor

Extrusion processing technology is mainly concerned with processes which lead to either consumer products (such as extrusion-textured food products resulting from GEP III) or intermediate products (such as extrusion-pressed agro-food materials resulting from GEP IV, or extrusion-processed fiber pulps resulting from the association of GEP II and GEP IV). According to their process-intensifying characteristics, GEP III and IV are able to provide food in poor economies or developing countries. In fact, in these countries very robust single screw extruder-cookers are already used to produce precooked flours from local raw materials (e.g. Moussa et al., 2011). Also, low-capacity single screw extrusion-pressing machines are used to extract vegetable oil from local oilseeds. However, very compact and flexible miniaturized extrusion processing lines need to be designed in order to convert local raw materials (starch-rich and protein-rich raw materials such as wheat, corn, cassava, soy, etc.) into a wide range of energy-dense and nutritive textured food products, while minimizing the consumption of utilities, thus helping to feed poor populations and encouraging integrated microeconomies in developing countries. In addition, compact and flexible extrusion-pressing units must be modernized to fit with the future needs of developing countries, to be able to process various agro-food raw materials (solid-liquid fractionation) and contribute to the development of basic food and feed products. The integrated twin screw extrusion pulping process (refer to Chapter 5, section 5.3.3) is particularly well adapted to provide printing-writing papers in poor economies from local non-wood raw materials, owing to its highly intensifying character (high fiber pulp yield, low consumption of utilities such as energy, water and chemicals, clean processing, low footprint, in particular), thus enhancing the education and culture of poor populations in developing countries.

10.2.3.1.2 Social acceptance; safety; smell; occupational health

Social acceptance of extrusion processing plants is not an issue in the food processing industry. In other industries, continuous extrusion processing technology allows existing or new processes to be intensified, which effectively raises the social acceptance of the processing plants they are integrated into. For instance, in chemical plants where extrusion technology is eligible, processing volumes can be reduced considerably, thus reducing drastically the size of the plant as well as the amount of chemically hazardous, reactive material and thus the risks of explosion. Diffusive emissions of volatile components are also reduced concomitantly, hence reducing smell. In that context, social acceptance and safety of reactive processes can be improved significantly by the introduction of reactive extrusion. Twin screw extrusion pulping processing of cellulose pulps provides a relevant example. Another example would be the processing of high-energy materials (such as explosives, propergols, etc.) where the use of continuous extrusion mixing units as an alternative to conventional, large-volume batch mixing allows significant advantages to be exploited (Tauzia, 1997), such as lower security surfaces of the production facilities owing to the small quantities of material used, lower capital costs, less direct labor, wider range of eligible material, preservation of the vital requirements of quality and safety.

10.2.3.1.3 Plot area impact; skyline impact

Continuous extrusion processing units are generally very compact owing to process intensification which improves acceptance with regard to plot area and skyline impacts.

10.2.3.2 Planet/ecological dimension

In general, extrusion process-intensifying methods allow material and energy intensity to be reduced significantly, hence reducing environmental impact.

10.2.3.2.1 Depletion of abiotic resources

Extrusion technology leads to clean processing with very low emissions, hence preserving air, surface water, and soil.

10.2.3.2.2 Global warming; photochemical ozone pollutants

Extrusion processing technology, via process intensification (such as integration of several functions in a single processing equipment, high dry solids processing, liquid expression, in particular), results in lower energy requirements and therefore lower carbon dioxide emissions (if energy is provided by fossil fuels), reducing the impact on global warming. This has been clearly shown in the previous subsections of this chapter when examining the characteristics of the three extrusion process-intensifying methods. For instance, energy requirements can be achieved by improving conversion selectivity in reactive extrusion (twin screw extrusion pulping process), by increasing the dry solids content in extrusion-cooking (food extrusion processing), or by avoiding liquid evaporation in solid-liquid fractionation (extrusion-pressing of solid-liquid media).

The introduction of extrusion process-intensifying methods (such as reactive extrusion) in chemical processing plants can contribute to reducing the number of pieces of equipment and therefore the number of flanges and seals, leading to a reduction of volatile organic compound emissions which cause the formation of photochemical oxidants in the environment.

In addition, extrusion processing technology allows environmentally friendly semi- or end-products to be produced, as alternatives to environmentally damaging products. It is worth mentioning the example of extrusion-processed wet fibrated proteins (based on vegetable proteins; refer to Chapter 6, section 6.3.3) which can realistically replace meat. It is well known that meat production (through intensive stock breeding, in particular) is one of the main sources of anthropogenic greenhouse gases; besides, meat production has a much higher environmental footprint (through water consumption, extensive use of land, deforestation, loss of biodiversity, energy consumption) than cereal production. For instance, the average energy consumption (from fossil fuel) per kg of animal protein is eight times larger than that required for the production of 1 kg of vegetable protein. Hence, the production of meat analogs from vegetable proteins via a high moisture extrusion-cooking process allows an important reduction of the environmental footprint to be achieved, while providing healthy and pleasant food to the population.

10.2.3.2.3 Acidification and thermal pollution

Extrusion processing technology has an indirect impact on these environmental issues owing to the consumption of electricity and steam, which result from the conversion of fossil fuel to energy. By reducing energy requirements, extrusion processing technology should have a positive impact on acidification and thermal pollution.

10.2.3.3 Profit/economic dimension

10.2.3.3.1 Scarce resource depletion; fossil fuel depletion; drinking water resource depletion

Non-renewable planetary resources are limited and development of alternative, renewable resources must be pursued actively in the coming decades. Through process-intensifying methods, extrusion processing can not only help to alleviate processing requirements through savings of important consumables (e.g. water, energy, possible chemical reagents) in existing processes, but also contribute to the development of new processes dedicated to the conversion of biomass and renewable resources into alternative products (such as 2G biofuels, fiber-reinforced composites, bio-sourced thermoplastics, bio-sourced chemicals, etc.).

10.2.3.3.2 Capital and operational expenditures; profitable over total lifecycle

As already seen in previous subsections of this chapter, when compared with conventional processes (batch or semi-continuous), extrusion process-intensifying processes allow significant reduction of capital and operational expenditures to be achieved; based on processes which are in operation in the food and non-food sectors, a cost reduction factor of 1.5–3 can be reasonably expected.

As for profitability over total lifecycle, extrusion processing shows long-term profitability owing to its high level of flexibility and adaptability and to the fact that there is no concern with external costs; life duration of extrusion equipment is typically in the range of 20–30 years, as it can be regularly upgraded and therefore benefit from incremental technical improvements.

10.3 Case studies: exercises

10.3.1 Exercise 1: Residence time distribution

Determine the residence time distribution function $F(t)$ as a function of reduced time, $t/\bar{t}$, for the two following cases (section 10.2.2.1.2).

1. The ideal continuous stirred tank reactor (CSTR).

2. The plane Poiseuille flow (or flow between infinite parallel plates); refer to Figure 10.9a. An isothermal, laminar, fully developed flow down z direction is considered while

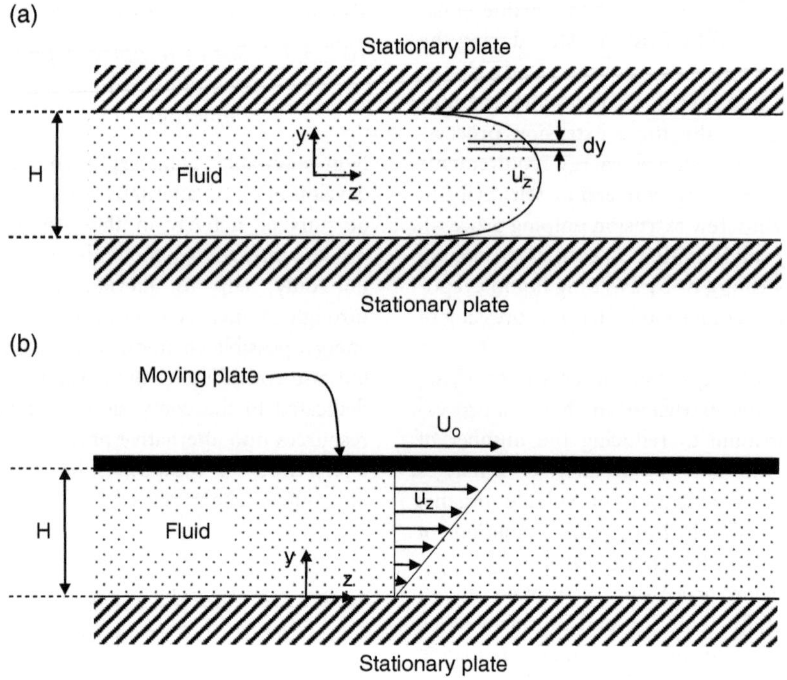

Figure 10.9 (a) Geometry for plane Poiseuille flow. (b) Geometry for parallel plate drag flow.

the fluid is assumed to be Newtonian. Hence, the velocity profile in z direction is the following:

$$u_z = \frac{3}{2}U\left[1-\left(\frac{2y}{H}\right)^2\right] \tag{10.45}$$

where u_z and U are the velocity in z direction and the average velocity. u_z is independent of the axial position z but it depends on y. H is the distance between the parallel plates.

10.3.1.1 Solution

1. Consider a steady continuous flow through the stirred tank of volume V. A non-reactive fluid of volumetric flow rate Q and volume concentration C_i of some component i enters the tank. As the system is at steady state, the contents of the tank are uniform and the component i enters and leaves the system at a rate of C_iQ. Suppose that at time $t = 0$, all the component i entering the tank is marked (say radioactive, for instance), so that the marked component can be identified from the unmarked material. At time $t > 0$, the component of interest leaving the tank comprises both the unmarked and marked components, while the concentration of the marked component C_i^* increases versus time.

Let the function $F(t)$ be the fraction of marked component leaving the tank at time t, then:

$$F(t) = \frac{C_i^*(t)}{C_i} \tag{10.46}$$

The function $F(t)$ can be determined from a material balance on the marked component in the tank, as follows:

$$QC_i - QC_i^* = V\frac{dC_i^*}{dt} \tag{10.47}$$

with $C_i^* = 0$ at $t \leq 0$.

Since perfect mixing is assumed, C_i^* represents the concentration of marked material in the tank and in the effluent. The solution of Eq. 10.47 is given by:

$$\frac{C_i^*}{C_i} = 1 - e^{-Qt/V} = F(t) \tag{10.48}$$

Hence, the function $F(t)$ for an ideal CSTR becomes:

$$F(t) = 1 - e^{-t/\bar{t}} \tag{10.49}$$

where $\bar{t} = V/Q$ is the mean residence time. This function $F(t)$ is shown in Figure 10.4.

2. With reference to Figure 10.9a, let us define a function $f(t)$ such that $f(t)dt$ is the fraction of the output of the flow which had resided in a fluid element at position y for a time comprised between t and $t + dt$. If Q is the volumetric flow rate, then it follows:

$$f(t)dt = \frac{dQ}{Q} = \frac{2u_z dy}{UH} \tag{10.50}$$

At position y, the fluid element moves at velocity u_z and requires a time t to cover a distance Z such as:

$$t = \frac{Z}{u_z} \tag{10.51}$$

And the average residence time of the whole fluid to cover the same distance is given by:

$$\bar{t} = \frac{Z}{U} \tag{10.52}$$

From Eqs 10.51 and 10.52, it follows:

$$\frac{u_z}{U} = \frac{\bar{t}}{t} \tag{10.53}$$

From $u_z = \frac{3}{2}U\left[1 - \left(\frac{2y}{H}\right)^2\right] = \frac{Z}{t}$, one can deduce:

$$\frac{12y dy}{H^2} = \frac{\bar{t} dt}{t^2} \tag{10.54}$$

and

$$\frac{2y}{H} = \left(1 - \frac{2\bar{t}}{3t}\right)^{1/2} \tag{10.55}$$

By use of Eq. 10.55, Eq. 10.54 allows the quantity dy/H to be determined:

$$\frac{dy}{H} = \frac{1}{6}\left(1 - \frac{2\bar{t}}{3t}\right)^{-1/2} \frac{\bar{t} dt}{t^2} \tag{10.56}$$

Finally, from Eqs 10.53 and 10.56, Eq. 10.50 takes the following form:

$$f(t)dt = \frac{\bar{t}^2}{3}\left(1 - \frac{2\bar{t}}{3t}\right)^{-1/2} \frac{dt}{t^3} \tag{10.57}$$

The residence time distribution function $F(t)$ can be defined by:

$$F(t) = \int_{t_{min}}^{t} f(t)dt \tag{10.58}$$

In Eq. 10.58, the lower limit relates to the shortest time of any fluid element in the volume, which corresponds to the fluid moving at position $y = 0$ (or along the center plane), where the velocity is greatest. Hence, $t_{min} = 2\bar{t}/3$ and $F(t)$ function takes the following form:

$$F(t) = \frac{\bar{t}^2}{3} \int_{\frac{2\bar{t}}{3}}^{t} \left(1 - \frac{2\bar{t}}{3t}\right)^{-1/2} \frac{dt}{t^3} \tag{10.59}$$

Equation 10.59 can be easily solved, hence leading to Eq. 10.60:

$$F(t) = \left(1 - \frac{2\bar{t}}{3t}\right)^{1/2} \left(1 + \frac{1\bar{t}}{3t}\right) \tag{10.60}$$

$F(t)$ function of the plane Poiseuille flow is shown in Figure 10.4.

10.3.2 Exercise 2: Polymer melt coupling in reactive extrusion

Two miscible polymers (say, polymer A and polymer B) have different terminal functional groups and are subjected to reactive extrusion. The polymers are initially in powder state with an initial particle size of approximately 200 μm. After feeding into the single screw extruder-reactor, the mix of polymers A and B is melted in the melting section and the resulting melt is continuously mixed in the reaction section to perform the coupling reaction between polymer A and B. The reaction section of the screw-barrel assembly is characterized by a L/D ratio of 42, a L_{ch}/H ratio of 100, a helix angle of 17.66°, a shear rate of $100\ s^{-1}$, and an average residence time of 40 seconds.

1. Suppose that a static layer of polymer A is introduced into a large volume of polymer B. Calculate the time required to homogenize the mix through pure diffusion.

2. If the mixing of polymer A and polymer B is realized in a single screw extruder-reactor, calculate the time to homogenize the mix. Is the average residence time of the reaction section large enough to complete the reactive coupling?

3. Determine the specific contribution of the extruder-reactor to the reduction of the segregation number for mixing, S_m, compared to that for the laminar flow in a pipe. An initially vertical interface will be assumed.

For this polymer melt coupling: $Da >> 1$ and the diffusion coefficient $D = 2.10^{-12} cm^2/s$.

10.3.2.1 Solution

1. If a static layer of polymer A of thickness $2l_o$ is introduced into a large volume of polymer melt B, the diffusion of polymer B into A is governed by Fick's second law which allows the diffusion time to be predicted. For instance, for a concentration difference in A layer less than 1%, the time required is given by:

$$t = \frac{2l^2}{D} \quad (10.61)$$

Accounting for the initial size of the powder (≅ 200 μm), thus $l \cong 100\ \mu m$, and the diffusion time to homogenize the mix is equal to 10^8 seconds, which is excessively long.

2. This time can be drastically reduced when using an extruder-reactor owing to the shear strain of the rotating screw(s) and the subsequent reduction of the striation thickness l (Eq. 10.18). Hence, combining Eqs 10.18 and 10.61, the time required to homogenize the mix in the single screw extruder-reactor is given by:

$$t = \sqrt[3]{\frac{8l_o^2}{D(\dot{\gamma})^2}} \cong 34 s \quad (10.62)$$

As $Da >> 1$, polymer coupling is not reaction limited. Thus, the average residence time in the reaction section (that is, 40 seconds) is long enough to complete the reactive coupling.

3. The specific contribution of the extruder-reactor to the reduction of S_m is given by the quantity $(m_i.\ Rc)^2$ (Eq. 10.21), that is: 5362.

10.3.3 Exercise 3: Weighted average total strain

1. Show that the WATS for a laminar flow of a Newtonian fluid in a long circular pipe (or Poiseuille flow) is $16L/3D$ (L and D are the length and diameter of the pipe, respectively). Calculate WATS for $L/D = 42$.

2. Determine the WATS of a continuous parallel plate drag flow (Figure 10.9b). An isothermal fully developed pure drag flow down z direction is considered while the fluid is assumed to be Newtonian. Compare the WATS obtained with that of a single screw extruder (see section 10.2.2.1.3).

3. Determine the WATS of the reactive extrusion process considered in Exercise 2 (reaction section); it is supposed that the flow behavior in this extruder-reactor is ideal plug flow. Assuming that an adequate mixing required for that process corresponds to a $WATS_{target}$ of 18, calculate the process-intensifying capability of the extruder-reactor.

10.3.3.1 Solution

1. WATS was defined by Eq. 10.24 (see section 10.2.2.1.3) which shows that the residence time distribution for a laminar flow in the circular pipe must be determined. It is worth recalling that the velocity profile of a Poiseuille flow is given by:

$$u_z = 2U\left[1 - \left(\frac{r}{R}\right)^2\right] \quad (10.63)$$

where R and U are the radius of the pipe and the average velocity of the fluid. Following the same demonstration as that for the plane Poiseuille flow (see Exercise 1), we can easily demonstrate that:

$$f(t)dt = \frac{\bar{t}^2\, dt}{2\ t^3} \quad (10.64)$$

where $\bar{t}$ is the average residence time; hence the expression for WATS is given by:

$$WATS = \int_{t_o}^{\infty} \gamma \frac{\bar{t}^2}{2t^3} dt = \int_{t_o}^{\infty} \dot{\gamma}_r t \frac{\bar{t}^2}{2t^3} dt \quad (10.65)$$

The local shear rate, $\dot{\gamma}_r$, varies across the pipe and is easily expressed as:

$$\dot{\gamma}_r = \frac{4Ur}{R^2} \quad (10.66)$$

Hence Eq. 10.65 becomes:

$$WATS = \frac{2U\bar{t}^2}{R^2} \int_{t_o}^{\infty} r \frac{dt}{t^2} \quad (10.67)$$

In this integral, keep in mind that r and t are dependent variables in the sense described by Eq. 10.51 and Eq. 10.63. Using these equations, one can obtain:

$$\frac{r}{R}=\left(1-\frac{\bar{t}}{2t}\right)^{1/2} \quad (10.68)$$

and Eq. 10.67 becomes:

$$WATS=\frac{2U\bar{t}^{2}}{R}\int_{t_o}^{\infty}\left(1-\frac{\bar{t}}{2t}\right)^{1/2}\frac{dt}{t^2} \quad (10.69)$$

Equation 10.69 can be easily solved knowing that the minimum residence time, t_o, corresponds to the fluid moving along the pipe axis ($r=0$) where the velocity is the greatest; hence, $t_o=\bar{t}/2$. Thus, one can easily find that:

$$WATS=\frac{16L}{3D} \quad (10.70)$$

And for $L/D=42$, $WATS=224=WATS_{ref}$ (as defined in section 10.2.2.1.4).

2. WATS can be determined by Eq. 10.24. With reference to Figure 10.9b, the velocity distribution for a fully developed, isothermal drag flow between parallel plates is:

$$u_z=\frac{U_o}{H}y \quad (10.71)$$

where U_o is the velocity of the moving plate (constant velocity). For continuous parallel plate drag flow, one can easily demonstrate that $f(t)dt$ is given by Eq. 10.64. Hence, knowing that the shear rate is uniform throughout the system ($\dot{\gamma}=U_o/H$), Eq. 10.24 becomes:

$$WATS=\frac{\bar{t}^{2}}{2}\frac{U_o}{H}\int_{\bar{t}/2}^{\infty}\frac{dt}{t^2}=\frac{2Z}{H} \quad (10.72)$$

where Z is the distance down the channel defined by the two parallel plates.

When comparing continuous parallel plate drag flow and drag flow achieved in a single screw extruder both characterized by identical geometry ratio (or Z/H = L/H), WATS achieved in a single screw extruder (in the range of 20 L/H to 30 L/H) is at least 10 times larger than that of the parallel plate drag flow, due to the effect of the helix angle of the screw channel which induces transverse convective mixing.

3. Accounting for the plug flow assumption, the WATS of the reactive extrusion process considered in Exercise 2 can be approximated by $WATS=\dot{\gamma}.t=4000=WATS_{process}$ (as defined in section 10.2.2.1.4). Hence, if $WATS_{target}=18$, the PIC of the corresponding extruder-reactor would be close to 1; this is in agreement with the conclusion of Exercise 2 which stipulates that the average residence time in the reaction section is long enough to complete the reaction.

10.3.4 Exercise 4: Energy saving in extrusion-cooking

Based on their process characteristics, calculate and compare the energy consumptions of both process alternatives to produce cereal flakes (refer to section 10.2.2.4, Table 10.4).

1. The conventional hydrothermal cooking process.
2. The pellet-to-flake extrusion-cooking process.

10.3.4.1 Solution

The specific energies, SCE, SDE and GSE, are given by Eqs 10.33, 10.34 and 10.35, respectively (see section 10.2.2.2.2). Using these equations, specific energies of both alternative processes can be calculated.

1. For the conventional hydrothermal cooking process, the following average cooking characteristics are considered: solids concentration of 62.5% (wwb) and temperature of 107.5°C. Hence, the specific energies are the following: SCE ≅ 314 kJ/kg, SDE ≅ 1677 kJ/kg and GSE ≅ 1991 kJ/kg.

2. For the pellet-to-flake extrusion-cooking process, the following average cooking characteristics are considered: solids concentration of 76% (wwb) and temperature of 137.5°C. Hence, the specific energies are the following: SCE ≅ 345 kJ/kg, SDE ≅ 806 kJ/kg and GSE ≅ 1151 kJ/kg. When compared with the conventional cereal flakes process (based on hydrothermal cooking), these rough estimations of the specific energies show that the pellet-to-flake process allows an important energy saving to be obtained (approximately 31%), due to process intensification (high solids concentration processing).

10.3.5 Exercise 5: Water saving in solid-liquid extrusion-pressing

A chemical pulping plant produces cellulose pulp from non-wood fiber crops. After alkaline cooking and bleaching, the resulting pulp (660 kg/h dry fiber; 25% dryness; concentration of solutes in retained solution: 2% weight basis) needs to be washed with fresh water, to remove

95% of the solutes (residual chemicals solubilized in water).

1. Calculate the consumption of fresh water for a single batch, contact-equilibrium washing process (standard gravity settling corresponds to 25% dryness).

2. Calculate the consumption of fresh water for a continuous, multistage contact-equilibrium washing process (standard gravity settling in each contacting stage corresponds to 25% dryness). Estimate the PIC of the continuous process.

3. Calculate the consumption of fresh water for a continuous, multistage contact-equilibrium washing process carried out in a twin screw extruder-press (extrusion-pressing conditions correspond to a filter cake of 40% dryness in each contacting stage). Estimate the PIC of the continuous screw extrusion-pressing washing process.

The consumption of fresh water will be expressed in metric tonnes per tonne of dry pulp. For questions 1 and 2, the ratio of fresh water to retained solution, k, will satisfy the condition $k \leq 2$.

10.3.5.1 Solution

The material balance at the inlet of any washing process is defined by:

- dry fiber: 660 kg/h
- retained solution (water + solutes): 1980 kg/h (25% dryness)
- solutes in retained solution: 39.6 kg/h (2% weight basis).

1. The consumption of fresh water is given by Eq. 10.36, where $x_o = 0.02$ and $x_1 = 0.001$ ($\rho_s = 0.95$). Hence, Eq. 10.36 leads to $k = 19$, which gives 37.62 tonnes/h of fresh water, or 57 tonnes/tonne of dry pulp.

2. Accounting for the conditions $k \leq 2$ and $\rho_s = 0.95$, Eq. 10.38 shows that the minimum number of contacting stages of the washing process is three, and Eq. 10.40 allows k to be determined; hence $k = 1.714$. The consumption of fresh water is then given by $n.\, k.\, m_{sol}$, that amounts to 10.18 tonnes/h, or 15.4 tonnes/tonne of dry pulp.

The PIC of the continuous, multistage contact-equilibrium washing process can be obtained by Eq. 10.42; hence $PIC \cong 0.73$. The high PIC of the three-stage washing process alternative shows clearly that switching from batch to continuous processing allows an important process intensification to be performed.

3. Accounting for the conditions $k \leq 2$ and $\rho_s = 0.95$, it can be shown that the minimum number of contacting stages of the washing process is three, and Eq. 10.43 allows k to be determined, with $f = 0.6/0.75 = 0.8$; hence $k = 1.52$. The consumption of fresh water is then given by $f.\, n.\, k.\, m_{sol}$, that amounts to 7.22 tones/h, or 10.9 tonnes/tonne of dry pulp.

The PIC of the continuous screw extrusion-pressing washing process can be obtained by Eq. 10.44; hence $PIC \cong 0.81$.

When compared with the conventional solid-liquid washing process (based on continuous multistage contact-equilibrium), the extrusion-pressing washing process allows important water saving to be obtained (water saving of approximately 29%), due to process intensification through pressure-induced solid-liquid separation.

It must be noted that the calculated consumption of fresh water may slightly differ from experimental observations, depending upon the distribution of the solutes between the solid and liquid phases.

10.4 Conclusion: future trends

Extrusion technology has substantially modified the technological environment of processing industries in recent decades. Since its first introduction in the 1930s in the polymer industry, extrusion processing has never stopped expanding its technical contribution, owing to the establishment of four fundamental generic extrusion processes: polymer plasticating (Chapter 4), reactive extrusion (Chapter 5), extrusion-cooking (Chapter 6), and solid-liquid extrusion-pressing (Chapter 8). Hence today, extrusion processing technology is able to provide relevant and cost-effective processing solutions in the polymer and plastics industry, the chemical industry, the food and feed industry, and the paper milling industry. It is well recognized that this industrial success is based on key technological characteristics (such as continuous short time high solids concentration processing, outstanding quality of mixing with viscous complex media, flexible and versatile processing, low processing volumes, in particular) which add real value to investors' businesses. It is worth noting that considerable amounts of innovative B-to-B and B-t-C products (such as thermoplastics materials, formulated chemicals, textured cereals and proteins, etc.) have been successfully created and marketed in various sectors, to satisfy the needs of society (such as food, health, well-being, pleasure, etc.). In addition, the technological characteristics of extrusion processing have allowed existing batch processes to be replaced by continuous extrusion processes in order to upgrade the technical and economical performance of conventional batch processing in numerous industrial sectors.

Of course, the potential of extrusion processing technology for product and process innovation, as well as for process substitution, is still being explored.

The key technological specificity of extrusion processing technology results from its outstanding quality of mixing and its ability to handle continuously inputs of mechanical and thermal energies while processing viscous complex media (homogeneous as well as heterogeneous). The remarkable mixing capability of extrusion processing technology has been described, discussed and illustrated extensively throughout this book, in particular when presenting the generic extrusion processes. For a given formulated medium, technical and economical performances of generic extrusion processes all depend primarily on the mixing efficiency of screw extruders, which then governs physical transfer phenomena (such as mass and heat transfer) and chemical reactions. Therefore, it is of paramount importance to apply our knowledge to the design of extrusion hardware (the screw-barrel assembly, in particular), to optimize the mixing efficiency in single screw and twin screw extruders. When comparing single screw and intermeshing co-rotating twin screw extruders, mixing efficiency is definitely much higher in the latter, which offer outstanding ability to intensify the mixing with viscous complex media such as multi-ingredient homogeneous melts and heterogeneous media (solid-liquid systems, emulsions, gas-liquid media, for instance). In such cases, significant scientific inputs and technical innovations will undoubtedly emerge in the coming decades to upgrade mixing capability of screw extruders according to upcoming process and product requirements.

The future of extrusion processing technology will be more and more driven by the development of process intensification, owing to principles of sustainable development and the subsequent need for sustainable processing technologies. As seen in this chapter, extrusion processing already belongs to the class of sustainable processing technology due to the four existing extrusion process-intensifying methods (polymer plasticating, reactive extrusion, extrusion-cooking, and solid-liquid extrusion-pressing), which significantly contribute to smaller, cleaner, safer, and cheaper processing. The need for sustainability will promote and boost the development of extrusion processing technology in two main directions.

• Potential replacement of conventional non-sustainable processing (such as batch processing, high energy-consuming processing, unsafe processing with regard to operators and plant substructures, environment-impacting processing, etc.) by extrusion processing technology. The ongoing development of the innovative extrusion-porosification process (as presented in Chapter 9), as an alternative to conventional high energy-consuming and high-CAPEX processes (such as spray drying, drum drying, freeze drying), provides a relevant example of emerging technology substitution.

• Emergence of innovative processing avenues owing to the development of green chemistry (synthesis of chemicals and biomaterials from renewable resources), biorefineries (production of biofuels from biomass such as second-generation biofuels), and biotechnology (intensification of enzymatic conversions), etc.

Our collective awareness of sustainable development and consequently of sustainable processing technology will increasingly reveal the need for new product and process concepts to accommodate "*the needs of the present without compromising the ability of future generations to meet their own needs*" (Brundtland, 1987). This should motivate product, process and equipment designers to innovate and develop appropriate sustainable processing technologies. Following these fundamental premises and long-term trends, existing as well as new extrusion process-intensifying processes should play a significant role in upcoming technology changes, leading to important developments in extrusion processing technologies in the coming years.

References

Biesenberger JA (1992) Principles of reaction engineering. In: Xanthos M (ed) *Reactive Extrusion – Principles and Practice*. Munich: Carl Hanser Verlag, pp. 227–255.

Bouzrara H, Vorobiev E (2003) Solid-liquid expression of cellular materials enhanced by pulsed electric field. Chemical Engineering and Processing 42(4): 249–257.

Brodkey RS, Hershey HC (1988) *Transport Phenomena. Part I – Basic Concepts in Transport Phenomena*. Columbus, Ohio: Brodkey Publishing, Columbus, pp. 330–333.

Brundtland GH (1987) *Our Common Future*. United Nations World Commission on Environment and Development (WCED) Report. Oxford: Oxford University Press.

Chella R, Ottino JM (1985) Fluid mechanics of mixing in a single-screw extruder. Industrial and Engineering Chemistry Fundamentals 24(2): 170–180.

Commenge JM, Falk L, Corriou JP, Matlosz M (2005) Analysis of microstructured reactor for process miniaturization and intensification. Chemical Engineering Technology 28(4): 446–458.

Davis WM (1992) Heat transfer in extruder reactors. In: Xanthos M (ed) *Reactive Extrusion – Principles and Practice.* Munich: Carl Hanser Verlag, pp. 257–282.

Della Valle G, Vergnes B (1994) Propriétés thermophysiques et rhéologiques des substrats utilisés en cuisson-extrusion. In: *Colonna P,* Della Valle G (eds) *La Cuisson-Extrusion.* Paris: Lavoisier, Tec & Doc, pp. 439–467.

Diamond J (2005) *Collapse – How Societies Choose to Fail or Succeed.* New York: Penguin Books.

Dorfman AS (2010) *Conjugate Problems in Convective Heat Transfer.* Boca Raton, Florida: CRC Press, pp. 275–277.

Ehrlich PR, Holdren JP (1971) Impact of population growth. Science 171: 1212–1217.

Graedel TE, Allenby BR (1995) *Industrial Ecology.* New York: Prentice-Hall.

Grimi N, Vorobiev E, Lebovka N, Vaxelaire J (2010) Solid-liquid expression from denatured plant tissue: filtration-consolidation behaviour. Journal of Food Engineering 96: 29–36.

Harmsen GJ, Korevaar G, Lemkowitz SM (2004) Process intensification contributions to sustainable development. In: Stankiewicz A, Moulijn JA (eds) *Re-engineering the Chemical Processing Plant – Process Intensification.* New York: Marcel Dekker, pp. 495–529.

Hoffert MI, Caldeira K, Jain AK, et al. (1998) Energy implications of future stabilization of atmospheric CO_2 content. Nature 395: 881–884.

Jackson T (1993) *Clean Production Strategies: Developing Preventive Environmental Management in the Industrial Economy.* Florida: Lewis Publishers.

Lanoisellé JL, Vorobyov EI, Bouvier JM, Piar G (1996) Modeling of solid/liquid expression for cellular materials. AIChE Journal 42(7): 2057–2068.

Martelli FG (1983) *Twin Screw Extruders – A Basic Understanding.* New York: Van Nostrand Reinhold.

Matlosz M, Falk L, Commenge JM (2009) Process intensification. In: Dietrich TR (ed) *Microchemical Engineering in Practice.* New Jersey: John Wiley, pp. 325–348.

Meadows DH, Meadows DL, Randers J, Behrens WW (1972) *The Limits to Growth.* A report for the Club of Rome's project on the predicament of mankind. New York: Universe Books.

Mottaz J, Bruyas L (2001) Optimized thermal performance in extrusion. In: Guy R (ed) *Extrusion Cooking – Technologies and Applications.* Cambridge: Woodhead Publishing, pp. 51–82.

Moussa M, Qin X, Chen LF, Campanella OH, Hamaker BR (2011) High-quality instant sorghum porridge flours for the West African market using continuous processor cooking. International Journal of Food Science & Technology 46(11): 2344–2350.

Ramshaw C (1983) "Higee" distillation – an example of process intensification. Chemical Engineer 389: 13–14.

Rauwendaal C (2001) *Polymer Extrusion,* 4th edn. Munich: Hanser Publishers.

Reay D, Ramshaw C, Harvey A (2008) *Process Intensification – Engineering for Efficiency, Sustainability and Flexibility.* Oxford: Butterworth-Heinemann.

Shirato M, Murase T, Iwata M, Nakatsuka S (1986) The Terzaghi–Voigt combined model for constant-pressure consolidation of filter cakes and homogeneous semi-solid materials. Chemical Engineering and Science 41(12): 3213–3218.

Stankiewicz A, Drinkenburg AAH (2004) Process intensification: history, philosophy, principles. In: Stankiewicz A, Moulijn JA (eds) *Re-engineering the Chemical Processing Plant – Process Intensification.* New York: Marcel Dekker, pp. 1–32.

Stankiewicz A, Moulijn JA (2000) *Process intensification: transforming chemical engineering.* Chemical Engineering Progress, Process Design Trends January: 22–34.

Stankiewicz A, Moulijn JA (2004) *Re-engineering the Chemical Processing Plant – Process Intensification.* New York: Marcel Dekker.

Tadmor Z, Gogos C.G (2006) *Principles of Polymer Processing,* 2nd edn. New York: John Wiley, pp. 322–408.

Tadmor Z, Klein I (1970) *Engineering Principles of Plasticating Extrusion.* Polymer Science and Engineering Series. Florida: Robert E. Krieger Publishing.

Tauzia JM (1997) L'évolution de la technologie de production des matériaux énergétiques. Revue Scientifique et Technique de la Défense 3(37): 133–138.

Vadke VS, Sosulski FW (1988) Mechanics of oil expression from canola. Journal of the American Oil Chemists' Society 65(7): 1169–1176.

Vauchel P, Kaas R, Arhaliass A, Baron R, Legrand J (2008) A new process for extracting alginates from Laminaria digitata: reactive extrusion. Food Bioprocessing Technology 1: 297–300.

Vauchel P, Leroux K, Kaas R, Arhaliass A, Baron R, Legrand J (2009) Kinetics modeling of alginate alkaline extraction from Laminaria digitata. Bioresource Technology 100: 1291–1296.

Vorobiev E, Lebovka NI (2006) Extraction of intercellular components by pulsed electric fields. In: Raso J, Heinz V (eds) *Pulsed Electric Fields Technology for the Food Industry – Fundamentals and Applications.* Food Engineering Series. New York: Springer Science + Business Media, pp. 153–195.

Wang SS, Chiang WC, Zhao B, Zheng XG, Kim IH (1991) Experimental analysis and computer simulation of starch–water interactions during phase transition. Journal of Food Science 56(1): 121–124 & 142.

Wang SS, Chiang WC, Zheng XG, Zhao B, Cho MH, Yeh AI (1992) Application of an energy equivalent concept to the study of the kinetics of starch conversion during extrusion. In: Kokini JL, Ho CT, Karwe MV (eds) *Food Extrusion Science and Technology.* New York: Marcel Dekker, pp. 165–176.

Westenbroek APH (2000) *Extrusion pulping of natural fibres.* PhD dissertation, Twente University, The Netherlands.

Index

Extrusion Processing Technology: Food and Non-Food Biomaterials, First Edition. Jean-Marie Bouvier and Osvaldo H. Campanella.

www.ingramcontent.com/pod-product-compliance
Ingram Content Group UK Ltd.
Pitfield, Milton Keynes, MK11 3LW, UK
UKHW051022091026
15624UKWH00012B/633